De-Shuang Huang   Changjun
Vitoantonio Bevilacqua
Juan Carlos Figueroa (Eds.)

T0238666

Intelligent
Computing Technology

8th International Conference, ICIC 2012
Huangshan, China, July 25-29, 2012
Proceedings

Springer

Volume Editors

De-Shuang Huang
Changjun Jiang
Tongji University
School of Electronics and Information Engineering
4800 Caoan Road, Shanghai 201804, China
E-mail: {dshuang, cjjiang}@tongji.edu.cn

Vitoantonio Bevilacqua
Polytechnic of Bari
Electrical and Electronics Department
Via Orabona, 4, 70125 Bari, Italy
E-mail: vitoantonio.bevilacqua@gmail.com

Juan Carlos Figueroa
District University Francisco José de Caldas
Faculty of Engineering
Cra. 7a No. 40-53, Fifth Floor, Bogotá, Colombia
E-mail: jcfigueroag@udistrital.edu.co

ISSN 0302-9743   e-ISSN 1611-3349
ISBN 978-3-642-31587-9   e-ISBN 978-3-642-31588-6
DOI 10.1007/978-3-642-31588-6
Springer Heidelberg Dordrecht London New York

Library of Congress Control Number: 2012941229

CR Subject Classification (1998): I.2, I.4, I.5, H.4, J.3, F.1, F.2, C.2, H.3, H.2.8, G.2.2, K.6.5

LNCS Sublibrary: SL 3 – Information Systems and Applications, incl. Internet/Web and HCI

© Springer-Verlag Berlin Heidelberg 2012
This work is subject to copyright. All rights are reserved, whether the whole or part of the material is concerned, specifically the rights of translation, reprinting, re-use of illustrations, recitation, broadcasting, reproduction on microfilms or in any other way, and storage in data banks. Duplication of this publication or parts thereof is permitted only under the provisions of the German Copyright Law of September 9, 1965, in its current version, and permission for use must always be obtained from Springer. Violations are liable to prosecution under the German Copyright Law.
The use of general descriptive names, registered names, trademarks, etc. in this publication does not imply, even in the absence of a specific statement, that such names are exempt from the relevant protective laws and regulations and therefore free for general use.

Typesetting: Camera-ready by author, data conversion by Scientific Publishing Services, Chennai, India

Printed on acid-free paper

Springer is part of Springer Science+Business Media (www.springer.com)

Lecture Notes in Computer Science 7389

Commenced Publication in 1973
Founding and Former Series Editors:
Gerhard Goos, Juris Hartmanis, and Jan van Leeuwen

Editorial Board

David Hutchison
Lancaster University, UK

Takeo Kanade
Carnegie Mellon University, Pittsburgh, PA, USA

Josef Kittler
University of Surrey, Guildford, UK

Jon M. Kleinberg
Cornell University, Ithaca, NY, USA

Alfred Kobsa
University of California, Irvine, CA, USA

Friedemann Mattern
ETH Zurich, Switzerland

John C. Mitchell
Stanford University, CA, USA

Moni Naor
Weizmann Institute of Science, Rehovot, Israel

Oscar Nierstrasz
University of Bern, Switzerland

C. Pandu Rangan
Indian Institute of Technology, Madras, India

Bernhard Steffen
TU Dortmund University, Germany

Madhu Sudan
Microsoft Research, Cambridge, MA, USA

Demetri Terzopoulos
University of California, Los Angeles, CA, USA

Doug Tygar
University of California, Berkeley, CA, USA

Gerhard Weikum
Max Planck Institute for Informatics, Saarbruecken, Germany

Preface

The International Conference on Intelligent Computing (ICIC) was started to provide an annual forum dedicated to the emerging and challenging topics in artificial intelligence, machine learning, pattern recognition, image processing, bioinformatics, and computational biology. It aims to bring together researchers and practitioners from both academia and industry to share ideas, problems, and solutions related to the multifaceted aspects of intelligent computing.

ICIC 2012, held in Huangshan, China, July 25–29, 2012, constituted the 8th International Conference on Intelligent Computing. It built upon the success of ICIC 2011, ICIC 2010, ICIC 2009, ICIC 2008, ICIC 2007, ICIC 2006, and ICIC 2005 that were held in Zhengzhou, Changsha, China, Ulsan, Korea, Shanghai, Qingdao, Kunming, and Hefei, China, respectively.

This year, the conference concentrated mainly on the theories and methodologies as well as the emerging applications of intelligent computing. Its aim was to unify the picture of contemporary intelligent computing techniques as an integral concept that highlights the trends in advanced computational intelligence and bridges theoretical research with applications. Therefore, the theme for this conference was "Advanced Intelligent Computing Technology and Applications." Papers focusing on this theme were solicited, addressing theories, methodologies, and applications in science and technology.

ICIC 2012 received 753 submissions from 28 countries and regions. All papers went through a rigorous peer-review procedure and each paper received at least three review reports. Based on the review reports, the Program Committee finally selected 242 high-quality papers for presentation at ICIC 2012, of which 242 papers are included in three volumes of proceedings published by Springer: one volume of *Lecture Notes in Computer Science* (LNCS), one volume of *Lecture Notes in Artificial Intelligence* (LNAI), and one volume of *Communications in Computer and Information Science* (CCIS).

This volume of *Lecture Notes in Computer Science* (LNCS) includes 84 papers.

The organizers of ICIC 2012, including Tongji University, made an enormous effort to ensure the success of the conference. We hereby would like to thank the members of the Program Committee and the referees for their collective effort in reviewing and soliciting the papers. We would like to thank Alfred Hofmann, executive editor from Springer, for his frank and helpful advice and guidance throughout and for his continuous support in publishing the proceedings. In particular, we would like to thank all the authors for contributing their papers.

Without the high-quality submissions from the authors, the success of the conference would not have been possible. Finally, we are especially grateful to the IEEE Computational Intelligence Society, the International Neural Network Society, and the National Science Foundation of China for their sponsorship.

May 2012

De-Shuang Huang
Changjun Jiang
Vitoantonio Bevilacqua
Juan Carlos Figueroa

ICIC 2012 Organization

General Co-chairs	Changjun Jiang, China
	Gary G. Yen, USA
Steering Committee Chair	De-Shuang Huang, China
Program Committee Co-chairs	Jianhua Ma, Japan
	Laurent Heutte, France
Organizing Committee Co-chairs	Duoqian Miao, China
	Yang Xiang, China
	Jihong Guan, China
Award Committee Chair	Kang-Hyun Jo, Korea
Publication Chair	Vitoantonio Bevilacqua, Italy
Workshop/Special Session Chair	Juan Carlos Figueroa, Colombia
Special Issue Chair	Michael Gromiha, India
Tutorial Chair	Phalguni Gupta, India
International Liaison Chair	Prashan Premaratne, Australia
Publicity Co-chairs	Kyungsook Han, Korea
	Ling Wang, China
	Xiang Zhang, USA
	Lei Zhang, China
Exhibition Chair	Qiong Wu, China
Organizing Committee Members	Zhijun Ding, China
	Hanli Wang, China
	Yan Wu, China
	Guo-Zheng Li, China
	Fanhuai Shi, China
Conference Secretary	Zhi-Yang Chen, China

Program Committee

Khalid Mahmood Aamir, Italy
Vasily Aristarkhov, Russian Federation
Costin Badica, Romania
Vitoantonio Bevilacqua, Italy
Shuhui Bi, China
Danail Bonchev, USA
Stefano Cagnoni, Italy
Chin-Chih Chang, Taiwan, China
Pei-Chann Chang, Taiwan, China
Jack Chen, Canada
Shih-Hsin Chen, Taiwan, China
Wen-Sheng Chen, China
Xiyuan Chen, China
Yang Chen, China
Ziping Chiang, Taiwan, China
Michal Choras, Poland
Angelo Ciaramella, Italy
Milan Cisty, Slovakia
Jose Alfredo F. Costa, Brazil
Loganathan D., India
Eng. Salvatore Distefano, Italy
Mariagrazia Dotoli, Italy
Karim Faez, Iran
Jianbo Fan, China
Minrui Fei, China
Wai-Keung Fung, Canada
Jun-Ying Gan, China
Xiao-Zhi Gao, Finland
Dunwei Gong, China
Valeriya Gribova, Russia
M. Michael Gromiha, Japan
Kayhan Gulez, Turkey
Anyuan Guo, China
Ping Guo, China
Phalguni Gupta, India
Fei Han, China
Kyungsook Han, Korea
Nojeong Heo, Korea
Laurent Heutte, France
Martin Holena, Czech Republic
Wei-Chiang Hong, Taiwan, China
Yuexian Hou, China
Sanqing Hu, China

Guangbin Huang, Singapore
Peter Hung, Ireland
Li Jia, China
Zhenran Jiang, China
Kang-Hyun Jo, Korea
Dah-Jing Jwo, Taiwan, China
Yoshiaki Kakuda, Japan
Vandana Dixit Kaushik, India
Muhammad Khurram Khan,
    Saudi Arabia
Bora Kumova, Turkey
Yoshinori Kuno, Japan
Takashi Kuremoto, Japan
Vincent C. S. Lee, Australia
Bo Li, China
Dalong Li, USA
Guo-Zheng Li, China
Shi-Hua Li, China
Xiaoou Li, Mexico
Hualou Liang, USA
Honghuang Lin, USA
Chunmei Liu, USA
Chun-Yu Liu, USA
Ju Liu, China
Ke Lv, China
Jinwen Ma, China
Igor V. Maslov, Japan
Xiandong Meng, USA
Filippo Menolascina, Italy
Pabitra Mitra, India
Ravi Monaragala, Sri Lanka
Tarik Veli Mumcu, Turkey
Primiano Di Nauta, Italy
Ben Niu, China
Sim-Heng Ong, Singapore
Vincenzo Pacelli, Italy
Shaoning Pang, New Zealand
Francesco Pappalardo, Italy
Young B. Park, Korea
Surya Prakash, India
Prashan Premaratne, Australia
Hong Qiao, China
Daowen Qiu, China

K. R. Seeja, India
Ajita Rattani, Italy
Angel D. Sappa, Spain
Simon See, Singapore
Akash K. Singh, USA
Jiatao Song, China
Qiankun Song, China
Zhan-Li Sun, Singapore
Stefano Squartini, Italy
Evi Syukur, Australia
Hao Tang, China
Chuan-Kang Ting, Taiwan, China
Jun Wan, USA
Bing Wang, USA
Jeen-Shing Wang, Taiwan, China
Ling Wang, China
Shitong Wang, China
Xuesong Wang, China
Yong Wang, China
Yufeng Wang, China
Zhi Wei, China
Xiaojun Wu, China

Junfeng Xia, USA
Shunren Xia, China
Bingji Xu, China
Shao Xu, Singapore
Zhenyu Xuan, USA
Yu Xue, China
Tao Ye, China
Jun-Heng Yeh, Taiwan, China
Myeong-Jae Yi, Korea
Zhi-Gang Zeng, China
Boyun Zhang, China
Chaoyang Joe Zhang, USA
Lei Zhang, Hong Kong, China
Rui Zhang, China
Xiaoguang Zhao, China
Xing-Ming Zhao, China
Zhongming Zhao, USA
Bo-Jin Zheng, China
Chun-Hou Zheng, China
Fengfeng Zhou, China
Waqas Haider Khan Bangyal, Pakistan
Yuhua Qian, China

Reviewers

Kezhi Mao
Xin Hao
Tarik Veli Mumcu
Muharrem Mercimek
Selin Ozcira
Ximo Torres
BinSong Cheng
Shihua Zhang
Yu Xue
Xiaoping Luo
Dingfei Ge
Jiayin Zhou
Mingyi Wang
Chung Chang Lien
Wei-Ling Hwang
Jian Jia
Jian Wang
Zhiliu Zuo
Sajid Bashir

Faisal Mufti
Hafiz Muhammad
  Farooq
Bilal Ahmed
Maryam Gul
Gurkan Tuna
Hajira Jabeen
Chandana Gamage
Prashan Premaratne
Chathura R. De Silva
Manodha Gamage
Kasun De Zoysa
Chesner Desir
Laksman Jayaratne
Francesco Camastra
Rémi Flamary
Antoninostaiano Alessio
  Ferone
Raffaele Montella

Nalin Karunasinghe
Vladislavs Dovgalecs
Pierrick Tranouez
Antonio Maratea
Giuseppe Vettigli
Ranga Rodrigo
Chyuan-Huei Yang
Rey-Sern Lin
Cheng-Hsiung Chiang
Jian-Shiun Hu
Yao-Hong Tsai
Hung-Chi Su
J.-H. Chen
Wen Ouyang
Chong Shen
Yuan Xu
Cucocris Tano
Tien-Dung Le
Hee-Jun Kang

Hong-Hee Lee
Ngoc-Tung Nguyen
Ju Kunru
Vladimir Brusic
Ping Zhang
Renjie Zhang
Alessandro Cincotti
Mojaharul Islam
Marzio Pennisi
Haili Wang
Santo Motta
Keun Ho Ryu
Alfredo Pulvirenti
Rosalba Giugno
Ge Guo
Chih-Min Lin
Yifeng Zhang
Xuefen Zhu
Lvzhou Li
Haozhen Situ
Qin Li
Nikola Paunkovic
Paulo Mateus
Jozef Gruska
Xiangfu Zou
Yasser Omar
Yin-Xiang Long
Bjoern Schuller
Erikcam Bria
Faundez-Zanuy Marcos
Rui Zhang
Yibin Ye
Qinglai Wei
Guangbin Huang
Lendasse Amaury
Michele Scarpiniti
Simone Bassis
Morabito Carlo
Amir Hussain
Li Zhang
Emilio Soria
Sanqing Hu
Hossein Javaherian
Veselin Stoyanov
Eric Fock

Yao-Nan Lien
Liangjun Xie
Nong Gu
Xuewei Wang
Shizhong Liao
Zheng Liu
Bingjun Sun
Yuexian Hou
Shiping Wen
Ailong Wu
Gang Bao
Takashi Kuremoto
Amin Yazdanpanah
Meng-Cheng Lau
Chi Tai Cheng
Jayanta Debnath
Raymond Ng
Baranyi Peter
Yongping Zhai
Baoquan Song
Weidi Dai
Jiangzhen Ran
Huiyu Jin
Guoping Lu
Xiaohua Qiao
Xuemei Ren
Mingxia Shen
Hao Tang
Zhong-Qiang Wu
Zhenhua Huang
Junlin Chang
Bin Ye
Yong Zhang
Yanzi Miao
Yindi Zhao
Jun Zhao
Mei-Qiang Zhu
Xue Xue
Yanjing Sun
Waqas Haider Khan
  Bangyal
Ming-Feng Yang
Guo-Feng Fan
Asma Nani
Xiangtao Li

Hongjun Jia
Yehu Shen
Tiantai Guo
Liya Ding
Dawen Xu
Jinhe Wang
Xiangyu Wang
Shihong Ding
Zhao Wang
Junyong Zhai
Haibo Du
Haibin Sun
Jun Yang
Chin-Sheng Yang
Jheng-Long Wu
Jyun-Jie Lin
Jun-Lin Lin
Liang-Chih Yu
S.H. Chen
Chien-Lung Chan
Eric Fan
X.H. Cloud
Yue Deng
Kun Yang
Badrinath Srinivas
Francesco Longo
Santo Motta
Giovanni Merlino
Shengjun Wen
Ni Bu
Changan Jiang
Caihong Zhang
Lihua Jiang
Aihui Wang
Cunchen Gao
Tianyu Liu
Pengfei Li
Jing Sun
Aimin Zhou
Ji-Hui Zhang
Xiufen Zou
Lianghong Wu
H. Chen
Jian Cheng
Zhihua Cui

Xiao-Zhi Gao

Guosheng Hao

Quan-Ke Pan

Bin Qian

Xiaoyan Sun

Byungjeong Lee

Woochang Shin

Jaewon Oh

Jong-Myon Kim

Yung-Keun Kwon

Mingjian Zhang

Xiai Yang

Lirong Wang

Xi Luo

Weidong Yang

Weiling Liu

Lanshen Guo

Yunxia Qu

Peng Kai

Song Yang

Xianxia Zhang

Min Zheng

Weiming Yu

Wangjun Xiang

Qing Liu

Xi Luo

Ali Ahmed Adam

Ibrahim Aliskan

Yusuf Altun

Kadir Erkan

Ilker Ustoglu

Levent Ucun

Janset Dasdemir

Xiai Yan

Stefano Ricciardi

Daniel Riccio

Marilena De Marsico

Fabio Narducci

Atsushi Yamashita

Kazunori Onoguchi

Ryuzo Okada

Naghmeh Garmsiri

Lockery Dan

Maddahi Yaser

Kurosh Zareinia

Ramhuzaini Abd
   Rahman

Xiaosong Li

Lei Song

Gang Chen

Yiming Peng

Fan Liu

Jun Zhang

Li Shang

Chunhou Zheng

Jayasudha John Suseela

Soniya Balram

K.J. Shanti

Aravindan Chandrabose

Parul Agarwal

Deepa Anand

Ranjit Biswas

Nobutaka Shimada

Hironobu Fujiyoshi

Giuseppe Vettigli

Francesco Napolitano

Xiao Zhang

Torres-Sospedra Joaquín

Kunikazu Kobayashi

Liangbing Feng

Fuhai Li

Yongsheng Dong

Shuyi Zhang

Yanqiao Zhu

Lei Huang

Yue Zhao

Yunsheng Jiang

Bin Xu

Wei Wang

Jin Wei

Kisha Ni

Yu-Liang Hsu

Che-Wei Lin

Jeen-Shing Wang

Yingke Lei

Jie Gui

Xiaoming Liu

Dong Yang

Jian Yu

Jin Gu

Chenghai Xue

Xiaowo Wang

Xin Feng

Bo Chen

Jianwei Yang

Chao Huang

Weixiang Liu

Qiang Huang

Yanjie Wei

Ao Li

Mingyuan Jiu

Dipankar Das

Gianluca Ippoliti

Lian Liu

Mohammad Bagher
   Bannae Sharifian

Hadi Afsharirad

S. Galvani

Chengdong Wu

Meiju Liu

Aamir Shahzad

Wei Xiong

Toshiaki Kondo

Andrea Prati

Bai Li

Domenico G. Sorrenti

Alessandro Rizzi

Raimondo Schettini

Mengjie Zhang

Gustavo Olague

Umarani Jayaraman

Aditya Nigam

Hunny Mehrotra

Gustavo Souza

Guilherme Barreto

Leandrodos Santos
   Coelho

Carlos Forster

Fernando Von Zuben

Anne Canuto

Jackson Souza

Carmelo Bastos Filho

Daniel Aloise

Sergio P. Santos

Ricardo Fabbri

Fábio Paiva
S.H. Chen
Tsung-Che Chiang
Cheng-Hung Chen
Shih-Hung Wu
Zhifeng Yun
Yanqing Ji
Kai Wang
Je-Ho Park
Junhong Wang
Jifang Pang
Thiran De Silva
Nalin Badara
Shaojing Fan
Chen Li
Qingfeng Li
Liangxu Liu
Rina Su
Hua Yu
Jie Sun
Linhua Zhou
Zhaohong Deng
Pengjiang Qian
Jun Wang
Puneet Gupta
Salim Flora
Jayaputera James
Sherchan Wanita
Helen Paik
Mohammed M. Gaber
Agustinus B. Waluyo
Dat Hoang
Hamid Motahari
Eric Pardede
Tim Ho
Jose A.F. Costa
Qiang Fan
Surya Prakash
Vandana Dixit K.
Saiful Islam
Kamlesh Tiwari
Sandesh Gupta
Zahid Akhtar
Min-Chih Chen
Andreas Konstantinidis

Quanming Zhao
Hongchun Li
Zhengjie Wang
Chong Meng
Lin Cai
Aiyu Zhang
Yang-Won Lee
Young Park
Chulantha Kulasekere
Akalanka Ranundeniya
Junfeng Xia
Min Zhao
Hamid Reza Rashidi
  Kanan
Mehdi Ezoji
Majid Ziaratban
Saeed Mozaffari
Javad Haddadnia
Peyman Moallem
Farzad Towhidkhah
Hamid
  Abrishamimoghaddam
Mohammad Reza
  Pourfard
M.J. Abdollahi Fard
Arana-Arexolaleiba
  Nestor
Carme Julià
Boris Vintimilla
Daniele Ranieri
Antonio De Giorgio
Vito Gallo
Leonarda Carnimeo
Paolo Pannarale
López-Chau Asdrúbal
Jair Cervantes
Debrup Chakraborty
Simon Dacey
Wei-Chiang Hong
Wenyong Dong
Lingling Wang
Hongrun Wu
Chien-Yuan Lai
Md.Kamrul Hasan
Mohammad Kaykobad

Young-Koo Lee
Sungyoung Lee
Chin-Chih Chang
Yuewang
Shinji Inoue
Tomoyuki Ohta
Eitaro Kohno
Alex Muscar
Sorin Ilie
Cosulschi Mirel
Min Chen
Wen Yu
Lopez-Arevalo Ivan
Sabooh Ajaz
Prashan Premaratne
Weimin Huang
Jingwen Wang
Kai Yin
Hong Wang
Yan Fan
Niu Qun
Youqing Wang
Dajun Du
Laurence T. Yang
Laurence Yang
Seng Loke
Syukur Evi
Luis Javier García
  Villalba
Tsutomu Terada
Tomas Sanchez Lopez
Eric Cheng
Battenfeld Oliver
Yokota Masao
Hanemann Sven
Yue Suo
Pao-Ann Hsiung
Kristiansen Lill
Callaghan Victor
Mzamudio Rodriguez
  Victor
Sherif Sakr
Rajiv Ranjan
Cheong Ghil Kim
Philip Chan

Wojtek Goscinski
Jefferson Tan
Bo Zhou
Huiwei Wang
Xiaofeng Chen
Bing Li
Wojtek Goscinski
Samar Zutshi
Rafal Kozik
Tomasz Andrysiak
Marian Cristian
  Mihaescu
Michal Choras
Yanwen Chong
Jinxing Liu
Miguel Gonzalez
  Mendoza
Ta-Yuan Chou
Hui Li
Chao Wu
Kyung DaeKo
Junhong Wang
Guoping Lin
Jiande Sun
Hui Yuan
Qiang Wu
Yannan Ren
Dianxing Liu
M. Sohel Rahman
Dengxin Li
Gerard J. Chang
Weidong Chang
Xulian Hua
Dan Tang
Sandesh Gupta
Uma Rani
Surya Prakash
Narendra Kohli
Meemee Ng
Olesya Kazakova
Vasily Aristarkhov
Ozgur Kaymakci
Xuesen Ma
Qiyue Li
Zhenchun Wei

Xin Wei
Xiangjuan Yao
Ling Wang
Shujuan Jiang
Changhai Nie
He Jiang
Fengfeng Zhou
Zexian Liu
Jian Ren
Xinjiao Gao
Tian-Shun Gao
Han Cheng
Yongbo Wang
Yuangen Yao
Juan Liu
Bing Luo
Zilu Ying
Junying Zeng
Guohui He
Yikui Zhai
Binyu Yi
Zhan Liu
Xiang Ji
Hongyuan Zha
Azzedine Boukerche
Horacio A.B.F. Oliveira
Eduardo F. Nakamura
Antonio A.F. Loureiro
Radhika Nagpal
Jonathan Bachrach
Daeyoung Choi
Woo Yul Kim
Amelia Badica
Fuqing Duan
Hui-Ping Tserng
Ren-Jye Dzeng
Machine Hsie
Milan Cisty
Muhammad Amjad
Muhammad Rashid
Waqas Bangyal
Bo Liu
Xueping Yu
Chenlong Liu
Jikui Shen

Julius Wan
Linlin Shen
Zhou Su
Weiyan Hou
Emil Vassev
Anuparp Boonsongsrikul
Paddy Nixon
Kyung-Suk Lhee
Man Pyo Hong
Vincent C.S. Lee
Yee-Wei Law
Touraj Banirostam
Ho-Quoc-Phuong
Nguyen
Bin Ye
Huijun Li
Xue Sen
Mu Qiao
Xuesen Ma
Weizhen Chun
Qian Zhang
Baosheng Yang
Xuanfang Fei
Fanggao Cui
Xiaoning Song
Dongjun Yu
Bo Li
Huajiang Shao
Ke Gu
Helong Xiao
Wensheng Tang
Andrey Vavilin
Jong Eun Ha
Mun-Ho Jeong
Taeho Kim
Kaushik Deb
Daenyeong Kim
Dongjoong Kang
Hyun-Deok Kang
Hoang-Hon Trinh
Andrey Yakovenko
Dmitry Brazhkin
Sergey Ryabinin
Stanislav Efremov
Andrey Maslennikov

Oleg Sklyarov
Pabitra Mitra
Juan Li
Tiziano Politi
Vitoantonio Bevilacqua
Abdul Rauf
Yuting Yang
Lei Zhao
Shih-Wei Lin
Vincent Li
Chunlu Lai
Qian Wang
Liuzhao Chen
Xiaozhao Zhao
Plaban Bhowmick
Anupam Mandal
Biswajit Das
Pabitra Mitra
Tripti Swarnkar
Yang Dai
Chao Chen
Yi Ma
Emmanuel Camdes
Chenglei Sun
Yinying Wang
Jiangning Song
Ziping Chiang
Vincent Chiang
Xingming Zhao
Chenglei Sun
Francesca Nardone
Angelo Ciaramella
Alessia Albanese
Francesco Napolitano
Guo-Zheng Li
Xu-Ying Liu
Dalong Li
Jonathan Sun
Nan Wang
Yi Yang
Mingwei Li
Wierzbicki Adam
Marcin Czenko
Ha Tran
Jeroen Doumen

Sandro Etalle
Pieter Hartel
Jerryden Hartog
Hai Ren
Xiong Li
Ling Liu
Félix Gómez Mármol
Jih-Gau Juang
He-Sheng Wang
Xin Lu
Kyung-Suk Lhee
Sangyoon Oh
Chisa Takano
Sungwook S. Kim
Junichi Funasaka
Yoko Kamidoi
Dan Wu
Dah-Jing Jwo
Abdollah Shidfar
Reza Pourgholi
Xiujun Zhang
Yan Wang
Kun Yang
Iliya Slavutin
Ling Wang
Huizhong Yang
Ning Li
Tao Ye
Smile Gu
Phalguni Gupta
Guangxu Jin
Huijia Li
Xin Gao
Dan Liu
Zhenyu Xuan
Changbin Du
Mingkun Li
Haiyun Zhang
Baoli Wang
Giuseppe Pappalardo
Huisen Wang
Hai Min
Nalin Bandara
Lin Zhu
Wen Jiang

Can-Yi Lu
Lei Zhang
Jian Lu
Jian Lu
Hong-Jie Yu
Ke Gu
Hangjun Wang
Zhi-De Zhi
Xiaoming Ren
Ben Niu
Hua-Yun Chen
Fuqing Duan
Jing Xu
Marco Falagario
Fabio Sciancalepore
Nicola Epicoco
Wei Zhang
Mu-Chung Chen
Chinyuan Fan
Chun-Wei Lin
Chun-Hao Chen
Lien-Chin Chen
Seiki Inoue
K.R. Seeja
Gurkan Tuna
Cagri Gungor
Qian Zhang
Huanting Feng
Boyun Zhang
Jun Qin
Yang Zhao
Qinghua Cui
Hsiao Piau Ng
Qunfeng Dong
Hailei Zhang
Woochang Hwang
Joe Zhang
Marek Rodny
Bing-Nan Li
Yee-Wei Law
Lu Zhen
Bei Ye
Jl Xu
Pei-Chann Chang
Valeria Gribova

Xiandong Meng
Lasantha Meegahapola
Angel Sappa
Rajivmal Hotra

George Smith
Carlor Ossi
Lijing Tan
Antonio Puliafito

Nojeong Heo
Santosh Bbehera
Giuliana Rotunno

Table of Contents

Evolutionary Learning and Genetic Algorithms

Fuzzy Theory and Models

Swarm Intelligence and Optimization

Kernel Methods and Supporting Vector Machines

Nature Inspired Computing and Optimization

Systems Biology and Computational Biology

Knowledge Discovery and Data Mining

Graph Theory and Algorithms

Machine Learning Theory and Methods

Biomedical Informatics Theory and Methods

Complex Systems Theory and Methods

Pervasive/Ubiquitous Computing Theory and Methods

Intelligent Computing in Bioinformatics

Intelligent Computing in Pattern Recognition

Intelligent Computing in Image Processing

Intelligent Computing in Robotics

Intelligent Computing in Computer Vision

Intelligent Computing in Petri Nets/Transportation Systems

Intelligent Data Fusion and Information Security

Intelligent Sensor Networks

Knowledge Representation/Reasoning and Expert Systems

Special Session on Hybrid Optimization: New Theories and Developments

Special Session on Bio-inspired Computing and Application

PSO Assisted NURB Neural Network Identification

Xia Hong[1] and Sheng Chen[2]

[1] School of Systems Engineering, University of Reading, UK
[2] School of Electronics and Computer Science, University of Southampton, UK, and Faculty of Engineering, King Abdulaziz University, Jeddah 21589, Saudi Arabia
x.hong@reading.ac.uk

Abstract. A system identification algorithm is introduced for Hammerstein systems that are modelled using a non-uniform rational B-spline (NURB) neural network. The proposed algorithm consists of two successive stages. First the shaping parameters in NURB network are estimated using a particle swarm optimization (PSO) procedure. Then the remaining parameters are estimated by the method of the singular value decomposition (SVD). Numerical examples are utilized to demonstrate the efficacy of the proposed approach.

Keywords: B-spline, NURB neural networks, De Boor algorithm, Hammerstein model, pole assignment controller, particle swarm optimization, system identification.

1 Introduction

The Hammerstein model, comprising a nonlinear static functional transformation followed by a linear dynamical model, has been widely researched [2,15,1,8]. The model characterization/representation of the unknown nonlinear static function is fundamental to the identification of Hammerstein model. Various approaches have been developed in order to capture the *a priori* unknown nonlinearity by use of both parametric [16] and nonparametric methods [12,7]. The special structure of Hammerstein models can be exploited to develop hybrid parameter estimation algorithms [1,4].

Both the uniform/nonrational B-spline curve and the non-uniform/rational B-spline (NURB) curve have also been widely used in computer graphics and computer aided geometric design (CAGD) [5]. These curves consist of many polynomial pieces, offering much more versatility than do Bezier curves while maintaining the same advantage of the best conditioning property. NURB is a generalization of the uniform, non-rational B-splines, and offers much more versatility and powerful approximation capability. The NURB neural network possesses a much more powerful modeling capability than a conventional non-rational B-spline neural network because of the extra shaping parameters. This motivates us to propose the use of NURB neural networks to model the nonlinear static function in the Hammerstein system. The PSO [10] constitutes a

D.-S. Huang et al. (Eds.): ICIC 2012, LNCS 7389, pp. 1–9, 2012.
© Springer-Verlag Berlin Heidelberg 2012

population based stochastic optimisation technique, which was inspired by the social behaviour of bird flocks or fish schools. It has been successfully applied to wide-ranging optimisation problems [13,14].

This paper introduces a hybrid system identification consisting two successive stages. In the proposed algorithm the shaping parameters in NURB neural networks are estimated using the particle swarm optimization (PSO) as the first step, in which the mean square error is used as the cost function. In order to satisfy the shaping parameters constraints, the normalisation are applied in PSO as appropriate. Once the shaping parameters are determined. The remaining parameters can be estimated by Bai's overparametrization approach [1] which is used in this study.

2   The Hammerstein System

The Hammerstein system consists of a cascade of two subsystems, a nonlinear memoryless function $\Psi(\bullet)$ as the first subsystem, followed by a linear dynamic part as the second subsystem. The system can be represented by

$$y(t) = \hat{y}(t) + \xi(t)$$
$$= -a_1 y(t-1) - a_2 y(t-2) - ... - a_{n_a} y(t-n_a)$$
$$+ b_1 v(t-1) + ... + b_{n_b} v(t-n_b) + \xi(t) \tag{1}$$
$$v(t-j) = \Psi(u(t-j)), \quad j = 1, ..., n_b \tag{2}$$

where $y(t)$ is the system output and $u(t)$ is the system input. $\xi(t)$ is assumed to be a white noise sequence independent of $u(t)$ with zero mean and variance of σ^2. $v(t)$ is the output of nonlinear subsystem and the input to the linear subsystem. a_j's, b_j's are parameters of the linear subsystem. n_a and n_b are assumed known system output and input lags. Denote $\mathbf{a} = [a_1, ..., a_{n_a}]^T \in \Re^{n_a}$ and $\mathbf{b} = [b_1, ..., b_{n_b}]^T \in \Re^{n_b}$. It is assumed that $A(q^{-1}) = 1 + a_1 q^{-1} + ... + a_{n_a} q^{-n_a}$ and $B(q^{-1}) = b_1 q^{-1} + ... + b_{n_b} q^{-n_b}$ are coprime polynomials of q^{-1}, where q^{-1} denotes the backward shift operator. The gain of the linear subsystem is given by

$$G = \lim_{q \to 1} \frac{B(q^{-1})}{A(q^{-1})} = \frac{\sum_{j=1}^{n_b} b_j}{1 + \sum_{j=1}^{n_a} a_j} \tag{3}$$

The two objectives of the work are that of the system identification and the subsequent controller design for the identified model. The objective of system identification for the above Hammerstein model is that, given an observational input/output data set $D_N = \{y(t), u(t)\}_{t=1}^N$, to identify $\Psi(\bullet)$ and to estimate the parameters a_j, b_j in the linear subsystems. Note that the signals between the two subsystems are unavailable.

Without significantly losing generality the following assumptions are initially made about the problem.

Assumption 1: $\Psi(\bullet)$ is a one to one mapping, i.e. it is an invertible and continuous function.

Assumption 2: $u(t)$ is bounded by $U_{min} < u(t) < U_{max}$, where U_{min} and U_{max} are assumed known finite real values.

3   Modelling of Hammerstein System Using NURB Neural Network

In this work the non-uniform rational B-spline (NURB) neural network is adopted in order to model $\Psi(\bullet)$. De Boor's algorithm is a fast and numerically stable algorithm for evaluating B-spline basis functions [3]. Univariate B-spline basis functions are parameterized by the order of a piecewise polynomial of order k, and also by a knot vector which is a set of values defined on the real line that break it up into a number of intervals. Supposing that there are d basis functions, the knot vector is specified by $(d + k)$ knot values, $\{U_1, U_2, \cdots, U_{d+k}\}$. At each end there are k knots satisfying the condition of being external to the input region, and as a result the number of internal knots is $(d - k)$. Specifically

$$U_1 < U_2 < U_k = U_{min} < U_{k+1} < U_{k+2} < \cdots <$$
$$U_d < U_{max} = U_{d+1} < \cdots < U_{d+k}. \tag{4}$$

Given these predetermined knots, a set of d B-spline basis functions can be formed by using the De Boor recursion [3], given by

$$\mathcal{B}_j^{(0)}(u) = \begin{cases} 1 \text{ if } U_j \leq u < U_{j+1} \\ 0 \qquad \text{otherwise} \end{cases} \tag{5}$$

$$j = 1, \cdots, (d+k)$$

$$\left. \begin{array}{l} \mathcal{B}_j^{(i)}(u) = \frac{u - U_j}{U_{i+j} - U_j} \mathcal{B}_i^{(i-1)}(u) \\ \qquad + \frac{U_{i+j+1} - u}{U_{i+j+1} - U_{j+1}} \mathcal{B}_{j+1}^{(i-1)}(u), \\ j = 1, \cdots, (d+k-i) \end{array} \right\} i = 1, \cdots, k \tag{6}$$

We model $\Psi(\bullet)$ as the NURB neural network in the form of

$$\Psi(u) = \sum_{j=1}^{d} \mathcal{N}_j^{(k)}(u)\omega_j \tag{7}$$

with

$$\mathcal{N}_j^{(k)}(u) = \frac{\lambda_j \mathcal{B}_j^{(k)}(u)}{\sum_{j=1}^{d} \lambda_j \mathcal{B}_j^{(k)}(u)} \tag{8}$$

where ω_j's are weights, $\lambda_j > 0$'s the shaping parameters that are to be determined. Denote $\boldsymbol{\omega} = [\omega_1, \cdots, \omega_d]^T \in \Re^d$. $\boldsymbol{\lambda} = [\lambda_1, \cdots, \lambda_d]^T \in \Re^d$. For uniqueness we set the constraint $\sum_{j=1}^{d} \lambda_j = 1$. Our algorithm involves estimating the weights and the shaping parameters in the NURB model. With specified knots and over the estimation data set D_N, $\boldsymbol{\lambda}, \boldsymbol{\omega}, \mathbf{a}, \mathbf{b}$ may be jointly estimated via

$$\min_{\boldsymbol{\lambda}, \boldsymbol{\omega}, \mathbf{a}, \mathbf{b}} \left\{ J = \sum_{t=1}^{N} (y - \hat{y}(t, \boldsymbol{\lambda}, \boldsymbol{\omega}, \mathbf{a}, \mathbf{b}))^2 \right\} \tag{9}$$

subject to

$$\lambda_j \geq 0, \forall j, \quad \boldsymbol{\lambda}^T \mathbf{1} = 1 \text{ and } G = 1 \tag{10}$$

in which $G = 1$ is imposed for unique solution. We point out that this is still a very difficult nonlinear optimization problem due to the mixed constraints, and this motivates us to propose the following hybrid procedure. It is proposed that the shaping parameters λ_j's are found using the PSO, as the first step of system identification, followed by the estimation of the remaining parameters.

4   The System Identification of Hammerstein System Based on NURB Using PSO

4.1   The Basic Idea

Initially consider using NURB approximation with a specified shape parameter vector $\boldsymbol{\lambda}$, the model predicted output $\hat{y}(t)$ in [??] can be written as

$$\hat{y}(t) = -a_1 y(t-1) - a_2 y(t-2) - ... - a_{n_a} y(t - n_a) + b_1 \sum_{j=1}^{d} \omega_j \mathcal{N}_j^{(k)}(t-1) + ...$$

$$+ b_{n_b} \sum_{j=1}^{d} \omega_j \mathcal{N}_j^{(k)}(t - n_b) \tag{11}$$

Over the estimation data set $D_N = \{y(t), u(t)\}_{t=1}^{N}$, [1] can be rewritten in a linear regression form

$$y(t) = [\mathbf{p}(\mathbf{x}(t))]^T \boldsymbol{\vartheta} + \xi(t) \tag{12}$$

where $\mathbf{x}(t) = [-y(t-1), ..., -y(t-n_a), u(t-1), ..., u(t-n_b)]^T$ is system input vector of observables with assumed known dimension of $(n_a + n_b)$, $\boldsymbol{\vartheta} = [\mathbf{a}^T, (b_1\omega_1), ..., (b_1\omega_d), ...(b_{n_b}\omega_1), ..., (b_{n_b}\omega_{n_b})]^T \in \Re^{n_a + d \cdot n_b}$,

$$\mathbf{p}(\mathbf{x}(t)) = [-y(t-1), ..., -y(t-n_a),$$
$$\mathcal{N}_1^{(k)}(t-1), ..., \mathcal{N}_d^{(k)}(t-1), ...\mathcal{N}_1^{(k)}(t-n_b)$$
$$, ..., \mathcal{N}_d^{(k)}(t-n_b)]^T \tag{13}$$

[12] can be rewritten in the matrix form as

$$\mathbf{y} = \mathbf{P}\boldsymbol{\vartheta} + \boldsymbol{\Xi} \tag{14}$$

where $\mathbf{y} = [y(1), \cdots, y(N)]^T$ is the output vector. $\boldsymbol{\Xi} = [\xi(1), ..., \xi(N)]^T$, and $\mathbf{P}$ is the regression matrix

$$\mathbf{P} = \begin{bmatrix} p_1(\mathbf{x}(1)) & p_2(\mathbf{x}(1)) & \cdots & p_{n_a + d \cdot n_b}(\mathbf{x}(1)) \\ p_1(\mathbf{x}(2)) & p_2(\mathbf{x}(2)) & \cdots & p_{n_a + d \cdot n_b}(\mathbf{x}(2)) \\ \cdots\cdots\cdots\cdots\cdots\cdots\cdots\cdots\cdots\cdots\cdots \\ p_1(\mathbf{x}(N)) & p_2(\mathbf{x}(N)) & \cdots & p_{n_a + d \cdot n_b}(\mathbf{x}(N)) \end{bmatrix} \tag{15}$$

Denote $\mathbf{B} = \mathbf{P}^T\mathbf{P}$. Performing the eigenvalue decomposition $\mathbf{BQ} = \mathbf{Q\Sigma}$, where $\mathbf{\Sigma} = \text{diag}[\sigma_1, ...\sigma_r, 0, \cdots, 0]$ with $\sigma_1 > \sigma_2 > ... > \sigma_r > 0$. $\mathbf{Q} = [\mathbf{q}_1, \cdots, \mathbf{q}_{n_a+d \cdot n_b}]$, followed by truncating the eigenvectors corresponding to zero eigenvalues, we have

$$\boldsymbol{\vartheta}_{LS}^{svd} = \sum_{i=1}^{r} \frac{\mathbf{y}^T\mathbf{P}\mathbf{q}_i}{\sigma_i} \mathbf{q}_i \tag{16}$$

Thus the mean square error can be readily computed from

$$J(\boldsymbol{\lambda}) = [\mathbf{y} - \mathbf{P}\boldsymbol{\vartheta}_{LS}^{svd}]^T[\mathbf{y} - \mathbf{P}\boldsymbol{\vartheta}_{LS}^{svd}]/N. \tag{17}$$

for any specified $\boldsymbol{\lambda}$. Note that it is computationally simple to evaluate $J(\boldsymbol{\lambda})$ due to the fact that the model has a linear in the parameter structure for a given $\boldsymbol{\lambda}$. This suggests that we can optimize $\boldsymbol{\lambda}$ as the first task.

4.2   Particle Swarm Optimisation for Estimating the Shaping Parameters λ_j's

In the following we propose to apply the PSO algorithm [10,11], and aim to solve

$$\boldsymbol{\lambda}_{\text{opt}} = \arg\min_{\boldsymbol{\lambda} \in \prod_{j=1}^d \Lambda_j} J(\boldsymbol{\lambda}), \quad \text{s.t.} \quad \boldsymbol{\lambda}^T\mathbf{1} = 1 \tag{18}$$

where $\mathbf{1}$ denotes a vector of all ones with appropriate dimension.

$$\prod_{j=1}^{d} \Lambda_j = \prod_{j=1}^{d} [0, \ 1] \quad \text{s.t.} \quad \boldsymbol{\lambda}^T\mathbf{1} = 1 \tag{19}$$

defines the search space. A swarm of particles, $\{\boldsymbol{\lambda}_i^{(l)}\}_{i=1}^S$, that represent potential solutions are "flying" in the search space $\prod_{j=1}^d \Lambda_j$, where S is the swarm size and index l denotes the iteration step. The algorithm is summarised as follows.

a) *Swarm initialisation.* Set the iteration index $l = 0$ and randomly generate $\{\boldsymbol{\lambda}_i^{(l)}\}_{i=1}^S$ in the search space $\prod_{j=1}^d \Lambda_j$. These are obtained by randomly set each element of $\{\boldsymbol{\lambda}_i^{(l)}\}_{i=1}^S$ as $rand()$ (denoting the uniform random number between 0 and 1), followed normalizing them by

$$\boldsymbol{\lambda}_i^{(0)} = \boldsymbol{\lambda}_i^{(0)} / \sum_{j=1}^{d} \boldsymbol{\lambda}_i^{(0)}|_j \tag{20}$$

where $\bullet|_j$ denotes the jth element of $\bullet$, so that $\{\boldsymbol{\lambda}_i^{(0)}\}^T\mathbf{1} = 1$ is valid.

b) *Swarm evaluation.* The cost of each particle $\boldsymbol{\lambda}_i^{(l)}$ is obtained as $J(\boldsymbol{\lambda}_i^{(l)})$. Each particle $\boldsymbol{\lambda}_i^{(l)}$ remembers its best position visited so far, denoted as $\mathbf{pb}_i^{(l)}$, which provides the cognitive information. Every particle also knows the best position visited so far among the entire swarm, denoted as $\mathbf{gb}^{(l)}$, which provides the social information. The cognitive information $\{\mathbf{pb}_i^{(l)}\}_{i=1}^S$ and the social information $\mathbf{gb}^{(l)}$ are updated at each iteration:

For $(i = 1; \ i \le S; \ i++)$
    If $(J(\boldsymbol{\lambda}_i^{(l)}) < J(\mathbf{pb}_i^{(l)}))$ $\mathbf{pb}_i^{(l)} = \boldsymbol{\lambda}_i^{(l)}$;
End for;
$i^* = \arg \min_{1 \le i \le S} J(\mathbf{pb}_i^{(l)})$;
If $(J(\mathbf{pb}_{i^*}^{(l)}) < J(\mathbf{gb}^{(l)}))$ $\mathbf{gb}^{(l)} = \mathbf{pb}_{i^*}^{(l)}$;

c) *Swarm update.* Each particle $\boldsymbol{\lambda}_i^{(l)}$ has a velocity, denoted as $\boldsymbol{\gamma}_i^{(l)}$, to direct its "flying". The velocity and position of the ith particle are updated in each iteration according to

$$\boldsymbol{\gamma}_i^{(l+1)} = \mu_0 * \boldsymbol{\gamma}_i^{(l)} + rand() * \mu_1 * (\mathbf{pb}_i^{(l)} - \boldsymbol{\lambda}_i^{(l)})$$
$$+ rand() * \mu_2 * (\mathbf{gb}^{(l)} - \boldsymbol{\lambda}_i^{(l)}), \tag{21}$$
$$\boldsymbol{\lambda}_i^{(l+1)} = \boldsymbol{\lambda}_i^{(l)} + \boldsymbol{\gamma}_i^{(l+1)}, \tag{22}$$

where μ_0 is the inertia weight, μ_1 and μ_2 are the two acceleration coefficients. In order to avoid excessive roaming of particles beyond the search space [9], a velocity space

$$\prod_{j=2}^d \Upsilon_j = \prod_{j=2}^d [-\Upsilon_{j,\max}, \ \Upsilon_{j,\max}] \tag{23}$$

is imposed on $\boldsymbol{\gamma}_i^{(l+1)}$ so that

If $(\boldsymbol{\gamma}_i^{(l+1)}|_j > \Upsilon_{j,\max})$ $\boldsymbol{\gamma}_i^{(l+1)}|_j = \Upsilon_{j,\max}$;
If $(\boldsymbol{\gamma}_i^{(l+1)}|_j < -\Upsilon_{j,\max})$ $\boldsymbol{\gamma}_i^{(l+1)}|_j = -\Upsilon_{j,\max}$;

Moreover, if the velocity as given in equation [21] approaches zero, it is reinitialised proportional to $\Upsilon_{j,\max}$ with a small factor ν

If $(\boldsymbol{\gamma}_i^{(l+1)}|_j == 0)$ $\boldsymbol{\gamma}_i^{(l+1)}|_j = \pm rand() * \nu * \Upsilon_{j,\max}$; \tag{24}

In order to ensure each element of $\boldsymbol{\lambda}_i^{(l+1)}$ that it satisfies the constraint and stays in the space, we modified constraint check in the PSO as follows;

If $(\boldsymbol{\lambda}_i^{(l+1)}|_j < 0)$ $\boldsymbol{\lambda}_i^{(l+1)}|_j = 0$;

then

$$\boldsymbol{\lambda}_i^{(l+1)} = \boldsymbol{\lambda}_i^{(l+1)} / \sum_{j=1}^d \boldsymbol{\lambda}_i^{(l+1)}|_j \tag{25}$$

Note that the normalization step that we introduced here does not affect the cost function value, rather it effectively keeps the solution stay inside the bound.

d) Termination condition check. If the maximum number of iterations, $I_{\max}$, is reached, terminate the algorithm with the solution $\mathbf{gb}^{(I_{\max})}$; otherwise, set $l = l + 1$ and go to Step *b)*.

Ratnaweera and co-authors [14] reported that using a time varying acceleration coefficient (TVAC) enhances the performance of PSO. We adopt this mechanism, in which μ_1 is reduced from 2.5 to 0.5 and μ_2 varies from 0.5 to 2.5 during the iterative procedure:

$$\mu_1 = (0.5 - 2.5) * l/I_{\max} + 2.5,$$
$$\mu_2 = (2.5 - 0.5) * l/I_{\max} + 0.5. \tag{26}$$

The search space as given in equation [19] is defined by the specific problem to be solved, and the velocity limit $\Upsilon_{j,\max}$ is empirically set. An appropriate value of the small control factor ν in equation [24] for avoiding zero velocity is empirically found to be $\nu = 0.1$ for our application.

4.3   Estimating the Parameter Vectors $\omega, \mathbf{a}, \mathbf{b}$ Using ϑ_{LS}^{svd}

In this section we describe the second stage of Bai's two stage identification algorithm [1] which can be used to recover $\omega, \mathbf{a}, \mathbf{b}$ from $\vartheta_{LS}^{svd}(\lambda_{\mathrm{opt}})$ based on the result of PSO above. Our final estimate of $\hat{\mathbf{a}}$, which is simply taken as the subvector of the resultant $\vartheta_{LS}^{svd}(\lambda_{\mathrm{opt}})$, consisting of its first n_a elements.

Rearrange the $(n_a + 1)^{th}$ to $(n_a + (d+1) \times n_b)^{th}$ elements of $\vartheta_{LS}^{svd}(\lambda_{\mathrm{opt}})$ into a matrix

$$\mathbf{M} = \begin{bmatrix} b_1\omega_0 & b_1\omega_1 & \cdots & b_1\omega_d \\ b_2\omega_0 & b_2\omega_1 & \cdots & b_2\omega_d \\ \cdots\cdots\cdots\cdots\cdots\cdots \\ b_{n_b}\omega_0 & b_{n_b}\omega_1 & \cdots & b_{n_b}\omega_d \end{bmatrix} = \mathbf{b}\omega^T \in \Re^{n_b \times (d+1)} \tag{27}$$

The matrix $\mathbf{M}$ has rank 1 and its singular value decomposition is of the form

$$\mathbf{M} = \boldsymbol{\Gamma} \begin{bmatrix} \delta_{\mathbf{M}} & 0 & \cdots & 0 \\ 0 & 0 & \cdots & 0 \\ \cdots\cdots\cdots\cdots \\ 0 & 0 & \cdots & 0 \end{bmatrix} \boldsymbol{\Delta}^T \tag{28}$$

where $\boldsymbol{\Gamma} = [\boldsymbol{\Gamma}_1, ..., \boldsymbol{\Gamma}_{n_b}] \in \Re^{n_b \times n_b}$ and $\boldsymbol{\Delta} = [\boldsymbol{\Delta}_1, ..., \boldsymbol{\Delta}_{d+1}] \in \Re^{(d+1) \times (d+1)}$, where $\boldsymbol{\Gamma}_i$ $(i = 1, .., n_b)$ and Λ_i $(i = 1, ..., (d+1))$ are orthonormal vectors. $\delta_{\mathbf{M}}$ is the sole non-zero singular value of $\mathbf{M}$. $\mathbf{b}$ and ω can be obtained using

$$\hat{\mathbf{b}} = \delta_{\mathbf{M}}\boldsymbol{\Gamma}_1, \quad \hat{\omega} = \boldsymbol{\Delta}_1 \tag{29}$$

followed by

$$\hat{\mathbf{b}} \longleftarrow \beta\hat{\mathbf{b}}, \quad \hat{\omega} \longleftarrow \hat{\omega}/\beta \tag{30}$$

where $\beta = (1 + \sum_{j=1}^{n_a} \hat{a}_j)/(\sum_{j=1}^{n_b} \hat{b}_j)$.

Note that the standard Bai's approach as above may suffer a serious numerical problem that the matrix $\mathbf{M}$ turns out to have rank higher than one, resulting in the parameters estimator far from usable. The more stable modified SVD approach [6] is used in our simulations.

5   An Illustrative Example

A Hammerstein system is simulated, in which the linear subsystem is $A(q^{-1}) = 1-1.2q^{-1}+0.9q^{-2}$, $B(q^{-1}) = 1.7q^{-1}-q^{-2}$, and the nonlinear subsystem $\Psi(u) = 2\mathrm{sign}(u)\sqrt{|u|}$. The variances of the additive noise to the system output is set as 0.01 (low noise) and 0.25 (high noise) respectively. 1000 training data samples $y(t)$ were generated by using [1] and [2], where $u(t)$ was uniformly distributed random variable $u(t) \in [-1.5, 1.5]$. The signal to noise ratio are calculated as 36dB and 22dB respectively. The polynomial degree of B-spline basis functions was set as $k = 2$ (piecewise quadratic). The knots sequence U_j is set as

$$[-3.2, \ -2.4, \ -1.6, \ -0.8, \ -0.05, \ 0, \ 0.05, \ 0.8, \ 1.6, \ 2.4, \ 3.2].$$

The proposed system identification algorithm was carried out. In the modified PSO algorithm, we set $S = 20$, $I_{max} = 20$, $\Upsilon_{j,max} = 0.025$. The resultant 8 NURB basis functions for the two data sets are plotted in Figure 1(a)(b). The nonlinear subsystem obtained with $\sigma^2 = 0.01$ data set was shown in Figure 1(c).(The plot obtained with $\sigma^2 = 0.25$ data set has the same appearance except for the external knots sequences). The modelling results are shown in Table 1 for the linear subsystem.

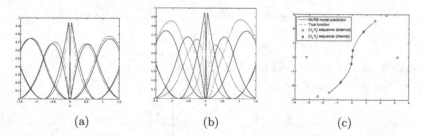

(a)   (b)   (c)

Fig. 1. The resultant B-spline (solid line) and NURB (dotted line) basis functions formed using PSO; (a) $\sigma^2 = 0.01$ and (b) $\sigma^2 = 0.25$, and the modelling result for the nonlinear function $\Psi(u)$ ($\sigma^2 = 0.01$)

Table 1. Results of linear subsystem parameter estimation for two systems

	a_1	a_2	b_1	b_2
True parameter	−1.2	0.9	1.7	−1
Estimate parameters ($\sigma^2 = 0.01$)	−1.2004	0.9004	1.7077	−1.0076
Estimate parameters ($\sigma^2 = 0.25$)	−1.2015	0.9027	1.7424	−1.0412

6   Conclusions

This paper introduced a new system identification algorithm for the Hammer-stein systems based on observational input/output data, using the non-uniform rational B-spline (NURB) neural network. We propose utilising PSO for the estimation of the shaping parameters in NURB neural networks. The efficacy of the proposed approach is demonstrated using illustrative example.

References

1. Bai, E.W., Fu, M.Y.: A Blind Approach to Hammerstein Model Identification. IEEE Transactions on Signal Processing 50(7), 1610–1619 (2002)
2. Billings, S.A., Fakhouri, S.Y.: Nonlinear System Identification Using the Hammerstein Model. International Journal of Systems Science 10, 567–578 (1979)
3. de Boor: A Practical Guide to Splines. Springer, New York (1978)
4. Chaoui, F.Z., Giri, F., Rochdi, Y., Haloua, M., Naitali, A.: System Identification Based Hammerstein Model. International Journal of Control 78(6), 430–442 (2005)
5. Farin, G.: Curves and Surfaces for Comnputer-aided Geometric Design: a Practical Guide. Academic Press, Boston (1994)
6. Goethals, I., Pelckmans, K., Suykens, J.A.K., Moor, B.D.: Identification of MIMO Hammerstein Models Using Least Squares Support Vector Machines. Automatica 41, 1263–1272 (2005)
7. Greblicki, W.: Stochastic Approximation in Nonparametric Identification of Hammerstein Systems. IEEE Transactions on Automatic Control 47(11), 1800–1810 (2002)
8. Greblicki, W., Pawlak, M.: Identification of discrete Hammerstein Systems Using Kernel Regression Estimate. IEEE Transactions on Automatic Control AC-31(1), 74–77 (1986)
9. Guru, S.M., Halgamuge, S.K., Fernando, S.: Particle Swarm Optimisers for Cluster Formation in Wireless Sensor Networks. In: Proc. 2005 Int. Conf. Intelligent Sensors, Sensor Networks and Information Processing, Melbourne, Australia, pp. 319–324 (2005)
10. Kennedy, J., Eberhart, R.: Particle Swarm Optimization. In: Proc. of 1995 IEEE Int. Conf. Neural Networks, Perth, Australia, vol. 4, pp. 1942–1948 (1995)
11. Kennedy, J., Eberhart, R.C.: Swarm Intelligence. Morgan Kaufmann (2001)
12. Lang, Z.Q.: A Nonparametric Polynomial Identification Algorithm for the Hammerstein System. IEEE Transactions on Automatic Control 42, 1435–1441 (1997)
13. van der Merwe, D.W., Engelbrecht, A.P.: Data Clustering Using Particle Swarm Optimization. In: Proc. CEC 2003, Cabberra, Australia, pp. 215–220 (2003)
14. Ratnaweera, A., Halgamuge, S.K., Watson, H.C.: Self-organizing Hierarchical Particle Swarm Optimizer with Time-varying Acceleration Coefficients. IEEE Trans. Evolutionary Computation 8, 240–255 (2004)
15. Stoica, P., Söderström, T.: Instrumental Variable Methods for Identification of Hammerstein Systems. International Journal of Control 35, 459–476 (1982)
16. Verhaegen, M., Westwick, D.: Identifying Mimo Hammerstein Systems in the Context of Subspace Model Identification. International Journal of Control 63(2), 331–349 (1996)

2D Discontinuous Function Approximation
with Real-Valued Grammar-Based Classifier System

Lukasz Cielecki and Olgierd Unold

Institute of Computer Engineering, Control and Robotics
Wroclaw University of Technology
Wybrzeze Wyspianskiego 27
50-370 Wroclaw, Poland
{lukasz.cielecki,olgierd.unold}@pwr.wroc.pl

Abstract. Learning classifier systems (LCSs) are rule-based, evolutionary learning systems. Recently, there is a growing interest among the researchers in exploring LCSs implemented in a real-valued environment, due to its practical applications. This paper describes the use of a real-valued Grammar-based Classifier System (rGCS) in a task of 2D function approximation. rGCS is based on Grammar-based Classifier System (GCS), which was originally used to process context free grammar sentences. In this paper, we propose an extension to rGCS, called Simple Accept Radius (SAR) mechanism, that filters invalid and unexpected input real values. Performance evaluations show that the additional Simple Accept Radius mechanism enables rGCS to accurately approximate 2D discontinuous function. Performance comparisons with another real-valued LCS show that rGCS yields competitive performance.

Keywords: Learning classifier systems, Grammatical inference, Context-free grammar, GCS, 2D function estimation.

1   Introduction

A Learning Classifier System (LCS) is an evolutionary algorithm that operates on a population of rules used to classify an environmental situation. The first learning classifier system was introduced by Holland [6] shortly after he created Genetic Algorithms (GAs) [5]. Many real-world problems are not conveniently expressed using the ternary representation typically used by LCSs (true, false, and don't care symbol). To overcome this limitation, Wilson [11] introduced real-valued XCS classifier system for problems which can be defined by a vector of bounded continuous real-coded variables.

In [2] we introduced a new model of real-valued LCS - the rGCS - to classify real value data. rGCS is based on Grammar-based Classifier System (GCS), which was used to process context free grammar (CFG) sentences [7]. In rGCS terminal rewriting rules were replaced with so-called environment probing rules. That enabled rGCS

D.-S. Huang et al. (Eds.): ICIC 2012, LNCS 7389, pp. 10–17, 2012.
© Springer-Verlag Berlin Heidelberg 2012

. to explore environment described by real values. First rGCS experiments utilized the checkerboard problem test sets. This simple benchmark proved an ability to solve "real-value input" problems and made new kind of data sets available for our system. Following these preliminary experiments some more complex benchmarks were tested, including 2D data, and function classification but prior to that some improvements to the system were applied.

To improve 2D function classification a covering technique was developed. This procedure, implemented earlier in the GCS system, replenishes the population of system's classifiers with new ones created on the y to complete current state of grammar evolution. As a result there is no need to launch a GA and wait for its results [3].

Current improvement to the rGCS - Simple Accept Radius or SAR - brought an ability to filter invalid and unexpected input values during the testing phase of an experiment. In this paper we describe the use of rGCS with SAR mechanism in a task of 2D discontinuous function approximation. The improved system was tested on benchmark functions introduced by Butz in [1], and compared with XCSR system.

2 The rGCS

The main idea behind the rGCS was to extend the space of eligible problems for grammar-based classifier system. Our original system (GCS - described in [7]) was virtually limited to natural and formal language based tasks. Improved rGCS is able to accept any data stored as a vector of real values which enables us to work with wide range of datasets.

Our rGCS operates in an environment formed by CYK table and evolves a CNF grammar. An experiment is divided into some cycles. Each cycle includes evolving new grammar, testing every single example from the training set and modifying existing rules' population. To fit to the real value input vectors we introduced environment probing rules which replace GCS standard terminal rewriting ones. They cooperate with regular rules during the parsing phase of the learning process.

Automatic learning of CFG is realized with so-called grammatical inference from text [4]. According to this technique system learns using a training set that in this case consists of sentences both syntactically correct and incorrect (training or learning phase). Grammar which accepts correct sentences and rejects incorrect ones is able to classify unseen so far sentences from a test set (during the test phase). Cocke-Younger-Kasami (CYK) parser, which operates in $\Theta(n^3 \cdot |G|)$ time (where n is the length of the parsed string and $|G|$ is the size of the CNF grammar G) [12], is used to parse sentences from the sets. Environment of classifier system is substituted by an array of CYK parser. Classifier system matches the rules according to the current environmental state (state of parsing) and generates an action (or set of actions in GCS) pushing the parsing process toward the complete derivation of the sentence.

The discovery component of the rGCS differs from the standard. It extends LCS with "covering" procedure that adds some rules useful in the current state of parsing (see [8]) and a "fertility" technique which preserves rules' dependencies in the parsing tree [10].

Environment Probing Rules

Environment probing rules are to put into the very first row of the CYK table as they sense the input data and translate real numbers into the non-terminal symbols. Each rule has the form of:

$$A \rightarrow f,$$

where:

A is a non-terminal symbol, f is a real number value.

f value (rule's factor or position) is used during the matching process and the non-terminal A is to put into the first row of the CYK table.

Regular Grammar Rules

These rules are identical to the ones used in the classic GCS. They are used in the CYK parsing process and the GA phase. They are in the form of:

$$B \rightarrow BC,$$

where A, B and C are non-terminal symbols.

Matching Phase

During the matching phase a bundle of non-terminals is defined to put into the first row of the CYK table, achieving the goal of translating real input values into the string of symbols capable of parsing.

First a list of distances between the element of an input vector (real number value) and each rule's factor is created and then simply the nearest rule is selected. As the result always one rule is put into the CYK cell. This scheme used during training process uses only a single real value from the rule - a point with no accepting range around it explicitly labeled. That approach differs from the one adopted in Wilson's XCSR [10] where several interval predicates are defined to "catch" input values located inside its boundaries. The main difference is that in the rGCS learning the single value method always chooses at least one rule - no matter where the input value is located. This means that even with limited number of environment probing rules there are no input values that are left unrecognized. This can be referred as some kind of interpolation of an input space.

Strategy where each rule can accept any real value without looking at the distance between them causes some problems however. In many problems only some specific subset of points in the input space (or in the certain dimension of an input space) can exist in the positive examples. Therefore we need to modify the matching phase to stop input values not included in that subset from being accepted. To achieve that we set up a Single Accept Radius (SAR) parameter which describes every environment probing rule. It is set up and updated (and then not yet used) during the matching phase of the learning process in the rGCS. Simple accept radius is the longest distance between rule's factor and any real value accepted by it that belongs to the positive example. During the test phase we accept only these real values that are located within the learned accept radius. This protects us from accepting values that are outside the area which our terminal rewriting rule is supposed to be employed for.

Matching regular rules used in the CYK parsing process follows the CYK algorithm procedure and is same as in GCS.

Rules Adjustments and Evolution

Environment probing rules adjustments were widely discussed in [2]. As the environment probing rules match the input vector, some data about the environment is collected to find the best possible value of each rule's real-value factor. The aim is to distribute environment probing rules over the input space. Modified Kohonen neural network adaptation techniques were implemented in the rGCS to allow classifiers' real-value factors to automatically tune to the environment.

Regular grammar rules are evolved just like in the classic GCS system during the evolutionary process. GA is launched at the end of each learning cycle. Fitness evaluation uses a fertility measurement technique (see [10] for discussion), for the rules that were present in any complete parsing tree generated during the cycle. This technique tries to keep all dependencies between rules in the parsing tree by promoting classifiers with large number of descendants.

3 Function Estimation Experiments

In [1] Butz presented some challenging two dimensional discontinuous functions to compare his modification of the XCSR with the original system [11]. The task was to decide whether a point at given coordinates belongs to the function or not. In this paper we test two Butz functions with rGCS classifier.

$$f_1(x,y) = \frac{(x*3)\%3}{3} + \frac{(y*3)\%3}{3} \qquad f_2(x,y) = \frac{((x+y)*2)\%4}{6} \qquad (1)$$

where the % is a modulo operator.

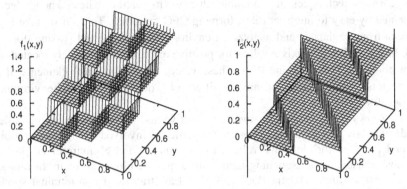

Fig. 1. Butz functions: the axis-parallel step function f_1 and the axis-diagonal step function f_2

During these experiments Simple Accept Radius (SAR) technique was employed in the rGCS for the very first time. For instance for function f_1 and the Z axis all values from positive examples are set exactly at five certain points (0, 1/3, 2/3, 1, 4/3) - therefore accept radius for perfectly adjusted (located at these points during learning) rules is equal to 0. That means that during the testing stage only these five real values will be translated into the corresponding non-terminal symbols while all other values will be simply rejected. This filtering in an environment probing step is a key to success in the correct classification but to achieve a final recognition an additional task must be done by regular rules of the rGCS and its CYK table.

To compare with Butz results achieved by his improved XCSR we performed a set of experiments with different population sizes. Fig. 2 shows minimum, mean and maximum fitness error (percentage of incorrectly recognized examples) averaged over 20 individual runs. Estimated error rate (average 0.03 at 150 regular classifiers, best result: 0.01) is rewarding and comparable with Butz results [1] with XCSR (best results: 1.00 with original system and 0.01 with modified one). Interestingly we have not noticed any important change to the grammar evolution time (always between 170 and 220 cycles) nor the best results (minimum error rate equal to 0.01) of the single experiment while changing the population size. On the other hand mean averaged over 20 runs is much better when employing more regular set of classifiers. That means that it is still possible to evolve robust grammars even with small classifier population but having it bigger raises the chances of rGCS to succeed. There is however always a minimum population size below which it is absolutely impossible to evolve a proper grammar. For instance during f_2 function estimation we observed that 50 regular rules was not enough to evolve grammar with 0.03 error rate (Fig. 2.a). For bigger grammars (Fig. 2.b and 2.c) it was still possible to achieve 0.01 error rate which again corresponds to Butz's results.

3.1   Induced Grammar Analysis

Let's have a closer look at how rGCS actually works. In contrast to many other machine learning techniques it's possible due to its nature where knowledge is represented by easy to interpret rules, forming CFG grammar. To do it we take some sentences from f_2 dataset and analyze an environment probing and parsing done by rGCS. In the f_2 function axis z values for positive examples may be only located at 4 certain points: 0, 0.25, 0.5 and 0.75. These values form third (last) elements of all input vectors in the dataset and are classified by a dedicated class of environment probing rules.

Fig. 3 shows rGCS actions for two true positive and one true negative classifications depicted by parsing trees, important properties of involved environment probing rules and input vector elements' values. In examples 1 and 8 environment probing rules have translated corresponding input values that are closest to theirs factors and fit into learned Simple Accept Radius (SAR). Last translated non-terminal symbol (presented most to the right in the bottom row of each parsing tree) always

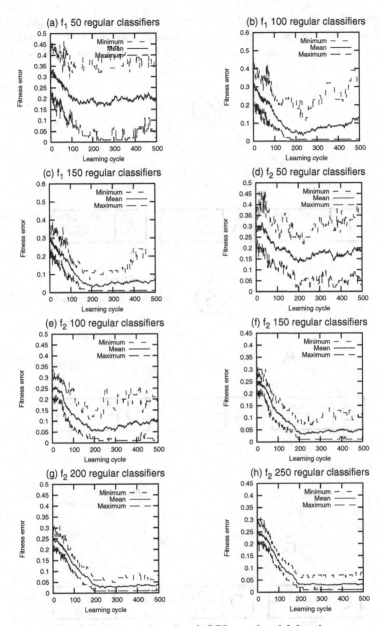

Fig. 2. Experimental results of rGCS over f_1 and f_2 functions

corresponds to the axis z value of an input vector and therefore dedicated rules' accept radius is set strictly to 0 (in the learning phase for f2 function all positive examples were strictly located as 4 points mentioned above, giving us zero-tolerance rules for z-axis). There is however no rule in sentence 6 that accepts last input value (0.2237 on axis z surely doesn't belong to f2 and SAR won't allow any rule to accept the value).

Consequently there is no rule which can let the parsing to complete and put a starting symbol S at the root of the parsing tree. That classification is therefore considered as (true) negative. Classifications of examples 1 and 8 with starting symbol S at the root are (true) positive ones.

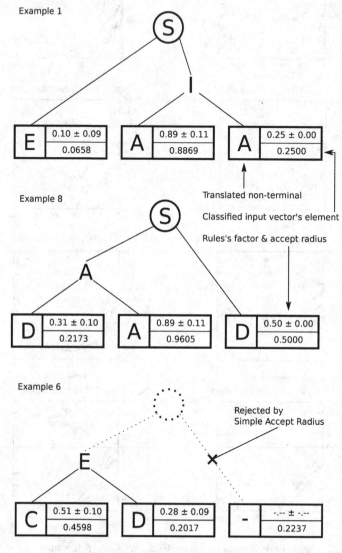

Fig. 3. Parsing trees for function f_2

4   Conclusion

In this paper real-valued Grammar-based Classifier System was extended by Simple Accept Radius mechanism. Thanks to use SAR the system is able to reject input real values that are located outside the learned accept radius. Experimental results showed that the improvement enabled rGCS to solve discontinuous function approximation problems with low fitness error.

However there are some areas where future development seems to be necessary. That includes building more sophisticated environment probing rules with complex accept conditions working in more than one dimensions and therefore checking more then one element of an input vector simultaneously. A number of other mechanisms are being considered to improve grammar evolution.

References

1. Butz, M.: Kernel-based, Ellipsoidal Conditions in the Real-valued XCS Classifier System. In: Proceedings of the 2005 Conference on Genetic and Evolutionary Computation, pp. 1835–1842. ACM (2005)
2. Cielecki, L., Unold, O.: Real-valued GCS Classifier System. International Journal of Applied Mathematics and Computer Science 17(4), 539–547 (2007)
3. Cielecki, L., Unold, O.: Modified Himmelblau Function Classification with RGCS System. In: Proc. of 8th International Conference on Hybrid Intelligent Systems, HIS 2008, pp. 879–884. IEEE (2008)
4. Gold, E.: Language Identification in the Limit. Information and Control 10(5), 447–474 (1967)
5. Holland, J.: Adaptation in Natural and Artificial Systems: An Introductory Analysis with Applications to Biology, Control, and Artificial Intelligence. U. Michigan Press (1975)
6. Holland, J.: Adaptation. In: Rosen, R., Snell, F. (eds.) Progress in Theoretical Biology. Academic Press (1976)
7. Unold, O.: Context-free Grammar Induction with Grammar-based Classifier System. Archives of Control Science 15(4), 681–690 (2005)
8. Unold, O., Cielecki, L.: Grammar-based Classifier System. In: Issues in Intelligent Systems: Paradigms, pp. 273–286 (2005)
9. Unold, O., Cielecki, L.: How to use Crowding Selection in Grammar-based Classifier System. In: Proceedings of 5th International Conference on Intelligent Systems Design and Applications, ISDA 2005, pp. 124–129. IEEE (2005)
10. Unold, O.: Playing a Toy-Grammar with GCS. In: Mira, J., Álvarez, J.R. (eds.) IWINAC 2005, Part II. LNCS, vol. 3562, pp. 300–309. Springer, Heidelberg (2005)
11. Wilson, S.W.: Get Real! XCS with Continuous-Valued Inputs. In: Lanzi, P.L., Stolzmann, W., Wilson, S.W. (eds.) IWLCS 1999. LNCS (LNAI), vol. 1813, pp. 209–219. Springer, Heidelberg (2000)
12. Younger, D.: Recognition and Parsing of Context-free Languages in Time n3. Information and Control 10(2), 189–208 (1967)

A Scatter Search Methodology
for the Aircraft Conflict Resolution Problem

Zhi-Zeng Li, Xue-Yan Song, Ji-Zhou Sun, and Zhao-Tong Huang

School of Computer Science and Technology, Tianjin University, Tianjin, China
lzz0613@126.com

Abstract. Accurate conflict resolution is important to improve traffic capacity and safety. In this paper, it shows how the evolutionary approach called scatter search(SS) can be used to solve the problem of aircraft conflict under the free flight conditions. The mathematic model of the conflict resolution problem is based on the path optimization problem. In order to justify the choice of SS, the paper describes the improvements that were used for solving the conflict resolution problem after a brief description of SS algorithm. A large number of simulation results show that: comparing with genetic algorithm, SS can reduce the running time about 15s to get the equivalent effect. In the same time, the total cost is reduced and the fairness between the airlines is taken into account.

Keywords: conflict resolution, free flight, scatter search, air traffic.

1   Introduction

The increasing development of air traffic in the limited and crowd airspace stimulates people to explore the new way to increase the utilization of airspace. Recent advances in navigation and data communication technologies make it feasible for individual aircraft to plan and fly their trajectories in the presence of other aircraft in the airspace. This way, individual aircraft can take advantage of the atmospheric and traffic conditions to optimally plan their paths[3][4]. Therefore, how to solve the conflicts fast and accuracy is the key technology to ensure air safe and achieve free flight.

The problem of conflict resolution has been researched for a long time. In total, over 60 different methods have been proposed by various researchers to address the conflict detection and resolution[1][2]. Karl D proposes the geometric optimization approach to aircraft conflict resolution[5]. Using the geometric characteristics of aircraft trajectories and information on aircraft position, speed and heading, the intuitive geometric can achieve more accurate conflict resolution. In response to the large-scale conflict, Stefan and Nicolas propose the multi-agent concept to solve the conflict[6][7]. The technology of multi-agent is suitable for the large-scale, complex structures and real-time problems. The key feature is the negotiation mechanism between agents which is effective to the multi-aircraft conflict.

Optimization algorithms, like genetic algorithm[8], are applied in the conflict resolution problem. It is based on the path planning and has its efficiency to solve

D.-S. Huang et al. (Eds.): ICIC 2012, LNCS 7389, pp. 18–24, 2012.
© Springer-Verlag Berlin Heidelberg 2012

aircraft conflicts. The authors note that the problem is simplified, but it still has common constraints and objectives.

In this paper we propose a novel, effective and efficient approach to solve the problem of conflict resolution, namely, the SS algorithm[9][10]. Like GA, SS is also based the population. In addition, SS algorithm is less dependent on the randomness of the search process, but the use of the dispersion-convergence-gather intelligence iteration mechanism. In comparison with the existing genetic algorithm, the proposed approach improves both running time and fairness between airlines.

In the next section the paper introduces the aircraft conflict resolution problem. Section 3 describes the SS implementation and section 4 contains the results of the simulation examples. Finally it concludes in section 5 with some views on the success of this research.

2   Problem Definition

Conflict resolution can be described as: when flights fly into the detection sector, according to the comparison of intended tracks, there may be appearing the risk of conflict in a moment. The Optimization algorithm makes the re-planning of the track to avoid conflict, as show in Fig1:

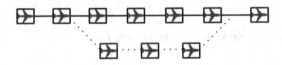

Fig. 1. Track changing diagram

2.1   Flight Safety Constraints

Due to the particularity of the air traffic, the relevant departments have expressly provided flight safety interval. In this paper, aircrafts take the free flight strategy in the cruise phase and fly in the same altitude. Aircraft can only make the flight angle be variation to achieve conflict resolution in case of emergency. The horizontal interval must be not less than $20\,km$.

Safety requirements:

$$d = \sqrt{\left(x_i - x_j\right)^2 + \left(y_i - y_j\right)^2} \geq 20km \tag{1}$$

2.2   The Objective Function

1. Minimum of the gap of distance between the point where the aircraft leave the detection sector scheduled and the point where the aircraft leave the detection sector actually

$$S_1 = e^{\frac{\sum_{i=1}^{n} d_i^2}{K}} \tag{2}$$

$d_i = \sqrt{(x_i - x_j)^2 + (y_i - y_j)^2}$ is the distance between the two points. K is a constant, namely the size of the initial population.

2. Minimum of the total voyage.

$$S_2 = \min \sum_{i=1}^{n} \sum_{j=1}^{m} s_{ij} \tag{3}$$

n expresses the number of aircraft. m is the number of point. s_{ij} expresses the more voyage of the i aircraft at the point j.

3. Minimum of the gap of total voyage between the aircrafts.

$$S_3 = \min |s_i - s_j| \tag{4}$$

$s_{i(j)} = \sum_{i(j)=1}^{19} \sqrt{(x_{i(j)+1} - x_{i(j)})^2 + (y_{i(j)+1} - y_{i(j)})^2}$, s_i is the total voyage of the aircraft i and s_j is the total voyage of the aircraft j.

3   Scatter Search Algorithm

The SS algorithm is a kind of optimization algorithm to solve the problem of decision planning and constraint policy. The basic idea: build the reference set in accordance with the principle of diversity and get the better solution set through the operations of the reference set, such as selection, combination, mutation and so on.

The SS algorithm consists of five main systematic sub methods: diversification generation method; reference set generation method; subset generation method; solution combination method; reference set update method.

3.1   Population Initialization

The following are the strategies of gene encoding: in a 200*200 km detection sector, the aircraft's flight path is divided into 20 sections. At each node position, the aircraft will take action; taking into account the passengers' comfort and safety, the paper sets the aircraft movements of three kinds:

1.turn left 30 degrees and be coded as 01;
2.turn right 30 degrees and be coded as 10;
3.keep the original direction and be coded as 00 or 11.
Therefore, the length of each binary string of code is 2*20=40. Like Fig2

| 01 | 11 | 00 | .... | 10 | 11 | 11 |

Fig. 2. The gene encoding of aircraft track

3.2   Reference Set Generation

In order to achieve the safety constrains, the randomly generated solution set shall be screened by the equation (1). Then it can get the feasible solution set to achieve the population initialization.

In the SS algorithm, the diversity solution set $R1$ and the high-quality solution set $R2$ together constitute the reference set R .

The diversity solution set generation: in order to maintain the diversity of the solution, the initial solution set is divided into n groups averagely according to the corresponding decimal value of solution. Then select one solution randomly from each group and the final n solutions that have been selected constitute the diversity solution set.

The high-quality solution set: First of all, according to the economy constrains, formula (2),(3), select $2n$ solutions. Then from the $2n$ solutions that have been chosen, according to the fairness constrains (formula (4)), select the n best solution to be the high-quality solution set.

3.3   Subset Generation

The size of the subset is 2 and every two solutions should combine in one subset. So if the number of reference set is B, the number of subset is B*(B-1)/2.

3.4   Solutions Combination Method

(1) Crossover: it takes the single-point crossover method. Select the point where the binary code of the first aircraft track is end as the cross point. Like Fig3
(2) Mutation: select the mutation point randomly and change the code. Like Fig4

Fig. 3. Crossover   **Fig. 4.** Mutation

3.5   Reference Set Update Method

It can determine that whether the new solutions can replace the solutions in the high-quality set. If it can replace, return to the step of reference set generation. Until there are no new solutions can replace the solutions in the high-quality set.

4   Performance Evaluation

In the Fig5, there is a $200*200km$ detection sector. If there is no any intervention, in the black dashed box the horizontal interval of the two aircrafts will be less than $20km$ and have a risk of collision.

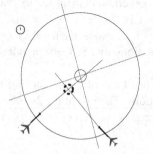

Fig. 5. Aircraft collision model

In this paper, establish the simulation system of aircraft conflict resolution using Matlab. The PC for simulation: CPU: 2.4GHz Core(TM)4; memory: 2GB. In SS, the size of reference set is 20. the initial population is 50,100,200 respectively. each instance we run 10 times and select the best results, as shown in Table1.

From the Table1, we can see that

Table 1. Result comparison of different methods'

Instance	The size of initial population	The number of iteration	GA			SS		
			Economic constraints	Fairness gap	Running time	Economic constraints	Fairness gap	Running time
A-P50-G20	50	20	7.8371	4.3	4s	7.2384	3.8	4
A-P50-G40	50	40	8.6231	3.7	7s	8.1363	4.9	9
A-P50-G60	50	60	8.7540	4.6	11s	8.6542	4.2	13
A-P50-G80	50	80	8.6913	3.9	14s			
B-P100-G20	100	20	6.3417	2.4	6s	7.3677	3.2	7s
B-P100-G40	100	40	7.6243	2.7	11s	9.7273	2.9	13s
B-P100-G60	100	60	9.5392	2.6	17s			
B-P100-G80	100	80	9.7365	2.3	24s			
C-P200-G20	200	20	6.5461	2.4	8s	7.7134	2.1	8s
C-P200-G40	200	40	7.5470	2.3	13s	8.9387	2.6	14s
C-P200-G60	200	60	8.7412	2.1	19s	9.9011	2.0	20s
C-P200-G80	200	80	9.1915	1.9	26s			
C-P200-G100	200	100	9.8624	2.6	34s			

(1)GA will run until reach the number of iteration, but the SS will stop when the new solutions can't replace the solutions in the reference set.

(2)The goal of the economy constrains we achieve using the SS algorithm is almost equal with the effect using the GA. It can fully demonstrate the effectiveness of the SS algorithm.

(3)The fairness between airlines not only decreases but also tends to balance. The fairness is improved.

(4)According to the running time, it just spend 13s to get the best solution (genetic algorithm requires 24s). It fully demonstrates the high efficiency of the algorithm in this paper.

Based on the instance of B-P-100, the Fig6 is the track using the SS algorithm and the Fig7 is the track using the genetic algorithm. The Figs can directly perceive the problem of aircraft conflict resolution through the senses.

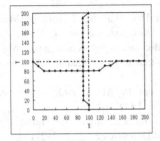   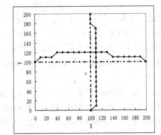

Fig. 6. The track after relief using SS   **Fig. 7.** The track after relief using GA

In the final, simulate the conflict resolution of three-aircraft using the SS.

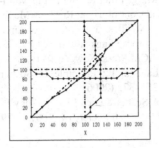

Fig. 8. The three aircrafts' track after relief

From the Fig8, the SS is suitable for the conflict resolution of many aircrafts. So there are some reasons to say SS has some advantages for the conflict resolution problem.

5   Conclusion

A new approach to aircraft conflict resolution, called scatter search algorithm, has been presented. Using the approach in this paper, the conflict resolutions are optimal in the sense that they not only achieve the equal effect that is got by using the genetic algorithm, but also spend less time and achieve the balance between airlines. Especially, the less time can meet the need the requirement of real-time in air traffic. The performance of the algorithm has been illustrated with simulation examples.

Acknowledgements. This study is supported by grants from National Natural Science Foundation of China and Civil Aviation Administration of China(61039001) and Tianjin Municipal Science and Technology Commission(11ZCKFGX04200).

References

1. Kuchar, J.K., Yang, L.C.: A Review of Conflict Detection and Resolution Modeling Methods. IEEE Transactions on Intelligent Transportation Systems 1(4), 179–189 (2000)
2. Kuchar, J.K., Yang, L.C.: Survey of Conflict Detection and Resolution Modeling Methods. In: AIAA Guidance, Navigation, and Control Conference, New Orleans, LA, August 11-13 (1997)
3. Schulz, R., Shaner, D., Zhao, Y.: Free Flight Concept. In: AIAA Guidance, Navigation, and Control Conference, New Orleans, LA, August 11-13 (1997)
4. Menon, P.K., Sweriduk, G.D.: Optimal Strategies for Free-flight Air Traffic Conflict Resolution. AIAA Journal of Guidance, Control, and Dynamics 22(2), 202–211 (1999)
5. Bilimoria, K.D.: A Geometric Optimization Approach to Aircraft Conflict Resolution. In: AIAA Guidance, Navigation, and Control Conference, Denver, CO, August 14-17 (2000)
6. Tomlin, C., Pappas, G.J., Sastry, S.S.: Conflict Resolution for Air Traffic Management: A Study in Multiagent Hybrid Systems. IEEE Transactions on Automatic Control 43(4), 509–521 (1998)
7. Resmerita, S., Heymann, M.: Conflict Resolution in Multi-Agent Systems. In: 42nd IEEE Conference on Decision and Control, Hyatt Regency Maui, December 9-12 (2003)
8. Durand, N., Alliot, J.M., Noailles, J.: Automatic Aircraft Conflict Resolution Using Genetic Algorithms. In: Proceedings of the 1996 ACM Symposium on Applied Computing, pp. 289–298. ACM Press, New York (1998)
9. Marti, R., Laguna, M., Laguna, F.: Principle of Scatter Search. European Journal of Operationl Research 169, 359–372 (2006)
10. Burke, E.K., Curtois, T.: A Scatter Search Methodology for the Nurse Rostering Problem. Journal of the Operational Research Society 61, 1667–1679 (2010)

Self-adaptive Differential Evolution Based Multi-objective Optimization Incorporating Local Search and Indicator-Based Selection

Datong Xie[1,2], Lixin Ding[1], Shenwen Wang[1], Zhaolu Guo[1], Yurong Hu[1],
and Chengwang Xie[3]

[1] State Key Laboratory of Software Engineering, School of Computer, Wuhan University,
Wuhan, 430000 P.R. China
[2] Department of Information Management Engineering, Fujian Commercial College, Fuzhou,
350012 P.R. China
[3] East China Jiao Tong University, Nanchang, 330000 P.R. China
{2010,lxding,wangshenwen}@whu.edu.cn,
{gz1990137,yuronghu118}@gmail.com, chengwangxie@163.com

Abstract. As an efficient and effective evolutionary algorithm, Differential evolution (DE) has received ever-increasing attention over recent years. However, how to make DE suitable for multi-objective optimization is still worth further studying. Moreover, various means from different perspectives are promising to promote the performance of the algorithm. In this study, we propose a novel multi-objective evolutionary algorithm, ILSDEMO, which incorporates indicator-based selection and local search with a self-adaptive DE. In this algorithm, we also use orthogonal design to initialize the population. In addition, the k-nearest neighbor rule is employed to eliminate the most crowded solution while a new solution is ready to join the archive population. The performance of ILSDEMO is investigated on three test instances in terms of three indicators. Compared with NSGAII, IBEA, and DEMO, the results indicate that ILSDEMO can approximate the true Pareto front more accurately and evenly.

Keywords: multi-objective optimization, self-adaptive differential evolution, orthogonal design, indicator-based selection, local search, k-nearest neighbor.

1   Introduction

In multi-objective optimization, a single method usually cannot perform well in terms of all performance indicators [1] such as convergence, uniformity, time costs, stability, etc. In order to get an overall promotion in multiple indicators, various means from different perspectives are worth trying. It is preferable to employ simple and efficient variation operators, such as Differential evolution (DE) [2], to improve the convergence performance. Several well-known DE-based multi-objective evolutionary algorithms such as PDE, DEMO, GDE3, ε-MyDE [3] had been developed. However, most of them use "DE/rand/n/bin". E. Mezura-Montes [4]

D.-S. Huang et al. (Eds.): ICIC 2012, LNCS 7389, pp. 25–33, 2012.
© Springer-Verlag Berlin Heidelberg 2012

experimentally validated that the most competitive one of 8 DE variants based on the quality and robustness of the results, regardless the characteristics of the problem to be solved, was "DE/best/1/bin". In addition, using different fitness assignment strategies and selection operators to adjust the selection pressure or employing local search operators to improve the accuracy of the results are also popular in recent years [5,6]. Recently, self-adaptive parameters and strategies have also received increasing attention due to the problem-dependent and stage-dependent parameters and strategies.

Based on the above considerations, we extend the two variants of DE ("DE/Best/1/bin" and "DE/Best/1/exp") and present a novel method, ILSDEMO, to solve multi-objective problems. More specifically, the proposed approach incorporates orthogonal design, local search, indicator-based selection [7], k-nearest neighbor rule with a self-adaptive DE.

The remainder of the paper is organized as follows. In section 2, we describe the proposed algorithm. Then, in section 3, the performance of the algorithm is investigated experimentally on three test instances in terms of three indicators, and the results are compared with those of NSGAII, IBEA, and DEMO. Finally, conclusions are drawn in section 4.

2   The Proposed Method

2.1   Population Initialization

The state of the initial population usually has an effect on the results. Experimental design methods such as uniform design and orthogonal design (OD) are statistically sound [8] and popular in many areas. OD had been used successfully in evolutionary algorithms [8,9,10]. In our method, OD is utilized to generate an initial population where the individuals are scattered uniformly over the feasible solution space, so that the algorithm can scan the search space evenly and locate the promising solutions quickly.

2.2   The Variation Operators

Exploration and Exploitation capabilities mainly propelled by variation operators are equally important in multi-objective evolutionary algorithms. They are realized via self-adaptive DE and a local search operator respectively in our method.

1) The Self-adaptive Differential Evolution

In our self-adaptive DE, not only the two control parameters but also the strategies are determined self-adaptively to avoid early stagnation. More precisely, the control parameters and the strategy are encoded into the individuals and undergo variation. Initially, the scaling factor and crossover probability are randomly initialized within [0.2, 0.8] and [0.7, 1.0] respectively. Meanwhile, one of the strategies, DE/best/1/bin or DE/best/1/exp, is randomly assigned to every individual. At the end of each iteration, the means and variances of two parameter values collected from the improved individuals are calculated to build two Gaussian models. The unimproved individuals will be assigned new parameter values by sampling the Gaussian models. Moreover, their strategies will be reinitialized according to the probabilities of the two strategies improving the individuals.

2) Local Search

The primary purpose of local search methods is to enhance the exploitation capability of the algorithm. Previous researches [1,11] had been focused on local search in multi-objective optimization. Different from other local search methods in the literature of multi-objective optimization, we use q-Gaussian mutation [12] to perform exploitation once the trial individual is better than the target individual. In q-Gaussian probability density function, the parameter q can control the shape of the q-Gaussian distribution. As described in [12], larger values of q result in longer tails of the q-Gaussian distribution. Thus, q-Gaussian mutation can adjust the range of local search with different values of q.

In our algorithm, if the individual obtained by local search dominates the target individual, the latter will be replaced by the former. Otherwise, the target individual remains unchanged. In either case, the trial individual should be used to update the archive population in order to avoid neglecting any possible improvement on the distribution of the archive population.

2.3   The Selection Operator

In this study, we utilize indicator-based selection to generate a new parent population based on the union of the old parent population and the archive population. First, the duplicated individuals are eliminated from the union. If the number of the remaining individuals is not beyond the parent population size, all the individuals will be propagated into the new parent population. Moreover, the new parent population may need to be supplemented by some new initialized individuals. Otherwise, the indicator-based selection will work. Specifically, indicator values of the remaining individuals will be worked out, and one extreme point for each objective will be ensured to survive to the new parent population via assigning small enough indicator values to them. Then, an individual with a maximum indicator value is deleted from the union, and the indicator values of the other individuals are updated concurrently. The procedure is repeated until the number of the remaining individuals is equal to the parent population size.

2.4   The Updating Rule of the Archive Population

Although indicator-based selection integrating with the preservation of extreme solutions is an effective method in favor of finding a well-distributed solution set, it still cannot obtain ideal uniformity as well as some other algorithms such as SPEA2. Therefore, we will introduce the k-nearest neighbor rule to update the archive population. In detail, when the number of non-dominated solutions exceeds the archive population size, the cumulative Euclidean distance to its k nearest neighbors for every non-dominated solution will be calculated. Here k is set to $2 * (m - 1)$ where m is the objective number. Then, the individual with the smallest distance is removed.

2.5   The Framework of the Proposed Algorithm

Based on the above design, the procedure of the proposed ILSDEMO algorithm for solving multi-objective problems is summarized in Algorithm 1.

Algorithm 1 ILSDEMO

```
1: Set n=0. Initialize Pⁿ{(Xᵢ, Fᵢ, CRᵢ, Sᵢ),i∈[1, NP]}
using OD and Evaluate Pⁿ
2: Copy the non-dominated solutions to the archive
population PA.
3: While the termination criteria are not satisfied do
4:    Set n = n + 1
5:    for i=1 to NP do
6:        select r1≠r2 ∈{1, 2, …, NP} randomly
7:        best∈{1, 2, …, fsize} //fsize is the size of the
non-dominated set in PA.
8:        Perform program segment 1 or 2 using Sᵢ, Fᵢ, CRᵢ
9:        Evaluate Vᵢ
10:       If Vᵢ is non-dominated by Xᵢ then
11:           If Vᵢ dominates Xᵢ then
12:              Mark Xᵢ as an improved individual
13:           End if
14:           Perform q-Gaussian mutation on Vᵢ
15:           If Vᵢ is non-dominated with the individuals in PA
16:              Add Vᵢ to PA using the k-nearest neighbor rule
17:           Else
18:              Remove the dominated individuals from PA and
add Vᵢ to PA
19:           End if
20:       End if
21:       If Vᵢ dominates Xᵢ then
22:           Xᵢ = Vᵢ
23:       End if
24:    Endfor
25:    Build two Gaussian Models using the parameters of
the improved individuals
26:    Calculate the probabilities of the strategies based
on the improved individuals
27:    Generate Pⁿ⁺¹ based on Pⁿ and PA using indicator-based
selection.
28:    Reinitialize F, CR, S for the unimproved individuals
according to the models
29: Endwhile
```

The variation operators of DE/best/1/bin and DE/best/1/exp are shown as follows.

Program segment 1
$j_{rand} = randint\,(1, D)$
for j=1 to D do
  if $rand(0,1)<CR_i \parallel j == j_{rand}$ then
    $V^j_i = X^j_{best} + F_i*(X^j_{r1}-X^j_{r2})$
  else
    $V^j_i=X^j_i$
  endif
endfor

Program segment 2
$j=randint\,(1, D)$
k=0
do
  $V^j_i = X^j_{best} + F_i*(X^j_{r1}-X^j_{r2})$
  $j = (j+1)\ \%\ D$
  k++
while $(rand(0,1)<CR_i$ && $k<D)$

In the above pseudo-codes, P^n is the parent population of the *n-th* generation; F_i, CR_i, and S_i are the two control parameters and the strategy corresponding to the individual P^n_i; D is the dimension of decision variables; randint(1, D) denotes generating a random integer number between 1 and D; $r1$ and $r2$ are two randomly generated integers with uniform distribution and their values are lower than or equal to the population size, and mutually different; *best* is a randomly generated integer with uniform distribution and less than the size of the non-dominated solution set.

3   Experiments

In this section, we experimentally investigate the performance of ILSDEMO compared with NSGAII, IBEA and DEMO/parent [13] on DTLZ1, DTLZ3, and DTLZ7 using three performance metrics. All the experiments are implemented in VC6.0, Matlab 7.1, Sigmaplot 12.0 and run on Intel i5 760 2.8GHz machines with 4GB DDR3 RAM.

3.1   Performance Indicators

We employ three performance metrics, namely, hypervolume (HV) [14], inverted generational distance (IGD) [15], and Spacing [16], to compare the non-dominated solutions obtained by the four algorithms. It must be pointed out that we use the HV difference between the Pareto front and the obtained non-dominated front to substitute the HV indicator. In the following experiment, the HV difference is denoted as HV*. Then, a small value of HV* corresponds to a high quality. Also noticeable is that we use an improved version of spacing, where the Euclidean distances between the two extreme ends of non-dominated front and the corresponding ends of the Pareto front are considered.

3.2   Parameters Settings

According to our preliminary experiments, the common parameters are set as follows: NP=50, NA=150, $F\in[0.2, 0.8]$, $CR\in[0.7,1.0]$, $RUNs$=30, Max_Evals=100000. Here, NP represents the parent population size, NA denotes archive population size, F and CR are the scaling factor and crossover rate of DE respectively, $RUNs$ and Max_Evals indicate the number of runs and evaluations in one run respectively. In addition, the factor k is set to 0.02 in indicator-based selection. In IBEA and NSGAII, the offspring population size is set to 50; the probabilities of simulated binary crossover (SBX) and polynomial mutation are set to 0.8 and 0.1 respectively; the distribution indices of SBX and polynomial mutation are set to 20. In the calculation of performance indicators, we use the objective values normalized by *arctan* function and finally mapped onto [0, 1], and each objective value of the reference point is set to 1.0.

3.3   Experimental Results

In this section, we present and discuss the experimental results obtained by ILSDEMO, NSGAII, IBEA and DEMO. In DTLZ1, DTLZ3, and DTLZ7, the number of decision variable is equal to M+k−1, where M is the number of objective functions and k can be set by the user. In this experiment, M is set to 3, and k is set to 5 for DTLZ1, 10 for DTLZ3, 20 for DTLZ7 respectively.

It can be seen from **Fig. 1** that ILSDEMO approaches to the true Pareto fronts of the DTLZ test instances more accurately and more evenly. In addition, For DTLZ1, the solutions obtained by IBEA are more evenly distributed while NSGAII maintains a worst distribution. For DTLZ3, the approximate front obtained by NSGAII is the worst owing to some outliers on f1 and f2, and DEMO is better than IBEA. Although the centers of the fronts obtained by IBEA are evenly distributed to a certain extent, not only large gaps exist between the centers and the edges, but also some extreme points of the three objectives are lost. For DTLZ7, the non-dominated front obtained by NSGAII is similar to the one of DEMO while IBEA obtains the worst result.

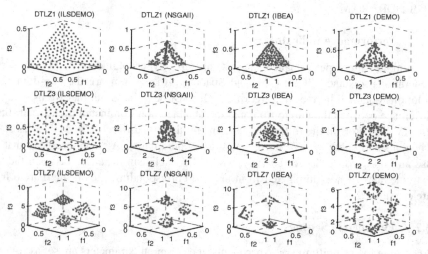

Fig. 1. The best approximate Pareto fronts of 25 runs obtained by the four algorithms on DTLZ1, DTLZ3, DTLZ7 according to HV* value

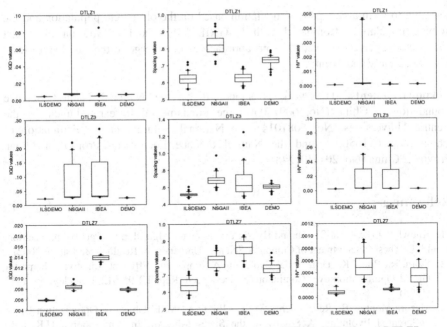

Fig. 2. The box plots of the three indicator values for DTLZ1, DTLZ3, and DTLZ7

Besides, the box-plots in **Fig. 2** statistically reflect the differences of the four algorithms by three performance indicators. The overall performance of ILSDEMO is highly competitive with the other three methods. Meanwhile, seen from the medians on DTLZ1, IBEA exhibits better performance than NSGAII and also slightly outperforms DEMO. For DTLZ3, although there exist some worse outliers for IBEA, NSGAII only shows better performance than IBEA regarding IGD while it has worse spacing and HV* values. For DTLZ7, the four algorithms have consistent performance ranks on IGD and Spacing, i.e., ILSDEMO has the best indicator values and IBEA obtains the worst indicator values while DEMO is better than NSGAII. However, IBEA is better than NSGAII and DEMO in terms of HV* metric.

4   Conclusions

In this paper, we have presented an efficient multi-objective evolutionary algorithm based on differential evolution. Multiple approaches from different perspectives have been applied to boost the performance of the algorithm. To generate a statistically sound population, orthogonal design is used in this algorithm. Self-adaptive parameters and strategies are employed to tune parameters automatically in order to adapt them to different problems or different stages. In the greedy selection of differential evolution, we use q-Gaussian mutation to enhance the local search capability. Moreover, the k-nearest neighbor rule is utilized to eliminate the crowded individual so as to maintain an evenly distributed archive population with a fixed size. To keep the parent population distributed evenly, we also employ indicator-based selection to form a new parent population based on the old parent population and the archive population. Compared with NSGAII, IBEA, and DEMO on three test instances, we have observed that our approach can converge to the true Pareto front more accurately and evenly.

Acknowledgments. The work is supported by the National Natural Science Foundation of China (No. 60975050), the Fundamental Research Funds for the Central Universities (No. 6081014), the National Natural Science Foundation of China (No. 61165004), and the Natural Science Foundation Project of Fujian Province, China (No. 2012J01248).

References

1. Knowles, J.D.: Local-Search and Hybrid Evolutionary Algorithm for Pareto Optimization. Ph.D. Thesis. Department of Computer Science, University of Reading, Berkshire (2002)
2. Storn, R., Price, K.: Differential Evolution–A Simple and Efficient Adaptive Scheme for Global Optimization over Continuous Spaces. Tech. Rep. TR-95-012, Berkeley, pp. 1–12 (1995)
3. Mezura-Montes, E., Reyes-Sierra, M., Coello, C.A.: Multi-Objective Optimization Using Differential Evolution: A Survey of the State-of-the-Art. In: Chakraborty, U.K. (ed.) Advances in Differential Evolution. SCI, vol. 143, pp. 173–196. Springer, Heidelberg (2008)
4. Mezura-Montes, E., Velázquez-Reyes, J., Coello, C.A.: A Comparative Study of Differential Evolution Variants for Global Optimization. In: Proceedings of the 8th Annual Conference on Genetic and Evolutionary Computation (GECCO 2006), Seattle, pp. 485–492 (2006)
5. Das, S., Suganthan, P.N.: Differential Evolution: A Survey of the State-of-the-Art. IEEE Trans. Evol. Comput. 15(1), 4–31 (2011)
6. Noman, N., Iba, H.: Accelerating Differential Evolution Using an Adaptive Local Search. IEEE Trans. Evol. Comput. 12(1), 107–125 (2008)
7. Zitzler, E., Künzli, S.: Indicator-Based Selection in Multiobjective Search. In: Yao, X., Burke, E.K., Lozano, J.A., Smith, J., Merelo-Guervós, J.J., Bullinaria, J.A., Rowe, J.E., Tiňo, P., Kabán, A., Schwefel, H.-P. (eds.) PPSN 2004. LNCS, vol. 3242, pp. 832–842. Springer, Heidelberg (2004)
8. Zhang, Q.F., Leung, Y.W.: An Orthogonal Genetic Algorithm for Multimedia Multicast Routing. IEEE Trans. Evol. Comput. 3(1), 53–62 (1999)
9. Leung, Y.W., Wang, Y.: An Orthogonal Genetic Algorithm with Quantization for Global Numerical Optimization. IEEE Trans. Evol. Comput. 5(1), 41–53 (2001)
10. Zeng, S.Y.: An Orthogonal Multi-objective Evolutionary Algorithm for Multi-objective Optimization Problems with Constraints. Evol. Comput. 12(1), 77–98 (2004)
11. Sindhya, K., Sinha, A., Deb, K., Miettinen, K.: Local Search Based Evolutionary Multi-objective Optimization Algorithm for Constrained and Unconstrained Problems. In: Proceedings of the 2009 IEEE Congress on Evolutionary Computation (CEC 2009), Trondheim, pp. 2919–2926. IEEE Press (2009)
12. Tinos, R., Yang, S.X.: Self-Adaptation of Mutation Distribution in Evolutionary Algorithms. In: Proceedings of the 2007 IEEE Congress on Evolutionary Computation (CEC 2007), Singapore, pp. 79–86 (2007)

13. Robič, T., Filipič, B.: DEMO: Differential Evolution for Multiobjective Optimization. In: Coello Coello, C.A., Hernández Aguirre, A., Zitzler, E. (eds.) EMO 2005. LNCS, vol. 3410, pp. 520–533. Springer, Heidelberg (2005)
14. Zitzler, E.: Evolutionary Algorithms for Multi-objective Optimization: Methods and Applications. PhD Thesis, ETH Zurich (1999)
15. Veldhuizen, D.A.: Multiobjective Evolutionary Algorithms: Classifications, Analyses, and New Innovations. Ph.D. Thesis. Department of Electrical and Computer Engineering, Graduate School of Engineering, Air Force Institute of Technology, Wright-Patterson AFB, Ohio (1999)
16. Deb, K., Pratap, A., Agarwal, S., Meyarivan, T.: A Fast and Elitist Multiobjective Genetic Algorithm: NSGAII. IEEE Trans. Evol. Comput. 6(2), 182–197 (2002)

Application of Data Mining in Coal Mine Safety Decision System Based on Rough Set

Tianpei Zhou

Xuzhou College of Industrial and Technology, Xuzhou, China
Zhoutianpei_001@163.com

Abstract. The coal mine safety decision systems, such as ventilation safety monitoring system, underground water inrush monitoring system, underground coal and gas emission monitoring system, have been established in many large and medium-sized coal mines. A large amount of original data had accumulated in these systems. How to transform data into information for scientific decision was a problem worth to consider for coal mine safety production. The rough set theory, quantitative analysis of incomplete, imprecise and uncertainty knowledge, provided a new method and tool for data mining. A kind of heuristic genetic algorithm for continuous attributes discretization was put forward to solve the problem of continuous attribute discretization of decision table; a kind of heuristic immune algorithm for attribute reduction was presented to conquer the shortage of existing attribute reduction algorithm; in order to solve the problem of reasoning and decision in incomplete and imprecise information, a kind of default rule mining model based on reduction lattice was proposed. Finally, data mining system based on rough set was designed, which was applied to data mining analysis of underground gas emission, good results were achieved.

Keywords: rough set, data mining, attribute reduction, genetic algorithm, Immure algorithm, reduction lattice, default rule.

1   Introduction

Coal and gas outburst was a complex geological disaster occurred in coal mines, the complexity mainly as follows: 1) the factors, affecting coal and gas outburst, itself was inaccurate and fuzzy; 2) the relationship between outburst factors and outburst events was fuzzy; 3) for the occurrence mechanism of coal and gas outburst, that was not yet fully clear; 4) all prediction index and prediction method of coal and gas outburst was presented based on the experience, which was difficult to generally applicable to different gas geological conditions [1].

Due to the above reasons, the practical applications of prediction method of coal and gas outburst were subject to greater restrictions, the effect was worst. The different prediction methods using different parameters, has used seven or eight, but also choose more than a dozen, two dozen, for many penetration workers, the problem that how to choose better predict effect index is a headache. To solve the problems, uncertainty, ambiguity, nonlinear, of existing prediction methods, a feasible way is to find

D.-S. Huang et al. (Eds.): ICIC 2012, LNCS 7389, pp. 34–41, 2012.
© Springer-Verlag Berlin Heidelberg 2012

new theory and method[2]. Data mining system based on rough set was designed, potential, valuable information, rules and knowledge were mined and discovered from a large number of original data, thus to effectively guide and predict safety, production, operation and management of coal mining enterprises, good results were achieved.

2   Model for Default Rule Mining

Default rule, reflecting general relationships of data, a certain degree of generality, relatively simple expression form, can deal the problems with some degree of incompleteness, the so-called incompleteness are shown that some attributes have no value in decision system[3]. Model for default rule mining based on reduction lattice was presented, and its idea is: for given confidence threshold μ_C and support threshold μ_S of target rules, starting from the sample data, according to rough set theory, reduction lattice was firstly built and searched through some search strategy, the rule set that meet the requirements of credibility and support was calculated according to searched reduction nodes, default rule was applied to reasoning or decision, due to the incomplete information, it was matched layer by layer in the lattice in accordance with existing information, the optimal solution of problem was obtained according to a priority decision algorithm.

$N_i^{(q)}$ was represented as the i-th node of the q-th layer, the corresponding decision subsystem for $(U, C_i^{(q)} \cup D)$, hich corresponded to a set of simplified rules, set $U / IND(C_i^{(q)})) = \{X_1,...,X_i\}$, $U / IND(D) = \{Y_1, ...Y_j\}$, if the credibility and support meet:

$$\mu_C(X_i,Y_j) > \mu_C, \ (\mu_C(X_i,Y_j) = card(X_i \cap Y_j)/card(X_i)) \tag{1}$$

$$\mu_C(X_i,Y_j) > \mu_C, \ (\mu_C(X_i,Y_j) = card(X_i \cap Y_j)/card(X_i)) \tag{2}$$

A default rule can be generated:

$$Des(X_i) \rightarrow Des(Y_j) \quad (\mu_C(X_i,Y_j),\mu_S(X_i,Y_j)) \tag{3}$$

And the rules were stored in the rule set of the node.

For node in reduction lattice, a reduction node may be found from top to bottom and from left to right, rule set was generated, if confidence of the node relative to a particular decision class Yj rule set was less than confidence threshold, all subsequent nodes relative to the decision class Yj rule set was not calculated, then a reduction node may be found from t bottom o top and from right to left, rule set was generated, if support of the node relative to a particular decision class Yj rule set was less than support threshold, all predecessor nodes relative to the decision class Yj rule set was not calculated, and so forth, finally all reduction node was traversed, the rules set of all nodes were obtained[4].

3   Data Mining System Based on Rough Set

Data mining system based on rough set achieved the main process: data preprocessing, data mining, model interpretation and model evaluation, as shown in Fig. 1, the whole system was composed of four modules: data pre-processing, attribute reduction, rule extraction ,rule evaluation and interpretation[5].

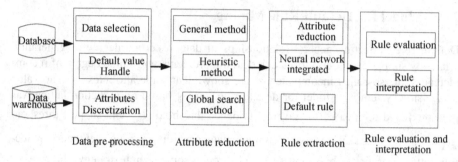

Fig. 1. Frame of data mining system based on rough set

3.1   Data Pre-processing Module

Data pre-processing was very important in data mining, the success or failure of a data mining was affected not only by algorithm model selection, but also by data pre-processing. The results of data analysis were directly affected by incomplete, noisy and inconsistent data. Data pretreatment technology can improve the quality of the data, and help to improve accuracy and performance of the mining. Data pre-processing module included three functions: data selection, data default value processing and data discretization. Data selection, including the selection of data tables and the selection of data object in tables; data default value processing, including several filling and ignore methods of default value; discretization of continuous attribute value was a generalization process of decision table. Heuristic genetic algorithm for discretization of continuous attributes was adopted[6], the algorithm was as follows:

Step 1: the genetic algorithm parameters, maximum evolution algebra G, population size Psize, initial temperature T0,Pc1,Pc2,Pm1,Pm2 were set;
Step 2: evolutionary algebraically counts t=0, according to the rules of initial candidate cut dots selection, initial candidate cut sets were selected, the population P(t) was initialized;
Step 3: each individual's fitness value of the initial population P(t) was calculated;
Step 4: population P'(t) was obtained by selection operation for population P(t);
Step 5: the population P''(t) was obtained by crossover and mutation operation for population P'(t);
Step 6: the next population P (t +1) was obtained through corrective operation for population P''(t);

Step 7: the calculation was end if t = G or optimal individual satisfying the requirements, otherwise each individual's fitness value of the initial population P(t+1) was calculated, went to Step 4.

3.2   Attribute Reduction Module

Attribute reduction module, the framework based on rough set theory, the core of all the applications of rough set theory, main function was reduction of data attributes and concentration and generalization of data. Attribute reduction methods included general reduction, heuristic reduction methods and global search attribute reduction method, where heuristic immune algorithm for attribute reduction was adopted[7], the algorithm was as follows:

Step 1: the relative difference comparison table of decision table was firstly solved, then core attribute of attribute reduction in decision table was calculated according to the relative difference comparison table;

Step 2: the algorithm parameters were initialized: antibody group size for N, the memory size for M, iterations for epochs, crossover probability for Pc, mutation probability for Pm, weighting factor for γ;

Step 3: antibody Ag was defined as attribute number of attribute reduction set;

Step 4: attribute reduction was encoded: each antibody contained core properties, only the non-core properties of antibody was encoded, the size of antibody group A1was set to N+M;

Step 5: corrective operation: new antibody in the antibody group A1 was corrected by correction strategy, so that each antibody was effective antibody;

Step 6: affinity calculation: affinity fj and concentration cj of each antibody Abj in Ak(N+M) was calculated;

Step 7: antibody evaluation: expected reproduction rate of antibody Abj for Prj = Prf j+ γ Prcj;

Step 8: Memory update: the poor antibody was replaced by M excellent antibodies with more affinity in antibody group Ak, which can ensure that the average affinity of actual antibody in the memory is greater than or equal to the prevenient antibody;

Step 9: termination condition judgment: whether it reached iterations, antibody affinity function value whether did not change after a certain number of successive iterations. all attribute reduction sets of memory were output, operation end. Otherwise, continue to the next step;

Step 10: genetic operation: for antibody Ak, comprehensive evaluation methods based on affinity and concentration were adopted, as parent group Bk, N antibodies were selected by roulette wheel selection method in Ak. For Bk, the corresponding antibodies were selected respectively in accordance with the crossover probability Pc and mutation probability Pm, Bk+1 was obtained through the crossover and mutation operations;

Step 11: corrective operation: new antibody in the antibody Bk+1was corrected by correction strategy, so that each antibody was effective antibody, effective antibody B'k+1 was obtained;

Step 12: a new antibody group generation: a new antibody group Ak+1 included antibodies in memory antibody group and offspring antibodies B'k+1 with corrective operation, then went to Step6.

3.3   Rule Extraction Module

The function of rule extraction module was to extract rules from the data set, attribute value reduction method based on rough set theory was adopted[8].

3.4   Rule Evaluation and Interpretation Module

The module included the prediction accuracy of classification rule, confidence and support analysis, the number of rules and the final interpretation of the rules[9].

4   Application

Gas emission in underground mines as an example, the relationship between gas emission and its affecting factors was found for scheduling decision, help mine leader to timely analyze and guide coal mine production. These data were collected from July 18, 2010 to September 20th, some were obtained by measure of underground gas detection system and mine coal belt system, some were calculated, a total of 1560 records, 20 data were randomly selected from records, as shown in table 1. S_1, S_2, S_3, S_4, S_5, S_6 as condition attributes, S_1 for the burial depth, S_2 for coal seam thickness, S_3 for gas content, S_4 for daily progress, S_5 for coal seam pitch, S_6 for daily output; D for the gas emission,1 for non-emergent, 2 for general, 3 for emergent[10].

Table 1. Data set of gas emission

	S_1(m)	S_2 (m)	S_3 (m^3.t^{-1})	S_4 (d.t^{-1})	S_5 (m)	S_6 (d.t^{-1})	D
1	409	2.1	1.92	4.42	21	1824	1
2	421	1.9	2.14	4.13	18	1750	1
3	434	2.4	2.50	4.71	16	2082	1
4	449	2.3	2.43	4.32	17	1995	1
5	528	2.6	2.80	3.25	10	1978	2
6	543	2.8	3.16	3.81	12	2206	2
7	562	3.1	3.68	3.53	11	2409	2
8	589	6.0	4.21	2.85	19	3238	3
9	606	6.2	4.34	2.77	16	3086	3
10	625	6.3	4.03	2.64	17	3353	3
11	639	6.4	4.67	2.75	15	3411	3
12	415	2.1	2.15	4.16	23	1526	1
13	431	2.4	2.58	4.67	18	2077	1
14	457	2.3	2.40	4.51	21	2102	1
15	515	2.9	3.22	3.45	13	2241	2
16	518	2.7	3.10	3.51	19	2089	2
17	530	3.0	3.35	3.68	12	2287	2
18	550	2.9	3.61	4.02	14	2325	2
19	628	6.5	4.62	2.80	20	3455	3
20	633	6.6	4.80	2.92	16	3649	3

Table 2. Data set of gas emission after discretization

	S_1	S_2	S_3	S_4	S_5	S_6	D
1	1	1	1	4	4	1	1
2	1	1	2	3	4	1	1
3	1	2	3	4	3	2	1
4	2	2	3	3	2	1	1
5	4	2	4	2	1	1	2
6	4	3	4	2	1	2	2
7	5	3	5	2	1	3	2
8	5	6	6	1	3	5	3
9	5	6	6	1	3	5	3
10	5	6	5	1	2	5	3
11	6	6	7	1	2	6	3
12	1	1	2	3	5	1	1
13	1	2	3	4	3	2	1
14	2	2	3	4	4	2	1
15	4	2	4	2	1	1	2
16	4	3	4	2	3	2	2
17	4	3	4	2	1	2	2
18	5	3	5	3	2	3	2
19	6	7	7	1	4	6	3
20	6	7	7	1	2	6	3

4.1   Discretization of Continuous Attributes

Due to condition attributes as the continuous, discretization of continuous firstly was carried out before extracting rule, 500 records that randomly selected from the gas emission data set were analyzed, discretization of the condition attributes, such as burial depth, coal seam thickness, gas content, daily progress, coal seam pitch and daily output was made by heuristic genetic algorithm, to get the minimum discrete points.

Burial depth data was discretized into 6 intervals, coal seam thickness data was discretized into 7 intervals, gas content data was discretized into 7 intervals, daily progress data was discretized into 4 intervals, coal seam pitch data was discretized into 5 intervals, daily output data was discretized into 6intervals, data set of instance of gas emission after discretization from table 1 as shown in Table 2.

4.2   Attribute Reduction

The discrete data sets with records for attribute reduction by using heuristic immune algorithm, attribute reduction was burial depth, coal seam thickness, gas content, daily progress, coal seam pitch and daily output.

4.3   Default Rules Mining

Gas emission data set was analyzed and mined by using model for default rule mining based on rough set. The classification rules were obtained in condition that the minimum reliability was 0.45 and the minimum support degree was 0.02, f which five rules were interpreted as follows:

Table 3. Test data set of gas emission

	S_1(m)	S_2 (m)	S_3 (m^3.t^{-1})	S_4 (d .t^{-1})	S_5 (m)	S_6 (d .t^{-1})	D
1	419	2.0	1.52	4.32	22	1722	1
2	467	2.2	2.91	4.21	22	2004	1
3	498	2.8	3.20	3.41	19	2083	1
4	545	2.8	3.31	4.22	15	2227	2
5	580	2.9	3.36	3.81	13	2108	2
6	615	6.4	4.45	2.84	18	3259	3
7	638	6.4	4.42	2.70	21	3357	3
8	425	2.5	2.35	4.26	24	1423	1
9	431	1.8	2.62	4.33	19	1653	1
10	444	2.3	2.70	4.31	17	2084	1
11	459	2.5	2.33	4.42	18	1896	1
12	525	3.0	3.12	3.35	14	2144	1
13	540	3.3	3.65	3.78	13	2188	2
14	620	6.2	4.35	2.95	16	3312	3
15	650	5.6	4.40	2.42	15	3546	3

S_1(burial depth) $=1 \wedge S_2$(coal seam thickness)$=1 \wedge S_3$(gas content)$=1 \rightarrow$ D(gas emission) $=1$ (non-emergent) (0.73, 020);

S_1(burial depth)$=4 \wedge S_2$(coal seam thickness)$=3 \wedge S_3$(gas content)$=4 \wedge S_4$(daily progress)$=2 \wedge S_5$(coal seam pitch)$=1 \wedge S_6$(daily output)$= 2 \rightarrow$ D(gas emission)$=2$(general) (0.70,032);

S_1(burial depth)$=4 \wedge S_2$(coal seam thickness)$=2 \wedge S_3$(gas content)$=4 \wedge S_4$(daily progress)$=2 \wedge S_5$(coal seam pitch)$=1 \wedge S_6$(daily output)$=1 \rightarrow$ D(gas emission)$=2$(general) (0.82,0.24);

S_1(burial depth)$=5 \wedge S_2$(coal seam thickness)$=6 \wedge S_3$(gas content)$=6 \wedge S_5$(coal seam pitch)$= 3 \wedge S_6$(daily output)$=5 \rightarrow$ (gas emission)$=3$(emergent) (0.73,0.02);

S_1(burial depth)$=6 \wedge S_2$(coal seam thickness)$=7 \wedge S_3$(gas content)$=7 \wedge S_5$(coal seam pitch)$= 2 \wedge S_6$(daily output)$=6 \rightarrow$ D(gas emission)$=3$(emergent) (0.85,0.021).

The gas emission data were tested after obtaining the gas emission classification rules, 15 samples of test data set as shown in Table 3. Firstly, the test data set was discretized, then discretized test data set was predicted for classification through the classification rules, prediction accuracy rate reached 98%, indicating that it was effective to data mining based rough set on for extracting the gas emission classification rules. Finally, the rules were evaluated and interpreted, and the rules were stored in knowledge base for the prediction of the gas emission or auxiliary decision.

5   Summary

The rough set theory, quantitative analysis of incomplete, imprecise and uncertainty knowledge, has been a research hotspots in recent years. In order to solve the problem of continuous attribute discretization of decision table, a kind of heuristic genetic algorithm for continuous attributes discretization was put forward, the performance of the algorithm was improved. A kind of heuristic immune algorithm for attribute reduction was presented to conquer the shortage of existing attribute reduction algorithm. In order to solve the problem of reasoning and decision in incomplete and imprecise information, a kind of default rule mining model based on reduction lattice was proposed, which was used the research method of Top-down and Bottom-up in reduction lattice alternately to mining default rule. By setting rule credible degree and rule support degree based on RS, the algorithm introduced some research strategies which were used to optimize the process of mining default rule and reduce the time complexity of the algorithm. Finally, data mining system based on rough set was designed, which was applied to data mining analysis of underground gas emission, good results were achieved.

References

1. Niu, L.D.: Mine Gas Linkage Monitoring Method Based on Data Mining Approach. China Safety Science Journal 21, 62–68 (2011)
2. Ni, L.Q., Zhou, H.T., Gao, S.S.: A More Effective Data Mining Approach That Adroitly Combines Rough Set Theory with Evidence Theory. Journal of Northwestern Polytechnical University 28, 927–931 (2010)
3. Zhao, Z.P., Yin, Z.M., Chen, J.C.: Mine Hidden Danger Data Digging Model and Applicative Digging Algorithm. Coal Science and Technology 38, 67–69 (2010)
4. Zheng, H.Z., Liu, Y., Zhan, D.C.: Default Rules Frame of Non-monotonous Problems Based on Data Mining. Computer Science 33, 181–182 (2006)
5. Wang, Y.Y.: Knowledge Discovery Methods Research Based on Rough Set Theory. Shanghai Jiao Tong University, Shanghai (2006)
6. Xu, X., Zhai, J.M.: Multiscale Genetic Algorithms for Discretization in Rough Set on Trees. Modern Manufacturing Engineering 10, 1–4 (2009)
7. Liu, Y., Li, W.H., Chen, Y.L.: Research on Intelligent Fault Diagnosis Based on Artificial Immune System. Computer Measurement & Control 18, 2694–2696 (2010)
8. Zhao, L.S., Shi, J.H.: Real Value Attribute Reduction Method Based on Rough Sets. Journal of Inner Mongolia University 41, 97–101 (2010)
9. Liu, B., Pan, J.H., Liu, P.S.: Rule Evaluation Method and Data Quality Mining System. Computer Integrated Manufacturing Systems 15, 1436–1441 (2009)
10. Hou, G.Z.: Focus on Gas Safety Management Face Ventilation and Gas Management Means. Coal Technology 28, 199–200 (2009)

Study on Web Text Feature Selection Based on Rough Set

Xianghua Lu and Weijing Wang

Department of Computer and Information Engineering, Luoyang Institute of Science
and Technology, Luoyang, P.R. China
xh_xianghua@sohu.com

Abstract. This paper uses vector space model as the description of the Web text, analyses the feature of the Web pages which are written in HTML, and improves the traditional formula of TF-IDF. The feature weight is calculated according to the term location in the document. In addition, a text classification system based on Vector Space Model is studied. In the article, feature selection and text classification is connected and feature terms are selected depending on the term's importance to classification, and then the paper proposes a feature selection algorithm based on rough set. Experiments show that this method can effectively improve the classification accuracy. It can not only reduce the dimension of feature space, but also improve the accuracy of classification.

Keywords: feature weight, feature selection, rough set, text classification.

1   Introduction

With the rapid development of WWW technology, Internet has become the largest information gathering place. As these large, heterogeneous Web information resources with great potential value, people urgently need tools to find the resources and knowledge from the Web rapidly and efficiently. Web text mining is a kind of method and tool to find the implicit knowledge and pattern from a large number of Web documents [1]. Web text classification is an important technology in Web text mining, and it refers to classify each of the Web documents into a pre-defined category [2]. Through the Web text classification, a lot of web pages can be classified automatically to help the users get their interested text information quickly and accurately, reduce the search space, accelerate the retrieval speed, and improve the query precision, etc.

The development of Web text classification is based on traditional text classification among which some technology are applicable in the Web text classification as well. However, there is a fundamental difference between the two kinds of classification. The former deal with Web texts and the latter do with plain texts. Web text is mainly described in the form of webpage which contains a lot of marker symbols; those symbols provide much richer information for Web text classification. In addition, the Web document also has high dimension features, sparse sample and less obvious characteristics [3]. The classification performance of some frequently-used classifier with superior performance usually dropped sharply in high dimension

D.-S. Huang et al. (Eds.): ICIC 2012, LNCS 7389, pp. 42–48, 2012.
© Springer-Verlag Berlin Heidelberg 2012

feature spaces, while the huge computation cost makes it difficult for various algorithms to deal with large-scale text collections. In view of the features above, the calculation method of feature weight has been improved, and then this paper proposed a feature selection method based on attribute reduction of rough set. Experiments proved that when reduced to the same feature number, classification can get higher accuracy using training subset extracted by the method.

2   The Improvement of Weight Calculation Formula

Vector Space Model (VSM) is usually used to describe the texts in Web text classification. In VSM, each text is described as a feature vector of n dimension:

$V(d) = (t_1, w_1(d); ...; t_i, w_i(d); ...; t_n, w_n(d))$ .Where, t_i stands for feature item, $w_i(d)$ is the weight of term t_i in text d, which is generally defined as the function of the frequency $tf_i(d)$ that t_i occurs in text d, namely $w_i(d) = \varphi(tf_i(d))$. Weight of feature describes the importance for expressing text content, so feature selection and weight calculation turns out important particularly. In VSM, TF-IDF is usually used to calculate feature weight, and the following is a commonly-used formula for weight calculation:

$$w_i(d) = \frac{tf_i(d) * \log(N/n_i + 0.01)}{\sqrt{\sum_{i=1}^{n} tf_i^2(d) * \log^2(N/n_i + 0.01)}} \tag{1}$$

N is the total number of training text, n is the character number, n_i is the document frequency of the item t_i, and $tf_i(d)$ is the frequency of t_i occurs in the text d:

$$tf_i(d) = \frac{freq_{t_i}}{\max_1 freq_{t_i}}, t_i \in \{t_1, t_2, ..., t_n\} \tag{2}$$

The corresponding part IDF measures the inverse document frequency:

$$idf_i = \log(N/n_i + 0.01) \tag{3}$$

When text feature weight calculated, the TF-IDF function only aims at the information about the times that the feature item appears in the text and the distribution of item in the whole text space. It doesn't consider the information of the feature location in the text and the uneven distribution. In a specific text, the description ability of feature which appears in different location is different. Its contribution to text classification is also obviously different. On the other hand, in the entire training set, the different position of feature in each text can also reflect the different performance to describe text [4]. So the item location should be considered when its weight calculated. At present, in most of text classifications based on position weight, absolute weighting method is the common method to calculate the weight of feature according to the main structure features. In the weight calculation based on position, what commonly used is the absolute value weighted method; it is a matter of experience. Firstly, the feature items in different positions are given different weighting coefficient; secondly, final weight is calculated together with the feature items in the main body of the text [5]. This method can improve the classification effect, but only to a certain extent. The disadvantage of the method is that weighting coefficient is fixed value, which has different effect to long text and short text. So the influence of structure feature to body text will weaken with the length of body text increase. To solve the problem, the reference [6] proposed an improved weighting method, namely relative weighting method. The method considers both the length of HTML document's body and the length of the text which is written in labels. Then, the method expands words of each element in proportion according to the length of the document, so as to reduce the disadvantages that the effect each element plays on the text is always influenced by the document length.

Web document contains a large number of HTML tags. Accordingly the term included in different tags has different ability to describe text's content or distinguish text categories. Therefore, the traditional TF-IDF algorithm is not so effective on the Web text. So, according to HTML structural feature, the weight of feature in different position should be calculated using different methods.

Based on the above analysis, each HTML document will be divided into three sections according to the Web document structure in this paper: title, anchor text and body. The same sections are included in a collection respectively, so that the entire training set can be divided into three "pseudo text set" [7]: $S_i, i = 1, 2, 3$; where: S_i is made of the text segments corresponding to the training set of the i section. In each" pseudo text set" S_i, formula (4) and (5) are used to calculate the feature frequency corresponding to the feature item and the inverted text frequency:

$$f_{i,j}(d) = \frac{freq_{t_i,j}}{\max_1 freq_{t_i,j}}, t_i \in \{t_1, t_2, ..., t_n\}, j = 1, 2, 3 \tag{4}$$

$$idf_{i,j} = \log\left(N / n_{i,j} + 0.01\right) \tag{5}$$

Where, $n_{i,j}$ is the number of texts containing the feature t_i in the text set $j, j = 1, 2, 3$.

In the entire training set the weight function of the feature t_i can be improved as:

$$w_i(d) = \frac{\sum\limits_{j=1}^{3} \lambda_j * tf_{i,j}(d) * \sum\limits_{j=1}^{3} \log \lambda_j (N/n_{i,j} + 0.01)}{\sqrt{\sum\limits_{i=1}^{n} \left[\sum\limits_{j=1}^{3} \lambda_j * tf_{i,j}(d) * \sum\limits_{j=1}^{3} \log \lambda_j (N/n_{i,j} + 0.01) \right]^2}} \qquad (6)$$

where, $\lambda_1 > \lambda_2 > \lambda_3$, $\sum\limits_{j=1}^{3} \lambda_j = 1$, it's determined by the experiment as the follow-

ing $\lambda_1 = 0.46, \lambda_2 = 0.36, \lambda_3 = 0.18$. Then make use of formula (6) to weight each feature item.

3   Feature Selection Based on Rough Set

A part of features, all of which have larger weight than those unselected, is selected to construct the feature vector. Where, the number of the selected features is far litter than the total of all items in text. And then, texts are classified using KNN automatically. However, the experiment shows that the classification efficiency is not very ideal because it is not easy to determine an appropriate value of m. it's would affect the classification accuracy whether m is too big or too small. To solve the problem, this paper put forward a method of feature selection based on rough set, which is a math tool to deal with incertitude and uncertain data [8]. The main idea of rough set theory is to extract decision rules by attribute reduction and value reduction in the premises of keeping the ability of classification. Attribute reduction based on attribute significance can remove those feature items which have little effect on the classification or even no effect, and reduce the dimension of feature vector. However, it exists obvious shortcomings when directly using the attribute reduction of rough sets for feature selection because that feature dimension of text is too large and the mathematics amount of attribute reduction is too high [9]. Extremely uneven attribute value may cause null able core of some attributes. This paper proposes a method of feature selection based on rough set. Firstly, making use of the previously-mentioned TF-ID algorithm, a certain number of features are selected to constitute primitive feature sets; secondly, beneficial features to classification are selected from the primitive feature sets by attribute reduction of rough sets in the premise of keeping the accuracy of classification. Further feature selection can reduce the feature dimension to a few hundred dimensions.

   Web text feature selection based on rough set theory is founded on the vector space model VSM obtained in the previous stage. Firstly, the weighted features are discredited and a decision table is established , the entire training sets are   regarded as U, feature subset as conditional   attribute and text categories set as decision attribute D. Secondly, conditional   attributes are filtered properly depending on its significance which   is evaluated through dependence of conditional attribute. So the feature subset including minimum features is obtained and the vector space model is reconstructed. The specific steps are as follows [10]:

- Calculate the positive region $Pos_c(D)$ that decision attributes D is relative to condition attribute C in the two-dimensional decision table:

$$Pos_C = \bigcup_{X \in U/D} \underline{C}X \tag{7}$$

- According to the dependence function of rough set:

$$K = \gamma_c(D) = \frac{card\left(Pos_c(D)\right)}{card(U)} \tag{8}$$

K reflects the degree that decision attributes D depend on attribute C, the larger the value of K, the higher the dependence degree.

- To each attribute t_i in the two-dimensional decision table, calculate its importance $IM_{t_i}(D)$ to decision attribute D:

$$IM_{t_i}(D) = \gamma_c(D) - \gamma_{c-\{t_i\}}(D) \tag{9}$$

The larger the value of $IM_{t_i}(D)$ is, the more important the attribute t_i is to classification. If the value of $IM_{t_i}(D)$ is zero, attribute t_i has no effect on classification. And it should be removed.

- Set a threshold value δ and remove all attributes that meet the condition $IM_{t_i} < \delta$. This method won't result in a loss of information, and can delete vast redundant attributes, so as to reduce the original dimension to a certain degree.

4   Experiments and Analysis of Eight Categories

This paper collects 800 webpage documents from Sohu website as dataset; those documents are divided into eight categories: cars finance and economics, health, travel, entertainment, sports, real estate and military, and each category contain 100 webpage documents. 600 of them as training set, 200 as the test set. Those documents are classified by the most commonly-used classification: k-NN. In the experiment, the classification results are evaluated by a general comprehensive classification rate, namely F1, which comprehensively considers the precision and recall of text classification. Its specific calculation formula is as follows:

$$F1 = \frac{precision * recalll * 2}{precision + recall} \tag{10}$$

Web pages are cleaned to extract the title, anchor text and body, those constitute plain texts. And then the Chinese plain texts are segmented into words with ICTCLAS, which is a high precision Chinese segmentation tool from CAS. The high frequency words, low-frequency words and stop words are eliminated. Weight of every term is calculated respectively using the improved weighting method and the traditional TF-IDF method.

The experiments are divided into two levels. In order to verify the effect of the method based on position weighting, the experiments in the first level compared the improved TF-IDF weighting algorithm and the traditional. Seven hundred and eighty terms are selected to construct the feature vector respectively; all of those selected terms have larger weight than others. And then they are classified using KNN classifier.

The experiments in the second level are based on the improved weighting method before and after using rough set theory, and then by the experiment comparison validate the effect of method based on rough set theory for further feature selection. Features are selected preliminary from the terms whose weight is calculated using the improved weighting method; there, seven hundred and eighty terms are selected as before. Feature dimensions are reduced by mean of the attribute reduction of rough sets, finally one hundred and eighty dimensions is determined.F1 test value of the classification results is shown in the following table:

Table 1. F1 test value

Category	weight formula		
	TF-IDF	Improved TF-IDF	
		Direct	reduction of rough sets
Cars	0.781	0.829	0.865
Military	0.759	0.796	0.839
Finance and Economics	0.672	0.768	0.812
Health	0.703	0.759	0.821
Travel	0.802	0.856	0.866
Entertainment	0.798	0.812	0.854
Sports	0.751	0.823	0.846
Real Estate	0.697	0.737	0.805

The experimental results prove that this improved algorithm of term weighting based on position in k-NN classifier can improve the precision of text classification. The method of feature selection based on rough set can not only achieve the purpose of dimension reduction, but also increase the final text classification accuracy.

5   Conclusion

With the extensive application of Web text classification technology in search engine, digital library, information filtering, information retrieval, Internet information monitoring and other fields, the research of Web text classification has been an advanced issue in the field of information processing. Searching for effective text feature reduction method will be the key technology to improve the efficiency of automatic text classification. Feature selection method based on evaluation function can reduce feature to over a thousand dimensions. The quantity of the dimensions is still very large to machine learning; further reducing the dimensions will affect classification accuracy.

The feature selection method based on rough was proposed in this paper. Firstly the improved weight algorithm based on location to select a certain number of features is used, and then the attribute reduction algorithm of rough set theory to further reduce the feature dimensions was used. The experiments showed the method is effective. It could get higher classification accuracy when the same number of features is reduced and the method to extract the feature subset training is used.

Acknowledgement. The project was supported by Key Project on Science and Technology of the Education Department of Henan Province (NO.12A520031), Youth Foundation of Luoyang Institute of Science and Technology (NO.2011QZ15).

References

1. Wang, J.C., Pan, J.G., Zhang, F.Y.: Research on Web Text Mining. Journal of Computer Research and Development 37, 513–520 (2007) (in Chinese)
2. Liu, H.: Research on Some Problems in Text Classification. Jilin University, Jilin (2009)
3. Liu, L.: The Research and Implementation of Automatic Classification for Chinese Web Text. University of Changchun for Science and Technology, Changchun (2007)
4. Chu, J.C., Liu, P.Y., Wang, W.L.: Improvement Approach to Weighting Terms in Web Text. Computer Engineering and Applications 43, 192–194 (2007) (in Chinese)
5. Tai, D.Y., Xie, F., Hu, X.G.: Text Categorization Based on Position Weight of Feature Term. Journal of Anhui Technical College of Water Resources and Hydroelectric (3), 64–66 (2008) (in Chinese)
6. Tan, J.B., Yang, X.J., Li, Y.: An Improved Approach to Term Weighting in Automatic Web Page Classification. Journal of the China Society for Scientific and Technical Information 27, 56–61 (2008) (in Chinese)
7. Liu, H.F., Zhao, H., Liu, S.S.: An Improved Method of Chinese Text Feature Selection Based on Position. Library and Information Service 53, 102–105 (2009) (in Chinese)
8. Wang, G.Y.: Rough Sets Theory and Knowledge Acquisition. Xi'an JiaoTong University Press, Xi'an (2001) (in Chinese)
9. Zhang, B.F., Shi, H.J.: Improved Algorithm of Automatic Classification Based on Rough Sets. Computer Engineering and Applications 47, 129–131 (2011) (in Chinese)
10. Chen, S.R., Zhang, Y., Yang, Z.Y.: The Research of the Feature Selection Method Based on Rough Set. Computer Engineering and Applications 42, 159–161 (2006) (in Chinese)

A Generalized Approach for Determining Fuzzy Temporal Relations

Luyi Bai and Zongmin Ma

College of Information Science & Engineering
Northeastern University
Shenyang, China
mazongmin@ise.neu.edu.cn

Abstract. Fuzzy temporal relations have been defined to support temporal knowledge representation and reasoning in the presence of fuzziness, which are still open issues. In this paper, we propose a generalized approach for determining fuzzy temporal relations assuming that fuzzy temporal intervals are all fuzzy. We firstly present the basics of representation of fuzzy temporal relations from two aspects: fuzzy time point and fuzzy time interval, and then give definitions of their fuzzy relations. On this basis, correspondences between fuzzy and crisp temporal relations are investigated. Finally, a general formalized algorithm for determining fuzzy temporal relations is proposed.

Keywords: Fuzzy time point, Fuzzy time interval, Fuzzy temporal relations.

1 Introduction

Temporal relations is a topic of great importance in various application domains such as scheduling and planning [7][8], natural language understanding [17], question answering [10], etc. Starting from Allen's seminal work on temporal interval relations [1], increasingly more expressive efforts have been proposed both in qualitative and quantitative aspects of reasoning about temporal data: combining qualitative and metric temporal data [12][14], specifying metric constraints on points and durations [15], and studying disjunctions of temporal constraints [11][13], etc.

However, the fact that notion of time is usually fuzzy in the real world, has led to emergence of various kinds of representation and reasoning approaches for dealing with fuzzy temporal information. The fuzzy set theory [20] and the possibility theory [21] seem to provide a efficient framework for dealing with fuzzy temporal information and are applied in [4] and [6]. The two theories allow relating the qualitative linguistic terms to a quantitative interpretation, and providing a sort of interface between the qualitative and quantitative levels of descriptions [9]. Moreover, it has been proven [5] the fuzzy methodologies can account for both preference and fuzziness.

Since fuzzy temporal relations play a fundamental role in the real world applications, the problems that emerge are how fuzzy temporal relations should be formalized to determine their fuzzy relations. To fill this gap, a lot of efforts

D.-S. Huang et al. (Eds.): ICIC 2012, LNCS 7389, pp. 49–56, 2012.
© Springer-Verlag Berlin Heidelberg 2012

[2][3][16][18][19] have been done in developing fuzzy temporal relations in the recent years. In [3], Badaloni and Giacomin integrate the ideas of flexibility and fuzziness into Allen's interval-based temporal framework, defining a new formalism called IA^{fuz}, which extends classical Interval Algebra (IA). Furthermore, Badaloni and Falda [2] study the problem of representing different forms of fuzzy temporal relations in order to develop a more general way to represent it. The work of Ribarić et al. [16] introduce a geometrical approach to validation of the possibility and necessity measures for temporal relations between two fuzzy time points and a fuzzy time point and a fuzzy time interval. Unfortunately, one of the most import limitations of their efforts for practical applications is that the fuzzy temporal relations are computationally rather expensive to evaluate. In order to solve the problem, Schockaert et al. [19] investigate how the evaluation of the fuzzy temporal interval relations for piecewise linear fuzzy intervals boils down to the evaluation for linear fuzzy intervals, and provide a characterization for linear fuzzy intervals that is both efficient to evaluate and easy to implement. What's more, Schockaert and De Cock [18] present a framework for reasoning about qualitative and metric temporal relations between fuzzy time intervals, which can be drawn upon well-established results for solving disjunctive temporal reasoning problems.

Following the previous ideas of building a model capable to effectively reason about fuzzy temporal relations, we propose a generalized approach for determining fuzzy temporal relations in this paper, assuming that fuzzy temporal intervals are all fuzzy. We firstly present the basics of representation of fuzzy temporal relations from two aspects: fuzzy time point and fuzzy time interval, and then give definitions of their fuzzy relations. On this basis, correspondences between fuzzy and crisp temporal relations are investigated, which has been received little attention although fuzzy temporal relations have been extensively studied. Finally, a general formalized algorithm for determining fuzzy temporal relations is proposed.

The rest of the paper is organized as follows. After presenting basics of representation of fuzzy temporal relations and giving definitions of them in Section 2, Section 3 investigates correspondences between fuzzy and crisp fuzzy temporal relations. Section 4 proposes a general formalized algorithm for determining fuzzy temporal relations and Section 5 concludes the paper.

2   Representation of Fuzzy Temporal Relations

In this section, we present the basics of representation of fuzzy temporal relations from two aspects, which are fuzzy time point and fuzzy time interval. Among each aspect, fuzzy temporal data and their relations are defined.

2.1   Fuzzy Time Point

The time of occurrence of an instantaneous event may be vaguely known and called a fuzzy time point (*FTP*). A fuzzy time point is interpreted as a representation of a set of possible time points along with the degrees of possibility that an event occurs.

Definition 1 (Fuzzy time point). For a fuzzy time point $FTP = (T, \delta)$ we have:

- T is a fuzzy time point
- δ is the membership degree of the fuzzy time point, where $0 \leq \delta \leq 1$.

The fuzzy relations of two fuzzy time points contain three cases, which are fuzzy equal, fuzzy disjoint, and fuzzy overlap (denoted as *FTPequal*, *FTPdisjoint*, *FTPoverlap*, respectively). Here, we introduce a mathematic symbol *supp*, where *supp* $A = \{u \mid u \in U, A(u) > 0\}$.

Definition 2 (Fuzzy relations of fuzzy time points). For two fuzzy time points $FTP_1 = (T_1, \delta_1)$ and $FTP_2 = (T_2, \delta_2)$, $(t_{1i}, \delta_{1i}) \in FTP_1$ and $(t_{2i}, \delta_{2i}) \in FTP_2$, we have:

- *FTPequal* (FTP_1, FTP_2): min $(supp\ (t_{1i}, \delta_{1i})) =$ min $(supp\ (t_{2i}, \delta_{2i})) \vee$ max $(supp\ (t_{2i}, \delta_{2i})) =$ max $(supp\ (t_{1i}, \delta_{1i}))$.
- *FTPdisjoint* (FTP_1, FTP_2): max $(supp\ (t_{1i}, \delta_{1i})) <$ min $(supp\ (t_{2i}, \delta_{2i})) \vee$ max $(supp\ (t_{2i}, \delta_{2i})) <$ min $(supp\ (t_{1i}, \delta_{1i}))$.
- *FTPoverlap* (FTP_1, FTP_2): (min $(supp\ (t_{1i}, \delta_{1i})) <$ min $(supp\ (t_{2i}, \delta_{2i}))) \wedge$ (min $(supp\ (t_{2i}, \delta_{2i})) <$ max $(supp\ (t_{1i}, \delta_{1i})) <$ max $(supp\ (t_{2i}, \delta_{2i}))) \vee$ (min $(supp\ (t_{2i}, \delta_{2i})) <$ min $(supp\ (t_{1i}, \delta_{1i}))) \wedge$ (min $(supp\ (t_{1i}, \delta_{1i})) <$ max $(supp\ (t_{2i}, \delta_{2i})) <$ max $(supp\ (t_{1i}, \delta_{1i}))) \vee$ max $(supp\ (t_{1i}, \delta_{1i})) =$ min $(supp\ (t_{2i}, \delta_{2i})) \vee$ min $(supp\ (t_{1i}, \delta_{1i})) =$ max $(supp\ (t_{2i}, \delta_{2i})) \vee$ min $(supp\ (t_{1i}, \delta_{1i})) <$ min $(supp\ (t_{2i}, \delta_{2i})) <$ max $(supp\ (t_{2i}, \delta_{2i})) <$ min $(supp\ (t_{1i}, \delta_{1i})) \vee$ min $(supp\ (t_{2i}, \delta_{2i})) <$ min $(supp\ (t_{1i}, \delta_{1i})) <$ max $(supp\ (t_{1i}, \delta_{1i})) <$ min $(supp\ (t_{2i}, \delta_{2i}))$.

Theorem 1. For two fuzzy time points $FTP_1 = (T_1, \delta_1)$ and $FTP_2 = (T_2, \delta_2)$, $\propto$ denotes precedence, we have:

i) $FTP_1 = FTP_2 \Leftrightarrow T_1 = T_2 \wedge \delta_1 = \delta_2$

ii) $FTP_1 \propto FTP_2 \Leftrightarrow supp\ (T_1, \delta_1) < supp\ (T_2, \delta_2)$

Proof. Because fuzzy time point represent all the possible time point so that it is straightforward that two fuzzy time points $FTP_1 = FTP_2$ if and only if each $t_i \in T_1$ and $t_j \in T_2$ that $t_i = t_j$ and $\delta_i = \delta_j$. In the case of precedence, similarly, if and only if each $t_i \in T_1$ and $t_j \in T_2$ with positive possibility that $t_i < t_j$ (in other words, there is no $t_i \in T_1$ and $t_j \in T_2$ that $t_j \propto t_i$), $FTP_1 \propto FTP_2$. $\square$

Theorem 2. Let $FTP_1 = (T_1, \delta_1)$ and $FTP_2 = (T_2, \delta_2)$ be two fuzzy time points and *poss* (E) denotes the possibility distribution of event E is true.

i) max $(poss\ (FTP_1 = FTP_2), poss\ (FTP_1 \propto FTP_2), poss\ (FTP_2 \propto FTP_1)) = 1$.

ii) min $(poss\ (FTP_1 = FTP_2), poss\ (FTP_1 \propto FTP_2), poss\ (FTP_2 \propto FTP_1)) = 0$.

Proof. It is completely possible that either $FTP_1 = FTP_2$, or $(FTP_1 \propto FTP_2)$, or $(FTP_2 \propto FTP_1)$, and impossible that $FTP_1 = FTP_2$, and $(FTP_1 \propto FTP_2)$, and $(FTP_2 \propto FTP_1)$. This interpretation is intuitive and understandable. $\square$

2.2 Fuzzy Time Interval

A fuzzy time interval (*FTI*) is a time interval with fuzzy bounds, which is interpreted as representation of a set of possible time intervals along with degrees of possibility.

Definition 3 (Fuzzy time interval). For a fuzzy time interval, we have $FTI = (T_s, \delta, T_e, \delta')$, including:

- T_s and T_e are fuzzy starting and ending time point respectively, where $T_s \propto T_e$.
- δ and δ' are the membership degrees of the fuzzy starting and ending time points being the time T_s and T_e, respectively.

Theorem 3. Let $FTI = (T_s, \delta, T_e, \delta')$ be a fuzzy time interval, the membership degree of the fuzzy time interval δ_{FTI} is:

i) if $\delta = 1$, $\delta_{FTI} = \delta'$.
ii) if $\delta' = 1$, $\delta_{FTI} = \delta$.
iii) if $0 < \delta < 1$ and $0 < \delta' < 1$, $\delta_{FTI} = \delta \times \delta'$.

Proof. The membership degree of the fuzzy time interval computation can be divided into three cases:

- Case 1: If the starting time point of the fuzzy time interval is crisp, we have $\delta_{FTI} = 1 \times \delta' = \delta'$.
- Case 2: If the ending time point of the fuzzy time interval is crisp, we have $\delta_{FTI} = \delta \times 1 = \delta$.
- Case 3: If the two ending time points of the fuzzy time interval are all fuzzy, we have $\delta_{FTI} = \delta \times \delta'$ because the two membership degrees of the two ending time points are assumed to be independent.   □

The fuzzy relations of two fuzzy time intervals contain seven cases, which are fuzzy before, fuzzy equal, fuzzy meet, fuzzy overlap, fuzzy during, fuzzy start, and fuzzy finish (denoted as *FTIbefore*, *FTIequal*, *FTImeet*, *FTIoverlap*, *FTIduring*, *FTIstart*, *FTIfinish*, respectively).

Definition 4 (Fuzzy relations of fuzzy time intervals). For two fuzzy time intervals $FTI_1 = (T_{1s}, \delta_1, T_{1e}, \delta_1')$ and $FTI_2 = (T_{2s}, \delta_2, T_{2e}, \delta_2')$, we have:

- *FTIbefore* (FTI_1, FTI_2): $supp (T_{1e}, \delta_1') < supp (T_{2s}, \delta_2)$.
- *FTIequal* (FTI_1, FTI_2): $supp (T_{1s}, \delta_1) = supp (T_{2s}, \delta_2) \wedge supp (T_{1e}, \delta_1') = supp (T_{2e}, \delta_2')$.
- *FTImeet* (FTI_1, FTI_2): $supp (T_{1e}, \delta_1') = supp (T_{2e}, \delta_2')$.
- *FTIoverlap* (FTI_1, FTI_2): $supp (T_{2s}, \delta_2) < supp (T_{1e}, \delta_1') < supp (T_{2e}, \delta_2') \wedge supp (T_{1s}, \delta_1) < supp (T_{2s}, \delta_2) < supp (T_{1e}, \delta_1')$.
- *FTIduring* (FTI_1, FTI_2): $supp (T_{2s}, \delta_2) < supp (T_{1s}, \delta_1) < supp (T_{2e}, \delta_2') \wedge supp (T_{2s}, \delta_2) < supp (T_{1e}, \delta_1') < supp (T_{2e}, \delta_2')$.
- *FTIstart* (FTI_1, FTI_2): $supp (T_{1s}, \delta_1) = supp (T_{2s}, \delta_2)$.
- *FTIfinish* (FTI_1, FTI_2): $supp (T_{1e}, \delta_1') = supp (T_{2e}, \delta_2')$.

3   Correspondences between Fuzzy Temporal Relations and Temporal Relations

In this section, we present correspondences between fuzzy and crisp temporal relations on the basis of the studies in the above section. The correspondences come from two cases: between fuzzy time points and between fuzzy time intervals.

Fig. 1 shows the three fuzzy relations of two fuzzy time points, and their corresponding crisp relations. Take *FTPequal* for example, the possible fuzzy time intervals of FTP_1 and FTP_2 are the same (from t_1 to t_2). However, the corresponding crisp relations can be either disjoint or equal because the time point is fuzzy and may be exist in any time point between the possible fuzzy time interval (from t_1 to t_2). Then we consider *FTPdisjoint* that the possible fuzzy time intervals of FTP_1 and FTP_2 are disjoint. The corresponding crisp relation must be disjoint because the possible fuzzy time intervals of FTP_1 and FTP_2 are disjoint $((FTP_1 = [t_1, t_2]) \cap (FTP_2 = [t_3, t_4]) = \varnothing)$. Finally, we define the relation of fuzzy time points is *FTPoverlap* if there is common area in the possible time area. The corresponding crisp relations of *FTPoverlap* can also be either disjoint or equal as *FTPequal*. Note that, *FTPoverlap* is different from *FTPequal* although the corresponding crisp relations are either disjoint or equal. It can be considered from two points of view: one is similarity as *FTPdisjoint* if $t \in (t_1, t_2) \subset FTP_1 \wedge t' \in (t_3, t_4) \subset FTP_2$ and the other is similarity as *FTPequal* if $t \in [t_2, t_3] \subset FTP_1 = t' \in [t_2, t_3] \subset FTP_2$. It is noted that it is only illustrated in one of the cases that T_1 and T_2 are overlap visually. Other cases like meet and contain can be extended in a similarly way.

Fig. 2 shows fuzzy relations of fuzzy time intervals, and their corresponding crisp relations. The dashed area represents fuzzy time area of each fuzzy time interval, which can be any sub-interval of the fuzzy area. *FTIbefore* is the fuzzy relation if any possible time interval of one fuzzy time interval stays before the other fuzzy time interval; *FTIequal* is the fuzzy relation if the minimum and the maximum possible

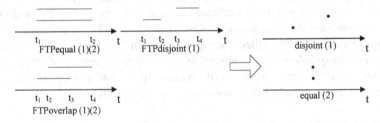

Fig. 1. Fuzzy relations of fuzzy time points

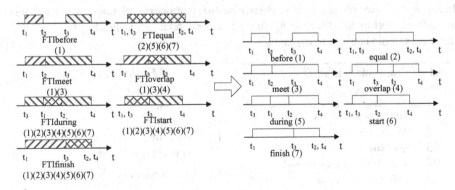

Fig. 2. Fuzzy relations of fuzzy time intervals

1 2 3 4 / 5 6 7 8	FTPdisjoint	FTPequal	FTPoverlap	FTIbefore	FTIequal
disjoint (1)	√ × ×	√ √ × ×	√ √ × ×	× √	× √
equal (2) overlap (3)	× × × ×	× × × ×	× × × ×		√ √ √
	FTImeet	**FTIoverlap**	**FTIduring**	**FTIstart**	**FTIfinish**
before (4) meet (5) during (6) start (7)	× √ √ √ ×	√ √	× √ √ √	× √ √ √	× √ √ √
finish (8)	√	√ √ √ √	√ √ √ √	√ √ √ √	√ √ √ √

Fig. 3. Correspondences between fuzzy and crisp temporal relations

Time intervals of the two fuzzy time intervals are equal; *FTImeet* is the fuzzy relation if the minimum possible time point of one fuzzy time interval equals to the maximum possible time point of the other fuzzy time interval; *FTIoverlap* is the fuzzy relation if the minimum possible time point of one fuzzy time interval stays between another possible time interval of the fuzzy time interval and the maximum possible time point of the other fuzzy time interval stays between this possible time interval of the fuzzy time interval; *FTIduring* is the fuzzy relation if any possible time intervals of one fuzzy time interval are contained by any possible time intervals of the other fuzzy time interval; *FTIstart* is the fuzzy relation if the minimum possible time points of the two fuzzy time intervals are equal; *FTIfinish* is the fuzzy relation if the maximum possible time points of the two fuzzy time intervals are equal.

The corresponding crisp temporal relations of fuzzy relations between fuzzy time points and between fuzzy time intervals are shown in parentheses in Fig. 1 and Fig. 2.

In order to present the correspondences between fuzzy and crisp temporal relations more intuitively, Fig. 3 shows their correspondences. In Fig. 3, it is denoted as cross if there are no correspondences; it is denoted as tick if there are correspondences and the fuzzy relation has the corresponding crisp relation; it is blank if there are correspondences but the fuzzy relation has no corresponding crisp relation.

4   Determination of Fuzzy Temporal Relations

In this section, we present how to determine fuzzy temporal relations. A general formalized algorithm for determining fuzzy temporal relations is firstly proposed, and then an example is given to explain it.

```
Algorithm Frelation Y, Z

01    for (k = 1; k <= X; k++)
02    let δ_k = 0
03    end for
04    for (m = 1; m <= i; i++)
05       for (n = 1; n <= m; n++)
06          for (r = 1; r <= X; r++)
07             if F_r relation (Y, Z)
08                δ_r = δ_r + δ_X
09             end for
10          end for
11    end for
12    if true (F_T relation)
13       return δ_T / Σ_{i=1}^{x} δi
```

The algorithm *Frelation* is a general algorithm for determining fuzzy temporal relations. It contains two loops in order to compare all possible fuzzy temporal relations. The membership degrees of the fuzzy temporal relation employ cumulative way to compute. The finally returned value should be transformed into relative one since it is obtained by two membership degrees which composed of two fuzzy time points

For the algorithm, Y and Z indicate two fuzzy temporal data (it can be further explained by their representing time points); X indicates number of fuzzy relations between Y and Z. $F_r relation$ (Y, Z) compares fuzzy relations X times according to the definition 2 or 4; T indicates the number satisfying fuzzy relation ranging from 1 to X, and the returned value is divided by all the possible membership degrees.

Consider $FTI_1 = \{[2, 0.6, 5, 0.7], [3, 0.4, 5, 0.7], [3, 0.4, 6, 0.3]\}$ and $FTI_2 = \{[3, 0.8, 4, 0.6], [4, 0.2, 5, 0.2], [4, 0.2, 6, 0.2]\}$. Because there are two fuzzy time intervals, Y and Z are FTI_1 and FTI_2, and their representing time points are $(T_{1si}, \delta_{1i}, T_{1ei}, \delta'_{1i})$, and $(T_{2sj}, \delta_{2j}, T_{2ej}, \delta'_{2j})$. There are seven fuzzy relations between two fuzzy time intervals so that X equals to 7. According to the definition 4, we can get the fuzzy relation of FTI_1 and FTI_2 is $FTIfinish$. Then we get the *during, finish, overlap, start* relation pair. Accordingly, we compute the membership degrees of each relation: $\delta_5 = 0.1392$; $\delta_7 = 0.0328$; $\delta_4 = 0.028$; $\delta_6 = 0.192$. Finally, the returned value is divided by all the possible membership degrees: the possibility that the relation is *during* approximately amounts to 0.355, is *finish* approximately amounts to 0.084, is *overlap* approximately amounts to 0.071, and is *start* approximately amounts to 0.490.

5   Conclusion

In the scope of this paper, we propose a general characterization of representing and determining fuzzy temporal relations. Definitions of fuzzy temporal data and their fuzzy relations are defined from mathematical point of view assuming that fuzzy temporal intervals are all fuzzy. On this basis, correspondences between fuzzy and crisp fuzzy temporal relations are investigated. Finally, a general formalized algorithm for determining fuzzy temporal relations is proposed. Future work is extending fuzzy temporal relations to fuzzy spatiotemporal relations.

Acknowledgments. This work is supported by the National Natural Science Foundation of China (60873010, 61073139) and the Fundamental Research Funds for the Central Universities (N090504005), and in part by the Program for New Century Excellent Talents in University (NCET-05-0288).

References

1. Allen, J.F.: Maintaining Knowledge about Temporal Intervals. Communications of ACM 26(11), 832–843 (1983)
2. Badaloni, S., Falda, M.: Classical and Fuzzy Neighborhood Relations of the Temporal Qualitative Algebra. In: Proceedings of the 16th International Symposium on Temporal Representation and Reasoning, pp. 147–154 (2009)

3. Badaloni, S., Giacomin, M.: The Algebra IA^{fuz}: A Framework for Qualitative Fuzzy Temporal Reasoning. Artificial Intelligence 170(10), 872–908 (2006)
4. Barro, S., Marin, R., Mira, J., Paton, A.R.: Model and Language for the Fuzzy Representation and Handling of Time. Fuzzy Sets and System 61(2), 153–175 (1994)
5. Dubois, D., Fargier, H., Prade, H.: Possibility Theory in Constraint Satisfaction Problems: Handling Priority, Preference and Uncertainty. Applied Intelligence 6(4), 287–309 (1996)
6. Dubois, D., Prade, H.: Processing Fuzzy Temporal Knowledge. IEEE Transaction on System, Man and Cybernetics 19(4), 729–744 (1989)
7. El-Kholy, A., Richards, B.: Temporal and Resource Reasoning in Planning: the parcPLAN Approach. In: Proceedings of the 12th European Conference on Artificial Intelligence (ECAI 1996), pp. 614–618 (1996)
8. Fox, M., Long, D.: PDDL2.1: An Extension to PDDL for Expressing Temporal Planning Domains. Journal of Artificial Intelligence Research 20, 61–124 (2003)
9. Freksa, C.: Spatial and Temporal Structures in Cognitive Processes. In: Freksa, C., Jantzen, M., Valk, R. (eds.) Foundations of Computer Science. LNCS, vol. 1337, pp. 379–387. Springer, Heidelberg (1997)
10. Harabagiu, S., Bejan, C.: Question Answering Based on Temporal Inference. In: AAAI 2005 Workshop on Inference for Textual Question Answering, Pittsburgh, PA (2005)
11. Jonsson, P., Bäckström, C.: A Unifying Approach to Temporal Constraint Reasoning. Artificial Intelligence 102(1), 143–155 (1998)
12. Kautz, H., Ladkin, P.: Integrating Metric and Qualitative Temporal Reasoning. In: The 9th National Conference on Artificial Intelligence, pp. 241–246 (1991)
13. Koubarakis, M.: Tractable Disjunctions of Linear Constraints: Basic Results and Applications to Temporal Reasoning. Theoretical Computer Science 266(1), 311–339 (2001)
14. Meiri, I.: Combining Qualitative and Quantitative Constraints in Temporal Reasoning. Artificial Intelligence 87(1), 343–385 (1996)
15. Navarrete, I., Sattar, A., Wetprasit, R., Marin, R.: On Point-Duration Networks for Temporal Reasoning. Artificial Intelligence 140(1-2), 39–70 (2002)
16. Ribarić, S., Bašić, B.D., Maleš, L.: An Approach to Validation of Fuzzy Qualitative Temporal Relations. In: Proceedings of the 24th International Conference on Information Technology Interfaces, pp. 223–228 (2002)
17. Sanampudi, S.K., Kumari, G.V.: Temporal Reasoning in Natural Language Processing: A Survey. International Journal of Computer Applications 1(4), 53–57 (2010)
18. Schockaert, S., De Cock, M.: Temporal Reasoning about Fuzzy Intervals. Artificial Intelligence 172(8-9), 1158–1193 (2008)
19. Schockaert, S., De Cock, M., Kerre, E.E.: An Efficient Characterization of Fuzzy Temporal Interval Relations. In: Proceedings of the IEEE International Conference on Fuzzy Systems, pp. 1894–1901 (2006)
20. Zadeh, L.A.: Fuzzy Sets. Information and Control 8(4), 338–353 (1965)
21. Zadeh, L.A.: Fuzzy Sets as a Basis for Theory of Possibility. Fuzzy Sets and Systems 1(1), 3–28 (1978)

Applications on Information Flow and Biomedical Treatment of FDES Based on Fuzzy Sequential Machines Theory

Hongyan Xing[1,3] and Daowen Qiu[1,2]

[1] Department of Computer Science, Sun Yat-sen University, Guangzhou, 510006, P.R. China
issqdw@mail.sysu.edu.cn
[2] SQIG--Instituto de Telecomunicações, Departamento de Matemática, Instituto Superior Técnico, TULisbon, Av. Rovisco Pais 1049-001, Lisbon, Portugal
[3] Faculty of Applied Mathematics, Guangdong University of Technology, Guangzhou, 510090, P.R. China

Abstract. In order to more effectively cope with the real world problems of vagueness, impreciseness, and subjectivity, fuzzy discrete event systems (FDES) were proposed and developed in recent ten years. In this paper, we study the applications of FDES to information flow inference and then, to biomedical control treatment planning and decision making based on fuzzy sequential machines (FSM) theory. Through modeling security system and biomedical decision problem with FDES as an FSM, we extend propositions and procedures to decide the equivalent states, display the ideas for checking the observation label based on event-state approach to decide whether hidden massage flow exists.

Keywords: fuzzy sequential machines, fuzzy discrete event systems, information flow, supervisory control.

1 Introduction

Discrete event systems (DES) [1] are dynamical systems whose evolution in time is governed by the abrupt occurrence of physical events at possibly irregular time intervals. In most of engineering applications, the states of a DES are crisp. However, this is not the case in many other applications such as biomedical systems and economic systems, in which vagueness, impreciseness, and subjectivity are typical features. Notably, Lin and Ying [2] initiated significantly the study of fuzzy discrete event systems (FDES) by combining fuzzy set theory with classical DES. Then, Qiu [3] established a supervisory control theory of FDES and designed a test-algorithm for checking the existence of supervisors. As was known, the main task of supervisory control is to find a supervisor that restricts the plant behavior modeled by a machine in order to comply with a specification behavior. In classical DES, under certain condition we can check within finite number of steps whether the controllability condition holds by utilizing the finiteness of states in crisp finite automata, but the infiniteness of fuzzy states in fuzzy DESs modeled by max-product automata gives

D.-S. Huang et al. (Eds.): ICIC 2012, LNCS 7389, pp. 57–64, 2012.
© Springer-Verlag Berlin Heidelberg 2012

rise to considerable complexity for formulating a uniform fashion to check the fuzzy controllability conditions. However, when FDESs are modeled by max-min automata, the author in [3] derived a uniform criterion to check this condition.

With a desire to provide further applications of FSM theory in FDES, in this paper, we first reformulate DES using state vectors and event transition matrices and then extend to fuzzy vectors and matrices by allowing their elements to take values in [0,1]. We deal with information flow analysis, which is closely related to confidentiality -- an important security requirement in many secure systems.

As is known, in practice, information safety directly decides the safety of the whole systems, in other words, secrecy is an often-appearing problem in finding and preventing the spread of secret or other important messages. It also plays an important role in describing the exact synthesis in communication networks. Theoretically, information flow analysis of the security systems is a fundamental way to label and analyze the hidden information flow and its skill and relative study have attracted attention for long times (see [4] and its references). For example, Goguren and Mesegurer proposed the first theoretical model in 1984 to describe the system confidentiality and security [5] called as noninterference-model. This model considered the condition of the conservation of the safety as forbidding high security part from executing influence to low security part. In [6], Zakintihinos put forward a general theory and gave out a unified form on the speciality of information flow.

However, we still cannot completely make sure that confidential information never leaks out, because the mechanisms cannot prevent hidden information flow and confidential information can stealthily leak out. In real systems, many safety strategies are described as state-transition rules, and also, they are formally described as a relative model of state-transition rules easily [7,8]. This process puts the system as a model firstly: taking the real safe system as a discrete event system. Motivated by this procedure, we can formalize the safe strategy as a fuzzy sequential machine (FSM), a state-based transition system driven by events in analyzing hidden information flow. As it is well known, sequential machines are simple but important models of computation such as in software engineering and lexical analysis [9], and so do their fuzzy-counterparts, which have still more practical importance in description of natural languages.

The rest of the paper is organized as follows. Section II recalls some required notations, properties, and methods deciding the equivalence between states and FSM. Then in Section III we describe the idea to analyze security systems modeled as FSM. In this model, security policies in actual secure systems are described, a state and event-based approach is displayed to decide whether hidden information flow exists, and a computation procedure to decide the equivalence is also proposed in this Section. Afterwards, we discuss an example in Section IV to illustrate the applications of the derived results and methods in fuzzy supervisory control of FDES. Finally, concluding remarks and further considerations are made in Section V.

2   Preliminaries

This section serves as recalling some preliminaries concerning fuzzy sequential machines (FSM). In what follows, N denotes the set of natural numbers, $|X|$ the cardinality of set X, Σ an alphabet and Σ^* the free monoid generated from Σ with the operation of concatenation; empty string ε is identified with the identity of Σ^*; $F(X)$ (or $\tilde{X}$, $[0,1]^X$) and $P(X)$ are used to represent the classes of all $[0,1]$-valued fuzzy subsets on X and classical subsets of X, respectively.

Definition 1. A quadruple $M = (Q, \Sigma_0, \Sigma_1, \delta)$ is called a fuzzy sequential machine, simply FSM, where Q, Σ_0, Σ_1 are finite sets, and $\delta : Q \times \Sigma_0 \times \Sigma_1 \times Q \to [0,1]$, a fuzzy subset of $Q \times \Sigma_0 \times \Sigma_1 \times Q$, is the transition function of FSM M.

As usual, Σ_0, Σ_1 stand for input and output alphabets, respectively; and $Q = \{q_1, q_2, ..., q_n\}$ a set of internal states. We can regard $\delta(q_j, u_i, v_r, q_k) = a_{jk}^{ir} \in [0,1]$ as the grade of the proposition that FSM M enters state q_k and produces output v_r given that the present state is q_j and input is u_i.

We call finite sequences of elements of $\Sigma_0(\Sigma_1)$ input (output) as tapes or strings. The collection of all input (output) tapes will be denoted by $\Sigma_0^*(\Sigma_1^*)$. The length of tape θ is denoted by $|\theta|$, so it is clear that $|\varepsilon| = 0$. Moreover, $(\Sigma_0 \times \Sigma_1)^* = \{(\theta, \gamma) | \theta \in \Sigma_0^*, \gamma \in \Sigma_1^*, |\theta| = |\gamma|\}$ and $\theta/\gamma \in (\Sigma_0 \times \Sigma_1)^*$. Furthermore, we assume all machines have the same input event set Σ_0, output event set Σ_1, and the input-output pair (θ, γ) satisfying $\theta/\gamma \in (\Sigma_0 \times \Sigma_1)^*$.

Definition 2. Let $Q = \{q_1, q_2, ..., q_n\}$ be a set, $|Q| = n$. A vector set pertaining to Q, denoted by vQ, is defined as:

$$VQ = \{q = (a_1, a_2, ..., a_n)^T | a_i \in [0,1], i = 1, 2, ..., n\}$$

i.e., column matrix $q = (a_1, a_2, ..., a_n)^T$ corresponds to Q with which elements a_i stands for the membership degree that the current state is $q_i, i = 1, 2, ..., n$.

In an FSM $M = (Q, \Sigma_0, \Sigma_1, \delta)$, δ can be defined by $\delta(q_j, u_i) = (a_{jk}^{ir})$, i.e.,

$$\delta : Q \times \Sigma_0 \to V(\Sigma_1 \times Q),$$

and, corresponding to δ, matrix $M_{u_i/v_r} = (a_{jk}^{ir})$ (also written as $M(u_i/v_r)$) characterizes the work of M. Note that the transition function δ can be extended to δ^*, from $Q \times (\Sigma_0 \times \Sigma_1)^* \times Q$ into $[0,1]$ defined inductively in the natural way.

Let us consider the words $\theta = u_1 u_2 ... u_m \in \Sigma_0^*, \gamma = v_1 v_2 ... v_m \in \Sigma_1^*$, , and the matrice $M_{u_i/v_r} = (a_{jk}(u_i/v_r))$, where $a_{jk}(u_i/v_r) = a_{jk}^{ir} = \delta(q_j, u_i, v_r, q_k) \in [0,1]$. With the operation of matrices, $\max - \min$ or $\max$-product, denoted by $\circ$, we obtain the expression

$$M_{\theta/\gamma} = M_{u_1/v_1} \circ M_{u_2/v_2} \circ ... \circ M_{u_m/v_m}, \text{ where} \tag{1}$$

$$M_{u_1/v_1} \circ M_{u_2/v_2} \underset{def}{=} \left(a_{jk}\left(u_1/v_1\right)\right) \circ \left(a_{jk}\left(u_2/v_2\right)\right) = \left(a_{jk}\left(u_1 u_2/v_1 v_2\right)\right), \text{ and} \tag{2}$$

$$a_{jk}\left(u_1 u_2/v_1 v_2\right) = \max_{q_s \in Q} a_{js}\left(u_1/v_1\right) \circ a_{sk}\left(u_2/v_2\right)$$
$$= \max_{q_s \in Q} \delta\left(q_j, u_1, v_1, q_s\right) \circ \delta\left(q_s, u_2, v_2, q_k\right). \tag{3}$$

Assume $Q = \{q_1, q_2, ..., q_n\}$, i.e. $|Q| = n$. Then

$$M_{\theta/\gamma} = a_{jk}\left(\theta/\gamma\right) = \left(a_{jk}\left(u_1 u_2 \cdots u_m/v_1 v_2 \cdots v_m\right)\right), \text{ where}$$

$$a_{jk}\left(u_1 u_2 \cdots u_m/v_1 v_2 \cdots v_m\right)$$
$$= \max_{1 \le i_1, i_2, ..., i_{m-1} \le n} a_{ji_1}\left(u_1/v_1\right) \circ a_{i_1 i_2}\left(u_2/v_2\right) \circ \cdots \circ a_{i_{m-1}k}\left(u_m/v_m\right). \tag{4}$$

Definition 3. An initial distribution of FSM $M = (Q, \Sigma_0, \Sigma_1, \delta)$ is a mapping η from Q to $[0,1]$, i.e., η is concentrated at $q \in Q$, denoted as η_q, if $\eta(q) = 1$ and 0 elsewhere. We will also use symbol η to denote the row vector whose i-th entry is $\eta(q_i)$. The ordered pair (M, η) is called an initialized FSM. If η is concentrated at $q \in Q$, we write (M, q) for (M, η).

Definition 4. Let $M = (Q, \Sigma_0, \Sigma_1, \delta)$, and let (M, η) be an initialized FSM. We set

$$S_\eta(\theta/\gamma)_M = \eta \circ M(\theta/\gamma) \circ P^1 \tag{5}$$

an entry indicating the maximal membership degree for the input-output pair θ/γ.

For the language theory, the generated and marked languages of (M, η), denoted by $L(M, \eta)$ and $L_m(M, \eta)$ respectively, or just simply $L(M)$ and $L_m(M)$, respectively, can be described as

$$L(M, \eta) = \left\{\theta/\gamma \in \left(\Sigma_0 \times \Sigma_1\right)^* \middle| S_{\eta,m}(\theta/\gamma) > 0\right\} \tag{6}$$

$$L_m(M, \eta) = \left\{\theta/\gamma \in \left(\Sigma_0 \times \Sigma_1\right)^* \middle| S_{\eta,m}(\theta/\gamma) = \max_{q_m \in Q_m} \eta \circ M_{\theta/\gamma} \circ q_m, q_m \in Q_m\right\} \tag{7}$$

Definition 5. Two initialized FSM (M, η) and (M', η'), where $M = (Q, \Sigma_0, \Sigma_1, \delta)$, $M' = (Q', \Sigma_0, \Sigma_1, \delta')$, are equivalent, denoted by $(M, \eta) \sim (M', \eta')$, if $L(M, \eta) = L(M', \eta')$, or more clearly, for all $\theta/\gamma \in \left(\Sigma \times \Sigma_1\right)^*$,

$$S_\eta(\theta/\gamma)_M = S_{\eta'}(\theta/\gamma)_{M'}. \tag{8}$$

In particular: If $(M, q) \sim (M', q')$, that is, η is concentrated at q and η' at q', i.e., $S_q(\theta/\gamma)_M = S_{q'}(\theta/\gamma)_{M'}$, then state q and q' are equivalent, denoted by $q \sim q'$.

Definition 6. (Observation-equivalence) Let $M = (Q, \Sigma_0, \Sigma_1, \delta)$ be an FSM. M is in reduced form if for each $q, q' \in Q$, relation $q \sim q'$ implies $q = q'$. Suppose two states $q, q' \in Q$ are in reduced FSM M. We call q, q' to be observation-equivalent if $(M, q) \sim (M', q')$.

Definition 7. (Synchronous composition of two FSMs) Let (M_i, η_i) be two initialized FSMs, where $M_i = (Q_i, \Sigma_0, \Sigma_1, \delta_i), i = 1,2$. The synchronous composition of (M_1, η_1) and (M_2, η_2) is a new machine (M, η), here

$$M = M_1 \| M_2 = (Q_1 \times Q_2, \Sigma_0, \Sigma_1, \delta_1 \| \delta_2), \eta = \eta_1 \| \eta_2, \tag{9}$$

where $Q_1 \times Q_2 = \{(q_1, q_2): q_i \in Q_i, i = 1,2\}$, and, for any $(q_1, q_2) \in Q_1 \times Q_2$ and any $\sigma \in \Sigma_0, \sigma' \in \Sigma_1$, $(\delta_1 \| \delta_2)((q_1, q_2), \sigma, \sigma', (q_1', q_2')) = \min(\delta_1(q_1, \sigma, \sigma', q_1'), \delta_2(q_2, \sigma, \sigma', q_2'))$.

3   Message Analyses Based on Equivalence Relations

In reality, information flow conforms to the demand of security and the secrecy of the system demands as follows: there exist two class of users, high-class user and low-class user (for simplicity, written as H. user and L. user, respectively). Every user has its own special operation and observation.

Definition 8. For initialized FSM (M, η), $M = (Q, \Sigma_0, \Sigma_1, \delta)$, the projection of M from H. users onto L. users, is also an initialized FSM (M_p, η), where $M_p = (Q, \Sigma_{0p}, \Sigma_{1p}, \delta_p)$, satisfying $\Sigma_{0p} \subseteq \Sigma_0$, $\Sigma_{1p} \subseteq \Sigma_1$, which denotes input and output events executable for L. users, respectively; and the transition function $\delta_p: Q \times \Sigma_{0p} \times \Sigma_{1p} \times Q \to [0,1]$, defined as :

$$\delta_p(q, a, y, q') = \begin{cases} \delta(q, a, y, q'), & a \in \Sigma_{0p}, \\ \delta_p(q, \varepsilon, \varepsilon, q') = 1, & a \notin \Sigma_{0p}. \end{cases} \tag{10}$$

i.e., $\delta_p = \delta|_{\Sigma_{0p} \times \Sigma_{1p}}$.

Definition 9. Given projection (M_p, η) of (M, η), where

$$M_p = (Q, \Sigma_{0p}, \Sigma_{1p}, \delta_p), M = (Q, \Sigma_0, \Sigma_1, \delta). \tag{11}$$

We say there exists hidden information flow if $\Sigma_0 \times \Sigma_1 - \Sigma_{0p} \times \Sigma_{1p} \neq \phi$.

Remark 1. In DES theory, usually a mask m is defined as a function $m: Q \to \Gamma$ that maps elements from the machine state space Q to the observation space Γ. In an FDES, we can also assume mask m in system $M = (Q, \Sigma_0, \Sigma_1, \delta)$ constructed as $m: Q \to \Gamma$ such that if $\forall q_i, q_j \in Q$, satisfy $\eta \circ M(\theta/\gamma)_M \circ P_i = \eta \circ M(\theta/\gamma)_M \circ P_j$ for some $\theta/\gamma \in (\Sigma_0 \times \Sigma_1)^*$, then $m(q_i) = m(q_j)$ and q_i, q_j to be indistinguishable relative to θ/γ. In other words, two states that are reached by some input/output string having same membership degree are relative-indistinguishable and the mask m may be constructed timely to map these states the same observation.

Procedure 1. Computing Steps decide distinguished states and distinct buttons.

Step 1: For two states q_i, q_j waiting for differentiate in FSM M, set two initial distribution $\eta = \eta_{q_i}, \eta' = \eta_{q_j}$ of FSM M;

Step 2: Choose $\theta/\gamma \in (\Sigma_0 \times \Sigma_1)^*$, write out the transition matrix $M(\theta/\gamma)$; decide whether $M(\theta/\gamma) \circ P_i = M(\theta/\gamma) \circ P_j$; for ``yes'', go to **step 3**; for ``no'', go to **step 4**;

Step 3: Compute $\eta \circ M(\theta/\gamma)$ and $\eta' \circ M(\theta/\gamma)$, decide whether $\eta \circ M(\theta/\gamma) = \eta' \circ M(\theta/\gamma)$; for answer ``yes'', $q_i \sim q_j$, otherwise, go to **step 4**;

Step 4: Note down the input/output symbol θ/γ as the distinct-button of q_i, q_j.

4   Application to Medical Decision Systems

When a DES is modeled by a finite automaton $G = (Q, \Sigma, \delta, q_0, Q_m)$, language $L(G)$ generated by G may be interpreted as the physically possible behavior. In order to alter the behavior of G, a supervisor S is introduced. Formally, S is defined as a function from $L(G)$ to $P(\Sigma)$. It is interpreted that for each $s \in L(G)$, $S(s) \cap \{\sigma : s\sigma \in L(G)\}$ represents the set of enabled events after the occurrence of s. Furthermore, it is required that for any $s \in L(G)$,

$$\Sigma_{uc} \cap \{\sigma \in \Sigma : s\sigma \in L(G)\} \subseteq S(s). \tag{12}$$

As before, we use FSM M to model FDES in this section. Suppose each fuzzy event σ is associated with a degree of controllability, so, the uncontrollable set Σ_{uc} and controllable set Σ_c are two fuzzy subsets of Σ, i.e., $\Sigma_{uc}, \Sigma_c \in F(\Sigma)$, and satisfy: for any $\sigma \in F(\Sigma)$,

$$\Sigma_{uc}(\sigma) + \Sigma_c(\sigma) = 1. \tag{13}$$

A supervisor $\tilde{S}$ of FDES M is defined as a function:

$$\tilde{S} : F(\Sigma_0 \times \Sigma_1)^* \to F(\Sigma_0 \times \Sigma_1). \tag{14}$$

Similar to the admissibility condition (12), for crisp supervisors, $\tilde{S}$ is usually required to satisfy that for any $s \in \tilde{\Sigma}^*$ and $\sigma \in \tilde{\Sigma}$,

$$\min\{\Sigma_{uc}(\sigma), L(M)(s\sigma)\} \le \tilde{S}(s)(\sigma). \tag{15}$$

Notations concerning prefix-closed property in the sense of FDES are: for any fuzzy string $s \in \tilde{\Sigma}^*$, $pr(s) = \{t \in \tilde{\Sigma}^* : \exists r \in \tilde{\Sigma}^*, tr = s\}$. For any fuzzy language L over $\tilde{\Sigma}^*$, its prefix-closure $pr(L): \tilde{\Sigma}^* \rightarrow [0,1]$ is defined as: $pr(L)(s) = \sup_{s \in pr(t)} L(t)$. So $pr(L)(s)$ denotes the possibility of string s belonging to the prefix-closure of L. By means of the formulation of the above concepts, now we can present the controllability theorem concerning FDESs. The following proposition is changed from Theorem 1 in [3].

Proposition 1. Let an FDES be modeled by FSM $M = (Q, \Sigma_0, \Sigma_1, \delta)$. Suppose fuzzy uncontrollable subset $\Sigma_{uc} \in \tilde{\Sigma} = F(\Sigma_0 \times \Sigma_1)$, and fuzzy legal subset $K \in \tilde{\Sigma}^*$ that satisfies: $K \subseteq L(M)$, and $K(\varepsilon) = 1$. Then there exists supervisor $\tilde{S} \in \tilde{\Sigma}^* \rightarrow \tilde{\Sigma}$, such that $\tilde{S}$ satisfies the fuzzy admissibility condition Eq. (15) and $L(\tilde{S}/M) = pr(K)$ if and only if for any $s \in \tilde{\Sigma}^*$ and any $\sigma \in \tilde{\Sigma}$,

$$\min\{pr(K)(s), \Sigma_{uc}(\sigma), L(M)(s\sigma)\} \leq pr(K)(s\sigma), \tag{16}$$

where Eq. (16) is called fuzzy controllability condition of K with respect to M and Σ_{uc}.

5   Concluding Remarks

We further developed FDES by dealing with information flow analysis based on sequential machine theory. Through modeling a security system by using a fuzzy sequential machine, the idea of a state-event-based approach to decide whether hidden massage flows exist was displayed. Furthermore, we study the supervisory control theory in FDES and discussed a medicine issue modeled by FSM. The technical contributions are mainly as follows: (i) we reformulated the parallel composition of crisp DES, and defined the parallel composition of FDES; (ii) we gave a state-event-based approach to decide whether hidden massage flows exist and a computing process was introduced to get the equivalent states and, by means of this, we can search for all possible distinguishable states and their distinct-buttons.

With the results obtained in [3] and this paper, it is worth further considering to apply the supervisory control theory of FDES to practical control issues, particularly in economic and traffic control systems. Moreover, dealing with FDES modeled by fuzzy petri nets [10] is of interest, as the issue of DES modeled by Petri nets [1], and we deem it a significant research direction.

Acknowledgments. This work is supported in part by the National Natural Science Foundation (Nos. 60873055, 61073054), the Natural Science Foundation of Guangdong Province of China (No. 10251027501000004), the Research Foundation for the Doctoral Program of Higher School of Ministry of Education of China (No. 20100171110042), the Fundamental Research Funds for the Central Universities

(No. 10lgzd12), and the project of SQIG at IT, funded by FCT and EU FEDER projects QSec PTDC/EIA/67661/2006, FCT project PTDC/EEA-TEL/103402/2008 QuantPriv-Tel, FCT PEst-OE/EEI/LA0008/2011, AMDSC UTAustin/MAT/0057/2008, IT Project QuantTel, Network of Excellence, Euro-NF.

References

1. Cassandras, C.G., Lafortune, S.: Introduction to Discrete Event Systems. Kluwer, Boston (1999)
2. Lin, F., Ying, H.: Modeling and Control of Fuzzy Discrete Event Systems. IEEE Transactions on System, Man, Cybern., B 32(4), 408–415 (2002)
3. Qiu, D.W.: Supervisory Control of Fuzzy Discrete Event Systems: A Formal Approach. IEEE Transactions on System, Man, Cybern., B 35(2), 72–88 (2005)
4. Zi, X.C., Yao, L.H., Li, L.: A State-based Approach to Information Flow Analysis. J. of Computers 29(8), 1460–1467 (2006) (in Chinese)
5. Goguren, J.A., Messegurer, J.: Security Policies and Security Models. In: Proceedings of the IEEE Symposium on Security and Privacy, California, USA, pp. 75–85 (1984)
6. Zakintihinos, A., Lee, E.S.: A General Theory of Security Properties. In: Proceedings of the IEEE Symposium on Security and Privacy, California, USA, pp. 94–102 (1997)
7. Shayman, M.A., Kumar, R.: Supervisory Control of Nondeterministic Systems with Driven Events via Prioritized Synchronization and Trajectory Models. SIAM J. Control Optim. 33(2), 469–497 (1995)
8. Thomason, M.G., Marinos, P.N.: Deterministic Acceptors of Regular Fuzzy Languages. IEEE Transactions on System, Man and Cyber. 4(1), 228–230 (1974)
9. Kumar, R., Heymann, M.: Masked Prioritized Synchronization for Interaction and Control of Discrete Event Systems. IEEE Transactions on Automatic Control 45(11), 1970–1982 (2000)
10. Looney, C.G.: Fuzzy Petri Nets for Rule-based Decision making. IEEE Trans. Syst., Man, Cybern. 18(1), 178–183 (1988)

Discontinuous Fuzzy Systems
and Henstock Integrals of Fuzzy Number Valued
Functions

Yabin Shao[1] and Zengtai Gong[2]

[1] College of Mathematics and Computer Science
Northwest University for Nationalities, Lanzhou 730030, China
[2] College of Mathematics and Information Science
Northwest Normal University, Lanzhou 730070, China
yb-shao@163.com

Abstract. In this paper, using the properties of the strong Henstock integrals of fuzzy-number-valued functions and controlled convergence theorem, we prove the existence theorem for the discontinuous fuzzy system $x' = \tilde{f}(t, x)$ in fuzzy number space, where $\tilde{f}$ is strong fuzzy Henstock integrable.

Keywords: fuzzy number, strong fuzzy Henstock integrals, cauchy problem, existence of solution.

1 Introduction

The Henstock integral is designed to integrate highly oscillatory functions which the Lebesgue integral fails to do. It is known as nonabsolute integration and includes the Riemann, improper Riemann, Lebesgue and Newton integrals [7]. The Riemann-type definition of nonabsolute integration was introduced more recently by Henstock in 1963 and also independently by Kurzweil, the definition is now simple and furthermore the proof involving the integral also turns out to be easy.

It is well known that the theory of fuzzy sets provides an effective means of describing the behavior of system which are too complex or too ill-defined to admit precise mathematical analysis by classical methods and tools. We have combined the above theories and discussed the fuzzy Henstock integrals of fuzzy-number-valued functions which extended Kaleva integration. In order to complete the theory of fuzzy calculus and to meet the solving need of transferring a fuzzy differential equation into a fuzzy integral equation, we also have defined the strong fuzzy Henstock integrals and discussed some of their properties and the controlled convergence theorem.

On the other hand, the characterization of the derivatives, in both real and fuzzy analysis, is an important problem. Bede and Gal [1] have subsequently

D.-S. Huang et al. (Eds.): ICIC 2012, LNCS 7389, pp. 65–72, 2012.
© Springer-Verlag Berlin Heidelberg 2012

introduced a more general definition of a derivative for fuzzy-number-valued function enlarging the class of differentiable of fuzzy number valued functions.

The Cauchy problems for fuzzy differential equations have been studied by several authors [4,6,8,11] on the metric space (E^n, D) of normal fuzzy convex set with the distance D given by the maximum of the Hausdorff distance between the corresponding level sets. In [8], the author has been proved the Cauchy problem has a uniqueness result if f was continuous and bounded. Wu and Song [11] changed the initial value problems of fuzzy differential equations into a abstract differential equations on a closed convex cone in a Banach space by the operator j which is the isometric embedding from (E^n, D) onto its range in the Banach space X. They obtained the existence theorems under the compactness-type conditions. In 2002, Xue and Fu [12] established solutions to fuzzy differential equations with right-hand side functions satisfying Caratheodory conditions on a class of Lipschitz fuzzy sets. However, the study on fuzzy discontinuous systems is insufficient.

In this paper, according to the idea of [12] and the operator j which is the isometric embedding from (E^n, D) onto its rang in the Banach space X, we shall deal with the Cauchy problem of discontinuous systems as following

$$\begin{cases} x'(t) = \tilde{f}(t, x(t)) \\ x(0) = x_0 \in E^n, \end{cases} \tag{1}$$

where E^n is a fuzzy number space, and $\tilde{f} : [0, \gamma] \times B \to E^n$ is strong fuzzy Henstock integrable, and $B = \{x : D(x, \tilde{0}) \le D(x_0, \tilde{0}) + b, b > 0\}$.

2   Preliminaries

Let $P_k(R^n)$ denote the family of all nonempty compact convex subset of R^n and define the addition and scalar multiplication in $P_k(R^n)$ as usual. Let A and B be two nonempty bounded subset of R^n. The distance between A and B is defined by the Hausdorff metric [2]:

$$d_H(A, B) = \max\{\sup_{a \in A} \inf_{b \in B} |a - b|, \sup_{b \in B} \inf_{a \in A} |b - a|\}.$$

Denote $E^n = \{u : R^n \to [0,1], u$ satisfies $(1) - (4)$ below$\}$ is a fuzzy number space. Where

(1) u is normal, i.e. there exists an $x_0 \in R^n$ such that $u(x_0) = 1$;

(2) u is fuzzy convex, i.e. $u(\lambda x + (1 - \lambda)y) \ge \min\{u(x), u(y)\}$ for any

$x, y \in R^n$ and $0 \le \lambda \le 1,$

(3) u is upper semi-continuous;

(4) $[u]^0 = cl\{x \in R^n \mid u(x) > 0\}$ is compact.

Define $D : E^n \times E^n \to (0, +\infty)$

$$D(u, v) = \sup\{d_H([u]^\alpha, [v]^\alpha) : \alpha \in [0,1]\},$$

where d_H is the Hausdorff metric defined in $P_k(R^n)$. Then it is easy see that D is a metric in E^n. Using the results [3], we know that the metric space (E^n, D) has a linear structure, it can imbedded isomorphically as a cone in a Banach space of function $u^* : I \times S^{n-1} \to R$, where S^{n-1} is the unit sphere in R^n, which an imbedding function $u^* = j(u)$ defined by $u^*(r, x) = \sup_{\alpha \in [u]^\alpha} \langle \alpha, x \rangle$.

It is well know that the H-derivative for fuzzy functions was initially introduced by Puri and Ralescu [2] and it is based in the condition (H) of sets. In this paper, we consider a more general definition of a derivative for fuzzy number valued functions enlarging the class of differentiable fuzzy number valued functions, which has been introduced in [1].

Definition 1[1]. Let $\tilde{f} : (a,b) \to E^n$ and $x_0 \in (a,b)$. We say that $\tilde{f}$ is differentiable at x_0, if there exist an element $\tilde{f}'(t_0) \in E^n$, such that (1) for all $h > 0$ sufficiently small, there exist $\tilde{f}(x_0 + h) -_H \tilde{f}(x_0)$ and $\tilde{f}(x_0 + h) -_H \tilde{f}(x_0)$, the limits

$$\lim_{h \to 0} \frac{\tilde{f}(x_0 + h) -_H \tilde{f}(x_0)}{h} = \lim_{h \to 0} \frac{\tilde{f}(x_0) -_H \tilde{f}(x_0 - h)}{h} = \tilde{f}'(x_0)$$

(2) for all $h > 0$ sufficiently small, there exist $\tilde{f}(x_0) -_H \tilde{f}(x_0 + h)$ and $\tilde{f}(x_0 - h) -_H \tilde{f}(x_0)$, the limits

$$\lim_{h \to 0} \frac{\tilde{f}(x_0) -_H \tilde{f}(x_0 + h)}{-h} = \lim_{h \to 0} \frac{\tilde{f}(x_0 - h) -_H \tilde{f}(x_0)}{-h} = \tilde{f}'(x_0)$$

(3) for all $h > 0$ sufficiently small, there exist $\tilde{f}(x_0 + h) -_H \tilde{f}(x_0)$ and $\tilde{f}(x_0 - h) -_H \tilde{f}(x_0)$, the limits

$$\lim_{h \to 0} \frac{\tilde{f}(x_0 + h) -_H \tilde{f}(x_0)}{h} = \lim_{h \to 0} \frac{\tilde{f}(x_0 - h) -_H \tilde{f}(x_0)}{-h} = \tilde{f}'(x_0)$$

(4) for all $h > 0$ sufficiently small, there exist $\tilde{f}(x_0) -_H \tilde{f}(x_0 + h)$

and $\tilde{f}(x_0) -_H \tilde{f}(x_0 - h)$, the limits

$$\lim_{h \to 0} \frac{\tilde{f}(x_0) -_H \tilde{f}(x_0 + h)}{-h} = \lim_{h \to 0} \frac{\tilde{f}(x_0) -_H \tilde{f}(x_0 - h)}{h} = \tilde{f}'(x_0).$$

3   The Strong Henstock Integrals of Fuzzy Number Valued Functions in E^n

In this section we define the strong Henstock integrals of fuzzy number valued functions in fuzzy number space [5] and we give some properties of this integral.

Definition 2[9]. A function $\tilde{f} : [a,b] \to E^n$ is strong fuzzy Henstock integrable on $[a,b]$ ($\tilde{f} \in SFH([a,b], E^n)$) if there exist a function $\tilde{F} : [a,b] \to E^n$ with the following property: given $\varepsilon > 0$ there exist a positive function δ on $[a,b]$ such that $P = \{[c_i, d_i]; s_i\}$ is a tagged partition of $[a,b]$, then

$$\sum_{i=1}^{n} D(\tilde{f}(s_i)(d_i - c_i), \tilde{F}([c_i, d_i])) < \varepsilon.$$

where D is the Hausdorff metric of the distance between fuzzy numbers.

Theorem 1. Let $\tilde{f} : [a,b] \to E^n$ be a strong fuzzy Henstock integrable on $[a,b]$ and let $\tilde{F}(t) = \int_a^t \tilde{f}(s)ds$ for each $s \in [a,b]$. Then

(1) F is continuous on $[a,b]$,

(2) F is differentiable almost everywhere on $[a,b]$,

(3) f is measurable.

Proof. Because of the fuzzy number space (E^n, D) is a complete metric space, the proof is similar to the real valued function in Ref. [10, 15], we omit it here.

Theorem 2 [5]. Suppose $\{\tilde{f}_n\}$ is a sequence of SFH integrable functions on $[a,b]$ satisfying the following conditions:

(1) $\tilde{f}_n(x) \to \tilde{f}(x)$ a. e. in $[a,b]$ as $n \to \infty$;

(2) the primitives $\tilde{F}_n$ of $\tilde{f}_n$ are ACG^* uniformly in n;

(3) the primitives $\tilde{F}_n$ converge uniformly on $[a,b]$;

Then $\tilde{f}$ also SFH integrable on $[a,b]$ and

$$\lim_{n \to \infty} \int_a^b \tilde{f}_n(x)dx = \int_a^b \tilde{f}(x)dx.$$

4 The Existence of Solutions for Discontinuous Fuzzy Systems

In this section we will prove the existence theorem for the problem (1). For any bounded subset A of the Banach space X we denote $\alpha(A)$ the Kuratowski measure of noncompactness of A, i.e the infimum of all $\varepsilon > 0$ such that there exist a finite covering of A by sets of diameter less than ε. For the properties of α we refer to [3] for example.

Lemma 1 [3]. Let $H \subset C(I_\gamma, X)$ be a family of strong equicontinuous functions. Then

$$\alpha(H) = \sup_{t \in I_\gamma} \alpha(H(t)) = \alpha(H(I_\gamma))$$

where $\alpha(H)$ denote the Kuratowski measure of noncompactness in $C(I_\gamma, X)$ and the function $t \to \alpha(H(t))$ is continuous.

Let $C(x_0, \gamma) = \{x \in C(I_\gamma) : x(0) = x_0, D(x, \tilde{0}) \le D(x_0 \tilde{0}) + b\}$ (b and γ are some positive numbers). Obviously, the fuzzy set $C(x_0, \gamma)$ is closed and convex. Let $\tilde{F}_x$ be defined by

$$\tilde{F}_x(t) = x_0 + \int_0^t \tilde{f}(s, x(s))ds \quad \text{or} \quad \tilde{F}_x(t) = x_0 + (-1) \cdot \int_0^t \tilde{f}(s, x(s))ds$$

for $t \in I_\gamma$ and $x \in C(x_0, \gamma)$ where the integrals in the sense of SFH.

Definition 3. A fuzzy number valued function $\tilde{f}$ is a Caratheodory function, if

(1) $\tilde{f}$ is measurable for any $x \in E^n$;

(2) $\tilde{f}$ is continuous for any $t \in I_\gamma$.

Lemma 2. Let V be equicontinuous bounded set in $C(I_\gamma, E^n)$, $\tilde{f}$ is a Caratheodory function and $\tilde{f}(\cdot, x(\cdot))$ be a SFH integrable function for each $x \in V$. Let $\tilde{F} = \{\tilde{F}_x : x \in C(x_0, \gamma)\}$ be equicontinuous and uniformly ACG^* on I_γ. Then

$$\alpha(j \circ \int_0^t \tilde{f}(s, V(s))ds) \le \int_0^t \alpha(j \circ \tilde{f}(s, V(s)))ds,$$

whenever $\alpha(j \circ \tilde{f}(s, V(s))) \le \varphi(s)$, for $s \in I_\gamma$ a.e. $\varphi(s)$ is a Lebesgue integrable function, and

$$\int_0^t \tilde{f}(s, V(s))ds = \{\int_0^t \tilde{f}(s, x(s))ds, x(s) \in V(s)\}.$$

Proof. Because $\tilde{f}(\cdot, x(\cdot))$ be a *SFH* integrable function, then $\tilde{f}_\alpha^-$ and $\tilde{f}_\alpha^+$ is Henstock integrable for all $\alpha \in [0,1]$. According to the theory of real analysis, the conclusion is hold true.

Theorem 3 [3]. Let D be a closed convex subset of X, and Let F be a continuous function from D into itself. If for $x \in D$ the implication

$$\overline{V} = \overline{con}(\{x\} \cup F(V)) \Rightarrow V \qquad (2)$$

is relatively compact, then F has a fixed point.

Definition 4[3]. A nonnegative function $(t, r) \rightarrow h(t, r)$ is a Kamke function on $I \times R^+$ if

(1) $h(t, r)$ satisfies the Caratheodory conditions;

(2) $h(t,0) = 0$ and the function indentically equal to zero is the unique continuous solution of equation $u(t) = \int_0^t h(s, u(s))ds$, for $t \in I$ satisfying $u(0) = 0$.

Next, we give the main results for this paper.

Theorem 4. If for each continuous $x: I_\gamma \rightarrow E^n$, $\tilde{f}(\cdot, x(\cdot))$ be a *SFH* integrable, $\tilde{f}$ is Caratheodory function and

$$\alpha(j \circ \tilde{f}(t, X)) \le h(t, \alpha(j \circ X)), \qquad (3)$$

for each bounded subset $X \subset E^n$, where h is a Kamke function. Let $\tilde{F} = \{\tilde{F}_x : x \in C(x_0, \gamma)\}$ be equicontinuous and uniformly ACG^* on I_γ. Then there exist a solution of problem (1) on I_β for some $0 < \beta \le \gamma$.

Proof. By equicontinuous of $\tilde{F}$, there exist a number β and $0 < \beta \le \gamma$ such that $D(\int_0^t \tilde{f}(s, x(s))ds, \tilde{0}) \le b$ for $t \in I_\beta$ and $x \in C(x_0, \gamma)$. By assumption, the operaor $\tilde{F}_x$ is well defined and maps $C(x_0, \beta)$ into $C(x_0, \beta)$. Using Theorem 2 for the *SFH* integral we deduce that $\tilde{F}$ is continuous.

Suppose that $\overline{V} = \overline{con}(\{x\} \cup \tilde{F}(V))$ for some bounded $V \subset C(x_0, \beta)$. We shall prove that V is relatively compact, thus (2) is satisfied. In fact, $\tilde{F}(V)$ is equicontinuous, the function $v(t) \rightarrow \alpha(j \circ V(t))$ is continuous on I_β, and

$$\tilde{F}(V(t)) = \{x_0 + \int_0^t \tilde{f}(s,x(s))ds\} \qquad \text{or}$$

$$\tilde{F}(V(t)) = \{x_0 + (-1)\cdot \int_0^t \tilde{f}(s,x(s))ds\}.$$

By Lemma 2 and (3), we have

$$\alpha(j\circ\tilde{F}(V(t)) \le \alpha(j\circ\{\int_0^t \tilde{f}(s,x(s))ds, x(s)\in V(s)\})$$

$$\le \int_0^t \alpha(j\circ\tilde{f}(s,V(s)))ds \le \int_0^t h(s,\alpha(j\circ V(s)))ds.$$

Since $\overline{V} = \overline{con}(\{x\}\cup\tilde{F}(V))$, by the property of measure of noncompactness, we have $\alpha(j\circ V(t)) \le \alpha(j\circ\tilde{F}(V(t)))$ and

$$v(t) = \alpha(j\circ V(t)) \le \int_0^t h(s,v(s))ds.$$

Hence, we have $v(t) = \alpha(j\circ V(t))$. By Lemma 1, V is relatively compact. So, by Theorem 3 has a fixed point which is a solution of (1). The proof is completed.

5   Conclusions

In this paper, we deal with the Cauchy problem of discontinuous fuzzy differential equations involving the strong fuzzy Henstock integral in fuzzy number space. The function governing the equations is supposed to be discontinuous with respect to some variables and satisfy nonabsolute fuzzy integrablility. Our result improves the result given in Ref. [4, 6, 8] and [12] (where uniform continuity was required), as well as those referred therein.

Acknowledgements. The authors wish to thanks the referees for the careful reading and valuable remarks which improved the presentation of the paper. This work is supported by the National Natural Science Fund of China (No.11161041 and No.7106013).

References

1. Bede, B., Gal, S.: Generalizations of the Differentiability of Fuzzy-Number-Valued-Functions with Applications to Fuzzy Differential Equation. Fuzzy Sets and Syst. 151, 581–599 (2005)
2. Diamond, P., Kloeden, P.: Metric Space of Fuzzy Sets: Theory and Applications. World Scientific, Singapore (1994)
3. Banas, J., Goebel, K.: Measure of Noncompactness in Banach Space. Lecture Notes in Pure and Appl. Math., vol. 60. Mercel Dekker, New York (1980)
4. Gong, Z., Shao, Y.: Global Existence and Uniqueness of Solutions for Fuzzy Differential Equations under Dissipative-type Conditions. Computers & Math. with Appl. 56, 2716–2723 (2008)

5. Gong, Z., Shao, Y.: The Controlled Convergence Theorems for the Strong Henstock Integrals of Fuzzy-Number-Valued Functions. Fuzzy Sets and Syst. 160, 1528–1546 (2009)
6. Kaleva, O.: Fuzzy Differential Equations. Fuzzy Sets and Syst. 24, 301–319 (1987)
7. Lee, P.: Lanzhou Lectures on Henstock Integration. World Scientific, Singapore (1989)
8. Nieto, J.J.: The Cauchy Problem for Continuous Fuzzy Differential Equations. Fuzzy Sets and Syst. 102, 259–262 (1999)
9. Puri, M.L., Ralescu, D.A.: Differentials of Fuzzy Functions. J. Math. Anal. Appl. 91, 552–558 (1983)
10. Wu, C.X., Ma, M.: On Embedding Problem of Fuzzy Number Spaces: Part 2. Fuzzy Sets and Syst. 45, 189–202 (1992)
11. Wu, C., Song, S.: Existence Theorem to the Cauchy Problem of Fuzzy Differential Equations under Compactness-type Conditions. Inf. Sci. 108, 123–134 (1998)
12. Xue, X., Fu, Y.: Caratheodory Solution of Fuzzy Differential Equations. Fuzzy Sets and Syst. 125, 239–243 (2002)

A Phased Adaptive PSO Algorithm for Multimodal Function Optimization

Haiping Yu[1,*] and Fengying Yang[2]

[1] Faculty of Information Engineering, City College Wuhan University of Science and Technology, Wuhan, China
[2] College of information engineering, Huanghuai University, Henan, China
yhp0308@yahoo.com.cn, ziying661@163.com

Abstract. Particle swarm optimization is a powerful algorithm that has been applied to various kinds of problems. However, it suffers from falling into local minimum and prematurity especially on multimodal function optimization problems. In this paper, a phased adaptive particle swarm optimization(PAPSO) is proposed to solve such problem. The process is divided into the initial particle pre-searching phase and the post-searching cooperative phase. In the post phase, the strategy of selecting randomly a certain number of particles for entering the reverse-learning is one of the most effective ways of escaping local stagnation. The illustrative example is provided to confirm the validity, as compared with the SPSO, Dynamic Inertia Weight PSO(PSO-W), and Tradeoff PSO(PSO-T) in terms of convergence speed and the ability of jumping out of the local optimal value. Simulation results confirm that the proposed algorithm is effective and feasible.

Keywords: particle swarm optimization, multimodal function, adaptive.

1 Introduction

Particle swarm optimization is a swarm intelligence technique developed by Kennedy and Eberhart in 1995[1], because it is a simple and effective, various improved algorithms have been proposed in recent years. For aspects of inertia weight, such as references[2-3]; for hybrid algorithms, such as references[4-6],etc. Particle swarm optimization has been used increasingly as an effective technique for solving complex optimization problems successfully, such as multi-objective optimization, training neural network and emergent system identification[7-9]. Researchers have found the algorithm suffers from premature convergence and falls into local minima. So it is of great significance to solve the above shortcoming. As a result, accelerating convergence rate and avoiding local optima have become the most important and appealing goals in PSO research[10].To overcome the above limitation, a phased adaptive particle swarm optimization(PAPSO) is proposed in this paper which can search global value for multimodal functions in high-speed of convergence.

This paper is organized as follows. Section 2 is devoted to the standard particle swarm optimization. In section 3, the proposed technique of PAPSO are described and analyzed. In section 4, six multimodal functions are tested by the PAPSO, and the results are analyzed in detail. Finally, the paper closes with conclusion and ideas for the further research about PSO in section 5.

D.-S. Huang et al. (Eds.): ICIC 2012, LNCS 7389, pp. 73–79, 2012.
© Springer-Verlag Berlin Heidelberg 2012

2   The Standard Particle Swarm Optimization

2.1   Algorithm Background

Many scientists have created computer simulations of various interpretations of the movement of organisms in a bird flock or fish school. Notably, Reynolds , Heppner and Grenander presented simulations of bird flocking choreography, and Heppner, a zoologist, was interested in discovering the underlying rules that enabled large numbers of birds to flock synchronously, often changing direction suddenly, scattering and regrouping, etc. Both of these scientists had the insight that local processes, such as those modeled by cellular automata, might underlie the unpredictable group dynamics of bird social behavior. Both models relied heavily on manipulation of inter-individual distances[1].

Particle Swarm optimizer(PSO) is an optimization algorithm first proposed by James Kennedy and Russell Beernaert[1], The original algorithm is based on the sociological behavior associated with bird flock or fish school. Because of the less algorithm parameters, the PSO algorithm is applied in many fields such as network training, optimization, and fussy control and so on.

2.2   The Theory of the Particle Swarm Optimization

In PSO, each swarm member called a particle represents a potential solution to an optimization problem in searching space. Each particle adjusts the search direction by learning from its own experience and the global particles' experiences. Specifically, each particle velocity is updated by following two optimum values. The first one is the best solution (fitness) that has been achieved so far. This value is called pbest. The second one is the global best value obtained so far by any particle in the swarm. This best value is called gbest[11]. Each particle updates its velocity by using formula (1), and each particle updates its position by using formula(2). At the same time, when a particle finds a better position than the previous one; its location is stored in memory. In other words , the algorithm works on the social behavior of particles in the swarm. In the end, particles can find the global-best position by simply adjusting their own position. The formula is as follows:

$$v_{ij}(t+1) = wv_{ij}(t) + c_1 r_1 (pbest_{ij}(t) - x_{ij}(t)) + c_2 r_2 (gbest_{gj}(t) - x_{ij}(t)) \tag{1}$$

$$x_{ij}(t+1) = x_{ij}(t) + v_{ij}(t+1) \tag{2}$$

This equation shows that the new velocity is not only related to the old velocity but also associated with the position of the particle itself and of the global best one. And the degree of influence depends on the inertia weight and the two cognitive factors. So reasonable choice of parameters has a great influence on the algorithm. In 2002, a constriction named λ coefficient is used to prevent each particle from exploring too far away in the range of min and max, since λ applies a suppression effect to the oscillation size of a particle over time. This method constricted PSO suggested by Clerc and Kennedy is used with λ set it to 0.7298, according to the formula[12-13].

However, PSO lacks the ability of global search in searching space, and it easily falls into local optimum especially in solving the multimodal function optimization. There are some reasons for this phenomenon. The multimodal function shape is more complicated. In addition, because of the parameter settings of the particle number and particle number for improper ,the diversity of particles gradually disappears, so the algorithm is easy to fall into local optimal solution. As the result, a new method will be demonstrated by the multimodal function optimization.

Table 1. Definition of the Parameters

Parameters	Descriptions
t	iteration number
w	inertia weight
r_1, r_2	random vectors the range is (0,1)
c_1	self-cognitive factor
c_2	social-cognitive factor
i	equals 1,2,3,...,N(N--the swarm size)
j	equals 1,2,3,...,D(D--the dimension of searching place)
v	the velocity of each particle
x	the position of each particle

3   A Phased Adaptive Particle Swarm Optimization(PAPSO)

According to the above description., it is very difficult to find the global optimum just by using the standard PSO. In order to overcome this limitation, we have proposed a new concept named a phased adaptive particle swarm optimization(PAPSO). The main idea is described as follows:

In primary phase, particles have stronger self-learning ability than the global one, therefore, first, let each particle search around its own personal best by setting a control parameter. If the number of failures reaches the control parameter, adjust the particle search strategy by way of setting the current optimum equals the global optimum. Then the searching enters an advanced phase. In this phase, the synergies among particles become stronger. We set the current value equals the global optimum, at the same time adjust the learning factor. If the number of failures reaches the control parameter, randomly select a certain number of particles for entering the reverse learning. The searching process iterates, until the particles find the global optimum or reach to the maximum number of iterations.

The algorithm flow diagram are as follows:

Step1. Define the Basic Conditions

Define the parameters: the parameters of the algorithm involved includes that population size ; the minimum and maximum ranges(according to the problem); the range of velocity; the dimension; the control parameter.

Step2. Initialize the Particle Velocity and Position.

Initialize the locations and velocities of all particles randomly in the searching space, and let the initial pbest of each particle equal its current position.

Step3. Update Velocity

In primary phase, the velocity of the particles changes as the formula (3), in advanced phase, update the velocity using formula (1), the c1 changes as the formula(4), c2=2.

$$v_{ij}(t+1) = wv_{ij}(t) + c_1 r_1 (pbest_{ij}(t) - x_{ij}(t)) \tag{3}$$

$$c_1 = \frac{iter}{\max iter} * 2 \tag{4}$$

where iter means the current times, maxiter means the maximum number of iterations.

Step4. Update Position

In primary phase: each particle modifies its position according to the formula (2). In advanced phase: When the particle' position has not changed in the range of the control parameter, it seems that the particles are likely to fall into local optimum. In order to avoid the phenomenon, we select a certain number of particles and calculate their opposite particles as follows[14,15]:

$$\overline{P}_{i,j} = \lambda(a_j + b_j) - P_{i,j}, \lambda = rand(0,1) \tag{5}$$

where $P_{i,j}$ is the jth vector of the ith candidate in the particle, $\overline{P}_{i,j}$ is the transformed one. the range of the population is between a_j and b_j in the jth dimension.

Step5. Update pbests and gbests.

Step6. Repeat and Check Convergence.

Steps3-5 are repeated until all particles are gathered around the global best or a maximum iteration is encountered.

4   Multimodal Function Test and Results

Six testing multimodal functions have been used to verify the performance of the proposed algorithm. A procedure description of the proposed algorithm is provided by Table 2, and the experimental settings is provided by Table 3 the results are showed in Table4.

In table 2, the six test function used in the experiments, where D is the dimension of the function, R is the definition domain, and F_{min} means the minimum value of the function.

Table 2. Definition of the Multimodal Function

Name	F	D	R	F_{min}
Sine	$f_1(x) = \sum_{i=1}^{n} \sin(x_i) \cos^{20}(\dfrac{i \times x_i^2}{\pi})$	30	[-PI,PI]	0
Schewefel	$f_2(x) = \sum_{i=1}^{n}(-x_i \sin(\sqrt{|x_i|}))$	30	[-500,500]	0
Griewank	$f_3(x) = \dfrac{1}{4000}\sum_{i=1}^{n} x_i^2 - \prod_{i=1}^{n}\cos(\dfrac{x_i}{\sqrt{i}}) + 1$	30	[-600,600]	0
Rastrigin	$f_4(x) = \sum [x_i^2 - 10\cos(2\pi x_i + 10)]$	30	[-5.12,5.12]	0
Rosenbrock	$f_5(x) = \sum_{i=1}^{n}(100(x_{i+1} - x_i^2)^2 + (x_i - 1)^2)$	30	[-30,30]	0
Ackley	$f_6(x) = 20 + e - 20e^{-\frac{1}{5}\sqrt{\frac{1}{n}\sum_{i=1}^{n} x_i^2}} - e^{\frac{1}{n}\sum \cos(}$	30	[-32,32]	0

Table 3. The Experimental Parameter Settings

Algorithm	P_S	c_1	c_2	w_{max}/w_{min}	$Iter_{max}$
SPSO	50	1.49618	1.49618	0.729843	150000
PSO-W	50	2.05	2.05	0.95/0.4	150000
PSO-T	50	2	2	0.9/0.4	150000
PA PSO	50	-	2	0.9/0.4	150000

Table 4. Results of the Experiment Comparing with Other PSOs

		f_1	f_2	f_3	f_4	f_5	f_6
SPSO	Avg	-2.11E+1	-7.75E+03	1.77E-02	4.57E+01	2.38E+01	1.18
	Std_D	1.75	6.03E+02	1.92E-02	1.11E+01	3.13E+01	6.60E-01
PSO-W	Avg	**-2.12E+1**	-8.39E-03	1.57E-02	3.62E+01	4.43E+01	1.71E-01
	Std_D	**2.42**	4.43E+02	2.06E-2	9.15	3.10E+1	4.81E-01
PSO-T	Avg	-1.35E+1	**-9.52E+03**	1.13E-2	3.13E+1	2.67E+1	7.84E-06
	Std_D	1.72E+01	**3.04E+02**	8.96E-3	4.90	5.37E+1	8.37E-6
PAPSO	Avg	-2.04E+1	-7.06E+03	8.92E-4	1.20E+01	2.26E+1	7.26E-15
	Std_D	1.86	4.02E+02	**3.27E-3**	5.90	**1.61**	**2.03E-15**

In table 3, where P_S is the size of population; Wmax/Wmin is the inertia weight gradient descent range; Itermax is the maximum number of iterations.

In table 4 Avg means the average of the best value, Std means the standard deviation. PSO_T is the reference 3 algorithm The bold number is the best one.

5   Conclusion

A Phased Adaptive PSO Algorithm for Multimodal function optimization is proposed in this paper. It keeps the basic concepts of the PSO, at the same time embeds the search mechanism by controlling parameter and reversing learning strategies. The testing results show that it avoids falling into local minimum and premature convergence compared to conventional PSO, PSO-W, and PSO-T. Moreover the PAPSO has a unique advantage in many multimodal functions optimization. However the PAPSO is not suitable for all kinds of problems. Further study is to investigate the effectiveness of PAPSO in the near future.

References

1. Eberhart, R., Kennedy, J.: A New Optimizer Using Particle Swarm Theory. In: Proc. 6th Int. Symp. Micromach. Hum. Sci., Nagoya, Japan, pp. 39–43 (1995)
2. Nie, P., Ji, G.Q., Zhi, G.: Self-adaptive Inertia Weight PSO Test Case Generation Algorithm Considering Prematurity Restraining. International Journal of Digital Content Technology and its Applications 5(9), 125–133 (2011)
3. Chen, F.: Tradeoff Strategy Between Exploration and Exploitation for PSO. In: Seventh International Conference on Natural Computation, pp. 1216–1222 (2011)
4. Abdel, K., Rehab, F.: An Improved Discrete PSO with GA Operators for Qos-multicase Routing. International Journal of Hybrid Information Technology, 223–238 (2011)
5. Wang, X.H., Li, J.J.: Hybrid Particle Swarm Optimization with Simulated Annealing. In: Proceedings of the Third International Conference on Machine Learning and Cybernetics, pp. 26–29 (2004)
6. Li, S.T., Tan, M.K., Ivor, W.T.: A Hybrid PSO-BFGS Strategy for Global Optimization of Multimodal Functions. IEEE Trans. on Systems, Man and Cybernetics 41(4) (2011)
7. Wang, Y.F., Zhang, Y.F.: A PSO-based Multi-objective Optimization Approach to the Integration of Process Planning and Scheduling. In: 8th IEEE International Conference on Control and Automation, pp. 614–619 (2010)
8. Hu, X., Eberhart, R.: Multiobjective Optimization Using Dynamic Neighborhood Particle Swarm Ptimization. In: Congress on Evolutionary Computation (CEC 2002), vol. 2, pp. 1677–1681. IEEE Service Center, Piscataway (2002)
9. Y, S.: Design of Neural Network Gain Scheduling Flight Control Law Using a Modified PSO Algorithm Based on Immune Clone Principle. In: Second International Conference on Intelligent Computation Technology and Automation, pp. 259–263 (2009)
10. Ho, S.Y., Lin, H.S., Liauh, W.H., Ho, S.J.: OPSO Orthogonal Particle Swarm Optimization and its Application to Task Assignment Problems. IEEE Trans. Syst., Man, Cybern. A, Syst. Humans 38(2), 288–298 (2008)

11. Sotirios, K., Goudos, V.M., Theodoros, S.: Application of a Comprehensive Learning Particle Swarm Optimizer to Unequally Spaced Linear Array Synthesis With Sidelobe Level Suppression and Null Control. IEEE Antennas and Wireless Propagation Letters 9, 125–129 (2010)
12. Clerc, M., Kennedy, J.: The Particle Swarm-explosion Stability and Convergence in a Multidimensional Complex Space. IEEE Trans. Evol. Comput., 58–73 (2002)
13. Li, X.D.: Niching Without Niching Parameters: Particle Swarm Optimization Using a Ring Topology. IEEE Transactions on Evolutionary Computation, 150–169 (February 2010)
14. Rahnamayan, S., Tizhoosh, H.R., Salama, M.M.A.: Opposition-based Differential Evolution. IEEE Trans. Evolut. Comput. 12, 64–79 (2008)
15. Wang, H., Zhi, J.W., Shahryar, R.: Enhancing Particle Swarm Optimization Using Generalized Opposition-based Learning. Information Sciences 181 (2011)

The Comparative Study of Different Number of Particles in Clustering Based on Three-Layer Particle Swarm Optimization[*]

Guoliang Huang, Xinling Shi, Zhenzhou An, and He Sun

School of Information Science and Engineering,
Yunnan University, Kunming 650091, China
hgl_wan.2008@163.com

Abstract. To study how the different number of particles in clustering affect the performance of three-layer particle swarm optimization (THLPSO) that sets the global best location in each swarm to be the position of the particle in the swarm of the next layer, ten configurations of the different number of particles are compared. Fourteen benchmark functions, being in seven types with different circumstance, are used in the experiments. The experiments show that the searching ability of the algorithms is related to the number of particles in clustering, which is better with the number of particles transforming from as little as possible to as much as possible in each swarm when the function dimension is increasing from less to more. Finally, the original algorithm and THLPSO are compared to illustrate the efficiency of the proposed method.

Keywords: Particle swarm optimization, hierarchy, cluster.

1 Introduction

Particle swarm optimization (PSO) algorithm is based on the evolutionary computation technique. It has been used increasingly as a novel technique for solving complex optimization problems. Many researchers have expanded on the original ideas of PSO, and some improving approaches such as the idea of hierarchy and cluster have been reported. In [1], a dynamically changing branching degree of the tree topology was proposed for solving intractable large parameter optimization problems. In [2], the particles have been clustered in each iteration and the centroid of the cluster was used as an attractor instead of using the position of a single individual. In [3], the population was partitioned into several sub-swarms, each of which was made to evolve based on the PSO. In [4], the PSO approach that used an adaptive variable population size and periodic partial increasing or declining individuals in the form of ladder function was proposed. In [5], a two-layer particle swarm optimization (TLPSO) was proposed for unconstrained optimization problems.

---

[*] This paper is supported by the National Natural Science Foundation of China No. 61062005.

D.-S. Huang et al. (Eds.): ICIC 2012, LNCS 7389, pp. 80–86, 2012.
© Springer-Verlag Berlin Heidelberg 2012

To study the influence on the ability to search function optimization among the configurations of the different number of particles in clustering, some previous work [6] based on a two-layer particle swarm optimization have been done. It has come to a conclusion that a good efficiency of searching ability is related to the number of particles transforming from as little as possible to as much as possible in each swarm when the function dimension is increasing from less to more. In order to further study the affection degree among different configurations of the different number of particles in clustering, a three-layer particle swarm optimization (THLPSO) is proposed for the comparison. Through the experiment, it confirms the accuracy of the results of [6] and deeper summarizes the conclusion. To illustrate the efficiency of THLPSO, the original algorithm is used to compare with the proposed method in the end.

The rest of this article is organized as follows. Section 2 describes the main idea and the basic process of THLPSO. Section 3 presents ten configurations of the different number of particles in clustering and seven classifications for the functions. In section 4, ten configurations of the different number of particles in clustering are compared in the benchmark functions existing in the seven types, then the following passages is the contrasting between the original algorithm and THLPSO. Finally, section 5 draws conclusions about the comparison among the ten configurations of the different number of particles testing in the seven types.

2   Three-Layer Particle Swarm Optimization(THLPSO)

In the THLPSO approach, there will be three layers of the structure: bottom layer, middle layer and top layer. M swarms of particles, B swarms of particles and one swarm of particles are generated in the three layers, respectively. In the initial population, M swarms of N particle, $x_b^{di}, d = 1, 2 \ldots M, i = 1, 2, \ldots N$, are randomly generated in the bottom layer, where x_b^{di} is the position of the ith particle in the dth swarm of the bottom layer. According to the fitness contrasting among the particles in the dth swarm of the bottom layer, the global best location of the dth swarm, $y_m^d, d = 1, 2, \ldots M$, is determined. Then the global best location of each swarm of the bottom layer is set to be in the middle layer, that is to say the number of particles in the middle layer has been determined, which is M. In the middle layer, B swarms of F particle, $y_m^{hk}, h = 1, 2 \ldots B, k = 1, 2, \ldots F$, are also randomly generated, where y_m^{hk} is the position of the kth particle in the hth swarm of the middle layer. According to the fitness contrasting among the particles in the hth swarm of the middle layer, the global best location of the hth swarm, $z_t^h, h = 1, 2, \ldots B$, is determined. Then the global best location of each swarm of the middle layer is set to be in the top layer, that is to say the number of particles in the top layer has also been determined, which is B. Therefore, the global best location of the swarm in the top layer will be determined according to the fitness contrasting among B particles in the top layer. Furthermore, two mutation operations are added into the particles of each swarm in the bottom layer and middle layer, respectively. Consequently, the diversity of the population in the THLPSO increases so that the THLPSO has the ability to avoid trapping into a local optimum. The structure of the proposed THLPSO is shown in Fig.1.

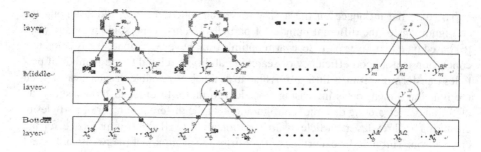

Fig. 1. The structure of the THLPSO

3   Classification and Analysis

3.1   Ten Configurations of the Number of Particles in Clustering and Computation Cost Analysis

In this experiment, the total number of particles is set to be one hundred and twenty, the number of swarms in the bottom layer (M) is set to be sixty. From the principle analysis of THLPSO in section 2, there will be sixty particles in the middle layer, which are chosen from each swarm with the global best location in the bottom layer, respectively. Ten different classifications about the sixty particles in the middle layer are shown in Table 1, where B means the total number of swarms in the middle layer and F stands for the average number of particles in clustering.

Table 1. The number of particles in clustering

B	2	3	4	5	6	10	12	15	20	30
F	30	20	15	12	10	6	5	4	3	2

Suppose the computation cost of one particle in the algorithms is c, then the total computation cost of the algorithms for one generation is $MNc+Mc+Bc$, where MNc stands for the computation cost of the bottom layer and Mc stands for the computation cost of the middle layer and Bc stands for the computation cost of the top layer. Therefore, the computation cost is increasing along with the increasing of the number of swarms in the middle layer. In other words, it corresponds to the decreasing of the number of particles in clustering.

3.2   Function Classification and Experiment Parameters Configuration

In [7], fourteen benchmark functions ($f_1 - f_{14}$), which were chosen for their variety, were generated and divided into three types. Functions $f_1 - f_3$ were simple unimodal problems, $f_4 - f_9$ were highly complex multimodal problems with many local minima, and $f_{10} - f_{14}$ were multimodal problems with few local minima. In this section, they are classified with the different dimensions in detail shown as follows:

- type 1. Unimodal function in two dimensions;
- type 2. Unimodal function in ten dimensions;
- type 3. Unimodal function in one hundred dimensions;
- type 4. Highly complex multimodal function with many local minima in two dimensions;
- type 5. Highly complex multimodal function with many local minima in ten dimensions;
- type 6. Highly complex multimodal function with many local minima in one hundred dimensions;
- type 7. Multimodal function with few local minima in two or four dimensions.

4   Experiments

4.1   Experimental Results Based on THLPSO

In this section, the seven types of the functions are employed to examine the efficiency of searching function optimization with the ten configurations of the different number of particles shown in Table 1, respectively. The simulation results about the ten configurations of the different number of particles in clustering testing in the seven types are shown in Fig.2. In the figure, the horizontal coordinate stands for the ten configurations of the different number of particles in clustering and the vertical coordinate stands for the average value of all the global optimum values produced in each cycle operation except for the maximum and the minimum values. And the figure of the left edge is the mark of the types from type 1 to type 7.

In Fig.2, type 1, type 4 and type 7 reveal the same results that the better searching ability is corresponding to the less number of particles in clustering in two or four dimensions, and type6 reveals the opposite that the more number of particles in clustering in one hundred dimensions shows better efficiency. Combined with type 2, type 3, type 5 and the results of [6], it comes to a conclusion that different types of functions will go through three stages of change along with the dimension of the functions transforming from less to more. The three stages of change are listed as follow:

- Stage 1: The less number of particles in clustering, the better searching ability for the function.
- Stage 2: The searching ability is better with the number of the particles transforming from as little as possible to as much as possible in clustering.
- Stage 3: The more number of particles in clustering, the better searching ability for the function.

4.2   The Contrast between Original Algorithm and THLPSO

Suppose the computation cost of one particle in the algorithms is c, then the total computation cost of THLPSO for one generation is $MNc+Mc+Bc$, which has been analysed above. On the other hand, the original algorithm (OPSO) has no the middle layer and top layer of the particles so that the computation cost for one generation is *MNc*. Therefore, the computation cost of the THLPSO is greater than that of the OPSO by *Mc+Bc*. However, the THLPSO spends *Mc+Bc* computation time on the movement of the particles in the middle layer and top layer for the global search. In order to illustrate the above results, Table 2 compares our results (the minimum average value among the ten configurations) of the proposed THLPSO approach with the results of the OPSO for the benchmark functions $f_1 - f_{14}$ with three different dimensions, respectively. In Table 2, *D* means the dimensions, *A* means the algorithms and *F* means the functions, all the values represents the average of all the global optimum values produced in each iteration.

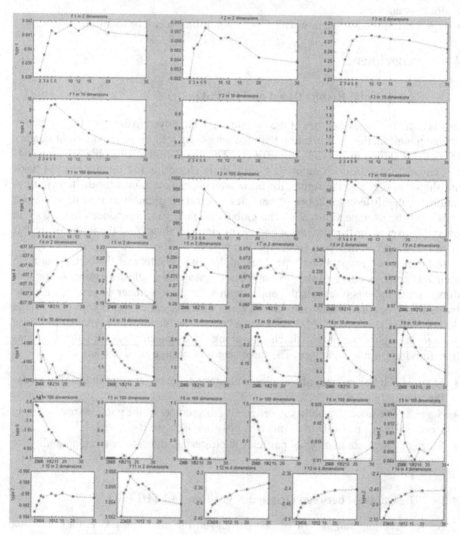

Fig. 2. Ten configurations of the number of particles compared from type 1 to type 7

Table 2. Comparison the Results between THLPSO and OPSO

D A F	D=2 or D=4 OPSO	THLPSO	D=10 OPSO	THLPSO	D=100 OPSO	THLPSO
f_1	0.2769	0.0385	115.6134	1.1289	3.22E+03	0.0922
f_2	0.2898	0.0521	4.6743	0.2496	3.11E+04	0.0404
f_3	1.3763	0.2357	5.2231	1.2576	280.2511	19.668
f_4	-809.9008	-837.8077	-3.64e+003	-4.19e+003	-2.69e+004	-4.09e+004
f_5	0.3345	0.1809	8.2818	1.1524	366.8465	3.45E-13
f_6	0.9323	0.2638	5.5007	1.3133	6.631	0.0186
f_7	0.1345	0.0689	1.0029	0.115	29.9999	0.0567
f_8	0.4475	0.3228	5.2959	0.304	7.39E+05	0.0102
f_9	0.0913	0.0709	11.6456	0.1617	4.71E+05	0.0043
f_{10}	0.0156	0.0104				
f_{11}	0.2023	0.1746				
f_{12}	-1.8778	-2.4379				
f_{13}	-1.9577	-2.4978				
f_{14}	-2.0649	-2.5444				

From the comparison, it can be seen that the accuracy of THLPSO are much better than OPSO in different dimensions with the same configuration parameters. It reveals that the OPSO might lead to the earlier convergence so that the result of the OPSO is trapped into the local optimum solution owing to the lack of swarm's diversity. On the other hand, the proposed THLPSO has more diversity such that it is hard to be trapped into the local optimum owing to having more searching choices for the particle swarm. Therefore, the THLPSO structure with three layers is necessary.

5   Conclusion

In this article, ten configurations of the different number of particles in clustering have been compared in the fourteen benchmark functions existing in different circumstance, respectively. According to the simulation results of the seven types, it can reach three general conclusions listed as follows:

- The less number of particles in clustering, the better searching ability for the function in less dimensions.
- The searching ability is better with the number of the particles transforming from as little as possible to as much as possible in clustering when the function dimension is increasing from less to more.
- The more number of particles in clustering, the better searching ability for the function in more dimensions.

References

1. Janson, S., Middendorf, M.: A Hierarchical Particle Swarm Optimizer and Its Adaptive Variant. IEEE Transactions on Systems, Man, and Cybernetics-Part B: Cybernetics 35, 1272–1282 (2005)
2. Kennedy, J.: Stereotyping: Improving Particle Swarm Performance with Cluster Analysis. In: Proceedings of the 2000 Congress on Evolutionary Computation, vol. 2, pp. 1507–1512 (2000)
3. Jiang, Y., Hu, T., Huang, C.C., Wu, X.: An Improved Particle Swarm Optimization algorithm. App. Math. Comp. 193, 231–239 (2007)
4. Chen, D.B., Zhao, C.X.: Particle Swarm Optimization with Adaptive Population Size and Its Application. App. Soft. Comp. 9, 39–48 (2009)
5. Chen, C.C.: Two-layer Particle Swarm Optimization for Unconstrained Optimization Problems. App. Soft. Comp. 11, 295–304 (2011)
6. Huang, G., Shi, X., An, Z.: The Comparative Study of Different Number of Particles in Clustering Based on Two-Layer Particle Swarm Optimization. In: Tan, Y., Shi, Y., Ji, Z. (eds.) ICSI 2012, Part I. LNCS, vol. 7331, pp. 109–115. Springer, Heidelberg (2012)
7. Bratton, D., Kennedy, J.: Defining a Standard for Particle Swarm Optimization. In: Proceedings of the 2007 IEEE Swarm Intelligence Symposium, pp. 120–127. IEEE Press, Honolulu (2007)

Implementation of Mutual Localization of Multi-robot Using Particle Filter

Yang Weon Lee

Department of Information and Communication Engineering,
Honam University, Seobongdong, Gwangsangu, Gwangju, 506-714, South Korea
ywlee@honam.ac.kr

Abstract. This paper describes an implementation of mutual localization of swarm robot using particle filter. Robots determine the location of the other robots using wireless sensors. Measured data will be used for determination of the robot itself moving method. It also effects on the other robot's formation such as circle and line type formation. We discuss the problem in circle formation enclosing target which moves. This method is the solution about enclosed invader in circle formation based on mutual localization of multi-robot without infrastructure. We use trilateration which does not need to know the value of the coordinates of reference points. So, specify enclosed point for the number of robots based on between the relative positions of the robot in the coordinate system. Particle filter is used to improve the accuracy of the robot's location. The particle filter is well operated for mutual location of robots than any other estimation algorithm. Through the experiments, we show that the proposed scheme is stable and works well in real environments

Keywords: swarm robot, particle filter, tracking.

1 Introduction

Swarm robotics is an approach to robotics that emphasizes many simple robots instead of a single complex robot. A robot swarm has much in common with an ant colony or swarm of bees. No individual in the group is very intelligent or complex, but combined, they can perform difficult tasks. Swarm robotics has been an experimental field, but many practical applications have been proposed. A traditional robot often needs complex components and significant computer processing power to accomplish its assigned tasks. In swarm robotics, each robot is relatively simple and inexpensive. As a group, these simple machines cooperate to perform advanced tasks that otherwise would require a more powerful, more expensive robot. Using many simple robots has other advantages as well. Robot swarms have high fault tolerance, meaning that they still will perform well if some of the individual units malfunction or are destroyed. Swarms also are scalable, so the size of the swarm can be increased or decreased as needed.

In this paper, we present a particle filter localization system for cooperative multiple robots in an environment where they can move automatically as a group and extract position information in a given map of the environment from stationary

D.-S. Huang et al. (Eds.): ICIC 2012, LNCS 7389, pp. 87–94, 2012.
© Springer-Verlag Berlin Heidelberg 2012

landmarks. For assessment of the system, we consider a scenario where some robots cannot detect landmarks. In this case, uncertainty in position estimates for all robots is expected to increase. We suppose that each mobile robot in the environment not only can detect and calculate the distance to other robots, but also can share its position information with other robots. Consequently, each mobile robot acts as an additional mobile landmark. However, it can introduce some amount of error to the position information from stationary landmarks, and localization could become unstable. Still, we desire self-localization performed by each robot to be improved by extra position information from other robots.

2   Localization Algorithm

Localization is the process of finding both position and orientation of a vehicle in a given referential system [20], Drumheller in [1],[2]. Navigation of mobile robots indoors usually requires accurate and reliable methods of localization. Many transportation systems now using wire-guided automated vehicles may benefit from the increased layout design flexibility provided by a wire-free localization method such as triangulation with active beacons.

2.1   Triangulation

Triangulation is based on the measurement of the bearings of the robot relatively to beacons placed in known positions. It differs from trilateration, which is based on the measurement of the distances between the robot and the beacons. These beacons are also called landmarks by some authors. According to [3], the term beacon is more appropriate for triangulation methods. When navigating on a plane, three distinguishable beacons - at least - are required for the robot to localize itself (Fig.1). λ_{12} is the oriented angle "seen" by the robot between beacons 1 and 2. It defines an arc between these beacons, which is a set of possible positions of the robot. An additional arc between beacons 1 and 3 is defined by λ_{31}. The robot is in the intersection of the two arcs. Usually, the use of more than three beacons results in redundancy.

2.2   Trilateration

Trilateration is a method to determine the position of an object based on simultaneous range measurements from three stations located at known sites. This is a common operation not only in robot localization [7,8,9], but also in kinematics, aeronautics, crystallography, and computer graphics. It can be trivially expressed as the problem of finding the intersection of three spheres that is, finding the solutions to the following system of quadratic equations:

$$(x - x_1)^2 + (y - y_1)^2 + (z - z_1)^2 = l_1^2$$
$$(x - x_2)^2 + (y - y_2)^2 + (z - z_2)^2 = l_2^2 \qquad (1)$$
$$(x - x_3)^2 + (y - y_3)^2 + (z - z_3)^2 = l_3^2$$

where x_1, x_2, x_3 are the coordinates of station $,i$ and l_i is the range measurement associated with it.

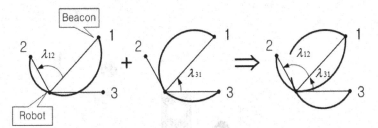

Fig. 1. Three-object triangulation

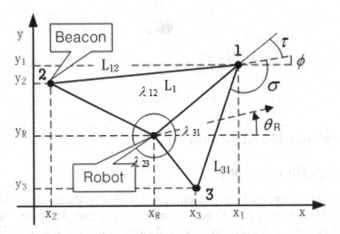

Fig. 2. Three-object triangulation

.In Figure 2, thick segments between stations define the base plane, and thin ones, those connecting the moving object and the stations, correspond to the range measurements.

2.3 Bounding Box

Bounding Box Method uses a simple box-shaped ranging area shown in Figure 3, which can be specified by lower left extreme coordinates, (x^l, y^l), and upper right extreme coordinates, (x^u, y^u). Figure 5 shows an example where node i is bounded by the ranging area of its neighbors, nodes 1, 2, and 3. Thus, the bounding area for the node i can be specified as follows:

$$(x^l(i), y^l(i)) = (\max(x^l(j)|j \in N(i)), \max(y^l(j)|j \in N(i)))$$
$$(x^u(i), y^u(i)) = (\max(x^u(j)|j \in N(i)), \max(y^u(j)|j \in N(i)))$$

(2)

Bounding Box Method can be implemented as a distributed algorithm; Simic and Sastry [9] suggest anchors broadcast their respective position estimates periodically, and nodes broadcast any changes in their estimates upon reception of any broadcast.

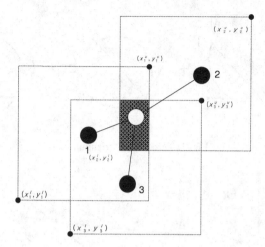

Fig. 3. Bounding box

3   Condensation Algorithm

3.1   Particle Filter Algorithms

The particle filter approach to track motion, also known as the condensation algorithm [4] and Monte Carlo localization, uses a large number of particles to 'explore' the state space. Each particle represents a hypothesized target location in state space. Initially the particles are uniformly randomly distributed across the state space, and each subsequent frame the algorithm cycles through the steps illustrated   as   follows :

1. Deterministic drift: particles are moved according to a deterministic motion model (a damped constant velocity motion model was used).
2. Update probability density function (PDF): Determine the probability for every new particle location.
3. Resample particles: 90 percent of the particles are resampled with replacement, such that the probability of choosing a particular sample is equal to the PDF at that point; the remaining 10 percent of particles are distributed randomly throughout the state space.
4. Diffuse particles: particles are moved a small distance in state space under Brownian motion.

This result in particles congregating in regions of high probability and dispersing from other regions, thus the particle density indicates the most likely target states. See[3] for a comprehensive discussion of this method. The key strengths of the particle filter approach to localization and tracking are its scalability (computational requirement varies linearly with the number of particles), and its ability to deal with multiple hypotheses (and thus more readily recover from tracking errors).However, the particle filter was applied here for several additional reasons:

- It provides an efficient means of searching for a target in a multi-dimensional state space.
- Reduces the search problem to a verification problem, i.e. is a given hypothesis face-like according to the sensor information?
- Allows fusion of cues running at different frequencies.

3.2   Application of Particle Filter for Multi-robots

In order to apply the particle filter algorithm to hand motion recognition, we extend the methods described by Black and Jepson [10,11]. Specifically, a state at time t is described as a parameter vector: $s^t = (\mu, \phi^i, \alpha^i, \rho^i)$ where μ is the integer index of the predictive model, ϕ^i indicates the current position in the model, α^i refers to An amplitude scaling factor and ρ^i is a scale factor in the time dimension.

1) Initialization

The sample set is initialized with N samples distributed over possible starting states and each assigned a weight of $\frac{1}{N}$. Specifically, the initial parameters are picked uniformly according to:

$$\left.\begin{array}{l} \mu \in [1, \mu_{max}] \\ \phi^i = \frac{1-\sqrt{y}}{\sqrt{y}}, y \in [0,1] \\ \alpha^i = [\alpha_{min}, \alpha_{max}] \\ \rho^i \in [\rho^{min}, \rho^{max}] \end{array}\right] \tag{3}$$

2) Prediction

In the prediction step, each parameter of a randomly sampled s_t is used to $t+1$ determine based on the parameters of that particular s_t. Each old state, s_t, is randomly chosen from the sample set, based on the weight of each sample. That is, the weight of each sample determines the probability of its being chosen. This is done efficiently by creating a cumulative probability table, choosing a uniform random number on [0, 1], and then using binary search to pull out a sample[14]. The following equations are used to choose the new state:

$$\left.\begin{array}{l} \mu_{t+1} = \mu_t \\ \phi^i_{t+1} = \phi^i_t + \rho^i_t + N(\sigma_\phi) \\ \alpha^i_{t+1} = \alpha^i_t + N(\sigma_\alpha) \\ \rho_{t+1} = \rho^i_t + N(\sigma_\rho) \end{array}\right] \tag{4}$$

where $N(\sigma_*)$ refers to a number chosen randomly according to the normal distribution with standard deviation σ_*. This adds an element of uncertainty to each prediction, which keeps the sample set diffuse enough to deal with noisy data. For a given drawn sample, predictions are generated until all of the parameters are within the accepted range. If, after, a set number of attempts it is still impossible to generate a valid prediction, a new sample is created according to the initialization procedure above.

3) Updating

After the Prediction step above, there exists a new set of N predicted samples which need to be assigned weights. The weight of each sample is a measure of its likelihood given the observed data $Z_t = (z_t, z_{t1}, \cdots)$. We define $Z_{t,i} = (z_{t,i}, z_{(t-1),i}, \cdots)$ as a sequence of observations for the ith coefficient over time; specifically, let $Z_{(t,1)}, Z_{(t,2)}, Z_{(t,3)}, Z_{(t,4)}$ be the sequence of observations of the horizontal velocity of the left robot, the vertical velocity of the left robot, the horizontal velocity of the right robot, and the vertical velocity of the right robot respectively. Extending Black and Jepson [5], we then calculate the weight by the following equation:

$$p(z_t|s_t) = \prod_{i=1}^{4} p(Z_{t,i}|s_t) \tag{5}$$

$$p(z_{t,i}|s_t) = \frac{1}{\sqrt{2\pi}} exp \frac{-\sum_{j=0}^{\omega-1}\left(z_{(t-j),i} - \alpha^* m^{\mu}_{(\phi-\rho^*j),i}\right)}{2(\omega-1)} \tag{6}$$

where ω is the size of a temporal window that spans back in time. Note that ϕ^*, α^* and ρ^* refer to the appropriate parameters of the model for the blob in question and that $\alpha^* m^{(\mu)}_{(\phi^*-\rho^*j),i}$ refers to the value given to the ith coefficient of the model μ interpolated at time $\phi^* - \rho^*j$ and scaled by α^*.

4   Experiment Result

To test the proposed particle filter scheme, we used MATLAB and visual studio. MATLAB is used for simulation of particle filter and visual studio is used to calculate the mutual localization of intruder and robot. Through experiment, we confirmed that accuracy of robot localization using particle filter is more than localization of using only sensor information. Therefore it is necessary to use particle filter to localize the robot when there is no more information except triangular measurement data. This information is shown in Figure [7] and [8].

Also, we evaluate the algorithm of intruder enclosed formation using trilateration. First of all, we assume that initially there are 6 robot and continue experiment alternatively swapped each robots position and intruder's position. Figure [9] and [10] is shown each robot make a circle formation to enclose intruder.

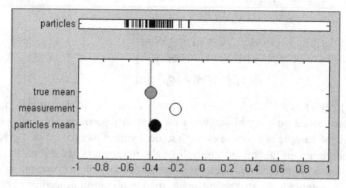

Fig. 4. Result 1 of experiment by particle filter

5   Conclusions

In this paper, we have developed the particle filter for the swarm robot localization and circle formation enclosing target which moves. It is method about enclosed invader in circle formation based on mutual localization of swarm robot without infrastructure. Therefore, we use trilateration that do not need to know the value of the coordinated of reference points. So, we specify the enclosed point for the number of robots base on between the relative positions of the robot in the coordinate system.

This scheme is important in providing a computationally feasible alternative to classify the robot motion in real system. We have proved that given an environment, particle filter scheme classify the robot location in real time.

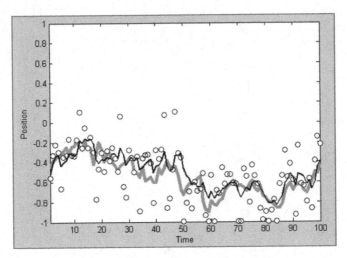

Fig. 5. Result 2 of experiment by particle filter

References

1. Huang, D.S., Jia, W., Zhang, D.: Palmprint Verification Based on Principal Lines. Pattern Recognition 41(4), 1316–1328 (2008)
2. Huang, D.S., Lawken, K., Ip, H., Chi, Z.: Zeroing Polynomials Using Modified Constrained Neural Network Approach. IEEE Trans. on Neural Networks 16(3), 721–732 (2005)
3. Huang, D.S., Ip, H., Chi, Z.: A Neural Root Finder of Polynomials Based on Root Moments. Neural Computation 16(8), 1721–1762 (2004)
4. Huang, D.S.: A Constructive Approach for Finding Arbitary Roots of Polyminals by Neural Networks. IEEE Trans. on Neural Networks 15(2), 477–491 (2004)
5. Huang, D.S.: Radial Basis Probabilistic Neural Networks: Model and Application. International Journal of Pattern Recognition and Artificial Intelligence 13(7), 1083–1101 (1999)
6. Huang, D.S.: The Local Minima Free Condition of Feedforward Neural Networks for Outer Supervised Learning. IEEE Trans. on Systems, Man and Cybernetics 28B(3), 477–480 (1998)

7. Yeo, T.K., Hong, S., Jeon, B.H.: Latest Tendency of Underwater multi-robots. Institute of Control, Robotics and Systems 16(1), 23–34 (2010)
8. Arai, T., Pagello, E., Parker, L.E.: Editorial: Advances in Multi-Robot Systems. IEEE Transactions on Robotics and Automation 18(5) (2002)
9. Isard, M., Blake, A.: CONDENSATION-conditional Density Propagation for Visual Tracking. International Journal of Computer Vision 29(1), 5–28 (1998)
10. Lee, Y.-W.: Adaptive Data Association for Multi-target Tracking Using Relaxation. In: Huang, D.-S., Zhang, X.-P., Huang, G.-B. (eds.) ICIC 2005, Part I. LNCS, vol. 3644, pp. 552–561. Springer, Heidelberg (2005)
11. Lee, Y.W., Seo, J.H., Lee, J.G.: A Study on The TWS Tracking Filter for Multi-Target Tracking. Journal of KIEE 41(4), 411–421 (2004)

Optimization of Orthogonal Poly Phase Coding Waveform Based on Bees Algorithm and Artificial Bee Colony for MIMO Radar

Milad Malekzadeh[1], Alireza Khosravi[1], Saeed Alighale[1],
and Hamed Azami[2]

[1] Department of Electrical and Computer Engineering,
Babol Industrial Univsity, Iran
[2] Department of Electrical Engineering, Iran University of Science and Technology, Iran
m.malekzade@stu.nit.ac.ir, akhosravi@nit.ac.ir,
saeed.alighale@gmail.com, hamed_azami@ieee.org

Abstract. Multi input multi output (MIMO) radars have multiple antenna radars that each of them can transmit signals. For that matter, to avoid interference, transmitted waveform must be mutually orthogonal. This paper has proposed a new approach by using bee algorithm (BA) and artificial bee colony (ABC) to design orthogonal discrete frequency coding waveforms (DFCWs) which have desirable autocorrelation and cross correlation characteristic for orthogonal MIMO radars. The results represent a various ability of these algorithms. The cross correlation and auto correlation are better designed by the BA and ABC, respectively.

Keywords: Bee algorithm, artificial bee colony, artificial bee colony, poly phase and MIMO radars.

1 Introduction

Multi input multi output radar (MIMO) is a great achievement in communication science in two past decades. The MIMO radar is a multiple antenna radar system which is capable of transmitting arbitrary waveform from each antenna element. Multiple transmitting antennas transmit orthogonal signals and multiple receiving antennas receive returns [1].

The MIMO radar technology has rapidly drawn considerable attention from many researchers. Several advantages of the MIMO radar have been discovered by many different researchers such as increased diversity of the target information, excellent interference rejection capability, improved parameter identify ability, and enhanced flexibility for transmit beam pattern design. However the ability to detect low speed target on the background of clutter and the detection ability of weak target on the background of strong clutter causes that MIMO radar performance to be considered in different projects [2].

D.-S. Huang et al. (Eds.): ICIC 2012, LNCS 7389, pp. 95–102, 2012.
© Springer-Verlag Berlin Heidelberg 2012

According to the MIMO radars' structure, there are different transmitted signal, so receiving the transmitted signal without interference is the most important purpose. Solving this problem, the transmitted signals must be mutually orthogonal [3].

The MIMO radar uses L orthogonal waveforms that are transmitted from different phase centers and N receiving phase centers. The received signals are matched filtered for each of the transmitted waveforms forming NL channels. This assumption denotes the necessity of low cross correlation properties between waveforms. In addition to have high resolution for multiple target detection, the low auto-correlation sidelobe peaks levels for transmitted signals is required [4].

Assume a MIMO radar system that compose of L transmitters, where each transmits a distinct signal from an orthogonal code set $\{s_l(t), l = 1, 2, 3, ..., L\}$ in which any two signals in the set are uncorrelated and the aperiodic autocorrelation function of any code $s_l(t)$ should be close to an impulse function.

The capability of being achieved of MIMO radars relates to the feasibility of a set of orthogonal signals with low autocorrelation and crosscorrelation properties. As a consequence, the acceptable design of such orthogonal code sets is very important for putting into effective MIMO radar systems [5]. MIMO radar systems can be coded with binary sequences, polyphase sequences, or frequency-hopped sequences. Polyphase code has some advantages over binary code; as might be expected, polyphase code is increasingly becoming a desirable alternative for radar signals [6].

In this paper, we intend to demonstrate impressive algorithms, bee algorithm (BA) and artificial bee colony (ABC), for the designing of orthogonal ployphase code sets that can be used in MIMO radars. The remainder of this paper is organized as follows: in the second section, the problem of the polyphase code set design will be presented. Then, we will introduce two kinds of swarm algorithms to numerically optimize polyphase code sets and the results from designing are presented. Finally, some conclusions are drawn in last section.

2   Orthogonal Polyphase Signal Design for MIMO

Consider that orthogonal polyphase code comprise L signals with each signal containing N subpulses represented by a complex number sequence, the signal set can be shown as follows [3]:

$$s_l(t) = e^{j\varphi_l(n)}, n = 1, 2, ..., N, \ l = 1, 2, ..., L \tag{1}$$

where $\varphi_l(n)(0 \le \varphi_l(n) \le 2\pi)$ is the phase of subpulse n of signal L in the signal set.

Assume a polyphase code set s with code length of N, set size of L, one can concisely represent the phase values of s with the following $L \times N$ phase matrix:

$$s(L, N) = \begin{bmatrix} \varphi_1(1), & \varphi_1(2), & ..., & \varphi_1(n) \\ \varphi_2(1), & \varphi_2(2), & ..., & \varphi_2(n) \\ & \vdots & & \\ \varphi_L(1), & \varphi_L(2), & ..., & \varphi_L(n) \end{bmatrix} \tag{2}$$

where the phase sequence in row l ($1 \leq l \leq L$) is the polyphase sequence of signal l, and all the elements in the matrix can only be chosen from the phase set in (2).

From the autocorrelation and cross correlation characteristic of orthogonal polyphase codes, we get:

$$A(\varphi_l, k) = \begin{cases} \dfrac{1}{N} \sum_{n=1}^{N-k} \exp j[\varphi_l(n) - \varphi_l(n+k)] = 0, \ 0 < k < N \\ \qquad\qquad\qquad\qquad\qquad l = 1, 2, ..., L \\ \dfrac{1}{N} \sum_{n=-k+1}^{N} \exp j[\varphi_l(n) - \varphi_l(n+k)] = 0, \ -N < k < 0 \end{cases} \tag{3}$$

$$C(\varphi_p, \varphi_q, k) = \begin{cases} \dfrac{1}{N} \sum_{n=1}^{N-k} \exp j[\varphi_q(n) - \varphi_p(n+k)] = 0, \ 0 < k < N \\ \qquad\qquad\qquad\qquad\qquad l = 1, 2, ..., L \\ \dfrac{1}{N} \sum_{n=-k+1}^{N} \exp j[\varphi_q(n) - \varphi_p(n+k)] = 0, \ -N < k < 0 \end{cases} \tag{4}$$

where $A(\varphi_l, k)$ and $C(\varphi_p, \varphi_q, k)$ are the aperiodic autocorrelation function of polyphase sequence s_l and the crosscorrelation function of sequences s_p and s_q, and k is the discrete time index. Therefore, designing an orthogonal polyphase code set is equivalent to the constructing a polyphase matrix in (2) with $A(\varphi_l, k)$ and $C(\varphi_p, \varphi_q, k)$ constraints in (3) and (4).

For the design of orthogonal polyphase code sets used in MIMO radar systems, an optimization criterion is not only to minimize the autocorrelation sidelobe peak (ASP) and the cross correlation peaks (CP), but also minimize the total autocorrelation sidelobe energy and crosscorrelation energy in (3) and (4). The peak and energy based cost function to be used for MIMO radar signals design is as follows [7]:

$$E = (1 - \mu) \sum_{l=1}^{L} \sum_{K=1}^{N-1} \max |A(\phi_l, k)|^2 + \mu \sum_{p=1}^{L-1} \sum_{q=p+1}^{L} \sum_{k=-(N-1)}^{N-1} |C(\phi_p, \phi_q, k)|^2 \tag{5}$$

3   Evolutionary Algorithms for DFCW

In past decades engineers have concentrated to present heuristic methods to solve optimization problems. In this way they have tried to inspire from nature and from this view they success to achieve different evolutionary algorithms. We purpose to present the BA and ABC in this paper. At the first step, the ABC algorithm is presented [8].

3.1   Artificial Bee Colony Algorithm Optimization of DFCW

ABC is one of the most recently defined algorithms proposed by Karaboga and Bas-tutk in 2006, is inspired by the intelligent behavior of honeybees. In the ABC algo-rithm, the colony of artificial bees contains three groups of bees: employed, onlookers and scouts. The performance of this algorithm has been shown at 9 steps as follows:

- Initialize the population of solutions
- Evaluate the population
- Produce new solutions for the employed bees
- Apply the greedy selection process
- Calculate the probability values
- Produce the new solutions for the onlookers
- Apply the greedy selection process
- Determine the abandoned solution for the scout, and   replace it with a new ran-domly produced   solution
- Memorize the best solution achieved so far

In this algorithm, first half of the colonies are selected as the employed artificial bees and the rest of them are chosen as the onlookers. For every food source, there is only one employed bee. In other words, the number of employed bees is equal to the num-ber of food sources around the hive. Scout is an employed bee whose food source has been abandoned. Generally, the ABC algorithm consists of local and global searches to find an optimum answer in desired space. In this algorithm, the position of a food source represents a possible solution to the optimization problem and the   nectar amount of a food source corresponds to the quality or fitness of the associated solu-tion. The number of the employed or the onlooker bees is equal to the number of solu-tions in the population. An onlooker bee assesses the information of the nectar taken from all employed bees and selects a food source with a probability p_i related to its fitness value [8].

$$p_i = \frac{fit_i}{\sum_{n=1}^{SN} fit_n}$$

(6)

where SN and fit_i are the size of population and the fitness value of the solution i which is proportional to the nectar amount of the food source in the position i, respec-tively. In order to produce a candidate food position from the old one in memory, the ABC uses the following expression:

$$v_{ij} = x_{ij} + \phi_{ij}(x_{ij} - x_{kj})$$

(7)

where ϕ_{ij} denotes a random number and selected between [-1,1] and $k \in \{1, 2,.., SN\}$ and $j \in \{1,2,...,D\}$ are randomly selected indexes. Each solution x_i ($i = 1, 2,..., N$) is represented by a D-dimensional vector, where D denotes the number of parameters to be optimized and each parameter is real coded.

4 Bees Algorithm Optimization for DFCW

The BA which imitates the food foraging behavior of honey bee colony is a novel swarm-based search algorithm developed by D.T. Pham [9]. This algorithm is based on a kind of neighborhood search combined with random search and can be used for multi-objective optimization. The performance of this algorithm has been shown at 8 steps as follows:

- Initialize population with random solutions
- Evaluate fitness of the population
- While (stopping criterion not met). Forming new population.
- Select sites for neighbourhood search.
- Recruit bees for selected sites (more bees for best e sites) and evaluate fitnesses.
- Select the fittest bee from each patch.
- Assign remaining bees to search randomly and evaluate their fitnesses.
- End While

The optimization of DFCW with Bee algorithm is summarized as follows:

At the first step the initial phase are produced by random solution. Then, according to the equation (5) we evaluate the fitness function. In order to minimize the fitness function; It is sorted in descending form. Now the m sites and best sites (e out of m) are selected with respect to fitness. The recruited bees investigate the selected sites. New values are produced with this equation:

$$u(i) = (x(i) - ngh) + (2 * ngh \cdot *rand(size(x(i),1),1)) \tag{8}$$

According to new value, the new fitness function would be derived. After comparing this fitness with old fitness and repeating this cycle, the BA processing will be done [10].

5 Simulation Results

In this section, according to described optimization algorithm we can design many different lengths of sequences. However, in this paper we will present only the correlation properties of sequences of the three code sequences with length N=128 and L=3. In the Figures 1 and 2, the results of ASPs and CPs for these sequences are shown for the BA and ABC, respectively. Also, Tables 1 and 2 list the three code sequences with length N=128 and L=3 for optimizing by BA and ABC, respectively. As can be seen in Tables 1 and 2, to design the CPs, applying of the BA is better than the ABC, while for designing the ASPs applying of the ABC is better than BA.

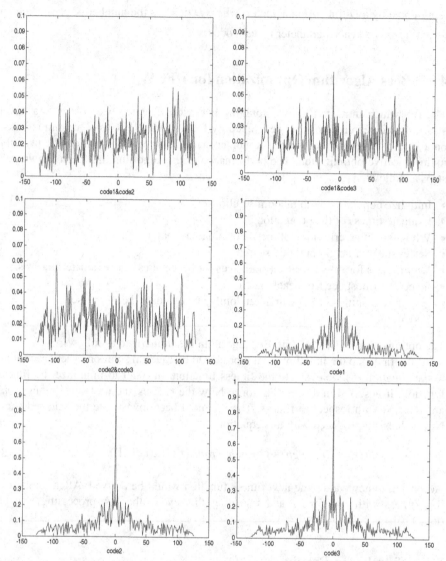

Fig. 1. Autocorrelation and cross-correlation functions of sequences with code length N=128 and set size L=3 (the BA results)

Table 1. ASP and CP of the designed DFCW set with N= 128 and L=3

	CODE 1	CODE 2	CODE3
CODE1	0.3835	0.0561	0.0649
CODE2	0.0561	0.3224	0.0573
CODE3	0.0649	0.0573	0.3032

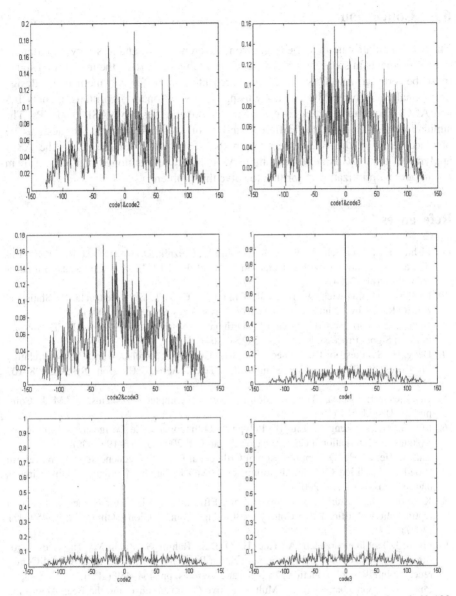

Fig. 2. Autocorrelation and cross-correlation functions of sequences with code length N=128 and set size L=3 (the ABC results)

Table 2. ASP and CP of the designed DFCW set with N=128 and L=3

	CODE 1	CODE 2	CODE3
CODE1	0.1254	0.1896	0.1565
CODE2	0.1896	0.1227	0.1676
CODE3	0.1565	0.1676	0.1657

6   Conclusion

The MIMO radars can track the target from different angles and it is very important to diagnose objects in space. In order to attain desirable properties, the transmitted signal must be orthogonal. This paper has presented an effective method to design orthogonal code for MIMO radars by using two evolutionary algorithms, namely BA and ABC. These approaches have tried to demonstrate desirable ASPs and CPs. The simulation results show the different ability of these algorithms. To design, the obtained CPs, by the BA is better than the ABC, while for designing the ASPs applying of the ABC is better than the BA. It should be mentioned that both of them are attractive optimization algorithm to solve this problem.

References

1. Fishler, E., Haimovich, A., Blum, R., Cimini, L., Chizhik, D., Valenzuela, R.: MIMO Radar: an Idea Whose Time Has Come. In: Proc. of the IEEE Int. Conf. on Radar, Philadelphia, PA (April 2004)
2. Fishler, E., Haimowich, A., Blum, R., Cimini, L., Chizhik, D., Valenzuela, R.: Statistical MIMO Radar. In: 12th Conf. on Adaptive Sensor Array Processing (2004)
3. Deng, H.: Polyphase Code Design for Orthogonal Netted Radar Systems. IEEE Transactions on Signal Processing 52, 3126–3135 (2004)
4. Deng, H.: Synthesis of Binary Sequences with Good Autocorrelation and Cross-correlation Properties by Simulated Annealing. IEEE Trans. Aerosp. Electron. Syst. 32, 98–107 (1996)
5. Grandjean, E.: Linear Time Algorithms and NP-complete Problems. SIAM J. Comput. 23(3), 573–597 (1994)
6. Liu, B., He, Z., Zeng, J., Liu, B.: Polyphase Orthogonal Code Design for MIMO Radar Systems. In: International Conference on Radar, CIE 2006, pp. 16–19 (2006)
7. Liu, B., He, Z., He, Q.: Optimization of Orthogonal Discrete Frequency-coding Waveform Based on Modified Genetic Algorithm for MIMO Radar. In: Communications, Circuits and Systems, ICCCAS (2007)
8. Karaboga, D., Bastürk, B.: A Powerful and Efficient Algorithm for Numerical Function Optimization: Artificial Bee Colony (ABC) Algorithm. J. Global Optim. 39(3), 459–471 (2007)
9. Pham, D.T., Ghanbarzadeh, A., Koc, E., Otri, S., Rahim, S., Zaidi, M.: The Bees Algorithm, a Novel Tool for Complex Optimization Problems. In: International Virtual Conference on Intelligent Production Machines and Systems, pp. 454–459 (2006)
10. Pham, D.T., Ghanbarzadeh, A.: Multi-objective Optimization Using the Bees Algorithm. In: International Virtual Conference on Intelligent Production Machines and Systems, Scotland (2007)

SVM Regularizer Models on RKHS vs. on R^{m*}

Yinli Dong[1] and Shuisheng Zhou[2,**]

[1] Foundation Department, Xian Eurasia University, P.R. China, 701165
[2] School of Science, Xidian University, P.R. China, 710071
sszhou@mail.xidian.edu.cn

Abstract. There are two types of regularizer for SVM. The most popular one is that the classification function is norm-regularized on a Reproduced Kernel Hilbert Space(RKHS), and another important model is generalized support vector machine(GSVM), in which the coefficients of the classification function is norm-regularized on a Euclidean space R^m. In this paper, we analyze the difference between them on computing stability, computational complexity and the efficiency of the Newton-type algorithms. Many typical loss functions are considered. The results show that the model of GSVM has more advantages than the other model. Some experiments support our analysis.

Keywords: representer theorem, regularizer, newton-type algorithm.

1 Introduction

Based on the Vapnik and Chervonenkis' structural risk minimization principle[1,2], SVMs are proposed as powerful machine learning methods for supervised learning. They are popular methods in the past 10 more years, and widely used in classification and regression problems, such as character identification, disease diagnoses, face recognition, the time series prediction etc. Historically, the optimal classification function obtained by SVMs has an offset. But the investigations of the generalization performance do not suggest that the offset offers any improvement for a large feature kernel space like Gaussian kernel[2,3]. For simplicity, here we study the SVMs without offset for the nonlinear classification problems with Gaussian kernel. Actually the offset can be considered with an extra attribute 1 added to every sample[4].

1.1 Representer Theory

Representer theorem is an important tool for the learning model with many samples as the inputs of the problem. It states that the learning solution is a linear combination of the kernel functions of input samples. It can transform the high or infinite dimensional learning problem into finite dimensional problems with the size of the input learning data, where the finite combination coefficients are solved according to the input data. Many kernel learning problems admit this, such as SVMs[1], PCA[5] etc.

* This work is supported by NNSFC under Grant No. 61072144, 61179040, 61173089 and 11101322.
** Corresponding author.

D.-S. Huang et al. (Eds.): ICIC 2012, LNCS 7389, pp. 103–111, 2012.
© Springer-Verlag Berlin Heidelberg 2012

Quantitatively speaking, given an input set $T=\{(x_1, y_1),..., (x_m, y_m)\}$ for samples $x_i \in R^d$ and labels $y_i \in \{-1, 1\}$, a kernel learning classification problem is to learn a classification function f in a reproduced kernel Hilbert spaces(RKHS) H corresponding to a kernel function $k:R^m \times R^m \to R$ with a good generalized capacity. RKHS has the reproducing property that admits $f(x)=<f,k(\cdot,x)>_H$ and $<k(\cdot,x), k(\cdot,z)>_H=k(x,z)$ for $f \in H$ and $x, z \in R^m$. Because of high dimensions the learning problem on H can not be solved efficiently. By the representer theory, the solution $f \in H$ can be represented as

$$f = \sum_{j=1}^{m} \alpha_j k(\cdot, x_j). \tag{1}$$

This is a finite linear combination of the basic hypotheses in H and the finite optimal combination coefficients are solved to represent the optimal hypothesis in H.

1.2   Two Regularized Models for SVMs

There are two regularization forms for SVMs. One model regularizes the classification function on H and is converted into a finite dimensional problem by representer theorem[4,6,7] or duality[1,8-10]. Another regularizes the combination coefficients on R^m, which is called Generalized Support Vector Machines(GSVMs)[11-13].

Regularization Model on RKHS. With the hypothesis f norm-regularized in H, a popular model is defined as:

$$M_1: \quad \min_{f \in H} \frac{\lambda}{2}\|f\|_H^2 + \sum_{i=1}^{m} L(1 - y_i < f, k(\cdot, x_i) >_H) \tag{2}$$

with the regularized parameter λ and the loss $L:R \to R^+$. Plugging (1) into (2), we have

$$\min_{\alpha \in R^m} \frac{\lambda}{2}\sum_{i,j} \alpha_i \alpha_j k(x_i, x_j) + \sum_{i=1}^{m} L\left(1 - y_i \sum_{j=1}^{m} \alpha_j k(x_i, x_j)\right) \tag{3}$$

Solving problem (3), we obtain m combination coefficients to represent the learning result of M_1 as (1), no matter how high the dimension of RKHS is.

Another way to convert (2) into a finite dimensional problem is the duality for a given loss [1,8-10]. We can derive a general form by *conjugate duality*[14]:

$$\max_{\gamma \in R^m} \frac{\lambda}{2}\sum_{i,j} y_i y_j k(x_i, x_j)\gamma_i \gamma_j + \sum_{i=1}^{m} (\gamma_i - L^*(\gamma_i)), \tag{4}$$

where $L^*:R \to R^+ \cup \{+\infty\}$ is the conjugate function of L. the formulation (3) is always more complicated than (4), which explains why so many researchers focus on the dual (4)[1,8,9,15 etc]. However, the dual has m dimensional, even the reduced representation as Eq. (5) is used to obtain an approximate solution [4,7,12,16].

$$f = \sum_{j \in J} \alpha_j k(\cdot, x_j). \tag{5}$$

where J is random chosen[12, 16] with $|J| \leq 0.1m$ or well-chosen[4] with less cardinal. Training the reduced SVMs by Newton methods in dual space is not a good choice.

Regularization Model on R^m. While the combinations coefficients (noted as β) for the hypothesis f in (1) are norm-regularized on R^m, the model called GSVMs sloves

$$\text{M}_2: \quad \min_{\beta \in R^m} \tfrac{\lambda}{2} \|\beta\|^2 + \sum_{i=1}^{m} L\Big(1 - y_i \sum_{j=1}^{m} \beta_j k(x_i, x_j)\Big) \tag{6}$$

It is first proposed by Mangasarian[11], and is used to design algorithms with squared hinge loss[12] and with least square loss[13]. Here β may be different from α in (3).

Relationship of Two Models. On the one hand, Mangasarian[11] points out that the dual of (6) can meet the same form with the dual of (2) for a well-chosen $g(\beta)$ instead of $\|\beta\|^2$. On the other hand, from (2) and (6), we observe that the difference of two models is on the first item of objective function, which is called the regularizer. The regularizer of the former is on L_2 norm of hypothesis f in RKHS $\mathcal{H}$, and the regularizer of the latter is on L_2 norm of the combination coefficients in Euclidean space R^m.

The rest of the paper is organized as follows. We analyze Newton-type algorithms for the reduced SVMs with different smooth loss in Section 2, give experimental results in Section 3 and conclude the paper in Section 4.

2   Newton Methods for Reduced Models with Different Losses

Let f in (1) be approximated as (5) with J is well-chosen or random chosen and $r=|J|$ be the reduced set size. For M_1 and M_2, the reduced form of (3) or (6) is:

$$\min_{\alpha \in R^m} \tfrac{\lambda}{2} z^\mathrm{T} A z + \sum_{i=1}^{m} L(1 - y_i K_{iJ} z) \tag{7}$$

where z can be α_J or β_J, and A is K_{JJ} for M_1 and $\mathbf{I}$ for M_2. The only difference between them is the regularizer $\alpha_J^\mathrm{T} K_{JJ} \alpha_J$ versus $\beta_J^\mathrm{T} \beta_J$. Commonly a kernel matrix $K \in R^{m \times m}$ has $\sigma_1 \geq 1 \geq \sigma_m \geq 0$, where $\sigma_1 (\sigma_m)$ is the largest(smallest) eigenvalue of K. We have $\kappa(\lambda I + Q) \leq \kappa(\lambda K + Q)$, where $\kappa(\cdot)$ is the condition number of a matrix and Q is a semi-positive definite matrix induced by the second part of (7). Roughly speaking, we have

Proposition.1 *The model of* M_2 *is stabler than the model of* M_1.

The most popular and meaningful loss function is the hinge loss function $L(u) = \max\{0, u\}$, but it is not differentiable and Newton-type methods do not work for it. Some losses are adopted to smooth it, including least squared loss[10, 13], squared hinge loss[4, 15], Huber loss[1, 7, 17] and logistic loss[12]. Next, those two regularizer problems with different loss functions are compared by Newton-type algorithms on performance of the classification and efficiency of the algorithms.

2.1   Setup of Newton-Type Algorithms

Given a smooth loss in (7), Newton-type algorithms[18] have quadratic convergence rate. Specifically, for a given z^t(can be α_J^t or β_J^t) at iteration t, let $\xi_i^t = 1 - y_i K_{iJ} z^t$ and $I_t = \{i \in M \mid L(\xi_i^t) > 0\}$, and $\bar{z}$ be the solution to Newton equations:

$$\Big(\lambda A + \sum_{i \in I_t} \nabla^2 L(\xi_i^t)\Big) z = \sum_{i \in I_t} \nabla^2 L(\xi_i^t) z^t - \sum_{i \in I^t} \nabla L(\xi_i^t) \tag{8}$$

If the full Newton step is acceptable, then $z^{t+1} = \bar{z}$, otherwise $z^{t+1} = \tau \bar{z} + (1-\tau)z^t$, where τ is chosen by line-search scheme. Experiments show the full step is always acceptable and Armijo line-search scheme is pretty well while the full step fails.

2.2   Specification of Algorithms with Smooth Loss Functions

Specification forms of (8) are analyzed according to different smooth loss, including least squared loss, squared hinge loss, Huber loss and logistic loss as follows.

Least Squared Loss. With least squared loss $L(u)=u^2$, Eq. (8) is independent to z^t and its solution is obtained by solving the following system of linear equations (9):

$$\left(\lambda A + K_{MJ}^T K_{MJ}\right)z = K_{MJ}\,y \tag{9}$$

The coefficients matrix in (9) for M2 is always positive definite while the coefficient matrix in (9) for M2 may be semi-positive definite need a ridge[4] added.

Squared Hinge Loss. With squared hinge loss $L(u)=\max\{0,u\}^2$, Eq. (8) is

$$\left(\lambda A + K_{I_t J}^T K_{I_t J}\right)z = K_{I_t J} y_{I_t} \tag{10}$$

In [4], they designed a complicated procedure to iteratively update the Cholesky factorization of $\lambda A + K_{I_t J}^T K_{I_t J}$ to reduce the computational complexity for solving (10). For simplicity, in this paper we use $B^t = B^{t-1} + K_{I_{in}J}^T K_{I_{in}J} - K_{I_{out}J}^T K_{I_{out}J}$ to update B^t iteratively as [15], then solve Newton equation (10) by "\" operator in Matlab, where $I_{in}:=\{i | i\in I_t,\ i\notin I_{t-1}\}$, $I_{out}:=\{i | i\notin I_t,\ i\in I_{t-1}\}$ and $|I_t|>|I_{in}|+|I_{out}|$ always holds. This scheme works well while the reduced set is randomly chosen in advanced.

Huber loss. With Huber loss $L_\delta(u) = \begin{cases} \max\{0,u\}, & |u|>\delta \\ (u+\delta)^2/4\delta, & |u|\le\delta \end{cases}$, (8) is simplified as

$$\left(\lambda A + \tfrac{1}{2\delta} K_{I_0^t J}^T K_{I_0^t J}\right)z = \tfrac{1+\delta}{2\delta} K_{I_0^t J}^T y_{I_0^t} + K_{I_+^t J}^T y_{I_+^t} \tag{11}$$

where $I_0^t:=\{i\in M|\ |\xi_i|<\delta\}$, $I_+^t:=\{i\in M|\ |\xi_i|\ge\delta\}$. At the beginning of the algorithm, we should start the algorithm with a big δ_0 like $\delta_0=1$, reduce it by $\delta_k:=0.1\delta_{k-1}$ while the current solution is good under some criterions (such as $\|g^t\|\le1$), and repeat the algorithm until $\delta_k\le10^{-4}$ and the final stop criterions are reached.

Logistical loss. With Logistical loss $L_p(u)=\max\{u,0\}+\log(1+\exp(-pu))/p$, (8) is

$$\left(\lambda A + K_{I_0 J}^T \Lambda_{I_0}^t K_{I_0 J}\right)z = K_{I_0 J}^T \Lambda_{I_0}^t K_{I_0 J} z^t + K_{I_+ J}^T L_p'(\xi_{I_+}^t) \tag{12}$$

where $I_+:=\{i\in M| L_p'(\xi_i^t)\ge v\}$ and $I_0:=\{i\in M|\ L_p''(\xi_i^t)\ge v\}$, $\Lambda^t:=\mathrm{diag}(L_p''(\xi_1^t), \cdots, L_p''(\xi_m^t))$ for a tiny number $v=10^{-10}$ or less. In order to make the Newton method working good, we should set p not very large such as $p=10$ at the beginning, then set $p:=10p$ if $\|g\|$ is small and repeat the algorithm until $p=10^4$ and $\|g\|\le\varepsilon$ for a given ε.

2.3   SMW Identity and Advantage of GSVM

SMW identity[18] $(A+U^T\Lambda U)^{-1}=A^{-1}-A^{-1}U^T(\Lambda^{-1}+UA^{-1}U^T)^{-1}UA^{-1}$ can be used to reduce the computational complexity for calculating $(A+U^T\Lambda U)^{-1}$ while A^{-1} is very simple and $\Lambda^{-1}+UA^{-1}U^T$ has a small size. Based on the analysis above, $\Sigma_{i\in I_t}L''(\xi_i^t)$

always has the form $U^T A U$ in Eq. (8), where U has the size of $r \times |I_t|$ and Λ is an $|I_t| \times |I_t|$ diagonal matrix. If $A=\mathbf{I}$ for GSVMs, the solution of Eq. (8) is

$$z = \lambda \mathbf{b} - \lambda U^T (\lambda \Lambda^{-1} + U U^T)^{-1} U \mathbf{b} \qquad (13)$$

where $\mathbf{b}$ is the right hand side of (8). So while $|I_t| < r$, which always happens for some problems that have a sparse solution, the solution to Eq. (8) is obtained by (13) with the computation complexity $O(r|I_t|^2)$, less than $O(r^3)$. This trick is not valid if $A=K_{JJ}$ for M_1. As a conclusion, we have

Proposition 2. *If $|I_t| < r$, Newton-type algorithm based on M_2 has less computational complexity than the same algorithm based on M_1.*

3   Experimental Results

In this section, we do two sets of experiments on some datasets to compare the reduced models with Newton-type methods. The first is a nonlinearly dataset called "tried and true" checkerboard[12,15] and the second is on practical datasets from site[19].

3.1   Artificial Data with Reduced Methods

In this section, we give some experiments on an artificial dataset to compare the performance of two kinds of regularization. The "tried and true" checkerboard datasets have often been used to show the effectiveness of nonlinear kernel methods[12,15]. It is generated by uniformly discretized the region $[0,199] \times [0,199]$ to $200^2 = 40,000$ points, and is labeled two classes spaced by 4×4 grids. The training set is random sampled from the total points, and the reminders are left in the testing set.

Over-Fitting of M_1 with Least Square Loss. Here we show that the resulted classification hyperplanes based on M_1 with least squared loss have strange phenomena but the hyperplanes based on M_2 haven't. The training results are plotted in Fig. 1. It shows that the classification lines of two methods are similar, but the high confidence area(satisfying $y_i f(x_i) \geq 1$) is very different. For M_2(right), the high confidence area is normal, i.e. the more central the grid, the higher the confidence, and the exception happens only on the four corners. But for M_1(left), the high confidence area is strange on nearly every grid, i.e., the central of the grids are not always in the high confidence area. These phenomena should be a kind of over fitting.

M_1 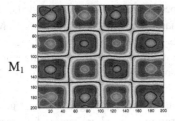 M_2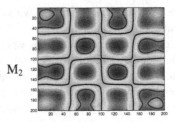

Fig. 1. Plots with least square loss($m=8000$, $r=7.5\%m$, $\lambda=0.01$). The blank lines are the classification $f(x)=0$ and the blue/green lines are support lines to $f(x)=\pm 1$ respectively.

Comparison M_1 and M_2 with Different Smooth Losses. The results in Table 1 are obtained by Newton-type algorithm with same stop criterion and $\lambda=0.01$, where the averaged test errors, averaged training time, averaged iterations of Newton step and numbers of the low confidence training samples that satisfying $y_i f(x_i)<1$ are listed with standard deviation. All the results are mean values on 20 random trials.

Table 1. Comparison of two regularizations with different smooth loss functions($\lambda=0.01$)

rithms	$m=2000$ $r=150$	$m=4000$ $r=300$	$m=8000$ $r=600$	$m=12000$ $r=800$	$m=20000$ $r=800$
	Test error(%) on the rest data with test 40000−m samples.				
M1+SH	1.52±0.23	0.70±0.11	0.23±0.06	0.11±0.04	0.06±0.03
M2+SH	**1.27±0.26**	**0.36±0.07**	**0.10±0.03**	**0.06±0.02**	**0.04±0.02**
M1+Log	1.92±0.23	0.97±0.13	0.35±0.07	0.22±0.05	0.16±0.05
M2+Log	1.65±0.25	0.48±0.11	0.18±0.05	0.10±0.03	0.07±0.03
M1+Hub	1.92±0.23	0.97±0.13	0.35±0.07	0.22±0.05	0.16±0.05
M2+Hub	1.65±0.25	0.48±0.11	0.18±0.05	0.10±0.03	0.07±0.03
	Training time(s)				
M1+SH	0.13±0.00	0.70±0.02	3.57±0.40	7.50±0.22	11.98±0.25
M2+SH	**0.11±0.08**	**0.57±0.03**	**2.89±0.06**	**6.38±0.17**	**10.67±0.32**
M1+Log	0.67±0.12	2.30±0.45	11.27±1.29	23.27±4.57	31.98±3.63
M2+Log	0.32±0.06	1.48±0.21	6.04±0.47	13.27±1.65	24.73±3.21
M1+Hub	0.51±0.02	2.10±0.11	9.83±0.86	19.77±1.11	26.52±1.37
M2+Hub	0.46±0.05	2.10±0.42	7.75±0.71	14.36±1.05	21.49±2.28
	Iterations of Newton step				
M1+SH	10.9±1.0	14.1±1.1	**15.7±0.7**	**16.9±0.7**	**19.1±0.7**
M2+SH	**10.4±0.8**	**11.7±0.7**	16.3±1.1	20.6±1.6	23.4±2.0
M1+Log	98.0±23.2	76.2±13.0	84.0±13.3	84.8±22.9	77.3±13.1
M2+Log	65.8±12.1	78.9±10.7	94.5±12.0	91.8±14.0	97.3±18.8
M1+Hub	112.8±5.3	110.7±8.3	113.3±9.5	117.5±6.0	117.3±8.6
M2+Hub	185.7±19.3	221.3±32.0	265.6±31.2	268.4±20.3	243.5±28.2
	Low-confidence training samples Number ($y_i f(x_i)<1$)				
M1+SH	223.8±10.1	351.9±9.4	592.4±16.7	842.0±14.0	1277.9±17.8
M2+SH	198.5±13.0	270.6±8.5	318.3±8.5	327.6±9.5	426.6±10.8
M1+Log	129.8±9.5	223.9±6.0	372.6±10.0	463.6±8.9	616.0±7.5
M2+Log	**123.2±10.2**	**142.8±7.8**	**156.6±7.7**	**176.3±7.8**	**243.2±7.7**
M1+Hub	130.3±9.7	224.3±5.7	372.6±10.0	463.6±8.6	616.2±7.4
M2+Hub	**123.2±10.2**	**142.9±7.9**	**156.6±7.8**	**176.3±7.7**	**243.2±7.7**

From the results in Table 1, we can get the following conclusions:

- It is observed that M_2 nearly wins most all aspects. Firstly, it is clear that the test errors corresponding to M_2 are always better than the results corresponding to M_1. Secondly, on the training time aspect, M_2 excels M_1 much for every loss functions, although the iterations of Newton step corresponding to M_2 are often longer than that of M_1. This is coherent with our conclusion in Proposition 2. Thirdly, for M_2 often has less low-confidence training samples than M_1 has.
- Compared three types of loss functions, the squared hinge loss gets the best generalization errors. No matter how large the training set, there are a few iterations(less than $2\log(m)$) of Newton steps needed.

- The advantages of the logistical loss and Huber loss are that they always get less low-confidence training samples(or called support vectors) than others. Especially for M_2, there always have a small number of low-confidence training samples.

By Table 1, we can conclude that M_2 has some advantages over M_1. With the same parameter settings, the algorithms based on M_2 always faster than those based on M_1. The former is stabler in computing than the latter who often needs a ridge on the Hessian matrix[4] to keep the Newton direction well-defined.

From those experimental results, M_2 with squared hinge loss is the best model to train reduced SVM with Newton method. In Section 3.2, we only do some experiments to compare M_1 and M_2 equipped with square hinge loss.

3.2   Benchmark Data Experiments Comparison

Six practical data sets from UCI repository of machine learning databases[19] are adopted to evaluate the related algorithms. The first five data are preprocessed in [20], but we exchange the training and the testing in order to get a large training set to implement the reduced technique. The sixth data is the Adult with a large training set. For simplicity, Gaussian kernel function $k(x, y)=\exp(-\gamma\|x-y\|^2)$ with different spread parameters γ is used for all dataset. Parameters γ and λ are roughly chosen by grid search method with $\gamma \in \{2^{-10},\ldots, 2^2\}$ and $\lambda \in \{10^{-5},\ldots, 10^0\}$. The test errors and the training time are listed in Table 2.

Table 2. Experiments on 6 data sets from UCI repository of machine learning databases

DataSet	Banana	Ringnorm	Splice	Twonorm	Waveform	Adult
(train test)	(4900, 400)	(7000, 400)	(2175, 100)	(7000, 400)	(4600, 400)	(30162, 15060)
Red. size	368	525	163	525	345	800
(γ, λ)	$(2^{-1}, 10^{-3})$	$(2^{-6}, 10^{-1})$	$(2^{-8}, 10^{-3})$	$(2^{-8}, 10^{-1})$	$(2^{-5}, 10^{-1})$	$(2^{-4}, 10^{-3})$
Error(%)						
M_1+SH	9.23±1.23	1.66±0.64	12.71±1.24	2.19±0.63	8.50±1.54	**15.50±0.03**
M_2+SH	9.25±1.28	1.70±0.67	**12.67±1.21**	**2.17±0.63**	**8.29±1.43**	15.56±0.04
Training Time(s)						
M_1+SH	2.19±0.17	3.77±0.18	0.28±0.03	3.70±0.13	1.86±0.12	65.45±7.71
M_2+SH	**1.64±0.18**	**3.17±0.10**	**0.20±0.02**	**3.20±0.09**	**1.68±0.13**	**22.59±2.84**

From the results in Table 2, the mean test errors are varied with the different methods with a little the difference. At the same time, it clearly sees that algorithms based on M_2 are always faster than those based on another. This is mainly because its Hessian matrix has a better condition number than those of M_1. In our experiments, a ridge is added to the Hessian matrix of M_1 to maintain the calculation stability.

4   Conclusion

There are two main regularization models of SVMs. One is norm-regularized the classification function in a reproduced kernel Hilbert space. The other is norm-regularized the coefficients of the classification function in a Euclidean space. All of them are converted to m dimension problems by representer theorem.

We compare M_1 and M_2 with the reduced methods by Newton-type algorithm. The Hessian matrix of M_2 is always positive and is simper than the Hessian matrix of M_1, and the SMW identity can be used to reduce the computation complexity of the corresponding algorithms for M_2 but cannot be used in algorithm for M_1. Our analysis and the experimental results support that M_2 has some advantages over M_1, such as simple in computing the Hessian matrix, stable in solving the Newton direction and the algorithm based on M_2 is also faster and stabler than the algorithm based on M_1 with the similar test errors. This work is valuable to extend the using of GSVMs.

References

1. Vapnik, V.N.: The Nature of Statistical Learning Theory. Springer, NY (2000)
2. Steinwart, I., Christmann, A.: Support Vector Machines. Springer (2008)
3. Steinwart, I.: Sparseness of Support Vector Machines. JMLR 4, 1071–1105 (2003)
4. Keerthi, S.S., Chapelle, O., Decoste, D.: Building Support Vector Machines with reduced classifier complexity. JMLR 7, 1493–1515 (2006)
5. Schölkopf, B., Smola, A., Müller, K.R.: Nonlinear component analysis as a kernel eigenvalue problem. Neural Comput. 10(5), 1299–1319 (1998)
6. Schölkopf, B., Herbrich, R., Smola, A.J.: A Generalized Representer Theorem. In: Helmbold, D.P., Williamson, B. (eds.) COLT 2001 and EuroCOLT 2001. LNCS (LNAI), vol. 2111, pp. 416–426. Springer, Heidelberg (2001)
7. Chapelle, O.: Training a Support Vector Machine in the primal. Neural Computation 19(5), 1155–1178 (2007)
8. Joachims, T.: Support Vector Machine (1998),
   http://www.cs.cornell.edu/people/tj/svmlight/
9. Platt, J.C.: Fast training of Support Vector Machines using Sequential Minimal Optimization. In: Schölkopf, B., et al. (eds.) Advances in Kernel Method-Support Vector Learning, pp. 185–208. MIT, Cambridge (1999)
10. Suykens, J., Vandewalle, J.: Least Square Support Vector Machine Classifiers. Neural Processing Letters 9(3), 293–300 (1999)
11. Mangasarian, O.L.: Generalized Support Vector Machine. In: Smola, A.J., et al. (eds.) Advances in Large Margin Classifiers, pp. 135–146. MIT Press (2000)
12. Lee, Y., Mangasarian, O.L.: RSVM: Reduced Support Vector Machines. Data Mining Institute, Computer Sciences Department, University of Wisconsin, pp. 1–7 (2001)
13. Fung, G., Mangasarian, O.L.: Proximal Support Vector Machine classifiers. In: Provost, Srikant, R. (eds.) Proceedings KDD 2001: Knowledge Discovery and Data Mining, San Francisco, CA, August 26-29, pp. 77–86. ACM, New York (2001)
14. Boyd, S.P., Vandenberghe, L.: Convex Optimization, 7th edn. Cambridge University Press, Cambridge (2009)
15. Zhou, S., Liu, H., Zhou, L., Ye, F.: Semi-smooth Newton Support Vector Machine. Pattern Recognition Letters 28, 2054–2062 (2007)
16. Lin, K.M., Lin, C.J.: A Study on Reduced Support Vector Machines. IEEE Trans. on Neural Networks 14(6), 1449–1459 (2003)

17. Ye, F., Liu, H., Zhou, S., Liu, S.: A Smoothing Trust-region Newton-CG method for Minimax Problem. Applied Mathematics and Computation 199(2), 581–589 (2008)
18. Golub, G.H., Loan, C.F.V.: Matrix Computations. The John Hopkins University Press, Baltimore (1996)
19. Blake, C.L., Merz, C.J.: UCI Repository of Machine Learning Databases (1998), http://www.ics.uci.edu/~mlearn/MLRepository.html
20. Rätsch, G.: Benchmark Repository (2003), http://users.rsise.anu.edu.au/~raetsch/data/index.html

Research on Performance Comprehensive Evaluation of Thermal Power Plant under Low-Carbon Economy

Xing Zhang

Department of Economy Management, North China Electric Power University,
Baoding 071003, Hebei, China
zx83316@126.com

Abstract. A performance evaluation index system of thermal power plant is established under low carbon economy, and a comprehensive evaluation model based on principal component analysis (PCA), support vector machine (SVM) and quick sort algorithm is presented. Then experiments are made by using the real data from 17 thermal power plants, and the sequence of them is obtained ultimately. The results show that the model proposed has high accuracy, and comparing with BP network, SVM shows better performance in the condition of few data.

Keywords: performance evaluation, thermal power plant, PCA, SVM, binary tree, quick sort algorithm.

1   Introduction

Low carbon and energy conservation has become a hot topic around the world since Copenhagen conference. Low carbon economy is the direction of economic development in China. Thermal power plant, as an industry which consumes large amounts of energy in low efficiency and damages the environment seriously, inevitably becomes the focus. Thermal power plant should transform its development mode to low carbon and energy saving. How to evaluating the performance of Thermal power plant synthetically under low carbon economy becomes an important topic.

SVM is a new and promising machine learning technique proposed by Vapnik [1], which can solve the nonlinear, small sample and high dimensional problems effectively. It has been attracting more and more researcher's attention in recent years because of its good generation performance. In this paper, SVM is introduced to compare two thermal power plants by vector connection, and then we obtain the sequence by means of quick sort algorithm. Experiment results show that the model has high accuracy; hence, it is effective to evaluate the performance of thermal power plant.

2   Performance Evaluation Index System under Low-Carbon Economy

According to the characteristic of thermal power plant and the requirement of low carbon economy, this paper designs the performance evaluation index system from economic performance, environmental performance, and social performance [2][3]:

D.-S. Huang et al. (Eds.): ICIC 2012, LNCS 7389, pp. 112–119, 2012.
© Springer-Verlag Berlin Heidelberg 2012

Table 1. Performance evaluation index system under low-carbon economy

The first level	The second level	The third level
Economic performance	Profit ability	Return on total asset (S_1)
		Return on net asset (S_2)
		Return on low carbon asset (S_3)
	Debt paying ability	Debt Asset ratio (S_4)
		Interest Protection Multiples (S_5)
	Turnover ability	Total Assets Turnover (S_6)
		Current Assets Turnover (S_7)
	Development ability	Rate of sales growth (S_8)
		Rate of Capital Accumulation (S_9)
Environmental performance	Pollutants discharge	CO_2 emission per unit power (S_{10})
		SO_2 emission per unit power (S_{11})
		NO_X emission per unit power (S_{12})
		Effluent emission per unit power (S_{13})
		Smoke emission per unit power (S_{14})
		Noise at the boundary of plant (S_{15})
	Resource utilization efficiency	Standard coal consumption rate (S_{16})
		Water consumption per unit power (S_{17})
		Plant water loss rate (S_{18})
		Plant power consumption rate (S_{19})
		Industrial water recycling rate (S_{20})
Social performance	Customer satisfaction	Contract compliance rate (S_{21})
		Annual number of complaints (S_{22})
	Staff satisfaction	Education funding Per capita (S_{23})
		Staff retention rate (S_{24})

3   Principal Component Analysis

Principal component analysis, which is also known as Karhunen-Loeve (KL) transform, Principal component analysis reorganizes a large number of originally indexes into a small amount of comprehensive indexes by linear transformation. Thereby it simplifies the data and reveals the relationship between variables [4]. The main process of PCA is as following:

(1) Convert the original matrix Y into standardized matrix Z.

$$z_{ij} = \frac{y_{ij} - \overline{y}_j}{s_j}, i = 1, 2, \cdots, n, \quad j = 1, 2, \cdots, p \tag{1}$$

where $\overline{y}_j = \frac{1}{n}\sum_{j=1}^{n} y_{ij}$, $s_j = \sqrt{\frac{1}{n-1}\sum_{i=1}^{n}(y_{ij} - \overline{y}_j)^2}$

(2) Obtain the correlation matrix R of the standardized matrix Z.

$$R = \left[r_{ij} \right]_{p \times p} = \frac{Z^T Z}{n-1} \quad i, j = 1, 2, \cdots, p \tag{2}$$

where $r_{ij} = \dfrac{\sum\limits_{k=1}^{n} z_{ki} z_{kj}}{n-1}$

(3) Compute the Eigen values λ_j and the corresponding eigenvectors b_j of the covariance matrix R respectively. The Eigen values λ_j are just the non-negative real roots of $|R - \lambda I_p| = 0$, order them from large to small $\lambda_1 \geq \lambda_2 \geq \cdots \geq \lambda_p \geq 0$, then the corresponding eigenvectors b_j can be obtained by $Rb_j = \lambda_j b_j$.

(4) Determine the principal components. p new variables are composed of the eigenvectors as follows:

$$\begin{cases} A_1 = b_{11} z_1 + b_{12} z_2 + \cdots + b_{1p} z_p \\ A_2 = b_{21} z_1 + b_{22} z_2 + \cdots + b_{2p} z_p \\ \vdots \\ A_p = b_{p1} z_1 + b_{p2} z_2 + \cdots + b_{pp} z_p \end{cases} \tag{3}$$

The first m principal components are selected as principal components when their accumulative contributive rate

$$w_m = \left. \sum_{j=1}^{m} \lambda_j \middle/ \sum_{j=1}^{p} \lambda_j \right. \geq 0.85 \tag{4}$$

4　Support Vector Machine

SVM is the theory based on statistical learning theory. It realizes the theory of VC dimension and principle of structural risk minimum [5]. The following is the brief introduction of SVM.

Given the training set in the case of linear separation, $(x_1, y_1), \cdots, (x_l, y_l)$, $x \in R^n$, $y \in \{+1, -1\}$, there exists a separating hyper plane with the target functions $w \cdot x + b = 0$ (w represents the weight vector and b the bias). The linear SVM for optimal separating hyper plane has the following optimization problem:

$$\begin{aligned} Min \quad & \phi(w) = \frac{1}{2}(w \cdot w) \\ s.t. \quad & y_i[w \cdot x_i + b] \geq 1, i = 1, \cdots l \end{aligned} \tag{5}$$

Considering the non-separable case, slack variable $\xi_i \geq 0$ is introduced and soft margin optimal hyper plane is constructed.

$$Min \quad \phi(w) = \frac{1}{2}(w \cdot w) + C \sum_{i=1}^{l} \xi_i$$

$$s.t. \quad y_i [w \cdot x_i + b] \geq 1 - \xi_i \tag{6}$$

$C \geq 0$ is the error penalty coefficient chosen by users.

The solution to above optimization problem can be converted into its dual problem. We can search the nonnegative Lagrange multipliers by solving the following optimization problem [6]:

$$Max \quad Q(\alpha) = \sum_{i=1}^{l} \alpha_i - \frac{1}{2} \sum_{i,j=1}^{l} \alpha_i \alpha_j y_i y_j x_i^T x_j$$

$$s.t. \quad \sum_{i=1}^{l} \alpha_i y_i = 0, 0 \leq \alpha_i \leq C \tag{7}$$

As to the non-linear case, SVM maps input vectors x into a high dimensional linear feature space by a nonlinear mapping $\varphi(\bullet)$, and searches the optimal separating hyper plane $w \cdot \varphi(x) - b = 0$ in the space. SVM use kernel function $K(x_i, x_j)$ of input space instead of the operation $\varphi(x_i) \cdot \varphi(x_j)$ of high dimensional feature space. Then the target function of (6) is changed to be

$$Max \quad Q(\alpha) = \sum_{i=1}^{l} \alpha_i - \frac{1}{2} \sum_{i,j=1}^{l} \alpha_i \alpha_j y_i y_j K(x_i \cdot x_j) \tag{8}$$

It has been proved that only the Lagrange coefficient of support vectors are not zero, that is, only support vectors can influence the result of classification. So, the optimal weight vectors are

$$w = \sum_{SV} y_i \alpha_i x_i \tag{9}$$

The optimal biases are

$$b = \frac{1}{2} \left[w \cdot x^*(1) + w \cdot x^*(-1) \right] \tag{10}$$

where $x^*(1)$ is one of the support vectors in the first class and $x^*(-1)$ is one of the support vectors in the second class.

Hence, the classification function is [7][8]

$$f(x) = sgn \left(\sum_{SV} y_i \alpha_i K(x_i \cdot x) + b \right) \tag{11}$$

5   Performance Comprehensive Evaluation Model

Comprehensive evaluation usually has two kinds of problems: one is classification, the other is ranking. This paper reduces the performance evaluation of thermal power plant to a ranking problem. The main process of the model is as follows:

(1) PCA preprocessing.

Because of a large number of performance evaluation indexes, the input dates of SVM are too complex; it is difficult to get good result in rapidity and accuracy. This paper introduces PCA into SVM, at first, the sample set is preprocessed by PCA, and several principal components are obtained instead of multiple original indexes. The preprocessed sample set is used to train and test SVM.

(2) Vector Connection.

SVM is established for classification. In order to introduce SVM to the model, vector connection is necessary. Given two thermal power plants A and B, the eigenvector of A is $(x_{A1}, x_{A2}, \cdots, x_{A5})$,and the eigenvector of B is $(x_{B1}, x_{B2}, \cdots, x_{B5})$, connect the eigenvectors of A and B $(x_{A1}, x_{B1}, x_{A2}, x_{B2}, \cdots, x_{A5}, x_{B5})$, and use it as the input vector of SVM, the output has three classes: higher, lower and equivalent.

(3) Establish two-layer SVM classifier based on SVM and binary tree.

The standard SVM is only used to classify two classes, but the output of the model have three classes, so a two-layer SVM classifier which combines SVM and binary tree is established. For the first SVM, if A is higher than B, the output is +1, otherwise the output is −1, and then the second SVM, if A is equivalent to B, the output is +1, otherwise the output is −1. It is illustrated by Fig. 2.

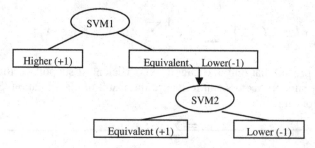

Fig. 1. Two-layer SVM classifier

(4) Sort by quick sort algorithm.

Quick sort algorithm is a recursive algorithm based on division. Given array S , $v \in S$ is chosen as pivot and $S - \{v\}$ is divided into two incompatible sets: $S_1 = \{x \in S - \{v\} \mid x \leq v\}$ and $S_2 = \{x \in S - \{v\} \mid x \geq v\}$, then repeat respectively the process for S_1 and S_2 until the array is ranked. The Time complexity of quick sort algorithm is $O(N \log N)$. Quick sort algorithm is considered as one of the best sorting algorithms

6   Experiments

6.1   Date Preprocessing by PCA

The credit data of 17 thermal power plants are obtained in this study. SPSS11.5 is utilized to simplify the index system. The accumulative contribution rate of the former five principal components can reach 85%, that is, the 24 indexes are reduced to five indexes.

Table 2. Contribution rates of principal components

Principal components	Contribution rate	Accumulative contribution rate
A_1	64.478%	64.478%
A_2	8.219%	72.697%
A_3	6.277%	78.974%
A_4	5.327%	84.301%
A_5	4.108%	88.409%

6.2   Two-Layer SVM Training and Testing

We randomly separate the dataset preprocessed by PCA into two parts: The data of 10 thermal power plants are used as training set, and then we obtain 90 input vectors by vector connection. The data of 7 thermal power plants are used as testing set.

Training set is used to train the two- layer SVM classifier, and LIBSVM (Version 2.6) is utilized in this paper. RBF function $k(x, y) = \exp[-\|x-y\|^2 /(2\delta^2)]$ is used as the kernel function, δ is the breadth of RBF function and its optimal value is 0.685313, C is the error penalty coefficient and its optimal value is 90.517457.

Testing set is tested by two-layer SVM classifier, and its main process is as follows: Suppose 7 thermal power plants P_1, P_2, P_3, P_4, P_5, P_6, P_7, P_4 is chosen as pivot and the rest is divided into two parts, compare P4 with the first part (P_1, P_2, P_3) and the second part (P_5, P_6, P_7) respectively, that is, input the vector connection of P_i $(i=1,2,3,5,6,7)$ and P_4 to two-layer SVM classifier, and adjust the position of P_i according to the output to make sure that the left of P_4 are all lower than P_4, and the right of P_4 are all higher than P_4, repeat the process for the two parts until all thermal power plants are ranked.

In order to verify the effectiveness of two-layer SVM classifier, BP network is also used to assess the same data. The neuron number of input layer, interlayer and output layer is 3, 6and 2, and sigmoid function is used as active function. Table 3 is the comparison of the results by using two-layer SVM classifier and BP network.

Table 3. Comparison of ranking results

Number	Original Rank	SVM	BP
P_1	1	1	2
P_2	3	3	3
P_3	6	6	6
P_4	7	7	7
P_5	5	5	4
P_6	2	2	1
P_7	4	4	5
Accuracy	—	100%	71.42%

From Table 3, we can see that the accuracy of two-layer SVM classifier is higher than BP network ,which because the theory of BP and SVM is different. SVM is established with principle of structural risk minimum, while BP is established with empirical risk minimum. In the condition of few data, SVM use few support vectors to represent the whole sample set and has good generalization performance, while BP has poor generalization performance because of over fitting. Only trained by sufficient data can BP network overcome the problem, however, this is difficult to realize, so SVM has greater use value than BP in practice.

7   Conclusions

A comprehensive index system of thermal power plant performance evaluation is established under low carbon economy, and a synthetic evaluation model based on PCA, SVM and quick sort algorithm is proposed. The method shows two merits in the research: ①After dimension reduction by PCA, the training set is reduced, the training time is shorten, and the prediction accuracy is improved. ② SVM, as the latter processor, has good generalization performance in the condition of small sample. Experiment results show that the model has great effectiveness for performance evaluation.

References

1. Vladimir, N.V., Zhang, X.G.: Nature of Statistics Theory. Tsinghua University Press, Beijing (2000)
2. Yao, X.Y., Li, Y., Wen, Q.: Low-carbon Economy Evaluation Index System of Thermal Power Industry. Journal of Ningxia University (Natural Science Edition) 12, 389–392 (2010)
3. Li, J.: Research on Performance Evaluation Index System for Thermo-Power Enterprises Based on Sustainable Development. Journal of Beijing Polytechnic College 1, 114–118 (2010)

4. Yang, K.R., Meng, F.R., Liang, Z.Z.: Adaptively Weighted PCA Algorithm. Computer Engineering and Applications 3, 189–191 (2012)
5. Wang, Y., Yang, J.A., Liu, H., Geng, Q.: A SVM Incremental Learning Algorithm Based on Inner Hull Vectors. Journal of Circuits and Systems 6, 109–113 (2011)
6. Ding, S.F., Sun, J.G., Chen, D.L., Li, Y., Jiang, X.L.: Improved SVM Decision-tree and Its Application in Remote Sensing Classification. Application Research of Computers 3, 1146–1148 (2012)
7. Feng, G.H.: Parameter Optimizing for Support Vector Machines Classification. Computer Engineering and Applications 3, 123–124 (2011)
8. Zhang, Z.Z., Dong, C.L., Chen, Z.Z., He, X.L.: Improved Fast Classifier Based on SVM and Density Clustering. Computer Engineering and Applications 2, 136–138 (2011)

Computing the Minimum λ-Cover in Weighted Sequences

Hui Zhang[1], Qing Guo[2,*] and Costas S. Iliopoulos[3]

[1] College of Computer Science and Technology,
Zhejiang University of Technology, Hangzhou, 310023, China
zhangh@zjut.edu.cn
[2] College of Computer Science, Zhejiang University, Hangzhou, 310027, China
13385718936@189.cn
[3] Department of Computer Science, King's College London Strand,
London WC2R 2LS, England
csi@dcs.kcl.ac.uk

Abstract. Given a weighted sequence X of length n and an integer constant λ, the minimum λ-cover problem of weighted sequences is to find the sets of λ factors of X each of equal length such that the set covers X, and the length of each element in the set is minimum. By constructing the Equivalence Class Tree and iteratively computing the occurrences of a set of factors in weighted sequences, we tackle the problem in $O(n^2)$ time for constant alphabet size.

Keywords: Weighted sequence, the minimum λ-cover problem, λ-combination, Equivalence Class Tree.

1   Introduction

It has long been an effort to investigate special areas in a biological sequence by their structure in biological sequence analysis, especially those repetitive genomic segments. The motivation comes from there exists high quantities of repetitive segments in the genome, some examples are tandem repeats, long interspersed nuclear sequences and short interspersed nuclear sequences[14].

A repeat of a string x is a substring that repeatedly occurs in x. A substring w of x is defined as a cover of x if and only if x can be constructed by concatenations and superpositions of w. For instance, aba is a cover of x=abababa. The study of covers dates back to the pioneering work of Apostilico etc.[1], who first introduced the notion of covers and presented a linear-time algorithm for finding the shortest cover of a string. A series of linear-time algorithms then followed to compute either the shortest cover or all the covers of a normal string [2, 8, 11].

Considering a set of repeated substrings "cooperatively" cover a given string, Iliopoulos and Smyth [10] introduced the k-covers and the minimum k-cover problem, that is, to compute a set w of substrings of x of minimum cardinality such that every element of w is of length k and every position of x lies within an occurrence of some

---

* Corresponding author.

D.-S. Huang et al. (Eds.): ICIC 2012, LNCS 7389, pp. 120–127, 2012.
© Springer-Verlag Berlin Heidelberg 2012

elements in w. Lots of work have been done on this problem. It was proved to be NP-Complete based on a reduction to 3-SAT[3]. Then by reducing to RVCPk, a minimum k-cover can be approximated to within a factor k in polynomial time[9].

Inspired by the k-covers, we introduced the idea of λ-covers [7] by restricting the size, instead of the length of each element of a generalized cover to be a given constant. Given a string x of length n and an integer λ, the λ-cover problem is to find all the sets $W=\{w_1, w_2,..., w_\lambda\}$ of substrings of x such that: (1) $|w_1|=|w_2|=...=|w_\lambda|$; and (2) the set W covers the string x. Through the construction of Equivalence Class Tree (ECT) and iterative computation of λ-combinations, the λ-covers of every possible length can be found in $O(n^2)$ time.

This paper focuses on the λ-covers in weighted sequences. A weighted sequence is a string that allows a set of characters to occur at each position of the sequence with respective probability. Weighted sequences are apt at summarizing poorly defined short sequences, e.g. transcription factor binding sites and the profiles of proteins[6].

The minimum λ-cover problem in weighed sequences is mainly investigated, that is, to find the λ-covers such that the length of each element in the set is minimum. If $\lambda=1$, this problem is reduced to compute the minimum cover of weighted sequences, which has been solved in $O(n^2)$ time [4, 12]. Thus, we simply consider the case $\lambda>1$.

2   Preliminaries

Definition 1. *Let an alphabet be $\Sigma = \{\sigma_1, \sigma_2, \ldots, \sigma_l\}$. A weighted sequence X over Σ, denoted by $X[1, n] = X[1]X[2] \ldots X[n]$, is a sequence of n sets $X[i]$ for $1 \leq i \leq n$, such that:*

$$X[i] = \{(\sigma_j, \pi_i(\sigma_j)) \mid 1 \leq j \leq l, \pi_i(\sigma_j) \geq 0, and \sum_{j=1}^{l} \pi_i(\sigma_j) = 1\}$$

Each $X[i]$ is a set of couples $(\sigma_j, \pi_i(\sigma_l))$, where $\pi_i(\sigma_l)$ is the non-negative weight of σ_j at position i,   representing the probability of having character σ_j at position i of X.

Let X be a weighted sequence of length n, σ be a character in Σ. We say that σ occurs at position i of X if and only if $\pi_i(\sigma)> 0$, written as $\sigma \in X[i]$. A nonempty non-weighted string $f[1, m]$ ($m \in [1, n]$) occurs at position i of X if and only if position $i + j - 1$ is an occurrence of the character $f[j]$ in X, for all $1 \leq j \leq m$. Then f is said to be a factor of X, and i is an occurrence of f in X. The probability of the presence of f at position i of X is called the weight of f at i, written as $\pi_i(f)$, can be obtained by using the cumulative weight, defined as the productive weight of the character at every position of f, that is, $\pi_i(f)=\prod_{j=1}^{m}\pi_{i+j-1}(f[j])$.

For instance, consider the following weighted sequence:

$$X = \left\{\begin{array}{l}(A,0.5)\\(C,0.25)\\(G,0.25)\end{array}\right\} G \left\{\begin{array}{l}(A,0.6)\\(C,0.4)\end{array}\right\} \left\{\begin{array}{l}(A,0.25)\\(C,0.25)\\(G,0.25)\\(T,0.25)\end{array}\right\} \left\{\begin{array}{l}(C,0.5)\\(T,0.5)\end{array}\right\}$$

The weight of a factor f =CGAT at position 1 of X is: $\pi_1(f) = 0.25 \times 1 \times 0.6 \times 0.25$ =0.0375. That is, CGAT occurs at position 1 of X with probability 0.0375.

Definition 2. *A factor f of a weighted sequence X is called a repeat in X if there exists at least two distinct positions of X that are occurrences of f in X.*

As scientists pay more attention to the pieces with high probabilities in DNA sequences, we fix a constant threshold $1/k$ (<1) for the presence probability of the motif. Then only those occurrences with probability not less than this threshold are counted, called real factors.

Definition 3. *Given a weighted sequence X of length n, and λ ($\lambda > 2$) factors of X such that $|w_1| = |w_2| = \ldots = |w_\lambda| = p$, we say that a set $W = \{w_1, w_2, \ldots, w_\lambda\}$ is a (λ, p)-cover of X if and only if every position of X lies within an occurrence of some w_i.*

To avoid trivialities we suppose $1 < p < n/\lambda$. Then the minimum λ-cover problem is to find such (λ, p)-covers of X with smallest p for a certain constant λ. Consequently, we need to iteratively record all the repeats of length p, then check if there exists combinations of these repeats that cover x until we find a (λ, p)-cover of X.

We have presented an algorithm for locating all the repeats in a weighted sequence, based on the following idea of equivalence relation on positions of a string [5]:

Definition 4. *Given a string x of length n over Σ, an integer $p \in \{1, 2, \ldots, n\}$, S be a set of positions of x: $\{1, 2, \ldots, n - p + 1\}$, then E_p is defined to be an equivalence relation on S such that: for two $i, j \in S$, $(i, j) \in E_p$ if $x[i, i + p - 1] = x[j, j + p - 1]$.*

Definition 5. *Consider a factor f of length p in a weighted sequence $X[1, n]$. An E_p-class associated with f is the set $C_f(p)$ of all position-probability pairs $(i, \pi_i(f))$, such that f occurs at position i with probability $\pi_i(f) \geq 1/k$.*

$C_f(p)$ is an ordered list that contains all the positions of X where a real factor f occurs. For this reason, the probability of each appearance of a factor should be recorded and kept for the next iteration. Briefly, we first partition all the n positions of X to build E_1, then iteratively compute E_p from E_{p-1} using Corollary 1 for $p \geq 2$. The computation stops at stage L, once no new E_{L+1}-classes can be created or each E_L-class is a singleton. For more information readers may refer to [13].

Corollary 1. *Let $p \in \{1, 2, \ldots, n\}$, $i, j \in \{1, 2, \ldots, n - p\}$. Then:*

$$((i, \pi_i(f)), (j, \pi_j(f))) \in C_f(p) \quad iff ((i, \pi_i(f')), (j, \pi_j(f'))) \in C_{f'}(p\text{-}1) \quad and$$
$$((i + p - 1, \pi_{i+p-1}(\sigma)), (j + p - 1, \pi_{i+p-1}(\sigma))) \in C_\sigma(1)$$

Where $\sigma \in \Sigma$, f and f' are two factors of length p and p-1 respectively, such that $f = f'\sigma$ and $\pi_j(f) \geq 1/k$, $\pi_j(f) \geq 1/k$.

3   Constructing the ECT

We follow to use the technique for computing λ-covers in non-weighted sequence. To help recording the λ-covers, we reinterpret the idea of the Equivalence Class Tree (ECT) for weighted sequences that was first introduced for non-weighted strings [7].

The ECT is a rooted tree built upon the building of equivalence classes, which expresses the relationship between each E_{p-1}-class and its corresponding E_p-classes by scanning characters from left to right in the ECT.

Let $\{f_1, f_2, ..., f_r\}$ be the real factors of length p-1 of X, $\{C_{f_1}(p-1),...,C_{f_r}(p-1)\}$ be the E_{p-1}-classes associated with these factors, denoted by $\{C_{f_1},...,C_{f_r}\}$ for simplification. The ECT is created as follows: The root has label 0. There are r nodes of depth p-1, each of which is a pair (C_{f_i}, f_i) ($i \in [1, r]$). To simplify, we label each node by f_i instead of the pair in the ECT. The children of f_i are the E_p-classes partitioned by C_{f_i} according to Corollary 1, corresponding to those real factors of length p produced by each f_i reading one character to the right. The construction of the ECT proceeds along with the computation of equivalence classes, until all the nodes are not repeats of X.

For example, consider the following weighted sequence and let the constant k=5: X = TAT[(A,0.5), (C, 0.3), (T, 0.2)]AT[(A,0.5), (C, 0.5)]A[(C,0.5), (T, 0.5)]A. Initially, there are three children of the root in the ECT: A, T, and C. When p=2, C_A is partitioned into three E_2-classes: C_{AA}, C_{AT}, C_{AC}, each of which has node A as the parent in the ECT. A subtree of the ECT of X rooted at A, is shown in Fig. 1.

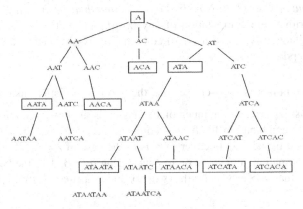

Fig. 1. A subtree of the ECT of X rooted at A

We label each factor that ends to position n in the ECT, called an end-aligned factor. It is easily observed that, in any branch of the ECT, every internal node represents a proper prefix of the leaf, of decreasing length from bottom to top.

The ECT provides a graphic model for the partitioning, which store the occurrences of all the repeats of X in different ways. The declaration the ECT is constructed along with the partitioning of equivalence classes implies that the construction takes time no more than the partitioning, that is $O(n^2)$ time.

4   Outline of Our Algorithm

Definition 6. *Let X be a weighted sequence of length n over Σ, λ be an integer. We say that a set $W=\{w_1, w_2,..., w_\lambda\}$ is a λ-combination of X if each w_j for $1 \le j \le \lambda$ is a factor of X. W is a (λ, p)-combination of X if each w_j is of length p.*

Consider a λ-combination $W=\{w_1, w_2,..., w_\lambda\}$ of X, we say that W occurs at position j of X if any w_i ($1 \le i \le \lambda$) occurs at position j. Obviously, a (λ, p)-combination W identifies a set of λ factors of length p, which contributes a candidate (λ, p)-cover of X.

Definition 7. *Suppose $W=\{w_1, w_2,..., w_\lambda\}$ is a λ-combination of X. A position list is a doubly linked list that stores all the occurrences of W in X in ascending numerical order, denoted by $L_W=\{j_1, j_1,..., j_r\}$, where for each element j in L_W, $j.prev$ and $j.next$ points to the previous and the next position respectively.*

Our method for finding the minimum λ-covers of a given weighted sequence X is then explicit. We begin with $p=2$, compute all the $(\lambda, 2)$-combinations of X if existed, then check each if it is a $(\lambda, 2)$-cover of X. If there exists at least a $(\lambda, 2)$-cover, the minimum λ-cover has been found, otherwise we increment p, to deduce all the (λ, p)-combinations from the corresponding $(\lambda, p-1)$-combinations and check if there exists a (λ, p)-cover of X by its position list for $p=3, 4, 5....$The iterative computation stops until at least a (λ, p)-cover was found at one stage.

Now we present a general algorithm for finding possible (λ, p)-covers W of a given weighted sequence X based on the ECT, for a certain $p \in (1, n/\lambda)$. Our algorithm mainly consists of two steps:

1.   Compute all the (λ, p)-combinations of X and their position lists.

In our algorithm, the computation of any (λ, p)-combination and its position list proceeds simultaneously. Once a (λ, p)-combination is obtained, the corresponding position list is updated.

(1).   Base case:

Let m_p be the number of E_p-classes. i.e, the number of distinct factors of length p of X. First consider $p=1$, the number of E_1-classes is at most $m_1= min (|\Sigma|, n)$. We store each $(\lambda, 1)$-combination for further computation. Note that the only possibility that there exists a trivial $(\lambda, 1)$-cover of X is in the case of $\lambda= m_1$. If $m_1 \leq \lambda$, we save this $(m_1, 1)$-combination, and move on to considering $p=2, 3, 4....$ the base case is the smallest p such that $m_p > \lambda$, which allows us to find qualified (λ, p)-combinations for iteration.

(2).   Inductive step:

Assume that we have created the ECT for X, and we have stored the position strings for all the (λ, p)-combinations at stage p. Consider a certain (λ, p)-combination $W=\{w_1, w_2,..., w_\lambda\}$, it might produce a series of $(\lambda, p+1)$-combinations as a result of the partitioning of each E_p-class $C_{w_i} (1\leq i\leq \lambda)$ according to the ECT. Let the number of sons of w_i in the ECT be δ_i. Then the relevant substrings of length $p+1$ can be denoted by $s_i^1, s_i^2,...s_i^{\delta_i}$, where $\delta_i \leq |\Sigma|$. Let the number of sons of w_i chosen to form the $(\lambda, p+1)$-combinations be τ_i, where $0 \leq \tau_i \leq \delta_i$.

From $i=1$ to $i=\lambda$, we successively substitute τ_i among these δ_i factors of length $p+1$ respectively for w_i. Every current combination obtained after w_i being replaced is saved to be further updated, denoted by S_i. In other words, we first compute $\{S_1\}$, then update each S_1 to obtain a set of S_2's, etc., until $\{S_\lambda\}$ is iteratively created. Obviously, $\{S_\lambda\}$ consists of all the $(\lambda, p+1)$-combinations associated with the given (λ, p)-combination W.

We simply describe the process of generating $\{S_1\}$ by considering w_1, since updating S_j to obtain a set of S_{j+1}'s follows the same method.

1. $\delta_1 < \lambda$: We update W by replacing w_1 with τ_1 among δ_1 sons of w_1, where $0 \leq \tau_1 \leq \delta_1$.

   a) $\tau_1 \neq 0$: We need to select λ-τ_1 elements from the remaining λ-1 w_j 's ($j \geq 2$) to compose λ-combinations, which leads to $\sum_{\tau_1=1}^{\delta_1} \binom{\delta_1}{\tau_1}\binom{\lambda-1}{\lambda-\tau_1}$ $\{S_1\}$'s. The cardinality of $\{S_1\}$'s is dependent on λ and $|\Sigma|$, which is a constant.

   b) $\tau_1 = 0$: Not choosing any s_i^j to substitute w_1, we simply keep the $(\lambda$-1, $p)$-combination $\{w_2, w_3,..., w_\lambda\}$ as an eligible S_1 to be further updated.

2. $\delta_1 \geq \lambda$: The algorithm runs almost the same with the case $\delta_1 < \lambda$, with the major distinction that $0 \leq \tau_1 \leq \lambda$. In this case, as $\delta_1 < |\Sigma|$, the number of S_1's is at most $\sum_{\tau_1=1}^{\lambda} \binom{|\Sigma|}{\tau_1}\binom{\lambda-1}{\lambda-\tau_1} + 1$, also a constant independent of p.

The position list L_{S_2} for any S_1 can be iteratively updated along with the induction of S_1 from W. Consider a certain S_1:

Case 1: w_1 is an unmarked internal node in the ECT.
As τ_1 sons of w_1 are chosen, those occurrences that are included in each of the τ_1 corresponding E_{p+1}-classes but not included in the class $C_{w_1}(p)$, and those of w_k's ($2 \leq$ k $\leq \lambda$) that are not included in S_1 should be eliminated from L_W. When an occurrence j needs to be removed, the node previous and next to it in the position list, ie. $j.prev$ and $j.next$ will be directly linked. After all the "removed" positions are examined, we get the position list for S_1.

Case 2: w_1 is an unmarked internal node in the ECT.
In this case, position n-p+1 should be removed. Besides this, the algorithm runs the same with Case 1.

Observe the fact that position 1 should be always the first element in the position list of a (λ, p)-cover, so during the iterative process, only those λ-combinations whose position list has position 1 as the first element are eligible to be stored, then to be further partitioned and updated. This will greatly decrease the amount of combinations that needs to be further processed.

2. *Check every (λ, p)-combination of X if it is a (λ, p)-cover.*

   (1) Filtering step:
   The last occurrence of a (λ, p)-cover must ends to n. Only those λ–combinations whose last element in its position list is n-p+1 will be remained as candidate (λ, p)-covers.

   (2) Comparison step:
   This step determines whether a candidate (λ, p)-cover is true or not based on the fact that any distance between adjacent occurrences of a (λ, p)-cover in X should not exceed the length of p. We attempt to maintain the maximum difference between adjacent occurrences of a λ-combination in X, denoted by MAX-GAP.

   Let MAX-GAP be g_0 for the given (λ, p)-combination $W=\{w_1, w_2,..., w_\lambda\}$. We can iteratively compute the value of MAX-GAP for a $(\lambda, p+1)$-combination S_λ at the iterative step along with the induction of S_λ from W. As mentioned earlier, when a position j is removed, the position previous and next to it in the position list will be directly linked, then the distance between two adjacent positions is correspondingly updated to $j.next$-$j.prev$. MAX-GAP is then maintained as the larger between this updated distance and the current MAX-GAP. When a S_λ is produced, MAX-GAP can also be achieved. Eliminating those candidates with MAX-GAP$>p$ from all those candidates, we obtain the true (λ, p)-covers.

Theorem 1. *The minimum λ-cover problem of a weighted sequence X of length n can be solved in $O(n^2)$ time.*

Proof. As we discussed in section 3, the ECT of weighted sequences is constructed along with the partitioning of equivalence classes, which costs $O(n^2)$ time[12].

Base step processes at most $O(n)$ λ-combinations. Consider the iterative step from p to $p+1$, $L_{\{w_1, w_2, ... w_\lambda\}}$ is updated according to the partitioning of each w_i. Computing the position list for any $(\lambda, p+1)$-combination simply removes some nodes of the list $L_{\{w_1, w_2, ... w_\lambda\}}$, which takes $O(1)$ time for each removed position.

As we have analyzed before, there are constant numbers of combinations produced after w_1 is processed, dependent on $|\Sigma|$ and λ. Hence, after all these λ w_j's are processed, we obtain constant numbers of $(\lambda, p+1)$-combinations related to a given (λ, p)-combination $\{w_1, w_2,..., w_\lambda\}$. Taking them all into account, every position is to be removed for $O(|\Sigma|)$ times. Therefore at stage $p+1$, Step 1 costs $O(n)$ time for all the n positions. That is, the total number of (λ, p)-combinations is $O(n)$.

The filtering step takes $O(1)$ time for each (λ, p)-combination since it simply checks the last element of its position list. The comparing step is performed along with the iterative step, thus computing maximal differences does not take more time than computing the position lists. Therefore, Step 2 runs in $O(1)$ time to determine whether a $(\lambda, p+1)$-combination is a $(\lambda, p+1)$-cover or not, hence $O(n)$ time for all $O(n)$ combinations.

To sum up, finding the $(\lambda, p+1)$-covers requires $O(n)$ time. Since $1< p < n/\lambda$, there are at most $O(n)$ stages, then the overall complexity for finding the minimum λ-cover is $O(n^2)$. Therefore, our algorithm needs $O(n^2)$ time in total.

5   Conclusions

In this paper we introduce the minimum λ-cover problem of a weighted sequence, and present an efficient solution to this problem with the time complexity $O(n^2)$. The main idea of our algorithm lies in the construction of the Equivalence Class Tree and iterative computations of the occurrences of a set of factors in weighted sequences. As opposed to its non-weighted version, the solution does not take more time.

Acknowledgments. This work was partially supported by Zhejiang Provincial Natural Science Foundation under Grant No: Y1101043 and Foundation of Zhejiang Provincial Education Department under Grant No: Y201018240 of China.

References

1. Apostolico, A., Farach, M., Iliopoulos, C.S.: Optimal Superprimitivity Testing for Strings. Information Processing Letters 39, 17–20 (1991)
2. Breslauer, D.: An On-line String Superprimitivity Test. Information Processing Letters 44, 345–347 (1992)
3. Cole, R., Iliopoulos, C.S., Mohamed, M., Smith, W.F., Yang, L.: Computing the Minimum k-cover of a String. In: Proc. of the 2003 Prague Stringology Conference (PSC 2003), pp. 51–64 (2003)
4. Christodoulakis, M., Iliopoulos, C.S., Mouchard, L., Perdikuri, K., Tsakalidis, A., Tsichlas, K.: Computation of Repetitions and Regularities on Biological Weighted Sequences. Journal of Computational Biology 13(6), 1214–1231 (2006)
5. Crochemore, M.: An Optimal Algorithm for Computing the Repetitions in a Word. Information Processing Letters 12(5), 244–250 (1981)
6. Gusfield, D.: Algorithms on Strings, Trees and Sequences: Computer Science and Computational Biology. Cambridge University Press (1997)
7. Guo, Q., Zhang, H., Iliopoulos, C.S.: Computing the λ-covers of a String. Information Sciences 177, 3957–3967 (2007)
8. Iliopoulos, C.S., Moore, D.W.G., Park, K.: Covering a String. Algorithmica 16, 288–297 (1996)
9. Iliopoulos, C.S., Mohamed, M., Smyth, W.F.: New Complexity Results for the k-covers Problem. Information Sciences 181, 251–255 (2011)
10. Iliopoulos, C.S., Smith, W.F.: An On-line Algorithm of Computing a Minimum Set of k-covers of a String. In: Proc. of the Ninth Australian Workshop on Combinatorial Algorithms (AWOCA), pp. 97–106 (1998)
11. Li, Y., Smyth, W.F.: Computing the Cover Array in Linear Time. Algorithmica 32(1), 95–106 (2002)
12. Zhang, H., Guo, Q., Iliopoulos, C.S.: Varieties of Regularities in Weighted Sequences. In: Chen, B. (ed.) AAIM 2010. LNCS, vol. 6124, pp. 271–280. Springer, Heidelberg (2010)
13. Zhang, H., Guo, Q., Iliopoulos, C.S.: Loose and Strict Repeats in Weighted Sequences. Protein and Peptide Letters 17(9), 1136–1142 (2010)
14. The Human Genome Project (HGP), http://www.nbgri.nih.gov/HGP/

A Novel Hybrid Evolutionary Algorithm for Solving Multi-Objective Optimization Problems

Huantong Geng[1], Haifeng Zhu[1], Rui Xing[2], and Tingting Wu[1]

[1] College of Computer and Software
[2] College of Atmospheric Science
Nanjing University of Information Science & Technology, Nanjing, China
htgeng@nuist.edu.cn

Abstract. This paper proposed a novel hybrid evolutionary algorithm for solving the multi-objective optimization problems (MOPs). The algorithm uses the idea of simulated annealing to combine co-evolution with genetic evolution, set several model sets according to the principle of "model student", and employs ε-dominant and crowding distance sorting to search the excellent population. In the co-evolution model, cluster analysis is used to classify the model set, and the estimation of distribution algorithm (EDA) is used to establish the probabilistic model for each class. The individuals are generated by sampling through the probabilistic model; in genetic evolution model, populations evolve based on the model set. This algorithm takes full advantage of the global and local search abilities, and makes comparison with the classical algorithm NSGA-II, the experimental results show that our algorithm for solving the multi-objective problem has better convergence and distribution.

Keywords: MOEAs, Co-evolution, EDA, Model Student.

1 Introduction

Many problems are affected by human activities and a number of factors in the natural environment, and many have multiple conflicting objectives, and can only get a group Pareto optimal solutions [1]. As a group search algorithm, evolutionary algorithm is very suitable for solving MOPs [2].

The classic multi-objective evolutionary algorithms (MOEAs), such as SPEA-II [3], and NSGA-II [4], are all based on Pareto dominance relationship. However, the MOEAs are mainly used for the two-objective optimization problems. As the number of objectives increases, the number of non-dominated individuals in population will increase exponentially, this can greatly weaken the selection and search capabilities based on Pareto sorting, and the problem of premature appears, so the traditional MOEAs based on Pareto sorting are needed to improve for the optimization problem. This paper combined ε-Pareto dominance [5] with Pareto dominance, and based on crowding distance sorting to achieve the purpose of selecting excellent individuals.

In recent years, our IS-MOEA [6] and SBMS-MOEA [7] are also proposed on the basis of previous work. In these evolutionary algorithms, the individual often

D.-S. Huang et al. (Eds.): ICIC 2012, LNCS 7389, pp. 128–136, 2012.
© Springer-Verlag Berlin Heidelberg 2012

represents only one solution of the problem, without taking into account the interaction between the individuals and the surrounding environment, that is, without considering the co-evolution. This may lead to lack of diversity of the solutions, and may affect the convergence and diversity. However, cooperative co-evolutionary genetic algorithm (CCGA) [8] solves the above problems to some extent. In this paper, the idea of CCGA is referred specific to the decision variable-independent MOPs, that is, if the function has V decision variables, uses V sub-populations, and each individual in the sub-populations only represents one part of the problem.

In the standard genetic algorithm, its evolution has excellent local optimization ability, but poor global search capability. At the late stages of the evolution, the evolution entirely dependents on the mutation, which makes the subsequent evolution become a completely aimlessly random search. For solving the complex problems, such as the problem of multiple local optima, the algorithm often takes on premature convergence phenomenon. EDA [9] is different from the traditional evolutionary algorithms. EDA has a good global search capability and can lead the population search toward the Pareto front. In order to converge to the Pareto front more effectively, a hybrid algorithm with better local and global search capability is needed.

This paper employs the idea of simulated annealing to combine the co-evolution al gorithm which is based on EDA with genetic evolution based on "model student", and proposes a new hybrid evolutionary algorithm for solving the multi-objective problem (CEDA-MSGA). This algorithm balances the relationship of the local search and the global search dynamically. In this paper, seven test functions are used to evaluate the performance of the algorithm, and comparison with the NSGA-II is also done.

2   CEDA-MSGA

CEDA-MSGA combines Pareto dominance with ε-dominance to find the model population, sorts the model population based on crowding distance to keep the diversity and distribution of the model set, uses the clustering to classify the model set, and establish the appropriate probabilistic model of each class. During the process of evolution simulated annealing is used to combine the co-evolution algorithm which is based on EDA with genetic evolution based on the "model student".

2.1   Individual Evaluation Method

Pareto dominance and ε-dominance are combined in this algorithm, compares p with q by Pareto dominated relations, that is, if p and q do not dominate mutually, and p has weaker performance on one objective only, but greatly improves on the other objectives, that is, p meets the condition of ε-dominance, then considering that p is better than q, i.e. p dominate q.

2.2   "Model Student" Set

"Model student": the assessment of the model student in real world is that the student who gets highest scores in one subject among the students who have good consolidated results will be chosen as the class representative of the subject, if the student acts excellently on two or three subjects, he can become the model undoubtedly. However, NSGA-II only simply uses the Pareto front and less model individuals.

In CEDA-MSGA, according to "model student" concept, a number of model sets are established correspondingly. Taking three objective functions as an example, set A, B, C that are used to store the individuals with outstanding performance in one, two, and three objectives respectively are set, and their capacities are set to M_1, M_2 and M_3, and if the problem has three objectives, $M_1 = M_2 = 3$, $M_3=200$.

Generation Method of Model Set. One individual from the initial population Pop is randomly selected to the set A, B and C in the first generation, then all individuals in Pop are compared with the individuals in C one by one, for $P_i \in Pop$,if P_i can't be dominated by any individual, add it to C, if the number of individuals in A and B is less than their maximal capacity, add P_i to A and B directly; Otherwise, only when P_i can't be dominated by any individual in C and is better than the corresponding objective of some individual, P_i can add to the set to replace the original location of the corresponding individual; if some individuals in C is dominated by P_i, these individuals will be removed from C. From the second generation, the individuals of populations through the evolution operations adjust C based on the domination relations, the update method used in A and B is same as in the first generation. The model sets are not needed to be sorted in the algorithm, thus the computing cost is greatly reduced.

The number of the excellent individuals in C is increasing as evolving. Taking the computational efficiency into account, once the number of individuals stored is more than M_3, the set C is needed to be trimmed, Whether each individual will be removed depends on its crowding distance, those poor individuals are removed to make the individuals maintain a certain distance in the set. In this way, the set not only maintain the diversity, but also have a better distribution.

Role of Model Sets. Firstly, take these model sets as the genetic evolutionary populations and their studying objects respectively, i.e., make the individuals which have three excellent objectives learn from the individuals which have one or two excellent objectives (make them mate with each other). The principle is shown as Fig.1, first, take C as the genetic evolutionary population, and select one individual randomly to the mating pool, then select one individual randomly from A or B as the studied object to the mating pool too, and finally, do the evolutionary operation to the individuals in the mating pool to achieve the purpose that the evolutionary population can learn from the model individual. Using this method can make the searched front evolve toward the true front direction and more completed.

Secondly, take C as a collection of excellent individuals, and use the clustering algorithm to classify these individuals according to the idea of EDA, and establish the appropriate probabilistic model for each dimension of individuals in each class (multi-probabilistic models can make the algorithm have better distribution), and this method can make all the co-evolutionary populations evolve toward a hopeful prospect, and find the complete Pareto front by this algorithm rapidly.

Considering the solutions of MOPs are an optimal solution set, EDA is used here, if only one probabilistic model is established, apparently this model can't make the population evolve towards all the fine directions, and the true Pareto front is most likely unable to find. In order to cover the Pareto optimal front better, the clustering algorithm---leader-follower [10] is used to classify C before the algorithm establishes the probabilistic model to obtain a number of different classes, and then establish the corresponding probabilistic model on each dimension for each class according to the idea of EDA. Considering C based on crowding distance sorting maintains the diversity, calculate the weight coefficient of each class according to the number of individuals in each class, and use the roulette wheel sampling algorithm based on the weight coefficient to select the corresponding model to generate new individuals, form next population, and it will also maintain this diversity distribution character.

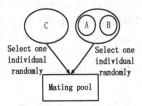

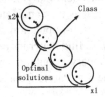

Fig. 1. Model set in genetic evolution   **Fig. 2.** Effect figure of the population cluster

Leader-follower can meet the requirement of uncertainty number of each generation probabilistic models without pre-determining the number of classes, and ensure that all searching areas are searched extensively and this can make the algorithm get better distribution. Fig.2 shows that four classes are searched in the two-dimensional search space.

Finally, the model sets save the excellent solutions, In the process of evolution, the algorithm always exploits and explores based on the best populations.

2.3   Genetic Evolution and Co-evolution Based on Simulated Annealing

The principle of the individuals mixed is the individuals come from different algorithms. In our algorithm, the individuals are generated from co-evolution and GA. The algorithm combined the co-evolution with GA using simulated annealing, in the early stages of evolution, global search is needed to be done by EDA of co-evolution to find the Pareto front quickly, and co-evolution plays a dominant role; while in the late stages of evolution, exploit need be strengthened around the Pareto front using the local search ability of GA, and GA plays a key role here. As shown in Fig.3, the proportion of the individuals generated by co-evolution gradually decreases as the number of evolutional generation increases, therefore, the algorithm not only maintains the local search ability, but also improves the global search ability.

In order to control the role played by co-evolution and GA at different period in the algorithm, simulated annealing method is used to control the scale factor p_t. p_t can be calculated by the equation set as follows:

$$\begin{cases} p_1 = p_{max} \\ p_t = p_{min} + \alpha(p_{t-1} - p_{min}) \end{cases} \tag{1}$$

In formula (1), p_{max} and p_{min} is the upper and lower limit scale factor respectively, and α is the annealing factor which interval is $0 \le \alpha \le 1$.

This algorithm generates new individuals by comparing random number *rand* with p_t: When *rand* $<p_t$, use the co-evolutionary approach to generate new individuals; when *rand* $\ge p_t$, use the crossover and mutation of GA to generate new individuals.

Generating Method of Parent Population. The random initialization is used to generate the first parent population, and the subsequent generation parent population is taking the excellent individuals entered the model sets, which have greater probability to be selected, as the candidate set of the next parent population. In this way, the successive new sub-population will be searched and explored based on the excellent populations to find better individuals to supply the model sets, therefore, the model sets and evolutionary population will be promoted mutually, and the algorithm will converge to the true Pareto front more rapidly.

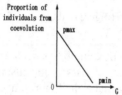

Fig. 3. Sketch map of proportion   **Fig. 4.** Co-evolutionary parent population

In the co-evolutionary algorithm, parent populations are generated by the approach shown as in Fig.4, that is, N complete individuals are randomly selected from model set C, and each complete individual has V components(In Fig.4, $V = 7$). Merging the components of the N complete individuals can reform the next generation parent populations which have V sub-populations(In Fig.4, Population 1 ... Population 7), and each size of the sub-population is N.

As shown in Fig.1, the parent population of GA comes from set A, B and C, and the parent individuals which are randomly selected from set A, B and C respectively.

3   Implementation Process of the Algorithm

The detailed process of CEDA-MSGA is described as follows:

Step1:**Initialization**: set the generation counter T to 1, co-evolutionary population size to N, the number of the sub-populations to V, and create model sets $A = B = C = \phi$, set the maximum capacity of A, B and C to M_1, M_2 and M_3, the population size of the genetic evolution to $V \times N$, and initialize the populations respectively.

Step2:**Population assessment**: Assess each *individual*$_{ij}$ ($i = 1, 2 ..., V; j = 1, 2 ..., N$) in the population, the individual which is assessed is chosen from co-evolution populations or GA population according to the relationship of random probability and simulated annealing scale factor. When *rand* <p_t, goto Step2.1; otherwise goto Step2.2. The step2 ends until all individuals are evaluated.

Step2.1:**Assessment in co-evolution**: evaluate the *j-th* individual of the *i-th* sub-population in co-evolution, and a complete solution is generated by the cooperation result from the individual and other sub-populations (understanding through Fig.4, this individual randomly combines with the individuals from other *V*-1 sub-populations to a complete individual. Different from CCGA, using this way, can greatly reduce the consumption of computing resources), and make a decision to update the model set or discard it according to the dominant concept, goto Step2.

Step2.2: **Assessment in genetic evolution**: when *T*=1, evaluate the *(i × j)*th individual in the population of genetic evolution, and decide to update the model set or discard it according to the dominant relationship; Otherwise, as shown in Fig.1, select a individual randomly from set *A*, *B* and *C* respectively, crossover and mutate in the mating pool to generate a new individual, and then decide to update the model set or discard it. Return to Step2.

Step3: **Trimming of set C** : If | *C* |> M_3, trim the *C*, and make the size of *C* equal to M_3, if | *C* | ≤M_3, goto Step4.

Step4: **Generating parent population**: the parent population of genetic evolution is generated in Step2.2, while the parent populations of co-evolution are generated by using the *N* complete individuals, the specific method is shown in section 2.3.

Step5: **Evolutionary operator**: genetic evolution is completed in Step2.2, about the co-evolution, classifying the parent populations of co-evolution to get different classes, and establishing the appropriate probabilistic model of each class. The new populations will be generated based on the probabilistic model.

Step6: **Termination**: $T = T +1$, If the termination conditions are satisfied then stop, and output the set *C* as the final solution set; otherwise goto Step2.

4   Experimental Results and Discussions

4.1   Test Problems

To verify the idea of CEDA-MSGA and its effectiveness, the performance of CEDA-MSGA is assessed using the experiments, and compared with NSGA-II, we adopted five standard test functions [4] (ZDT1~ZDT4, and ZDT6, minimization problems of two objectives) and two standard test functions in the references [11] (DTLZ1 ~DTLZ2, minimization problems of three objectives). They have different typical characteristics, that is, convex or non-convex and searching space which has a bias. The difficulties of using the EAs for solving these functions are representative.

4.2   Performance Metrics

(1) Approximation: the distance between the solutions set in the objective space and the true Pareto front is minimized. The coverage metrics *C* [12], the convergence measure *M1* [12] and distance indicators *GD* [13] are used;

(2) Distribution: the solution set in the objective space acquires the best distribution pattern, and this paper adopted the distribution measure SP [12] and SD [14].

In addition, the graphs of Pareto front are drawn to make the comparison.

4.3 Design of Experiments

CEDA-MSGA and NSGA-II are real-coded, and the mutation probability P_m is $1/V$ (V is the number of independent variables). The size of co-evolutionary sub-populations is N in CEDA-MSGA, and the amount of sub-populations is V. The GA population size is $V{\times}N$, and the scale factor $p_{max}{=}0.9$, $p_{min}{=}0.3$, $\alpha{=}0.9$, and the threshold b of clustering algorithm is 0.5. For each test function, two algorithms run 30 times independently.

To prove the idea of CEDA-MSGA and its effectiveness for solving MOPs, the classical NSGA-II is used to solve the same MOPs, and the results are compared with CEDA-MSGA. The sub-populations size N is set to 30 in CEDA-MSGA.

Experiment 1: In two objectives, the algorithm is terminated when the number of objective evaluation equals 90,000. The max-capacity M_3 of the model set C is set to 100, and crossover probability P_c is 0.9. Population size N in NSGA-II equals 100.

Experiment 2: In three objectives, when the algorithm is terminated when the number of objective evaluation equals 100,000, the max-capacity M_3 of the model set C is set to 200, and crossover probability P_c is 1. Size N in NSGA-II is equal to 200.

4.4 Experimental Results and Analysis

Experiment 1: Table 1 shows the average results of running CEDA-MSGA and NSGA-II 30 times independently on $M1$, C and SD metrics in each test function. Here $C(CEDA\text{-}MSGA, NSGA\text{-}II)$ is abbreviated as $C(C,N)$, and $C(NSGA\text{-}II, CEDA\text{-}MSGA)$ as $C(N, C)$; Fig.5 is the box plot which depicts the coverage metrics $C(C, N)$ and $C(N, C)$ results obtained from CEDA-MSGA and NSGA-II on 5 test functions. Fig.6 shows that the Pareto front searched by CEDA-MSGA and NSGA-II, and these figures indicate that the front searched by CEDA-MSGA is more completed and more approximation to the true front. The experimental results show that CEDA-MSGA is significantly better than NSGA-II on the test functions ZDT1~ ZDT4 and ZDT6 in the aspects of convergence and distribution.

Experiment 2: Table 2 and Fig.7 show that CEDA-MSGA is better than NSGA-II in the aspects of convergence (GD measure) and distribution (SP measure) in three-objective test optimization problems DTLZ1 and DTLZ2.

Table 1. Results obtained by NSGA-II and CEDA-MSGA on ZDT1-ZDT4, ZTD6

Problems	CEDA-MSGA			NSGA-II		
	$M1$	SD	$C(C,N)$	$M1$	SD	$C(N,C)$
ZDT1	**0.0001**	**0.0031**	**1.0000**	0.0179	0.0126	0.0000
ZDT2	**0.0001**	**0.0018**	**1.0000**	0.0222	0.0603	0.0000
ZDT3	**0.0003**	**0.0052**	**0.9424**	0.0133	0.0158	0.0000
ZDT4	**0.0002**	**0.0020**	**1.0000**	1.0123	0.0134	0.0000
ZDT6	**0.0004**	**0.0015**	**1.0000**	0.5859	0.0053	0.0000

Table 2. Results obtained by NSGA-II and CEDA-MSGA on DTLZ1 and DTLZ2

Problems	CEDA-MSGA		NSGA-II	
	GD	*SP*	*GD*	*SP*
DTLZ1	**0.000181**	**0.011234**	0.032412	0.016642
DTLZ2	**0.000313**	**0.052755**	0.122403	0.070044

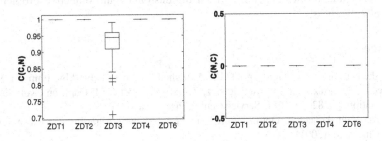

Fig. 5. C(C,N) *and* C(N,C) for two-objective optimization problems

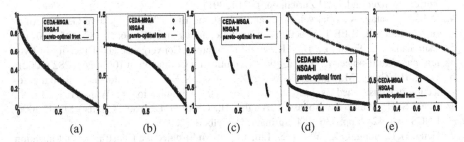

Fig. 6. Pareto fronts obtained by NSGA-II and CEDA-MSGA for ZDT1~ ZDT4, ZDT6 are shown in sub-figure (a)~(e)

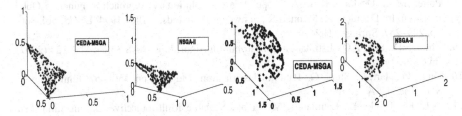

Fig. 7. Pareto fronts obtained by CEDA-MSGA or NSGAII for DTLZ1~DTLZ2

5   Conclusions and Future Works

This paper proposed a novel hybrid evolutionary algorithm to solve the multi-objective problem. The algorithm using the intrinsic character of diversity attached to CCGA, and combining co-evolution algorithm which is based on EDA with genetic evolution based on "model student" method, makes the new algorithm possess better capability of local and global search. Each strategy plays an important role during the different evolutionary period. In the seven standard test optimization problems, compared with NSGA-II, experimental results show that our algorithm has better search ability, and the Pareto solutions have better performance in convergence and distribution. The future work will further research the solving effect of our algorithm on other types of test functions, and make more comparisons with other classical algorithms(such as SPEA-II, etc.), and further apply it to high dimensional multi-objective problem to verify its effectiveness.

References

1. Horn, J., Nafpliotis, N., Goldberg, D.E.: A Niched Pareto Genetic Algorithm for Multi-objective Optimization. In: Michalewicz, Z. (ed.) Proc. 1st IEEE Conf. on Evolutionary Computation, pp. 82–87. IEEE Service Center, Piscataway (1994)
2. Deb, K.: Multi-Objective Optimization Using Evolutionary Algorithms. John Wiley & Sons, Chichester (2001)
3. Zitzler, E., Laumanns, M., Thiele, L.: SPEA2: Improving the Strength Pareto Evolutionary Algorithm. Technical Report 103, Computer Engineering and Networks Laboratory (TIK), Zurich, Switzerland, ETH Zurich, pp. 1–21 (2001)
4. Deb, K., Pratap, A., Agarwal, S., et al.: A Fast and Elitist Multi-objective Genetic Algorithm: NSGA-II. IEEE Transactions on Evolutionary Computation 6(2), 182–197 (2002)
5. Laumanns, M., Thiele, L., Deb, K., et al.: Combining Convergence and Diversity in Evolutionary Multi-objective Optimization. Evolutionary Computation 10(3), 263–282 (2002)
6. Geng, H., Zhang, M., Huang, L., Wang, X.: Infeasible Elitists and Stochastic Ranking Selection in Constrained Evolutionary Multi-objective Optimization. In: Wang, T.-D., Li, X., Chen, S.-H., Wang, X., Abbass, H.A., Iba, H., Chen, G.-L., Yao, X. (eds.) SEAL 2006. LNCS, vol. 4247, pp. 336–344. Springer, Heidelberg (2006)
7. Geng, H.T., Song, Q.X., Wu, T.T., Liu, J.F.: A Multi-objective Constrained Optimization Algorithm Based on Infeasible Individual Stochastic Binary-Modification. In: Proceedings of 2009 IEEE International Conference on Intelligent Computing and Intelligent Systems, pp. 89–93. IEEE, Shanghai (2009)
8. Potter, M.A., De Jong, K.A.: A Cooperative Coevolutionary Approach to Function Optimization. In: Davidor, Y., Männer, R., Schwefel, H.-P. (eds.) PPSN 1994. LNCS, vol. 866, pp. 249–257. Springer, Heidelberg (1994)
9. Zhou, S.D., Sun, Z.Q.: Estimation of Distribution Algorithms. AAS 33(2), 113–124 (2007) (in Chinese)
10. Duda, O., Hart, E., Stork, G.: Pattern Classification, 2nd edn., pp. 450–452. John Wiley & Sons, Inc., USA (2001)
11. Deb, K., Thiele, L., Laumanns, M., et al.: Scalable Multi-objective Optimization Test Problems. In: Proceedings of the Congress on Evolutionary Computation, vol. 1, pp. 825–830. IEEE Service Center, Piscataway (2002)
12. Zitzler, E., Deb, K., Thiele, L.: Comparison of Multiobjective Evolutionary Algorithms: Empirical Results. Evolutionary Computation 8(2), 173–195 (2000)
13. Coello, C.A.C.: A Comprehensive Survey of Evolutionary-based Multi-objective Optimization Techniques. Knowledge and Information Systems 1(3), 269–308 (1999)
14. Schott, J.R.: Fault Tolerant Design Using Single and Multicriteria Genetic Algorithm Optimization. Master's thesis, Department of Aeronautics and Astronautics, Massachusetts Institute of Technology, Cambridge, MA (1995)

Heuristic Algorithms for Solving Survivable Network Design Problem with Simultaneous Unicast and Anycast Flows

Huynh Thi Thanh Binh, Pham Vu Long, Nguyen Ngoc Dat, and Nguyen Sy Thai Ha

School of Information and Communication Technology, Hanoi University of Science and
Technology, Hanoi, Vietnam
binhht@soict.hut.edu.vn,
{long.opd,dat.thientai,nguyensythaiha}@gmail.com

Abstract. Given a connected, weighted, undirected graph G = (V, E), a set of
nodes, a set of links with modular cost based on ACMC model [2] and a set of
customers' demands. This paper proposes two heuristic algorithms for solving
the ACMC Survivable Network Design Problem (A-SNDP). The goal is to design connections based on customers' demands with the smallest network cost
to protect the network against all failures. This problem is NP-hard. The experimental results are reported to show the efficiency of proposed algorithm
comparing to the Tabu Search algorithm [2].

Keywords: survivable network design, anycast, unicast, A-SNDP.

1   Introduction

In the recent years, there are many kinds of transmissions have been used, included
unicast and anycast flow, which are two of the most popular types. A transmission
connected one host to another is called unicast. An anycast is defined as one-to-one-
of-many transmission to deliver a packet to one of many host, which is applied in
Domain Name Service (DNS), Web Service, Overlay Network, peer-to-peer (P2P)
systems, Content Delivery Network (CDN), software distribution...[2]. The populari-
ty of anycast technology will increase in the near future, since many new services that
can use anycast paradigm are developed [1].

The impact of Internet and other computer networks can be seen in almost all areas
of our lives, including business, science, technology, so on, and their importance will
grow. Thus, any network failure can cause serious consequences affect to many
people and corporations. Therefore, designing the survivable network is a crucial
problem. Because we solve the SNDP problem based on ACMC (All Capacities
Modular Cost) model which is defined in [2], we call this problem by A-SNDP
(ACMC Survivable Network Design Problem).

A-SNDP problem is defined as the following: Given an undirected graph G = (V,
E). Each link divides into some bandwidth levels, each level has a corresponding cost.
Link cost is the cost of total of corresponding bandwidths using in this link. And

D.-S. Huang et al. (Eds.): ICIC 2012, LNCS 7389, pp. 137–144, 2012.
© Springer-Verlag Berlin Heidelberg 2012

network cost is total of link cost. A connection from a node to another is a unicast transmission. An anycast is also a connection from a node to another, but the difference is that the destination node has a replica server which backs up for it. In this problem, a demand required by customer is a connection between two nodes with corresponding bandwidth and type of connection. The goal of problem is to find a set of connections for all demands such that the network cost is minimal.

We can formulate the problem as following:

Find an appropriate connection for each customer's demand so as to minimize:

$$NCost = \sum_i c_i \tag{1}$$

With condition is that if ($B_{k-1} < \sum_j R_{ij} < B_k$) then $c_i = C_k$, where c_i is cost of link i,

B_k, C_k is bandwidth and corresponding cost with level k.

To solve A-SNDP, we have to build a set of path for all demands to minimize the network cost. To guarantee the survivability of connections, we also use backup path approach [2] [3] [4] [5]. In particular, each connection (anycast or unicast) is divided into working path and backup path (they can use some same links, but not completely). If the working path is broken, the backup path must be restored and conversely.

In order to minimize the network cost, we propose a new approach called Free Bandwidth based Heuristic Algorithm (FBB). This approach is based on utilizing the redundant bandwidth corresponding with paid cost level in each link. We propose two heuristic algorithms: FBB1 and FBB2. The FBB1 applies the main idea of FBB algorithm. Whereas, the FBB2 combines the idea of FBB1 and Local Search algorithm [14] for solving A-SNDP. We hope that the results found by FBB1 and FBB2 are better than the one found by Tabu Search [2].

The rest of this paper is organized as following: Section II describes the related works. In section III, we present the proposed algorithms to solve A-SNDP. Our experiments and computational and comparative results are given in section IV. The paper concludes with section V with some discussions on the future extension of this work.

2   Related Works

The SNDP is generally presented in [10], when considering both economics and reliability in telecommunication network design. Thus, there are two problems requested in the A-SNDP. Those are to guarantee the survivability of network system and to minimize the network cost.

The most popular way mentioned in many researches is the single backup path approach. The main idea of this method is as following: Each connection has a working path and a backup path. The working path is used for transmitting data in normal, failure-free state of the network. After a failure of the working path, the failed connection is switched to the backup path [2], [3], [4], [5].

In the literature, there are several papers research on SNDP problem ([2], [8], [10]). They use branch – and – bounds or branch – and – cut methods to find optimal precisely solutions. These methods can only use for small networks with 30 nodes and 100 edges. For larger network, they may propose evolutionary algorithms, tabu search

[2], and simulated annealing [8]. In [10], Vissen and Gold apply the evolution strategy (ES). Following the discussion of [10], the quality of result is very flexible and effective. The graph instance which has m nodes, n edges is called by (m, n) – ES. Overall, the (30, 200) – ES with discrete recombination delivers the best solution. The (10, 50) – ES operates with a smaller population size, uses much less resources and still delivers good cost results, but it is not as good as the (30, 200) – ES. Clearly, when using ES, a larger population helps to achieve a better result than a smaller one by avoiding or delaying convergence on local suboptimal. However, this algorithm is useful in the network which has only unicast flows.

With the network which has both anycast and unicast flows, K.Walkowiak and Jacub Gladysz [13] presented a heuristic algorithm for solving A-SNDP. The main idea of this algorithm is based on Flow Deviation method [8] and Local Search algorithm [14]. They achieve the quite good result with Polska (12 nodes, 36 links, 65 unicast, 12 anycast) network, the detail is that the average gap of the proposed heuristic algorithm to optimal results is 7.11%. Furthermore, K. Walkowiak and Jakub Gładysz [2] have built Tabu search algorithm based on hill climbing algorithm with some heuristics to solve this problem. They experimented this algorithm with three large instances which are Polska (12 nodes, 36 links, 65 unicast, 12 anycast), Germany (17 nodes, 52 links, 119 unicast, 13 anycast), Atlanta (26 nodes, 82 links, 234 unicast, 22 anycast) and received effective results. In particular with Polska network, they achieve the average gap to optimal results is 2.57% for 70% anycast/30% unicast case and 2.00% for 80% anycast/ 20% unicast case. However, their Tabu Search algorithm is quite simple and their results cannot be optimal completely.

In the next section, we introduce two algorithms for solving A-SNDP. We hope that our proposed algorithms will have a better result than previous algorithms.

3   Proposed Algorithms

Almost the current heuristic algorithms for solving A-SNDP use local search method. Each step, they try to minimize the cost of the network using different path finding algorithms, with the default link cost. But in those problem, where link cost is modular, there is another type of link cost can be used to minimize the network cost.

This type of link cost based on the characteristic of modular-link-cost: the cost for redundant bandwidth is free. For instance, we choose a 20 MB cable type for a link which needs 15 MB of bandwidth (consider there is only 20MB and 10 MB cable types) so 5MB of bandwidth will be unused. If we route another connection which needs no more than 5 MB of bandwidth through this link, the cost is totally free. We call this 5 MB of bandwidth the Free-Bandwidth. If we can route connections through those Free-Bandwidth, that would be a greatly decrease of network cost. That is the ideal of the Free-Bandwidth based heuristic algorithms (FBB).

This algorithm is based on the utilizing the redundant bandwidth. Consider we have an initial solution: a network with a set of identified links and a routing table for all requests. First, we remove the connection v of a demand d from the routing table and recalculate the cable type for all link of the network. And then, we recalculate the complementary cost of all links in the current network if the new connection of demand go through those as following: demand d requires bandwidth rb. For every link, we add rb to its bandwidth, if the result is greater than the max value of bandwidth level which is used in this link so:

$$Ccost = Cost(nextLevel) - Cost(currentLevel) \qquad (2)$$

Otherwise, Ccost = 0. This is a new way to calculate graph costs, and it helps the path finding algorithms easier find a better solution.

3.1   Free Bandwidth Based Heuristic Algorithm 1 (FBB1)

Base on the general idea, FBB1 can find a new connection for demand d, and the new link cost is recalculated. If the new network cost is lower than the old one, we have a new best current solution.

However if we consider cost of any link is totally free, the path finding algorithms may not work well. For instance, we can see in the graph below:

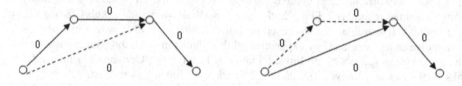

Fig. 1. A misleading situation for path finding algorithms

In this instance, we can see that the FBB1 cannot recognize the differences between two paths above. Although they both have free cost, the second path seems to be better. In some worse situations, the path finding algorithms may enter a loop. In order to overcome this drawback, instead of considering cost for redundant bandwidth is free, we set it a minor value (compare to the network link cost) so that the path finding algorithms still try to go through these links as possible, they are able to distinguish the better "free" path and will not enter any loop. The pseudo code for Free-Bandwidth based algorithm solving A-SNDP is shown below:

```
1.  Procedure FBB1
2.  //G represent graph with identified link
3.  //C represent connections for all request
4.  G,C <- initial Solution
5.  currentNetworkCost <- initial Solution Cost
6.  While !terminateCondition
7.      d <- randomRequest;
8.      c <- C.getConnections(d);
9.      V = V\{v}; //v belong to c
10.     For j = 1 to number_Of_Links do
11.         if ( requiredBandwidth(d) + usedBandwidth(j)<=
maxBandwidth(j))
12.             cost(j) = MINOR_VALUE;
13.         else cost(j) = costLevel(nextLevel(j))- costLe-
vel(currentLevel(j));
14.     Endfor
```

```
15.   c' ← findConnectionsForRequest(d);
16.   C=C∪c\c';
17.   newNetworkCost   ß recalculateSolutionCost;
18.   if (currentNetworkCost < newNetworkCost)
19.   C = C∪c\c';
20.   else C = C;
21. End While
22. End Procedure
```

3.2   Free Bandwidth Based Heuristic Algorithm 2 (FBB2)

In general, there are not much redundant bandwidths which are applied in FBB1. So, in order to improve, we could use some normal path finding algorithms, such as Dijkstra shortest path, DFS, etc. That is the idea of FBB2. First, we remove the connection v of a demand d from the routing table and recalculate the cable type for all link of the network. And then, we find two new paths, one using the same algorithm in FBB1, and the other using shortest path with normal link cost. Last, we choose the best of the current solutions and consider that is the current best solution.

4   Experimental Results

4.1   Problem Instances

In our experiments, we used three real world instances. They are Polska (12 nodes, 36 links, 65 unicast, 12 anycast), Germany (17 nodes, 52 links, 119 unicast, 13 anycast), Atlanta (26 nodes, 82 links, 234 unicast, 22 anycast). With each instance, we randomly create 10 test sets which are different from the content of customers' demands.

4.2   Experiment Setup

We experiment two our proposed algorithms independently and compare their performance with together and Tabu Search in [2].

4.3   System Setting

In the experiment, the system was run 50 times for each test set. All the programs were run on a machine with Intel Core 2 Duo U7700, RAM 2GB, Windows 7 Ultimate, and were installed by C++ and Java language.

4.4   Computational Results

The experiments show that:

- Figure 2 shows that on Polska network, the results found by FBB2 algorithm are the best. The best results found by Tabu Search and FBB1 are more equivalent, with the 5 first test sets, the ones of FBB1 are better than Tabu Search and conversely.

- Figure 3 and 4 show that on the larger instance (Germany and Atlanta network), the best results found by our proposed algorithms and Tabu Search [2] are quite equivalent. But in almost test sets, the results found by both FBB1 and FBB2 are better than the best results found by Tabu Search.
- Figure 5, 6 and 7 shows that the average results found by FBB1 and FBB2 are better than Tabu Search [2].

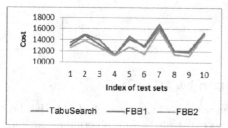
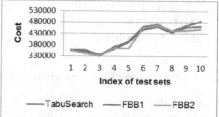

Fig. 2. The best result of Polska network found by Tabu Search [2], FBB1, FBB2 over 50 runing times

Fig. 3. The best result of Germany network found by Tabu Search [2], FBB1, FBB2 over 50 running times

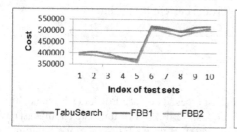
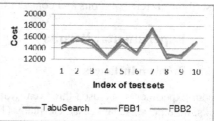

Fig. 4. The best result of Atlanta network found by Tabu Search [2], FBB1, FBB2 over 50 running times

Fig. 5. The average result of Polska network found by Tabu Search [2], FBB1, FBB2 over 50 running times

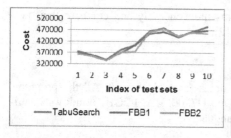
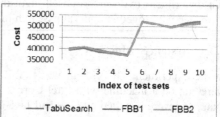

Fig. 6. The average result of Germany network found by Tabu Search [2], FBB1, FBB2 over 50 running times

Fig. 7. The average result of Atlanta network found by Tabu Search [2], FBB1, FBB2 over 50 running times

5   Conclusion

In this paper, we proposed two new heuristic algorithms for solving A-SNDP called FBB1 and FBB2. FBB1 algorithm is based on utilizing the redundant bandwidth corresponding with paid cost level in each link. FBB2 is the combination of FBB1 and Local Search algorithm [14]. We experimented on three instances which are Polska, Germany and Atlanta network [2], [13]. With each instance, we randomly create 10 test sets which are different from the content of customers' demands. The results show that our proposed approach is quite effective with A-SNDP. On all instances, FBB1 and FBB2 have better results than Tabu Search in most of test sets.

In the future, we are planning to improve the algorithm for solving larger instances. Moreover, we hope that we can find the other approach with better results for A-SNDP.

Acknowledgement. We would like to thank Prof. Jakub Gładysz, Prof. Krzysztof Walkowiak for providing us the A-SNDP problem instances, as well as sending us the materials related to their works on A-SNDP problem. This work was partially supported by the project "Models for next generation of robust Internet" funded by the Ministry of Science and Technology, Vietnam and the project "Some Advanced Statistical Learning Techniques for Computer Vision" funded by the National Foundation for Science and Technology Development under grant number 102.01-2011.17. The Vietnam Institute for Advanced Study in Mathematics provided part of the support funding for this work.

References

1. Walkowiak, K.: A Flow Deviation Algorithm for Joint Optimization of Unicast and Anycast Flows in Connection-Oriented Networks. In: Gervasi, O., Murgante, B., Laganà, A., Taniar, D., Mun, Y., Gavrilova, M.L. (eds.) ICCSA 2008, Part II. LNCS, vol. 5073, pp. 797–807. Springer, Heidelberg (2008)
2. Gładysz, J., Walkowiak, K.: Tabu Search Algorithm for Survivable Network Design Problem with Simultaneous Unicast and Anycast Flows. Intl Journal of Electronics and Telecommunications 56(1), 41–48 (2010)
3. Walkowiak, K.: A New Function for Optimization of Working Paths in Survivable MPLS Networks. In: Levi, A., Savaş, E., Yenigün, H., Balcısoy, S., Saygın, Y. (eds.) ISCIS 2006. LNCS, vol. 4263, pp. 424–433. Springer, Heidelberg (2006)
4. Grover, W.: Mesh-based Survivable Networks: Options and Strategies for Optical, MPLS, SONET and ATM Networking. Prentice Hall PTR, Upper Saddle River (2004)
5. Sharma, V., Hellstrand, F.: Framework for MPLS-based Recovery. RFC 3469 (2003)
6. Vasseur, J., Pickavet, M., Demeester, P.: Network Recovery: Protection and Restoration of Optical, SONET-SDH, IP and MPLS. Morgan Kaufmann, San Francisco (2004)
7. Kasprzak, A.: Algorithms of Flow, Capacity and Topology Structure in Computer Networks, Monography, Wroclaw, Polish (1989)
8. Pioro, M., Medhi, D.: Routing, Flow, and Capacity Design in Communication and Computer Networks. Morgan Kaufmann Publishers (2004)

9. Walkowiak, K.: Anycast Communication – A New Approach to Survivability of Connection-Oriented Networks. In: Gorodetsky, V., Kotenko, I., Skormin, V.A. (eds.) MMM-ACNS 2007. CCIS, vol. 1, pp. 378–389. Springer, Heidelberg (2007)
10. Nissen, V., Gold, S.: Survivable Network Design with An Evolution Strategy. In: Yang, A., Shan, Y., Bui, L.T. (eds.) Success in Evolutionary Computation. SCI, vol. 92, pp. 263–283. Springer, Heidelberg (2008)
11. Johnson, Deering: Reserved IPv6 Subnet Anycast Addresses. RFC 2526 (1999)
12. Anycast vs Unicast, http://communitydns.eu/Anycast.pdf
13. Gladysz, J., Walkowiak, K.: Optimization of Survivable Networks with Simultaneous Unicast and Anycast Flows. In: ICUMT, Poland, pp. 1–6 (2009)
14. Battiti, R., Brunato, M., Mascia, F.: Reactive Search and Intelligent Optimization. Springer, New York (2008)

Protein-Protein Binding Affinity Prediction Based on an SVR Ensemble

Xueling Li[1,*], Min Zhu[2,*], Xiaolai Li[1], Hong-Qiang Wang[1], and Shulin Wang[3]

[1] Intelligent Computing Lab, Hefei Institute of Intelligent Machines,
Chinese Academy of Sciences, Hefei, Anhui 230031, P.R. China
xlli@iim.ac.cn
[2] Robot Sensor and Human-Machine Interaction Laboratory, Hefei Institute of Intelligent
Machines, Chinese Academy of Sciences, Hefei, Anhui 230031, P.R. China
zhumin@iim.ac.cn
[3] School of Computer and Communication, Hunan University, Changsha,
Hunan 410082, P.R. China

Abstract. Accurately predicting generic protein-protein binding affinities (PPBA) is essential to analyze the outputs of protein docking and may help infer real status of cellular protein-protein interaction sub-networks. However, accurate PPBA prediction is still extremely challenging. Machine learning methods are promising to address this problem. We propose a two-layer support vector regression (TLSVR) model to implicitly capture binding contributions that are hard to explicitly model. The TLSVR circumvents both the descriptor compatibility problem and the need for problematic modeling assumptions. Input features for TLSVR in first layer are scores of 2209 interacting atom pairs within each distance bin. The base SVRs are combined by the second layer to infer the final affinities. Leave-one-out validation on our heterogeneous data shows that the TLSVR method obtains a very good result of R=0.80 and SD=1.32 with real affinities. Comparison experiment further demonstrates that TLSVR is superior to the previous state-of-art methods in predicting generic PPBA.

Keywords: Protein-protein interaction affinity, machine learning, two-layer support vector machine, potential of mean force.

1 Introduction

The affinity is the bridge of function and structure. Revealing the energetic characteristics of cellular multi-molecular complex is critical to understand the protein function. Four general scoring approaches have been developed to evaluate protein-protein docking results and to predict protein-protein binding affinity (PPBA). They are physical-based force fields [1], empirical scoring functions [2], knowledge-based statistical potentials, where volume correction is always considered to improve the prediction accuracy [3-6], and hybrid scoring functions [7, 8]. These scoring functions are successful in protein-protein docking evaluation and some are successful in protein binding affinity prediction. However, existing scoring functions for protein-protein

D.-S. Huang et al. (Eds.): ICIC 2012, LNCS 7389, pp. 145–151, 2012.
© Springer-Verlag Berlin Heidelberg 2012

docking usually do not hold capacity to predict the binding affinity of a complex [9]. Kastritis and Bonvin presented a protein–protein binding affinity benchmark consisting of binding constants (K_d) for 81 complexes to assess the performance of nine commonly used scoring algorithms. Their results revealed a poor correlation (R<0.3) between binding affinity and scores for all algorithms tested and concluded that accurate prediction of binding affinity remains beyond these methods reach. Therefore, improvement in protein-protein interaction affinity is still in great need.

In previous work, we developed a distance-independent residue level potential of mean force to predict PPBA [10] on a small data set of PPBA including 80 protein-protein complexes. Machine leaning methods can achieve satisfactory results without assuming any predefined model when used for affinity prediction. However, the generalization of methods based on one single classifier is often limited. Furthermore, a great number of features and small data set lead to the curse of dimensionality. Thus, researchers seek for classifier ensembles [11, 12] or multiple instance learning [13] to improve prediction accuracy and generalization and at the same time overcome the high feature dimensional disaster. For example, an ensemble learning method, random forest, as an ensemble is recently reported in diverse protein-ligand binding affinity prediction with higher prediction accuracy [14]. In our previous study, we used support vector regression models and rough set reduction model to predict TAP-peptide binding affinity and specificity [18, 19]. This method has a high interpretability.

In this work, we build a two-layer support vector regression (TLSVR) model with greater generality and prediction accuracy by capturing the non-linear combination effects on affinities of interacting atom pairs within each bin and between bins. Secondly, we construct a new data set by considering that structure diversity will greatly affect the accuracy of PPBA prediction. Finally, we evaluate our method with LOO cross validation.

2   Methods

2.1   Data Set

1056 heterogeneous protein complexes were obtained from PDBbind-CN, 2010 version, which includes complexes with a single residue mutation or multiple residue mutations [17, 18]. The dataset was then filtered with sequence similarity <50% by PDBculled (http://dunbrack.fccc.edu/Guoli/PISCES_InputB.php) with complex entities criteria and other default parameters. We finally integrated 49 proteins from [4] which did not exist in the dataset to get a heterogeneous larger data set. For simplicity, only complexes with two chains were held except 1CHD with four chains (EFG/I). Thus, 180 protein-protein interaction complexes were kept in our final data set.

2.2   Input Features

Machine learning methods require equal length of vectors. How to represent a protein complex structure in an equal-length vector as the input of TLSVR is the first prerequisite. 47 types of heavy atoms were defined as reported by Su, et al. [4]. The occurrence numbers of 2209 contact atom pairs within each 0.2 Å width bin were counted by Eq.1 at each protein complex interface. Our preliminary experiment shows that

TLSVR achieves best prediction results when the cutoff distance of a pair of contact atoms equals 16 Å. Thus, 71 distance bins, i.e., 80 bins minus 9 bins with distance threshold below 1.8 Å were kept. The final features consist of scores of 2209 contact atom pairs obtained with Eq.2 which are obtained by multiplying the occurrence number of each contact atom pair with the corresponding atom pair potential. The interacting atom pair's potentials were defined as the natural logarithm of the ratio of observed number of interacting atom pairs to those expected. The smoothed potentials were generated as illustrated in [4]. Thus, 2209 smoothed atom pair scores (features) were finally generated. Specifically,

$$f_{ij} = N_{obs}(i,j,r) \times P(i,j,r)$$ (1)

where $N_{obs}(i,j,r)$ is the observed number of interacting atom pairs i,j within the distance shell $r(r-\Delta r, r)$ in a given protein-protein binding structure. $\Delta r (\Delta r = 0.2$ Å) is the bin width for $1.8\text{Å} \le r_{cut} \le 16$ Å between two interface chains or proteins at n^{th} distance bin of width 0.2 Å.

$$P(i,j,r) = -\log(\frac{N_{obs}(i,j,r)}{N_{exp}(i,j,r)})f_{cor}, (i=1,2,3...47, j=1,2,3,..47)$$ (2)

The bins $r(r - \Delta r, r)$ range from 1.8 Å to 16 Å at 0.2 Å intervals. The interfacial atom pair potential: $N_{exp}(i,j,r) = X_i \times X_j \times N_{total}(r)$ is the expected number of interacting atom pairs of i,j between two interface chains or proteins if there are no preferential interactions between them. X_i is the mole fraction of atom type i and is calculated as N_i / N, where N_i and N are the total number of atom type i and all atoms, respectively, while $N_{total}(r)$ is the total number of interacting atom pairs derived from the reference database [4]. f_{cor} is the correction factor, derived from smoothing $N_{obs}(i,j,r) / N_{exp}(i,j,r)$ ranging from 1.8 Å to 16 Å, by a moving window of 3.0 Å width for bin of width 0.2 Å.

2.3   Two-Layer Support Vector Regression Model

Support Vector Machine (SVM) is introduced by Vapnik [19]. While SVM classification outputs binary results (binding or nonbinding), Support Vector Regression (SVR) model produces continuous values (affinity absolute value).

Protein-protein binding affinity was predicted by using two-layer SVR (TLSVR). Each input vector at first layer of TLSVR is 2209-dimensional. As depicted above, each real value of a vector represents a score of an atom pair in interface within each of 71 bins, i.e. 1.8 Å: 0.2 Å: 16 Å of a protein complex. Here the contact atom pairs with distance below 1.8 Å of atom clashes were disregarded. 71 individual SVR modes were included at first layer. As shown in Fig.1, the predicted values from the individual SVR modes of the first layer were input into the second layer SVR (the combiner). The output of the combiner was the final predicted affinity. Parameters were default in individual SVR models. All the computational experiments are carried out with LIBSVM that is available at http://www.csie.ntu.edu.tw/~cjlin/libsvm.

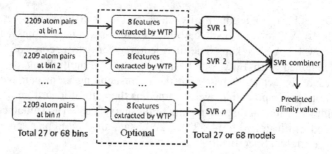

Fig. 1. Scheme of the proposed two-layer SVM prediction system

For comparison purpose, a one-layer SVR (OLSVR) model was also developed. Different from TLSVR, each input feature for OLSVR was the sum score of 2209 interacting atom pairs within each bin. Thus, total 71 scores from different distance bins were input into one-layer SVR after discarding the contact atom pairs with distance below 1.8 Å of atom clashes as described above. We then compared the prediction results of the one- with two-layer SVR.

2.4   Evaluation of TLSVR Model

The performance of SVR models was tested by the leave-one-out cross-validation (LOOCV) by testing of each protein complex structure of the data set due to the small size of the data set. The data set having n complexes is broken in n subsets, each having one example. The classifier was trained on $n-1$ subset and evaluated on n^{th} subset. The process was repeated n times using each subset as the testing set and the rest of complexes for training set. The correlation coefficient and standard derivation between the predicted values of all samples and the real ones are calculated by combining overall estimate of training procedure.

3   Results and Discussion

3.1   Results of OLSVR and TLSVR

TLSVR models were generated by using RBF kernel type in this study without extra specification. Prediction of binding affinities of protein complexes was done by using TLSVM. The final correlation coefficient (R) result of the TLSVR was obtained from LOOCV evaluation, which was recognized to be the most extreme and accurate type of cross-validation test.

Table 1. Results of the one- and two-layer radial base kernel SVM for protein-protein binding affinity prediction

SVR Models	Features	R^d	SD^e
PMF	--	0.18	2.18
One-layer	1 final score	0.24	2.16
One-layer	71 grid scores	0.50	1.92
Two-layer	1 final score	0.48	1.94
Two-layer	**2209 ×71 grids**	**0.80**	**1.32**

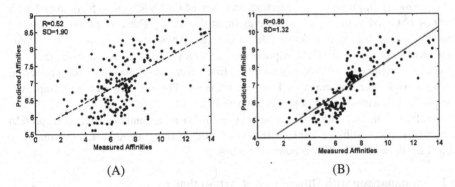

Fig. 2. Correlation of the predicted affinities and the observed ones. (A) Prediction results of one-layer SVR (OLSVR); (B) prediction results of two-layer SVR ensemble (TLSVR).

Table 1 shows the results of the OLSVR and TLSVR models with RBF kernel. The best OLSVR prediction obtained R=0.52 and SD=1.90 (Fig.2a) when input features were the sum over 2209 atom pair smoothed scores within each of 71 distance bins. TLSVR prediction obtained R=0.80 and SD=1.32 for 71 distance bins (Fig.2b), when input features were smoothed scores of each contact atom pair. All prediction results demonstrate that TLSVR has satisfactory performance and greatly improves the prediction accuracy of PPBA compared with OLSVR.

Why did TLSVR improve the prediction accuracy so much? In OLSVR, at each interval, 2208 atom pair scores were lineally summed up to one score value. Since each sum score of a distance bin constituted one feature, 71 features in total were input into OLSVR. The lower prediction accuracy of OLSVR shows that simply linearly summation of the scores of 2209 interacting atom pairs at each distance bin without consideration their non-linear effects on binding affinity results in significant deterioration in prediction. On the other hand, two-layer radial basis SVR improves the accuracy greatly by the consideration of the non-linear relationship between atom pair in different bins. Thus, the proposed TLSVR model leads to tremendous improvement in the prediction performance by filtering results of the first layer.

Table 2. Comparison of the prediction performance of TLSVM on dataset with slightly different sequence similarity cut

Heterogeneity	R	SD
50%_Similarity Cut Culled on Entity plus 49 Su's protein (180)	0.80	1.32
50%_Similarity Cut Culled on Entity (161)	0.84	1.22
40%_Similarity Cut Culled on Entity (91)	0.78	1.33

We notice that the non-linear combination of 71 scores captured by using radial kernel basis OLSVR did not improve the prediction accuracy by comparing one final sum score with 71 scores from different distance bins as input into OLSVR. As shown in Table 1, for OLSVR model, accounting for the non-linear effects of scores from 71 bins obviously improved the prediction accuracy of OLSVR to R=0.5 compared with R=0.24 obtained with one final score as input (Table 1). These results demonstrates that accounting for the non-linear relationship of 2209 atom pairs may be the important factor that improves TLSVR prediction accuracy. Using SVR combiner that correlated the results of models generated in first layer also contributes the prediction improvement. It can be concluded that the proposed TLSVR method highly improves the prediction accuracy and generalization ability.

Finally, on dataset with different sequence similarity cut and heterogeneous protein complexes, our results in Table 2 show that both heterogeneous and low similarity cutoff decrease the TLSVM prediction accuracy.

3.2   Comparison with Other State-of Art Methods

To our best knowledge, DFIRE and Su's methods are two representative recently proposed PMF-based methods for PPBA prediction. RF-score is recently developed method for protein-ligand affinity prediction. We compare our method with these three representative methods on our collected data set. Comparison results are shown in Table 3. We can see that DFIRE obtained a correlation of 0.12 only. Su's method only obtained a correlation coefficient of 0.18 in predicting the generic and heterogeneous PPBA. TLSVR model obtained much a higher correlation coefficient of 0.80 between the predicted and experimental affinities than both methods.

Table 3. Comparison of TLSVM with two state of art methods, DFIRE and Su's method with LOO cross-validation on our data set

On independent test set	R	SD
DFIRE on training set	0.12	2.21
Su's Method on training set	0.18	2.18
RF-score	0.18	2.18
TLSVM	0.80	1.32

Acknowledgements. This work was supported by the Knowledge Innovation Program of Chinese Academy of Sciences, No. 0823A16121, Anhui Provincial Natural Science Foundation No. 1208085MF96, and National Natural Science Foundation of China, Nos. 31071168, 30900321, 60973153 and 61133010.

References

1. Kollman, P.A., Massova, I., Reyes, C., Kuhn, B., Huo, S., Chong, L., Lee, M., Lee, T., Duan, Y., Wang, W., Donini, O., Cieplak, P., Srinivasan, J., Case, D.A., Cheatham, T.E.: 3rd: Calculating Structures and Free Energies of Complex Molecules: Combining Molecular Mechanics and Continuum Models. Acc. Chem. Res. 33, 889–897 (2000)
2. Bohm, H.J.: Prediction of Binding Constants of Protein Ligands: a Fast Method for the Prioritization of Hits Obtained from De Novo Design or 3D Database Search Programs. J. Comput. Aided Mol. Des. 12, 309–323 (1998)

3. Melo, F., Feytmans, E.: Novel Knowledge-based Mean force Potential at Atomic Level. J. Mol. Biol. 267, 207–222 (1997)
4. Su, Y., Zhou, A., Xia, X., Li, W., Sun, Z.: Quantitative Prediction of Protein-protein Binding Affinity with a Potential of Mean Force Considering Volume Correction. Protein Sci. 18, 2550–2558 (2009)
5. Lu, H., Lu, L., Skolnick, J.: Development of Unified Statistical Potentials Describing Protein-protein Interactions. Biophysical Journal 84, 1895–1901 (2003)
6. Muegge, I.: PMF Scoring Revisited. J. Med. Chem. 49, 5895–5902 (2006)
7. Englebienne, P., Moitessier, N.: Docking Ligands into Flexible and Solvated Macromolecules. 4. Are Popular Scoring Functions Accurate for this Class of Proteins? Journal of Chemical Information and Modeling 49, 1568–1580 (2009)
8. Oda, A., Tsuchida, K., Takakura, T., Yamaotsu, N., Hirono, S.: Comparison of Consensus Scoring Strategies for Evaluating Computational Models of Protein-ligand Complexes. Journal of Chemical Information and Modeling 46, 380–391 (2006)
9. Kastritis, P.L., Bonvin, A.M.J.J.: Are Scoring Functions in Protein-Protein Docking Ready To Predict Interactomes? Clues from a Novel Binding Affinity Benchmark. Journal of Proteome Research 9, 2216–2225 (2010)
10. Li, X.-L., Hou, M.-L., Wang, S.-L.: A Residual Level Potential of Mean Force Based Approach to Predict Protein-Protein Interaction Affinity. In: Huang, D.-S., Zhao, Z., Bevilacqua, V., Figueroa, J.C. (eds.) ICIC 2010. LNCS, vol. 6215, pp. 680–686. Springer, Heidelberg (2010)
11. Wolpert, D.H.: Stacked Generalization. Neural Network 5, 241–259 (1992)
12. Xia, J.-F., Zhao, X.-M., Huang, D.-S.: Predicting Protein-protein Interactions from Protein Sequences Using Meta Predictor. Amino. Acids 39, 1595–1599
13. Teramoto, R., Kashima, H.: Prediction of Protein-ligand Binding Affinities Using Multiple Instance Learning. Journal of Molecular Graphics and Modelling 29, 492–497
14. Ballester, P.J., Mitchell, J.B.O.: A Machine Learning Approach to Predicting Protein-ligand Binding Affinity with Applications to Molecular Docking. Bioinformatics 26, 1169–1175 (2010)
15. Li, X.-L., Wang, S.-L.: A Comparative Study on Feature Selection in Regression for Predicting the Affinity of TAP Binding Peptides. In: Huang, D.-S., Zhang, X., Reyes García, C.A., Zhang, L. (eds.) ICIC 2010. LNCS, vol. 6216, pp. 69–75. Springer, Heidelberg (2010)
16. Li, X.L., Wang, S.L., Hou, M.L.: Specificity of Transporter Associated with Antigen Processing Protein as Revealed by Feature Selection Method. Protein and Peptide Letters 17, 1129–1135 (2010)
17. Wang, R.X., Fang, X.L., Lu, Y.P., Wang, S.M.: The PDBbind Database: Collection of Binding Affinities for Protein-ligand Complexes with Known Three-dimensional Structures. Journal of Medicinal Chemistry 47, 2977–2980 (2004)
18. Wang, R.X., Fang, X.L., Lu, Y.P., Yang, C.Y., Wang, S.M.: The PDBbind Database: Methodologies and Updates. Journal of Medicinal Chemistry 48, 4111–4119 (2005)
19. Vapnik, V.N.: Statistical learning theory. Springer, New York (1998)

A Novel Two-Stage Alignment Method for Liquid Chromatography Mass Spectrometry-Based Metabolomics[*]

Xiaoli Wei[1], Xue Shi[1], Seongho Kim[2], Craig McClain[3,4,5,6], and Xiang Zhang[1]

[1] Departments of Chemistry, University of Louisville, Louisville, KY 40292
{jujuxiao,xueshisx}@gmail.com, xiang.zhang@louisville.edu
[2] Bioinformatics and Biostatistics, University of Louisville, Louisville, KY 40292
biostatistician.kim@gmail.com
[3] Medicine, [4] Pharmacology & Toxicology, [5] Alcohol Research Center, [6] Robley Rex Louisville
VAMC, University of Louisville, Louisville, KY 40292
craig.mcclain@louisville.edu

Abstract. We report a novel two-stage alignment algorithm that contains full alignment and partial alignment, for the analysis of LC-MS based metabolomics data. The purpose of full alignment is to detect landmark peaks that present in all peak lists to be aligned. These peaks were first selected based on m/z value and isotopic peak profile matching. After removing peaks with large Euclidian distance of retention time from the potential landmark peaks, a mixture score was calculated to measure the matching quality of each landmark peak pair between reference peak list and a test peak list. After optimizing the weight factor in the mixture score, the value of minimum mixture score of all landmark peaks was used as the threshold for peak matching in the partial alignment. A local optimization based retention time correction method was used to correct the retention time changes between peak lists during partial alignment. The two-stage alignment method was used to analyze a spiked-in experimental data and further compared with literature reported algorithm RANSAC implemented in MZmine.

Keywords: LC-MS, two-stage peak list alignment, local optimization.

1 Introduction

Metabolomics is the study of low molecular weight molecules (i.e., metabolites) found within cells and biological systems. It aims to measure and interpret the complex time-related concentration, activity and flux of large sets of metabolites in biological samples. Several types of instruments have been utilized to analyze

---

[*] This work was supported by NIH grant 1RC2AA019385 through the National Institute on Alcohol Abuse and Alcoholism. This work was also partially supported by National Institute of Health (NIH) grant 1RO1GM087735 through the National Institute of General Medical Sciences (NIGMS).

D.-S. Huang et al. (Eds.): ICIC 2012, LNCS 7389, pp. 152–159, 2012.
© Springer-Verlag Berlin Heidelberg 2012

metabolites including nuclear magnetic resonance (NMR), gas chromatography-mass spectrometry (GC-MS), and liquid chromatography-mass spectrometry (LC-MS). Each type of instrumental analysis provides limited coverage of the metabolites, and therefore, only generates a partial metabolite profile of each sample. Currently, significant challenges remain in almost every aspect for the application of metabolomics to biomedical research. Among these, the lack of accurate and efficient bioinformatics tools for the processing of metabolomics data has become a critical bottleneck to the progress of metabolomics. Many data analysis steps are involved in deciphering the mass spectrometry data, including data preprocessing, metabolite identification, quantification, network and pathway analysis.

Peak alignment is a key step of data preprocessing in LC-MS based metabolomics. It recognizes peaks generated by the same metabolite occurring in different samples from the millions of peaks detected during the course of an experiment [1, 2]. A large volume of information-rich data can be generated in a LC-MS based metabolomics study. To carry out the alignment procedure, several bioinformatics tools have been developed including *XCMS2* [3], *centWave* [4], *MZmine2* [5], *MZedDB* [6], *OpenMS* [7]. However, the accuracy of peak alignment remains challenge in metabolomics.

The objective of this work was to develop a novel approach for high accuracy peak alignment. For this reason, we developed a two-stage alignment method to align the peak lists generated from high-resolution mass spectrometry for metabolomics study. The developed method has been implemented in *MetSign* [8] and used to analyze a set of spiked-in experimental data acquired on a LC-MS system. The performance of this method was compared with existing software packages *MZmine*.

2   Experimental Section

2.1   Spiked-in Samples

A mixture of 30 compound standards was prepared at a concentration of 100 μg/mL for each compound. The standards included 11 fatty acid (behenic acid, tricosanoic acid, stearic acid, myristic acid, nonadecanoic acid, heptadecanoic acid, adipic acid, heneicosanoic acid, nonanoic acid, butyric acid, linoleic acid), 5 triglycerides (trilauroyl-glycerol, trimyristin, tripalmitin, tricaprylin, tricaprin), 9 phospholipids PC(16:0/16:0), PC(16:0/14:0), PC(12:0/12:0), PC(6:0/6:0), LysoPC(16:0/0:0), LysoPC(10:0), PC(20:4(5Z,8Z,11Z,14Z)/16:0), PC(18:2(9Z,12Z)/18:2(9Z,12Z)), PC(24:1(15Z)/24:1(15Z)), and 5 other small molecules (caffeine, L-tryptophan, lidocaine, creatine, trans-4hydroxyl-L-proline). 10 μL of the standard mixture was added to a 100 μL sample of metabolite extract of mouse liver, and dichloromethane: methanol (v/v = 2:1) was then added to the sample vials to make the total volume up to 200 μL.

2.2   LC-MS Analysis

A LECO Citius LC-HRT high resolution mass spectrometer equipped with an Agilent 1290 Infinity UHPLC with a Waters Acquity UPLC BEH hydrophilic interaction chromatography (HILIC) 1.7 μm 150 × 2.1 mm column was used in this work. The sample was loaded in H_2O + 5 mM NH_4OAc + 0.2% acetic acid (buffer A) and separated using a binary gradient consisting of buffer A and buffer B (90/10 acetonitrile/H_2O + 5 mM NH_4OAc + 0.2% acetic acid). Flow rate was set at 250 μL/min on the column with 100% B for 4 min, 45% B at 12 min holding to 20 min, 100% B at 21 min and holding to 60 min for the gradient. The Citius was operated with electrospray ionization in positive ion mode. The system was optimized in high resolution mode (R = 50,000 (FWHM)) and was mass calibrated externally using Agilent Tune Mixture (ATM). The mass spectrometry was operated in both full mass analysis and tandem MS/MS mode to acquire molecular m/z value and the corresponding MS/MS spectrum. The spiked-in sample was analyzed 6 times on the LC-MS system.

3   Theoretical Basis

The peak alignment method was developed as a two-stage algorithm: full alignment and partial alignment. The goal of full alignment is to recognize landmark peaks, which are defined as a set of metabolite peaks present in every sample. In the partial alignment stage, the peaks in a test sample that are not recognized as the landmark peaks are aligned.

Let $S = \{S_1, S_2, ..., S_i, ..., S_{n+1}\}$ be the sample set, and $n+1$ is the total number of samples to be aligned. After selecting the reference peak list (RPL) in a random manner, the rest of peak lists are considered as test samples, which can be written as $S = \{RPL, t_1, t_2, ..., t_i, ..., t_n\}$. Each of the test peak lists is aligned to the RPL, respectively.

Considering two peak lists $\{RPL, t_i\}$, all m/z value matched peak pairs between these two peak lists can be selected using a user defined m/z variation. If a peak can be matched to multiple peaks in the other peak list, the peak pair with the minimum retention time difference is selected as the most probable match and the other matches are discarded. Therefore, the m/z matched peak pairs can be recorded as $\{(r_1, s_1), (r_2, s_2), ..., (r_p, s_p)\}$, where r is a peak from RPL, s is a peak from t_i, and p is the total number of the m/z matched peak pairs. The m/z matched peak pairs are further filtered based on the Euclidean distance of retention time between r and s, i.e., $d_j = |r_j - s_j|$ with a confidence interval of 95%. The peak pairs filtered by retention time are represented as $\{(r_1, s_1), (r_2, s_2), ..., (r_m, s_m)\}$ and $m \le p$. This process is iteratively operated on all the test samples, respectively.

A mixture similarity score S_m was developed to measure the matching quality between two peaks as follows:

$$S_m(d_i, \Delta_i | w) = w * \exp\left(-1.6 * \frac{d_i - d_{min}}{d_{med} - d_{min}}\right) + (1-w) * \frac{1}{1+\Delta_i} \tag{1}$$

where d_i is the Euclidean distance of retention time between the i th matched peak pair, d_{min} and d_{med} are the minimum and median retention time distance among all matched peaks in the two peak lists, respectively, Δ is the absolute value of m/z difference between the i th matched peak pair, and w is a weight factor and $0 \leq w \leq 1$.

The peaks that are present in every test peak list and are matched to the same peak in the RPL are used to optimize the value of weight factor w for the alignment of a test peak list and the RPL by maximizing the value of $\sum_{i=1}^{k} S_m(d_i, \Delta_i \mid w)$, k is the number of matched peaks between the test peak list and the RPL, and w is set as 0.05, 0.1, 0.2, 0.3, 0.4, 0.5, 0.6, 0.7, 0.8, 0.9 and 0.95, respectively. After optimizing the weight factor w, the value of S_m is calculated for each matched peak pair between the test peak list and the RPL, followed by an outlier detection in S_m^j, $j = 1, \ldots, k$. By iteratively considering pair set $\{RPL, t_i \mid i = 1, \ldots, n\}$, the landmark peaks $\{(r_1, t_{11}, \ldots, t_{n1}), \ldots, (r_m, t_{1m}, \ldots, t_{nm})\}$ are obtained. The minimum mixture score S_m^{min} among all the test peak lists is then used as a threshold value in the partial alignment.

To perform the partial alignment, the retention time value of each landmark peak in the test peak list is assigned to the retention time value of the corresponding landmark peak in the RPL. A local polynomial fitting method is employed to correct the retention time of peaks present between two adjacent landmark peaks. Because multiple landmark peaks can be detected in a set of experimental data, adjusting retention time shifts using two adjacent landmark peaks can correct nonlinear retention time shifts. To correct the retention time of peaks not present between two landmark peaks, an iteratively optimization method is applied to the group of peaks eluted earlier than the first-eluted landmark peak and the group of peaks eluted later than the last-eluted landmark peaks, respectively. In each optimization process, 30% of landmark peaks are randomly selected from $\{(r_1, t_{11}), \ldots, (r_m, t_{1m})\}$ and a polynomial model fitting error is computed as follows

$$\varepsilon = \sum_{i=1}^{N} | t_{R,i}^o - t_{R,i}^f | \tag{2}$$

where $t_{R,i}^o$ is the original retention time of the i th peak, $t_{R,i}^f$ is the fitted retention time of the i th peak, N is the number of peaks in the test peak list at the region of interest. This process is repeated 1000 times and the model with minimum error is selected and used for retention time correction.

After the retention time correction, the partial alignment is applied to all the non-landmark peaks present in each of the test peak lists and aligns them to the peaks present in the RPL, where a mixture score S_m is calculated using equation (1) for each peak pair. A peak pair is considered to be a match if its mixture score is larger than S_m^{min}. It is possible that one peak in the test sample can be matched to multiple peaks in the RPL and *vice versa*. In these cases, the peak pair with the maximum mixture score is kept while the remaining matches are discarded. If there is a peak in the test peak list that cannot be matched to any peaks in the RPL, this peak is considered as a new peak to the RPL and is added to the RPL. The updated RPL is then used to align the peaks in the next test peak list, and this process is repeated until all the test peak lists are aligned.

4   Results and Discussion

The raw instrument data were first converted into *mz*ML format and further reduced to peak lists using *MetSign* software. [8] There are about 2300 peaks detected by *MetSign* software in each sample and about 1100 peaks assigned to a database compound. Of the 30 spiked-in compound standards, most of the compounds were detected based on the match of *m/z* values with a variation window of ≤ 5 ppm and the similarity of isotopic peak profile measured by Pearson's correlation coefficient ≤ 0.75. Table 1 shows the number of compound standards detected in each replicate injection. The variation of the number of detected compound standards was generated by the experimental variation. After *MetSign* processing, a total of six peak lists were generated, and these peak lists were subjected for peak alignment.

Table 1. The number of compound standards detected in six replicate injections

Index of replicate injection	1	2	3	4	5	6
No. of compound standards identified in each injection	25	23	23	24	24	24

During the full alignment, a total of 283 landmark peaks were detected with retention time ranges from 91.52 s to 643.14 s. Of the 283 landmark peaks, 18 were the peaks generated by the spiked-in compound standards while the remaining landmark peaks were generated by the metabolites extracted from mouse liver. Figure 1 shows the effectiveness of retention time correction to the non-landmark peaks during the partial alignment. Even though the experiments of the six replicate injections were performed under the identical experimental conditions, the retention time of each compound still drifted between injections and such a retention time drift is not linear. Therefore, the local polynomial fitting method is able to correct the retention time of peaks present between two adjacent landmark peaks.

To compare the alignment accuracy of the two-stage alignment method, the experiment data were also processed using publically available software *MZmine*. *MZmine* software has two alignment methods, Join aligner and RANSAC aligner. Join aligner is a simple alignment method, which aligns detected peaks in different samples

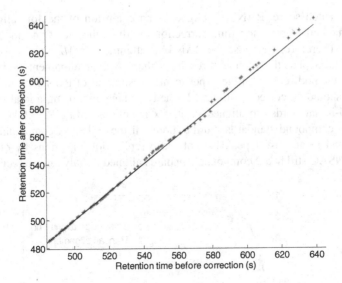

Fig. 1. Local optimization based retention time correction for compounds eluted between two landmark peaks. The value of each red star in x-axis is the retention time of a compound before correction and the value in y-axis is after retention time correction. The solid blue line is a guideline depicting a situation of no retention time correction.

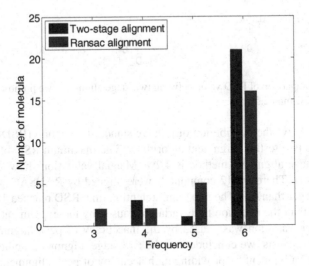

Fig. 2. Alignment results of spiked-in compound standards by two-stage method (blue) and RANSAC method of MZmine2.5 (red)

through a match score. RANSAC aligner is an extension of the Join aligner. It includes a method of retention time correction to adjust the retention time shift in all peak lists. Therefore, we chose the RANSAC alignment in *MZmine2.5* for comparison. Figure 2 depicts the alignment results of the two-stage alignment method and the RANSAC method. Based on the experimental design, all of the spiked-in compound standards should be correctly aligned. In the two-stage alignment, a total 21 peaks of the spiked-in standards are aligned in all six injections, while RANSAC only fully aligned 16 compound standards. Furthermore, all the spiked-in compound standards were aligned in at least 4 peak lists of the 6 replication injections by our method, while RANSAC still had 2 compound standards aligned in only three injections.

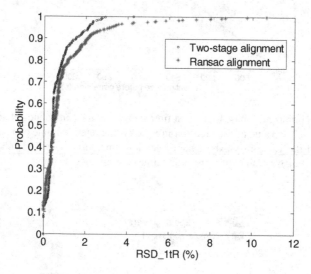

Fig. 3. The comparison of RSD value between two-stage alignment we proposed and RANSAC alignment in MZmine2.5

Figure 3 shows the distribution of relative standard deviation (RSD) of all aligned peaks by the two testing alignment algorithms. The maximum RSD of aligned peaks in the two-stage alignment method is 4.2%. Manual validation shows that this alignment is correct. There are 12 compounds were aligned by RANSAC with a retention time RSD larger than 4%. The maximum retention time RSD reached 10.7%, which is much larger than the retention time variation caused by the experiments. Such a large retention time variation was caused by the inaccuracy of peak alignment. From the comparative analysis, we conclude that the two-stage alignment method outperforms the RANSAC alignment by providing high accuracy of peak alignment for the analysis of LC-MS based metabolomics data.

5   Conclusions

A novel two-stage alignment algorithm, containing full alignment and partial alignment, was developed for high accuracy peak list alignment for LC-MS based metabolomics. The full alignment detects landmark peaks that are present in all the peak lists. During this process, the potential landmark peaks were first selected based on the m/z and isotopic peak profile matching. After removing outliers based on the Euclidian distance of retention time from the potential landmark peaks, a mixture score method was employed to evaluate the match quality of each landmark peak pair between the reference and the test sample peaks. The value of minimum mixture score of all landmark peaks was used as the threshold of peak matching during partial alignment, in which local optimization based retention time correction was employed to correct the retention time changes between peak lists. The performance of the two-stage alignment method was tested by analyzing a spiked-in experimental data and further compared with literature reported algorithm RANSAC implemented in *MZmine2.5*. The comparison demonstrates that our two-stage alignment method out-performs the RANSAC algorithm for high accuracy of peak alignment.

References

1. Zhang, X., et al.: Data pre-processing in Liquid Chromatography-mass Spectrometry-based Proteomics. Bioinformatics 21(21), 4054–4059 (2005)
2. Wang, B., et al.: DISCO: Distance and Spectrum Correlation Optimization Alignment for two-dimensional Gas Chromatography time-of-flight Mass Spectrometry-based Metabolomics. Anal. Chem. 82(12), 5069–5081 (2010)
3. Benton, H.P., et al.: XCMS2: Processing Tandem Mass Spectrometry Data for Metabolite Identification and Structural Characterization. Anal. Chem. 80(16), 6382–6389 (2008)
4. Tautenhahn, R., Bottcher, C., Neumann, S.: Highly Sensitive Feature Detection for High Resolution LC/MS. BMC Bioinformatics 9, 504 (2008)
5. Pluskal, T., et al.: MZmine 2: Modular Framework for Processing, Visualizing, and Analyzing Mass Spectrometry-based Molecular Profile Data. BMC Bioinformatics 11, 395 (2010)
6. Draper, J., et al.: Metabolite Signal Identification in Accurate Mass Metabolomics Data with MZedDB, an Interactive m/z Annotation Tool Utilising Predicted Ionisation Behaviour 'rules'. BMC Bioinformatics 10, 227 (2009)
7. Sturm, M., et al.: OpenMS-An Open-source Software Framework for Mass Spectrometry. BMC Bioinformatics 9 (2008)
8. Wei, X.L., et al.: MetSign: A Computational Platform for High-Resolution Mass Spectrometry-Based Metabolomics. Analytical Chemistry 83(20), 7668–7675 (2011)

Reconstruction of Metabolic Association Networks Using High-throughput Mass Spectrometry Data[*]

Imhoi Koo[1,2], Xiang Zhang[2], and Seongho Kim[1]

[1] Department of Bioinformatics and Biostatistics,
University of Louisville, Louisville KY, 40292, USA
[2] Department of Chemistry, University of Louisville, Louisville KY, 40292, USA
{imhoi.koo,xiang.zhang,s0kim023}@louisville.edu

Abstract. Graphical Gaussian model (GGM) has been widely used in genomics and proteomics to infer biological association networks, but the relative performances of various GGM-based methods are still unclear in metabolomics. The association between two nodes of GGM is calculated by partial correlation as a measure of conditional independence. To estimate the partial correlations with small sample size and large variables, two approaches have been introduced, which are arithmetic mean-based and geometric mean-based methods. In this study, we investigated the effects of these two approaches on constructing association metabolite networks and then compared their performances using partial least squares regression and principal component regression along with shrinkage covariance estimate as a reference. These approaches then are applied to simulated data and real metabolomics data.

Keywords: metabolomics, graphical Gaussian model, partial correlation, partial least squares regression, principal component regression, false discovery rate.

1   Introduction

Metabolomics is a rapidly emerging field to systemically analyze small-molecule metabolites in a biological organism [1]. It is equally important in systems biology as other "-omics" such as genomics, transcriptomics, and proteomics. One of the important approaches to integrating the individual "-omics" data for system level analysis is the reconstruction of cellular networks, which is collection and visualization of all physiologically relevant cellular processes.

In metabolomics, a relatively small number of studies have been reported for metabolic network construction. For instance, Arkin et al. [2] predicted interactions within reaction networks over time for the glycolytic pathway, where Pearson's correlation coefficient was used to construct the interaction networks. A major drawback of Pearson's correlation-based networks is unable to distinguish between

---

[*] This work was also partially supported by National Institute of Health (NIH) grant 1RO1GM087735 through the National Institute of General Medical Sciences (NIGMS).

D.-S. Huang et al. (Eds.): ICIC 2012, LNCS 7389, pp. 160–167, 2012.
© Springer-Verlag Berlin Heidelberg 2012

direct and indirect associations. On the other hand, graphical Gaussian models (GGMs) reveal direct associations with conditional independences/dependences among variables, using partial correlation coefficients that are calculated by the correlation of two variables after removing affection of other variables [3]. GGMs have been employed in metabolomics for several studies [4, 5]. Note that the size of samples (experiments) was larger than the number of variables (metabolites) for these studies so that network construction was straightforward.

If the number of samples is much smaller than number of variables, it is difficult to directly estimate partial correlation due to singularity. To overcome this difficulty, several methods have been developed by either reducing the number of given variables or a regularized estimation [6, 7]. Another alternative is to use dimension-reduced regression such as partial least squares regression (PLSR) and principal component regression (PCR) approaches. When calculating the partial correlations using regression coefficients, arithmetic and geometric means of regression coefficients were employed in Kramer et al. [8] and Pihur et al. [9], respectively. The partial correlation coefficients estimated by these two methods are not the same to each other and, it is important to investigate the effects of the different calculation methods on network reconstruction. Therefore, we evaluated the performance of PLSR and PCR using shrinkage covariance estimate as a reference in terms of network construction.

2   Methods and Materials

The graphical Gaussian model (GGM) is a statistical multivariate analysis to infer the direct relationship among variables using nodes and edges [3], where the nodes correspond to the variables under consideration, and the edges represent the conditional independence between two variables as measured by partial correlation coefficient.

Suppose a data matrix X consists of n observed samples and p metabolites with a mean of zero. Then the partial correlation coefficient matrix $P = (\pi_{ij})$ is calculated by the inverse of the covariance matrix $\Sigma = \frac{1}{n}X^\top X$ as follows:

$$\pi_{ij} = -\frac{w_{ij}}{\sqrt{w_{ii}w_{jj}}}, \tag{1}$$

where $\Sigma^{-1} = (w_{ij})$.

The covariance matrix Σ becomes singular when the sample size n is smaller than the number p of variables. To deal with singularity, several methods have been introduced [8]. In this study, the following three methods are considered.

2.1   Shrinkage Covariance Estimation

Schafer and Strimmer [10] introduced shrinkage covariance estimator (SCE) for the partial correlation estimation when the covariance matrix $\Sigma = \frac{1}{n-1}X^\top X$ is singular. Under singularity of covariance matrix, the SCE is to trade off the unbiased sample covariance Σ and low dimensional shrinkage target matrix T:

$$\hat{\Sigma} = \lambda T + (1 - \lambda)\Sigma, \tag{2}$$

where $\lambda \in (0, 1]$ is shrinkage intensity. The optimal value of the tuning parameter λ is analytically determined and estimated from the data.

2.2   Regression with Dimension Reduction

Partial Least Squares Regression and Principal Component Regression. The common property of both partial least squares regression (PLSR) and principal component regression (PCR) is to use dimension reduction method to avoid the singularity for the "small n, large p" paradigm. PLSR finds orthogonal vector $\mathbf{w}$ to maximize the covariance between $\mathbf{t} = X\mathbf{w}$ and dependent (response) variable $\mathbf{y}$, while PCR searches for orthogonal vector $\mathbf{w}$ to maximize the variance of $\mathbf{t} = X\mathbf{w}$.

Consider linear regression of dependent variable $\mathbf{y}$ on data matrix X as follows:

$$\mathbf{y} = X\beta + \epsilon, \tag{3}$$

where β is a vector of regression coefficients and ϵ is error. The estimation of coefficient β using PLSR consists of two steps [11]. The first step is to extract a latent variable set $T = (\mathbf{t}_1, \mathbf{t}_2, \cdots, \mathbf{t}_k)$ of orthogonal components $(k < p)$, which maximizes a covariance with dependent variable $\mathbf{y}$. The second step is to estimate the coefficient of regression of $\mathbf{y}$ on the new latent variable set T and then to transform it into space spanned by data X. The first PLSR component $\mathbf{t}_1 = X\mathbf{w}_1$ is obtained by maximizing the covariance as follows:

$$\mathbf{w}_1 = \arg\max_{\|\mathbf{w}\|=1} \mathrm{Cov}^2(X\mathbf{w}, \mathbf{y}). \tag{4}$$

The next components $\mathbf{t}_i$, $i = 2, \cdots, k$, are satisfied with maximizing the squared covariance to $\mathbf{y}$ and are mutually orthogonal to each other. Consider the orthogonal part X_k of X on all components $\mathbf{t}_1, \mathbf{t}_2, \cdots, \mathbf{t}_{k-1}$:

$$X_k = X - \mathcal{P}^{\perp}_{\mathbf{t}_1, \cdots, \mathbf{t}_{k-1}}(X), \tag{5}$$

where $\mathcal{P}^{\perp}_{\mathbf{t}_1, \cdots, \mathbf{t}_{k-1}}$ is the projection operator related to $\mathbf{t}_1, \cdots, \mathbf{t}_{k-1}$. The kth latent variable, $\mathbf{t}_k = X_k\mathbf{w}_k$, is then obtained by solving the optimization problem:

$$\mathbf{w}_k = \arg\max_{\|\mathbf{w}\|=1} \mathrm{Cov}^2(X_k\mathbf{w}, \mathbf{y}). \tag{6}$$

Using the following equations, the vector of regression coefficients $\hat{\beta}^{PLSR}(k)$ is determined to predict the output of a model including k components:

$$\hat{y}_k^{PLSR} = X\hat{\beta}^{PLSR}(k) \text{ and } \hat{\beta}^{PLSR}(k) = (\mathbf{w}_1, \cdots, \mathbf{w}_k)T^{\top}\mathbf{y}. \tag{7}$$

For PCR, the equations (4) and (6) are replaced with the following equations, respectively:

$$\mathbf{w}_1 = \arg\max_{\|\mathbf{w}\|=1} \text{Var}(X\mathbf{w}), \text{ and } \mathbf{w}_k = \arg\max_{\|\mathbf{w}\|=1} \text{Var}(X_{k-1}\mathbf{w}). \tag{8}$$

Then, the predicted output $\hat{y}^{PCR}$ and regression coefficients $\hat{\beta}^{PCR}$ can be calculated by

$$\hat{y}^{PCR} = X\hat{\beta}^{PCR}(k) \text{ and } \hat{\beta}^{PCR}(k) = (\mathbf{w}_1, \cdots, \mathbf{w}_k)T^\top y, \tag{9}$$

Once the regression coefficients in equations (7) and (9) are computed, the partial correlation coefficients are estimated by using either geometric or arithmetic mean of regression coefficients.

Method 1: Geometric mean approach. In this approach, the partial correlation coefficient π_{ij} of X in the equation (1) is estimated by

$$\pi_{ij} = \text{sign}(\hat{\beta}_{ij})\sqrt{\hat{\beta}_{ij}\hat{\beta}_{ji}} \tag{10}$$

Method 2: Arithmetic mean approach. Arithmetic mean approach of the association/interaction scores was introduced by Pihur et al. [9]. The partial correlation is calculated by

$$\hat{\pi}_{ij} = \frac{\sum_{l=1}^{k} \hat{\beta}_{il}c_{jl}^{(i)} + \sum_{l=1}^{k} \hat{\beta}_{jl}c_{il}^{(j)}}{2}. \tag{11}$$

In this equation (11), the coefficients are obtained from

$$x_i = \sum_{j=1}^{k} \beta_{ij}t_j^{(i)} + \epsilon, \text{ and } t_j^{(i)} = \sum_{l \neq i}^{p} c_{jl}^{(i)}X_l^{(i)}, \tag{12}$$

where $t_j^{(i)}$ is a jth latent variable of PLSR and PCR , and k is the number of latent variables which is pre-determined by user.

2.3   False Discovery Rate

After estimating partial correlation coefficients, statistical hypothesis test is performed to select the significant edges indicating strong association between two variables. To do this, false discovery rate (FDR) is applied to control the expected proportion of incorrectly rejected null hypotheses by using the q-value method in R software package *fdrtool* [12].

2.4   Data

Simulation Data. The simulated data were generated using two conditions, sample size and network complexity. The number of variables p was always set to 100. We used three different densities, 5%, 15%, and 25%, to describe the complexity of the network. Given each density, we considered five different sample sizes, 25, 50, 100, 150, and 200, to generate simulated data. For each case, we generated 100 data sets and then compared the performance of each method with their averages. The R software package *GeneNet* was used to generate the simulated data [13].

Experimental Data. We also investigated the performance of each method using experimental data of metabolites extracted from mouse liver. The experimental data consist of all compounds detected from mouse samples on a linear trap quadruple-Fourier transform ion cyclotron resonance mass spectrometer (LTQ-FTICR MS) via direct infusion electrospray ionization (DI-ESI)-mass spectrometry. For the association network study, we used 99 compound peaks that were detected in all 40 samples by *MetSign* software [14].

2.5   Performance Evaluation

We evaluated the five estimation methods, shrinkage covariance estimation (SCE), geometric/arithmetic mean-based partial least squares regression (PLSR.G and PLSR.A, respectively), and geometric/arithmetic mean-based principle component regression (PCR.G and PCR.A, respectively), in this study. In order to evaluate their performance, the following three criteria were considered:

1. The true positive rate is the proportion of true positives which are correctly predicted; $TPR = \frac{TP}{TP+FN}$,
2. The positive predictive value is the proportion of subjects with positive output results which are correctly predicted; $PPV = \frac{TP}{TP+FP}$,
3. F1 score is a measure of accuracy, which is the harmonic average of TPR and PPV; $F1 = 2 \cdot \frac{TPR \cdot PPV}{TPR+PPV}$.

3   Results and Discussion

Fig. 1 (a)-(c) show the F1 scores of each method in terms of network construction based on simulated data. It can be seen that the performance of geometric mean-based and arithmetic mean-based approaches relies on the estimation methods (PLSR and PCR). As for PCR, geometric mean-based approach performs better than arithmetic mean-based approach when the network is complex regardless of the sample size. However, arithmetic mean-based approach has the larger F1 score than geometric mean-based approach with PLSR (PLSR.G) when the sample size is large. In particular, when the sample size is less than 50, PLSR.G performs the best with density of 5%, while PCR.G is the best method if the density is 15% or 25% based on F1 score, as shown in the figure.

Interestingly, in case of PPV as shown in Fig. 1 (d)-(f), arithmetic mean-based approach with PCR outperforms geometric mean-based approach regardless of sample size and network complexity. On the other hand, as for TPR in Fig. 1 (g)-(i), geometric mean-based approach performs better than arithmetic mean-based approach when PCR is applied, while arithmetic mean-based approach with PLSR (PLSR.A) is better when the density is 15% or 20%.

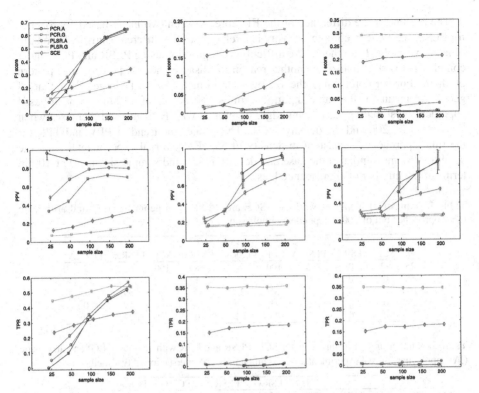

Fig. 1. Performance plots. (a), (b), and (c) show the F1 scores. (d), (e), and (f) show the positive predictive value. (g), (h), and (i) show the true positive rate. (a), (d), and (g) correspond to density 5%. (b), (e), and (h) correspond to density 15%. (c), (f), and (i) correspond to density 25%. SCE, PLSR.G, PLSR.A, PCR.G and PCR.A stand for shrinkage covariance estimate, geometric mean-based partial least squared regression, arithmetic mean-based partial least squared regression, geometric mean-based principle component regression and arithmetic mean-based principle component regression, respectively. Error bar stands for average value and 95% confidence interval of F1 score, PPV and TPR over 100 runs.

Table 1 shows the numbers of empty network estimated by SCE, PLSR.G and PLSR.A out of 100 independent runs. Note that PCR methods generated no empty network. When the true network becomes more complex, those methods generated more estimated empty network. Furthermore, the number of estimated empty networks is decreased as the sample size goes to 200. However, the trend of PLSR.G is different with other methods. For example, the empty network for PLSR.G with density of 25% is increased as the sample size is increased.

The results of network construction using real experimental data are shown in Table 2. The number of significant edges and the number of intersection of edges of pair of two methods are reported. The geometric mean-based approaches, PLSR.G and PCR.G, generate larger significant edges than arithmetic mean-based approaches. Namely, PLSR.A and PCR.A selected at least 5.4 times more edges. Most edges (90% and 89%) of PLSR.A and PCR.A are overlapped with these of PLSR.G and PCR.G, respectively.

The reason for the difference of the F1 score between two mean-based approaches for PLSR and PCR in complex density is likely due to the different statistical property of latent variables from them. The difficulty of regression using PLSR and PCR under complex network can also be another reason for disagreement of performance pattern of them. Furthermore, since the output of arithmetic mean is larger than that of geometric mean for the same input, discrimination power of arithmetic mean-based approach to estimating significant edges combined with FDR method increases when $n = 100, 150, 200$ and the density is 5%. This makes the trend of PPV and TPR for the final approaches consistent in density of 5%. For the real experimental data, the condition seems similar to the case that density is 5% and sample size is 50 or 100 in terms of the number of significant edges.

Table 1. Number of empty networks for SCE and PLSR with geometric (.G) and arithmetic (.A) approaches out of 100 independent simulations

	5%			15%			25%		
n	SCE	PLSR.G	PLSR.A	SCE	PLSR.G	PLSR.A	SCE	PLSR.G	PLSR.A
25	92	0	0	100	6	8	100	5	7
50	0	0	0	97	8	1	100	12	2
100	0	0	0	85	39	0	93	59	1
150	0	0	0	41	26	0	86	69	0
200	0	0	0	24	11	0	82	72	0

Table 2. Number of significant edges for SCE, PLSR and PCR with geometric (.G) and arithmetic (.A) approaches for real experimental results and number of intersection of two methods

	SCE	PLSR.G	PLSR.A	PCR.G	PCR.A
SCE	259	150	131	137	85
PLSR.G		1228	172	608	136
PLSR.A			191	133	79
PCR.G				1292	188
PCR.A					211

4   Conclusion

We evaluated the performance of two estimation methods, arithmetic mean-based and geometric mean-based approaches, using regression coefficients to construct association networks. We observed that the performances of geometric mean-based and arithmetic mean-based approaches are dependent on the dimension-reduced regression methods (PLSR and PCR) and simulation settings such as sample size and density. Arithmetic estimation outperforms geometric mean when it is incorporated with PLSR and the sample size is larger, while the geometric mean-based approach performs better when the true network is complex and it is used with PCR in terms of F1 score.

References

1. Watkins, S.M., German, J.B., Hammock, B.D.: Metabolomics: Building on a Century of Biochemistry to Guide Human Health. Metabolomics 1(1), 3–9 (2005)
2. Arkin, A., Shen, P.D., Ross, J.: A Test Case of Correlation Metric Construction of a Reaction Pathway from Measurements. Science 277(5330), 1275–1279 (1997)

3. Whittaker, J.: Graphical Models in Applied Multivariate Statistics. Wiley series in probability and mathematical statistics. Wiley, Chichester (1990)
4. Theis, F.J., Krumsiek, J., Suhre, K., Illig, T., Adamski, J.: Gaussian Graphical Modeling Reconstructs Pathway Reactions from High-throughput Metabolomics data. BMC Systems Biology 5 (2011)
5. Chan, E., Rowe, H., Hansen, B., Kliebenstein, D.: The Complex Genetic Architecture of the Metabolome. PLoS Genetics 6(11), e1001198 (2010)
6. Dobra, A., Hans, C., Jones, B., Nevins, J.R., Yao, G., West, M.: Sparse Graphical Models for Exploring Gene Expression Data. Journal of Multivariate Analysis 90(1), 196–212 (2004)
7. de la Fuente, A., Bing, N., Hoeschele, I., Mendes, P.: Discovery of Meaningful Associations in Genomic Data using Partial Correlation Coefficients. Bioinformatics 20(18), 3565–3574 (2004)
8. Kramer, N., Schafer, J., Boulesteix, A.L.: Regularized Estimation of Large-scale Gene association Networks Using Graphical Gaussian Models. BMC Bioinformatics 10 (2009)
9. Pihur, V., Datta, S., Datta, S.: Reconstruction of Genetic Association Networks from Microarray Data: a Partial Least Squares Approach. Bioinformatics 24(4), 561–568 (2008)
10. Schafer, J., Strimmer, K.: A Shrinkage Approach to Large-scale Covariance Matrix Estimation and Implications for Functional Genomics. Statistical Applications in Genetics and Molecular Biology 4 (2005)
11. Houskuldsson, A.: Pls Regression Methods. Journal of Chemometrics 2(3), 211–228 (1988)
12. Strimmer, K.: fdrtool: a Versatile r Package for Estimating Local and Tail Area-based False Discovery Rates. Bioinformatics 24(12), 1461–1462 (2008)
13. Schafer, J., Strimmer, K.: An Empirical Bayes Approach to Inferring Large-scale Gene Association Aetworks. Bioinformatics 21(6), 754–764 (2005)
14. Wei, X., Sun, W., Shi, X., Koo, I., Wang, B., Zhang, J., Yin, X., Tang, Y., Bogdanov, B., Kim, S., Zhou, Z., McClain, C., Zhang, X.: Metsign: A Computational Platform for High-resolution Mass Spectrometry-based Metabolomics. Analytical Chemistry 83(20), 7668–7675 (2011)

Predicting Protein Subcellular Localization by Fusing Binary Tree and Error-Correcting Output Coding

Lili Guo[1,2] and Yuehui Chen[1,2]

[1] Computational Intelligence Lab, School of Information Science and Engineering, University of Jinan, 106 Jiwei Road, 250022 Jinan, P.R.China
[2] Shandong Provincial Key Laboratory of Network Based Intelligent Computing
gfcguoguo@163.com

Abstract. In this paper, a new method was applied to predict the protein subcellular localization. The features used in the paper were the Distance frequency (DF), the Physical and chemical composition (PCC) and the Pseudo Amino Acid composition (PseAA). The classifier was integrated by Binary tree and Error-Correcting Output Coding (ECOC) based six Artifical neural networks (ANN). The prediction ability was evaluated by 5-jackknife cross-validation. By comparing its results with other methods, such as Lei-SVM and ESVM, the experimental result demonstrated that our method outperformed their predictions, and indicate the new approach is feasible and effective.

Keywords: subcellular localization, feature extraction, Binary tree, ECOC, ANN, ensemble classifier.

1   Introduction

We can realize from the biology that the proteins play a vital function on cell survival. The cells are highly ordered structure. We usually divide intracellular regions into different organelles or cell-areas depending on different functions and space distributions, as nucleus, Golgi body, endoplasmic reticulum, chondriosome,   endochylema and cytomembrane etc, which are called subcellular organelles. After protein synthesis in ribosome, they are transported to specific position.

Precise knowledge of a protein's function requires appropriate subcellular localization [1], as it has been observed that a protein may lose its functions if not properly localized [2]. In addition, the information about subcellular localization may provide understanding about the study of proteins' functions, also about protein interaction, evolutionary analysis.

2   Materials and Methods

2.1   Dataset

We choose the SNL6 [3] dataset to validate the availability of our classifier. This dataset is founded by Lei and Dai and more commonly used in subcellular

D.-S. Huang et al. (Eds.): ICIC 2012, LNCS 7389, pp. 168–173, 2012.
© Springer-Verlag Berlin Heidelberg 2012

localization. SNL6 contains 504 protein sequences and 6 subcellular positions. Among the 504 sequences, 61 chromatin, 55 nuclear lamina, 56 nuclear speckle, 219 nucleolus, 75 nucleoplasm, 38 PML body.

2.2   Representation of Protein Sequence

Distance Frequency (DF)

Relevant researches indicate that the physical and chemical properties of amino acids are much consequence to the subcellular localization [4], so we think that the hydropathy / hydrophobic distribution of amino acids in sequence closely relate with subcellular localization. So the 20 native amino acids (20 letters in alphabet) are divided into 6 classes [4] according to the properties. Table1 shows the result of classification.

Table 1. Amino acids hydration property classification [5]

Classification	Abbreviation	Amino acids
hydrophily	L	R,D,E,N,Q,K,H
hydrophobicity	B	L,I,V,A,M,F
neutral	W	S,T,Y,W
proline	P	P
glycocoll	G	G
cysteine	C	C

After the classification, any protein sequence can be represented by combination of the six letters [6]. For every type of amino acids, we separately calculate the distance-value's occurrence number of two letters which belong to same type. One sequence thus gets one vector V_i on the basis of distance frequency:

$$V_i = \left[v_1^L, v_2^L, \cdots, v_s^L, v_1^B, v_2^B, \cdots, v_s^B, \cdots, v_1^C, v_2^C, \cdots, v_s^C \right] \qquad (1)$$

The value of s is a key problem. If it is too small, we can not extract enough feature information; conversely, it can produce lots of noise data. Experimental results show that we can get sufficient information when s is 11. So this method consists of 66 descriptor values. $v_j^\xi (j = 1, 2 \cdots s, \xi = L, B, W, P, G, C)$ has two parts: 1) when j<s, v_j^ξ is occurrence number of distance value j of two letters which belong to ξ type; 2) when j=s, v_j^ξ is occurrence number of distance value H which equal or greater than s.

Physical and Chemical Composition (PCC).

The 20 native amino acids are divided into three groups for their physicochemical properties, including seven types [7] of hydrophobicity, normalized van der Vaals volume, polarity, polarizibility, charge, secondary structures and solvent accessibility. For instance, using hydrophobicity attribute all amino acids are divided into three groups: polar, neutral and hydrophobic. A protein sequence is then transformed into a sequence of hydrophobicity attribute. Therefore, the composition descriptor consists of three values: the global percent compositions of polar, neutral and hydrophobic residues in the new sequence. For seven types of attributes, PCC consists of a total of 3×7=21 descriptor values.

Pseudo Amino Acid composition (PseAA)

According to the concept of Chou's PseAA composition [8], a sample of protein sequence is a point in (20+ λ)-D space.

$$X = \left[x_1, x_2, \cdots, x_{20}, x_{21}, \cdots, x_{20+\lambda}\right]^T \in \Re^{(20+\lambda)} \tag{2}$$

$$x_i = \begin{cases} \dfrac{f_i}{\sum\limits_{j=1}^{20} f_j + w \sum\limits_{j=1}^{\lambda} P_j} & 1 \le i \le 20 \\[4ex] \dfrac{w \mu_i}{\sum\limits_{j=1}^{20} f_j + w \sum\limits_{j=1}^{\lambda} P_j} & 21 \le i \le 20 + \gamma \end{cases} \tag{3}$$

Where, the $f_i(1 \le i \le 20)$ in Ep.(3) is the occurrence frequencies of 20 amino acids in sequence. $P_i(21 \le i \le 20 + \lambda)$ is the additional factors that incorporate some sort of sequence order information. The parameter w is weight factors.

Features Fusion

Putting the characteristic data attached to another and this is called the fusion of two features extraction methods. DF consists of 66 descriptor values, PCC consists of a total of 21 descriptor values and PseAA we used consists of 40 values, so the fusion of DF, PCC and PseAA consist of 127 values.

2.3   Ensemble Classifier Prediction System

The biggest problem of studying subcellular localization is data imbalance [9], as the sequences's number of Nucleolus is 219 which is much greater than other types'. So we introduce the Binary tree to classify the Nucleolus and others firstly. The following flowchart shows the prediction process.

Binary Tree Structure

Actually the Binary tree is not a concrete classifier but a kind of classification structure and it classifies the subcellular position Nucleolus and others in this paper. Here we use the simplest Artifical neural network to classify two categories.

ECOC Framework

ECOC [10] trains single classifier respectively according to encoding matrix. In testing process, every single classifier outputs a predicted value which forms a output vector $H(x) = (h_1(x), h_2(x), \cdots, h_n(x))$. It uses Hamming distance function or Euclidean distance function to calculate the distance between the output vector H(x) and each row of encoding matrix, the corresponding class label of the shortest coding is the output of the test sample [11].

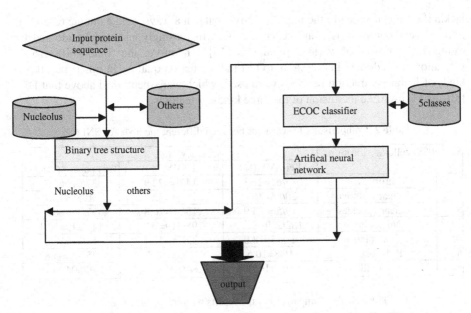

Fig. 1. A flowchart to show the prediction process of ensemble classifier

The encoding matrix is defined as $M_{K \times n}$ and each element of the matrix is $\{0, 1\}$. K refers to class number of the dataset and n stands for the number of the single classifier. Each row in M corresponds to one class, while each column one single classifier. For instance, the all single classifiers are $h_1(x), h_2(x), \cdots, h_n(x)$; if the M(i,j)=0, that means that classifier j regards all the samples of class label i as positive samples, or the samples are regarded as negative ones. The encoding matrix takes many forms [12], as one-to-many matrix, one-to-one matrix, random sparse coding matrix and dense random coding matrix etc. In this paper, our encoding matrix is $\{(0,0,0,0,0,0,),(1,0,1,1,0,1),(0,1,1,0,1,1),(1,1,0,1,1,0),(0,0,0,1,1,1)\}$. The five rows stands for the five categories except for Nucleolus and each column corresponds to one single classifier.

Artifical Neural Network (ANN)
The single classifier of ECOC framework is ANN. ANN has obtained very good application in many fields of pattern recognition and is such a algorithm which imitates the message processing of people's neurons [13]; it has strong robustness and tolerance and can study uncertain system. So the ANN has been applied to the subcellular location. In this paper the Particle swarm optimization (PSO) is adopted to optimize the parameters (weights and thresholds) of the ANN.

3   Experimental Results

In the prediction and classification problems, we often use the following three methods to examine the quality of a predictor: independent dataset test, sub-sampling test, and jackknife test [14]. But among the three cross-validation methods, the jackknife test is deemed to the most objective that can always yield a unique result for a given benchmark dataset and hence has been increasingly and widely adopted to examine the power of various predictors [15]. Therefore, the 5-jackknife cross-validation was adopted in our study to test the prediction quality. In this paper, three kinds of feature extraction methods were used, which were mentioned above that DF, PCC and PseAA, i.e the fusion of the three kinds of features.

Table 2. Comparison of accuracies between different methods for SNL6

Subset	Subcellular location	Accuracy rate		
		Lei-SVM(%)	ESVM(%)	This paper [1](%)
S1	Chromatin	13/61=21.3	14/61=22.9	37/61=60.7
S2	Nuclear-Lamina	20/55=36.4	18/55=32.7	40/55=72.7
S3	Nuclear-speckles	19/56=33.9	15/56=26.8	37/56=66.0
S4	Nucleolus	182/219=83.1	198/219=90.3	147/219=67.1
S5	Nucleoplasm	21/75=28.0	32/75=42.7	54/75=72.0
S6	PML body	4/38=10.5	7/38=18.4	25/38=65.8
	Overall	259/504=51.4	284/504=56.4	340/504=67.5

Table 3. The comparison of the results with my prior research

Compartments	Different methods (%)		
	PseAA+PCC+ECOC	PseAA + PCC +HpAA[2]+ECOC	This paper
Chromatin	35/61=57.3	32/61=52.5	37/61=60.7
Nuclear lamina	25/55=45.4	31/55=56.3	40/55=72.7
Nuclear speckles	30/56=53.5	27/56=48.2	37/56=66.0
Nucleolus	163/219=74.4	165/219=75.3	147/219=67.1
Nuclear diffuse	38/75=50.7	39/75=52.0	54/75=72.0
PML body	13/38=34.2	20/38=52.6	25/38=65.8
Overall	304/504=60.2	314/504=62.6	340/504=67.5

As for how to calculate the overall success rate for a statistical system, we count classification accuracy of every class and the overall accuracy. Accuracy is refers to the ratio of the number of proteins which are correctly classified and the all proteins. Listed in Table 2 are the results obtained, respectively, with other methods and ours on the benchmark dataset SNL6. The comparison of results with my prior research which also use the same dataset is also showed in Table 3.

4   Conclusion

In this paper, we used the features fusion and integrated classifiers to predict the sub-cellular localization. The overall accuracy rate achieved by this paper was 67.5%, which was much better than that by Lei-SVM and ESVM. The result was better than

---

[1] The method of this paper is DF + PCC + PseAA + Binary tree + ECOC.

[2] HpAA Notes that Amino acids hydration properties composition which is a method of feature extraction.

others and my prior work because we analyzed the imbalance of data, so we first used the Binary tree to classify. This was also an innovation of this paper. A lot of people have studied this subject, so the next step of research, I hope to make innovation in feature extraction and classifier to improve the accuracy of classification.

Acknowledgments. This research was partially supported by the Natural Science Foundation of China (61070130), the Key Project of Natural Science Foundation of Shandong Province (ZR2011FZ001), the Key Subject Research Foundation of Shandong Province and the Shandong Provincial Key Laboratory of Network Based Intelligent Computing.

Reference

1. Deng, M., Zhang, K., Mehta, S., Chen, T., Sun, F.: Prediction of Protein Function Using Protein–protein Interaction Data. Journal of Computational Biology 10, 947–960 (2003)
2. Boden, M., Teasdale, R.D.: Determining Nucleolar Association from sequence by Leveraging Protein-protein Interactions. Journal of Computational Biology 15, 291–304 (2008)
3. Lei, Z., Dai, Y.: An SVM-based System for Predicting Protein Subnuclear Localizations. BMC Bioinformatics 6, 291–298 (2005)
4. Pánek, J., Eidhammer, I., Aasland, R.: A New Method for Identification of Protein (Sub) Families in a Set of Proteins Based on Hydropathy Distribution in Proteins. Proteins: Struct. Funct. Bioinformatics 558, 923–934 (2005)
5. Chen, Y.L., Li, Q.Z.: Prediction of the Subcellular Location of Apoptosis Proteins. J. Theor. Biol. 245, 775–783 (2007)
6. Zhang, L., Liao, B., Li, D.C., Zhu, W.: A Novel Representation for Apoptosis Protein Subcellular Localization Prediction Using Support Vector Machine. J. Theor. Biol. 259, 361–365 (2009)
7. Shi, J.Y., Zhang, S.W., Pan, Q., Cheng, Y.M., Xie, J.: SVM-based Method for Subcellular Localization of Protein Using Multi-scale Energy and Pseudo Amino Acid Composition. Amino Acids 33(1), 69–74 (2007)
8. Chou, K.C.: Prediction of Protein Cellular Attributes Using Pseudo-amino Acid Compositio. Proteins. Struct. Funct. Genet. 43(3), 246–255 (2001)
9. Zhang, S., Huang, B., Xia, X., et al.: Bioinformatics Research in Subcellular Localization of Protein. Prog. Biochem. Biophys. 34(6), 573–579 (2007)
10. Huang, Y., Li, Y.D.: Prediction of Protein Subcellular Locations Using Fuzzy K-NN method. Bioinformatics 20(1), 21–28 (2004)
11. Dietterich, T.G., Bakiri, G.: Solving Multiclass Learning Problems via Error-Correcting Output Codes. Artificial Intelligence Research (2), 263–286 (1995)
12. Luo, D., Xiong, R.: Distance Function Learning in Error-Correcting Output Coding Framework. In: King, I., Wang, J., Chan, L.-W., Wang, D. (eds.) ICONIP 2006, Part II. LNCS, vol. 4233, pp. 1–10. Springer, Heidelberg (2006)
13. Masulli, F., Valentini, G.: Effectiveness of Error Correcting Output Codes in Multiclass Learning Problems. In: Kittler, J., Roli, F. (eds.) MCS 2000. LNCS, vol. 1857, pp. 107–116. Springer, Heidelberg (2000)
14. Chou, K.C., Zhang, C.T.: Review: Prediction of Protein Structural Classes. Crit. Rev. Biochem. Mol. Biol. 30, 275–349 (1995)
15. Chen, C., Chen, L., Zou, X., Cai, P.: Prediction of Protein Secondary Structure Content by Using the Concept of Chou's Pseudo-amino Acid Composition and Support Vector Machine. Protein Pept. Lett. 16, 27–31 (2009)

Exponential Stability of a Class of High-Order Hybrid Neural Networks[*]

Qian Ye, Baotong Cui, Xuyang Lou, and Ke Lou

Key Laboratory of Advanced Process Control for Light Industry (Ministry of Education),
Jiangnan University, Wuxi 214122, China
yeqian85@gmail.com

Abstract. This paper considers a generalized model of high-order hybrid neural networks with time-varying delays and impulsive effects is considered. By establishing an impulsive delay differential inequality and using the method of Lyapunov functions, we investigate the global exponential stability of high-order dynamical neural networks with time-varying delays and impulsive effects. Our sufficient conditions ensuring the stability are dependent on delays and impulses and show delay and impulsive effects on the stability of neural networks.

Keywords: Exponential stability, high-order hybrid neural networks, Lyapunov function, impulse effects.

1 Introduction

Dynamical neural networks are often used to describe dynamic systems due to its practical importance and wide applications in many areas such as optimization, industry, biology, economics and so on. In such applications, it is of prime importance to ensure the stability of the equilibrium points of the designed network. Recently, the stability of delayed neural networks has also been studied extensively (see, e.g. [1-3]). Since the existence of delays is frequently a source of instability, there has been a considerable attention given in the literature on Hopfield-type neural networks with time delays. High-order neural networks have been investigated recently in [4,5]. It is known that high-order neural networks have stronger approximation property, greater storage capacity, and higher fault tolerance than lower-order neural networks.

On the other hand, most neural networks can be classified as either continuous or discrete. However, there are many real-world systems and natural processes that behave in a piecewise continuous style interlaced with instantaneous and abrupt changes (impulses) [6]. Correspondingly, there is not much work dedicated to investigate the stability of impulsive neural networks [5, 7-9]. In particular, Li and Hu [8] gave a criterion for the existence and global exponential stability of a periodic solution in a class of Hopfield neural networks with impulses but without delays. As far as we know, few results have been reported in literature on the exponential stability of high-order dynamical neural networks with both impulsive effects and time-varying delays.

---

[*] This work is partially supported by National Natural Science Foundation of China (No.61174021, No.61104155), and the 111 Project (B12018).

D.-S. Huang et al. (Eds.): ICIC 2012, LNCS 7389, pp. 174–181, 2012.
© Springer-Verlag Berlin Heidelberg 2012

In this paper, we introduce a new class of high-order dynamical neural networks with impulsive effects. By establishing an impulsive delay differential inequality, we obtain the sufficient conditions ensuring the global exponential stability of impulsive high-order dynamical neural networks with time-varying delay. The conditions are dependent on the greatest delay value and the strength values of the impulsive effects, and are less restrictive. Our condition is widely applicable because it does not need the differentiability or the monotonicity of the activation functions. An example is given to demonstrate the effectiveness of the results.

Notation. For $x \in R^n, A \in R^{n \times n}$, we denote

$$| x | = (| x_1 |, \cdots, | x_n |)^T, | A | = (| a_{ij} |)_{n \times n}, \| x \| = \sum_{i=1}^{n} |x_i|, \| A \| = \max_{1 \le j \le n} \sum_{i=1}^{n} |a_{ij}|.$$

For $\varphi: R \to R$, denote

$$\varphi(t^+) = \lim_{s \to 0^+} \varphi(t+s), \varphi(t^-) = \lim_{s \to 0^-} \varphi(t+s),$$

$$[\varphi(t)]_\tau = \sup_{-\tau \le s \le 0} \{\varphi(t+s)\}, [\varphi(t)]_{\tau^-} = \sup_{-\tau \le s < 0} \{\varphi(t+s)\}.$$

$PC := \{\psi \mid \psi: R \to R^n$ is bounded variation function and right-hand continuous on any subinterval $[t - \tau, t]\}$. Let R denote the set of real numbers. R^+ denotes the set of nonnegative real numbers. Denote E an identity matrix. The notation $P > 0$, (respectively, ..) means that is symmetric and positive definite (respectively, negative definite) matrix. We use P^T, P^{-1}, λ_m, and λ_M to denote, respectively, the transpose of, the inverse of, the smallest and the largest eigenvalues of a square matrix P. The norm $\| \cdot \|$ is either the Euclidean vector norm or the induced matrix norm.

2   Preliminaries

Consider an impulsive high-order dynamical neural network with time-varying delays described by

$$D^+ y_i(t) = -c_i y_i(t) + \sum_{j=1}^{n} A_{ij} g_j(y_j(t)) DU_j + \sum_{j=1}^{n} B_{ij} g_j(y_j(t - \tau_j(t))) DV_j$$

$$+ \sum_{j=1}^{n} \sum_{l=1}^{n} T_{ijl} g_j(y_j(t - \tau_j(t))) g_l(y_l(t - \tau_l(t))) DW_{jl} + I_i, \tag{1}$$

where $i \in \{1, 2, \cdots, n\}$, $t \ge t_0$, $y_i(t)$ is the neuron state; c_i is positive constant, it denotes the rate with which the cell i resets its potential to the resting state; A_{ij}, B_{ij} are the first-order synaptic weights of the neural networks; T_{ijl} is the second-order synaptic weights of the neural networks; $\tau_j(t)$ $(j = 1, 2, \cdots, n)$ is the transmission delay of the j th neuron such that $0 < \tau_j(t) \le \sigma$, where σ is a constant; the activation function g_j is continuous on $[t_0 - \sigma, +\infty)$; I_i is the external input; the operator D represents the distribution derivative, bounded variation functions U_j, V_j, $W_{jl}: [t_0, +\infty) \to R$ are right continuous on any compact subinterval of

$[t_0, +\infty)$. DU_j, DV_j and DW_{jl} depict the impulsive effects of neural networks, $i \in \{1, 2, \cdots, m\}$, $j \in \{1, 2, \cdots, n\}$, $l \in \{1, 2, \cdots, n\}$. We assume that

$$DU_j = 1 + \sum_{k=1}^{\infty} u_{jk} \delta(t - t_k), j = 1, 2, \cdots, n, k \in N \equiv \{1, 2, \cdots\},$$

$$DV_j = 1 + \sum_{k=1}^{\infty} v_{jk} \delta(t - t_k), j = 1, 2, \cdots, n, k \in N \equiv \{1, 2, \cdots\},$$

$$DW_{jl} = 1 + \sum_{k=1}^{\infty} w_{jk} \delta(t - t_k), j = 1, 2, \cdots, n, k \in N \equiv \{1, 2, \cdots\},$$

where the fixed impulsive moments t_k satisfy $t_{k-1} < t_k$ and $\lim_{k \to \infty} t_k = \infty$, $\delta(t)$ is the Dirac function, u_{jk}, v_{jk} w_{jk} represent the strength of impulsive effects of the j th neuron at time t_k, $t - t_k$ and $t - t_k$, respectively.

In order to obtain our results, we need establishing the following definitions and lemmas:

Definition 1. The function $\{y(t) : [t_0 - \sigma, +\infty) \to R^n\}$ is called a solution of Eq.(1) with the initial condition given by

$$y(t) = \varphi(t), t \in [t_0 - \sigma, t_0], \varphi \in PC, \tag{2}$$

if $y(t)$ is continuous at $t \neq t_k$ and $t \geq t_0$, $y(t_k) = y(t_k^+)$ and $y(t_k^-)$ exists, $y(t)$ satisfies Eq.(1) for $t \geq t_0$ under the initial condition. Especially, a point $y^* \in R^n$ is called an equilibrium point of (1), if $y(t) = y^*$ is a solution of (1).

Lemma 1 [7]. Suppose $p > q \geq 0$ and $V(t)$ satisfies scalar impulsive differential inequality

$$\begin{cases} D^+ V(t) \leq -pV(t) + q[V(t)]_\tau, t \neq t_k, t \geq t_0, \\ V(t^+) \leq b_k V(t_k^-) + d_k [V(t_k)]_{\tau^-}, k \in N, \\ V(t) = \phi(t), t \in [t_0 - \tau, t_0] \end{cases} \tag{3}$$

where $V(t)$ is continuous at $t \neq t_0$, $V(t_k) = V(t_k^+)$ and $V(t_k^-)$ exists, $\phi \in PC$ with $n = 1$. Then

$$V(t) \leq (\prod_{t_0 \leq t_k \leq t} \delta_k) e^{-\lambda(t - t_0)} \| \phi(t_0) \|_\tau, t \geq t_0,$$

where $\delta_k := \max\{1, |b_k| + |d_k| e^{\lambda \tau}\}$ and $\lambda > 0$ is a solution of the inequality $\lambda - p + q e^{\lambda \tau} \leq 0$.

Lemma 2 [9]. Let $B \in R^{n \times n}$. Then:

1) $\rho(B) < \| B \|$, where $\rho(\cdot)$ denotes the spectral radius;

2) $\| (E - B)^{-1} \| \leq (1 - \| B \|)^{-1}$ if $\| B \| < 1$;

3) $(E-B)^{-1}$ exists and $(E-B)^{-1} \geq 0$ if $\rho(B) < 1$ and $B \geq 0$, where $B \geq 0$ means B is a nonnegative matrix;

4) $\lambda_m x^T x \leq x^T Bx \leq \lambda_M(B)x^T x$ for any $x \in R^n$ if B is a symmetric matrix, where $\lambda_m(\cdot)$ and $\lambda_M(\cdot)$ denote the minimum eigenvalue of the matrix and the maximum one, respectively.

Assume that the neuron activation function g_i is continuously differentiable and satisfies the following conditions

$$(H_1) \quad |g_i(u_i)| \leq \chi_i, 0 \leq \frac{g_i(u_i) - g_i(v_i)}{u_i - v_i} \leq L_i, \forall u_i \neq v_i, u_i, v_i \in R, i = 1,2,\cdots,n.$$

Notice that the activation function g_i satisfy Lipschitz condition, are not necessarily linear and differentiable. Define $\chi = (\chi_1, \cdots, \chi_n)^T, L = diag(L_1, \cdots, L_n)$. For any $\varphi \in PC$, we assume that there exists at least one solution of (1) with the initial condition (2). Let y^* be an equilibrium point of (1), $y(t)$ be any solution of (1) and $x(t) = y(t) - y^*$.

Set $f_j(x_j(t)) = g_j(y_j(t)) - g_j(y_j^*)$, $f_j(x_j(t - \tau_j(t))) = g_j(y_j(t - \tau_j(t))) - g_j(y_j^*)$. Then, for each $i = 1,2,\cdots,n$, $|f_j(z)| \leq L_j |z|$, and $zf_j(z) \geq 0, \forall z \in R$.

Substituting them into (1), we get

$$D^+ x_i(t) = -c_i x_i(t) + \sum_{j=1}^n A_{ij} f_j(x_j(t))DU_j + \sum_{j=1}^n B_{ij} f_j(x_j(t - \tau_j(t)))DV_j$$

$$+ \sum_{j=1}^n \left(\sum_{l=1}^n (T_{ijl} + T_{ilj})\zeta_l\right) f_j(x_j(t - \tau_j(t)))DW_{jl}, n \tag{4}$$

where $i = 1,2,\cdots,n$; ζ_l is between $g_l(y_l(t - \tau_l(t)))$ and $g_l(y_l^*)$.

3   Main Results

It is clear that the stability of the zero solution of Eq.(4) is equivalent to the stability of the equilibrium point y^* of Eq.(1). Therefore, we mainly discuss the stability of the zero of Eq.(4). For convenience, we denote

$$C = diag(c_1, c_2, \cdots, c_n), \quad A = (A_{ij})_{n \times n}, \quad B = (B_{ij})_{n \times n}$$

$$\hat{A} = (\hat{A}_{ij}) = \begin{cases} c_j - L_j A_{jj}^+, & \text{if } j = i, \\ -(|A_{ij}| L_j + |A_{ji}| L_i)/2, & \text{if } j \neq i, \end{cases}$$

$$S = \begin{bmatrix} 0 & PL \\ LP^T & 0 \end{bmatrix}, \quad D_k = \begin{bmatrix} 0 & P_k^T Q_k \\ Q_k^T P_k & 0 \end{bmatrix},$$

$$A_{jj}^+ = \max\{0, A_{jj}\}, \quad A_k = (A_{ij}^{(k)})_{n \times n} = (A_{ij}u_{jk})_{n \times n}, \quad B_k = (B_{ij}^{(k)})_{n \times n} = (B_{ij}v_{jk})_{n \times n},$$

$$\zeta = [\zeta_1, \zeta_2, \cdots, \zeta_n]^T, \quad \Gamma = diag(\zeta, \zeta, \cdots, \zeta)^T,$$

$$T_i = (T_{ijl})_{n \times n}, \quad T_H = [(T_1 + T_1^T), (T_2 + T_2^T), \cdots, (T_n + T_n^T)]^T,$$

$$T_{Hk} = [(T_1 + T_1^T) w_{1k}, (T_2 + T_2^T) w_{2k}, \cdots, (T_n + T_n^T) w_{nk}]^T,$$

$$P_k = (E - |A_k| L)^{-1}, \quad Q_k = (E - |A_k| L)^{-1} (|B_k| L + \chi \| T_{Hk} | L),$$

$$a = \min_{1 \le j \le n} \{c_j - s_j L_j\}, \quad b = \max_{1 \le j \le n} \left\{ \sum_{i=1}^n |B_{ij}| L_j + \sum_{i=1}^n \sum_{l=1}^n T_{ijl} + T_{ilj} |\chi_l L_j \right\},$$

$$s_j = \max \{0, A_{jj} + \sum_{j \ne i} |A_{ij}| \}, \quad Q = (|A_{ij}|)_{n \times n},$$

$$p_{ij} = |B_{ij}| + \sum_{l=1}^n T_{ijl} + T_{ilj} |\chi_l, P = (p_{ij})_{n \times n}, \quad P = (p_{ij})_{n \times n},$$

$$\xi_k = \lambda_M (P_k^T P_k) + \lambda_M (D_k), \quad \varsigma_k = \lambda_M (Q_k^T Q_k) + \lambda_M (D_k).$$

Theorem 1. Assume that, in addition to (H_1), the following conditions are satisfied for $k \in N$

$(H_2) \quad \rho(|A_k| L) < 1;$

$(H_3) \quad \mu < \lambda_m(\hat{A});$

(H_4) let $\lambda > 0$ satisfy $\lambda - \lambda_m(\hat{A}) + \mu e^{\lambda \sigma} \le 0, \quad \theta_k := \max\{1, \xi_k + \varsigma_k e^{\lambda \sigma}\},$

$$\theta := \sup_{k \in N} \left\{ \frac{\ln \theta_k}{t_k - t_{k-1}} \right\} \text{ where } \mu = \lambda_M(S).$$

Then the zero solution of (4) is globally exponentially stable if $\theta < \lambda$.

Proof. From $(H_3) - (H_4)$, the inequality $\lambda - \lambda_m(\hat{A}) + \mu e^{\lambda \sigma} \le 0$ has at least one solution $\lambda > 0$. Consider the Lyapunov functional

$$V(t) = \frac{1}{2} \sum_{i=1}^n x^2_i(t). \tag{5}$$

From (3) and (H_1) and Lemma 2, for $t \ne t_k$ we can get

$$D^+ V(t) = \sum_{i=1}^n x_i(t) | x_i(t) |'$$

$$\le -\sum_{i=1}^n c_i x_i^2(t) + \sum_{i=1}^n \left(A_{ii}^+ L_i x_i^2(t) + \sum_{j \ne i} |A_{ij}| L_j | x_i(t) \| x_j(t) | \right)$$

$$+ \sum_{i=1}^n \sum_{j=1}^n L_j | B_{ij} \| x_i(t) \| x_j(t - \tau_j(t)) |$$

$$+ \sum_{i=1}^n \sum_{j=1}^n \sum_{l=1}^n T_{ijl} + T_{ilj} |\chi_l L_j | x_i(t) \| x_j(t - \tau_j(t)) | \tag{6}$$

$$= -|x(t)|^T \hat{A} |x(t)| + \frac{1}{2} \begin{bmatrix} |x(t)| \\ |x(t - \tau(t))| \end{bmatrix}^T \begin{bmatrix} 0 & ML \\ LM^T & 0 \end{bmatrix} \begin{bmatrix} |x(t)| \\ |x(t - \tau(t))| \end{bmatrix}$$

$$\le -2\lambda_m(\hat{A}) V(t) + \mu V(t) + \mu V(t - \tau(t))$$

$$\le -(2\lambda_m(\hat{A}) - \mu) V(t) + \mu V(t),$$

where
$$V(t) = \sup_{t-\sigma \le s \le t} V(s), \; |x(t)| = (|x_1(t)|, |x_2(t)|, \cdots, |x_n(t)|)^T,$$

$$|x(t-\tau(t))| = (|x_1(t-\tau_1(t))|, |x_2(t-\tau_2(t))|, \cdots, |x_n(t-\tau_n(t))|)^T.$$

On the other hand, by using the properties of Dirac measure, we have

$$x_i(t_k) - x_i(t_k^-) = \sum_{j=1}^n A_{ij} f_j(x_j(t_k)) u_{jk} + \sum_{j=1}^n B_{ij} f_j(x_j(t_k - \tau_j(t_k))) v_{jk}$$

$$+ \sum_{j=1}^n \Big(\sum_{l=1}^n (T_{ijl} + T_{ilj}) \zeta_l \Big) f_j(x_j(t_k - \tau_j(t_k))) w_{jk}. \tag{7}$$

that is, $x(t_k) = x(t_k^-) + A_k f(x(t_k)) + B_k f(x(t_k - \tau(t_k))) + \Gamma^T T_{Hk} f(x(t_k - \tau(t_k))).$

From (H_2) and Lemma 2, $(E - |A_k| L)^{-1} \ge 0$. Then,

$$|x(t_k)| = |x(t_k^-)| + |A_k| L |x(t_k)| + |B_k| L |x(t_k - \tau(t_k))| + |\chi| \|T_{Hk}| L |x(t_k - \tau(t_k))|$$

$$\le (E - |A_k| L)^{-1} |x(t_k^-)| + (E - |A_k| L)^{-1} (|B_k| L + |\chi| \|T_{Hk}| L) |x(t_k - \tau(t_k))|, \tag{8}$$

yielding

$$V(t_k) = \frac{1}{2} |x(t_k)|^T |x(t_k)| \le \frac{1}{2} (P_k |x(t_k^-)| + Q_k |x(t_k - \tau(t_k))|)^T (P_k |x(t_k^-)| + Q_k |x(t_k - \tau(t_k))|)$$

$$= \frac{1}{2} |x(t_k^-)|^T P_k^T P_k |x(t_k^-)| + \frac{1}{2} |x(t_k - \tau(t_k))|^T Q_k^T Q_k |x(t_k - \tau(t_k))|$$

$$+ \frac{1}{2} \begin{bmatrix} |x(t_k^-)| \, (10) \\ |x(t_k - \tau(t_k))| \end{bmatrix}^T \begin{bmatrix} 0 & P_k^T Q_k \, (11) \\ Q_k^T P_k & 0 \end{bmatrix} \begin{bmatrix} |x(t_k^-)| \, (12) \\ |x(t_k - \tau(t_k))| \end{bmatrix}$$

$$\le (\lambda_M (P_k^T P_k) + \lambda_M (D_k)) V(t_k^-) + (\lambda_M (Q_k^T Q_k) + \lambda_M (D_k)) (V(t_k))_\sigma$$

$$= \xi_k V(t_k^-) + \varsigma_k [V(t_k)]_\sigma. \tag{9}$$

Employing Lemma 1, from (6), (9), (H_3) and (H_4), we have

$$V(t) \le \theta_1 \ldots \theta_{k-1} e^{-\lambda(t-t_0)} \| V(t_0) \|_\sigma$$

$$\le e^{\theta(t_1-t_0)} \ldots e^{\theta(t_{k-1}-t_{k-2})} e^{-\lambda(t-t_0)} \| V(t_0) \|_\sigma$$

$$\le e^{\theta(t-t_0)} e^{-\lambda(t-t_0)} \| V(t_0) \|_\sigma \tag{10}$$

$$= e^{-(\lambda-\theta)(t-t_0)} \| V(t_0) \|_\sigma, t_{k-1} \le t < t_k, k \in N.$$

So, for all $t \ge t_0$, $V(t) \le e^{-(\lambda-\theta)(t-t_0)} \| V(t_0) \|_\sigma$.

Then the conclusion holds and the proof is complete.

Theorem 2. Assume that, in addition to (H_1) and (H_2), the following conditions are satisfied for $k \in N$

(H_1') $b < a$;

(H_2') let $\lambda \ge 0$ be a solution of $\lambda - a + b e^{\lambda \sigma} \le 0$ and

$$\eta_k := \max \{ 1, \| P_k \| + \| Q_k \| \, e^{\lambda \sigma} \}, \eta := \sup_{k \in N} \{ \frac{\ln \eta_k}{t_k - t_{k-1}} \}.$$

Then the zero solution of (4) is globally exponentially stable if $\eta < \lambda$ $(\eta \le \lambda)$.

Proof. Since $b < a$, the inequality $\lambda - a + be^{\lambda\sigma} \le 0$ has at least one solution $\lambda > 0$. Consider the Lyapunov functional

$$V(t) = \sum_{i=1}^{n} |x_i(t)| = \| x(t) \|.$$ (11)

Calculating the upper right derivative D^+V along the solutions of (4), from the conditions (H_1) and (H_2), we have

$$D^+V(t) = \sum_{i=1}^{n} \text{sgn}(x_i(t)) \, |\, x_i(t)\,|'$$

$$\le -\sum_{j=1}^{n}(c_j - s_j L_j) |\, x_j(t)\,| + \sum_{j=1}^{n}\sum_{i=1}^{n} |B_{ij}| \, |\, L_j\,| \, |\, x_j(t - \tau_j(t))\,|$$

$$+ \sum_{j=1}^{n}\sum_{i=1}^{n}\sum_{l=1}^{n} |T_{ijl} + T_{ilj}| \, |\, \chi_l L_j\,| \, |\, x_j(t - \tau_j(t))\,|$$ (12)

$$\le -aV(t) + b[V]_\sigma.$$

On the other hand, by (H_2) and Lemma 2, then the inequality (8) holds, which implies $V(t_k) = \| x(t_k) \| \le \| P_k \| V(t_k^-) + \| Q_k \| [V(t_k)]_{\sigma^-}$. From this inequality, (13), (H_1') and (H_2'), we can get by Lemma 1 $V(t) \le e^{-(\lambda-\eta)(t-t_0)} \| V(t_0) \|_\sigma, t \ge t_0$. Therefore, the conclusion holds and the proof is complete.

4   Example

Consider the hybrid neural network (1) with parameters $i = 1, 2$, $g_1(y_1) = \tanh(0.53 y_1)$, $g_2(y_2) = \tanh(0.67 y_2)$, $\tau_j(t) = 0.5 e^{-t}$, $j = 1, 2$, $t_0 = 0$, $t_k = t_{k-1} + 0.6k$, $u_{jk} = 0.2(-1)^k$, $v_{jk} = 0.2 e^{0.2k}$, $w_{jk} = 0.1 e^{0.2k}$, for $j, l = 1, 2$, $k \in N$.

$$C = \begin{bmatrix} 1.90 & 0 \\ 0 & 1.89 \end{bmatrix}, A = \begin{bmatrix} 0.05 & 0.14 \\ 0.20 & 0.31 \end{bmatrix}, B = \begin{bmatrix} 0.09 & 0.25 \\ 0.21 & 0.45 \end{bmatrix},$$

$$T_1 = \begin{bmatrix} 0.05 & 0.14 \\ -0.06 & 0.05 \end{bmatrix}, T_2 = \begin{bmatrix} 0.29 & 0.10 \\ 0.23 & -0.14 \end{bmatrix}, T_3 = \begin{bmatrix} -0.23 & 0.07 \\ 0.09 & -0.02 \end{bmatrix}.$$

In terms of the parameters defined in the section 3, we have

$\sigma = 0.5$, $\chi_i = 1$, $\hat{A} = \begin{bmatrix} 1.85 & -0.17 \\ -0.17 & 1.58 \end{bmatrix}$, $\lambda_m(\hat{A}) = 1.4979$, $\mu = 0.5837$,

$\| P_k \| = 1.0410$, $\| Q_k \| = 0.1190 e^{0.2k}$, $a = 1.45$, $b = 0.7$, $\xi_k = 1.2350$, $\varsigma_k = 0.1725$,

$t_k - t_{k-1} = 0.6k$ and $\rho(|A_k| L) = 0.0784 < 1$, $\mu = 0.5837 < 1.4979 = \lambda_m(\hat{A})$,

$\ln \theta_k / (t_k - t_{k-1}) \le 0.6502 < \lambda = 0.6785$, where λ is an unique solution of equation: $\lambda - \lambda_m(\hat{A}) + \mu e^{\lambda \sigma} = 0$. It follows from Theorem 1 that the equilibrium point $(0,0)^T$ is globally exponentially stable and the exponentially convergent rate is approximately equal to 0.0283. To use Theorem 2, we note that $\eta_k = 1.2731, a = 1.45 > 0.7 = b$, and $\ln \theta_k / (t_k - t_{k-1}) \le 0.4024 < \lambda = 0.5352$, where $\lambda = 0.5352$ is the unique solution of the equation $\lambda = a - b e^{\lambda \sigma}$. Thus from Theorem 2 that the equilibrium point $(0,0)^T$ is globally exponentially stable. Moreover, when $T_{ijl} \equiv 0$, the system (1) reduces to the model studied in [7], it is easy to check all the conditions of Theorem 1 in [7] are satisfied, and $\ln \theta_k / (t_k - t_{k-1}) \le 0.6429 < \lambda = 0.6794$, where λ is an unique solution of equation: $\lambda - \lambda_m(\hat{A}) + \lambda_M(S) e^{\lambda \sigma} = 0$. Thus the example demonstrate the effectiveness of the results.

5   Conclusions

By means of an impulsive delay differential inequality and several Lyapunov functions, we have analyzed the global exponential stability of high-order dynamical neural networks with time-varying delays and impulsive effects. Several delay-dependent sufficient conditions ensuring the stability have been proposed, which is illustrated in the given numerical example.

References

1. Ozcan, N.: A new sufficient condition for global robust stability of delayed neural networks. Neural Processing Letters 34(3), 305–316 (2011)
2. Di Marco, M., Grazzini, M., Pancioni, L.: Global robust stability criteria for interval delayed full-range cellular neural networks. IEEE Transactions on Neural Networks 22(4), 666–671 (2011)
3. Faydasicok, O., Arik, S.: Equilibrium and stability analysis of delayed neural networks under parameter uncertainties. Applied Mathematics and Computation 218(12), 6716–6726 (2012)
4. Liao, X.X., Liao, Y.: Stability of Hopfield-type neural networks (II). Sci. China (Series A) 40(8), 813–816 (1997)
5. Liu, X.Z., Teo, K.L.: Exponential stability of impulsive high-order Hopfield-type neural networks with time-varying delays. IEEE Transactions on Neural Networks 16(6), 1329–1339 (2005)
6. Liu, X.Z., Ballinger, G.: Uniform asymptotic stability of impulsive delay differential equations. Comput. Math. Applicat. 41, 903–915 (2001)
7. Yang, Z., Xu, D.: Stability analysis of delay neural networks with impulsive effects. IEEE Transactions on Circuits and Systems II: Express Briefs 52(8), 517–521 (2005)
8. Li, Y.K., Hu, L.H.: Global exponential stability and existence of periodic solution of Hopfield-type neural networks with impulses. Phys. Lett. A 333, 62–71 (2004)
9. Berman, A., Plemmons, R.J.: Nonnegative Matrices in Mathematical Sciences. Academic Press, New York (1979)

Mining Google Scholar Citations: An Exploratory Study

Ze Huang and Bo Yuan

Intelligent Computing Lab, Division of Informatics,
Graduate School at Shenzhen, Tsinghua University, Shenzhen 518055, P.R. China
workthy@hotmail.com, yuanb@sz.tsinghua.edu.cn

Abstract. The official launch of Google Scholar Citations in 2011 opens a new horizon for analyzing the citations of individual researchers with unprecedented convenience and accuracy. This paper presents one of the first exploratory studies based on the data provided by Google Scholar Citations. More specifically, we conduct a series of investigations on: i) the overall citation patterns across different disciplines; ii) the correlation among various index metrics; iii) the personal citation patterns of researchers; iv) the transformation of research topics over time. Our results suggest that Google Scholar Citations is a powerful data source for citation analysis and provides a solid basis for performing more sophisticated data mining research in the future.

Keywords: Google Scholar Citations, Citation Analysis, Tag Cloud, Clustering.

1 Introduction

Citation analysis refers to the investigation of the frequency and patterns of citation records (i.e., references to published or unpublished sources) in scholarly literature. It has been widely used as a method of bibliometrics to evaluate the quality of journals and to establish the links among works and authors. For example, it is possible to identify groups of people that collaborate frequently or how a piece of work is related to existing studies [1]. It is also important for researchers to choose the right journals as well as track the development of specific research topics [2]. In many academic institutes, citation record is also being used as one of the major selection criteria in the process of recruiting and promotion.

There have been some critics [3, 4] on the potential misinterpretation of citation in evaluating journals and researchers: i) the number of citations can be manipulated to deliberately increase the impact of a journal; ii) the influence of self-citation and negative citation needs to be taken into account; iii) it may be difficult to find a good tradeoff between the number of citations and the number of publications; iv) different research disciplines may have significantly different typical citation numbers. Nevertheless, citation record provides a practical and quantitative performance measure, which has been accepted widely in academia.

Online bibliographic databases such as Web of Science (published by Thomson ISI), SciVerse Scopus (published by Elsevier) and Google Scholar (a freely accessible

D.-S. Huang et al. (Eds.): ICIC 2012, LNCS 7389, pp. 182–189, 2012.
© Springer-Verlag Berlin Heidelberg 2012

web search engine released in 2004 by Google) have brought tremendous benefits to researchers across different disciplines [5]. In nowadays, Web of Science covers over 12,000 journals and 150,000 conference proceedings but its most famous citation index Science Citation Index Expanded only covers around 8,000 journals. In the meantime, SciVerse Scopus contains nearly 19,500 titles from 5,000 publishers worldwide with 46 million records most of which are journal articles.

Different from subscription-based commercial databases, Google Scholar retrieves bibliographic data of academic literature by automatically crawling over the Web and uses a ranking algorithm, which relies heavily on citation counts, to display the search results. Although its exact coverage is not known to the public (some publishers may not allow Google Scholar to crawl their databases), some studies show that Google Scholar usually returns the highest number of citations compared to other similar services [6, 7]. Note that conference papers are extensively indexed in Google Scholar and their citation counts are calculated in combination with journal articles. This is particularly important for disciplines such as computer science where many high quality research outcomes are published in conferences.

Although most bibliographic databases provide comprehensive search functions, there is an inherent issue making the accurate evaluation of individual researchers a very challenging task: the disambiguation of authors. In many occasions, different researchers share exactly the same name (e.g., even different names may appear to be identical in terms of spelling when translated into English) and it is necessary to group the publications corresponding to the same author before doing any further analysis. Despite of some recent progress in this area, it is still not a fully reliable procedure [8]. After all, a researcher may have different affiliations and collaborate with different people and publish papers in seemingly irrelevant disciplines.

The official launch of a new service in the name of Google Scholar Citations[1] (GSC) in late 2011 provides a different solution to the above issue. Built on the top of Google Scholar, it allows researchers to create personal accounts and add, manually or automatically, papers published by them. By doing so, each registered researcher has a mini-homepage (similar to a blog) with a list of papers and citation counts. In other words, GSC gives researchers the freedom to maintain the list of publications to ensure the highest accuracy and integrity of data. For example, the bibliographic information of each paper is user-editable, which means that errors accidentally introduced during the crawling of web pages can be corrected in a straightforward manner. Also, similar to many social networking services, researchers can follow new articles and citations of any registered researchers.

In this paper, we present one of the first exploratory studies on the analysis of the structured data in GSC. We will investigate: i) the overall citation patterns across different disciplines; ii) the correlation among various index metrics; iii) the personal citation patterns of researchers; iv) the transformation of research topics over time. Section 2 describes the data collection procedure including the content structure of GSC. Section 3 presents the major results of citation analysis while Section 4 focuses on the analysis of topic trends. This paper is concluded in Section 5 with some discussion and directions for future work.

---

[1] http://scholar.google.com/citations

2   Data in GSC

In GSC, researchers can manually label their research disciplines. In many cases, each registered researcher has three to four discipline labels. To explain how we collected the data, all web pages in GSC are divided into two types in this paper: *Discipline Level* (DL) pages and *Author Level* (AL) pages.

2.1   Web Page Description

The major information in GSC is shown in Table 1. DL pages display authors in a certain discipline (e.g., data mining). For example, with "label: data_mining" as the keyword, GSC returns a name list of 10 authors who have identified themselves as in the *data mining* discipline, sorted based on the total citation number in descending order. By clicking the "Next" link, the next 10 authors (if any) will be displayed.

Each author in the list has an URL linking to his/her personal page (AL page), which shows detailed publication information. The AL page can be divided into 3 sections from top to bottom: i) author profile; ii) citation indexes table; iii) a list of papers that the author has published, sorted by each paper's citation number in descending order. The paper list shows maximally 20 or 100 papers and additional papers (if any) may be accessed by clicking the "Next" link. This action was simulated in our program to gather the information of all papers corresponding to an author.

Table 1. The content description of two types of web pages

Page Type	Content	Details
DL Page	Author List	URL links to each author's personal page
AL Page	Author Profile	Affiliation, Disciplines, Home Page
	Citation Index Table	Citations, h-index, i10-index (All & Recent)
	Paper List	Title, Author, Year, Citation Number

Note that, the citation indexes table has two columns. The first column consists of statistical values based on an author's entire publication records, and the other one is based on recent papers published within 5 years (e.g., since 2007 as of 2012).

2.2   Data Collection

Since GSC does not provide APIs to the public, we collected the data by analyzing the web page source code with a crawler program. The crawler extracted required information through pattern match using regular expression.

For DL pages, we focused on 6 related disciplines: Data Mining (DM), Artificial Intelligence (AI), Bioinformatics (Bio), Information Retrieval (IR), Machine Learning (ML) and Pattern Recognition (PR). So far, many disciplines in GSC did not have sufficient number of registered authors (e.g., 200 authors in Sociology). Note that the same author may appear in multiple disciplines and authors whose papers had zero citation were not counted. For AL pages, the 6 indexes in the citation index table and publication information (title, year of publication and citation count) were retrieved. Papers without any citation were excluded.

Totally, we collected up to 1000 authors in each discipline and no more than 100 papers for each author. It took less than 30 minutes for the crawler to collect the data, and the dataset used in the following experiments was collected on March 11, 2012.

3   Citation Analysis

3.1   Index Metrics

The number of total citations is a most commonly used index metric to quantify the impact of an individual's research output. However, it is far from sufficient to compare and evaluate research work comprehensively. Recently, many new metrics were designed and enhanced, such as the h-index [9] and the g-index [10]. In GSC, only three metrics are adopted: the total number of citations (TC), the h-index and the i10-index. The h-index attempts to address both the productivity and the impact factors and is defined as the maximum number h so that there are h papers each with citation number $\geq h$. The i10-index was introduced in July 2011, which indicates the number of academic papers of an author that have received at least 10 citations.

Fig. 1 shows the TC values of the top 30 scholars in each of the 6 disciplines. It is clear that researchers especially the most eminent ones in AI tend to have much larger TC values compared to disciplines such as IR and PR. This fact suggests that TC is not a reliable metric for evaluating the impact of individuals in different disciplines.

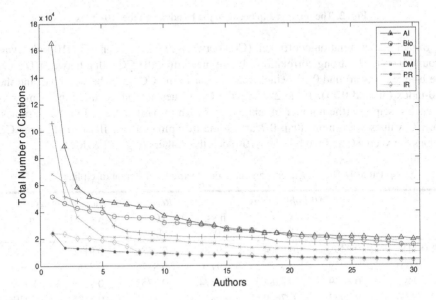

Fig. 1. A comparison of the total number of citations of researchers in different disciplines

As mentioned above, currently three index metrics are calculated in GSC, which are divided into two groups based on all publications and recent publications respectively. The Pearson Correlation Coefficient was used to measure the dependences among these metrics. Fig. 2 shows a scatter plot based on the values of the h-index and the i10-index, over the entire dataset (i.e., all disciplines and all publications). Intuitively, there was strong linear correlation between the two index metrics.

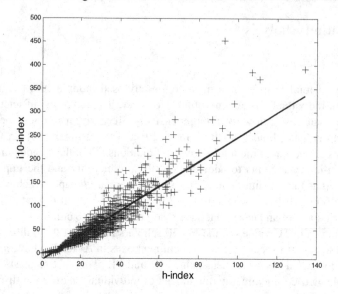

Fig. 2. The relationship between the h-index and the i10-index

In fact, the correlation coefficient (CC) between the h-index and the i10-index was around 0.95 (very strong correlation). In the meantime, the CC value between TC and the h-index was around 0.77, which was similar to the CC value between TC and the i10-index (around 0.75). Table 2 shows the CC values among the three index metrics in each discipline (the number of authors is shown in parentheses). In all disciplines, the CC values were more than 0.7 and some disciplines (e.g., IR) had stronger CC values between TC and the h-index/i10-index than others (e.g., AI & ML).

Table 1. The coefficients among index metrics in different disciplines

Discipline	All Publications			Recent Publications (Since 2007)		
	TC vs. h	TC vs. i10	h vs. i10	TC vs. h	TC vs. i10	h vs. i10
IR (511)	0.8495	0.9004	0.9389	0.8583	0.9185	0.9289
AI (1000)	0.7829	0.7738	0.9710	0.8276	0.8366	0.9636
Bio (999)	0.8019	0.7282	0.9374	0.7752	0.7205	0.9180
DM (879)	0.7862	0.7669	0.9327	0.8052	0.8225	0.9306
ML (1000)	0.7690	0.7240	0.9512	0.7604	0.7498	0.9511
PR (539)	0.8434	0.8583	0.9306	0.8404	0.8613	0.9334

3.2   Personal Citation Pattern

One of the major benefits of GSC is that it provides the most accurate citation profiles for individual researchers. Among many potential research questions that can be addressed, we focused on the personal citation patterns at this stage. Regardless of the TC value of a researcher, it is interesting to investigate how the citations are distributed among his/her publications. For example, some authors may have a small number of highly cited documents (e.g., review papers and books are often cited heavily) while other authors may have citations distributed relatively evenly.

To testify this hypothesis, cluster analysis was conducted as follows. Firstly, only researchers with more than 200 citations and at least 10 papers were selected (3539 authors in total). Secondly, the papers of each researcher were sorted based on the citation numbers in descending order. Thirdly, a 10D vector was created with the first element corresponding to the proportion of the citations of the top 10% papers among all citations of the specific researcher. Similarly, the second element corresponded to the citation proportion of the second 10% papers and so on. Note that the actual number of papers was rounded to the nearest integer towards minus infinity for the first 9 subsets and all rest papers were assigned to the 10th subset (the number of papers in it may be higher than average).

As a result, each researcher was represented by a data point in the 10D space, specified by the distribution of citations. Fig. 3 shows the results of clustering using the K-Means method (K=2) and the Euclidean distance as the similarity measure.

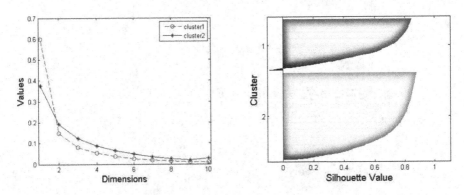

Fig. 3. Personal citation pattern: cluster centroids (left); clustering evaluation (right)

Fig. 3 (left) shows the two cluster centroids after clustering. It is evident that authors in *cluster 1* tended to have the majority of citations concentrated on the few very best papers. In fact, the top 10% papers contributed around 60% of the total citations. By contrast, papers of authors in *cluster 2* received citations in a more uniform manner. Fig. 3 (right) demonstrates the quality of clustering using the Silhouette method. There were more authors in *cluster 2* (2225 authors or 62.87%) than in *cluster 1* (1314 or 37.13%) in *cluster 2*. Moreover, only few data points had negative Silhouette values (the average Silhouette value was 0.6717) and it is clear that the two clusters were reasonably well structured.

4   Topic Trend

The titles of papers may provide some vital clues on how topics or keywords evolved in a discipline. One solution is to show the title texts in the form of tag cloud using IBM Word Cloud Generator[2] where the font size of a word in the cloud is proportional to its frequency in the text. We divided the papers into 2 subsets: before 2007 and since 2007 to observe the change over time. Fig. 4 shows an example of the keywords in the field of IR. For better visual effect, some common title words such as *based*, *approach*, *systems*, *analysis*, *using* were ignored as these words appeared frequently across all disciplines. Additionally, we filtered out *information*, *retrieval* and *search*, due to their large number of occurrences in both clouds.

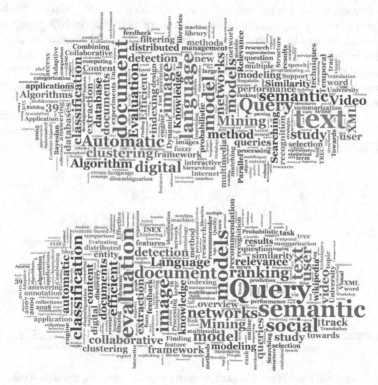

Fig. 4. Tag clouds of titles in IR: top (before 2007); bottom (since 2007)

By comparing the two clouds corresponding to two consecutive time periods, it is possible to find some interesting clues. For example, the font size of *text* dropped down while the font size of *semantic* scaled up, which may suggest that semantic retrieval has been growing rapidly as the mainstream research topic in IR. Meanwhile, the word *social* appeared only in the bottom cloud, indicating that a new research direction related to social networks has emerged in recent years.

---

[2] http://www.wordle.net/

5   Conclusion

Citation analysis has attracted significant attentions from researchers in all aspects. With the availability of online searchable citation databases such as Web of Science and Google Scholar, it is now possible to conduct decent analysis using data mining and knowledge discovery techniques. In this paper, we presented one of the first studies on Google Scholar Citations, which provides well organized citation records in terms of discipline and authorship. We found that different disciplines had different numbers of typical citations and there were strong correlations among the h-index, the i10-index and the total number of citations. For individual researchers, we identified two distinct groups of authors in terms of the distribution of citations. Finally, we demonstrated the effectiveness of tag cloud in discovering the topic trend in certain research field. In the future, with the increasing number of registered researchers, we will be able to collect more comprehensive data sets and conduct more thorough analysis. It would also be interesting to combine the domain knowledge with the results of data analysis to provide more insights into each discipline.

Acknowledgements. This work was supported by the National Natural Science Foundation of China (No. 60905030).

References

1. White, H.D., McCain, K.W.: Visualizing a Discipline: An Author Co-Citation Analysis of Information Science, 1972-1995. Journal of the American Society for Information Science and Technology 49(4), 327–355 (1998)
2. Chen, C.: CiteSpace II: Detecting and Visualizing Emerging Trends and Transient Patterns in Scientific Literature. Journal of the American Society for Information Science and Technology 57(3), 359–377 (2006)
3. Gisvold, S.E.: Citation Analysis and Journal Impact Factors – Is the Tail Wagging the Dog? Acta Anaesthesiol. Scand. 43(10), 971–973 (1999)
4. MacRoberts, M.H., MacRoberts, B.R.: Problems of Citation Analysis: A Critical Review. Journal of the American Society for Information Science and Technology 40(5), 342–349 (1989)
5. Bakkalbasi, N., Bauer, K., Glover, J., Wang, L.: Three Options for Citation Tracking: Google Scholar, Scopus and Web of Science. Biomedical Digital Libraries 3(7) (2006)
6. Harzing, A., Wal, R.: Google Scholar as a New Source for Citation Analysis. Ethics in Science and Environmental Politics 8(1), 61–73 (2008)
7. Meho, L.I., Yang, K.: Impact of Data Sources on Citation Counts and Rankings of LIS Faculty: Web of Science versus Scopus and Google Scholar. Journal of The American Society for Information Science and Technology 58(13), 2105–2125 (2007)
8. Torvik, V., Smalheiser, N.: Author Name Disambiguation in MEDLINE. ACM Transactions on Knowledge Discovery from Data 3(3), Article 11 (2009)
9. Hirsch, J.: An Index to Quantify an Individual's Scientific Research Output. Proceedings of the National Academy of Sciences 102(46), 16569–16572 (2005)
10. Egghe, L.: Theory and Practise of the g-index. Scientometrics 69(1), 137–152 (2006)

Knowledge Acquisition of Multiple Information Sources Based on Aircraft Assembly Design

Liang Xia [1], Lizhi Zhang [2], and Zhenguo Yan[2]

[1] Xian University of Science and Technology, Xi'an, China
Zlz365@126.com
[2] The Ministry of Education Key Laboratory of Contemporary Design and Integrated
Manufacturing Technology, Northwestern Polytechnical University, Xi'an, China
593658826@qq.com

Abstract. This paper analyzes the importance of knowledge acquisition in aircraft assembly design process and introduces briefly how to create an assembly knowledge base. The knowledge involved in the aircraft assembly design knowledge-based system is classified into three types: fact knowledge, rule knowledge and instance knowledge. The acquisition methods and the corresponding examples of these three kinds of knowledge are described in detail.

Keywords: aircraft assembly design, fact knowledge acquisition, rule knowledge acquisition, instance knowledge acquisition

1 Introduction

Aircraft assembly design is a complex and long period work, need to carry out a lot of researches and tests, meanwhile a large number of formulas must be used and produce a mass of data. With the development of E-learning and network technique, human will possess of more data. But it is difficult that designers find useful knowledge from the resource library. To solve the problem, we must transform the resource information to useful knowledge to help people make a decision [1]. The aircraft assembly project is one of the CoPS(Complex Product Systems), and it needs inter-unit cooperation, so aircraft assembly design is actually a problem solving process under the guidance of multi-type, multi-disciplinary expertise. But the current state of the various information points scattered in the different sources of information, lacking of effective organization, management and guidance, we need to consider how knowledge extraction and form a unified description. Now, there are some methods about knowledge acquisition, for example: by in-training exams [2], based on scenario model and repository grid and attribute ordering table technology [3], distributed knowledge acquisition based on semantic grid [4] and so on. Because the different applications, the different knowledge characteristics, we can not copy the other knowledge acquisition methods.

D.-S. Huang et al. (Eds.): ICIC 2012, LNCS 7389, pp. 190–197, 2012.
© Springer-Verlag Berlin Heidelberg 2012

2 Building Process of Assembly Knowledge Base

Assembly rapid design knowledge is the basis of the assembly design; the whole assembly design process revolves around the assembly design knowledge. Divided by locations and functions in conceptual design process, the assembly knowledge can be divided into assembly meta-knowledge and assembly instance-knowledge. Meta-knowledge can be divided into domain knowledge (knowledge of demand domain, knowledge of functional domain, knowledge of process domain, and knowledge of physical domain), mapping rules, rules of interpretation, evaluation rules [5].

Building processes of assembly knowledge base include the introduction of assembly design knowledge, knowledge classification, acquisition, representation, creating a database, and knowledge base maintenance techniques. It can be summarized in four stages: conceptualization stage, formalization stage, implementation stage and testing phase. These four stages contact closely but indivisible. The build process is shown in Fig. 1.

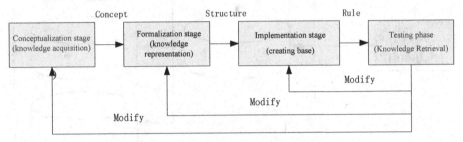

Fig. 1. The building process of Knowledge Base

Conceptualization stage is the knowledge acquisition phase; the task at this stage is to identify the overall mission and relationship. Such as" How these are formed"," what is the relationship between the child objects " and" what is the relationship between the child object and parent object". These questions as a starting point to obtain the knowledge, so as to achieve a comprehensive analysis.

3 Knowledge Acquisition

Knowledge acquisition is the most important and difficult part of assembly knowledge base construction; it is directly related to the quality and quantity of knowledge base. Knowledge acquisition methods can be divided into two types, which are direct access to knowledge and indirect access to the knowledge. Based on the phenomenon and the data, direct access to knowledge summed up the knowledge of the conclusion or decision into the knowledge base. This method attempts to solve the difficulties of automatic acquisition of knowledge. Due to induction is an integrated process, so it is more difficult than the interpretation.

At present, there is no such knowledge of the program. This paper discusses the method of indirect access, and imparts the knowledge to knowledge processing systems with the knowledge editor and other tools. Aircraft assembly design knowledge base involves three types of knowledge, namely the fact knowledge, rule knowledge and instance knowledge. The knowledge acquisition mode is shown in Fig. 2.

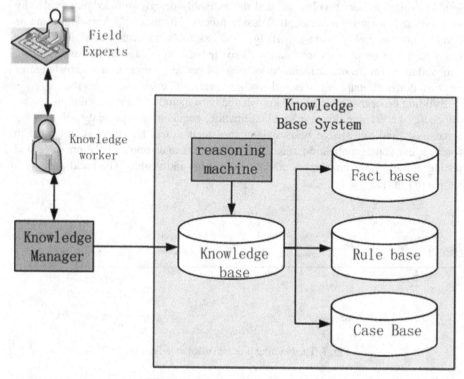

Fig. 2. Knowledge acquisition mode

3.1   Acquisition of Rule Knowledge

The rule knowledge are some experiences, standards and so on, they can be used in the rule-based reasoning process, derived from practical experience of the experts and books theory. Assembly design is the realization of the design requirements, such as selection and design, layout design, as well as to determine the performance requirements. Thus the main task of rule knowledge acquisition in assembly design is to extract the related knowledge of selection and design, the related knowledge of the layout design and the relevant knowledge to determine the performance requirements.

The acquisition problem of rule knowledge is to solve the design process in which knowledge is required. For example, when we select the rules for the strengthening of the molding structure in the type design of the assembly design, the crucial question is how to determine the strengthening of the molding structure. Answers to the questions: First, identify ways of strengthening the structure according to the molding shape curvature. The second is to identify ways of strengthening the structure according to the relative size of profile molding appearance. Then according to the characteristics of rule knowledge of the assembly design, to further refine the knowledge of rules.

①For the acquisition of unity knowledge, as to whether the use of flanging strengthening structure of assembly design as an example, we can analyze from two aspects. According to the first aspect from the forming shape curvature defining reinforcing structure, and gets a rule A: If the molding shape with big curvature radius will choose flanging strengthening structure. According to the second aspect from the molding shape profile relative to the size of the angle consideration, and gets a rule B: If the molding shape outline is relatively small, select flanging strengthening structure.

②For the acquisition of overall knowledge, knowledge due to its large capacity, first is to arrangement the experience knowledge into that form, and then transform into rules from the assertion.

For example, the following paragraph is expert knowledge about the fixture sample together in the form selection model:

"For the choice of fixture model together form, model appearance and involution accuracy requirement is that the decisive condition of the fixture model together way. In general, simple shape and co-error model using the pair together in the form; involution accuracy higher assembly, component model using triangle pairs together in the form; high accuracy of the co-template hole involution forms; if it is the block structure of the larger plane model, the fixture sample is used to dovetail together in the form."

Organized into assertions as:

"If the model shape is simple, and the co-error, using the pair together in the form"

"If the template used for assembly, components, and higher on the co-accuracy, the use of triangular form of involution"

"If the model is large, and the high accuracy, the co-hole pairs together in the form"

"If the model plane larger and sub-block structure, the dovetail together in the form"

According to the above assertion can create the following fixture model selected set of rules together in the form:

(**RS** Fixture model selection together in the form

(**Features** : involutive forms)

(**Rules**)

(**Rules1** Using the pair together in the form

IF (simple shape) AND (co-accurate low)

THEN (Object name by the line together to form))

(**Rules2** Using triangle pairs together in the form

IF (the model shape for assembly, parts) AND (higher accuracy)

THEN (Object name by triangle to form))

(**Rules3** Using holes together in the form

IF (model of larger size) AND (high accuracy)

THEN (Object name by hole to form))

(**Rules4** Using dovetail together in the form
IF (larger model plane and block shape)
THEN (Object name by dovetail to form))

3.2   Acquisition of Fact Knowledge

In object-oriented knowledge base, the fact knowledge can be obtained by the human-computer interaction, characteristics inheritance, rule-based reasoning. The characteristics inheritance obtains fact value from the parent class as the fact property value. Based on existing conditions, the rule-based reasoning invoked the rules set of object class, the reasoning results as the eigenvalue of the fact knowledge. All the three methods to achieve the knowledge acquisition is recording the "ask" "inherit" "deduce" on the corresponding position of characteristic value.

The following is an acquisition example of fact knowledge. The vacuum fixture is typically used on a CNC (Computerized Numerical Control) milling machine, it can machining the large area parts such as the aircraft integral panel, the overall beam, the overall rib and honeycomb core.

```
(Object vacuum fixture: class:Intensify force
calculations
(Sealing coefficient: Ask)
(Residual pressure of Vacuum chamber :Ask)
(Effective vacuum area: Ask)
(Seal bar compression, rebound per meter length: Ask)
(The length of Seal bar: Ask)
(Atmospheric pressure: Inherit)
(Pressure difference: Deduce)
```

The logic algorithm processes of fact knowledge acquisition are shown in Fig. 3.

3.3   Acquisition of Instance Knowledge

Instances, as the name suggests are the classic typical examples, methods or solutions used in the assembly design. The task to obtain instance knowledge is extracted success stories from the previous assembly design as an example, stored in instance library after the knowledge engineer finish work.

Understanding stage in the knowledge acquisition of instance, the first step is to understand where the problem is in assembly design instance. The main features include those elements, how to communicate with assembly design experts and research.

Instance knowledge of assembly design, includes the asking of instance problem, the solution to the problem, the evaluation of instance. The asking of instance problem is the demand analysis for assembly design. According to the design needs the result obtained is the solution to the problem. The evaluation of instance is the evaluation of the practicality and reliability of instance. Then according to the feature of the

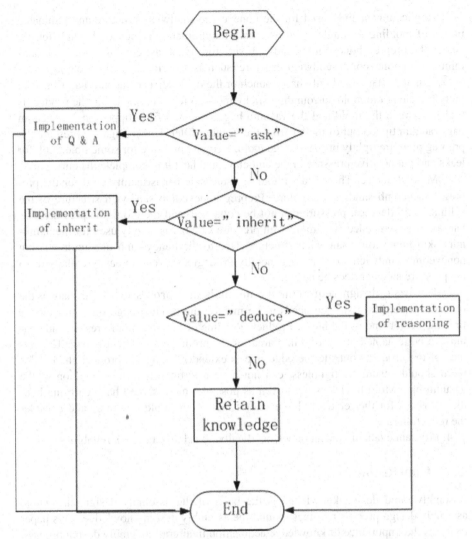

Fig. 3. Logic algorithm processes of the fact knowledge acquisition

characteristics problem of the assembly design instance, to further refine the content of the instance knowledge. Generally, the assembly design example includes the following aspects:

1. Instance number: the number of technology and equipment in actual production.

2. The problem of instance requirements to solve (i.e. design requirements): assembly type, purpose, structure, strength, functional requirements, user requirements, and development costs.

3. Solution to the problem (i.e. specific design)

Vacuum molding, for example, its design includes the following four aspects:

①Application: mainly used for the honeycomb sandwich structure and laminated pieces of molding of single-sided assembly requirements. Honeycomb sandwich includes the prepreg honeycomb sandwich structure (such as: cowlings, etc.) and aluminum skin honeycomb sandwich structure (such as: flooring, tail boom, etc.).

②Principle: The mold only has a punch or the die side. The vacuum bag film with putty bar along the mold surrounding sticky motif to form a sealed bag, the product is sealed between the mold and the vacuum bag, a vacuum filter tipped on the vacuum bag, vacuum tipped out of the air from the vacuum filter. Under negative pressure the prepreg plies are tightly pressed to the mold. Keeping pressure for some time, look for leaks and put positive pressure in the autoclave and heat it to complete the curing.

③Mold structure: The mold without pressure side bar is mainly used for the prepreg honeycomb sandwich structure. Prepreg honeycomb sandwich structure of the cellular is 45 degrees, pressure can not be collapsed, and therefore, the pressure side bar does not be needed. The mold with pressure side bar is mainly used for the aluminum skin honeycomb sandwich structure. Due to the honeycomb of aluminum skin honeycomb sandwich structure is generally 90 degrees, pressure easily collapsed, so the pressure side bar does be needed.

④The basic design requirements, methods and processes: Mold shape is the product shape. Mold surface needs to be marked the product edge line, the cellular line, the center line of the holes. Product edge line or the outside of pressure side bar must to be remained at least 80 mm area for the production of vacuum bag. The surface of the products outside the edge line is extended along the product shape. The mold should ensure air tightness, can not have a through-hole in the region of the vacuum bag. Mold thickness is as even as possible; mold should have carrying handles or rings; for the relatively large size of the pieces, mold design should consider the reduction ratio.

4. Program evaluation: accuracy (specifically, to achieve targets), reliability.

4   Conclusion

Assembly rapid design knowledge is the basis of the assembly design; the whole assembly design process revolves around the assembly design knowledge. This paper analyzes the importance of knowledge acquisition in aircraft assembly design process. Introduce briefly how to create an assembly knowledge base. On this basis, the acquisition methods of fact knowledge, rule knowledge and instance knowledge are introduced in detail. Assembly design knowledge-based systems full use of the above three kinds of knowledge representation. When the user query, the system can display the corresponding knowledge according to the multiple types of feature information given by the user.

Acknowledgment. Project supported by the Defense Industrial Technology Development Program (Grant No. A0520110003) and the Shaanxi Science and Technology projects (Grant No. 2010K09-07). The author would also like to gratefully acknowledge the useful remarks and kind support of Dr. Yu jian-feng and Dong-liang.

References

1. Romero, P.C., Ventura, P.S.: Educational data mining: A Survey from 1995 to 2005. Expert Systems with Applications: an International Journal 33(1), 135–146 (2007)
2. Furman, S.M., Scott, L.Z., Joseph, C.K.: Factors Associated with Medical Knowledge Acquisition During Internal Medicine Residency. Journal of General Internal Medicine 22(7), 962–968 (2007)
3. Tai, L., Guo, H., Zhong, T., Li, D.: Journal of Harbin Institute of Technology. Journal of General Internal Medicine 5 (2007)
4. Wang, H., Nie, G., Fu, K.: Distributed Knowledge Acquisition Based on Semantic Grid. In: Proceedings of 2009 Asia-Pacific Conference on Information Processing, vol. 1 (2009)
5. Lin, J., Liao, W., Li, Y.: Object-oriented-based Knowledge Representation for Aircraft Assembly Assembly Conceptual Design. Machine Building Automation 38(5), 80–82 (2009)

Generalizing Sufficient Conditions and Traceable Graphs

Kewen Zhao

Institute of Information Science and Mathematics, Qiongzhou University,
Sanya, Hainan, 572022, P.R. China
kwzqzu@yahoo.cn

Abstract. In 2005, Rahman and Kaykobad proved that if G is a 2-connected graph with n vertices and $d(u)+d(v)+\delta(u,v)\geq n+1$ for each pair of distinct non-adjacent vertices u,v in G, then G is traceable [*Information Processing Letters,* 94 (2005), 1, 37-41]. In 2006, Li proved that if G is a 2-connected graph with n vertices and $d(u)+d(v)+\delta(u,v)\geq n+3$ for each pair of distinct non-adjacent vertices u,v in G, then G is Hamiltonian-connected [*Information Processing Letters,* 98 (2006), 4, 159-161]. In this present paper, we prove that if G is a 2-connected graph with n vertices and $d(u)+d(v)+\delta(u,v)\geq n$ for each pair of distinct non-adjacent vertices u,v in G, then G has a Hamiltonian path or G belongs to a class of exceptional graphs. We also prove that if G is a 2-connected graph with n vertices and $d(u)+d(v)+\delta(u,v)\geq n+2$ for each pair of distinct non-adjacent vertices u,v in G, then G is Hamiltonian-connected or G belongs to a classes of exceptional graphs. Thus, our the two results generalize the above two results by Rahman et al. and Li, respectively.

Keywords: rahman-kaykobad condition, new sufficient condition, traceable graphs, hamiltonian graphs, Hamiltonian-connected.

1 Introduction

We consider only finite undirected graphs without loops or multiple edges. For a graph G, let $V(G)$ be the vertex set of G and $E(G)$ the edge set of G. The complete graph of order n is denoted by K_n. For two vertices u and v, let $\delta(u,v)$ be the length of a shortest path between vertices u and v in G, that is, $\delta(u, v)$ is the distance between u and v. If H and S are subgraphs of G or subsets of $V(G)$, let $N_H(S)$ be the set of vertices in H that are adjacent to some vertex in S and let the cardinality of $N_H(S)$ be $|N_H(S)| = d_H(S)$. In particular, if $H = G$ and S is a vertex u, then $N_G(S)$ is the neighborhood of u in G. Furthermore, let G-H and G[S] denote the subgraphs of G induced by $V(G)-V(H)$ and S, respectively. For each integer $m\geq 3$, let $P_m(x_1, x_m)=x_1x_2\ldots x_m$ denote a path of order m whose two end-vertices are x_1, x_m and define

$$N^+_{Pm}(u)=\{x_{i+1}\in V(P_m):x_i\in N_{Pm}(u)\}$$

$$N^-_{Pm}(u)=\{x_{i-1}\in V(P_m):x_i\in N_{Pm}(u)\},$$

D.-S. Huang et al. (Eds.): ICIC 2012, LNCS 7389, pp. 198–205, 2012.
© Springer-Verlag Berlin Heidelberg 2012

$$N^{\pm}{}_{Pm}(u) = N^{+}{}_{Pm}(u) \cup N^{-}{}_{Pm}(u),$$

where subscripts are expressed as integers modulo m.

A path in a graph G that contains every vertex of G is called a Hamiltonian path. A graph G is said to be Hamiltonian-connected if its any two vertices can be connected by a Hamiltonian path. Hamiltonian path and Hamiltonian-connected are important subjects in interconnection networks.

If no ambiguity can arise we sometimes write $N(u)$ instead of $N_G(u)$, V instead of $V(G)$, etc. We refer to [3][5] for graph theory notation and terminology not described in this paper.

It is well-known that the Hamiltonian graph problem is NP-complete [1]. In 2005, Rahman and Kaykobad [4] established a sufficient condition for Hamiltonian path graphs.

Theorem 1.1(Rahman and Kaykobad [4]). If G is a connected graph with n vertices and $d(u)+d(v)+\delta(u,v) \geq n+1$ for each pair of distinct non-adjacent vertices u,v in G, then G has a Hamiltonian path.

In this present paper, we study further condition $d(u)+d(v)+\delta(u,v) \geq n$, which improves the above condition of Theorem 1.1.

Theorem 1.2. If G is a 2-connected graph with n vertices and $d(u)+d(v)+\delta(u,v) \geq n$ for each pair of distinct non-adjacent vertices u,v in G, then G has a Hamiltonian path or $G \in G_{(n-2)/2} \vee K^{-}{}_{(n+2)/2}$.

Where $G_{(n-2)/2}$ is graphs of order (n-2)/2 and n is even, $K^{-}{}_{(n+2)/2}$ is empty graph of order (n+2)/2. For graphs A and B the join operator "$A \vee B$" of A and B is the graph constructed from A and B by adding all edge joining the vertices of A and the vertices of B.

Theorem 1.3. If G is a 2-connected graph with n vertices and $d(u)+d(v) \geq n-2$ for each pair of distinct non-adjacent vertices u,v in $\delta(u,v)=2$ in G, then G has a Hamiltonian path or $G \in G_{(n-2)/2} \vee K^{-}{}_{(n+2)/2}$.

In [5], we studied the pancyclic with condition $d(u)+d(v)+\delta(u,v) \geq n+1$, and in [6] we studied Hamiltonian-connected, which is also the following topic.

In 2006 Li [3] investigated Hamiltonian-connected with the following Rahman-Kaykobad condition $d(x)+d(y)+\delta(x,y) \geq n+3$.

Theorem 1.4(Li[3]). If G is a 3-connected graph with n vertices and $d(x)+d(y)+\delta(x,y) \geq n+3$ for each pair of nonadjacent vertices x,y in G, then G is Hamiltonian-connected.

In this paper, we also present the following two results, which improve the above Theorem 1.4.

Theorem 1.5. If G is a 2-connected graph of order $n \geq 3$ such that $d(x)+d(y)+\delta(x,y) \geq n+2$ for each pair of nonadjacent vertices x,y in G, then G is Hamiltonian-connected or $G \in G_{n/2} \vee K^{-}{}_{n/2}$.

Theorem 1.6. If G is a 2-connected graph of order $n \geq 3$ such that $d(x)+d(y) \geq n$ for each pair of nonadjacent vertices x,y with $d(x,y)=2$ in G, then G is Hamiltonian-connected or $G \in G_{n/2} \vee K^-_{n/2}$.

2   Proof of Main Results

Clearly, Theorem 1.2 can be obtained easily from Theorem 1.3, so in this section, first, we will prove Theorem 1.3.

Proof of Theorem 1.3. Suppose G satisfies the hypothesis of Theorem 1.3 and contains a longest path $P_m = x_1 x_2 \ldots x_m$ that is not a Hamiltonian path. Let H be a component of $G - P_m$. First, we consider the following two claims.

Claim 1. $d(x_{i+1}, x_{j+1}) = 2$ for each pair $x_{i+1}, x_{j+1} \in N^+_{Pm}(H)$.

Proof of Claim 1. Suppose, to the contrary, that $d(x_{i+1}, x_{j+1}) \neq 2$. Then if $d(x_{i+1}, x_{j+1}) = 1$, let P be a path in H which two end-vertices adjacent to x_i, x_j, respectively. Without loss of generality, assume $i < j$, then $x_1 x_2 \ldots x_i P x_j x_{j-1} \ldots x_{i+1} x_{j+1} x_{j+2} \ldots x_m$ is **a** path longer than P_m, a contradiction. Thus, under the hypothesis $d(x_{i+1}, x_{j+1}) \neq 2$, we have $d(x_{i+1}, x_{j+1}) \geq 3$. Without loss of generality, assume $u \in V(H)$ and $x_i \in N_{Pm}(u)$, since P_m is a longest path, so clearly both u and x_{i+1} do not have any common neighbor in $G - P_m$. Hence we have

$$|N_{G-Pm}(u)| + |N_{G-Pm}(x_{i+1})| \leq |V(G-P_m-u)| \tag{1}$$

Also, since P_m is a longest path, so clearly x_{i+1} is not adjacent to any of $N^+_{Pm}(u)$, thus $|N_{Pm}(x_{i+1})| \leq |V(P_m)| - |N^+_{Pm}(u)|$, and since $d(x_{i+1}, x_{j+1}) \geq 3$ so x_{i+1} is also not adjacent to x_j, x_{j+2} and clearly $x_j, x_{j+2} \notin N^+_{Pm}(u)$. Hence we further have inequality (i) $|N_{Pm}(x_{i+1})| \leq |V(P_m)| - |N^+_{Pm}(u)| - |\{x_j, x_{j+2}\}|$, together with that P_m is a longest path, then u is not adjacent to x_m, so $|N^+_{Pm}(u)| = |N_{Pm}(u)|$, then from inequality (i) $|N_{Pm}(x_{i+1})| \leq |V(P_m)| - |N^+_{Pm}(u)| - |\{x_j, x_{j+2}\}|$, we have

$$|N_{Pm}(u)| + |N_{Pm}(x_{i+1})| \leq |V(P_m)| - |\{x_j, x_{j+2}\}| \tag{2}$$

Combining inequalities (1) and (2), we have

$$d(u)+d(x_{i+1}) \leq |V(G-P_m-u)| + |V(P_m)| - |\{x_j, x_{j+2}\}| = n-3$$

a contradiction. Therefore, the above Claim 1 that $d(x_{i+1}, x_{j+1}) = 2$ holds.

Claim 2. $d(x_{i+1}) + d(x_{j+1}) \leq n - |V(H)|$.

Proof of Claim 2. Without loss of generality, assume $i < j$ and let S_1 denote the path $x_1 x_2 \ldots x_i$ that is a section of P_m and S_2 denote the path $x_{i+1} x_{i+2} \ldots x_j \setminus \{x_j\}$ and S_3 denote the path $x_{j+1} x_{j+2} \ldots x_m$. Since P_m is a longest path of G, so clearly none of $N^-_{S1}(x_{j+1})$ are adjacent to x_{i+1}. (Otherwise, if $x_h \in N^-_{S1}(x_{j+1})$ is adjacent to x_{i+1}. Let P be a path in H which two end-vertices adjacent to x_i, x_j, respectively, then $x_1 x_2 \ldots x_h x_{i+1} x_{i+2} \ldots x_j P x_i x_{i-2} \ldots x_{h+1} x_{j+1} x_{j+2} \ldots x_m$ is a path longer than P_m, a contradiction.). Similarly, none of $N^+_{S2}(x_{j+1})$ are adjacent to x_{i+1}, and none of $N^-_{S3}(x_{j+1})$ are adjacent to x_{i+1}. Also clearly $N^-_{S1}(x_{j+1}) \cap N^+_{S2}(x_{j+1}) = \varnothing$ and $N^+_{S2}(x_{j+1}) \cap N^-_{S3}(x_{j+1}) = \varnothing$, and $x_{i+1} \notin N^-_{S1}(x_{j+1}) \cup N^+_{S2}(x_{j+1}) \cup N^-_{S3}(x_{j+1})$. Hence we have

$|N_{\underline{Pm}}(x_{i+1})| \leq |V(P_m)| - (|N^-_{S1}(x_{j+1})| + |N^+_{S2}(x_{j+1})| + |N^-_{S3}(x_{j+1})|) - |\{x_{i+1}\}| \leq |V(P_m)| - (|N_{Pm}(x_{j+1})| -$
$|\{x_j\}|) - |\{x_{i+1}\}| = |V(P_m)| - |N_{Pm}(x_{j+1})|$, this implies

$$|N_{Pm}(x_{i+1})| + |N_{Pm}(x_{j+1})| \leq |V(P_m)| \qquad (3)$$

Also, both x_{i+1}, x_{j+1} do not have any common neighbor in $G\text{-}P_m\text{-}H$ and both x_{i+1}, x_{j+1} all are not adjacent to any vertex of H, hence

$$|N_{G\text{-}Pm}(x_{i+1})| + |N_{G\text{-}Pm}(x_{j+1})| \leq |V(G\text{-}P_m\text{-}H)| \qquad (4)$$

Combining inequalities (3) and (4), we have

$$d(x_{i+1}) + d(x_{j+1}) = N(x_{i+1}) + |N(x_{j+1})| \leq |V(P_m)| + |V(G\text{-}P_m\text{-}H)| \leq n - |V(H)| \qquad (5)$$

Therefore, the above Claim 2 that $d(x_{i+1}) + d(x_{j+1}) \leq n - |V(H)|$ holds.

Then together with the assumption of Theorem that $d(x_{i+1}) + d(x_{j+1}) \geq n-2$, we have $|V(H)| \leq 2$. Now we consider the following cases on $|V(H)| \leq 2$.

Case 1. $|V(H)| = 2$.

In this case, since P_m is a longest path and $|V(H)| = 2$, so clearly $|\{x_{i+1}, x_{i+2}, \ldots x_{j-1}\}| \geq 2$ for each pairs $x_i \in N_{Pm}(u), x_j \in N_{Pm}(v)$, where $\{u,v\} = V(H)$. Also, since $|V(H)| = 2$, so u and v all are not adjacent to x_1, x_2, x_{m-1}, x_m, thus, we can check $\min\{d(u), d(v)\} \leq (|V(P_m)| - |\{x_1, x_2, x_{m-1}, x_m\}|)/3 + 1 + |V(H-u)| \leq (n-6)/3 + 2 = n/3$. Without loss of generality, assume $d(u) = \min\{d(u), d(v)\}$, then by the assumption of Theorem that $d(u) + d(x_{i+1}) \geq n-2$, we have $d(x_{i+1}) \geq (n-2) - n/3 = 2n/3 - 2$.

(i). If there exist two distinct vertices $x_i, x_j \in N_{Pm}(u)$. Then we have

$$d(x_{i+1}) + d(x_{j+1}) \geq 4n/3 - 4 \qquad (6)$$

When $n \geq 7$, inequality (6) contradicts inequality (5).

When $n \leq 6$. Since u is not adjacent to x_1, x_2, x_{m-1}, x_m, then we can obtain a path that is longer than P_m, a contradiction

(ii). If $\{x_i\} = N_{Pm}(u)$, i.e., $|N_{Pm}(u)| = 1$, so we have $d(u) = 2$. By the assumption of Theorem $d(u) + d(x_{i+1}) \geq n-2$, we have $d(x_{i+1}) \geq n-4$. Since G is 2-connected, so $|N_{Pm}(v)| \geq 1$. If there exist $x_h, x_j \in N_{Pm}(v) \backslash \{x_i\}$, by $d(x_{h+1}) + d(x_{j+1}) \geq n-2$, $\max\{d(x_{h+1}), d(x_{j+1})\} \geq (n-2)/2$, so we can check $d(x_{i+1}) + \max\{d(x_{h+1}), d(x_{j+1})\} > n - |V(H)|$, which contradicts the inequality (5). If there only exist $\{x_j\} = N_{Pm}(v) \backslash \{x_i\}$, then we also can check $d(x_{i+1}) + d(x_{j+1})\} > n - |V(H)|$, which contradicts the inequality (5).

Case 2. $|V(H)| = 1$.

In this case, we claim $d(u) \geq (n-3)/2$. Otherwise, if $d(u) < (n-3)/2$, by that G is 2-connected, let $x_i, x_j \in N_{Pm}(u)$, and by assumption of Theorem that $d(u) + d(x_{i+1}) \geq n-2$ and $d(u) + d(x_{j+1}) \geq n-2$, $d(x_{i+1}) > (n-1)/2$ and $d(x_{j+1}) > (n-1)/2$, so $d(x_{i+1}) + d(x_{j+1}) > n-1$, which contradicts inequality (5).

Thus $d(u) \geq (n-3)/2$ holds.

When n is even. Since $d(u)$ is integer, by $d(u) \geq (n-3)/2$, so $d(u) \geq (n-2)/2$.

When n is odd, we **claim.** $d(u) \geq (n-2)/2$.

Otherwise, suppose to the contrary that $d(u) < (n-2)/2$, together with the above claim $d(u) \geq (n-3)/2$, this implies $d(u) = (n-3)/2$. Consider the following

When $|V(G-P_m)| \geq 2$, let $u, v \in V(G-P_m)$. Since $|V(H)| = 1$ for each component H of $G-P_m$ and by the above consequence that $d(u) = (n-3)/2$ and $d(v) = (n-3)/2$, so both u,v have at least a common neighbor in P_m, this implies $d(u) + d(v) \leq (2n-6)/2 < n-2$, which contradicts the condition of Theorem.

When $|V(G-P_m)| = 1$. (i).If $d(u) \geq 3$, since $d(u) = (n-3)/2$ so there must exist two vertices x_i, x_{i+2} of P_m that are adjacent to u (Otherwise, if there does not exist this case, then we can check $d(u) \leq (|V(P_m)| - |\{x_1, x_m\}|)/3 + 1 = n/3$, which contradicts $d(u) = (n-3)/2$), then by $d(u) = (n-3)/2$, $|\{x_{i+1}, x_{i+2}, \ldots x_{j-1}\}| \leq 2$ for any two consecutive neighbor vertices each pairs $x_i, x_j \in N_{Pm}(u)$, so clearly x_{i+1} is not adjacent to every vertex of $V(P_m) \backslash N_{Pm}(u)$ (Otherwise, we easily obtain a path longer than P_m, a contradiction), so $d(x_{i+1}) \leq |N_{Pm}(u)|$. Hence we have $d(u) + d(x_{i+1}) \leq (n-3)/2 + (n-3)/2 = n-3$, a contradiction. (ii). If $d(u) = 2$. Since $d(u) = (n-3)/2$, so n=7. When $N(u) = \{x_i, x_{i+2}\}$, then clearly $N(x_{i+1}) = \{x_i, x_{i+2}\}$, hence $d(u) + d(x_{i+1}) = 4 \leq n-3$, a contradiction. When $N(u) = \{x_2, x_5\}$, then $x_1 x_3, x_4 x_6 \notin E(G)$ (Otherwise, $x_1 x_3 x_2 u x_5 x_4 x_6$ is Hamiltonian path, a contradiction). Without loss of generality, assume $x_1 x_3 \notin E(G)$, then clearly $d(x_1) \leq 2$, hence $d(u) + d(x_1) \leq n-3$, a contradiction..

Therefore, $d(u) \geq (n-2)/2$ holds for all even and odd n in Case 2.

Then, since P_m is a longest path of G, so u is not adjacent to x_1, x_m and u is at most adjacent to a vertex of $\{x_i, x_{i+1}\}$ for each pair of consecutive vertices x_i, x_{i+1} on P_m. By $d(u) \geq (n-2)/2$, clearly we have $N(u) = \{x_2, x_4, \ldots x_{m-1}\}$. Then, also since P_m is a longest path of G, so clearly, vertex set $\{x_1, x_3, \ldots x_m, u\}$ is a independent set. By the assumption of theorem that $d(u) + d(v) \geq n-2$ for any two $u, v \in \{x_1, x_3, \ldots x_m, u\}$, this implies $d(u) = (n-2)/2$ for each $u \in \{x_1, x_3, \ldots x_m, u\}$. Thus, we easily obtain $G \in G_{(n-2)/2} \vee K^-_{(n+2)/2}$, where $V(G_{(n-2)/2}) = G[\{x_2, x_4, \ldots x_{m-1}\}]$, $K^-_{(n+2)/2} = G[\{x_1, x_3, \ldots x_m, u\}]$, and "$\vee$" is the join operator.

Therefore, this completes the proof of Theorem 1.3.

The Proof of Theorem 1.2. Clearly, if graph G satisfies the condition of Theorem 1.2, then G also satisfies the condition of Theorem 1.3. And clearly $G_{(n-2)/2} \vee K^-_{(n+2)/2}$ satisfies the condition of Theorem 1.2, so Theorem 1.2 can be obtained from Theorem 1.3 immediately.

Similarly, Theorem 1.5 can be obtained easily from Theorem 1.6, so in this following, we only need to prove Theorem 1.6.

Proof of Theorem 1.6. Assume that G is not Hamiltonian-connected with satisfying the condition of Theorem. Then, there exist two distinct vertices x,y such that the longest path $P_m(x,y)$ is not Hamiltonian path. Let H be a component of $G-P_m$. Since G is

3-connected so there exist two vertices $u,v \in V(H)$ satisfying $x_{i+1} \in N^+_{Pm}(u)$ and $x_{j+1} \in N^+_{Pm}(v)$. Then we claim $d(x_{i+1}, x_{j+1}) = 2$. Otherwise, if $d(x_{i+1}, x_{j+1}) \neq 2$, by $P_m(x,y)$ is a longest (x,y)-path, so $x_{i+1}x_{j+1} \notin E(G)$, this implies $d(x_{i+1}, x_{j+1}) \geq 3$. Then also since $P_m(x,y)$ is a longest (x,y)-path, so both u and x_{i+1} do not have any common neighbor in $G-P_m$. Hence we have

$$|N_{G-Pm}(u)| + |N_{G-Pm}(x_{i+1})| \leq |V(G-P_m-u)|. \tag{2}$$

Let $P_{m-1} = P_m - x_m$, so $|N_{Pm}(u)| \leq |N^+_{Pm-1}(u)| + 1$. Clearly x_{i+1} is not adjacent to any of $N^+_{Pm-1}(u)$ and since $d(x_{i+1}, x_{j+1}) \geq 3$ then x_{i+1} is not adjacent to x_j, x_{j+2} and $x_j, x_{j+2} \notin N^+_{Pm}(u)$. Thus, when $x_{j+1} \neq x_m$, we can check $|N_{Pm}(x_{i+1})| \leq |V(P_m)| - (|N^+_{Pm-1}(u)| + |\{x_j, x_{j+2}\}|)$, so we have

$$|N_{Pm}(u)| + |N_{Pm}(x_{i+1})| \leq |N_{Pm}(u)| + (|V(P_m)| - |N^+_{Pm-1}(u)| - |\{x_j, x_{j+2}\}|) \leq |V(P_m)| - 1 \tag{2 *}$$

When $x_{j+1} = x_m$. By P_m is a longest (x,y)-path so u is not adjacent to x_{m-1}, and $d(x_{i+1}, x_{j+1}) \geq 3$ so x_{i+1} is not adjacent to x_m. So similarly, we can check $|N_{Pm}(x_{i+1})| \leq |V(P_m)| - (|N^+_{Pm-1}(u)| + |\{x_j, x_m\}|)$, so we have

$$|N_{Pm}(u)| + |N_{Pm}(x_{i+1})| \leq |N_{Pm}(u)| + (|V(P_m)| - |N^+_{Pm}(u)| - |\{x_j, x_m\}|) \leq |V(P_m)| - 1 \tag{2**}$$

Combining inequality (1) and one of (2)*, (2)**, we have $d(u) + d(x_{i+1}) \leq |V(G-P_m-u)| + |V(P_m)| - 1 = n-2$, a contradiction. Thus, $d(x_{i+1}, x_{j+1}) = 2$ holds.

Then, we let path $x_1x_2 \ldots x_i = P_1$, path $x_{i+1}x_{i+2} \ldots x_j \setminus \{x_j\} = P_2$ and $x_{j+1}x_{j+2} \ldots x_m = P_3$. Since $P_m(x,y)$ is a longest (x,y)-path, so clearly none of $N^-_{P1}(x_{j+1})$ are adjacent to x_{i+1}, none of $N^+_{P2}(x_{j+1})$ are adjacent to x_{i+1}, and none of $N^-_{P3}(x_{j+1})$ are adjacent to x_{i+1}, and x_{i+1} is not adjacent to itself, with $|N^-_{P1}(x_{j+1})| + |N^+_{P2}(x_{j+1})| + |N^-_{P3}(x_{j+1})| \geq |N_{Pm}(x_{j+1})| - |\{x_j\}| - |\{x_1\}|$, hence we have

$|N_{Pm}(i+1)| \leq |(m - (|N^-_{P1}(x_{j+1})| - |N^+_{P2}(x_{j+1})| - |N^-_{P3}(x_{j+1})|) - |\{x_{i+1}\}|)| \leq m - (|N_{Pm}(x_{j+1})| - |\{x_j\}| - |\{x_1\}|) - |\{x_{i+1}\}| = m - |N_{Pm}(x_{j+1})| + 1$, then, by plus $|N_{Pm}(x_{j+1})|$ in two side of the inequality, we have

$$|N_{Pm}(i+1)| + |N_{Pm}(x_{j+1})| \leq (m - |N_{Pm}(x_{j+1})|) + |N_{Pm}(x_{j+1})| = m+1 \tag{3}$$

Also, by $P_m(x,y)$ is a longest (x,y)-path, so both x_{i+1}, x_{j+1} do not have any common neighbor vertex in $G-P_m-H$, and x_{i+1}, x_{j+1} are not adjacent to any vertex of H. Hence we can check

$$|N_{G-Pm}(i+1)| + |N_{G-Pm}(x_{j+1})| \leq |V(G-P_m-H)| \tag{4}$$

Combining (3) and (4), we have

$$d(x_{i+1})+d(x_{j+1})\leq n-|V(H)|+1 \tag{5}$$

Since $d(x_{i+1})+d(x_{j+1})\geq n$, this implies $|V(H)|\leq 1$. Then we consider the following cases.

In this case, we claim $d(u)>(n-1)/2$ (Otherwise, if $d(u)\leq(n-1)/2$, by condition of Theorem that $d(u)+d(x_{i+1})\geq n$, we have $d(x_{i+1})\geq(n+1)/2$ and $d(x_{j+1})\geq(n+1)/2$, then we have $d(x_{i+1})+d(x_{j+1})\geq n+1$, this contradicts the inequality (5) that $d(x_{i+1})+d(x_{j+1})\leq n-|V(H)|+1)$.

We also claim that there exist at most two components $H=\{u\}$ and $R=\{v\}$ in $G-P_m$. Otherwise, if there at least exist three components in $G-P_m$ and let H, R, T are three components of $G-P_m$, by Case 2, we let $H=\{u\}$, $R=\{v\}$ and $T=\{w\}$, since G is 3-connected, we may let $x_{i+1},x_{j+1}\in N^+_{Pm}(H)$. Since $P_m(x,y)$ is a longest (x,y)-path, then x_{i+1} or x_{j+1} is not adjacent to vertex $V(R)$(Otherwise, if x_{i+1} and x_{j+1} are all adjacent to vertex $V(R)$, then we obtain a (x,y)-path $x_1x_2...x_iux_jx_{j-1}...x_{i+1}vx_{j+1}x_{j+2}...x_m$ is a longer (x,y) path, a contradiction). Without loss of generality, assume x_{i+1} is not adjacent to vertex $V(R)$. Since $|V(H)|=1$ for each component H of $G-P_m$, we claim that vertex $V(H)=\{u\}$ and $V(R)=\{v\}$ must be adjacent to x_1 and x_m(Otherwise, for example, if u is not adjacent to x_m, then we have $|N^+_{Pm}(u)|=|N_{Pm}(u)|$, $d(u)=|N_{Pm}(u)|$ and clearly $d(x_{i+1})\leq|V(P_m)|-|N^+_{Pm}(u)|+|V(G-P_m-H-T)|$. Hence we have $d(u)+d(x_{i+1})\leq|N_{Pm}(u)|+(|V(P_m)|-|N^+_{Pm}(u)|+|V(G-P_m-H-T)|)=n-2$, a contradiction. Similarly, we can prove that u is adjacent to x_1, and $V(R)=\{v\}$ is adjacent to x_1 and x_m). Thus, $d(u,v)=2$, then clearly $d(u)\leq(m+1)/2$ and $d(v)\leq(m+1)/2$, so we have $d(u)+d(v)\leq m+1\leq n-2$, a contradiction. The contradiction shows that there at most exist two components in $G-P_m$. By $d(u)\geq(n-1)/2$ then we consider the following two cases.

In this case there only exist a components $H=\{u\}$ in $G-P_m$(Otherwise, if there exist two components $H=\{u\}$ and $R=\{v\}$ in $G-P_m$, then m $\leq n-2$, so there must exist two x_i and x_{i+1} that are adjacent u, so we obtain a (x,y)-path longer than $P_m(x,y)$, a contradiction)

By $d(u)>(n-1)/2$ and $P_m(x,y)$ is a longest (x,y)-path, we have $d(u)=n/2$ and $N(u)=\{x_1,x_3,...,x_{2r-1},x_{2r+1},...x_m\}$, and clearly $\{x_2,x_4,...,x_{2r}, x_{2r+2},...x_{m-1}\} \cup \{u\}$ is an independent set. Thus, we have $G\in G_{n/2}\vee K^-_{n/2}$.

3   Conclusion

Recently, Hasan, Kaykobad, Lee et al.[2] presented a detailed analysis of the recent works on Hamiltonian graphs under Rahman-Kaykobad conditions. They showed each classes of exceptional graphs under each distances of graphs, Rahman-Kaykobad condition $(d(u)+d(v)+\delta(u,v)\geq n+1)$ is sufficient to make a graph Hamiltonian. In this present paper, we study further condition $(d(u)+d(v)+\delta(u,v)\geq n)$ for the existence of Hamiltonian paths. Naturally, it is interesting that there are what classes of non-Hamiltonian graphs under the condition $d(u)+d(v)+\delta(u,v)\geq n$.

Acknowledgements. The authors are very grateful to the anonymous reviewer for his very helpful remarks and comments.The work of the first author was supported by the NSF of Hainan Province (no. 10501).

References

1. Garey, M.R., Johnson, D.S.: Computers and Intractability: A Guide to the Theory of NP-Completeness. W. H. Freeman and Company, New York (1979)
2. Kamrul, M., Hasan, M., Kaykobad, Y.K., Lee, S.Y.: Note: A Comprehensive Analysis of Degree Based Condition for Hamiltonian Cycles. Theoretical Computer Science 411(1), 285–287 (2010)
3. Li, R.: A New Sufficient Condition for Hamiltonicity of Graphs. Information Processing Letters 98(4), 159–161 (2006)
4. Rahman, M., Kaykobad, M.: On Hamiltonian Cycles and Hamiltonian Paths. Inform. Process. Lett. 94(1), 37–41 (2005)
5. Zhao, K.W., Lin, Y., Zhang, P.: A Sufficient Condition for Pancyclic Graphs. Information Processing Letters 109(16), 991–996 (2009)
6. Zhao, K.W., Lao, H.J., Zhou, J.: Hamiltonian-connected Graphs. Computers Math. 55(12), 2707–2714 (2008)

Note on the Minimal Energy Ordering
of Conjugated Trees[*]

Yulan Xiao[1] and Bofeng Huo[1,2]

[1] Department of Mathematics, Qinghai Normal University
[2] Key Lab of Tibetan Information Processing (Qinghai Normal University),
Ministry of Education and Qinghai Province
Xining 810008, P.R. China
hbf@qhnu.edu.cn

Abstract. For a simple graph G, the energy $E(G)$ is defined as the sum of the absolute values of all eigenvalues of its adjacency matrix $A(G)$. Gutman proposed two conjectures on the minimal energy of the class of conjugated trees (trees having a perfect matching). Zhang and Li determined the trees in the class with the minimal and second-minimal energies, which confirms the conjectures. Zhang and Li also found that the conjugated tree with the third-minimal energy is one of the two graphs which are quasi-order incomparable. Recently, Huo, Li and Shi found there exists a fixed positive integer N_0, such that for all $n > N_0$, the energy of the graphs with the third-minimal through the sixth-minimal are determined. In this paper, the N_0 is fixed by a recursive method, and the problem is solved completely.

Keywords. extremal graph, minimal energy, quasi-order incomparable, conjugated tree.

1   Introduction

For a given simple graph G of order n, denote by $A(G)$ the adjacency matrix of G. The characteristic polynomial of $A(G)$ is usually called the characteristic polynomial of G. It is well-known [3] that the characteristic polynomial of a tree T can be expressed as

$$\phi(T,x) = \sum_{k=0}^{\lfloor \frac{n}{2} \rfloor} (-1)^k m(T,k) x^{n-2k}$$

where $m(T,k)$ denotes the number of the k-matchings of T.

The energy is one graph parameter stemming from the Hückel molecular orbital (HMO) approximation for the total π-electron energy, see [9]. If $\lambda_1, \lambda_2, \ldots, \lambda_n$ denote the eigenvalues of adjacency matrix $A(G)$, the energy of G is defined as

---

[*] Supported by NSFC and Youth Innovation foundation of Qinghai Normal University.

D.-S. Huang et al. (Eds.): ICIC 2012, LNCS 7389, pp. 206–213, 2012.
© Springer-Verlag Berlin Heidelberg 2012

$$E(G) = \sum_{i=1}^{n} \lambda_i$$

Coulson [2] obtained

$$E(G) = \frac{1}{\pi} \int_{-\infty}^{+\infty} \left[n - \frac{ix\phi'(G,ix)}{\phi(G,x)} \right] dx . \tag{1}$$

In particular the energy of a tree T [6] is

$$E(T) = \frac{2}{\pi} \int_{0}^{+\infty} x^{-2} \log \left[1 + \sum_{k=1}^{\lfloor \frac{n}{2} \rfloor} m(T,k) x^{2k} \right] dx ,$$

where $m(T,k)$ is the number of the k-matchings of T. Now we show one well-known result due to Gutman [9].

Lemma 1. If G_1 and G_2 are two graphs with the same number of vertices, then

$$E(G_1) - E(G_2) = \frac{1}{\pi} \int_{-\infty}^{+\infty} \log \frac{\phi(G_1,ix)}{\phi(G_2,ix)} dx .$$

If T_1 and T_2 are two trees with the same number of vertices, it is clear that $E(T_1) \le E(T_2)$ if $m(T_1,k) \le m(T_2,k)$ for all $k = 1,\ldots,\lfloor \frac{n}{2} \rfloor$. So there exists a quasi-order $\prec$ in the set of trees with the same order. For two trees T_1 and T_1 with n vertices, if $m(T,k) \le m(T_2,k)$ holds for all $k \ge 0$, then we define $T_1 \prec T_2$. Thus $T_1 \prec T_2$ implies $E(T_1) \le E(T_2)$ [5, 7, 18]. Similarly, a quasi-order can be defined in bipartite graphs [16] and unicyclic graphs[11], these relations have been established for numerous pairs of graphs [4, 5, 8, 11, 12, 14–19]. We will compare the energy of trees of order n having a perfect matching, and determine the third- , forth-, fifth- and sixth-minimal energy in this class. For more results on graph energy, we refer to [9, 10], and for terminology and notation not defined here, we refer to Bondy and Murty [1].

2   Preliminaries

Denote by X_n the star $K_{1,n-1}$, Y_n the tree obtained by attaching a pendent edge to a pendent vertex of the star $K_{1,n-2}$, Z_n by attaching two pendent edges to a pendent vertex of $K_{1,n-3}$ and W_n by attaching a P_3 to a pendent vertex of $K_{1,n-3}$. In [5], Gutman gave the following

Lemma 2. For any tree T of order n, if $T \ne X_n, Y_n, Z_n, W_n$, then

$$X_n \prec Y_n \prec Z_n \prec W_n \prec T .$$

Denoted by Φ_n the class of trees with n vertices which have a perfect matching. For the minimal energy tree in Φ_n, Gutman proposed two conjectures in [8]. Zhang and Li [17] confirmed that both conjectures are true by using quasi-ordering relation $\prec$.

Lemma 3. [17]   In the class Φ_n , $E(T)$ is minimal for the graph F_n , and $E(T) = E(F_n)$ if and only if $T = F_n$, where F_n is obtained by attaching a pendent edge to each vertex of the star $K_{1,\frac{n}{2}-1}$.

For the trees of the second-, third-, and forth-minimal energies in Φ_n , they obtained the following

Lemma 4. [17] In the class Φ_n , the graph attained the second-minimal energy is B_n , where B_n is the graph obtained from F_{n-2} by attaching a P_3 to the 2-degree vertex of a pendent edge, and $E(T) = E(B_n)$ if and only if $T = B_n$.

Lemma 5. [17] In the class Φ_n , the graphs attained the third- and forth-minimal energy are in $\{ L_n, M_n \}$, where L_n is the graph obtained from F_{n-4} by attaching two P_3's to the 2-degree vertex of a pendent edge, and M_n is obtained from F_{n-2} by attaching a P_3 to a 1-degree vertex to form a path of length 6. Furthermore, L_n and M_n are quasi-order incomparable. Denote by I_n the graph obtained by attaching a P_3 to a 1-degree vertex of X_n and attaching a pendent edge to each other 1-degree vertex of $X_{\frac{n}{2}-1}$; $W_{\frac{n}{2}}^*$ the graph obtained by attaching a pendent edge to each vertex of $W_{\frac{n}{2}}$.

Li and Li [15] proved the following results.

Lemma 6.   In the class Φ_n , the graph of the third-, forth-, fifth- and sixth-minimal energy are in $\{ L_n, M_n, I_n, W_{\frac{n}{2}}^* \}$. Furthermore, $M_n \prec I_n$ and $L_n \prec W_{\frac{n}{2}}^*$, but I_n and L_n are quasi-order incomparable.

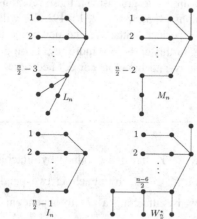

3   Main Results

In [13], Huo et al. gave the following consequence.

Theorem 7.   There exists a fixed positive integer N_0, for all $n > N_0$, $E(I_n) < E(L_n)$.

Corollary 8. There exists a fixed positive integer N_0, for all $n > N_0$,

$$E(M_n) < E(I_n) < E(L_n) < E(W^*_{\frac{n}{2}}).$$

In the following theorem, by a recursive method, we will give the accurate value of N_0 in Theorem 7 and Corollary 8.

Theorem 9. For all even integers $n \geq 84$, $E(M_n) < E(I_n) < E(L_n) < E(W^*_{\frac{n}{2}})$.

Proof. By using Lemma 1, it is easy to get

$$E(I_n) - E(L_n) = \frac{2}{\pi} \int_0^{\infty} \log \frac{x^8 + (\frac{n}{2} + 3)x^6 + (\frac{3}{2}n + 2)x^4 + (n+1)x^2 + 1}{x^8 + (\frac{n}{2} + 3)x^6 + (2n - 4)x^4 + (\frac{n}{2} + 3)x^2 + 1} dx.$$

In the proof of Theorem 7, the uniformly convergence of the sequence $g(n) = E(I_n) - E(L_n)$ has been proved. Denote by $f(x, n)$ the integrand. Since for all $X \geq -1$,

$$\frac{X}{1 + X} \leq \log(1 + X) \leq X, \tag{2}$$

we can express $f(x, n)$ as $f(x, n) = 1 + \dfrac{(-\frac{n}{2} + 6)x^4 + (\frac{n}{2} - 2)x^2}{x^8 + (\frac{n}{2} + 3)x^6 + (2n - 4)x^4 + (\frac{n}{2} + 3)x^2 + 1}$.

For all real number x and all positive integer n,

$$f(x, n) \leq \frac{(-\frac{n}{2} + 6)x^4 + (\frac{n}{2} - 2)x^2}{x^8 + (\frac{n}{2} + 3)x^6 + (2n - 4)x^4 + (\frac{n}{2} + 3)x^2 + 1},$$

$$f(x, n) \geq \frac{(-\frac{n}{2} + 6)x^4 + (\frac{n}{2} - 2)x^2}{x^8 + (\frac{n}{2} + 3)x^6 + (\frac{3}{2}n + 2)x^4 + (n+1)x^2 + 1}.$$

It follows that

$$|f(x, n)| \leq \frac{(-\frac{n}{2} + 6)x^4 + (\frac{n}{2} - 2)x^2}{x^8 + (\frac{n}{2} + 3)x^6 + (2n - 4)x^4 + (\frac{n}{2} + 3)x^2 + 1}, \text{if } |x| \leq \sqrt{\frac{n - 4}{n - 12}},$$

$$|f(x, n)| \geq \frac{(-\frac{n}{2} + 6)x^4 + (\frac{n}{2} - 2)x^2}{x^8 + (\frac{n}{2} + 3)x^6 + (\frac{3}{2}n + 2)x^4 + (n+1)x^2 + 1}, \text{if } |x| \geq \sqrt{\frac{n - 4}{n - 12}}.$$

Notice that $f(x, n)$ is a pointwise convergent sequence, since $\lim_{n \to \infty} f(x, n)$ exists, and is a piecewise continuous function

$$\varphi(x) = \begin{cases} \log \dfrac{\frac{1}{2}x^6 + \frac{3}{2}x^4 + x^2}{\frac{1}{2}x^6 + 2x^4 + \frac{1}{2}x^2} & x \neq 0 \\ 0 & x = 0 \end{cases}.$$

Theorem 7 has proved that $f(x, n)$ converge uniformly to $\varphi(x)$ in a interval $I = [\delta, +\infty] \subset [0, +\infty]$ and δ will be introduced later. For $x \neq 0$,

$$f(x, n) - \varphi(x) =$$

$$\log\left[1 + \frac{x^{10} + 8x^8 + 15x^6 + 5x^4 - 4x^2 - 1}{x^{12} + (\frac{n}{2} + 6)x^{10} + (\frac{7}{2}n + 7)x^8 + (\frac{15}{2}n - 3)x^6 + (\frac{11}{2}n + 2)x^4 + (n+9)x^2 + 2}\right].$$

If $x^{10} + 8x^8 + 15x^6 + 5x^4 - 4x^2 - 1 \geq 0$,

$$|f(x,n) - \varphi(x)| \leq \frac{x^{10} + 8x^8 + 15x^6 + 5x^4 - 4x^2 - 1}{x^{12} + (\frac{n}{2} + 6)x^{10} + (\frac{7}{2}n + 7)x^8 + (\frac{15}{2}n - 3)x^6 + (\frac{11}{2}n + 2)x^4 + (n+9)x^2 + 2};$$

if $x^{10} + 8x^8 + 15x^6 + 5x^4 - 4x^2 - 1 \leq 0$,

$$|f(x,n) - \varphi(x)| \leq \frac{-(x^{10} + 8x^8 + 15x^6 + 5x^4 - 4x^2 - 1)}{x^{12} + (\frac{n}{2} + 7)x^{10} + (\frac{7}{2}n + 15)x^8 + (\frac{15}{2}n + 12)x^6 + (\frac{11}{2}n + 7)x^4 + (n+5)x^2 + 1}.$$

It is not hard to verify that $x^{10} + 8x^8 + 15x^6 + 5x^4 - 4x^2 - 1 \geq 0$, for $n \geq 3$ and any $x \in [\delta, +\infty]$, where δ is the only positive root of polynomial $x^{10} + 8x^8 + 15x^6 + 5x^4 - 4x^2 - 1$ and $\delta \in (0.6750, 0.6751)$. For $x \in [0, \delta]$, we have $f(x,n) \leq \varphi(x)$. Therefore, in the interval $[\delta, +\infty]$,

$$\lim_{n \to +\infty} \int_\delta^{+\infty} f(x,n)dx = \int_\delta^{+\infty} \lim_{n \to +\infty} f(x,n)dx = \int_\delta^{+\infty} \varphi(x)dx.$$

That is, for arbitrarily small $\varepsilon > 0$, there exists positive integer N, such that for $n > N$,

$$\int_\delta^{+\infty} \varphi(x)dx - \varepsilon < \int_\delta^{+\infty} f(x,n)dx < \int_\delta^{+\infty} \varphi(x)dx + \varepsilon.$$

Consequently,

$$\int_0^{+\infty} f(x,n)dx = \int_0^\delta f(x,n)dx + \int_\delta^{+\infty} f(x,n)dx$$

$$< \int_0^\delta \varphi(x)dx + \int_\delta^{+\infty} \varphi(x)dx + \varepsilon = \int_0^{+\infty} \varphi(x)dx + \varepsilon.$$

Since $\int_0^{+\infty} \varphi(x)dx = \alpha \approx -0.118023$, if take $\varepsilon = |\alpha|$, there exists N_0 such that for $n > N_0$,

$$\int_0^{+\infty} f(x,n)dx < \int_0^{+\infty} \varphi(x)dx + \varepsilon = 0.$$

For the integrand f(x, n), we have

$$f(x,n+1) - f(x,n) = \log\left(1 - \frac{h(x)}{g(x,n)}\right),$$

where $h(x) = \frac{1}{2}x^2(x^{10} + 8x^8 + 15x^6 + 5x^4 - 4x^2 - 1)$ and

$$g(x,n) = x^{16} + (n + \frac{13}{2})x^{14} + (\frac{1}{4}n^2 + \frac{27}{4}n + \frac{21}{2})x^{12} + (\frac{7}{4}n^2 + \frac{51}{4}n + \frac{11}{2})x^{10}$$
$$+ (\frac{15}{4}n^2 + \frac{33}{4}n + 12)x^8 + (\frac{11}{4}n^2 + \frac{29}{4}n + \frac{23}{2})x^6$$
$$+ (\frac{1}{2}n^2 + \frac{15}{2}n + \frac{7}{2})x^4 + (\frac{3}{2}n + \frac{7}{2})x^2 + 1.$$

As the only positive root of the polynomial $x^{10} + 8x^8 + 15x^6 + 5x^4 - 4x^2 - 1$, the number δ is also the only positive root of $h(x)$. Thus for $x \in [0, \delta]$,

$$f(x,n) \leq f(x,n+1) \leq \varphi(x) \tag{3}$$

and for $x \in [\delta, +\infty]$,

$$\varphi(x) \leq f(x, n+1) \leq f(x, n). \tag{4}$$

Naturally, it follows that for $x \in [0, \delta]$, $f(x, n)$ is monotonically increasing on n while for $x \in [\delta, +\infty]$, $f(x, n)$ is monotonically decreasing on n

Notice that $f(x, n) > 0$ for $x \in \left(0, \sqrt{\frac{n-4}{n-12}}\right)$ and $f(x, n) < 0$ for $x \in \left(\sqrt{\frac{n-4}{n-12}}, +\infty\right)$.

Moreover, $\delta \in (0.6750, 0.6751)$ and $\sqrt{\frac{n-4}{n-12}} \in [1, 3]$ for $n \geq 13$. Let $d_1 = 0.6750$, $d_2 = 0.6751$, we deduce that for any $n \geq 13$, $f(d_1, n)$, $f(d_2, n)$ and $f(\delta, n)$ are all positive. According to the geometrical interpretation on the integral of real-value function, the following inequalities are plain.

$$\int_{d_2}^{+\infty} f(x, n) dx < \int_{\delta}^{+\infty} f(x, n) dx < \int_{d_1}^{+\infty} f(x, n) dx, \quad (n \geq 13) \tag{5}$$

Similarly, since $\varphi(x) > 0$ for $x \in [0, 1)$ and $\varphi(x) < 0$ for $x \in (1, +\infty)$, according to the geometrical interpretation on the integral of real-value function, we have

$$\int_0^{d_1} \varphi(x) dx < \int_0^{\delta} \varphi(x) dx < \int_0^{d_2} \varphi(x) dx. \tag{6}$$

Let $N_1 \geq 13$ be a positive integer, by inequalities from (3) to (6), the following inequalities hold for $n \geq N_1$.

$$\int_0^{+\infty} f(x, n) dx = \int_0^{\delta} f(x, n) dx + \int_{\delta}^{+\infty} f(x, n) dx$$

$$\leq \int_0^{\delta} \varphi(x) dx + \int_{\delta}^{+\infty} f(x, N_1) dx < \int_0^{d_2} \varphi(x) dx + \int_{d_1}^{+\infty} f(x, N_1) dx.$$

We find the critical point of N_1 at which the expression 2 switches from negative to positive. The point is $N_1 = 377$. In fact, by running a computer with Maple program, we get

$$\int_0^{d_2} \varphi(x) dx + \int_{d_1}^{+\infty} f(x, 377) dx = -0.0001 \tag{7}$$

It implies that for $n \geq 377$, $\int_0^{+\infty} f(x, n) dx < 0$.

For $94 \leq n \leq 377$, we use a recursive method that evaluate the integral piecewise. Concretely, let n_1 and n_2 be positive integers with $n_i \geq 13$, $i = 1, 2$. By means of the inequalities from (3) to (6), for positive integer n with $n_1 \leq n \leq n_2$, we have

$$\int_0^{+\infty} f(x, n) dx = \int_0^{\delta} f(x, n) dx + \int_{\delta}^{+\infty} f(x, n) dx$$

$$\leq \int_0^{\delta} f(x, n_2) dx + \int_{\delta}^{+\infty} f(x, n_1) dx < \int_0^{d_2} f(x, n_2) dx + \int_{d_1}^{+\infty} f(x, n_1) dx. \tag{8}$$

By running a computer with Maple program, we get

$$\int_0^{d_2} f(x,377)dx + \int_{d_1}^{+\infty} f(x,152)dx \approx -0.0004;$$

$$\int_0^{d_2} f(x,152)dx + \int_{d_1}^{+\infty} f(x,108)dx \approx -0.0001; \qquad (9)$$

$$\int_0^{d_2} f(x,108)dx + \int_{d_1}^{+\infty} f(x,94)dx \approx -0.0004;$$

According to inequalities (8) and (9), we deduce $\int_0^{+\infty} f(x,n)dx < 0$ for $94 \leq n \leq 377$.
It implies $E(I_n) < E(L_n)$ for $94 \leq n \leq 377$. For $8 \leq n \leq 94$, we give Table 1 which
is also attained by employing Maple program. From the table one can see that
$E(I_n) > E(L_n)$ for the even integers n from $n = 8$ to 82, while $E(I_n) < E(L_n)$
for even integers n from $n = 84$ to 94. The theorem is a summary of the discussion above, in which the conjugated trees with the third-minimal through the sixth-minimal energies for $n \geq 84$ are determined.

Table 1. Values of $E(I_n) - E(L_n)$ for even integers n with $8 \leq n \leq 94$

n	$E(I_n) - E(L_n)$	n	$E(I_n) - E(L_n)$
$n=8$	0.154368	$n=10$	0.133281
$n=12$	0.117005	$n=14$	0.103962
$n=16$	0.093213	$n=18$	0.084161
$n=20$	0.076404	$n=22$	0.069664
$n=24$	0.063737	$n=26$	0.058475
$n=28$	0.053762	$n=30$	0.049510
$n=32$	0.045649	$n=34$	0.042123
$n=36$	0.038887	$n=38$	0.035904
$n=40$	0.033143	$n=42$	0.030577
$n=44$	0.028185	$n=46$	0.025949
$n=48$	0.023852	$n=50$	0.021881
$n=52$	0.020023	$n=54$	0.018269
$n=56$	0.016609	$n=58$	0.015035
$n=60$	0.013541	$n=62$	0.012118
$n=64$	0.010763	$n=66$	0.009470
$n=68$	0.008235	$n=70$	0.007052
$n=72$	0.005920	$n=74$	0.004834
$n=76$	0.003791	$n=78$	0.002788
$n=80$	0.001824	$n=82$	0.000895
$n=84$	-0.6×10^{-8}	$n=86$	-0.000864
$n=88$	-0.001697	$n=90$	-0.002503
$n=92$	-0.003282	$n=94$	-0.004036

References

1. Bondy, J.A., Murty, U.S.R.: Graph Theory. Springer, Berlin (2008)
2. Coulson, C.A.: On the Calculation of the Energy in Unsaturated Hydrocarbon Molecules. Proc. Cambridge Phil. Soc. 36, 201–203 (1940)
3. Cvetković, D.M., Doob, M., Sachs, H.: Spectra of Graphs-Theory and Application. Academic Press, New York (1980)
4. Chen, A., Chang, A., Shiu, W.C.: Energy Ordering of Unicyclic Graphs. MATCH Commun. Math. Comput. Chem. 55, 95–102 (2006)
5. Gutman, I.: Acylclic Systems with Extremal Hückel π-electron energy. Theor. Chim. Acta 45, 79–87 (1977)
6. Gutman, I., Polansky, O.E.: Mathematical Concepts in Organic Chemistry. Springer, Berlin (1986)
7. Gutman, I., Zhang, F.: On the Ordering of Graphs with Respect to Their Matching Numbers. Discrete Appl. Math. 15, 25–33 (1986)
8. Gutman, I.: Acylclic Conjugated Molecules, Trees and Their Energies. J. Math. Chem. 1, 123–143 (1987)
9. Gutman, I.: The Energy of a Graph: Old and New Results. In: Betten, A., Kohnert, A., Laue, R., Wassermann, A. (eds.) Algebraic Combinatorics and Applications, pp. 196–211. Springer, Berlin (2001)
10. Gutman, I., Li, X., Zhang, J.: Graph Energy. In: Dehmer, M., Emmert-Streib, F. (eds.) Analysis of Complex Networks: From Biology to Linguistics, pp. 145–174. Wiley-VCH Verlag, Weinheim (2009)
11. Hou, Y.: Unicyclic Graphs with Minimal Energy. J. Math. Chem. 29, 163–168 (2001)
12. Hou, Y.: On Trees with the Least Energy and a Given Size of Matching. J. Syst. Sci. Math. Sci. 23, 491–494 (2003)
13. Huo, B., Li, X., Shi, Y., Wang, L.: Determining the Conjugated Trees with the Third-through the Sixth-minimal Energies. MATCH Commun. Math. Comput. Chem. 65, 521–532 (2011)
14. Lin, W., Guo, X., Li, H.: On the Extremal Energies of Trees with a Given Maximum Degree. MATCH Commun. Math. Comput. Chem. 54, 363–378 (2005)
15. Li, S., Li, N.: On Minimal Energies of Acyclic Conjugated Molecules. MATCH Commun. Math. Comput. Chem 61, 341–349 (2009)
16. Li, X., Zhang, J., Wang, L.: On Bipartite Graphs with Minimal Energy. Discrete Appl. Math. 157, 869–873 (2009)
17. Zhang, F., Li, H.: On acyclic Conjugated Molecules with Minimal Energies. Discrete Appl. Math. 92, 71–84 (1999)
18. Zhang, F., Lai, Z.: Three Theorems of Comparison of Trees by Their Energy. Science Exploration 3, 12–19 (1983)
19. Zhou, B., Li, F.: On Minimal Energies of Trees of a Prescribed Diameter. J. Math. Chem. 39, 465–473 (2006)

An Ensemble Method Based on Confidence Probability for Multi-domain Sentiment Classification

Quan Zhou, Yuhong Zhang, and Xuegang Hu

School of Computer & Information Hefei University of Technology
Hefei, 230009, Anhui, China
yuan_zhouquan@163.com,
{zhangyh,jsjxhuxg}@hfut.edu.cn

Abstract. Multi-domain sentiment classification methods based on ensemble decision attracts more and more attention. These methods avoid collecting a large amount of new training data in target domain and expand aspect of deploying source domain systems. However, these methods face some important issues: the quantity of incorrect pre-labeled data remains high and the fixed weights limit accuracy of the ensemble classifier. Thus, we propose a novel method, named CEC, which integrates the ideas of self-training and co-training into multi-domain sentiment classification. Classification confidence is used to pre-label the data in the target domain. Meanwhile, CEC combines the base classifiers according to classification confidence probabilities when taking a vote for prediction. The experiments show the accuracy of the proposed algorithm has highly improved compared with the baseline algorithms.

Keywords: ensemble, multi-domain sentiment classification, co-training.

1 Introduction

Nowadays, expressing ones' ideas and feelings on the Web has become a trend. Product reviews, blogs and forum posts are everywhere. How to identify ones' emotion expressed in these texts is of growing importance for practical applications (e.g. e-commerce, business intelligence, information monitoring, et al). Thus, there are increasingly interests in the study of sentiment classification.

However, with the differences among domains, the ways to show affection are rich and varied. This affects and limits sentimental classification. For example, "durable and delicate" is often used to express positive sentiment for electronics review, but, hardly used for movie remark. When a traditional classification scenario is carried over into a new sentimental domain, a mass of labeled training data in the new domain are needed. Obviously, it is costly and time-consuming. Thus, multi-domain sentiment classification is proposed. Matthew Whitehead et al. [1] and Li [2] proposed multi-domain sentiment classification methods based on the ensemble classification, using the labeled data in source domains. Nevertheless, there are some issues needing to be promoted: The quantity of incorrect pre-labeled data remains high and the fixed weights limit accuracy of the ensemble classifier.

D.-S. Huang et al. (Eds.): ICIC 2012, LNCS 7389, pp. 214–220, 2012.
© Springer-Verlag Berlin Heidelberg 2012

To address these issues, we introduce the classification confidence into the ensemble strategy, and propose CEC (Confident Ensemble Classifier), in which, classification confidence is used to pre-label data in the target domain and to weight the voting of ensemble classification. It successfully decreases the differences between the source domains and the target domain and gets a more accurate classifier for the target domain.

The rest of this paper is organized as follows: Section 2 describes the proposed algorithm. Section 3 presents experimental results. Finally, section 4 concludes this paper.

2   An Ensemble Method Based on Confidence Probability

This paper attempts to address the problem of sentiment classification from multiple sources. Assume that there are k source domains, in which data are labeled and each domain is denoted as S_i. Let T be the target domain and L_i be the data pre-labeled from S_i. Labels are denoted as $y \in \{1, -1\}$, where "1" represents a positive review and "-1" represents a negative review. Our objective is to maximize the accuracy of assigning a label in y to the data in T utilizing the training data S_i.

2.1   The CEC Description

The formal description of the proposed algorithm is as follows:

Input: source domains $S_1, S_2, ...S_k$, target domain T, confidence proportion r, Maximum number of loop m

Output: $TL=\{ (x_1, y_1), (x_2, y_2), ..., (x_n, ..., y_n) \}$

1) Set the initial number of iteration $t=1$, $T_i=\{\}$, where $i \in \{1, 2, ..., k\}$.
2) Train MaxEnt classifier on $S_i \cup T_i$ and predict T to get $p_i(y/x)$ respectively.
3) If $t=1$
   Sort all the $p_i(y/x)$ by descending order and choose the α^{-th} ones as the threshold, where $\alpha = |r*n|$.
4) Add instance x to T_i, if it satisfies
   $P_j(y|x) \geq threshlod$, where $j=1, ..., i-1, i+1, ..., k$ and $i \in \{1, 2, ..., k\}$.
5) If $avg(p_i(y|x))$ decreases or $t==m$
   go to 6);
   else go to 2);

6) Predict each instance in the target domain T according to the

Formula as follows: $p(y|x) = \dfrac{1}{k}\sum_{i=1}^{k} p_i(y|x)$

---

1) The first step is initialization, where t is the initial number of loop. 2) Get base classifiers with Maxent model. 3) Confidence probability threshold is determined according to the parameter r. 4) Pick out the pre-labeled data whose confidence probability is higher than threshold and add them to the training sets of other base

classifiers. 5) Judge whether the average of $p_i(y|x)$ declines. If the average falls down, turn to 6). Otherwise, go to 2) and enter the next loop. 6) Predict each data in the target domain T, through combining the base classifiers.

2.2   The Pre-labeling Method Based on Confidence Probability

Avrim[3], Cardie [4] et al. tended to set the number of pre-labeled data in advance for each base classifier. However, because of the different data distributions, most base classifiers can not achieve good classification performance [1]. Setting the number of pre-labeled data blindly can not guarantee pre-labeled data with high confidence probabilities. Meanwhile, as the similarities between source and target domains are different, setting uniform number is bad for the more similar sources and setting specific number for each source domain is difficult.

To solve these problems, unlike traditional methods, we combine and sort the classification results of base classifiers and get the threshold of classification confidence probability utilizing r and n, where r is a parameter and n is the number of data in the target domain. The threshold is calculated as follows:

$$\alpha = |r*n|$$

Only the data, of which classification confidence probability is above the α^{-th} top confidence probability, can be pre-labeled and added into the training data sets of other base classifiers. This strategy ensures the accuracy of pre-labeling data and allows the source domain which is more similar to the target domain to pre-label more data. Thus, it can reduce the interferences caused by less similar source domains.

2.3   The Number of Iteration

Avrim[3], Stephen[5] et al. utilized a simple method, setting the maximum number of iteration, to end their algorithms. Hence, their methods can not exit the loop in a flexible way.

The confidence probability is also used to define the number of iteration. When the average confidence probability of the pre-labeled data declines, the loop is ended. It can be understood intuitively. Through the experiments, this criterion of ending the loop not only ensures pre-labeling data adequately and efficiently, but voids a mount of incorrect pre-labeled data by exiting the loop timely.

2.4   The Ensemble Method

The ensemble classifiers can be arranged into two categories: majority voting and weighted voting. Majority voting methods predict the labels of data simply, utilizing the principle of that the minority is subordinate to the majority. They can achieve great performance when the accuracy of base classifiers are roughly equal[6]. However, because the base classifiers in this paper come from different source domains and the similarities between these source domains and the target domain are different, the accuracies of base classifiers are different and majority voting methods are ineffective.

Weighted voting methods[7] adjust the weights of training data with the error of each base classifier on T. They are not applicable too. As data distributions of target domain and source domain are different, it is useless to calculate errors on source domains. Furthermore, because of the lack of labels in the target domain, it is an impossible mission to calculate the error on target domain in advance. Thus, the CEC demonstrates it in a different path: Summarize the classification confidence probabilities from all base classifiers to label the instances in target domain T. The formula is shown as follows:

$$p(y \mid x) = \frac{1}{k} \sum_{i=1}^{k} p_i(y \mid x)$$

Here, y corresponds to the label of instance x, k is the number of base classifiers, $p_i(y \mid x)$ denotes the confidence probability of base classifier c_i labeling y to instance x.

The classification results are influenced by the probability distribution of labels through each base classifier. This method predicts the data according to ensemble probability distribution rather than labels predicted by base classifier directly.

3   Experimental Evaluation

In order to determine how well the proposed algorithm performs, we investigate performance over a variety of review-plus-rating English data sets[1] and Chinese data sets[2]. The English data sets include 11 domains, such as camera, camp and so on. The Chinese data sets come from three domains including hotel, electronics and stock.

The proposed framework is compared with two algorithms: one combines the base classifiers directly and uses a simple majority vote of its component models for each new classification, recorded as SEC(Simple Ensemble Classifier)[2]; the other is the weighted ensemble method mentioned above, recorded as CWEC[1]. For all of our experiments involving MaxEnt model we used the NLTK library, which could be download at www.sourceforge.net.

Parameter r represents the share of classification results accounted for by the convinced correct results. Because of different data distributions, base classifiers could not get high accuracy and it is not appropriate to set a high value for r. Fig. 1 illustrates the accuracy results as a function of r on the tv and stock data sets. It suggests that we can get the best accuracy when r falls in [0.05, 0.2].

In order to evaluate how the proposed ensemble algorithm performs on different feature spaces, we utilized four feature selection criteria – OR[1], MI[8], DB2[9] and LLRTF. Table 1 shows experimental results of the proposed CEC and the baselines CWEC[1], SEC[2] on 11 English datasets. In the experiments, we set r=0.2 and maximum number of loop m=10. Furthermore, we also added some constraints to the classifiers to avoid the case of pre-labeling data be unbalanced.

---

[1] The English data sets are available online: http://www.cs.indiana.edu/~mewhiteh/html/opinion_mining.html.

[2] The Chinese data sets download address: http://www.searchforum.org.cn/tansongbo/senti_corpus.jsp.

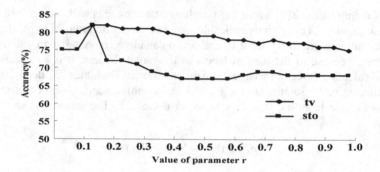

Fig. 1. Accuracy curves of CEC for different parameter *r*

Table 1. Accuracy comparison of different methods on English data sets (%)

Dataset	OR			MI			DB2			LLRTF		
	SEC	CWEC	CEC	SEC	CWEC	CEC	SEC	CWEC	CEC	SEC	CWEC	CEC
camp	51	52	55	55	55	56	55	55	56	80	81	87
camera	53	59	63	63	63	64	63	63	64	75	76	81
doctor	54	58	50	50	54	50	50	54	50	82	82	86
drug	51	51	51	51	50	52	51	50	52	60	59	66
lawyer	52	52	57	57	58	57	57	58	57	80	81	86
radio	50	50	50	50	50	54	50	50	54	70	72	76
restaurant	51	57	58	55	58	60	63	66	66	84	83	88
music	53	56	58	55	56	59	61	65	65	69	69	72
laptop	53	52	55	55	55	60	52	51	54	70	68	79
tv	52	53	63	63	63	57	61	61	58	76	77	82
movie	50	50	50	50	50	50	50	50	50	57	56	74

Tab. 1 reveals several observations: 1) the proposed algorithm outperforms baselines in all different feature spaces. 2) Especially in LLRTF, the classification accuracy of camera data set can achieve to 80%. Compared to baselines, the accuracy can be improved as much as 17%. Meanwhile, on all of data sets, CEC can improve the accuracy by almost 7% and achieve to 80% on average. 3) It is noticed that the method and the baselines give low performance on the drug data set. The accuracy is low, which is due to the similarities between drug domain and other domains are low. 4) Meanwhile, we note that the proposed algorithm and baselines can not get pretty high accuracies in other three feature spaces including OR, MI and DB2. This suggests that the method of feature selection is an important factor in influencing cross-domain classification accuracy.

Table 1. Accuracy comparison of different methods on Chinese data sets (%)

Dataset	OR			MI			DB2			LLRTF		
	SEC	CWEC	CEC	SEC	CWEC	CEC	SEC	CWEC	CEC	SEC	CWEC	CEC
hotel	50	50	57	51	51	54	50	51	54	73	73	75
electronics	54	46	54	50	50	50	52	50	52	67	67	77
stock	50	60	50	57	57	55	50	54	55	68	69	82

Tab.2 presents the comparison results of different methods on three Chinese data-sets. It is illustrated that CEC can get better performance than the baselines, even though the source domain datasets is scarce.

4   Conclusions and Future Work

In this paper, a novel ensemble method is proposed for transferring knowledge from multiple sources to the target domain in sentiment classification. The basic idea is providing believable label for unlabeled data in the target domain and combining the base classifiers by the confidence probability to solve the problem of domain adaptation. However, the method trains base classifiers on all source domains, without automatically choosing the useful ones. It makes the method space-time consuming and lacking efficiency. In the future, we will extend relational theory to measure the similarity between domains to address this issue.

Acknowledgments. This research is supported by the National Natural Science Foundation of China (NSFC) under grants 60975034, the Fundamental Research Funds for the Central Universities under grant 2011HGBZ1329 and 2011HGQC1013.

References

1. Whitehead, M., Yaeger, L.: Building a General Purpose Cross-Domain Sentiment Mining Model. In: Proceedings of the 2009 WRI World Congress on Computer Science and Information Engineering, pp. 472–476 (2009)
2. Li, S., Zong, C.: Multi-domain Sentiment Classification. In: Proceedings of ACL 2008: HTL. Short Papers (Companion Volume), pp. 257–260 (2008)
3. Avrim, B., Tom, M.: Combining Labeled and Unlabeled Data with Co-Training. In: Proceeding of The Eleventh Annual Conference on Computational Learning Theory, pp. 92–100 (1998)
4. Ng, V., Cardie, C.: Weakly Supervised Natural Language Learning without Redundant Views. In: Proceedings of the 2003 Conference of the North American Chapter of the Association for Computational Linguistics on Human Language Technology, Edmonton, Canada, pp. 94–101 (2003)
5. Clark, S., Curran, J., Osborne, M.: Bootstrapping Pos Taggers Using Unlabelled Data. In: Proceedings of the Seventh Conference on Natural Language Learning at HLT-NAACL 2003, Edmonton, Canada, pp. 49–55 (2003)

6. Dietterich, T.G.: Ensemble Methods in Machine Learning. In: Kittler, J., Roli, F. (eds.) MCS 2000. LNCS, vol. 1857, pp. 1–15. Springer, Heidelberg (2000)
7. Freund, Y., Schapire, R.E.: A Decision Theoretic Generalization of On-line Learning and an Application to Boosting. Journal of Computer and System Sciences 55(1), 119–139 (1997)
8. Yang, Y., Pedersen, J.: A Comparative Study on Feature Selection in Text Categorization. In: ICML, pp. 412–420 (1997)
9. Tan, S., Wang, Y., Cheng, X.: An Efficient Feature Ranking Measure for Text Categorization. In: SAC 2008, Fortaleza, Ceará, Brazil, pp. 407–413 (2008)

Robust ISOMAP Based on Neighbor Ranking Metric

Chun Du, Shilin Zhou, Jixiang Sun, and Jingjing Zhao

School of Electronic Science and Engineering, National University of Defense Technology,
Changsha, Hunan, P.R. China
yxduchun@gmail.com

Abstract. ISOMAP is one of classical manifold learning methods that can discover the low-dimensional nonlinear structure automatically in a high-dimensional data space. However, it is very sensitive to the outlier, which is a great disadvantage to its applications. To solve the noisy manifold learning problem, this paper proposes a robust ISOMAP based on neighbor ranking metric (NRM). Firstly, NRM is applied to remove outliers partially, then a two-step strategy is adopted to select suitable neighbors for each point to construct neighborhood graph. The experimental results indicate that the method can effectively improve robustness in noisy manifold learning both on synthetic and real-world data.

Keywords: ISOMAP, noisy manifold learning, neighbor ranking metric.

1   Introduction

Recently, there has been growing concern about manifold learning, which is effective for nonlinear dimensionality reduction. Some manifold learning algorithms such as isometric mapping (ISOMAP) [1], locally linear embedding (LLE) [2] and local tangent space alignment (LTSA) [3] have been successfully used in pattern recognition and data visualization. Most existing manifold learning algorithms try to explore the low-dimensional nonlinear manifold structure of high-dimensional data. However, they are not usually robust in presence of noisy data. The noisy data lie outside the manifold (outliers) and can easily change the local linear structure of the manifold.

In order to resolve the noisy manifold learning problem, several extended algorithms have been proposed. For example, Choi, et al [4] presented robust kernel ISOMAP, which could handling with critical outliers (which result in short-circuit edges) to improve the topological stability. But it may be not effective when the noise is uniform or gaussian distribution. Chang and Yeung [5] proposed Robust locally linear embedding (RLLE) based on the robust PCA. RLLE can reduce the undesirable effect of outliers largely. Nevertheless, the computational requirement of this method is significantly higher than LLE. Moreover, the difficulty of parameter selection is another impediment for its wide application.

In this paper, we address the noisy manifold learning problem in the context of ISOMAP. First of all, a new and intuitive neighborhood similarity measure called

D.-S. Huang et al. (Eds.): ICIC 2012, LNCS 7389, pp. 221–229, 2012.
© Springer-Verlag Berlin Heidelberg 2012

neighbor ranking metric (NRM) is presented. Secondly, NRM is respectively used for removing outliers and selecting suitable neighbors of each point to construct neighborhood graph. Lastly, we apply above techniques to make ISOMAP robust.

2   A Brief Review of ISOMAP

ISOMAP works by assuming isometry of geodesic distances in the manifold. It can be considered as a special case of the classical multidimensional scaling(MDS)[6], where dissimilarities is measured by geodesic distance instead of Euclidean distance. Given a m-dimensional data set $X = \{x_1, x_2, ..., x_N\} \in R^{m \times N}$, sampled from a d-dimensional manifold M ($d < m$), the signal model relevant to noisy manifold learning is given as follows:

$$x_i = f(\tau_i) + \varepsilon_i \tag{1}$$

where f is mapping function, $f : \Omega \subset R^d \rightarrow R^m$, Ω is an open subset, and ε_i denotes noise. ISOMAP assumes that intrinsic dimension d is known and proceeds in the following steps:

1.   Set Neighborhood. Two simple methods are commonly used: one is to select the k nearest neighbors($k - NN$), the other is to choose all neighbors within a fixed radius($\varepsilon - $hypersphere).
2.   Estimate the geodesic distances. Supposed x_i and x_j are a pair of samples, the geodesic distances $d_G(x_i, x_j)$ can be approximated by the shortest path distances between node $x_i^{'}$ and $x_j^{'}$ (corresponding to data point x_i and x_j) in the graph G.
3.   Compute the low-dimensional coordinates. One can use the classical MDS algorithm with the matrix of graph distances $D_G = \{d_G(x_i, x_j)\}$ to map the data set into a low-dimensional embedded space.

Unlike the linear dimensionality reduction techniques such as Principal Component Analysis (PCA), ISOMAP can discover the nonlinear degrees of freedom that underlie complex natural observations. However, it still has some shortcomings: (1) ISOMAP is not robust in presence of outliers, (2) the resulting output is extremely sensitive to parameters that affect the selection of neighbors.

3   Robust ISOMAP Method

3.1   Neighbor Ranking Metric

At the first step of the classical ISOMAP algorithms, we often select the neighborhood of each point using $k - NN$ method. But in some cases, the k nearest neighbors based on Euclidean may not be that in the sense of manifold. For example, as Figure 1 depicts, the 4 nearest neighbors of data 'H' according to the ranking by Euclidean distance is {G, I, F, J} and that of data 'A' is {H, I, G, J}. Obviously, 'A' is an outlier near to dataset based on Euclidean distance but far from the manifold based on geodesic distance. If 'A' is inserted into graph G as a node while 'A-H' as an edge, the structure of manifold will be changed. So it is necessary to find a measure on how likely a data is a neighbor of another data.

Fig. 1. Schematic plot of suitable neighborhood selection

The core idea of our method is inspired from the practical experiences that neighbors should be admitted each other. Similar to the relationship between two individuals in real world, person A considers B as his best friend, but B may not think so. Only if A and B all regard each other as his best friend, they are real best friends.

Motivated by the ideas mentioned above, we present a new neighborhood similarity measure called neighbor ranking metric(NRM). For each sample x_i, let $U_{x_i} = \{x_{i_1}, x_{i_2}, ..., x_{i_N}\}$ be the neighborhood set of x_i sorted by Euclidean distance in ascending order. Then NRM is defined as follows:

$$NRM(x_i, x_j) = |O_{ij} - O_{ji}| \tag{2}$$

where $|.|$ is the absolute value operator, O_{ij} denotes the order where x_j lies in U_{x_i}, O_{ji} denotes the order where x_i lies in U_{x_j}. It can be clearly seen from equation (2) that if x_i and x_j have small NRM value, the possibility of x_j to be a neighbor of x_i is very high. So, let η be the threshold to judge the value of NRM. $NRM(x_i, x_j) < \eta$ indicates that x_j belongs to the neighborhood of x_i and vice versa.

3.2   Outlier Removing

NRM can be easily applied to outlier removing, which is important to noisy manifold learning. Supposed x_i belongs to the k nearest neighborhood of some other samples, whether x_i is an outlier or not largely depends on the number of samples who consider x_i as their k nearest neighbor. Specifically, we can detect outliers by the criterion $S_{x_i} < T$, where S_{x_i} is the number of neighbors belongs to x_i and T is the threshold. In this paper, we set $T = 3$ by experience and remove the samples whose neighborhood size is less than 3.

3.3   Neighborhood Selection

After removing outliers, neighborhood selection is needed. The k-NN method is widely used because it can guarantee the sparseness of resulting structures[7]. However, this method is not without its shortcomings. To begin with, the choice of an appropriate k is difficult. If k is too small, the local neighborhoods may estimate the correlation between the points and their neighbors falsely, even divided the manifold into several disconnected sub-manifolds. In contrast, too large k will make the local neighborhoods include points in different patches of data manifold and result in short circuits. Furthermore, the k-NN method sets parameter the same value for all points. This strategy is not reasonable because local structures of each point are usually different. In general, the selection of neighborhood size should be data-driven and depend on such factors as density, curvature and noise.

As indicated in some previous works[8-9], there are several requisitions for neighborhood selection: (1) the selected neighbors for each sample should reflect the local geometric structure of the manifold, (2) the number of sample in each neighborhood should be large enough to make sure the neighborhood graph is connected.

In this paper, we mainly focus on neighborhood selection in presence of noisy data and further extend the k-NN method based on NRM. The basic idea of our method is to expand each neighborhood while removing its non-neighbor elements. The detail of our method is described as follows:

Input: data set X, the initial parameter k, the tolerable threshold η.

Output: the neighborhood size k_{x_i}, the neighborhood N_{x_i}.

Step1: compute Euclidean distance matrix D_X between any pair wise points, then determine the initial k-NN neighborhood $N_{x_i} = \{x_i^1, x_i^2, ..., x_i^k\}$.

Step2: expand neighborhood N_{x_i}. For each element of N_{x_i}, obtain its k nearest neighborhood $N_{x_i^l}$, $l = 1, 2, ... k$, then expand $N_{x_i} = N_{x_i} \cup N_{x_i^l}$ and delete the repetitive elements. At last, sort the updated N_{x_i} by Euclidean distance to x_i in ascending order and set $k_{x_i} = S(N_{x_i})$, where $S(N_{x_i})$ denotes the number of the elements in N_{x_i}.

Step3: remove non-neighbor elements of N_{x_i}. For each element x_i^l, $l = 1, 2, ... k_{x_i}$, compute the $NRM(x_i, x_i^l)$ using Equation (2), then judge whether x_i^l is a suitable neighbor or not. If $NRM(x_i, x_i^l) < \eta$ and $\min(O_{il}, O_{li}) < k$, retain x_i^l in N_{x_i}, otherwise, consider it as a non-neighbor and remove it from N_{x_i}, then let $k_{x_i} = k_{x_i} - 1$.

Step4: For each $x_i \in X$ that has not been processed, repeat step2 to step3.

Obviously, the method described above is a two-step strategy. In order to construct a connected neighborhood graph, the method firstly expands initial k-NN neighborhood by merging the set of N_{x_i} and N_{x_j} . The expansion process is easy to implement and can make N_{x_i} comprise candidate neighbors as large as possible. However, not all elements in expanding N_{x_i} are suitable neighbors in the sense of manifold. So, in the second step of our method, NRM is applied to detect and delete these non-neighbor elements. The remaining elements would be considered as the suitable neighbors of x_i in neighborhood graph G .

3.4   Robust ISOMAP Algorithms

Based on the discussion mentioned above, the outline of the proposed algorithm can be concluded as follows:

Input: data set X , the initial parameter k , the tolerable threshold η
Output: the low-dimensional embedding result Y
Step1: Identify and remove the outliers from the original data X by performing our outlier removing method in section 3.2.
Step2: Construct the neighborhood graph G based on our neighborhood selection method in section 3.3.
Step3: Carry out the classical ISOMAP algorithm to G and build the low-dimensional embedding Y that best preserves the manifold's intrinsic structure.

4   Experiment

Two examples are presented to illustrate the performance of the robust ISOMAP method. The test data sets include Swiss roll and USC-SIPI dataset. In order to evaluate the embedding result, we use the residual variance which is defined as below:

$$Residual \_ variance = 1 - \rho(D_X, D_Y)^2 \tag{3}$$

where D_X and D_Y are two vectors which express the geodesic distances in the high-dimensional data space and Euclidean distances in the low-dimensional embedded space respectively, ρ is the standard linear correlation coefficient. The lower residual variance is, the better the performance of the manifold learning method is.

4.1   Experiments on Swiss Roll Data

At first, the robust ISOMAP method and classical ISOMAP method are applied to Swiss roll data without noises. 2000 clean points are randomly sampled from the Swiss roll manifold. The classical ISOMAP method uses $k-NN$ method to select neighborhood and k is assigned within the range from 4 to 30. For the robust ISOMAP method, we set the parameter $\eta = k$ by experience. The embedding results with $k = 10$ are shown in Fig. 2(a) and Fig. 2(b). The change of the residual variances with different k is shown in Fig.3. From Fig. 2 and Fig. 3, we can see that both methods achieve almost the same embedding results with $k \in [7,15]$. More importantly, the robust ISOMAP method can attain lower residual variances than classical ISOMAP method with the increase of k. That is to say, the resulting output of the robust ISOMAP is not sensitive to parameter k and the robust ISOMAP method seems superior to classical ISOMAP method when the original Swiss roll data is clean.

(a)   (b)

Fig. 2. Embedding results on 2000 clean Swiss roll data (a) by applying the classical ISOMAP algorithm, (b) by applying the robust ISOMAP

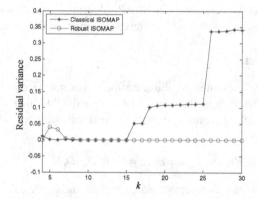

Fig. 3. The change of the residual variances with different k

Furthermore, we test robust ISOMAP and classical ISOMAP on the noisy Swiss roll data. As shown in Fig. 4(a), 2000 clean points are randomly sampled from the Swiss roll manifold and 200 uniform distributed outliers are added as noises (marked as '*'). Shown in Fig.4(c) is the outlier removing results. We can find that the vast majority of clean points are preserved while almost half of the labeled outliers are removed. The embedding result of classical ISOMAP is shown in Fig. 4(b). As can be seen, there is a strong distortion between the embedding result in presence of outliers and that in absence of outliers (see Fig. 2(a)). The embedding result of robust ISOMAP is shown in Fig. 4(d). Comparing Fig. 4(b) with Fig. 4(d), we find that the robust ISOMAP method is more topologically stable than the classical ISOMAP,

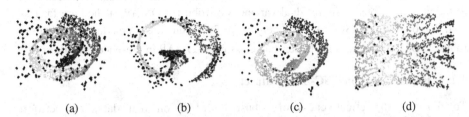

Fig. 4. (a) Noisy Swiss roll data (200 outliers). (b) Embedding results by classical ISOMAP. (c) The result of outlier removing. (d) Embedding results by robust ISOMAP.

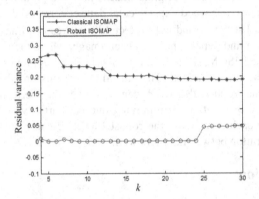

Fig. 5. The change of the residual variances with different k

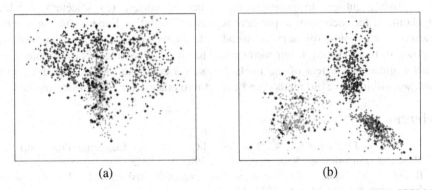

Fig. 6. Embedding results on wood texture data (a) by applying the classical ISOMAP algorithm, (b) by applying the robust ISOMAP

since it adopt an effective strategy to suppress the effect of noises and select suitable neighborhood for each point. Fig.5 shown the change of the residual variances with different k .Obviously, the residual variances of robust ISOMAP method are far less than that of classical ISOMAP for every $k \in [4,30]$. Thus it can be concluded that the robust ISOMAP can perform better than classical ISOMAP in presence of outliers.

4.2   Experiments on USC-SIPI Dataset

To illustrate the effectiveness of robust ISOMAP on real data, we perform experiments on wood texture data from the USC-SIPI image database[1]. The wood texture data consists of rotated texture images of seven different rotation angles. The size of each image is 512×512 . We select three rotated texture images ($0°, 60°, 90°$) and divide each original images into 400 partially overlapping blocks of size 32×32. Thus, our wood experimental dataset consists of 1200 images each of 1024 dimensions. Then, we randomly select 120 images (40 images for each rotated angle) and add 'salt and pepper' noise to each image, where the noise density is 0.05. Lastly, the classical ISOMAP and robust ISOMAP are applied to the wood texture data set with noisy images added. The embedding results are shown in Fig. 6. It is obviously that the robust ISOMAP can improve the performance of classical ISOMAP on noisy data set. With removing outliers partially and selecting suitable neighborhood for each data point, the robust ISOMAP can restrain noise effect and preserve the separation between clusters well.

5   Conclusion

This paper investigates the noisy manifold learning problem and proposes a robust version of ISOMAP method. The proposed method presents a new neighborhood similarity measure called NRM and makes ISOMAP more robust from two aspects: (1) removing outliers to suppress the effect of noises, (2) selecting suitable neighborhood for each point to preserve topological stability. Experiments on several data sets such as Swiss roll data and wood texture data, verified the robustness of our proposed method. However, our method still has its shortcomings. For example, how to determinate parameters of the method is still a complicated work. In the future work, we will improve our method and extend it to other manifold learning methods.

References

1. Tenenbaum, J.B., de Silva, V., Langford, J.C.: A Global Geometric Framework for Nonlinear Dimensionality Reduction. Science 290, 2319–2323 (2000)
2. Roweis, S.T., Saul, L.K.: Nonlinear Dimensionality Reduction by Locally Linear Embedding. Science 290, 2323–2326 (2000)
3. Zhang, Z., Zha, H.: Principal Manifolds and Nonlinear Dimension Reduction via Local Tangent Space Alignment. SIAM J. Scientific Computing 26, 313–338 (2005)
4. Choi, H., Choi, S.: Robust kernel Isomap. Pattern Recognition 40, 853–862 (2007)

---

[1] http://sipi.usc.edu/services/database

5. Chang, H., Yeung, D.Y.: Robust locally Linear Embedding. Pattern Recognition 39, 1053–1065 (2006)
6. Cox, T., Cox, M.: Multidimensional Scaling. Chapman and Hall, Boca Raton (1994)
7. Mekuz, N., Tsotsos, J.K.: Parameterless Isomap with Adaptive Neighborhood Selection. In: Franke, K., Müller, K.-R., Nickolay, B., Schäfer, R. (eds.) DAGM 2006. LNCS, vol. 4174, pp. 364–373. Springer, Heidelberg (2006)
8. Zhan, Y., Yin, J., Liu, X., Zhang, G.: Adaptive Neighborhood Select Based on Local Linearity for Nonlinear Dimensionality Reduction. In: Cai, Z., Li, Z., Kang, Z., Liu, Y. (eds.) ISICA 2009. LNCS, vol. 5821, pp. 337–348. Springer, Heidelberg (2009)
9. Zhang, Z., Wang, J., Zha, H.: Adaptive Manifold Learning. IEEE Trans. Pattern Anal. Mach. Intell. 34, 253–265 (2012)

Modeling by Combining Dimension Reduction and L2Boosting

Junlong Zhao

School of Mathematics and System Science, Beihang University, LMIB of the Ministry of Education, Beijing, China
zjlczh@126.com

Abstract. Dimension reduction techniques are widely used in high dimensional modeling. The two stage approach, first making dimension reduction and then applying existing regression or classification method, is commonly used in practice. However, an important issue is that when two stage approach can lead to consistent estimate. In this paper, we focus on $L2$boosting and discuss the consistency of the two stage method-dimension reduction based $L2$boosting (briey DR$L2$B). We establish the conditions under which DR$L2$B method results in consistent estimate. This theoretical finding provides some useful guideline for practical application. In addition, we propose an iterative DR$L2$B approach and make some simulation study. Simulation results shows that iterative DR$L2$B method has good performance

Keywords: dimension reduction, $L2$boosting, consistency.

1 Introduction

Dimension reduction methods have been widely used in high dimensional modeling to extract the main information and to reduce the noise in the data. Suppose that $X \in R^{p \times 1}$ is p-dimensional vector. The aim of the many dimension reduction methods is to find the low dimension linear combination $B^T X$ where $B \in R^{p \times k}$ with $k < p$. There are many dimension reduction methods have been developed. For example, the Principal Component Analysis (PCA), Linear Discriminant Analysis (LDA) and Independent Component Analysis (ICA) are the most widely used methods(Jolliffe, 2002). In regression setting, the well-known methods include the MAVE (Xia et al, 2002), SIR (Li, 1991), SAVE (Cook and Weisberg, 1991), KDR(Fukumizu, et al, 2009) etc.

The two stage approach applies dimension reduction first and then builds the model by the existing regression or classification methods developed in low dimensional setting. Two stage method is widely used in the different fields, such as, image processing and face recognition(Zhao et al, 2003). However, for two stage method, the error generated in estimating B will inevitably has impact on the following regression or classification procedure. Therefore, an important issue is that whether such impact is negligible, or equivalently, wether two stage method can lead

D.-S. Huang et al. (Eds.): ICIC 2012, LNCS 7389, pp. 230–235, 2012.
© Springer-Verlag Berlin Heidelberg 2012

to consistent estimate. L2Boosting (Bühlmann and Yu, 2003) is an effective method for regression and classification and has been widely used in different settings. Many authors had studied the theoretical and numerical property of L2Boosting (Zhang and Yu, 2005; Bühlmann and Horthorn, 2007; Chang et al 2010). In this paper, we focus on two stage method dimension reduction based L2Boosting method (briefly DRL2B) and obtain the conditions that guarantee the consistency of the DRL2B approach. This condition suggests that, CART as the most widely used base learner in boosting approach may lead to inconsistent estimate in DRL2B method. This provides some useful guideline in the application.

The main contents of this paper are arranged as follows. The DRL2B algorithm is reviewed in section 2.1 and its consistency is established in section 2.2 and the iterative DRL2B algorithm is proposed in section 2.3. Simulation results in section 3 show that iterative DRL2B and DRL2B method have good performance.

2   Dimension Reduction Based $L2$Boosting and Its Extension

2.1   Algorithm of Dimension Reduction Based $L2$Boosting

Let S denote the set of the real value functions and define span(S) $= \{ \sum_{k=1}^{m} \omega^k f^k$ $\omega^k \in R, f^k \in S, m \in Z^+ \}$ being the linear function space spanned by S, where Z^+ is the set of positive integer. For any $f \in \text{span}(S)$, define the 1-norm as $\| f \|_1 = \inf \{ \| \omega \|_1 :$ $f = \sum_{i=1}^{m} \omega^i f^k \ f^k \in S, m \in Z^+ \}$. In high dimensional setting, dimension reduction methods usually assume the true model has the low dimensional structure $Y = f(B^T X, \varepsilon)$ (Li, 1991), where $B \in R^{p \times k}$ and $B^T B = I_k$ and k is usually much small. Many dimension reduction method has been developed to estimate the subspace span(B), such as, LDA, OLS, SIR (Li, 1991), MAVE (Xia, et al., 2002) etc.

Let $A(f) = E(Y - f)^2 / 2$, the L_2 loss of the function f. Define the base function space $S_B = \{ g(B^T x) : g$ being some kinds of base learner$\}$. Throughout the paper, we further require that S_B satisfies that, for any $k \times k$ orthogonal matrix H,

$$S_B = S_{BH} \tag{1}$$

that is, for any $g(B^T x) \in S_B$, we have $g((HB)^T x) \in S_B$. Therefore, it suffices to find one of the orthonormal base of span(B). Suppose that $\{ (X_i, Y_i), i = 1, \cdots, n \}$ are i.i.d. observations of (X, Y). Let B_n be the estimate of B with $B_n^T B_n = I_k$.

Define $A_n(f) = \sum_{i=1}^{n} (Y_i - f(X_i))^2 / 2 := E_n(Y - f(X))^2 / 2$ and $S_{B_n} = \{ g(B_n^T x)$ being some kinds of base learner$\}$. Similar to greed boosting algorithm (Zhang and Yu,2005), the major steps of DRL2B algorithm is as follows.

1. Obtain the estimate B_n of some orthnormal basis B by dimension reduction methods.
2. Select $f_0 \in S_{B_n}$

3. For $k = 0, 1, 2 \cdots$ select a closed subset $\Lambda_k \subset R$, such that $0 \in \Lambda_k$ and $\Lambda_k = -\Lambda_k$

4. Find $\hat{\alpha}_k \in \Lambda_k$ and $\hat{g}_k (B_n^T x) \in S_{B_n}$ which approximately minimize the loss function

$A_n(\hat{f}_k + \hat{\alpha}_k \hat{g}_k) \leq \inf_{\alpha_k \in \Lambda_k, g_k \in S_{B_n}} A_n(\hat{f}_k + \alpha_k g_k) + \varepsilon_k$ where ε_k is a series of nonnegative real number towards 0.

5. Let $\hat{f}_{k+1} = \hat{f}_k + \hat{\alpha}_k \hat{g}_k$

2.2   Consistency of DRL2B

Our goal is to show the statistical asymptotic convergence, that is, as $n \to \infty$, $A_n(\hat{f}_k)$ $\to \inf\{A(f), f \in S_B\}$ in probability, where $k = k(n) \to \infty$, as $n \to \infty$. Many dimension reduction, such as SIR, SAVE, MAVE etc, can obtain $B_n \to B$ in probability, as $n \to \infty$. Without loss of generality, we assume that $\sup_g E[g(X)]^2 \leq 1$. Let h_k and s_k satisfy $|\hat{\alpha}_k| \leq h_k \in \Lambda_k$, $s_k = \| f_0 \|_1 + \sum_{i=0}^{k-1} h_i$ and suppose that (1) holds. Furhtermore, we make the following assumptions.

(C1) Lipschiz condition. There exists positive constant $L > 0$, such that, $|g(x_1) - g(x_2)| \leq L \| x_1 - x_2 \|_F$, for any $g \in S$ and any x_1, x_2.

(C2) There exists a basis B of span(B) and $s > 0$, such that $B_n - B = O_p(n^{-s})$.

(C3) $[(k(n)+1)]s_{k(n)+1} = o(n^s)$, as $k = k(n) \to \infty$.

(C4) The covering numbers $N(S_B, \varepsilon)$ of the space span(S_B) is finite.

(C5) $|f(X) - Y| < M_0$ for all $f \in S_B$, for some M_0 almost surely.

(C6) Assumption 3.1 of Zhang and Yu(2005) holds.

Assumption $N(S_B, \varepsilon) < \infty$ holds in many situations, for example, span(S_B) is finite dimensional function space, Sobolev spaces or spaces associated to a kernel(Cucker and Smale, 2001). Condition (C5) holds as Y and $\sup\{f : f \in$ span$(S_B)\}$ are finite almost surely. The following theorem presents the statistical asymptotic convergence of the DRL_2B algorithm.

Theorem 1. Under (C1)--(C6), as $n \to \infty$, we have $A_n(\hat{f}_{k(n)}) \xrightarrow{P} \inf_{f \in span(S_B)} A(f)$.

2.3   Further Refinement: Itrative DRL2B

Borrowing the idea of iteration method, such as the PLS, which iteratively apply the dimension reduction and the linear regression, we propose the following corrected algorithm, called iterative DRL2B. Taking the case of fixing step length as an example(that is $\alpha_k = r$ for some r), we describe the iterative DRL2B algorithm as follows.

1.   Initialize $\hat{f}_0$, denoted as $\hat{f}_0 = \bar{Y}$. Let $k = 0$.

2.   Increase k by 1, and compute the residuals $U_i^{[k]} = Y_i - \bar{f}_{k-1}(X_i)$, $i = 1, 2, \cdots, n$.

3. Apply dimension reduction method to $(U^{[k]}, X)$ to obtain the estimate $B_n^{[k]}$. Let $V^{[k]} = XB_n^{[k]}$.

4. Fit $U^{[k]}$ by $V^{[k]}$ using the base learner to obtain the estimate $\hat{g}_k(\cdot)$.

5. Update $\hat{f}_k = \hat{f}_{k-1} + r\hat{g}_k$, where $0 < r \le 1$ is the step length.

6. Repeat step 2–5, until $k = m_{stop}$.

3 Simulations and Real Data Analysis

3.1 Simulations

In this section, we conduct simulations to compare DRL2B, iterative DRL2B, and the component-wise L2Boosting (Bülhmann and Yu, 2003), briefly denoted as CL2B. Define the testing error as $error = (\hat{Y} - Y)^2 / |testset|$. Take at random 80% data as training set and the rest as testing set. We select optimal stopping number mopt by five-fold CV method based on the training data and compute the corresponding prediction error on the test data. Repeat the procedure for 100 times and compute the mean and standard deviation of prediction error under the corresponding mopt, denoted respectively as MEAN and STD in the following table 1 of this section. Let Mopt denote the mean of mopt over 100 replicas. We consider three setting of the values of n and p: Case I: $n = 200$, $p = 10$; Case II: $n = 100$, $p = 50$; Case III: $n = 100$, $p = 100$. As generally suggested in the literature, the small value of r, generally leading to better results (Friedman, 2001), is preferred, especially for large p. So, we take take $r = 0.1$ for three cases.

For Case II and III, in both DRL2B and iterative DRL2B, partial SIR method (Li, et al 2007), which is computationally simple, is used to estimate the index in Case II and III. For case I of small p but large n, we take MAVE to infer the index for DRL2B and SIR for iterative DRL2B. In the following models, we always assume that $\varepsilon \sim N(0,1)$. Let $b_1 = p^{-1/2}(1,...,1)^T \in R^p$, $b_2 = (1,2...,p)^T$, $b_3 = (1,1,...,1)^T \in R^p$. Consider the following non-linear models.

- Model 1 $Y = (b_1^T X + 1)^2 + 2\sin(b_2^T X / \|b_2\|) + \varepsilon$, where $X \sim N(0, I_p)$.

- Model 2 $Y = 2\mathrm{sign}(b_2^T X / \|b_2\|)(1 + b_3^T X / \|b_3\|)^2 + \varepsilon$ where, $X \sim N(0, I_p/4)\Sigma$, where Σ is the matrix with the (i, j) elements $\sigma_{ij} = 0.3^{|i-j|}$.

Here $B = (b_1, b_2)$ and $B = (b_2, b_3)$, for model 2 and 3, respectively. And the dim(B)=2. We compare simulation results of four methods. CL2B with spline base learner, denoted as SplL2; DRL2B with dim(B)=2 (that is, we use the true dimension of B) and thin-plate spline base learner, denoted as Ts-tpDR; iterative DRL2B method with dim($B^{[k]}$)=1 and spline base learner, denoted as It-splDR; iterative DRL2B with dim($B^{[k]}$)=2 and thin-plate spline base learner, denoted as It-tpDR.

Comparison of CL2B with DRL2B and iterative DRL2B. From Table 1, it follows that Ts-tpDR, It-splDR and It-tpDR are much much better than splL2 in all three cases.

Comparison between DRL2B with iterative DRL2B. From Table 1, three methods Ts-tpDR, It-splDR and It-tpDR have the nearly the same prediction performance for Case I, where p is small but n is large; and they also have the similar performance for Case II, where p is moderate ($p = n/2$). The performances of Ts-tpDR in case I and II coincide with our theoretical finding in section 2. And for Case III, where p is large ($p = n$), the performance It-splDR and It-tpDR are better than Ts-tpDR. Therefore for large p but small or moderate n, iterative DRL2B is a good choice.

3.2   Real Data Analysis

We also compare these methods on two real data sets: Ozone data and Boston Housing data. Simulation results confirm the better performance of DRL2B and iterative DRL2B than SplL2. Simulations are presented in table 1.

Table 1. Simulation results for model 1 and 2 and real data

			SplL2	It-splDR	Ts-tpDR	It-pDR
		MEAN	3.542	1.617	1.739	1.541
		STD	1.195	0.428	0.492	0.431
	CASE I	Mopt	202.600	76.800	22.650	38.700
		MEAN	12.034	9.359	9.450	8.901
model 1	CASE II	STD	5.175	5.886	6.450	5.966
		Mopt	241.550	86.300	42.600	28.350
		MEAN	15.564	11.748	11.871	10.542
	CASE III	STD	5.206	4.595	5.519	4.131
		Mopt	134.600	148.100	40.650	18.650
		MEAN	10.612	3.685	2.959	3.077
		STD	3.175	1.061	1.092	0.805
	CASE I	Mopt	179.200	115.800	36.600	45.500
		MEAN	18.867	13.973	14.327	13.129
model 2	CASE II	STD	6.165	5.358	5.395	5.512
		Mopt	22.700	124.200	18.800	47.100
		MEAN	19.044	12.615	12.436	10.735
	CASE III	STD	5.536	4.411	4.474	3.597
		Mopt	220.300	130	57.900	72.600
Ozone data		MEAN	20.134	18.837	17.077	16.864
		STD	2.521	3.326	2.252	2.433
		Mopt	103.450	19.700	18.450	43.100
Boston Housing data		MEAN	19.889	17.798	17.355	14.478
		STD	4.093	4.398	4.160	3.573
		Mopt	133.600	119.250	23.100	149.700

Ozone data (Breiman and Friedman, 1985). This data contains $n = 330$ observations, and the number of the covariates $p = 8$. In each simulation, 220 observations are selected at random as training data and rest observations as testing data.

Boston Housing data. This data is about the price of the house of the suburb of Boston (Cook and Weisberg, 1994), including $n = 506$ observations, each observation has $p = 13$ covariates. Select at random $[3n/5]$ (the integer part of $3n/5$) observations as training data and rest as testing data.

4   Conclusions

In this paper, we prove the consistency of the dimension reduction based L2Boosting, a simple two stage method. Moreover, we propose a refined algorithm--Iterative DRL2B. Simulation results and real data analysis confirm the effective ness of DRL2B.

Acknowledgements. This work is supported by National Natural Science Foundation of China (No. 11026049 and No.11101022) and Foundation of the Ministry of Education of China for Youths (No. 10YJC910013).

References

1. Breiman, L., Friedman, J.H.: Estimating Optimal Transformations for Multiple Regression and Correlation. J. Amer. Statist. Assoc. 80, 580–598 (1985)
2. Bühlmann, P., Hothorn, T.: Boosting Algorithms: Regularization, Prediction and Model Fitting. Statist. Sci. 22(4), 477–505 (2007)
3. Bühlmann, P., Yu, B.: Boosting with the L2-loss: Regression and Classification. J. Amer. Statist. Assoc. 98, 324–339 (2003)
4. Cook, R.D., Weisberg, S.: Sliced Inverse Regression for Dimension Reduction: Comment. J. Amer. Statist. Assoc. 86(414), 328–332 (1991)
5. Cook, R.D., Weisberg, S.: An Introduction to Regression Graphics. Wiley, New York (1994)
6. Chang, Y., Huang, Y., Huang, Y.: Early Stopping in L2 Boosting. Comp. Stat. & Data Anal. 54(11), 2203–2213 (2010)
7. Fukumizu, K., Bach, F.R., Jordan, M.I.: Kernel Dimension Reduction in Regression. Ann. Statist. 37(1), 1871–1905 (2009)
8. Friedman, J.: Greedy Function Approximation: A Gradient Boosting machine. Ann. Statist. 29, 1189–1232 (2001)
9. Li, K.C.: Sliced Inverse Regression for Dimension Reduction (with discussion). J. Amer. Statist. Assoc. 86, 316–327 (1991)
10. Li, L., Cook, R.D., Tsai, C.L.: Partial Inverse Regression Method. Biometrika 94, 615–625 (2007)
11. Jolliffe, I.T.: Principal Component Analysis, 2nd edn. Springer, New York (2002)
12. Xia, Y., Tong, H., Li, W.K., Zhu, L.X.: An Adaptive Estimation of Dimension Reduction Space. J. Roy. Statist. Soc. Ser. B 64, 363–410 (2002)
13. Zhang, T., Yu, B.: Boosting with Early Stopping: Convergence and Consistency. Ann. Statist. 33, 1538–1579 (2005)
14. Zhao, W., Chellappa, R., Phillips, P.J., Rosenfeld, A.: Face Recognition: A Literature Survey. ACM Computing Surveys 35(4), 399–459 (2003)

Geometric Linear Regression and Geometric Relation

Kaijun Wang[1] and Liying Yang[2]

[1] School of Mathematics & Computer, Fujian Normal University, Fuzhou 350108, P.R. China
wkjwang@gmail.com
[2] School of Computer Science and Technology, Xidian University, Xian 710071, P.R. China

Abstract. When a linear regression model is constructed by statistical calculation, all the data are treated without order, even if they are order data. We propose the Geometric regression and geometric relation method (GR2) to utilize the relation information inside the order of data. The GR2 transforms the order data of each variable to a curve (or geometric relation), and uses the curves to establish a geometric regression model. The prediction method using this geometric regression model is developed to give predictions. Experimental results on simulated and real datasets show that the GR2 method is effective and has lower prediction errors than traditional linear regression.

Keywords: relations between data, geometric regression, geometric relation.

1   Introduction

The regression analysis technique [1-4] is widely used to analyze dependency relationship between a dependent (or response) variable and a set of independent (or predictor) variables in multivariate data analysis, and the regression model learned from training data may be used for a prediction task. Let there be one dependent variable Y and q independent variables X_1, X_2, ..., X_q, the linear regression model about these variables is written as [4]: $Y = b_0 + b_1 X_1 + ... + b_q X_q + \varepsilon$, where b_j ($j=0,1,...,q$) are regression coefficients, random error ε is assumed to be normal distribution $N(0, \sigma^2)$, and $f(X) = b_0 + b_1 X_1 + ... + b_q X_q$ is the regression function. Usually, the coefficients b_j can be found by the least squares estimates from given data [4].

When a regression model is constructed with data, all the data are treated without order by statistical calculation, even if the data are time series. However, there might be relation information inside the order of data, and this information may be helpful to establish a regression model with better predictive ability for some applications.

To utilize the relation information inside data order, we propose Geometric regression and geometric relation method (GR2). GR2 transforms the order data of each variable to a curve, and then GR2 uses the curves instead of original data to establish a geometric regression model. For a prediction task, when some new order data of each independent variable are available, GR2 makes a curve with the new data for each independent variable, and these curves are sent to the geometric regression model to produce (batch) prediction values of dependent variable Y.

D.-S. Huang et al. (Eds.): ICIC 2012, LNCS 7389, pp. 236–243, 2012.
© Springer-Verlag Berlin Heidelberg 2012

The GR2 method is proposed in Section 2, and the experimental results are in Section 3. Finally, Section 4 is the discussion and Section 5 gives the conclusion.

2   Geometric Regression and Geometric Relation

We first discuss the projection of the data of each variable to a 2-dimensional curve manifold according to relation T. The geometric regression is designed to solve a regression model F that fits curves. Finally, the prediction by model F is discussed.

2.1   Making a Curve with Order Data

Let 1-dimensional continuous dependent variable Y have n data $\{y_i\}$ in order $\{y_1, y_2, y_3,..., y_n\}$. Typically, order relation T is the time order (e.g., T=$\{t_1, t_2, ..., t_n\}$ where the time $t_1<t_2<...<t_n$) when the n data are time series. The order relation is transformed to *geometric relation* when T is represented by a geometric curve.

It is ideal to make a *curve manifold*, one 2-dimensional smooth curve, to represent the n data and their order relations. Here we approximate the ideal curve manifold via piecewise curves that fit local data. This work includes the data division, the making of a piecewise curve fitting a data group, and the connection of piecewise curves. The following is a simple introduction (more detail refers to [5]):

The work will be discussed within the framework of manifold theory [6]. Let n data be divided equally into k groups along T, then there is $m=n/k$ data for each data group (called *local data*), e.g., the first data group is $\{y_1, y_2, ..., y_m\}$ and the second $\{y_{m+1}, y_{m+2}, ..., y_{m+m}\}$. Let there be a curve manifold M in Cartesian coordinate system tOy, and *local region* U_j of M correspond to the jth data group. The quadratic polynomial is selected on account of easy computation, and then the fitting curve $y=at^2+bt+c$ may be found by least-squares fit [7].

Every piecewise curve is made under local Cartesian coordinate system uOy, where coordinates u represents local T. In detail, a piece of quadratic polynomial curve C_j: $y=f_j(u)$ ($u\in[1,m]$) is made under uOy with the jth group of data --- m pairs of $\{u_i,y_i\}$, i.e., the jth data group is projected to a 2-dimensional local curve:

$$P_j: \{y_i\}_{i=(j-1)m+1\sim(j-1)m+m} \rightarrow C_j: y = f_j(u). \tag{1}$$

Then, geometric properties of C_i are taken as geometric properties of corresponding local region U_i of M. Thus, all the k local regions corresponding to k data groups are ready. A 2-dimensional curve manifold is constructed when k local regions are joined along T, i.e., the n data are projected to M in plane tOy:

$$P: \{y_i\}_{i=1\sim n} \rightarrow M : M =\bigcup U_j. \tag{2}$$

For a better approximation to a smooth curve manifold, we take the central 1/3 part of each local curve to form local region U_i, and use a same length for all local regions. It is similar to construct a curve manifold for each variable X_i by using above method.

2.2   Geometric Regression Method

A geometric regression method using integral calculation will be designed to fit a regression model to curve manifolds. Without losing generality, let there be linear regression equation $y=\phi_0+\phi x$ between variables X and Y, and let the same local coordinate system uOx be used for local regions of curve manifold M_X and the same uOy for M_Y. Then, any point on M_X is (u,x) and any point on M_Y is (u,y). The least-squares-fit method [7] added with the integral computation of curves for establishing a regression equation between curve manifolds is called *geometric regression* method.

Theorem 1 (Geometric regression). Let $y=\phi_0+\phi x$ be the linear regression equation between variables X and Y (their curve manifolds are M_X and M_Y respectively). Let every curve manifold comprise k local regions, and the local region of M_X be $x=g_l(u)$ and the one of M_Y be $y=f_l(u)$, where $u\in[u_1,u_2]$ and $l=1\sim k$. The means of x values of M_X and the means of y values of M_Y are set as follows:

$$\bar{g}=\frac{1}{k}\frac{1}{u_2-u_1}\sum_{l=1}^{k}\int_{u_1}^{u_2}g_l(u)du, \quad \bar{f}=\frac{1}{k}\frac{1}{u_2-u_1}\sum_{l=1}^{k}\int_{u_1}^{u_2}f_l(u)du.$$

The coefficients are found by the least-squares fit method with M_X and M_Y :

$$\phi=L_{fg}/L_{gg}, \tag{3}$$

$$\phi_0=\bar{f}-\phi\,\bar{g}, \tag{4}$$

where

$$L_{fg}=\sum_{l=1}^{k}\int_{u_1}^{u_2}f_l(u)g_l(u)du-k(u_2-u_1)\bar{f}\,\bar{g}, \tag{5}$$

$$L_{gg}=\sum_{l=1}^{k}\int_{u_1}^{u_2}g_l(u)^2du-k(u_2-u_1)\bar{g}^2, \tag{6}$$

$$L_{ff}=\sum_{l=1}^{k}\int_{u_1}^{u_2}f_l(u)^2du-k(u_2-u_1)\bar{f}^2. \tag{7}$$

2.3   Prediction by Geometric Regression Model

Regression model F is first constructed with the geometric regression method from a training dataset, and then model F is used for predictions. The prediction performance of F is usually tested with a test dataset. The GR2 procedure is in the following:

(1) Geometric treatment of training data
    Sort training data by order relation T, and make one curve manifold with training data under relation T for each variable (e.g. Y and X) by the method in subsection 2.1.

(2) Constructing geometric regression model F with curve manifolds

Construct geometric regression model F (e.g. $y = \phi_0 + \phi x$) with curve manifolds by the method in subsection 2.2.

(3) Geometric treatment of new observed data for independent variables

When some new data (in order T) of each independent variable (e.g. X) are available (e.g., $\{x1, x2, ..., xp\}$), make a curve manifold with these new data under order relation T for each independent variable.

If there are fewer new data, the new data and training data are combined under order relation T, and then the combined data of each independent variable are transformed to a curve manifold (e.g. M_X) under T.

Cut every curve manifold into parts (each part with a unit length of T) along coordinates t, and then take the starting points of each unit part as projection points (e.g., point (u,x') on M_X), and the projection points corresponding to the new data are kept (e.g., $\{x'1, x'2, ..., x'p\}$, but ui is useless and discarded).

(4) Prediction by geometric regression model F

The kept data (e.g., $\{x'1, x'2, ..., x'p\}$) from step (3) are as input of model F to produce prediction values (in batch) of dependent variable Y (e.g., $\{y'1, y'2, ..., y'p\}$).

Note: when the geometric regression process is regarded as a projection $(u,x') \rightarrow (u,y')$, the projection $u \rightarrow u$ is an invariable projection, so the u is meaningless and discarded for a prediction task.

2.4   Parameters of Curve Manifolds

Different parameter m will lead to different curve manifolds and different regression models. The optimal m (or m_0) may be obtained while we find the geometric regression model $y = \phi_0 + \phi x$ by minimizing residual sums of squares $RSS(\phi, m)$:

$$RSS(\phi, m) = L_{ff} - 2\phi L_{fg} + \phi^2 L_{gg}. \tag{8}$$

Generally, $m = 5$ to 12 for $n \geq 42$ is the proper searching scope of m_o according to our experiences; $m = m_1$ to m_2 for $n < 42$ and $n \geq 15$, where m_1 and m_2 are the rounding down of $\sqrt{n} - 1$ and $n/3$ respectively; and $m = 2$ to m_2 for $n < 15$.

3   Experimental Results

This section illustrates the prediction performance of GR2, which is compared with traditional linear regression (LR) [1,4]. Here the prediction performance is evaluated by prediction error (PE), the mean squared error between prediction values of a regression model on test data and true values of test data. For artificial data, the training and test data sets are randomly generated from same functions. For real data, 10-fold cross validation is used, i.e., the data are first divided into 10 non-overlapping subsets, and each subset is used as a test set in turn and the remainder as a training set. This procedure is repeated 20 times for artificial data and 10 times for real data, resulting in an average PE as the final performance.

The artificial dataset is generated from the linear function $Z(t)=2X(t)+Y(t)+2$, where $X(t)= 4\sin(0.08t) +10$, and $Y(t)= 4\cos(0.04t+4)+10$. 300 noisy data of every variable are generated at $t=1,2,3,\ldots,300$, and then random white Gaussian noise is added to the data. In the experiments, the order relation $T=\{1,2,3, \ldots,300\}$, the linear regression model is F: $z=ax+by+c$, and the optimal parameter m_o is searched from $m=5,6,\ldots,12$. The test process is: under m_o the 300 test data $\{x_i'\}$ of X are transformed to curve manifold Mx' and the 300 test data $\{y_i\}$ of Y to My'; for x_i and y_i, their corresponding points on Mx' and My' are (u_i,x_i') and (u_i,y_i') respectively; the x_i' and y_i' are taken as inputs of model F, and output z_i' of model F is the prediction; finally, prediction errors are calculated for performance evaluation.

The real data sets (the servo, auto-price, cpu and auto-mpg) are from the UCI database [8]. These data are scaled to zero mean and unit variance to prevent possible numeric problems, and the nominate attributes in these data sets are replaced by integers (e.g. 1,2,3,...) [9]. The linear regression model is F: $Y = b_0 + b_1 X_1 + \ldots + b_q X_q$, and the optimal parameter m_o is searched from $m=5,6,\ldots,12$ in the experiments.

As the data relation information of the four real datasets is not given, we mine potential small-to-big order as order relation T from data. The mining method is: find a regression model F_{train} with training data by LR method, use F_{train} to yield predictions of Y with training data of X (prediction set Sp_{train} based on the training set), and use F_{train} to yield predictions of Y with test data of X (prediction set Sp_{test} based on the test set), and then sort the combination of Sp_{train} and Sp_{test} by values to give small-to-big order relation T. (Note: it is reasonable to use test data of X here, since independent variable X is observed in real prediction tasks, while unobserved dependent variable Y needs to be predicted.)

The results versus different noise levels for artificial data are shown in Table 1. It can be seen that GR2 reduces much the prediction errors (about 44% - 69% PE reduction) compared with LR. The results for real data are shown in Table 2. One can find that GR2 reduces the prediction errors by about 11% - 21% compared with LR.

Table 1. Prediction errors (PE) in mean squared errors of LR and GR2 for artificial data sets. PR denotes proportional reduction of PE by GR2 compared with LR.

Noise va-riance	LR	GR2	PR(%)
0.2	1.19	0.66	44.8
0.6	3.47	1.23	64.7
1.0	5.55	1.78	68.0
1.4	7.42	2.38	68.0
1.8	9.08	2.80	69.2
2.2	10.88	3.46	68.2

Table 2. Prediction errors (PE) in mean squared errors of LR and GR2 for real data sets. PR denotes proportional reduction of PE by GR2 compared with LR.

Data sets	LR	GR2	PR(%)
servo	1.32	1.03	21.6
autoprice	1.188×10^7	1.056×10^7	11.1
cpu	6297	5402	14.2
autompg	12.48	10.66	14.6

4   Discussion

When order relations of data hold latent varying information of data (called *profitable case*), it would be helpful to mine and utilize this information for a regression-prediction task by GR2. Stronger the relations among data, more benefit the GR2 can obtain from data relations (e.g., the artificial data have strong relations of data in the experiments). In contrast, the traditional linear regression does not utilize this information. Therefore, when GR2 gains additional information about relations of data, it would have better prediction performance on regression models than the traditional regression technique (more analyses refer to [10]). This also implies that GR2 method usually needs prior knowledge about data relations.

The differences between GR2 and traditional linear regression method (LR) include: sum of n data (e.g., $\sum_{i=1}^{n} y_i^2$) is performed in LR, while the integral technique (e.g., $\int_{u_1}^{u_2} f_i(u)^2 du$) is used in GR2; LR uses data in regression directly, while GR2 uses curves transformed from data of each variable. It is the geometric curves that reflect additional information about data varying, so GR2 prefers more training data and more new data of independent variables to uncover the law of data varying.

The experimental results suggest that the data relation information is helpful for improving the predictive ability of linear regression models.

5   Conclusion

The relations between data are helpful for the regression analysis and predictions. We propose the GR2 method to mine and utilize the relations of data for the linear regression. GR2 will improve the prediction ability of linear regression methods when relatively strong relations of data are obtained and the predictions by traditional linear regression methods are bad.

Acknowledgements. This work is supported in part by Natural Science Found of China (No. 61175123) and a Key Project of Fujian Provincial Universities - Information Technology Research Based on Mathematics.

References

1. Wagner, M., Adamczak, R., Porollo, A., Meller, J.: Linear Regression Models for Solvent Accessibility Prediction in Proteins. Journal of Computational Biology 12(3), 355–369 (2005)
2. Hashimoto, E.M., Ortega, E.M.M., Paula, G.A., Barreto, M.L.: Regression Models for Grouped Survival Data: Estimation and Sensitivity Analysis. Computational Statistics & Data Analysis 55(2), 993–1007 (2011)
3. Cogger, K.O.: Nonlinear Multiple Regression Methods: a Survey and Extensions. Intelligent Systems in Accounting, Finance and Management 17(1), 19–39 (2010)
4. Yuan, Z.F., Song, S.D.: Multivariate Statistical Analysis. Science Press, Beijing (2009)
5. Wang, K., Zhang, J., Shen, F., Shi, L.: Adaptive Learning of Dynamic Bayesian Networks with Changing Structures by Detecting Geometric Structures of Time series. Knowledge and Information Systems 17(1), 121–133 (2008)
6. Chen, W.: An Introduction to Differential Manifold. High Education Press, Beijing (2001)
7. Sampaio Jr., J.H.B.: An Iterative Procedure for Perpendicular Offsets Linear Least Squares Fitting with Extension to Multiple Linear Regression. Applied Mathematics and Computation 176(1), 91–98 (2006)
8. Blake, C.L., Merz, C.J.: UCI Repository of Machine Learning Databases. University of California, Irvine (1998), http://mlearn.ics.uci.edu/MLRepository.html
9. Meyer, D., Leisch, F., Hornik, K.: The Support Vector Machine under Test. Neurocomputing 55, 169–186 (2003)
10. Wang, K., Zhang, J., Guo, L., Tu, C.: Linear and Support Vector Regressions based on Geometrical Correlation of Data. Data Science Journal 6, 99–106 (2007)

Appendices

Proof of Theorem 1

Proof. The coefficients a linear regression equation may be found by the least-squares-fit method that minimizes the sum of squared residuals. A curve is regarded as infinite data, and then the integral technique is used to deal with infinite data.

These are known: every curve manifold has k local regions, any point of a local region of M_X is (u,x) ($u \in [u_1, u_2]$) under local coordinate system uOx, and any point of a local region of M_Y is (u,y) under uOy.

For regression equation $y = \phi_0 + \phi x$, we sum and average its two sides for all the infinite points of curve manifolds, where $x_{li} = g_i(u_i)$ and $y_{li} = f_i(u_i)$ are used:

$$\frac{1}{k}\sum_{l=1}^{k}\left(\lim_{N\to\infty}\frac{1}{N}\sum_{i=1}^{N}y_{li}\right) = \phi_0 + \phi \times \frac{1}{k}\sum_{l=1}^{k}\left(\lim_{N\to\infty}\frac{1}{N}\sum_{i=1}^{N}x_{li}\right).$$

This work is done through the integral to local regions $x = g_l(u)$ and $y = f_l(u)$:

$$\frac{1}{k}\frac{1}{u_2-u_1}\sum_{l=1}^{k}\int_{u_1}^{u_2}f_l(u)du = \phi_0 + \phi \times \frac{1}{k}\frac{1}{u_2-u_1}\sum_{l=1}^{k}\int_{u_1}^{u_2}g_l(u)du .$$

When $\overline{g} = \dfrac{1}{k}\dfrac{1}{u_2-u_1}\sum\limits_{l=1}^{k}\int_{u_1}^{u_2}g_l(u)du$ and $\overline{f} = \dfrac{1}{k}\dfrac{1}{u_2-u_1}\sum\limits_{l=1}^{k}\int_{u_1}^{u_2}f_l(u)du$ are set, and

$\overline{f} = \phi_0 + \phi\overline{g}$, we can find out ϕ_0: $\phi_0 = \overline{f} - \phi\overline{g}$.

The sum of squared residuals $E(\phi)$ of $y=\phi_0+\phi x$ fitting M_X and M_Y is the sum of errors between $y_{li}=f_i(u_i)$ and $\hat{y}_{li}=\phi_0+\phi x_{li}$ $(i=1\sim\infty)$ in k local regions, so $E(\phi)$ is:

$$E(\phi) = \sum_{l=1}^{k}\left[\lim_{N\to\infty}\sum_{i=1}^{N}\left(y_{li}-\phi_0-\phi x_{li}\right)^2\right] = \sum_{l=1}^{k}\left[\lim_{N\to\infty}\sum_{i=1}^{N}\left(y_{li}-\overline{f}-\phi x_{li}+\phi\overline{g}\right)^2\right]$$

$$= \sum_{l=1}^{k}\left\{\lim_{N\to\infty}\sum_{i=1}^{N}\left[\left(y_{li}-\overline{f}\right)^2-2\phi\left(y_{li}-\overline{f}\right)\left(x_{li}-\overline{g}\right)+\phi^2\left(x_{li}-\overline{g}\right)^2\right]\right\}.$$

When the integral is applied to local regions $x=g_l(u)$ and $y=f_i(u)$ $(l=1\sim k)$ under local coordinate system uOx and uOy, we have

$$E(\phi) = L_{ff} - 2\phi L_{fg} + \phi^2 L_{gg},$$

where

$$L_{gg} = \sum_{l=1}^{k}\left[\lim_{N\to\infty}\sum_{i=1}^{N}\left(x_{li}^2+\overline{g}^2-2x_{li}\overline{g}\right)\right] = \sum_{l=1}^{k}\left(\int_{u_1}^{u_2}x_l^2du+\overline{g}^2\int_{u_1}^{u_2}du-2\overline{g}\int_{u_1}^{u_2}x_l du\right)$$

$$= \sum_{l=1}^{k}\left(\int_{u_1}^{u_2}g_l(u)^2du+\overline{g}^2\int_{u_1}^{u_2}du-2\overline{g}\int_{u_1}^{u_2}g_l(u)du\right)$$

$$= \sum_{l=1}^{k}\int_{u_1}^{u_2}g_l(u)^2du-k(u_2-u_1)\overline{g}^2,$$

and similarly,

$$L_{ff} = \sum_{l=1}^{k}\left[\lim_{N\to\infty}\sum_{i=1}^{N}\left(y_{li}^2+\overline{f}^2-2y_{li}\overline{f}\right)\right] = \sum_{l=1}^{k}\int_{u_1}^{u_2}f_l(u)^2du-k(u_2-u_1)\overline{f}^2,$$

$$L_{fg} = \sum_{l=1}^{k}\int_{u_1}^{u_2}f_l(u)g_l(u)du-k(u_2-u_1)\overline{f}\,\overline{g}.$$

For minimal fitting error, set $\partial E(\phi)/\partial\phi = 0$, and then the ϕ is:

$$\phi = L_{fg}/L_{gg}.$$

Optimal Control Strategies of a Tuberculosis Model with Exogenous Reinfection

Yali Yang[1,2,*], Xiuchao Song[2], Yuzhou Wang[2], and Guoyun Luo[3]

[1] College of Mathematics and Information Science, Shaanxi Normal University, Xi'an, 710062, China
[2] College of Science, Air Force Engineering University, Xi'an 710051, China
[3] Unit 94170 of the PLA, Xi'an, 710082, China
yylhgr@126.com, xiuchaosong@163.com,
{575884777,799513324}@qq.com

Abstract. For the tuberculosis (TB) model with exogenous reinfection, we study the impact of two control strategies: chemoprophylaxis and treatment. We focus primarily on controlling the disease using an objective function based on a combination of minimizing the number of TB infections and minimizing the cost of control strategies. By using Pontryagin's Maximum Principle, we derive the optimal levels of the two controls. Numerical simulations of the optimal system indicate that the strategies should be improved as the exogenous reinfection transmission coefficient increasing.

Keywords: tuberculosis model, exogenous reinfection, optimal control, pontryagin's Maximum Principle.

1 Introduction

Tuberculosis (TB) is an infectious disease caused by the bacillus Mycobacterium tuberculosis (Mtb), which is spread in the air when people who are sick with pulmonary TB expel bacteria, for example by coughing [1]. Two billion peoples, about one-third of the world's population are infected by it [2]. Fortunately, the vast majority (90%) of people infected with Mtb do not develop TB disease, 5% of people infected progress rapidly to active infection and die without treatment, while 5% progress slowly over their lifetime [3]. It is shown that infected individuals may remain in this latent stage for long and variable periods of time (in fact, many die without ever developing active TB), and latent individuals progression towards active TB may accelerate with re-exposure to TB bacilli through repeated contacts with individuals with active TB [4]. Hence, in our TB model, we must not only look at TB infection as the progression from primary infection but also include the possibility of exogenous reinfection. Treatment for the infectious TB is the basis control strategy [5], moreover, it has been reported that TB latent who with chemoprophylaxis are only 1/30 to 1/40 fraction less likely to develop active TB than those without chemoprophylaxis[6]. Therefore, in this paper, for the TB model with exogenous reinfection, we consider two optimal control strategies: chemoprophylaxis for the latent infections and treatment for the infectious.

D.-S. Huang et al. (Eds.): ICIC 2012, LNCS 7389, pp. 244–251, 2012.
© Springer-Verlag Berlin Heidelberg 2012

Our objective function balances the effect of minimizing the cases of infected TB and minimizing the cost of implementing the control strategies.

This paper is organized as follows. In Section 2, a TB model with exogenous reinfection and two control strategies is formulated, and our objective function is introduced. In Section 3, we derive the optimality system by using Pontryagin's Maximum Principle. At last, we use the Runge-Kutta fourth order scheme to numerically simulate the optimal control model.

2   ATB Model with Exogenous Reinfection

The total population is divided into five epidemiological compartments: susceptible individuals, S; latent individuals, infected with Mtb but not infectious, L; infectious with Mtb (that is, TB active cases), I; being treated for infectious with Mtb, T; chemoprophylaxis for latent with Mtb, V. Then $N = S + L + I + T + V$.

We assume that an individual may be infected only through contacting with infectious individuals. Furthermore, we formulate an $SLIT$ - V epidemic model with exogenous reinfection, seeking to reduce the latent and infectious TB groups, we use chemoprophylaxis for the latent infection and treatment for the infectious, respectively. Furthmore, system (1) parameters' definitions are tabulated in Table 1.

Then the model is given by the following system of differential equations:

$$\begin{cases} S' = m(A - S) - b_1 IS, \\ L' = b_1 IS + (1 - k)dT - (m + u_1(t) + e_1)L - b_2 IL, \\ I' = e_1 L + e_2 V + kdT - (m + a_1)I - (g + g_0 u_2(t))I + b_2 IL, \\ T' = (g + g_0 u_2(t))I - (m + d + a_2)T, \\ V' = u_1(t)L - (m + e_2 + a_3)V. \end{cases} \tag{1}$$

In system (1), $u_1(t)(0 \le u_1(t) \le 1)$ is the fraction which an individual leaves the latent compartment and enters the chemoprophylaxis compartment; $u_2(t)(0 \le u_2(t) \le 1)$ is the fraction which an individual leaves the infectious compartment and enters the treatment compartment during the treatment is strengthened. The control functions $u_i(t)(i = 1, 2)$ are bounded, Lebesgue Integrable functions. The "chemoprophylaxis for latent infection" control $u_1(t)$, represents the fraction of TB latent individuals that is identified and will be put under chemoprophylaxis (to reduce the number of individuals that may be infectious). When the rate $u_1(t)$ is higher, there is less latent and higher implementation costs. Correspondingly, the "treatment for infectious infection" control $u_2(t)$, represents the strengthened fraction of TB infectious individuals that is identified and will be treated. When the rate $u_2(t)$ is higher, there is less infectious and higher implementation costs.

Table 1. Definitions of used symbols

Parameter	Explanation
m	Natural death rate
mA	Recruitment rate of individuals
b_1	Transmission coefficient of the infectious cases for the susceptible individuals
b_2	Transmission coefficient of the infectious cases for the latent individuals
e_1	Rate of a latent individual becoming infectious
e_2	Rate of an individual from chemoprophylaxis to infectious compartment
g	Rate of an infective individual be treated before the treatment policy is strengthened
g_0	The maximal added rate of an infective individual be treated for during the treatment policy is strengthened
d	Rate of an individual complete the treatment
k	Fraction of disease-relapsed
a_1	Disease-induced death rate of infectious individuals
a_2	Disease-induced death rate of treatment individuals
a_3	Disease-induced death rate of chemoprophylaxis individuals

Our objective functional to be minimized is

$$J(u) = \int_0^{t_f} [B_1 L(t) + B_2 I(t) + B_3 T(t) + B_4 V(t) + C_1 u_1^2(t) + C_2 u_2^2(t)]dt \qquad (2)$$

subject to the state system given by (1). The control goal is to minimize the infected populations, include the being treated populations, and the cost of implementing the control measures. The cost result from a variety of sources, we assume that the cost of chemoprophylaxis is nonlinear, we take quadratic form here[7]. The coefficients, B_1, B_2, B_3, B_4 and C_1, C_2 represent the corresponding weight constant, respectively, and these weights are balancing cost factors due to size and importance of the parts of the objective functional.

We seek to find an optimal control function $(u_1^*(t), u_2^*(t))$ such that

$$J(u_1^*, u_2^*) = \min\{ J(u_1, u_2) : (u_1, u_2) \in \Omega \} \qquad (3)$$

where

$$\Omega = \{(u_1, u_2) : u_i \in L^1[0, t_f] \mid \alpha_i \le u_i \le \beta_i, i = 1, 2\} \qquad (4)$$

and $\alpha_i, \beta_i (i = 1, 2)$ are fixed positive constants.

3   Analysis of Optimal Controls

The necessary conditions that an optimal pair must satisfy come from Pontryagin's Maximum Principle[8]. This principle converts (1), (2) and (3) into a problem of minimizing pointwise a Hamiltonian, H, with respect to $(u_1(t), u_2(t))$:

$$
\begin{aligned}
H = &\, B_1 L(t) + B_2 I(t) + B_3 T(t) + B_4 V(t) + C_1 u_1^2(t) + C_2 u_2^2(t) \\
&+ l_1[m(A - S) - b_1 IS] \\
&+ l_2[b_1 IS + (1 - k)dT - (m + u_1(t) + e_1)L - b_2 IL] \\
&+ l_3[e_1 L + e_2 V + kdT - (m + a_1)I - (g + g_0 u_2(t))I + b_2 IL] \\
&+ l_4[(g + g_0 u_2(t))I - (m + d + a_2)T] \\
&+ l_5[u_1(t)L - (m + e_2 + a_3)V],
\end{aligned}
\tag{5}
$$

where $l_i (i = 1, \cdots, 5)$ are the adjoint variables. By applying Pontryagin's Maximum Principle and the existence result for the optimal control pairs from[9], we obtain the following result.

Theorem. There exists an optimal control (u_1^*, u_2^*) and corresponding solution, S^*, L^*, I^*, T^* and V^* that minimizes $J(u_1, u_2)$ over Ω. Furthermore, there exists adjoint functions, $l_1(t), \ldots, l_5(t)$, such that

$$
\begin{cases}
l_1' = (l_1 - l_2)b_1 I^* + l_1 m, \\
l_2' = -B_1 + l_2(m + u(t) + e_1) + (l_2 - l_3)b_2 I^* - l_3 e_1 - l_5 u_1(t), \\
l_3' = -B_2 + (l_1 - l_2)b_1 S^* + (l_2 - l_3)b_2 L^* + l_3(m + a_1) + l_3(g + g_0 u_2(t)) - l_4 g, \\
l_4' = -B_3 - l_2(1 - k)d - l_3 kd + l_4(m + d + a_2), \\
l_5' = -B_4 - l_3 e_2 + l_5(e_2 + m + a_3),
\end{cases}
\tag{6}
$$

with transversality conditions

$$
l_i(t_f) = 0 \ (i = 1, \ldots 5).
\tag{7}
$$

The following characterization holds

$$
u_1^* = \max\{\alpha_1, \min\{\beta_1, \frac{(l_2 - l_5)L^*}{2C_1}\}\},
$$

$$
u_2^* = \max\{\alpha_2, \min\{\beta_2, \frac{(l_3 - l_4)g_0 I^*}{2C_2}\}\}
\tag{8}
$$

Proof. Corollary 4.1[9] gives the existence of an optimal control due to the convexity of integrand of J with respect to $(u_1(t), u_2(t))$, a priori boundedness of the state solutions, and the Lipschitz property of the state system with respect to the state variables.

Applying Pontryagin's Maximum Principle [8], we obtain $l_1' = -\dfrac{\partial H}{\partial S}, l_2' = -\dfrac{\partial H}{\partial L}, \ldots, l_5' = -\dfrac{\partial H}{\partial V}$ and $l_i(t_f) = 0 \ (i = 1, \ldots 5)$ evaluated at the optimal control and corresponding states, which results in the stated adjoint system (6) with transversality conditions (7).

By considering the optimality conditions, $\dfrac{\partial H}{\partial u_i} = 0$ and solving for $u_i(t)(i = 1, 2)$, subject to the constraints, the characterizations (8) can be derived.

To illustrate the characterization of $u_i(t)(i=1,2)$, we have

$$\frac{\partial H}{\partial u_1} = 2C_1 u_1(t) - l_2 L + l_5 L = 0,$$

$$\frac{\partial H}{\partial u_2} = 2C_2 u_2(t) - l_3 g_0 L + l_4 g_0 L = 0 \tag{9}$$

at $u_i(t)$ on the set $\{t | \alpha_i \le u_i \le \beta_i\}(i=1,2)$. On this set,

$$u_1^* = \frac{(l_2 - l_5)L^*}{2C_1},$$

$$u_2^* = \frac{(l_3 - l_4)g_0 I^*}{2C_2} \tag{10}$$

Taking into account the bounds on $u_i(t)(i=1,2)$, we obtain the characterization of $(u_1^*(t), u_2^*(t))$ in (8).

The optimality system consists of the state system (1) with the initial conditions, the adjoint system with the terminal transversality conditions (6), (7) and the control characterization (8). The uniqueness of solutions to the optimality system can be obtained by standard results. Thus, the uniqueness of the optimal control follows from the uniqueness of the optimality system.

4   Numerical Simulation

In this section, we discuss the numerical solutions of the optimality system and the interpretations from various cases.

we study numerically two optimal control strategies of the TB model. The optimal control strategies are obtained by solving the optimality system, consisting of 10 ordinary differential equations from the state and adjoint equations, coupled with the control characterization. An iterative method is used for solving the optimality system. Given initial guess for the controls and initial conditions for the states, then, the state differential equations are solved forward in time using a fourth-order Runge-Kutta scheme. Using the current iteration solution of the state equations and the given transversality conditions (7), the adjoint system is solved backward in time, again employing a fourth-order Runge-Kutta scheme. Both state and adjoint values are used to update the control using the characterizations given by (8), and then the iterative process is repeated. This iterative process terminates when current state, adjoint, and control values converge sufficiently, in other words, the iteration is stopped if the values of unknowns at the previous iteration are very close to the ones at the present iteration.

In our simulation, we assume $A = 100000$, thus $L = mA = 1/70 \times 100000$. The bound on the control coming from estimating about the effectiveness of the intervention strategy. In choosing lower and upper bounds for the controls, for the control $u_1(t)$, it is reasonable to assume that upper bound is 0.5, a reasonable lower bound is

0; for the control $u_2(t)$, it is reasonable to assume that upper bound is 0.9, a reasonable lower bound is 0. The initial values for the states are given by $S(0) = 95000, L(0) = 2000, I(0) = 2000, T(0) = 1000, V(0) = 0$, respectively.

Table 2. Parameter values used in the Fig. 1

Parameter	Value	Source	Parameter	Value	Source
m	1/70.0	[10]	b_1	30/100000	Assumed
e_1	0.00368	[11]	e_2	0.001	[6]
g	0.50	[12]	d	1.5	[13]
k	0.01	[1]	a_1	0.3	[14]
a_2	0.05	[15]	a_3	0.03	Assumed
g_0	0.50	Assumed	$b_2 = s * b_1$	$s = 0.3$	Assumed

In Fig. 1, parameter values as the Table 2 list. Due to lack of data, some parameter values are assumed within realistic ranges for the purpose of simulation. The units where applicable are per year. The weights in the objective function are $B_1 = 1, B_2 = 100, B_3 = 10, B_4 = 0.5$ which mean that minimization of the number of latent, infectious, treated, and Chemoprophylaxis has different importance. Fig. 1 shows the optimal treatment strategy for the case of $C_1 = 400, C_2 = 2000$. The populations with optimal control actions are shown in dashed lines, to compare with the populations without control actions. Obviously, we can know that, from Fig. 1, the infected individuals are decreased sharply, when to use the optimal control strategies, comparing with the case without control actions. Then, the infected will be eliminated if we adopt the optimal control measures.

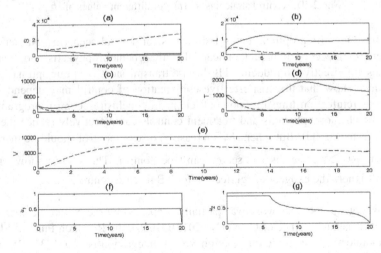

Fig. 1. Simulations of the model (1). Dashed line: Populations with control; Solid line: without control

In Fig. 2, we show how the control strategy $u_2(t)$ depends on the exogenous rein-fection transmission coefficient b_2, which is the exogenous reinfection transmission coefficient of the infectious cases for the latent individuals. We suppose $b_2 = s * b_1$, then let s choose $0, 0.5, 1, 1.5, 2$ respectively. s is shown that there doesn't exist exogenous reinfection for the TB; $s = 0.5$ is shown that the exogenous reinfection transmission coefficient b_2 is smaller than b_1; $s = 1.5$ is shown that the exogenous reinfection transmission coefficient b_2 is bigger than b_1; and so on. From Fig. 2, we find that the control strategy $u_2(t)$ should be increased as b_2 increasing.

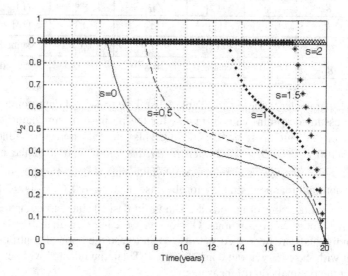

Fig. 2. The control strategies $u_2(t)$ for different values of b_2

We have identified optimal control strategies for several scenarios. Numerical si-mulations of the control model indicate that the chemoprophylaxis and treatment strategies are effective in reducing TB disease transmission. And our optimal control results also show that the cost-effective combination of control may depend on the exogenous reinfection transmission coefficient. In conclusion, control programs that follow the chemoprophylaxis and treatment controls can effectively reduce the num-ber of infected for TB. But when b_2 increasing, disease control should be increased, even when $b_2 \geq 2$, the disease seems can't be control. The studies show that we shouldn't ignore the exogenous reinfection for TB disease control.

Acknowledgements. This work was partially supported by the Nature Science Foun-dation of Shaanxi Province of China (No.2012JQ1019), the Research Fund of Shaanxi Key Laboratory of Electronic Information System Integration(No.201113Y12), and the Nature Science Foundation of China (No.11071256).

References

1. WHO: Global Tuberculosis Control, WHO Report 2011, Switzerland (2011)
2. WHO: Global Tuberculosis Control, WHO Report 2006, Geneva (2006)
3. Blower, S.M., Small, P.M., Howell, P.C.: Control Strategies for Tuberculosis Epidemics: New Models for Old Problems. Science 273, 497–500 (1996)
4. Feng, Z.L., Castillo-Chavez, C., Capurro, A.F.: A Model for Tuberculosis with Exogenous Reinfection. Theor. Popul. Biol. 57, 235–247 (2000)
5. Russell, D.G., Barry, C.E., Flynn, J.L.: Tuberculosis: What We Don't Know Can, and Does, Hurt Us. Science 328, 852–856 (2010)
6. Ziv, E., Daley, C.L., Blower, S.M.: Early Therapy for Latent Tuberculosis Infection. Am. J. Epidemiol. 153, 381–385 (2001)
7. Tchuenche, J., Khamis, S., Agusto, F., et al.: Optimal Control and Sensitivity Analysis of an Influenza Model with Treatment and Vaccination. Acta Biotheor. (2010)
8. Pontryagin, L.S., Boltyanskii, V.G., Gamkrelidze, R.V., et al.: The Mathematical Theory of Optimal Processes. Wiley, New York (1962)
9. Fleming, W.H., Rishel, R.W.: Deterministic and Stochastic Optimal Control. Springer, New York (1975)
10. Castillo-chavez, C., Song, B.J.: Dynamical Models of Tuberculosis and Their Applications. Math. Biosci. Eng. 1, 361–404 (2004)
11. Blower, S.M., McLean, A.R., Porco, T.C., et al.: The Intrinsic Transmission Dynamics of Tuberculosis Epidemics. Nature Med. 1, 815–821 (1995)
12. Bhunu, C., Garira, W.: Tuberculosis Transmission Model with Chemoprophylaxis and Treatment. Bulletin of Mathematical Biology 70, 1163–1191 (2008)
13. Rodriguesa, P., Gomes, M.G.M., Rebelo, C.: Drug Resistance in Tuberculosis-a Reinfection Model. Theor. Pop. Biol. 71, 196–212 (2007)
14. Ted, C., Megan, M.: Modeling Epidemics of Multidrug-Resisitant M.tuberculosis of Heterogeneous Fitness. Nature Medicine 10, 1117–1121 (2004)
15. Sanchez, M.A., Blower, S.M.: Uncertainty and Sensitivity Analysis of the Basic Reproductive Rate: Tuberculosis as an Example. American Journal of Epidemiology 145, 1127–1137 (1997)

Method of Smartphone Users' Information Protection Based on Composite Behavior Monitor

Hua Zha and Chunlin Peng

Research Institute of Electronic Science and Technology, University of Electronic Science and Technology of China, Chengdu,Si Chuan Province, P.R.China
{zhahua_orphen,peng_chunlin}@163.com

Abstract. For the users' information of the Smartphone disclosure issues, a method of users' information secure for the Smartphone has been researched. Based on hidden Markov model and fuzzy pattern recognition model, the paper proposes a compound monitoring approach which use the heuristic scanning-behavior block monitoring, to discriminate the behavior of the program whether is malicious. Finally, the paper introduces a method for testing Smartphone users' information protection. The results show that this method is able to well implement to monitor and block the malicious behavior of the program.

Keywords: heuristic scanning, behavior blocking, malicious behavior, Smartphones Information Protection.

1   Introduction

Today, according to market research agency IDC recently released [1], at the end of the fourth quarter of 2011, every sold three mobile phones in the global, there is a smartphone. It is becoming the people second data processing center. However, the Smartphone users' information data is more and more unsafe.

On the Smartphone, the users' information disclosure is particularly dangerous. The literature [2] indicates that malicious program will collect the Smartphone users' privacy information after these implant in the Smartphone. Then the Smartphone users' privacy information will be sent out. As these programs have hidden, cell firewall is also difficult to intercept. Therefore, for Smartphone users' information security, we need to find a suitable method of protection for the Smartphone characteristics.

Combining the characteristics of Smartphone in this paper, based on hidden Markov model and fuzzy pattern recognition model, we put forward the heuristic scan-behavior blocker composite monitoring methods, to achieve real-time security protection on mobile users' information.

2   Related Research

Today, Smartphone monitoring method is on mature, there is: feature code detection method, heuristic analysis method and behavioral blocking method [3, 4].

D.-S. Huang et al. (Eds.): ICIC 2012, LNCS 7389, pp. 252–259, 2012.
© Springer-Verlag Berlin Heidelberg 2012

2.1   Feature Code Detection Method

This detection technology is single which is not affective for the unknown malicious program. The literature [5] indicates that as the security threatened on Smartphones will become diverse, the malicious programs will render polymorphism, for the detection capability of a single feature code technology will gradually be replaced by other ways.

2.2   Heuristic Analysis Method

The literature [6, 7] indicates that the heuristic analysis is a priori program behavior prediction methods, through knowledge and experience, it analysis the intent of the program and determine the behavior of the program. By differences from the features of the traditional detection methods for detection of program code, it detects suspicious logic of the program so that detecting unknown malicious programs is functional. In the specific implementation, the heuristic analysis method exists to identify suspicious code instruction sequence requires time-consuming, and false negatives and false positives problem.

2.3   Behavior Blocking Method

The literature [8] indicates that the behavior blocking methods will define security policy initially. While the program is running, real-time monitor the program behavior. When the program is illegally, the security policy blocks it. There are two types of security policy: one is a smart approach, according to certain algorithm, set the factors and analyzes and judges the behavior of the program; another one is user direct involvement, determine whether the program is malicious behavior by the user. The literature [4] indicates, "The behavior of the different types of viruses can widely vary, and may cause the system behavior patterns have many kinds of infection", this will lead to the monitoring and analysis of the behavior of the program will take a lot of time and resources.

Today, Smartphone computing power and storage capacity increase substantially, the calculation time and resource consumption on heuristic analysis and behavior blocking constraints of the both of these two methods are no longer obvious. In this paper, we combine the heuristic analysis method and behavior blocking method, and propose a Smartphone users' information protection method based on heuristic scanning - behavior blocking composite monitoring.

3   Method of Composite Monitoring

3.1   Malicious Behavior Feature

According to feature analysis of the program of the malicious behavior for users' information, we mine the malicious behavior factors, as shown in Table 1.

Table 1. The program of malicious behavior feature

Number	Behavior Feature	Threat Degree
1	program self-delete	high
2	service auto start	high
3	copying in the system directory	high
4	process hidden	high
5	SMS interception	high
6	auto start of the memory card	high
7	reading address book	middle
8	reading SMS	middle
9	reading call records	middle
10	voice recording	high
11	mobile locating	high
12	abnormally calling	high
13	sending SMS	low
14	sending E-mail	middle
15	sending by network	high

3.2   Heuristic Scanning

We construct a model for the procedure focused using the double stochastic process of hidden Markov model. λ as a 5-tuple (S,V,A,B,Π).

Monitoring the behavior of procedures can be considered as a binary classification problem. S as a group of state set, define the set of procedure state $S = \{0, 1\}$, 0 is none-malice, 1 is malice, the state number $N = 2$. s as a behavior detected, has $s \in S$.

V is the set of procedures for malicious behavior, as shown in Table 1, $V=\{V_1,V_2,V_3,...,V_M\}$, V_M is the malicious behavior of M, M is the total number of currently known malicious behavior.

A is the state transition matrix, $A=\left[a_{ij}\right]_{N \times N} = \begin{bmatrix} a_{00} & a_{01} \\ a_{10} & a_{11} \end{bmatrix}$, $a_{ij} = P(q_{t+1}=j \mid q_t=i), 1 \leq i$, $j \leq N$, where q_t is the current program state, and q_{t+1} is the next program state.

B is the malicious programs probability distribution, $B = \{b_j(k)\}$, $bj(k)$ is the probability distribution for procedure state j, $b_j(k) = P(V_k \mid j), 1 \leq k \leq M, 1 \leq j \leq N$.

Π is the initial state probability distribution. The state of the system is normal or abnormal, so $\Pi = \{\pi_1, \pi_2\}$, $\pi_i = P(q_1=i)$, is the probability of choosing i when time 1.

B is the probability distribution of malicious programs, $B = \{b_j(k)\}$, $b_j(k)$ express the probability of malicious behavior while the program is in state j, $b_j(k) = P(V_k \mid j)$, $1 \leq k \leq M, 1 \leq j \leq N$.

Π is the initial state probability distribution. System state in a normal state or malicious state, so $\Pi = \{\pi_1, \pi_2\}$, $\pi_i = P(q_1=i)$, π_i is the probability of selecting state i in initial time.

HMM model as the malicious behavior of the generating device, according to a certain step, generate malicious behavior series: $O(t) = (O_1,O_2, O_3,..., O_T) \in V$, T is the number of malicious behavior monitoring to program.

Use the forward - backward algorithm:

Forward variable α_t (i) = $P(O_1 O_2 O_3 \ldots O_t$, $q_t = i \mid \lambda)$, $\alpha_t(i)$ refers to the probability of malicious behavior of the program sequence $O_1 O_2 O_3 \ldots O_t$ at the given model λ in time t under state i.

(1) Initialization: α_1 (i) = π_i b_i (o_1) ($1 \leq i \leq N$)

(2) Iterative calculation: $\alpha_{t+1}(j) = \left[\sum_{i=1}^{N} \alpha_t(i) a_{ij} \right] b_j(O_{t+1})$,$1 \leq t \leq T-1$, $1 \leq i, j \leq N$

(3) Distinguishing malicious behavior: $P\{O \mid \lambda\} = \sum_{i=1}^{N} \alpha_T(i)$

Backward variable $\beta_t(i)$ = $P(O_t O_{t+1} O_{t+2} \ldots O_T \mid q_t = i$, $\lambda)$, $\beta_t(i)$ refers to the probability of malicious behavior of the program sequence $O_t O_{t+1} O_{t+2} \ldots O_T$ at the given model λ in time t under state i.

(1) Initialization: $\beta_T(i)$ = 1 ($1 \leq i \leq N$)

(2) Iterative calculation: $\beta_t(i) = \sum_{i=1}^{N} a_{ij} b_j(O_{t+1}) \beta_{t+1}(j)$ $1 \leq t \leq T-1$, $1 \leq j \leq N$

(3) Distinguishing malicious behavior: $P\{O \mid \lambda\} = \sum_{i=1}^{N} \pi_i b_i(O_1) \beta_1(i)$

We derive the value $P\{O \mid \lambda\}$ by forward and backward algorithm, and set the value by weighted average to arrive at a final predictive value $P\{O \mid \lambda\}$. Where weight is simplified, we subject to the average.

We calculate the malicious behavior programs probability value $P\{O \mid \lambda\}$ and set the threshold ε. When calculating the probability value is greater than the threshold ε, we can be submitted to user determine whether the program is malicious behavior.

3.3   Behavior Blocking Monitor

Behavior blocking monitors focus on suspicious malicious behavior, such as listed in Table 1 a series of malicious behavior, and monitor the type of malicious behavior of the corresponding system API functions. We construct a model for the method of fuzzy pattern recognition. First, we use fuzzy sets to the program behavior is divided into the normal behavior and malicious behavior [9], then use the Optional Near principle for the classification monitoring program.

Fuzzy Pattern Recognition: Definitions of sets, the behavior of the program is divided into two categories, normal and malicious. Select a sample program from the sample space, and extract the sample programs in a set of feature set A, which related to user information security. Each program has a corresponding feature, that can be defined a fuzzy set on the feature class to describe an object or object class. Based on the domain Q={$q_1, q_2, \ldots, q_n$}, it composed by n feature. An executable program or class can be used to define a fuzzy set on the domain of Q to describe, that is $\tilde{M}$. The program has the feature q_i membership is defined as the real number μ_i , $\mu_i \in [0,1]$.

Training:

(1)   According to Table 1 corresponds to the access to the system API function, select t feature set of API functions to form a fuzzy feature set. t=15, feature set can increase with the increase of malicious behavior in the future.

(2)   According to the feature set, we establish the membership function of fuzzy set $\tilde{V}$ of malicious behavior: $\mu_V(E(A_i^V)) = \begin{cases} 0 & , E(A_i^V) \leq 0 \\ 1-e^{-(E(A_i^V))^2/\sigma^2} & , E(A_i^V) \geq 0 \end{cases}$

In the Formula, the $E(A_i^V)$ is the frequency average of the system API functions used in malicious behavior. $\sigma = Max\{E(A_1^V), E(A_2^V),...,E(A_t^V)\}/3$, t is number of feature quantities. With membership function $\mu_V(E(A_i^V))$, evaluate the fuzzy set $\tilde{V} = \{\mu_1^V/A_1, \mu_2^V/A_2,...,\mu_t^V/A_t\}$.

(3)   Establishment membership functions of the fuzzy set $\tilde{N}$ in the API function of the normal behavior: $\mu_V(E(A_i^N)) = \begin{cases} 0 & , E(A_i^N) \leq 0 \\ 1-e^{-(E(A_i^N))^2/\sigma^2} & , E(A_i^N) \geq 0 \end{cases}$

In the Formula, the $E(A_i^N)$ is the frequency average of the system API functions used in normal behavior. With membership function $\mu_V(E(A_i^N))$, evaluate the fuzzy set $\tilde{N} = \{\mu_1^N/A_1, \mu_2^N/A_2,...,\mu_t^N/A_t\}$.

Classification:

(1)   When the program calls a system API functions related to Smartphone users' information , get the API function information , statistics on its frequency of feature, obtain the set $\{A_1, A_2,...,A_t\}$.t is feature space dimension;

(2)   Membership function of the program is: $\mu_M(A_i) = \begin{cases} 0 & , A_i \leq 0 \\ 1-e^{-(A_i)^2/\sigma^2} & , A_i \geq 0 \end{cases}$

With membership function $\mu_M(A_i)$, evaluate the program fuzzy set $\tilde{M} = \{\mu_1/A_1, \mu_2/A_2,...,\mu_t/A_t\}$;

(3)   Get the fuzzy set $\tilde{V}$ and $\tilde{N}$ from the training, separately calculate the Euclid close degree $\Psi(\tilde{M},\tilde{V})$ and $\Psi(\tilde{M},\tilde{N})$.

The formula of the Euclid close degree: $\Psi(\tilde{A},\tilde{B}) = 1 - \frac{1}{\sqrt{t}}\left(\sum_{i=1}^{t}(\mu_i^A - \mu_i^B)^2\right)^{\frac{1}{2}}$;

(4)   Use the" optional near principles" to categorize detected program. The principle is: for the fuzzy sets $\tilde{A}_i, \tilde{B}$ (i= 1, 2,...,n),if $\exists i$ has $\Psi(\tilde{A}_i,\tilde{B}) = \bigvee_{1 \leq j \leq n} \Psi(\tilde{A}_j,\tilde{B})$,that consider the $\tilde{A}_i$ and $\tilde{B}$ closest, classify the both as one class.

Based on the above two methods, we propose the composite model. The model for the Smartphone users' information protection system: Use heuristic scanning methods, scan the program, determine whether the program behavior threat users' information; Use behavior blocking method, monitor these system API functions related to users' information in real-time, when the program use these system API functions, determine whether the program behavior threat users' information. The entire interception program is processed by user. Then we construct the confidence list of this complementary measure; store the users' selection in the list.

4   A Protection Model of Users' Information Based on Composite Monitoring

Based on composite monitor method, this paper design users' information protection model, which of architecture as shown in Figure 1.

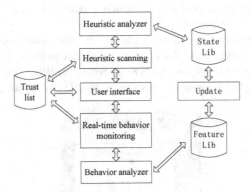

Fig. 1. The architecture of users' information protection model

Heuristic scanning module: it's initiated by user, scans the Smartphone system program which not in the trust list, detects the program behavior whether threat the users' information. For malicious behavior program, it submits user to select blocking or releasing the program, and adds the program into the trust list.

Heuristic analyzer: it uses hidden Markov rule, analyzes the program behavior which is submitted by the scanning module, determines the program whether has the malicious behavior and backs the result to the scanning module.

Real-time behavior monitoring module: it monitors system API functions which is related to users' information. If the program which is not in the trust list uses these system API functions, it determines the program behavior whether threat the users' information. For the malicious behavior of program, it submits user to select blocking or releasing the program, and adds the program into the trust list.

Behavior analyzer: it uses fuzzy pattern recognition module, classifies the program behavior which is submitted by real-time behavior monitoring module and backs the classified results to monitoring module.

User interface: the interface of user and module. It submits the malicious behavior of program which monitor from scanning module and real-time monitoring module, prompts the user to select block or release. Finally, it returns the result to their selected module and records into the trust list.

Trust list: it establishes the black and white list, stores separately the behavior program of normal and malicious.

5   Experimental Analyses

The experiment uses Windows Mobile Smartphone platform, for the spy software WinCE.Flexispy.A sample[10], writes a malicious behavior program named Alfred.exe. The malicious program enters a latent state after implantation. When the malicious program receives command, turn on the running state. It'll steal contacts, SMS messages, call records and phone physical geographic information, via text messaging, e-mail and networking send out. It's also as detectaphone, monitors user talk.

After the malicious behavior program Alfred.exe implanted in the phone, start the information program model. We run the heuristic scanning to detect the phone in the program. When it detects the program Alfred.exe which is not in the trust list, analyzes the behavior of program, draws a conclusion which Alfred.exe is a malicious program, and submits user to handle.

Open real-time behavior monitoring, and run the malicious behavior program Alfred.exe. When the program get the contacts to access to the relevant API function, the real-time monitoring block and analyze its behavior. While the program behavior classifies monitoring behavior, it submits user to handle. As shown in Figure 2.

Fig. 2. A tip from real-time monitoring of the behavior blocking

Table 2. User information monitor list

	Heuristic scanning	Real-time behavior monitoring
Contacts	√	√
SMS	√	√
Call records	√	√
Calling	√	√
Voice recording	√	√
geographic locating	√	√
SMS sending		√
E-mail sending		√
networking		√

In this experiment, we select Allow, and continue to test. Then the program will continue to get SMS messages, call records, geographic information and calling to access to the relevant system API function, the real-time monitoring module can be blocked, and   submit user to handle.

As the program Alfred.exe covers most of current malicious behavior which is related to Smartphone users' information disclosure, from this experiment can be seen that the heuristic scanning and real-time behavior monitoring can be good find the malicious behavior, thereby demonstrated the method of Smartphone users' information protection based on composite behavior monitor is effective.

Table 2 shows the model can protect the Smartphone users' information.

6   Conclusion

This paper analyzes the malicious behavior feature of the program which threatens the security of Smartphone users' information, and studies the method of the information protection. Based on Hidden Markov Models and fuzzy pattern recognition model, the paper proposes a compound monitoring approach which uses the heuristic scanning-behavior block monitoring. It designs the Smartphone users' information protection model based on the method mentioned above. Experiments show that, the model can be good to monitor the program which threatens the user's information security, and verify the correctness of the compound monitoring approach which uses the heuristic scanning-behavior block monitoring, achieved the expected goals. In the furniture, we need to further improve the malicious behavior library, reduce omissions and false, and increase the accuracy of the blocking of malicious behavior program.

References

1. Statistical report of the global mobile phone (OL) (2012),
   http://www.199it.com/archives/23609.html
2. Cheng, Z.: Mobile Malware: Threats and Prevention, McAfee Avert Labs (2008)
3. Yap, T.S., Ewe, H.T.: A Mobile Phone Malicious Software Detection Model with Behavior Checker. In: Proc. of HIS 2005, pp. 57–65 (2005)
4. Szor, P.: The Art of Computer Virus Research and Defense. Addison Wesley Professional (2005)
5. Wang, Z.H., Wang, H.F.: Study on Anti-Virus Engine Based on Heuristic Search of Polymorphic Virus Behavior. Research and Exploration in Laboratory 25(9), 1089–1091 (2006)
6. Wu, J.J., Fang, M.W., Zhang, X.F.: Research of Mobile Phone Virus Defense Based on Heuristic Behavior-Checking. Computer Engineering and Science 32(1), 35–38 (2010)
7. Cui, P.: Application of Formal Semantics to Heuristic Anti-virus Engine. Journal of Liaodong University (Natural Sciences) 15(3), 167–171 (2008)
8. Hu, Y.T., Chen, G., Zheng, N., Guo, Y.H.: Malicious Executable Detection Based on Runtime Behavior. Computer Engineering and Applications 45(17), 64–66 (2009)
9. Zhang, B.Y., Yin, J.P., Tang, W.S.: Study and Implementation Intelligent Detection System to Recognize Unknown Computer Virus. Computer Engineering and Design 27(11), 1936–1938 (2006)
10. Liu, X.L., Liu, K.: Principle of Malware Analysis and Study of Protective Measures on Windows Mobile Phone. Network Security Technology & Application 9, 41–43 (2008)

Improved Digital Chaotic Sequence Generator Utilized in Encryption Communication

Xiaoyuan Li[1], Bin Qi[2,*], and Lu Wang[3]

[1] Electronic Engineering Department, Harbin Vocational Technical College, Harbin, China
[2] College of Information and Communication Engineering College,
Harbin Engineering University, Harbin, China
[3] School of Overseas Education, Northeast Normal University, Changchun, China
{qibinwinter,wanglu.daily}@gmail.com
hzylxywl@163.com,

Abstract. Chaos has good statistical characteristics and is sensitive with initial value which is appropriate for encryption system. Since digital chaos is conducted under the condition of finite precision, which will lead to the degradation of the system and the output sequence will appear periodicity. This paper proposed an improvement for the short period of digital chaotic sequence and designs a new type of digital chaotic sequence generator based on FPGA. Experimental data show that this new cycle of the system output sequence can be set according to the encryption necessary which remedies the short-period phenomenon of digital chaos, so that digital chaotic sequence generator is safe to be applied to the encryption system.

Keywords: correlative peak interval, finite precision, correlation, chaotic pseudo-random sequence.

1 Introduction

In recent years, chaotic encryption is widely used in many research fields for its good characteristics that chaos is sensitive with initial value and is similar to random sequence [1]. Since random sequence can be used in many situations, the replacement of random sequence with chaos has become a hotpot and conducted by many researchers [2, 3]. However, using digital system to realize chaos has many difficulties that chaos is restrained by the finite precision of the system and even small errors introduced in each iteration will have a big effect on the implement of chaos [4]. Consequently, the accumulation of the error will result in the deviation of the orbit and greatly affect the characteristic of the system [5]. Using analog circuits to realize chaotic sequence had been used for a long time [6]. However, due to the sensitivity of chaos to initial values and parameters, chaotic key stream is easily affected by environment conditions [7]. Moreover, it is hard to set up the cooperation between the sending and receiving terminal [8]. And most research just focus on the simulation experiment, which can not discover the problems happened in the actual hardware [9, 10]. Theoretically, chaotic sequence in non-period, but it will turn to be a period

D.-S. Huang et al. (Eds.): ICIC 2012, LNCS 7389, pp. 260–268, 2012.
© Springer-Verlag Berlin Heidelberg 2012

sequence in the situation of finite precision. In addition, chaos will have strong correlation after discretization, which will greatly affect its performance [11, 12]. So this paper did some research on the correlative peak interval and presents a new type of digital chaotic sequence generator.

2   Chaos Theory

Bifurcation chart is the description of state variant based on parameter space. General mapping form is:

$$x_{n+1} = f(\mu, x_n), x_n \in R \tag{1}$$

Where $f(\cdot, \cdot)$ is the differentiable function, μ is a parameter. If initial value is fixed, the value of x_n repeats infinite cycle among $p(p \geq 1)$ states which is a periodic orbit. According to the method of linear stability, the conditions for stable periodic orbits is

$$\left| \prod_{t=1}^{p} f'(\mu, x'_t) \right| \leq 1 \tag{2}$$

where the periodic orbit p is called super-stable. In this paper, the logistic map is defined as:

$$x_{n+1} = \mu x(1-x), \mu \in [0,4] x_n \in [0,1] \tag{3}$$

We get the fixed point $O: x = 0; A: x = 1 - 1/\mu$ .The stability of fixed point is determined by $|f'(x)|$, which is:

$$f'(x) = \mu - 2x\mu \tag{4}$$

Consequently, the stability of fixed points depends on the parameter μ. From the behavior of iterative equation (3), it could obtained that when the parameter μ becomes larger from zero, the iterative process has different dynamic behavior. Nonlinear equations change the topology structure of system trajectory, which will cause the overall shape of the system changes suddenly and produce the phenomenon of bifurcation. However, it is the necessary process of the generation of chaos.

3   Logistic Sequence Generator and Key Selection

Logistic mapping mathematical expression is shown as equation (3), when μ is in the interval of [3.5699456, 4], logistic map enters into chaotic state and presents complex dynamics. We design logistic chaotic mapping digital circuit model according to the equation shown in Fig. 1, where Input is the initial value, Shift and Sampler are two binary output sequence, and the other module is the logistic chaotic mapping operations unit. Output is the output of the key sequence generator.

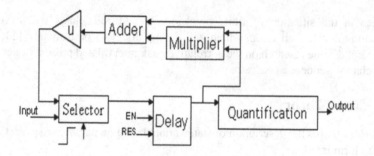

Fig. 1. Logistic chaotic map circuit based on FPGA

Initial value is the first consideration in the analysis of chaos. It can be seen that the statistical characteristics of the output sequence are affected by the computing accuracy, equation parameters and initial values. As the computing precision in the system is limited, so it is not suitable if we use the key input in chaotic encryption algorithm. According to the research of chaotic dynamics, when parameter μ is in the interval of [3.5699456, 4], logistic map is in the chaotic state and the output sequence is non-periodic and non-convergent. However, it can be seen from Lapunov index curve that the interval is not always in chaos state, and when $\mu = 4$, the map is a full shot in the unit interval [0, 1] that the chaotic sequence has the characteristic of periodicity. Therefore μ can not be the initial key input of chaotic encryption when the initial value has a tiny deviation, the orbit will separate with exponential speed. So it is impossible to have a long-term prediction on the behavior of the system. Just as the chaotic system is sensitive with the initial value, when the chaotic system is assigned with different initial value, we can get a series different and not related chaotic sequence. Therefore, we choose the initial value of chaotic systems as the chaotic key input. We select $\mu = 4$, precision is 38, and initial value X (0) as the initial key conditions in logistic equation under the condition of Matlab simulation environment. With different selected initial keys, auto-correlation tests were shown in Fig. 2.

It can be seen that with the same accuracy and parameters, when the initial value is different, the output sequences are not the same. 10^6 groups of data were selected to test relevance in this paper. When the initial value is 0.216, it does not have any relevant peak interval in the whole operation region, and when the initial value is 0.1378, there are a few of relevant peak intervals. After repeated test, any initial value will produce an unpredictable relevant peak interval. That is using initial value as the input key could not make the encryption system security when digital chaotic system is used for plaintext encryption.

Consequently, if the ciphertext sequence has cyclical relevant peak interval, it will directly threaten the safety of the key sequences. But the discretization of the chaotic relevant peak interval is inevitable, so how to get the ideal interval between the peak values becomes a key issue in the digital chaotic sequence encryption system. After a lot of relevant test we can see that relevant peak interval is equal or approximately equal, so in this paper, we do the analysis on relevant interval which is equal to the peak period.

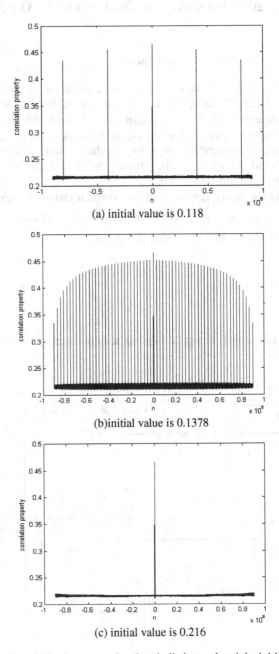

(a) initial value is 0.118

(b)initial value is 0.1378

(c) initial value is 0.216

Fig. 2. The relation between related periodic interval and the initial value

4   Digital Logistic Hardware Realization and Its Output Sequence Test

4.1   Digital Logistic Hardware Realization

Both the design and development process are completed in the chip of Altera development board EP2C8Q208C8N. The digital chaotic key stream generator was designed with the operating platform of Quartus II 8.1 based on logistic equation. Fig. 3 is the logistic hardware circuit diagram which includes obtaining the initial value, floating point unit, the implementation of the standard double-precision addition and multiplication, iteration, fixed point, floating-point conversion, and changing quantitative modules into binary sequences.

In order to generate binary chaotic sequence, original chaotic sequence $\{x(n)\}$ was changed to binary sequence $\{s(n)\}$ which subjects to $s(n) = T[x(n)]$, $n = 0,1,\cdots\cdots$, where $T[x(n)]$ is an irreversible function defined as

$$T[x(n)] = \begin{cases} 0 & x(n) \in \bigcup_{k=0}^{2^m-1} I_{2k}^m \\ 1 & x(n) \in \bigcup_{k=0}^{2^m-1} I_{2k=1}^m \end{cases} \tag{5}$$

As chaotic signal $\{x(n)\}$ has good random statistic characteristics, so $\{s(n)\}$ has balanced 0-1 ratio.

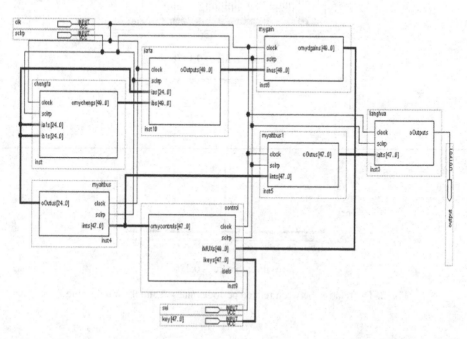

Fig. 3. Logistic chaotic map hardware circuit connection diagram

According to the equation (3), when $X = \{x(n) \mid n = 0,1,2,\cdots, x(n) \in [0,1]\}$, the binary sequence after conversion is $S = \{s(n) \mid n = 0,1,2,\cdots, s(n) \in \{0,1\}\}$. In order to make the design simple and the easy to be realized, equation (3) is done with the linear transformation.

$$T[x(n)] = \begin{cases} 0, & 2^m x(n) \in [2k \quad (2k+1)) \\ 1, & 2^m x(n) \in [(2k+1) \quad (2k+2)] \end{cases} \tag{6}$$

Where m is the arbitrary positive integer and $k = 0,1,2,\cdots\cdots,2^m -1$. Whether the output of sequence is 0 or 1 will be determined by the parity of bit. Bit extractor will complete the judgment of the bit parity that barrel shift register and the bit selector can complete the function of equation (3). In order to improve the period of chaotic sequence, digital chaotic key stream generator we designed consists of two parts. The first part is the output of the uncertain chaotic sequence where $\mu = 4$, the precision is 38, and the initial value could be any value. The simulation results were shown in Fig. 4. The default system clock period is 10ns, rising edge trigger, sclrp is the circuit reset signal. When sclrp is high, the whole circuit is valid to reset initialization. Sel is the initial key loading start signal. When sclrp is low and sel is high, the circuit will sends the initial key into the circuit, however, when sel changes from high to low, the circuit uses the input key as the initial value for the iteration logistic equation. This paper chooses 48 bits binary number as the initial key input, and output is the sequence output of Logistic iteration equation.

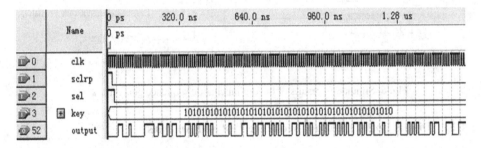

Fig. 4. Uncertain chaotic system output with arbitrary chaotic key

The second part is used for fixing initial value, $\mu = 4$, the reliable system simulation waveform under the precision of 38 (Fig. 5).

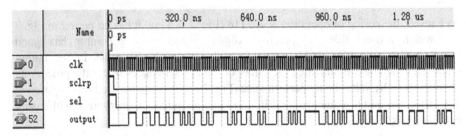

Fig. 5. Reliable system output sequence with fixed initial values, parameters and accuracy

The XOR output between the reliable systems with initial key chaotic sequence system is shown in Fig. 6. The reset signal is sclrp and the high value is effective. When sclrp is low and sel is high, the system will load initial key AAAAAAAAAAAA. N1 is the output sequence of arbitrary system. When the third rising edge of the clock signal is triggered, the system will do the iteration with initial key. N is a reliable system, and the output is the result after the operation with N1.

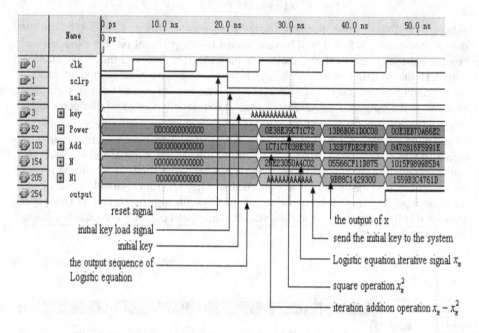

Fig. 6. Simulation results of the digital chaotic sequence generator key

4.2   Digital Logistic Output Sequence Test and Analysis

The length of the binary sequence is N, then the correlation coefficient of the binary sequence is defined as following, where m is the parameters for the step.

$$R(m) = \frac{1}{N} \cdot \sum_{i=1}^{N-m} x_i \cdot x_{i+m} \tag{7}$$

As a result, the period of operation is 402114 (Fig. 7 (a)) and it has no period in Fig. 7 (b), which proved that the practical digital chaotic key generator has good cryptographic properties. It is inevitable to have correlation peaks in the output sequence which directly threaten the safety of the key sequences. After the processing of operator XOR, it has better performance (Fig. 7(c)). So with the method proposed in this paper, the output sequence used in digital communication encryption will ensure better autocorrelation.

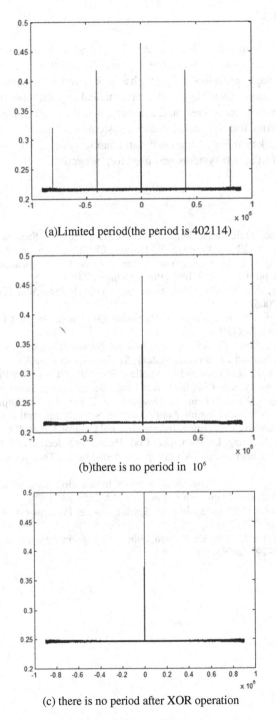

(a)Limited period(the period is 402114)

(b)there is no period in 10^6

(c) there is no period after XOR operation

Fig. 7. the relevance of the digital chaotic sequence key generator

5   Conclusion

Discrete chaotic sequence will generate chaos degradation, which will make the periodic and pseudo-random sequence worse, and result in the encryption system insecurity. This paper proposes a digital chaotic key sequence generator based on the short-period phenomena and digital chaotic practical applications. It consists of fixed initial value, equation parameters and accuracy, and the chaotic sequences after good statistical properties test are called reliable system. The initial values of the system were used as the key input of the uncertain chaotic system. The new sequence after XOR operation of the two systems was used for encryption.

References

1. Suneel, M.: Electronic Circuit Realization of the Logistic Map. Sadhana 31, 69–78 (2006)
2. Ding, Q., Zhu, Y., Zhang, F., Peng, X.: Discrete Chaotic Circuit and the Property Analysis of Output Sequence. In: International Symposium on Communications and Information Technologies, pp. 1009–1012. IEEE Press, Beijing (2005)
3. Li, Q., Yang, X.: 2D Chaotic Signals Generator Design. Acta Electronic Sinica 33, 1299–1302 (2005)
4. Yang, X., Tang, Y.: Horseshoes in Piecewise Continuous Maps. Chaos, Solutions and Fractals 19, 841–845 (2004)
5. Li, K., Soh, Y., Zhang, C.: A Frequency Aliasing Approach to Chaos-based Cryptosystems. IEEE Transactions on Circuits and Systems 51, 2470–2475 (2004)
6. Huang, R., Huang, H.: Chaos and Its Applications. Wuhan University Press (2005)
7. Lv, J., Lu, J.: Analysis and Application of Chaos Series. Wuhan University Press (2002)
8. Ding, Q., Pang, J., Fang, J., Peng, X.: Designing of Chaotic System Output Sequence Circuit Based on FPGA and its Possible Applications in Network Encryption Cards. International Journal of Innovative Computing, Information and Control 3, 449–456 (2007)
9. Ding, Q., Pan, J., Wang, L.: The Cipher Code Parameter Selection and its Impact on Output Cycles. In: 2009 International Workshop on Chaos-Fractals Theories and Applications, pp. 143–147 (2009)
10. Zhao, G., Fang, J.: The Progress of Modern Information Security and Chaotic Secure Communication. The Progress of Physics 23, 212–225 (2003)
11. Zhang, X.: Logistic Chaotic Mapping Section and its Performance Analysis. Electronic Journal 37, 720–725 (2009)
12. Chee, C., Xu, D.: Chaotic Encryption Using Discrete-time Synchronous Chaos. Physics Letters 3, 284–292 (2006)

Modeling and Adaptive Control for Flapping-Wing Micro Aerial Vehicle

Qingwei Li[1] and Hongjun Duan[2]

[1] Department of Environmental Science and Engineering,
Northeastern University at Qinhuangdao, Qinhuangdao, China
[2] Department of Automation Engineering, Northeastern University at Qinhuangdao,
Qinhuangdao, China
lqwday@126.com, dhj@mail.neuq.edu.cn

Abstract. Flight quality of flapping-wing micro aerial vehicle (FMAV) depends much upon efficient control of flight attitude. So, an accurate model of flight attitude is of utmost importance. The fly mechanism of birds and big insects, especially the motion rule of wings were investigated to establish a complete dynamic model and mathematical model for flight attitude of FMAV. The design of attitude controller is challenging due to the complexity of the flight process, and the difficulty is system uncertainty, nonlinearity, multi-coupled parameters, and all kinds of disturbances. To control the attitude movement effectively, a global adaptive H∞ control strategy was constructed that the controller synthesis was based on Lyapunov function instead of solving the Hamilton-Jacobi-Isaacs (HJI) partial differential equation. The method overcomes the impact of time-varying parameters and unknown disturbances to the system. Simulation results support the effectiveness of the dynamic model and the control strategy.

Keywords: flapping-wing micro aerial vehicle, dynamic model, nonlinearity, adaptive H∞ control, attitude control.

1 Introduction

Flapping-wing Micro Aerial Vehicle (FMAV) has a great technological potential in remote surveillance missions, inspection of infrastructures like dam walls and investigation of hazardous or dangerous environment [1-2]. Safe autonomous flight is essential for widespread acceptance of aircraft that must fly close to the ground and such capability is widely sought. For example, search and rescue operations in the setting of a natural disaster allow different vantage points at low altitude. Likewise, FMAV performing reconnaissance for the police, news or the military must fly low enough that the environment presents obstacles. Flying close to and among obstacles is difficult because of the challenges in sensing small obstacles, in three dimensions. Some aspects of collision avoidance are easier for air vehicles than ground vehicles. Any object close to the intended path of an air vehicle must be avoided as opposed to ground vehicles where deviations from the nominal ground plane indicate obstacles

D.-S. Huang et al. (Eds.): ICIC 2012, LNCS 7389, pp. 269–276, 2012.
© Springer-Verlag Berlin Heidelberg 2012

and are often not visible until they are close. The use of helicopters, rather than fixed wing aircraft also helps because in the worst case it is possible to come to a hover in front of an obstacle [3]. Significant research interest and energy has been directed towards the development of autonomous helicopters due to their high payload to power ratio. Helicopters, however, are extremely dangerous in practice due to the exposed rotor blades, and are only suitable for autonomous applications where there is no chance of unintended human-robot interaction. The small size, highly coupled dynamics and low cost implementation of indoor aerial robotic devices poses a number of significant challenges in both construction and control [4-5].

The flight circumstances of FMAV are similar to that of birds or big insects: fluid dynamics under low Reynolds numbers and unsteady aerodynamics. Flapping flight is possible with only two degrees of freedom: flapping ("up and down" motion) and feathering (angular movement about the wing longitudinal axis) [6]. In flight process, the propulsion of FMAV is aerodynamic force which is represented as the sum of instantaneous translational force and rotational force [7]. The flight control is accomplished by controlling the aerodynamic forces and moments generated by the airfoils during flapping. However, aerodynamic forces on airfoils are highly nonlinear and time-variable during a stroke cycle. As a result, the system dynamics cannot be approximated by a linear time-invariant model [9]. Paper [10] describes the model identification and attitude control system for a Micromechanical Flying Insect (MFI). In nonlinear optimal control theory, nonlinear H∞ control method is robust and potential approach to the attitude control problem. However, the practical applications of the nonlinear H∞ control method still remain open due to the difficulty in solving the associated Hamilton-Jacobi-Isaacs (HJI) partial differential equation. There have been some replaceable attempts for it [12-17]. In this paper, a global H∞ control strategy for nonlinear time-varying system is introduced where Lyapunov function is used to solve the H∞ problem instead of solving the HJI partial differential equation. This method is applied to H∞ attitude control for FMAV. The design process consists of two steps. First, without regard to the disturbances, a global asymptotically stable adaptive controller is contrived. Then a nonlinear H∞ controller synthesis is developed to overcome the external disturbances.

2   Dynamic Modeling of FMAV

In flight process, aerodynamic force, gravity, and propulsion from airfoils act normal to airframe. According to the transform of inertia coordinate system and airframe coordinate frame, we get

$$\begin{cases} \dot{\vartheta} = q\sin\psi + r\cos\psi \\ \dot{\theta} = (q\cos\psi - r\sin\psi)/\cos\vartheta \\ \dot{\psi} = p - \tan\vartheta(q\cos\psi - r\sin\psi) \end{cases} \tag{1}$$

where ϑ, θ , ψ are pitch, yaw, roll angles respectively; p, q, r are angular velocities respectively. In inertia coordinate system, the dynamic equations of FMAV rotating around its centroid are as follows

$$\begin{cases} \dot{p} = ((J_y - J_z)rq + M_{bx} + M_{wx})/J_x \\ \dot{q} = ((J_z - J_x)pr + M_{by} + M_{wy})/J_y \\ \dot{r} = ((J_x - J_y)qp + M_{bz} + M_{wz})/J_z \end{cases} \quad (2)$$

where M_{bx}, M_{by}, M_{bz} are roll, yaw, pitch moments of airframe respectively; M_{wx}, M_{wy}, M_{wz} are roll, yaw, pitch moments of airfoils respectively; J_x, J_y, J_z are the inertia moments of airframe. We can simulate airfoils motion by

$$\begin{cases} \varphi_l(t) = \varphi_L \sin(2\pi f t - \lambda) \\ \varphi_r(t) = -\varphi_R \sin(2\pi f t - \lambda) \\ \phi_l(t) = -\phi_L \cos 2\pi f t - \Delta\phi_l \\ \phi_r(t) = \phi_R \cos 2\pi f t + \Delta\phi_r \end{cases} \quad (3)$$

where $\varphi_l(t)$, $\varphi_r(t)$, $\phi_l(t)$, $\phi_r(t)$ are the real-time values of feathering angles and flapping angles, respectively; φ_L, φ_R, ϕ_L, ϕ_R are the maximum of feathering angles and flapping angles, respectively; $\Delta\phi_l$, $\Delta\phi_r$ are the emendation values of flapping angles; f is the wingbeat frequency; λ is the phase shifting between feathering and flapping; the subscripts l and L mean "left", r and R mean "right". The flight attitude control is accomplished by adjusting the angles $\varphi_l(t)$, $\varphi_r(t)$, $\phi_l(t)$, $\phi_r(t)$ real-timely to change airfoil aerodynamic moments M_{wz}, M_{wy}, M_{wx}. So the following mathematical model can be elicited.

$$m(x)\ddot{x} + f(x, \dot{x}) = u + w \quad (4)$$

where x is the state vector; u is the input vector; w is the unknown disturbance vector; $m(x)$ is the positive definite coefficient matrix; $f(x, \dot{x})$ is the system vector function, and

$$x = \begin{bmatrix} \vartheta \\ \theta \\ \psi \end{bmatrix}, f = \begin{bmatrix} f_1 \\ f_2 \\ f_3 \end{bmatrix}, u = \begin{bmatrix} M_{wz}/J_z \\ M_{wy}/J_y \\ M_{wx}/J_x \end{bmatrix}, m(x) = \begin{bmatrix} \cos\psi & -\cos\vartheta\sin\psi & 0 \\ \sin\psi & \cos\vartheta\cos\psi & 0 \\ 0 & \sin\psi & 1 \end{bmatrix},$$

$$f_1 = \dot{\vartheta}\dot{\psi}\sin\psi(1 - \sin\vartheta) + \dot{\psi}^2\cos\vartheta\cos\psi$$
$$+ \{(J_x - J_y)(\dot{\theta}\sin\vartheta(\dot{\vartheta}\sin\psi + \dot{\theta}\cos\vartheta\cos\psi)$$
$$+ \dot{\psi}(\dot{\vartheta}\sin\psi + \dot{\theta}\cos\vartheta\cos\psi)) + M_{bz}\}/J_z,$$

$$f_2 = -\dot{\psi}\cos\vartheta(\dot{\vartheta}+\dot{\theta}\sin\psi)-\dot{\vartheta}\dot{\theta}\sin\vartheta\cos\psi$$
$$+\{(J_z - J_x)(-\dot{\psi}\cos\vartheta\sin\psi(\dot{\theta}\sin\vartheta+\dot{\psi})$$
$$+\dot{\vartheta}\cos\psi(\dot{\theta}\sin\vartheta+\dot{\psi}))+M_{by}\}/J_y,$$
$$f_3 = -\dot{\vartheta}\dot{\theta}\cos\vartheta+\{(J_y - J_z)(\dot{\vartheta}\cos\psi(\dot{\vartheta}\sin\psi$$
$$+\dot{\theta}\cos\vartheta)-\dot{\vartheta}\dot{\psi}\cos\vartheta\sin\psi(\sin\psi+1))+M_{bz}\}/J_z.$$

3   Adaptive H∞ Control and Performance

The design process of nonlinear H∞ control law consists of two steps.
Step1, the following adaptive controller is proposed

$$u = m(x)\ddot{x}_r + f(x,\dot{x})+k_D\dot{e}+k_P e+v \tag{5}$$

where, x_r is the desired value of x, $e = x_r - x$; k_D, k_P are symmetric positive definite matrices; v is the new control input vector that will be designed.

When $w = 0$, the closed loop system (4) is global asymptotically stable with $v = 0$. When $w \neq 0$, we will design a controller v that still makes the system be global asymptotically stable.
Step2, an extended error vector is defined by

$$s := \dot{e} \tag{6}$$

Substitute (5)~(6) into (4), the following closed loop system can be obtained

$$\dot{s} = g(s,t)+g_v(s,t)v(t)+g_w(s,t)w(t) \tag{7}$$

where $g(s,t) = -m(x)^{-1}(k_D s+k_P e)$, $g_v(s,t) = g_w(s,t) = -m(x)^{-1}$.

Define the output vector to be controlled as follows

$$z = h_1(s,t)+h_2(s,t)v \tag{8}$$

where $h_1(s,t) = [e,s,0]^T$, $h_2(s,t) = [0,0,I]^T$.

Define the following positive definite function

$$V(s,t) := \frac{1}{2}s^T m(x)s \tag{9}$$

Theorem 1: Given the attitude control system (4), if the controller parameters matrix k_D and k_P are selected properly, that the following conditions are satisfied

$$s^T(\dot{m}/2+I-k_D)s \leq 0 \tag{10}$$

$$s^T m\dot{s} - s^T k_p e + e^T e \leq 0 \qquad (11)$$

Under the given control law (5), with

$$v = s/2 \qquad (12)$$

The closed loop system is globally asymptotic stable. Moreover, given any positive scalar $\gamma > 1$,

$$\int_0^t \|z(t)\|^2 \, dt \leq \gamma^2 \int_0^t \|w(t)\|^2 \, dt \qquad (13)$$

is satisfied for all $t > 0$ and all piecewise continuous functions $w(t)$.

Proof: Take the time derivative of $V(s,t)$

$$\dot{V} = dV / dt + (\partial V / \partial s)(g + g_v v + g_w w) \qquad (14)$$

Define

$$H(s,v,w,t) := \dot{V} + \|z(t)\|^2 - \gamma^2 \|w(t)\|^2 \qquad (15)$$

Substitute (14) and (8) into (15), the following result can be obtained

$$H = dV / dt + (\partial V / \partial s)(g + g_v v + g_w w) + h_1^T h_1 + v^T v - \gamma^2 w^T w \qquad (16)$$

Solve equations $\partial H / \partial w = 0$ and $\partial H / \partial v = 0$, the results can be obtained as follows

$$w_1 = \frac{1}{2\gamma^2} g_w^T(s,t)\left(\frac{\partial V}{\partial s}\right)^T = -\frac{s}{2\gamma^2}, v_1 = -\frac{1}{2} g_v^T(s,t)\left(\frac{\partial V}{\partial s}\right)^T = \frac{s}{2}.$$

$H(s,v,w,t)$ takes the maximum when $v = v_1$, $w = w_1$.

$$H(s,v,w,t) \leq H(s,v_1,w_1,t) = s^T m\dot{s} - s^T k_p e + e^T e + e^T(\dot{m}/2 + I - k_D)s$$
$$+ s^T((1/4\gamma^2) - 1/4)s \qquad (17)$$

Substitute (10)~(12) and $\gamma > 1$ into (17), we have

$$H(s,v,w,t) \leq 0 \qquad (18)$$

It follows from (15) that

$$\dot{V} + \|z(t)\|^2 - \gamma^2 \|w(t)\|^2 \le 0 \qquad (19)$$

When the system has an undisturbed motion, i.e. $w = 0$, the following result can be gotten

$$\dot{V} \le -\|z(t)\|^2 \qquad (20)$$

It indicates that the closed loop system is globally asymptotically stable when $w = 0$ and $v = v_1$. Integrate on both sides of the inequality (19) that

$$V(s) - V(s_0) \le -\int_0^s \|z(t)\|^2 \, dt + \gamma^2 \int_0^s \|w(t)\|^2 \, dt \qquad (21)$$

The initial condition of the closed loop system (4) is $s_0 = 0$, so $V(s_0) = 0$. Moreover, it follows from (9) that $V(s) \ge 0$. The result of (13) can be obtained, and the proof is end.

4   Simulation Study

The dynamic model parameters of FMAV are: $R = 10\text{cm}$; $\bar{c} = 5\text{cm}$; $\rho = 1.20\text{kg}/\text{m}^3$; $m = 120\text{g}$; $V = 5\text{m}/\text{s}$; $f = 12.5\text{Hz}$; $\phi_L = \phi_R = 60°$; $\varphi_L = \varphi_R = 30°$; $\Delta\phi_l = \Delta\phi_r = 20°$; $\lambda = 10°$; $\alpha = 5°$; $\beta = 0°$. The initial control conditions are: $\vartheta = \theta = \psi = 0°$, $m(0) = \text{diag}\{1,1,1\}$, $\Delta m = \text{diag}\{0.5, 0.4, 0.2\}$, $f(0) = [5.5, 4.5, 2.0]^T$, $\Delta f = [1.5, 1.5, 0.5]^T$, $w = [1.0° \sin t, 1.5° \sin t, 0.5° \sin t]^T$. The controller parameters are: $\gamma = 2$, $k_D = \text{diag}\{2.55, 2.25, 1.65\}$, $k_P = \text{diag}\{5.5, 5, 4\}$. The desired attitude angles of are given with step shapes.

The simulation results are in Fig.1~Fig.3. The real lines denote the tracking results of nominal system, and the attitude angles are with subscript "1"; the broken lines denote the tracking results of actual system with uncertainty and disturbance, and the attitude angles are with subscript "2". From the simulation results, we get to know that the desired attitude angles are tracked stably.

$$\vartheta_r = \begin{cases} 15°, & 0s \le t < 4s \\ 0°, & 4s \le t < 8s \\ -15°, & 8s \le t < 12s \\ 0°, & 12s \le t \le 16s \end{cases}$$

$$\theta_r = \begin{cases} 0°, & 0s \le t < 4s \\ 45°, & 4s \le t < 8s \\ -45°, & 8s \le t < 12s \\ 0°, & 12s \le t \le 16s \end{cases}$$

$$\psi_r = \begin{cases} -10°, & 0s \le t < 4s \\ 0°, & 4s \le t < 8s \\ 10°, & 8s \le t < 12s \\ 0°, & 12s \le t \le 16s \end{cases}$$

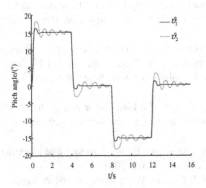

Fig. 1. The step response of pitch angle

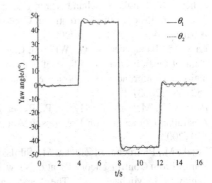

Fig. 2. The step response of yaw angle

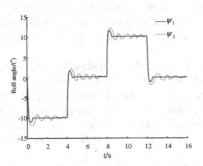

Fig. 3. The step response of roll angle

5   Conclusion

Through observing the wings motion of birds or insects in flight, we investigated the airframe dynamics, airfoil aerodynamic forces and moments of FMAV, then its dynamic model and mathematical model of flight attitude were developed. The attitude control system of FMAV embodies nonlinearity, parameter coupling, uncertainty, and all kinds of disturbances. A novel control scheme — global adaptive H∞ control scheme was proposed. Lyapunov function was used to solve the H∞ problem instead of solving the HJI partial differential equation. Simulation results indicate that the tracking errors of attitude angles reach into an acceptant bound in finite time.

References

1. Wzorek, M., Conte, G., Rudol, P., Merz, T., Duranti, S., Doherty, P.: From Motion Planning to Control – A Navigation Framework for an Autonomous Unmanned Aerial Vehicle. In: Proc. of the 21st Bristol International UAV Systems Conference (2006)
2. Le, B.F., Mahony, R., Hamel, T., Binetti, P.: Adaptive Filtering and Image Based Visual Servo Control of a Ducted Fan Flying Robot. In: Proc. of the 45th IEEE Conference on Decision & Control, San Diego, CA, USA, December 13-15, pp. 1751–1757 (2006)

3. Scherer, S., Singh, S., Chamberlain, L.J., Saripalli, S.: Flying Fast and Low Among Obstacles. In: Proc. of the International Conference on Robotics & Automation, Roma, Italy, April 10-14, pp. 2023–2029 (2007)
4. La, C.M., Papageorgiou, G., William, C.M., Kanade, T.: Integrated Modeling and Robust Control for Full-envelope Flight of Robotic Helicopters. In: Proc. of the 2003 IEEE International Conference on Robotics & Automation, Taipei, Taiwan, September 14-19, pp. 552–557 (2003)
5. Pounds, P., Mahony, R., Hynes, P., Roberts, J.: Design of a Four Rotor Aerial Robot. In: Australian Conference on Robotics & Automation, Auckland, November 27-29, pp. 145–150 (2002)
6. Lasek, M., Pietrucha, J., Zlocka, M., Sibilski, K.: Analogies Between Rotary and Flapping Wings From Control Theory Point of View. AIAA-2001–4002 (2001)
7. Sanjay, P.S., Michael, H.D.: The Aerodynamic Effects of Wing Rotation and A Revised Quasi-steady Model of Flapping Flight. Journal of Experimental Biology 205, 1087–1096 (2002)
8. Yan, J., Wood, R.J., Avadhanula, S., Sitti, M., Fearing, R.S.: Towards Flapping Wing Control for a Micromechanical Flying Insect. In: Proceedings of the 2001 IEEE International Conference on Robotics & Automation, Seoul, Korea, May 21-26, pp. 3901–3908 (2001)
9. Schenato, L., Campolo, D., Sastry, S.: Controllability Issues in Flapping Flight for Biomimetic Micro Aerial Vehicles (MAVs). In: Proceedings of the 42nd IEEE International Conference on Decision & Control, Maui, Hawaii, USA, pp. 6441–6447 (December 2003)
10. Deng, X.Y., Schenato, L., Sastry, S.S.: Model Identification and Attitude Control for a Micromechanical Flying Insect Including Thorax and Sensor Models. In: Proc. of IEEE International Conference on Robotics & Automation, Teipei, Taiwan, September 14-19, pp. 1152–1157 (2003)
11. Liu, Z.L., Svobode, J.: A New Control Scheme for Nonlinear Systems with Disturbances. IEEE Transactions on Control Systems Technology 14(1), 176–181 (2006)
12. Magni, L., Nijmeijer, H., van der Schaft, A.J.: A Receding-horizon Approach to the Nonlinear H∞ control problem. Automatica 37, 429–435 (2001)
13. Aliyu, M.D.S.: A Transformation Approach for Solving the Hamilton-Jacobi-Bellman equation in H_2 deterministic and stochastic optimal control of affine nonlinear systems. Automatica 39, 1243–1249 (2003)
14. Aliyu, M.D.S.: An Approach for Solving the Hamilton-Jacobi-Isaacs Equation (HJIE) in Nonlinear H∞ control. Automatica 39, 877–884 (2003)
15. Aguilar, L.T., Orlov, Y., Acho, L.: Nonlinear H∞-control of Nonsmooth Time-varying Systems with Application to Friction Mechanical Manipulators. Automatica 39, 1532–1542 (2003)
16. Chang, Y.C.: An Adaptive H∞ Tracking Control for a Class of Nonlinear Multiple-Input - Multiple-Output (MIMO) systems. IEEE Transactions on Automatic Control 46(9), 1432–1437 (2001)
17. Su, W., Souza, C., Xie, L.: H∞ Control for Asymptotically Stable Nonlinear Systems. IEEE Transactions on Automatic Control 44(5), 989–993 (1999)

Distributed Staff's Integral Systems Design and Implementation

Qing Xie, Guo-Dong Liu, Zheng-Hua Shu, Bing-Xin Wang, and Deng-Ji Zhao

Key Laboratory for Optoelectronics & Communication of Jiangxi Province
Jiangxi Science & Technology Normal University, Jiangxi, Nanchang, P.R. China
xqwyy163@163.com

Abstract. This article is based on the practical application of the communications industry, to make fully use of the characteristic of distributed systems such as it can solve the dispersed organization and the data that needs to be communicated, expand the organization with minimal impact, balance the load to find the right system architecture. So as to achieve the distribution application, centralized management, in the building characteristics of the management mechanism to achieve the different system needs of different parts.

Keywords: communications industry, distributed systems, minimal impact, distribution application, centralized management.

1 Introduction

With the rapid development of information technology, the enterprise will improve the timeliness requirements of the network system reliability and the cost in the future. But for a large telecommunications company, the principle of Centralized system is accounted for part of ratio, the problem of centralized system is also a very significant and outstanding issue. For example the host is leaded to an excessive burden on running the process. There is often appearing the speed so slowly or the phenomenon of the collapse. Although the centralized system was revaluating in the last century, some enterprises is still using the centralized system. Because they are using the original software and the software are often very expensive. Therefore, the principle of centralized systems has not been suited to the modern fast and efficient development of the communication.

So we bring up the distributed system, applying the principle of a distributed system to create an economic benefit, faster and stronger, any extension of the processing system. Data operations are scripting by Linux, and interface browser is achieving through the web.

2 The Concept

The distributed system made up of more independent calculating devices and they can be able to communicate with each other. Right users, the systems like a computer.

D.-S. Huang et al. (Eds.): ICIC 2012, LNCS 7389, pp. 277–283, 2012.
© Springer-Verlag Berlin Heidelberg 2012

This sentence contains two meaning: Firstly, the computer is autonomous in hardware. Secondly, users will make the entire system as a computer in software. Generally speaking, the goal of parallel processing uses all processors to perform a large task. But each processor of the distributed systems usually executes the quasi-sovereign program sequence. Due to resource sharing, availability and fault tolerance for various reasons, the processor needs to coordinate action with each other.

Distributed programming language is used to write the distributed program that is running on distributed computer. The distributed program made up of a number of programs which can perform independently. They are distributed in a distributed processing system and simultaneously executed on several computers. With compared to the centralized programming language, there are three characteristics: Distribution, Communication and Stability.

The integral system of distributed is the information management system ,which draw on the principles of distributed systems, use cities as a unit, design the clerk management subsystem integration individually for its own, in order to meet the different needs for various cities. Using distributed systems, realize distributed applications and centralized management. To establish the efficient and special management mechanism, as the same time convenient for the system expansion without intervention with each other, and coordinate each other.

3   Design and Realization

3.1   The General Framework

Supposing each city (that is A, B, C, D, E, F .etc) sees a separate computer, and the province company is user. All cities are combined to form a high operation without

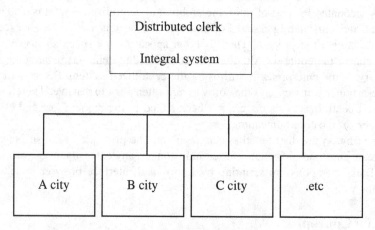

Fig. 1. The general framework

intervention each other. So users (that is the provinces) through the system platform, can directly observed the operation of each city. Therefore we make a simple distributed system which is simple and general.

3.2   The System Topology Schematic

Distributed system structures are divided into two parts: the surrounding system and the shop assistant integral system. At the same time two parts share different function. As is shown in the chart below.

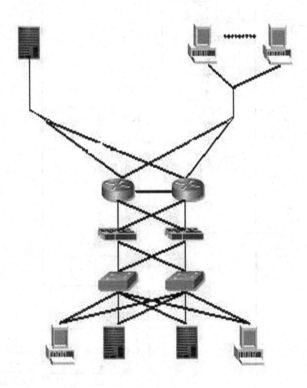

Fig. 2. The system topology schematic

The hardware of Clerk integral system including: double workstation, database server, business library server and Web server hardware performance (according to the hardware performance,several severs can be merged into a computer). Server shelves is seted in the support center.Server different user can use this system by the web, different users can use the system by the web.

3.3   System Integration Rules

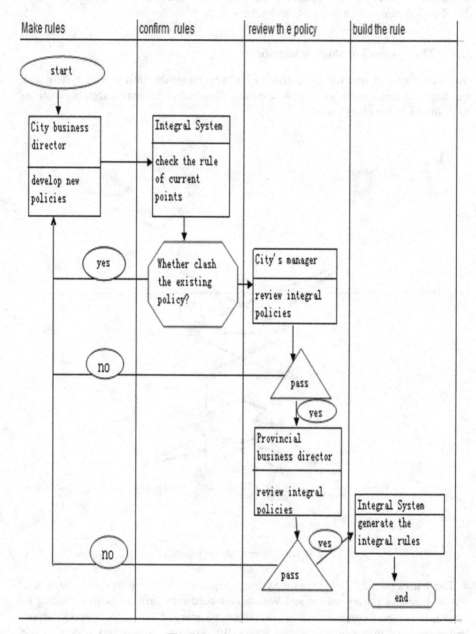

Fig. 3. System integration rules

3.4   The Core Algorithm

Integral template-based operation is to make the rules and main points program separated, No longer binding the complex points to the rules of the main program. Instead, it was singled out and formed some multiple templates. This has the advantage of flexible, fast, actuating operation of integral rules and management of dictionary, which reduces system maintenance costs and maintenance requirements, reflects the user-level data and the effective separation of system-level data.

The Integral Process Is as Follows. First, according to the local city market, carry on the personalized modification of the template rules for the province's uniform integral, and submitted to the management system. Second, the integral management system process the submission of points around the city, generate different Integral template, Store in the system database. Third, processing the main points program regularly, transferring to the relevant Integral template, generating a temporary integral program, processing the data from the boss by the program.

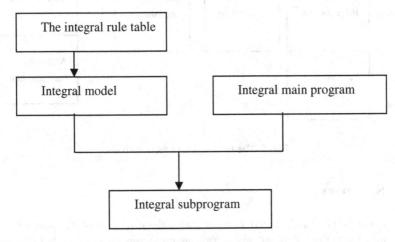

Fig. 4. The integral process

The Template Operation of Integration Disposal Has the Following Advantages. First, simplify the integration rule modification process, business directors only need to submit by themselves, then take effect, no longer will need the intervention of system Maintenance personnel, modify the main program code. Second, reduce maintenance costs, Because of the integration rules modification, there is no need to modify the main program's code, thus significantly reducing the pressure of the system maintenance personnel, while also reducing systemic risk. Third, make the integration rule configuration flexible, because of the dictionary management of the integration rules, business directors of every city can carry on all kinds of optimization configuration conveniently, greatly enhance the market response speed.

The Total System Function Charts

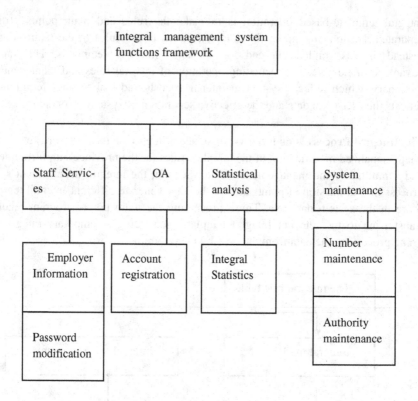

Fig. 5. The total system function charts

4   Summary

This system is using distributed systems under the direction of the principle, it not only can achieve the independently in various prefectures and cities, but also modify their template, create its own characteristics, conducive to head office, update policy adjustments more nearly and timely for the different market conditions, while significantly reducing the investment of resources and equipment. At present distributed systems is everywhere (commercial, academic, government and family). These systems usually provide to share resources (such as color printers or scanners, and other special equipment) means of sharing data. It is extremely important for our information-based economy. It is an example of the distributed system to calculation, at the same time it is more and more popular in providing computing resources and services. Distributed systems is more challenging to solve sub-problems through parallel computing to provide higher performance, and they also provide greater availability of certain components to prevent failure.

References

1. George, C., Jean, D., Tim, K.: Distributed Systems Concepts and Design, 3rd edn. Addison-Wesley, Pearson Education (2001)
2. Garofalakis, M.: Mining Sequential Patterns With Regular Expression Constraints. IEEE Transactions on Knowledge and Data Engineering 14(3), 530–552 (2002)
3. Franks, R.G.: Performance Analysis of Distributed Server Systems. Carleton University, Ottawa (1999)
4. Reiser, M., Lavenburg, S.: Mean Value Analysis of Closed Queuing Networks. ACM 28(3), 629 (1981)
5. Reiser, M.: Mean Value Analysis of Queuing Networks, A New Look at An Old Problem. In: Performance of Computer Systems, pp. 63–77. North-Holland, Amsterdam (1979)
6. John, D., Rich, F., Tai, J.: Web Server Performance Measurement and Modeling Techniques. Performance Evaluation (33), 5–26 (1988)
7. Caraca-Valente, J.P., Lopez-Chavarrias, I.: Discovering Similar Patterns in Time Series. In: Proceedings of the 6th ACM SIGKDD International Conference on Knowledge Discovery and Data Mining, USA, Boston, Massachusetts, pp. 497–505 (2000)

Research of QoC-aware Service Adaptation in Pervasive Environment

Di Zheng[1], Qingwei Xu[2], and Ke-rong Ben[1]

[1] Department of Computer Science, Naval University of Engineering,
Wuhan, Hubei, China 430033
[2] Communication Institute, Beijing, China 100085

Abstract. Existing context-aware middleware are designed to automatically adapt its behavior to changing environment such as effectively selecting services for adaptation according to the user's current context. However, with the increase of pervasive services and the kinds of sensors, the importance of information quality becomes more urgent. We should continuously monitor/capture and interpret the environment related information efficiently to assure efficient context awareness. Many attentions have been paid to the research of the context-aware pervasive applications and context-aware service adaptation. However, little of them pay attention to the quality based service adaptation. Therefore, we propose a quality management middleware to support QoC management for service adaptation. By using this framework, we can configure different strategies to refinery raw context, discard duplicate and inconsistent context so as to protect and provide QoS-enriched context information of users to support context-aware service adaptation.

Keywords: QoC, middleware, context-aware, pervasive, service adaptation.

1 Introduction

In recent years, context-awareness has become one of the core technologies in pervasive computing environment gradually and been considered as the indispensable function for pervasive applications[1].Many research efforts have been done for gathering, processing, providing, and using context information [2].

At the same time, with the help of context-ware middleware, we can efficiently select services for adaptation according to the user's current context. And one service can dynamic reconfigure the software components of it or request the help of other services can be a efficient approach for addressing context-awareness of applications. The service can add, replace, or remove the component of it, and also change the value of its variables. However, with the increase of the scale of applications, we can not make sure all the contexts from different sources are valid and certain. So Buchholz et al. [3] defined QoC "as any information describing the quality of information that is used as context" firstly. Then following researchers have characterized context information by certain well-defined QoC aspects, such as precision, accuracy, completeness, up-to-dateness, security and so on [4~8]. But these studies evaluate quality only on some

D.-S. Huang et al. (Eds.): ICIC 2012, LNCS 7389, pp. 284–292, 2012.
© Springer-Verlag Berlin Heidelberg 2012

aspects, i.e. they do not consider complex and comprehensive applications which need pay attention to different context configuration and adapt to the user as wells as to the environment, the devices and the complex relationships between them adaptively and efficiently.

In our previous works, we have put forward a middleware for the context-aware service-based applications so as to make these applications can be adapted more easily than traditional applications by simply adding and deleting services[9,10,11]. Based on this middleware, we use the agents to support the configuration of different quality factors. Furthermore the strategies used by these agents can help us evaluate raw context, discard duplicate and inconsistent context so as to protect and provide QoS-enriched context information of users to support quality based service adaptation.

2   Quality Based Service Adaptation

2.1   Agent Based Quality Management of Context

As depicted info figure 2, we divide the entire context –aware process into five layers including sensor layer, retriever layer, deal layer, distribution layer and application layer. Different from existing methods, we pay attention to the quality management of context through all these layers.

Firstly, in the sensor layer, we set agents supporting different threshold to implement auto context discarding. The agents can be configured with one or more threshold, by this way we can reduce the number of the raw contexts. Secondly, in the retriever layer, we use the context quality index to describe the quality of the contexts. All these factors are decided by user's demand and they may be different at all in various applications. Furthermore, we use the agents to complete the computation of these factors.

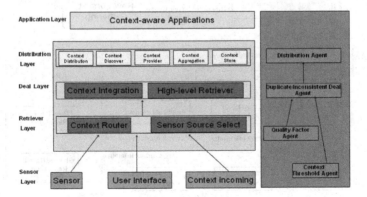

Fig. 1. Agent based Quality Management of Context

Then in the deal layer, we expand existing context dealing process with the duplicate context discarding and inconsistent context discarding to provide more accurate and more efficient contexts for the applications. All the algorithms are configured in the duplicate and inconsistent dealing agent.

At last, in the distribution layer and application layer, we expand traditional context-aware component/service adaptation/deployment algorithms with the help of the incoming contexts. This process is also helped by the distribution agent.

2.2   QoC-aware Context Processing Procedure

The detailed quality-aware context processing procedure is shown in Figure 2.The first step is the raw context gathering, in which raw contexts from various sensor sources are collected during a fixed short period. In this step we will use one or multiple thresholds to refinery the raw context.

The second step is the duplicate and inconsistency resolution during context interpretation and aggregation. If there are some contexts having the same identifier or the same name/value pairs, we will check the sources of context. If they have different sources then some errors may occur and we should check the gathering of the contexts. If these two context objects are from the same source then we check the time when these context objects are generated. If they have the same timestamp then it means that they are the exact duplicate of each other and anyone of them can be discarded as well as keeping the other one. If they have the different timestamps it means that these are the duplicate contexts and will be discarded.

Then, we apply rule-based reasoning to generate high-level contexts. The user-defined rules are in the form of Jena generic rules without negation and "or" operation. The two reasoners are configured as "traceable" in order to facilitate updating dependency graphs in context repository, though more memory is required. After that, we use inferred high-level contexts to update the context repository and notify applications which register context triggers.

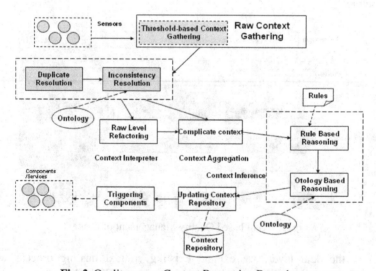

Fig. 2. Quality-aware Context Processing Procedure

2.3   Typical Quality Factors Used for Adaptation

In pervasive environment, there may be lots of factors affecting the adaptation. For example, we can use precision to reflect the granularity how the context can describes the real world. For numeric context information, the value described with three significant figures (e.g. 32.2) is more precise than with two significant figures (i.e. 32). There are many other factors like this and they can help us improve the efficiency of the middleware as well as make the applications be adaptive to the environment more accuracy. As follows are several typical quality factors:

- Precision. It is used to reflect the granularity how the context can describes the real world. For numeric context information, the value described with three significant figures (e.g. 32.2) is more precise than with two significant figures (i.e. 32).
- Completeness. It is used to describe the relative ration of the available context versus the entire collected context.
- Freshness. It is a tempo definition and it is used to define the time that elapses between the determination of context information and its delivery to a requester
- Certainty. It is described as the certainty of the context. When we have similar contexts from different sensors, we should choose the right context with the help of certainty.
- Relativity. It is used to describe the frequency of the context. Usually, the context will be reused in a short period. So we think the context being reused is relative.
- Usability. Usability of context information will be measured by comparing the granularity of context information presented by the context object and the granularity of context object that is required by a context consumer.
- Accuracy. It is used to describe the degree of the realness that the context can reflect the real environment. The lower accuracy may affected by many reasons such as sensors and wrong dealing of context.

Up-to-dateness. This quality measure indicates the degree to use a context object for a specific application at a given time.

The services including in different applications may pay attention to different factor, one or their combination. For example: Healthcare patient monitoring service: This service is an important part of HIS and it pays more attention to the time and the accuracy. Supposing the service get an input context that a patient having heart disease may be in trouble, then the doctors and the nurses may come to help him soon. If the context is wrong, the dealing is useless. But if the context is delayed then the life of the patient may be lost. So we can use the combination of delay and accuracy to help the service complete the adaptation. The percent of these two factors can be adjusted by the users according to their demand.

2.4   QoC-aware Adaptation

To different service, we use the configuration file of different factors to help the adaptation firstly. For example, we can configure the file as follows:

$$ADA(S_1) = Up(CxtObj)*W_1 + A(CxtObj)*W_2 + C(CxtObj)*W_3$$
$$W_1 + W_2 + W_3 = 1$$

These rules are used in the case the service is used to open the device automatically when it finds the user is walking near. In this equation W_1, W_2 and W_3 represent the weight of these three factors and we can configure them as we need. We choose the value as 0.3 , 0.4, 0.4 for accuracy and certainty are more important.

$Up(CxtObj)$ represents age of the context object $CxtObj$ by taking the difference between the current time and the measurement time of that context object.

$$Age(CxtObj) = t_{curr} - t_{measu}(CxtObj)$$

$$Up(CxtObj) = \begin{cases} 1 - \dfrac{Age(CxtObj)}{TimePeriod(CxtObj)} : \\ if \ Age(CxtObj) < TimePeriod(CxtObj) \\ 0 : otherwise \end{cases}$$

The value of up-to-dateness and hence the validity of context object $CxtObj$ decrease as the age of that context object increases.

✧ $A(CxtObj)$ represents accuracy as follows: $A(CxtObj) = \dfrac{CorrectnessProbability}{MinimumCorrectnessProbability}$

We use *CorrectnessProbability* to represent current correctness probability of context and use *MinimumCorrectnessProbability* to represent the minimum correctness probability defined by user. If the ratio is larger than 1, the freshness may be good. According to the difference of user's demand, we can use different $A(CxtObj)$ to select sensor source.

✧ $C(CxtObj)$ represents certainty as follows:

$$C(CxtObj) = \begin{cases} CO(CxtObj) \times \dfrac{NumberofAnsweredRequest + 1}{NumberofRequest + 1} \\ : if \ F(CxtObj) \neq 0 \ and \ CxtObj \neq null \\ CO(CxtObj) \times \dfrac{NumberofAnsweredRequest}{NumberofRequest + 1} : otherwise \end{cases}$$

$$F(CxtObj) = \dfrac{CurrentFreshInterval}{MinimunFreshInterval} \qquad CO(CxtObj) = \dfrac{\sum_{j=0}^{m} \omega_j(\sigma)}{\sum_{i=0}^{n} \omega_i(\sigma)}$$

We use *NumberofAnsweredRequest* to represent the number of the reply requests and use *NumberofRequest* to represent the sending requests. Furthermore, we use the ratio to represent the certainty with the help of completeness $CO(CxtObj)$ and freshness $F(CxtObj)$. If the ratio is more close to 1, the resolution may be better. According to the difference of user's demand, we can use different $C(CxtObj)$ to select sensor source.

By using the configuration like this, the service will compute the adaptation value and decide to react to the circumstance by itself.

Algorithm 1. Algorithm for QoC Adaptation
INPUT: New arrived context

1.   *Configure the service demanding for the quality factors*
2.   *Configure the distribution Agent*
3.   *Set the inconsistent context discarding algorithm*
4.   *While true*
5.   *get the identifier of contexts*
6.   *if There exists contexts with same ID*
7.   *then*
8.   *if sourceID of both context objects match*
9.   *then*
10.   *if timestamp of both context objects match*
11.   *then*
12.   *Find duplicate contexts and discard anyone*
13.   *end if*
14.   *else*
15.   *Check the context gathering*
16.   *end if*
17.   *else if There exists contexts with same name/value pairs*
18.   *Discard one according to the quality tuple*
19.   *end if*
20.   *end if*
21.   *get the new instance of context in the queue of matching patterns* pat_que
22.   *To all the patterns* $pat_1, pat_2, ... pat_n$
23.   *In the pat's trigger* tgr
24.   *if exists* ins_1 *in* pat_que_1 , ins_2 *in* pat_que_2 , ..., *and* ins_n *in*
        pat_que_n *and* *tgr satisfy the constraint of* ins_1, ins_2 , ..., ins_n
25.   *then*
26.   *if the constraint of tgr is satisfied*
27.   *then*
28.   *We get inconsistency*
29.   *delete all the inconsistent context instances*
30.   *add the remaining instances to the repository*
31.   *end if*
32.   *end if*
33.   *reasoning the low-level context and get the high-level context*
34.   *if the context asks the service to adapt the changing of the environment and act to user's activities*
35.   *Computing the* $ADA(S_1)$ *of the service*
36.   *if the* $ADA(S_1)$ *satisfy out configuration*
37.   *Adaptation*
38.   *end if*
39.   *end if*
40.   *goto 4*

3   Performance Results

In fact, the adaptation may be more costly for using the QoC management. They offer a quality measure (i.e. 'probability of correctness') which aids in increasing the utility of the decision to change/adapt context source. So we should balance the effect and the cost of the QoC based adaptation.

Firstly, we compare the overhead of the common context processing procedure as well as the procedure with QoC-aware dealing. As depicted in figure 3, the curve with minimum time represents traditional context dealing. This time is composed of the time of sensing, transferring, reasoning and distribution. The second curve above the bottom represents the dealing with replicate context discarding and the process may exhaust more time. And the third curve above the bottom represents the dealing with replicate context discarding and inconsistent context dealing. We use all inconsistent contexts discarding algorithm and we can find it may exhaust more time. The fourth curve above the bottom represents the dealing with replicate context discarding, inconsistent context dealing and quality factors computing.

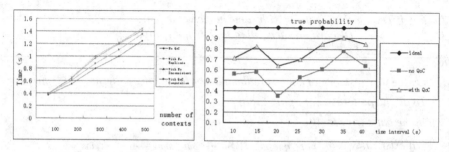

Fig. 3. The Overhead   and true probability of the QoC-aware Context Dealing

From the figure above we can see the QoC dealing may exhaust more time and the time is lower than 30 percent of the normal dealing time of the context. In fact, comparing to the effect of the QoC, this degree of extra time may be accepted.

Secondly, we will discuss the effect of the using of QoC. As the discussion above, we set several duplicate and wrong contexts in the context stream which may lead to wrong adaptation decisions. In fact, in our testing example the inconsistent context is more important for duplicate contexts may just lead to more time be used to get the high level context but the inconsistent context may lead to wrong results. We record the number of right or wrong adaptation and compute the true probability of them. We use the all inconsistent contexts discarding algorithm for ease.

We set different bad contexts for testing:

✧   A random percent between 10 and 30 wrong contexts;
✧   A random percent between 10 and 30 inconsistent contexts at the same time by the sensor pair;
✧   A random percent between 10 and 30 duplicate contexts at the same time by the sensor pair.

From the above figure we can see in all the procession the service adaptation with the help of QoC may have higher true probability. Most of the wrong adaptation for the reason of inconsistent contexts or duplicate contexts may be eliminated by the quality management framework. At the same time, though we can use the QoC mechanism to improve the true probability of adaptation, the effect of the wrong contexts can not be avoided. To resolve this problem, we need more complex algorithms. In fact, in most of the applications, the wrong contexts may come from the failure of the sensors and they may be inconsistent with the value of the sensors nearby. Therefore our framework can support the efficient adaptation in most circumstances.

4   Conclusions

The diversity of the sources of context information, the characteristics of pervasive environments, and the nature of collaborative tasks pose a stern challenge to the efficient management of context information by sensing a lot of redundant and conflicting information. Most of existing research just use the raw context directly or take just some aspects of the Quality of Context (QoC) into account. In this paper, we have proposed a middleware based context-aware framework that support QoC based service adaptation. By this framework we can evaluate raw context, discard duplicate and inconsistent context so as to protect and provide QoS-enriched context information of users for efficient service adaptation. In future work, we will complete more experiments to discuss more aspects of the framework.

References

1. Dey, K.: Understanding and Using Context. Personal and Ubiquitous Computing 5(1), 4–7 (2001)
2. Chen, G., Kotz, D.: A Survey of Context-aware Mobile Computing Research, Hanover, NH, USA, Tech. Rep. (2000)
3. Buchholz, T., Küpper, A., Schiffers, D.: Quality of Context: What It Is and Why We Need It. In: HPOVUA 2003, Geneva (2003)
4. Kim, Y., Lee, K.: A Quality Measurement Method of Context Information in Ubiquitous Environments. In: ICHIT 2006, pp. 576–581. IEEE Computer Society, Washington, DC (2006)
5. Razzaque, P.N.M., Dobson, S.: Categorization and Modelling of Quality in Context Information. In: Proceedings of the IJCAI 2005 (2005)
6. Preuveneers, D., Berbers, B.: Quality Extensions and Uncertainty Handling for Context Ontologies. In: Shvaiko, P., Euzenat, J., Léger, A., McGuinness, D.L., Wache, H. (eds.) Proceedings of (C&O 2006), Riva del Garda, Italy, pp. 62–64 (August 2006), http://www.cs.kuleuven.be/davy/publications/cando06.pdf
7. Sheikh, K., Wegdam, M., Sinderen, M.: Quality-of-context and Its Use for Protecting Privacy in Context Aware Systems. JSW 3(3), 83–93 (2008)
8. Manzoor, A., Truong, H.-L., Dustdar, S.: On the Evaluation of Quality of Context. In: Roggen, D., Lombriser, C., Tröster, G., Kortuem, G., Havinga, P. (eds.) EuroSSC 2008. LNCS, vol. 5279, pp. 140–153. Springer, Heidelberg (2008)

9. Zheng, D., Jia, Y., Zhou, P., Han, W.-H.: Context-Aware Middleware Support for Component Based Applications in Pervasive Computing. In: Xu, M., Zhan, Y.-W., Cao, J., Liu, Y. (eds.) APPT 2007. LNCS, vol. 4847, pp. 161–171. Springer, Heidelberg (2007)
10. Zheng, D., Yan, J., Wang, J.: Research of the Middleware based Quality Management for Context-aware Pervasive Applications. In: 2011 International Conference on Computer and Management, Wuhan (May 2011)
11. Zheng, D., Yan, J., Wang, J.: Research of the QoC-aware Service Selection for Middleware based Pervasive Applications. In: The 2nd International Conference on Biomedical Engineering and Computer Science, Wuhan (April 2011)
12. Zheng, D., Yan, J., Wang, J.: Research of the Middleware based Quality Management for Context-aware Pervasive Applications. In: 2011 International Conference on Computer and Management, Wuhan (2011)
13. Zheng, D., Yan, J., Wang, J.: Research of the QoC-aware Service Selection for Middleware based Pervasive Applications. In: The 2nd International Conference on Biomedical Engineering and Computer Science, Wuhan (April 2011)

Energy Efficient Filtering Nodes Assignment Method for Sensor Networks Using Fuzzy Logic*

Soo Young Moon and Tae Ho Cho

College of Information and Communication Engineering, Sungkyunkwan University
Suwon 440-746, Republic of Korea
{moonmous,taecho}@ece.skku.ac.kr

Abstract. Wireless Sensor Network (WSN) can enable context-aware services through sensing, processing, and reporting event information. Due to limited resources WSNs are vulnerable to various malicious attacks. In one of these attacks false event reports are generated to compromise the integrity of the sensor data. In most filtering schemes every sensor node on a path from an event source to a sink node operates as a filtering node which verifies received reports and determine whether they are valid or false. Hence, even the valid reports are verified multiple times as they are forwarded toward the sink node, causing unnecessary energy consumption. In this paper we propose a filtering nodes assignment method to reduce the energy consumption while verifying the event reports. The proposed method partitions the network into several areas and assigns filtering nodes for each area according to a fuzzy output value derived from the three inputs - the number of valid event reports received from the area, the elapsed time since the last valid event report received from the area, and the average hop count from the nodes in the area to the sink node. The experimental results show that the proposed method conserves sensor nodes' energy and increases the network lifetime with similar security level.

Keywords: false report injection attacks, filtering scheme, SEF, fuzzy logic.

1 Introduction

A wireless sensor network (WSN) is composed of sensor nodes, which are equipped with sensing, computation, and communication capabilities, and one or more sink node(s) that connects the WSN to another network [1, 2]. Sensor nodes detect interesting events and report them to the sink node. A sink node refines the data and provides it to users. WSNs are prone to many security attacks due to their resource constraints and wireless communication [3, 4]. In false report injection attacks [5], an attacker can capture a few sensor nodes in the field and inject forged event reports through them. The false reports are forwarded from the compromised nodes to the sink node using multi hop routing, causing the energy depletion at the intermediate nodes and unnecessary response for fake events.

---

* This research was supported by Basic Science Research Program through the National Research Foundation of Korea (NRF) funded by the Ministry of Education, Science and Technology (No. 2012-0002475).

D.-S. Huang et al. (Eds.): ICIC 2012, LNCS 7389, pp. 293–300, 2012.
© Springer-Verlag Berlin Heidelberg 2012

Filtering schemes [5-7] have been proposed to defend against false report injection attacks. Statistical en route filtering scheme (SEF) [5] was proposed by Ye et al. and its goal is to detect and drop false reports early in their phase. Multiple sensing nodes generate and authenticate event reports collaboratively using their authentication keys. Although SEF can detect and remove false reports early in their phase, energy inefficiency can occur, since a valid report is verified en-route many times. We propose an event report verification limiting method for energy efficiency in WSN using fuzzy logic. In the proposed method, a sensor network is divided into several areas. The sink node determines the reliability level of each area based on 1) the number of valid reports received from the area, 2) the elapsed time since the last valid report received from the area, and 3) the average hop count from the nodes in the area to itself. The sink node also controls the number of filtering nodes for event reports from each area based on its reliability level.

The remaining sections of this paper are as follows. In section 2, we summarize the operation of SEF. In section 3, we explain our method in detail. Section 4 shows the simulation results. Section 5 concludes the paper and plans future work.

2   Statistical En-route Filtering (SEF)

Statistical En-route Filtering (SEF) [5] defends against false report injection attacks, especially for large sensor networks. SEF achieves early detection and dropping of false reports by performing collective generation of reports and en-route filtering using globally-shared authentication keys. There are four phases in the operation of SEF: key assignment phase, report generation phase, en-route filtering phase, and sink verification phase. The key assignment phase is executed only once, but the other three phases are performed repeatedly. In the key assignment phase, every sensor node is associated with a key partition of the global key pool and loaded with some portion of keys in the partition. After the phase, the nodes are deployed in the sensor field randomly or manually. In the report generation phase, multiple nodes sense an event, elect the center of stimulus (CoS) node among themselves, and collaboratively generate and authenticate the final event report. The CoS collects message authentication codes (MACs) created by other sensing nodes, of distinct partitions, and attaches them to the event report. In the en-route filtering phase, a receiving node checks if one of its keys has been used to generate a MAC in the report. If so, the node verifies the corresponding MAC in the report using its own key, and forwards or drops it, depending on the verification result. In the sink verification phase, the sink node validates all the MACs in the received report, since it knows all the keys in the global key pool. It detects and removes the report if at least one MAC was forged. SEF can detect and drop false reports if the number of compromised nodes is smaller than the security parameter value [5]. Fig.1 shows the operation of SEF.

Each node (except the sink node) in fig. 1 is assigned two authentication keys from one of three key partitions in the global key pool. K_{ij} means j_{th} key in i_{th} key partition. An event report includes three MACs created by using K_{12}, K_{23}, and K_{31}, respectively. Every node on the forwarding path tries to verify a MAC in the received event report. However, it can verify one of the MACs only when it stores the corresponding key in its memory. For example, N_1 does not know any of the three keys and therefore it just

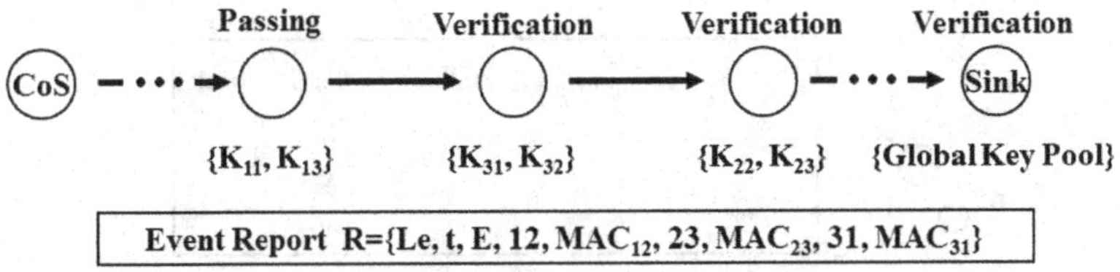

Fig. 1. SEF Operation

passes the received report to the next node N_2. On the other hand, each of N_2 and N_3 knows K_{23} and K_{31}, respectively, and verifies the corresponding MAC and conditionally forwards the event report to the next node based on the verification result. The sink node knows all the authentication keys in the global key pool and hence, it verifies every MAC in the received report and detects a false report.

In SEF, every node is assigned as a filtering node and performs verification of received reports without considering how reliable the source nodes are. The one hop verification probability (OVP) at each node is the same regardless of the event sources. As a result, valid reports are verified en-route repeatedly, and unnecessary energy consumption for verifying legitimate reports occurs.

3   Proposed Method

3.1   System Model and Assumptions

We target sensor networks for monitoring applications in which selected nodes periodically generate and send event reports to the sink node. We assume that sensor nodes are distributed randomly and their locations do not change. Sensor nodes are equipped with limited memory [8] and communicate via wireless links. Node compromise attack can occur, although the sink node is safe from the attack. The sink node can also authenticate its message to the sensor nodes [9].

3.2   Motivation

In SEF, every node on a forwarding path is a filtering node that verifies a MAC in a received report. Therefore, a valid report is verified many times as it travels from source to the sink node, especially when the path is long. The repetitive verification operations result in unnecessary energy consumption and shorten the lifetime of sensor networks. The proposed method controls the number of filtering nodes for each area based on its reliability, to limit the number of verification operations. If an area is more reliable than the others, then smaller numbers of filtering nodes are assigned to the event reports originating from the area.

3.3   System Operation

After key assignment and node deployment, the sink node partitions the sensor field into several areas based on geographic region. The sink node then manages a reliability level table for the areas. Figure 2 illustrates the partitioned field and the reliability level table.

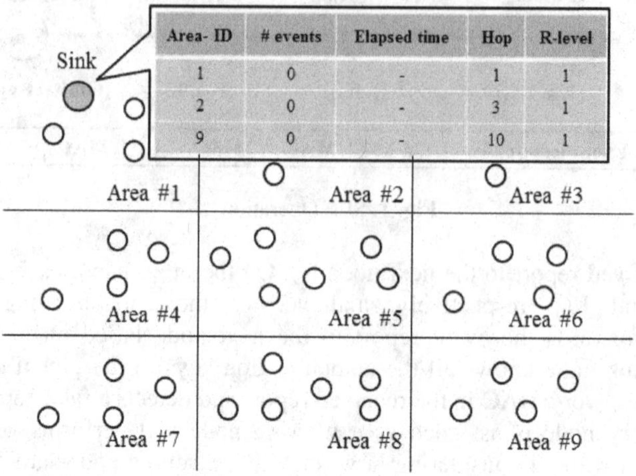

Area- ID	# events	Elapsed time	Hop	R-level
1	0	-	1	1
2	0	-	3	1
9	0	-	10	1

Fig. 2. Field partitioning and R-level initialization

There are nine areas in the sensor field. Each entry of the table contains the area ID, the number of valid event reports received from the area (# events), the elapsed time since the last valid report from the area, the average hop count (Hop) from the sink node to the nodes in the area and the reliability level (R-level) of the area. The R-level ranges from one (initial value) to five.

After field partitioning and R-level initialization, sensor nodes report events. Periodically, sensor nodes elect one of them as the CoS node which gathers MACs from other sensing nodes, creates event report, and sends the report to the next node. Every event report should include the information of event location.

When a node on the forwarding path receives an event report, it check whether it is a filtering node for the area of event source and whether it knows a key for generating one of MACs in the report. Only when both conditions are true, the node verifies the corresponding MAC and drops it if the verification fails. Each node manages the list of areas for which it is a filtering node. At first, every node is the filtering node for all the areas in the network. The list at each node can change independently as the network operates.

The sink node verifies all the MACs in the event reports received from areas in the field. If one of the MACs is incorrect, the sink node detects a false report, drops it and resets the reliability level of the corresponding area to the level one. In addition, the sink node periodically (e.g., every 10 reports received) computes the reliability levels of areas and updates the reliability level table. Whenever the sink node receives an event report, it records the factors to compute the reliability levels of areas. The relationship between the three factors – the number of valid event reports from an area, the elapsed time since the last valid report received from the area and the average hop count from the nodes in the area to the sink node - and the reliability level - is as follows. If the sink node receives many valid reports from an area and the elapsed time since the last valid report is short, it judges the area as reliable. Conversely, it considers an area unreliable if few valid reports have been received from the area and (or) the elapsed time is long. In addition, an area close to the sink node is assumed to be more reliable than another far from the sink node. When the reliability level of an area is changed, the sink node re-assigns filtering nodes for event reports from the area by determining the number of filtering nodes on the path from itself to the area. Figure 3 shows the reliability level update operation.

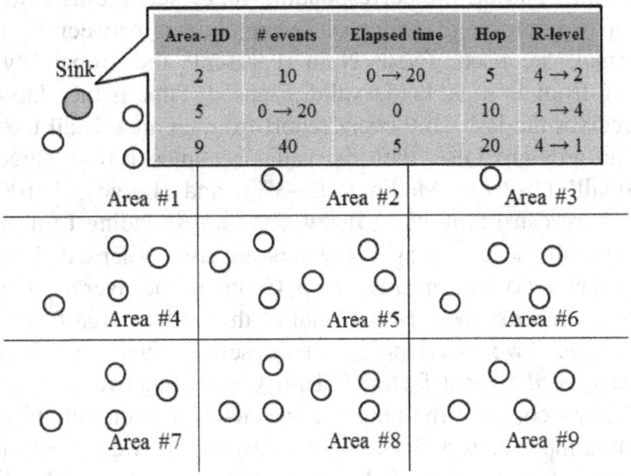

Area- ID	# events	Elapsed time	Hop	R-level
2	10	$0 \to 20$	5	$4 \to 2$
5	$0 \to 20$	0	10	$1 \to 4$
9	40	5	20	$4 \to 1$

Fig. 3. R-level update

In figure 3, the reliability levels of three areas change. The sink node does not receive a valid event report from area #2 for 20 time-units. Hence, the reliability level (R-level) for area #2 decreases from 4 to 1. In area #5, 20 event reports occur and they are all valid. As a result, its R-level increases from 1 to 4. Area #9 sends 30 false reports and one of them is detected by the sink node. Then the sink node resets the reliability level of area #9 to 1. In addition, the number of valid event reports from the area becomes zero. In the proposed method, the sink node controls the number of filtering nodes for each area based on its reliability level. An area with a high reliability level does not require many filtering nodes, and vice versa. In the proposed method, the sink node sends a control message to adjust the number of filtering nodes for an area when its reliability level changes. A control message includes an area ID, remaining hop count and the number of filtering nodes for event reports originating from the area. When a node on the path receives a control message, it reduces the remaining hop count by one. Then, it checks if the remaining hop count is no more than the number of filtering nodes for the area. If so, it inserts the area ID to its list of areas for which it is a filtering node.

3.4   Fuzzy Logic Design

The sink node generates the reliability level of the area based on fuzzy logic [10] from the three factors. The three factors may contain errors due to node and link failure or topology change of the WSN. Further, the same value for a factor may be considered high or low depending on the application. It is efficient to use fuzzy logic due to these uncertainties and ambiguity to reduce error and generate sub-optimal output without complex equations. All the fuzzy input factors are normalized. Event_Num is the number of valid event reports that the sink node received from an area. If all event reports originating from an area are valid, Event_Num becomes 100%. Event_Num is defined by three overlapping fuzzy sets - 'Few' (0~50%), 'Normal' (0~100%), and 'Many' (50~100%). A given value of Event_Num can be included in multiple fuzzy sets at the same time, and all the corresponding fuzzy sets are used to calculate the output value. For example, if the value of Event_Num is between 0 and 50%, both 'Few' and 'Normal' are used. Event_Num represents the probability of an event report, originating from an area, being valid. Elapsed_Time is the elapsed time since the sink node receives the last valid event report from an area. If all the event reports which occur at an area are false, then the value becomes 100%. Three overlapping fuzzy sets - 'Small' (1~2%), 'Medium' (2~3%), and 'Large' (3~100%) – define Elapsed_Time. There can be multiple fuzzy sets corresponding to a given value of Elapsed_Time, and all the matching fuzzy sets are used. Elapsed_Time is used for detecting false event report in an area. Hop_Count is the average hop count from nodes in an area to the sink node. Hop_Count of the farthest area from the sink node is defined as 100%. Two overlapping fuzzy sets - 'Near' (1~60%) and 'Far' (30~100%) – exist for the input factor. If Hop_Count is between 30 and 60%, both fuzzy sets are used to compute the output value. R-level is the reliability level of an area for given three input factors. Its value is between 1 (lowest level) and 5 (highest level). There are five fuzzy sets for R-level - 'Very Low', 'Low', 'Medium', 'High', and 'Very High'. Fig. 4 shows the fuzzy membership functions for the input and output variables.

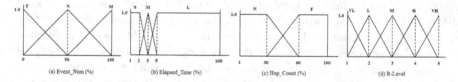

(a) Event_Num (%)   (b) Elapsed_Time (%)   (c) Hop_Count (%)   (d) R-Level

Fig. 4. Fuzzy Membership Functions

Given input values to the system, the sink node derives the output value R-Level using fuzzy if-then rules. Table 1 represents a portion of the if-then rules of the proposed method [10].

Table 1. Fuzzy if-then rules

Rule #	Event_Num	Elapsed_Time	Hop_Count	R-Level
1	F	S	N	H
4	F	M	F	L
8	N	S	F	M
13	M	S	N	VH
18	M	L	F	L

The strategy of the rules can be summarized as follows. If Event_Num is high, Elapsed_Time and Hop_Count are low, the sink node sets the R-level of the area high. On the other hand, if Event_Num is low, Elapsed_Time and Hop_Count are high, it sets the R-level of the area low. For example, in rule 13, if Event_Num is many, Elapsed_Time is small, and Hop_Count is near, then the R_Level is very high.

4   Simulation Results

Simulation parameters for the proposed method are as follows. The sensor field of 1200 m × 240 m is divided into 80 areas of 60 m × 60 m size. Both the sensing and transmission range of each sensor node are 30 m, and there are 1920 nodes in the field.

The global key pool comprises 100 keys organized with 10 partitions of 10 keys each. Each node is loaded with five keys from one of 10 partitions. Every event report contains 5 MACs. 16.25/12.5 µJ is consumed to transmit / receive a byte, and 75 µJ for one filtering operation [5]. We generated 300 event reports in one simulation run. Shortest path-based routing [11] was assumed and free fuzzy logic library [12] was used to calculate reliability values. Figure 5 shows the energy consumption and detection probabilities of SEF and the proposed method.

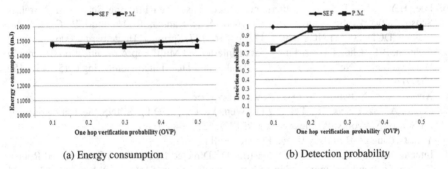

(a) Energy consumption   (b) Detection probability

Fig. 5. Simulation results (FTR = 5%, hop count ≒ 50)

OVP is the probability of verifying a MAC at each filtering node and determines security strength. It can be shown that the proposed method consumes less energy than SEF when OVP is > 0.1. The amount of energy saved increases as the OVP increases. The detection probability of false event reports of the proposed method is lower than SEF. However, the difference is very small when OVP > 0.1.

5   Conclusion

In this paper, we proposed the energy-efficient filtering nodes assignment method for sensor networks using fuzzy logic. Existing filtering schemes do not consider different reliability levels of areas in the sensor field and consume much energy to repeatedly verify valid reports. In the proposed method, the sink node manages the reliability levels of areas in the sensor field and adjusts the number of filtering nodes for the areas, based on their reliability levels. Simulation results show that our method reduces energy consumption for en-route filtering of existing schemes, without losing false report detection capability. Future work will enable each sensor node to manage the reliability levels of its neighbor nodes. This will control the filtering probabilities at each node for event reports from different nodes, based on their reliability levels.

References

1. Akyildiz, I.F., Su, W., Sankarasubramaniam, Y., Cayirci, E.: A Survey on Sensor Networks. IEEE Communications Magazine 40(8), 102–116 (2002)
2. Al-Karaki, J.N., Kamal, A.E.: Routing Techniques in Wireless Sensor Networks: a Survey. IEEE Wireless Communication Magazine 11(6), 6–28 (2004)
3. Djenouri, D., Khelladi, L., Badache, N.: A Survey of Security Issues in Mobile Ad-Hoc and Sensor Networks. IEEE Communications Surveys (2005)
4. Karlof, C., Wagner, D.: Secure Routing in Wireless Sensor Networks: Attacks and Countermeasures. Elsevier's Ad Hoc Networks Journal, Special Issue on Sensor Network Protocols and Applications 1(2-3), 293–315 (2003)
5. Ye, F., Luo, H., Lu, S., Zhang, L.: Statistical En-route Filtering of Injected False Data in Sensor Networks. IEEE JSAC 23(4), 839–850 (2005)
6. Lee, H.Y., Cho, T.H.: Key Inheritance-Based False Data Filtering Scheme in Wireless Sensor Networks. In: Madria, S.K., Claypool, K.T., Kannan, R., Uppuluri, P., Gore, M.M. (eds.) ICDCIT 2006. LNCS, vol. 4317, pp. 116–127. Springer, Heidelberg (2006)
7. Nghiem, P.T., Cho, T.H.: A Fuzzy-based Interleaved Multi-hop Authentication Scheme in Wireless Sensor Networks. Journal of Parallel and Distributed Computing 69(5), 441–450 (2009)
8. Xbow Sensor Networks, http://www.xbow.com
9. Perrig, A., Szewczyk, R., Tygar, J.D., Wen, V., Culler, D.E.: SPINS: Security Protocols for Sensor Networks. Wireless Networks 8(5), 521–534 (2002)
10. Yen, J., Langari, R.: Fuzzy Logic. Prentice Hall (1999)
11. Intanagonwiwat, C., Govindan, R., Estrin, D.: Directed Diffusion: A Scalable and Robust Communication Paradigm for Sensor Networks. In: Proc. of MOBICOM, pp. 56–67. ACM (2000)
12. FFLL, http://ffll.sourceforge.net

Sentiment Analysis with Multi-source Product Reviews

Hongwei Jin, Minlie Huang, and Xiaoyan Zhu

State Key Laboratory of Intelligent Technology and Systems
Tsinghua National Laboratory for Information Science and Technology
Department of Computer Science and Technology, Tsinghua University
Beijing, 100084, China
cs.jin.hongwei@gmail.com, {aihuang,zxy-dcs}@tsinghua.edu.cn

Abstract. More and more product reviews emerge on E-commerce sites and microblog systems nowadays. This information is useful for consumers to know the others' opinion on the products before purchasing, or companies who want to learn the public sentiment of their products. In order to effectively utilize this information, this paper has done some sentiment analysis on these multi-source reviews. For one thing, a binary classification framework based on the aspects of product is proposed. Both explicit and implicit aspect is considered and multiple kinds of feature weighing and classifiers are compared in our framework. For another, we use several machine learning algorithms to classify the product reviews in microblog systems into positive, negative and neutral classes, and find OVA-SVMs perform best. Part of our work in this paper has been applied in a Chinese Product Review Mining System.

Keywords: product review, sentiment analysis, microblog, SVM.

1 Introduction

With the development of Internet, more and more customers get used to purchasing products on E-commerce sites such as 360buy[1] and Newegg[2]. They also write reviews on the products after using them, which produce a large number of reviews on the Internet. In addition, microblog, a system that allows users to post messages of no more than 140 words, and share information instantaneously based on the relationship between users, is under rapid development, such as Twitter[3], Sina microblog[4] and Tencent microblog[5]. A lot of microblogs contain latest product reviews.

Reviews from the above two large data sources contain much useful information for users and companies. Users can make better purchasing decisions based on these reviews, while companies can also analyze customers' satisfaction according to these reviews, and further improve the quality of their products. Since there is a mass of

---

[1] http://www.360buy.com
[2] http://www.newegg.com.cn
[3] http://www.twitter.com
[4] http://weibo.com
[5] http://t.qq.cn

D.-S. Huang et al. (Eds.): ICIC 2012, LNCS 7389, pp. 301–308, 2012.
© Springer-Verlag Berlin Heidelberg 2012

product reviews and a single user cannot read all of them, automatically mining the reviews from multiple sources is particularly important.

Most reviews in Chinese E-commerce sites are labeled with advantage or disadvantage, which is naturally suitable for binary classification. The state-of-art research in Chinese sentiment analysis mainly focuses on the whole review classification, while customers often desire a more detailed understanding of products. For example, they want to know others' opinion on the battery of a cell phone. Therefore, we propose a framework of sentiment classification at aspect level to solve this problem. In our framework, not only explicit but also implicit aspects are taken into account. To our knowledge, no implicit aspect discovery work of product review in Chinese language has been reported before. For the reviews of each aspect, the unigram features of words are used as text features. We also compare the performance of three feature weighing strategies, three reduction dimension, and three classification approaches.

The sentiment analysis for reviews of products on microblogs is in its infancy. Besides the microblogs that express opinion on the products, some microblogs only give some statements relative to the products, which contain no sentiment polarity, or are neutral. Therefore, in this paper, we exploit linear regression, multi-class classification, two-stage classification and Mincut model optimization to classify the product related microblogs into three classes, and compare the performance of these methods.

2   Related Work

Sentiment classification mainly includes two methods: supervised learning and unsupervised learning. Turney[1] proposed a simple unsupervised method to classify reviews into positive and negative categories. Pang[2] used Naive Bayes (NB), Max entropy (ME) and SVM separately as supervised learning algorithms for binary classification of reviews. Besides these, Pang[3] exploited regression and One-Vs-All SVMs to predict the score for classification, namely the multi-classification, which can also be realized by combining binary classifiers in a two-stage manner. Pang[4] and Su[5] both optimized their multi-classification results using the Mincut model.

For sentiment classification in microblogs, Go[6] was the first to classify Twitter data into two classes (positive and negative). Barbosa[7] classified Twitter data into three classes in a two-stage manner. Jiang[8] analyzed topic related sentiment for Twitter text. Based on the two-stage binary classification result, and the forwarding relationship between users, the performance was promoted using the method of graph. But for Chinese microblogs systems, only text can be obtained instead of the relationship between users and microblogs.

3   Binary Sentiment Classification Based on Product Aspect

The overall flow chart about polarity classification (positive and negative) based on product aspect is shown in Fig.1.

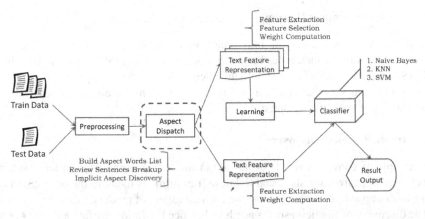

Fig. 1. The polarity classification (positive and negative) based on product aspect framework

Aspect, Aspect Word. Product aspect refers to some attribute, component, or function of the product. The word or phrase used to represent the aspect of product is called the aspect word. Aspect is actually the concept at semantic level, and the aspect word is the external presentation form of the aspect. Different categories of products generally have a different set of aspects. Product aspect can be divided into explicit aspect and implicit aspect. Explicit aspect refers to the aspect word which describes the product performance or function can be directly found in the product reviews. Implicit aspect may be found after the sentence semantic understanding because there are no aspect words in review sentence.

3.1 Preprocessing

In order to get relatively clean text about product reviews, messy code will be removed in the preprocessing stage according to a user-defined messy code list. Next we use the word segmentation software tools to make the sentence processed. There are some words that appear in almost all of the text which are called stop words. The text content is represented more accurately after deleting stop words. Some words in the traditional stop words list is retained because they have sentiment polarity which is useful information for sentiment classification.

3.2 Build Aspect Words List

We use a statistics-based algorithm to automatically build the vocabulary about the product aspects. Here is the algorithm:

1. Count occurrences number of all the nominal words.
We count the occurrences number of all the unigram whose part of speech(POS) tagging is noun in the reviews of same product category.
2. Filter high-frequency words
The final aspect words list is filtered by threshold of high frequency and product specifications.

3.3   Review Sentences Breakup

First of all, we break a review into sentences according to various sentence end punctuations, each of which is further cut into short sentences (SS) with comma. Then each SS is assigned to different sentiment categories of different aspects according to the aspect vocabulary made in advance. Finally, if an SS has implicit aspect, its words should also be correctly distributed.

3.4   Implicit Aspect Discovery

For example, if there is a word "beautiful" in the sentence and without any explicit aspect word appeared, we also need to know "beautiful" is likely in the description of the product's appearance which is just the implicit aspect of product.

The discovery of implicit aspects needs to balance the following factors:

1. Calculate mutual information (MI) scores between adjectives in the current SS and each aspect. Adjectives with most relevant aspects will be calculated and the aspect of the entire SS is determined by voting.

2. If the above rule doesn't work, the SS is just assigned to the previous SS's aspect.

3.5   Text Feature Representation

After the processing mentioned above, we get all the review sentences in each aspect in each category, and vector space model is employed to represent them.

Features of the review text using word-based unigram, but different word have the different ability and importance to represent the text which is called weight. In this paper, BOOL, TF, and TF-IDF are used as feature weights. BOOL weight means that if the count of the term appears in the current document greater than zero, the weight is 1, otherwise 0. TF (term frequency) refers to the frequency of the words occurrence in the file. IDF (inverse document frequency) measure the common importance of the word. In addition, we use chi-square statistics to reduce feature dimension.

3.6   Classifier

Naive Bayes. Navie Bayes is a simple model which calculates posterior probability of each class based on the priori probability and the likelihood. Class with the largest posterior probability is assigned as the class of the document.

KNN. The basic idea of the KNN algorithm is considering the K nearest (the most K similar) texts in the training set of the new given text. The label of the new text is determined by these K texts.

SVM. SVM is a popular classification algorithm which compresses the raw data set to support vector set and learn a decision hyperplane in the vector space. This hyperplane best splits the data points into the two classes.

In this paper, NB, KNN and SVM are implemented using data mining tool Weka[6].

---

[6] http://www.cs.waikato.ac.nz/ml/weka/

4   Sentiment Analysis on Product Reviews in Microblog System

4.1   Text Feature Extraction

Like mentioned in Section 3.1, stop words and messy code in the microblog will be removed, and then the Chinese word segmentation and POS tagging will be done. The vector space model is still employed to represent each microblog. The features are unigrams and the weight is BOOL value. However, we can take the advantage of characteristic of microblog to portray it better and reduce the feature dimension.

Emoticons. For example, the text form of emoticon 😊 in microblog text is "[ha ha]". All the positive emoticons will be converted into token "POS", and all negative into token "NEG" according to the manual defined emotions list.

Usernames. Forwarding mechanism of microblog systems makes microblog amazing transmitted. It is in the form of @ + username (such as @ryanking1219). So we use string "USERNAME" to instead all words beginning with @.

Links. Users like to include the URL which usually begins with "http" when they share videos or news, such as http://t.cn/aCKddG. All website links are normalized to the token "URL".

Topics. In microblog systems, topic starts with "#" and also ends with "#". We will replace all the string in this form with an equivalence class string "TOPIC".

4.2   Two-Stage SVM

Step 1: Subjectivity Classifier
Neutral samples in the training data is considered as positive examples, while positive and negative samples as negative cases to train the classifier. Do binary classification on the test data for subjective and objective detection.

Step 2: Polarity Classifier
Positive samples in the training data is considered as positive examples, while negative samples as negative cases to train the classifier. Do binary classification on the test data which is correctly divided into subject class in Step 1 for positive and negative detection.

4.3   Minimum Cut Model Optimization

Inspired by work of Pang[4] and Su[5], we also use Minimum cut (Mincut) model to optimize the Two-stage SVM result. Binary classification with Mincut in graph is based on the idea that similar items should be split in the same cut. We build an undirected graph G with vertices $\{s, t, v_1, v_2, ..., v_n\}$; s is the source and t is the sink, and all items in the test data are seen as vertices v_i. Each vertex v is connected to s and t via a weighted edge. The weight is the estimation of the probability converted from SVM classifier output. Each edges (v_i, v_k) with weight $assoc(v_i, v_k)$ expresses the similarity between v_i and v_k and how important they should be in the same class. Then we remove a set of edges to divide graph into two disconnected sub-graphs. The vertices via s are positive instances and the vertices via t are negative instances. We penalize when putting highly similar items into different classes, so the best split is one which removes edges with lowest weights. This is exactly the Mincut of graph. Because the capacity of the Mincut equals the max flow value, we can employ Edmonds-Karp algorithm to compute the Mincut.

In experiment, we set

$$assoc(v_i, v_j) = \begin{cases} \alpha * \cos(v_i, v_j), & if\ \alpha * \cos(v_i, v_j) > t \\ 0 & , otherwise \end{cases} \tag{1}$$

where $\cos(v_i, v_j)$ is the text similarity between v_i and v_j. Constant α is the scale factor of association scores. Only scores higher than threshold t will be taken into consideration.

4.4   Regression

Text classification is also a problem that utilizes text features to predict a category label, and similar entries should have similar category labels. The most classic model is the linear regression function which is actually a linear weighted sum of all the features. The final score is rounded as a category label.

4.5   OVA-SVMs

Although the original SVM is used to solve binary classification problem, we can combine several SVM classifiers to achieve multi-class classification, such as one-versus-all method (for short of OVA-SVMs). When Training each classifier we mark one category samples as positive examples, and all the rest samples as negative examples. So that k categories of samples can construct k SVM classifiers in turn. Class label of the test sample assigned to the class which has the largest classification function value.

5   Experiments

5.1   Sentiment Classification Based on Product Aspect

We downloaded 821 cell phone products data from the 360buy and Newegg, and finally get 663,537 advantages reviews and 314,529 disadvantage reviews.

CHI dimension reduction selects 5%, 20% and 100% words most relevant to each aspect of each product category, represented as CHI-5, CHI-20, and CHI-100. Of course, from the whole view of each product category, the most relevant words of each aspect will be overlapped with each other.

Performance evaluation methods commonly apply PRECISION, RECALL, and F-values, but these only represent local significance (i.e., the performance of each aspect in each product category). If you want to evaluate the overall classification performance, all the aspects of product category need to be taken into account to calculate the F-value of the MICRO AVERAGING and MACRO AVERAGING.

Micro-average gives the same weight to each document in each product category while Macro-average gives the same weight to each aspect in each product category. Results are illustrated in Fig.2.

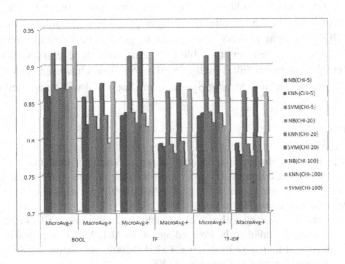

Fig. 2. Cell phone experiment result (17 aspects)

The following conclusions could be drawn from the experimental result:

First, the best combination is BOOL-SVM. Second, the consequent of TF is very close to the TF-IDF, but is slightly lower than BOOL. In addition, the SVM classification performance is better than NB and KNN. Last but not least, CHI-5 result is similar with CHI-20 and CHI-100, even sometimes CHI-5 perform best, so dimension reduction is proved effective.

5.2   Sentiment Analysis on Product Reviews in Microblog System

We select 5 popular items including "Nokia", "iphone 4s", "E72i", "Lenovo" and "Canon". For all of these items, we downloaded 2,100 microblogs through the API provided by Sina microblog and Tencent microblog from October 2011 to November 2011. We manually labeled all the microblogs and finally obtain 729 positive, 345 negative, and 1,026 neutral microblogs respectively. The results are listed in Table 1.

Table 1. Result for microblog classification

	Regression	OVA-SVMs	Two-stage SVM	Two-stage SVM + MinCut
Accuracy	51.6%	**64.5%**	62.6%	64.0%

As shown in Table 1, the OVA-SVMs classifier is best. Two-stage SVM classifier after optimization with the minimum cut model (when parameter $\alpha = 0.2$, t = 0.5) has 1.4% increase in performance, and reach almost the same effect of OVA-SVMs.

6   Conclusion

This paper makes sentiment analysis on product reviews from multiple data sources. Firstly, a binary sentiment classification framework based on the aspects of the product is proposed. On the granularity of aspect, we use unigram as review feature, and adopt BOOL weight and SVM classifier to get the best results. Secondly, we classify our product review in microblog systems into three classes. OVA-SVMs method offers the optimal result. Finally, Work in Section 3 has been partially applied in a Chinese Product Review Mining System[7], and Section 4 will be applied soon.

References

1. Turney, P.D.: Thumbs up or Thumbs down?: Semantic Orientation Applied to Unsupervised Classification of Reviews. In: Proceedings of the 40th Annual Meeting on Association for Computational Linguistics, ACL 2002, pp. 417–424. Association for Computational Linguistics, Stroudsburg (2002)
2. Pang, B., Lee, L., Vaithyanathan, S.: Thumbs up?: Sentiment Classification Using Machine Learning Techniques. In: Proceedings of the ACL 2002 Conference on Empirical Methods in Natural Language Processing, EMNLP 2002, vol. 10, pp. 79–86. Association for Computational Linguistics, Stroudsburg (2002)
3. Pang, B., Lee, L.: Seeing Stars: Exploiting Class Relationships for Sentiment Categorization with Respect to Rating Scales. In: Proceedings of the 43rd Annual Meeting on Association for Computational Linguistics, ACL 2005, pp. 115–124. Association for Computational Linguistics, Stroudsburg (2005)
4. Pang, B., Lee, L.: A Sentimental Education: Sentiment Analysis Using Subjectivity Summarization Based on Minimum Cuts. In: Proceedings of the 42nd Annual Meeting on Association for Computational Linguistics, ACL 2004. Association for Computational Linguistics, Stroudsburg (2004)
5. Su, F., Markert, K.: Subjectivity Recognition on Word Senses via Semi-supervised Mincut. In: Proceedings of Human Language Technologies: The 2009 Annual Conference of the North American Chapter of the Association for Computational Linguistics, NAACL 2009, pp. 1–9. Association for Computational Linguistics, Stroudsburg (2009)
6. Go, A., Bhayani, R., Huang, L.: Twitter Sentiment Classification Using Distant Supervision. Technical report, Stanford Digital Library Technologies Project (2009)
7. Barbosa, L., Feng, J.: Robust Sentiment Detection on Twitter from Biased and Noisy Data. In: Proceedings of the 23rd International Conference on Computational Linguistics: Posters, COLING 2010, pp. 36–44. Association for Computational Linguistics, Stroudsburg (2010)
8. Jiang, L., Yu, M., Zhou, M., Liu, X., Zhao, T.: Target-dependent Twitter Sentiment Classification. In: Proceedings of the 49th Annual Meeting of the Association for Computational Linguistics: Human Language Technologies, HLT 2011, vol. 1, pp. 151–160. Association for Computational Linguistics, Stroudsburg (2011)

---

[7] http://166.111.138.18/cReviewMiner/

Identifying CpG Islands in Genome Using Conditional Random Fields

Wei Liu [1,2], Hanwu Chen[1], and Ling Chen [2,3]

[1] Department of Computer Science and Engineering,
Southeast University, Nanjing 210096, China
[2] Department of Computer Science, Yangzhou University, Yangzhou 225127, China
[3] National Key Lab of Novel Software Tech, Nanjing University, Nanjing 210093, China
yzliuwei@126.com

Abstract. This paper presents a novel method for CpG islands location identification based on conditional random fields (CRF) model. The method transforms CpG islands location identification into the problem of sequential data labeling. Based on the nature of CpG islands location, we design the methods of model constructing, training and decoding in CRF accordingly. Experimental results on benchmark data sets show that our algorithm is more practicable and efficient than the traditional methods.

Keywords: conditional random fields model, CpG islands, sequential data labeling.

1 Introduction

During the large scale genome sequencing project, we probably can discover a new gene by detecting a CpG island [1-2]. In genetics, CpG islands [3]are genomic regions that contain a high frequency of CpG sites but to date objective definitions for CpG islands are limited. Therefore, the prediction of CpG islands [4]is theoretically and practically important in the mining and identifying new gene and early diagnosis of tumor.

In the past twenty years, extensive efforts have been devoted to developing algorithms and criterias for identifying CpG islands. Gardiner-Garden and Frommer first presented a technical standard for CpG islands detecting in 1987. Later Takai et al. [5] made some amendments and proposed a new technical standard of CpG islands which is helpful to remove Alu repeats. In 2006, Hachenberg et al. presented a new CpG islands recognition method named CpGCluster [6] based on objective physical distance. But since the criterias of CpG islands mentioned above are all artificially defined, the CpG islands identified have diminutive biological meaning, and they require tremendous computational time to get the results. Many biologists are actively searching for more perfect criteria of CpG islands [7] which are helpful to identify true CpG islands of biological significance. Wang Jin-long et al. [8]advanced a method to predict CpG islands based on fuzzy theory. Shi Ou-yan et al. [9]and Jiang Hong-jing et al. [10]respectively used HMM which has strict independence assumption to identify the location of CpG islands.

D.-S. Huang et al. (Eds.): ICIC 2012, LNCS 7389, pp. 309–318, 2012.
© Springer-Verlag Berlin Heidelberg 2012

In this paper, we present a new method for identifying CpG islands based on the model of conditional random fields [11]. The method transforms the problem of CpG islands location identifying into the problem of sequential data labeling. Based on the nature of CpG islands location, we design the methods of model constructing, training and decoding in CRF accordingly. Experimental results on benchmark data sets show that our algorithm is more practicable, precision and efficient than some traditional methods.

2   Conditional Random Fields

Definition1. Let $X = (X_1, X_2, ..., X_n)$ and $Y = (Y_1, Y_2, ..., Y_n)$ be the observation and labeling sequences respectively. Suppose the graph G = (V, E) of Y is a tree or, in the simplest case, a chain, where every edge is a cut of the graph. By the fundamental theorem of random fields [12](Hammersley & Clifford, 1971) , the joint distribution over the label sequence Y given X has the form:

$$p_\theta(y \mid x) \varnothing \exp\left(\left(\sum_{e \in E, K} \lambda_k f_k(e, y \mid e, x)\right) + \left(\sum_{v \in V, K} u_K g_K(v, y \mid v, x)\right)\right) \tag{1}$$

where V and E are the sets of vertexes and edges in G respectfully, f_k is a transitional feature function and g_k is a state feature function. The parameter $\theta = (\lambda_1, \lambda_2, ...; \mu_1, \mu_2, ...)$ is estimated from training data $D = \left\{\left(x^{(i)}, y^{(i)}\right)\right\}_{i=1}^{N}$.

In the CRFs model, the parameters f_k, g_k in (1) can be unified as the form of $f_k(y_{i-1}, y_i, x, i)$, and the joint distribution of chain-structured CRFs is written as:

$$p_\lambda(y \mid x) = \frac{1}{Z(X)} \exp\left(\sum_i \sum_j \lambda_j f_j(y_j - 1, y_j, x, i)\right) \tag{2}$$

Here, $Z(x)$ is an instance-specific normalization function as follows

$$Z(X) = \sum_y \exp\left(\sum_i \sum_j \lambda_j f_j(y_i - 1, y_i, x, i)\right) \tag{3}$$

In (2) $f_j(y_{i-1}, y_i, x, i)$ is a feature function with value 1 or 0. And the parameter λ_j in (2), which is obtained during the training procedure, is used to index the weight corresponding to each feature function [18]. A CRF model is specified by a group of feature functions f_i and the corresponding weights λ_i.

3   Identifying CpG Islands Using CRFs

3.1   Feature Selection

3.1.1 Defining the States of Each Position

We can see that there are eight states in the whole CRFs model illustrated in fig.1. We define that the state "X^+" indicates "the character X is in the CpG island"; otherwise the state "X^-" indicates "the character X is out of the CpG island", where $X \in \{A, C, G, T\}$. We can use the CRFs model mentioned above to give the most suitable labels for an observation sequence.

state:	A^+	C^+	G^+	T^+	A^-	C^-	G^-	T^-
The Released character:	A	C	G	T	A	C	G	T

Fig. 1. A CRFs Model for CpG Island Detecting

3.1.2   Feature Functions

Based on the above defined eight states and the nature of CpG islands, we set eight feature functions as follows:

$$f_1\left(y_{i-1}, y_i, x_i\right) = I_{\left(y_{i-1}='+'\right)}I_{\left(y_i='-'\right)}I_{\left(x_i='C'\right)} \qquad f_5\left(y_{i-1}, y_i, x_i\right) = I_{\left(y_{i-1}='+'\right)}I_{\left(y_i='-'\right)}I_{\left(x_i='G'\right)}$$

$$f_2\left(y_{i-1}, y_i, x_i\right) = I_{\left(y_{i-1}='+'\right)}I_{\left(y_i='+'\right)}I_{\left(x_i='C'\right)} \qquad f_6\left(y_{i-1}, y_i, x_i\right) = I_{\left(y_{i-1}='+'\right)}I_{\left(y_i='+'\right)}I_{\left(x_i='G'\right)}$$

$$f_3\left(y_{i-1}, y_i, x_i\right) = I_{\left(y_{i-1}='-'\right)}I_{\left(y_i='+'\right)}I_{\left(x_i='A'\right)} \qquad f_7\left(y_{i-1}, y_i, x_i\right) = I_{\left(y_{i-1}='-'\right)}I_{\left(y_i='+'\right)}I_{\left(x_i='T'\right)}$$

$$f_4\left(y_{i-1}, y_i, x_i\right) = I_{\left(y_{i-1}='-'\right)}I_{\left(y_i='-'\right)}I_{\left(x_i='A'\right)} \qquad f_8\left(y_{i-1}, y_i, x_i\right) = I_{\left(y_{i-1}='-'\right)}I_{\left(y_i='-'\right)}I_{\left(x_i='T'\right)}$$

3.2   Parameter Training

In the CRFs model, the parameter λ_j in (2) indexes the weight corresponding to each feature function, and its value can be obtained by training from the known samples of CpG. In the CRFs model, we use maximum likelihood to train the samples for parameter estimation.

Given a DNA sequence set $\left\{x^{(i)}\right\}$, the corresponding label sequence set $\left\{y^{(i)}\right\}$ $(i = 1, \cdots, N)$ and eight feature functions defined above, firstly we need to get the parameters λ_j $(j = 1, 2, \cdots, 8)$ of CRFs model so as to maximize the conditional probability of the log likelihood $L_{(\lambda)} = \sum_{i=1}^{N} \log P_\lambda \left(y^{(i)} \big| x^{(i)} \right)$, where $P_\lambda \left(y^{(i)} \big| x^{(i)} \right)$ is defined by (2).

Since Newton's Method costs large amount of time to compute the inverse of Hessian matrix and requires the Hessian matrix always being positive definite, we use Quasi-Newton method in the parameter optimization for CRFs model.

3.2.1   Quasi-Newton Method

Supposed the optimization problem is: $\max f(x)$ s.t. $x \in R^n$ .Assuming that $f(x)$ has continuous second partial derivatives on $x^{(k)}$, we would use

$$\nabla^2 f\left(x_{k+1}\right) = \frac{\nabla f\left(x_{k+1}\right) - \nabla f\left(x_k\right)}{x_{k+1} - x_k}$$ as the iterative formula instead of Hessian matrix.

Let the iterative formula be

$$x_{k+1} = x_k - H_k \nabla f\left(x_k\right) \tag{4}$$

supposed that the iterative formula is

$$H_k = H_{k-1} + A_k + B_k \tag{5}$$

Then we can get the values of A_k and B_k . Moreover A_k and B_k satisfy

$$\sum_{k=1}^{n} A_k = H^{-1}, \sum_{k=1}^{n} B_k = -H_0 = -I$$

After substituting A_k and B_k into (5), we can get the expression of H_k . And the iterative formula can be obtained after substituting H_k into (4).

3.2.2   Parameter Estimate

For the CRFs model mentioned above, the object function $f(x)$ of Quasi-Newton Method is

$$L(\lambda) = \sum_{i=1}^{N} \sum_{t=1}^{T} \sum_{k=1}^{K} \lambda_k f_k \left(y_t^{(i)}, y_{t-1}^{(i)}, x_t^{(i)} \right) - \sum_{i=1}^{N} \log Z(X^i)$$

Its first derivative is written as:

$$\frac{\partial L}{\partial \lambda_{K}} = \sum_{i=1}^{N}\sum_{t=1}^{T} f_{k}\left(y_{t}^{(i)}, y_{t-1}^{(i)}, x_{t}^{(i)}\right) - \sum_{i=1}^{N}\sum_{t=1}^{T}\sum_{k=1}^{K} f_{k}\left(y, y, x_{t}^{(k)}\right)p\left(y, y \mid x^{i}\right) \qquad (6)$$

The k-dimensional vector $\nabla L(\lambda) = \left(\dfrac{\partial L}{\partial \lambda_{1}}, \dfrac{\partial L}{\partial \lambda_{2}}, I \dfrac{\partial L}{\partial \lambda_{K}}\right)$ consists of elements $\dfrac{\delta L}{\delta \lambda_{k}}$,

($k = 1, 2, \cdots K$) , which is just the $\nabla f(x)$ in the iterative formula (4) of Quasi-Newton Method.

To calculate $p(y, y' \mid x^{(i)})$ in (6), we can use a dynamic programming method which is similar to the forward-backward algorithm for Hidden Markov Model.

3.3   Decoding Problem

The decoding problem is described as follows: Given DNA sequences $x^{(i)}$ $(i = 1, \cdots, N)$, feature functions $f_{1} \sim f_{8}$ the corresponding weights λ_{j} $(j = 1, 2, \cdots, 8)$, and input sequence $x = "x_{1}x_{2} \cdots x_{n}"$, our goal is to label each position of x, namely, finding out label sequence $y = "y_{1}y_{2} \cdots y_{n}"$, $y_{i} \in \{+, -\}$. $y_{i} = "-"$ indicates this character is out of the CPG island otherwise y_{i} is in the island.

Because the label sequence to be solved satisfies the condition of the optimal substructure, Viterbi algorithm, which is based on the dynamic programming technique, can be applied to solve the decoding problem of CPG islands. In the Viterbi algorithm, a Viterbi variable $\Phi(i, y)$ is defined to denote the possibility of optimal label sequence y' on the processed subsequence $x' = "x_{1}, x_{2}, \ldots, x_{i}"$ of the observation sequence $x = "x_{1}, x_{2}, \ldots, x_{l}"$. Accordingly, $\Psi(i, y)$ is defined to denote the label sequence with the maximal possibility. The formal definitions of $\Phi(i, y)$ and $\Psi(i, y)$ are described as follows:

$$\Phi(i, y) = \max_{y_{1}, y_{2}, \ldots, y_{i}} P_{\lambda}\left(y_{1}, y_{2}, \ldots, y_{i} \mid x_{1}, x_{2}, \ldots, x_{i}\right)$$

and

$$\Psi(i, y) = \arg\max_{y_{1}, y_{2}, \ldots, y_{i}} P_{\lambda}\left(y_{1}, y_{2}, \ldots, y_{i} \mid x_{1}, x_{2}, \ldots, x_{i}\right)$$

The algorithm gradually gets the optimal label in each position based on $\Phi(i, y)$ until the whole label sequence is obtained. The framework of the algorithm is depicted as follows:

Algorithm. CpG-Viterbi

Input: $f_k(y_{i-1}, y_i, x_i)$: feature functions;

        $\lambda_k \; k=1,..,8$: the corresponding weights;

        $X="x_1, x_2,..., x_l"$: Sequence to be labeled;

Output: $Y="y_1, y_2,..., y_l"$: The label sequence over x;

Begin

1、Initialization

    For all labels y, set $\Phi(1, y) = 1$;

2、**For** i=2 **to** l **do**

    For all labels y, calculate

$$\phi(i, y) = \max_{y'} \left\{ \phi(i-1, y') * \exp\left(\sum_k \lambda_k f_k(y', y, x, i) \right) \right\}$$

$$\psi(i, y) = \arg\max_{y'} \left\{ \phi(i-1, y') * \exp\left(\sum_k \lambda_k f_k(y', y, x, i) \right) \right\}$$

    Endfor

  Endfor

3、Termination:

    $y_l = \Psi(l, y)$;

4、Computing label sequence $y_1,..., y_l$:

  For i=l-1 **downto** 1 **do**

    $y_i = \Psi(i+1, y_{i+1})$

  Endfor

End

4 Experimental Results and Analysis

4.1 Experimental Environment

From the database of human genome CPG islands [13],we select 72 sequences included 80 islands as the training data set , and the remainders are used to test the precision of our method. We use CRF++0.54 [14] as the tool to train CRFs model. All the experiments were conducted on a 3.0GHzPentium with 1GB memory, Windows XP operating system. All codes were complied using Microsoft Visual C++ 6.0.

4.2 Data Files and Test File Formats

The training procedure of CRF++ needs several data files such as training files, model files, label files and a feature template file. These structures and formats of the data files on CPG islands database are illustrated as follows.

4.2.1 Training Files, Model Files and Label Files

The model file, which describes the CRFs model, is a binary file built by the training process and is used in the decoding. Based on model files, we use the algorithm

CpG-Viterbi to obtain the global optimal labeling sequences which is presented by the label files.

A training file is used to record the information of the training sample. It comprises multiple tokens. The data of the first column shown in Table 1 denotes the observation characters to be labeled. The data of the second column represents the basic attribute feature of the sequence. For characters A and T, the attribute feature values are 0 indicating that the two characters are out of CPG islands. Similarly, we assign 1 to the attribute feature value of characters C and G indicating that they are in CPG islands. The states (in or out of the CPG islands) corresponding to observation characters are listed in the third column. Table 1 is a part of a training file used in the experiment, where the observation sequence is "ATCGGACGAT". It can be seen from the third column, "CGGACG" is a CPG island.

Table 1. An example of training file

Observation character	Attribute feature	Label
A	0	A^-
T	0	T^-
C	1	C^+
G(current focusing token)	1	G^+
G	1	G^+
A	0	A^+
C	1	C^+
G	1	G^+
A	0	A^-
T	0	T^-

4.2.2   Feature Template Files

A feature template file is used to describe feature function which can define unary features, dualistic features, n-dimensional features and even compound features based on character level. As shown in table 2, a group of feature template is practically used in our experiments

Table 2. A feature template example

#Unigram
U00:%x[-2,0]
U01:%x[-1,0]
U02:%x[0,0]
U03:%x[1,0]
U04:%x[2,0]
U05:%x[0,1]
U06:%x[-1,0]/%x[0,0]
U07:%x[0,0]/%x[1,0]
#Bigram
B1

Each line in the template file denotes one *template*. Note also that there are two types of templates, that is, "Unigram" and "Bigram". The types are specified with the first character of templates. In the above feature template file, the first character of parameter *Uxx* is "*U*", which denotes a unigram template. "*xx*" means the number of feature templates. In each template, special macro *%x[row,col]* will be used to specify a token in the input data, where *%x* represents taking strings from the database, *row* specifies the relative position from the current focusing token, and *col* specifies the absolute position of the column. To test our algorithm, we use four groups of templates as shown in Table 3.

Table 3. The feature templates

#Unigram	#Unigram	#Unigram	#Unigram
U00:%x[-2,0]	U00:%x[-2,0]	U00:%x[-2,0]	U00:%x[-2,0]
U01:%x[-1,0]	U01:%x[-1,0]	U01:%x[-1,0]	U01:%x[-1,0]
U02:%x[0,0]	U02:%x[0,0]	U02:%x[0,0]	U02:%x[0,0]
U03:%x[1,0]	U03:%x[1,0]	U03:%x[1,0]	U03:%x[2,0]
U04:%x[2,0]	U04: %x[-2,0]/%x[-1,0]	U04:%x[2,0]	U04: %x[0,0]/%x[2,0]
U05:%x[0,1]	U05: %x[-1,0]/%x[0,0]	U05: %x[0,0]/%x[1,0]	U05: %x[-2,0]/%x[-1,0]
U06:%x[-1,0]/%x[0,0]	U06: %x[0,0]/%x[1,0]	U06:%x[1,0]/%x[2,0]	U06: %x[-1,0]/%x[0,0]
U07:%x[0,0]/%x[1,0]	U07: %x[-1,0]/%x[1,0]	U07:%x[0,0]/%x[1,0]/%x[2,0]	U07:%x[-2,0]/%x[-1,0]/%x[0,0]
#Bigram	#Bigram	#Bigram	#Bigram
B1	B2	B3	B4

4.3   Experimental Results and Analysis

We test our algorithm and compare its performance with other methods in terms of recall and precision, which are defined as follows:

$$Recall = \frac{\text{The number of CpG islands extracted correctly}}{\text{The total number of true CpG islands in the test set}}$$

and

$$Precision = \frac{\text{The number of CpG islands extracted correctly}}{\text{The total number of CpG islands extracted}}$$

To comprehensively evaluate the system performance, we used the *F*-score which is the weighted geometrical average of precision and recall. *F*-score is defined as:

$$F = \frac{P*R*(\beta^2+1)}{R+\beta^2*P}$$

Here we set $\beta=1$ in our experiments which indicates recall and precision have the same importance.

We also conduct experiments of CpG islands identification using our algorithm and other three methods. We compare their performance in terms of recall and precision, and F-score. In our experiments, four feature templates shown in Table3 are used and the results are listed in Table4.

Table 4. Comparison of the results among CRFs, MEMM and HMM

Model	Feature template	Precision(P)（%）	Recall® （%）	F-score(F)（%）
CRFs	Template1	93.14	91.21	92.16
	Template 2	90.20	89.95	90.07
	Template 3	87.33	88.20	87.76
	Template 4	86.36	87.96	86.80
MEMM	Template 1	90.10	89.70	89.90
	Template 2	87.79	87.40	87.70
	Template 3	84.27	85.40	84.83
	Template 4	82.62	84.51	83.47
HMM	Template 1	91.50	89.60	90.59
	Template 2	89.36	88.13	88.98
	Template 3	83.72	84.37	84.04
	Template 4	78.26	80.20	79.22

From Table4, we can see that our algorithm using CRFs outperforms the other methods in the aspects of both recall, precision and F-score for CpG islands identification. Our experimental results have fully highlighted the merit of CRFs in CpG islands identification.

5   Conclusions

In this paper, we present a novel method for CpG islands location identification based on the model of conditional random fields. In order to overcome the shortcomings such as the strong independence assumptions and the label-bias problem exhibited by other models, our method transforms the problem of CpG islands location identification into sequential data labeling. Based on the nature of CpG islands location, we define the corresponding feature functions and design the methods of model constructing, training and decoding in CRF accordingly. Experimental results on benchmark data sets show that our algorithm is more practicable and efficient than the methods based on hidden Markov model (HMM) or the maximum entropy Markov model (MEMM).

Acknowledgements. This research was supported in part by the Chinese National Natural Science Foundation under grant Nos. 61070047、61070133 and 61003180, Natural Science Foundation of Jiangsu Province under contracts BK2010318, BK21010134, and Natural Science Foundation of Education Department of Jiangsu Province under contract 09KJB20013.

References

1. Li, W.J., Li, M., Xin, R.H., Wei, L.H.: Promoter Recognition in Human Genome Based on KL Divergence and BP Neural Network. Journal of Liaoning Normal University (Natural Science Edition) 3(33), 42–45 (2010)
2. Huang, Y.K.: Promoter Recognition System Research from Gene Sequence Data. The Master's Thesis of Harbin Engineering University (2009)
3. Zhang, C.T.: The Current Status and The Prospect of Bioinformatics. The Journal of Liaoning Science and Technology 08, 25–26 (2001)
4. Tong, Q., Zhen, H.R., Ning, Y.: A Gene-prediction Algorithm Based on the Statistical Combination and the Classification in Terms of CpG Content. Journal of Beijing Biomedical Engineering 4(26), 178–181 (2007)
5. Takai, D., Jones, P.A.: Comprehensive Analysis of CpG Islands in Hhuman ChromoSomes 21 and 22. Proc. Natl. Acad. Sci. USA 99(6), 3740–3745 (2002)
6. Hackenberg, M., Previtil, C., Luis, L.E., et al.: CpG Cluster: a Distance-based Algorithm for CpG-island Detection. BMC Bioinformatics 7, 446 (2006)
7. Wang, Y., Leung, F.C.: An Evaluation of New Criteria for CpG Islands in the Human Genome as Gene Markers. Bioinformatics 20(7), 1170–1177 (2004)
8. Wang, J.L., Su, J.Z., Wang, F.C., Ying, Z.Y.: A New Method to Predict CpG-islands Based on Fuzzy Theory. China Journal of Bioinformatics 6(7), 91–94 (2009)
9. Shi, O.Y., Yang, J., Tian, X.: Hidden Markov Model for CpG Islands Prediction Based on Matlab. Computer Applications and Software 11(25), 214–215 (2008)
10. Jiang, H.J., Zhang, Z.L.: Discrimination of CpG Islands Location Based on HMM. Mathematical Theory and Applications 6(29), 113–116 (2009)
11. John, L., Andrew, M.C., Fernando, P.: Conditional Random Fields: Probabilistic Models for Segmenting and Labeling Sequence Data. In: Proc. of the 18th ICML, pp. 282–289. Morgan Kaufmann, San Francisco (2001)
12. Hammersley, J.M., Clifford, P.: Markov Field on Finite Graphs and Lattices (1971) (unpublished)
13. ftp://ftp.ebi.ac.uk/pub/databases/cpgisle
14. http://sourceforge.net/projects/crfpp/files/

A Novel Gene Selection Method
for Multi-catalog Cancer Data Classification

Xuejiao Lei[1], Yuehui Chen[2], and Yaou Zhao[3]

[1] School of Information science and Engineering, University of Jinan, PR China
aduosi@126.com
[2] School of Information science and Engineering, University of Jinan, PR China
yhchen@ujn.edu.cn
[3] School of Information science and Engineering, University of Jinan, PR China
ise_zhaoyo@ujn.edu.cn

Abstract. In this paper, a novel gene selection method which was merging the relevance score (BW ratio) and the Flexible Neural Tree (FNT) together was proposed for the multi-class cancer data classification. Firstly, the BW ratio method was adopted to select some informative genes, and then the FNT method was used to extract more characteristic genes from the gene subsets. FNT is a tree-structured neural network with input variables selection, over-layer connections and different activation functions for different nodes. Based on the pre-defined instruction/operator sets, a flexible neural tree model can be created and evolved. The FNT structure is developed by using probabilistic incremental program evolution (PIPE) algorithm, and the free parameters embedded in neural trees are optimized by particle swarm optimization (PSO) algorithm. Experiment on two well-known cancer datasets shows that the proposed method achieved better results compared with other methods.

Keywords: gene selection, BW ratio, Flexible Neural Tree, Probabilistic Incremental Program Evolution, Particle Swarm Optimization.

1 Introduction

In addition, reducing the number of genes can help to cut down the inputs for computation, so the classifiers are much more efficient for classification and run much faster. For this so-called "high-dimensional small sample" problem, a suitable gene selection method is very important.

Feature selection techniques can be divided into three categories, depending on how they interact with the classifiers [1]. Filter methods directly operate on the dataset, that is, they are independent for classification models. The output of these methods are also score–ranking values, according to these ranking values the informative genes are selected. There are many filter methods which are proposed by predecessors, such as BW ratio [2], Bscatter[3], MinMax[3] and so on. The advantage of these methods is fast, but has inferior results. Wrapper methods run a search in the space of feature subsets, guided by the outcome of the model (e.g. classification performance

D.-S. Huang et al. (Eds.): ICIC 2012, LNCS 7389, pp. 319–326, 2012.
© Springer-Verlag Berlin Heidelberg 2012

on a cross-validation of the training set). There are many researchers using these methods to select feature subsets, such as Particle Swarm Optimization (PSO) [4], genetic algorithm (GA) [5] and so on. They often obtain better results than filter methods, but have an increased computational cost [6]. Finally, embedded methods with internal information of the classification model was used to perform feature selection, for instance Genetic programming (GP) [7] and Flexible Neural Tree (FNT)[8]. They often provide a good trade-off between performance and computational cost [9].

In this paper, we proposed a novel gene selection method which was mixing the BW ratio and the flexible neural tree. Firstly, the BW ratio method was used to select feature subsets, and then the FNT was utilized to extract more informative gene subsets from the selected feature subsets which got from the first step. The FNT structure is developed by probabilistic incremental program evolution (PIPE) algorithm and the free parameters embedded in neural trees are optimized by particle swarm optimization (PSO) algorithm.

2   Gene Selection Methods

Generally, the micro-array data has very high dimensionality (in thousands) and small size of samples (in dozens). However, only small parts of genes have great impact on classification and most of them are useless. These irrelevant genes not only confuse learning algorithms, but also degrade their performance and efficiency. Moreover, the prediction model induced from irrelevant genes may prone to over-fitting. So in this paper we applied mixing BW ratio method and FNT to select some informative genes.

2.1   BW Ratio Gene Selection Method and Flexible Neural Tree

The BW ratio [2] gene selection method is firstly introduced by Sandrine Dudoit et al. It is a preliminary genes selection method based on the ratio of their between-group to within-group sums of squares. The highest BW ratio value is most informative.

The FNT[8] is a tree-structured neural network with input variables selection, over-layer connections and different activation functions for different nodes. Based on the pre-defined instruction and operator sets, a flexible neural tree model can be created and evolved

2.2   Mix Gene Selection Method

Depending on the advantage of filter methods and the embedded methods, we proposed a novel gene method with mixing the BW ratio method and the FNT.

In our paper, the FNT was used for the multi-class cancer data to select feature genes. Firstly, the BW ratio was applied to select some informative genes. And then, according to the pair-wise comparison classification strategy, the selected subsets which were got from the first step were divided into two-label subsets. Finally, the FNT was utilized for the two-label subsets to choose more characteristic genes (see Fig.2).

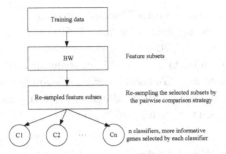

Fig. 1. BW indicates the feature sets which are generated by BW ratio method respectively. C1, C1, ..., Cn denote n FNT-base classifiers. Each FNT selects the more informative genes.

3   Classification Method

In this paper, the flexible neural tree (FNT) was also applied as the base classifier and the pair-wise comparison classification strategy. Depending on the extracted datasets, the FNT model can be created and evolved. The structure of FNT is developed by the Probabilistic Incremental Program Evolution (PIPE) [10] and the parameters are optimized by the Particle Swarm Optimization (PSO). The PIPE programs are encoded in n-array trees that are parsed depth first from left to right, with n being the maximal number of function arguments.

3.1   Procedure of the General Learning Algorithm

The general learning procedure for constructing the FNT model can be described as follows.

1. Create an initial population randomly (FNT trees and its corresponding parameters);
2. Structure optimization is achieved by using the PIPE algorithm;
3. If a better structure is found, then go to step 4, otherwise go to step 2;
4. Parameter optimization is achieved by the PSO algorithm as described in subsection chapter 3.2. In this stage, the architecture of FNT model is fixed, and it is the best tree developed during the end of run of the structure search. The parameters (weights and flexible activation function parameters) encoded in the best tree formulate a particle.
5. If the maximum number of local search is reached, or no better parameter vector is found for a significantly long time then go to step 6; otherwise go to step 4;
6. If the satisfactory solution is found, the algorithm is stopped; otherwise go to step 2.

4   Cancer Classification Using FNT

4.1   Data Sets

We performed extensive experiments on two benchmark cancer datasets, namely the MLL and Brain tumor (GLIOMA), which were downloaded from the website given in [11].

The MLL dataset [12] contains total 72 samples in three classes, acute lymphoblastic leukemia (ALL), acute myeloid leukemia (AML), and mixed-lineage leukemia gene (MLL), which have 24, 28, 20 samples, respectively.

The Brain tumor (GLIOMA) dataset [13] contains in total 50 samples in four classes, cancer glioblastomas (CG), non-cancer glioblastomas (NG), cancer oligodendrogliomas (CO) and non-cancer oligodendrogliomas (NO), which have 14, 14, 7, 15 samples, respectively. Each sample has 12625 genes.

4.2   Experiment Result and Analysis

In our experiment, firstly the BW ratio method was used to select 35 informative genes, and then the FNT method was employed for the selected gene subsets to choose more characteristic genes. Meanwhile, we utilized the pair-wise comparison strategy for multiclass classification. The FNT was employed to be the base classifier. The parameter of each FNT was adjusted by the PSO and the structure of each FNT was optimized by the PIPE.

For comparing with the other peoples' work, the classification performance was measured by 5-fold cross validation technique. The 5-fold cross validation technique: the samples were randomly divided into 5 equally sized subsets; each subset was used as a test set for a classifier, and the remaining 4 subsets were for training. The training data was used to select informative features. This process was repeated for 10 times to obtain the average results.

- MLL cancer

Firstly the BW ratio method was used to select 35 informative genes (see Fig.3 left). It was shown that the curve dot of ALL was close to the curve dot of MLL in the left picture of Figure 3. This was disadvantageous for the classification. But our method supplied the shortage of the BW ratio method. The selected subset was divided into a training set with 58 samples and a testing set with 14 samples by 5-fold cross validation technique. The MLL had three classes: 0(AML), 1(ALL), 2(MLL). According to the pair-wise comparison classification strategy, there were three classifiers. And then utilize the FNT to select genes and classify. The best FNT trees obtained by the proposed method were shown from Figure 4 to Figure 5 (left). It should be noted that the informative features for constructing the FNT model are formulated in the light of the procedure mentioned in the previous section. These final informative genes selected by FNT are shown in Table 2.

For comparison purpose, the classification performances of a KPCSR [14], SVM+F-test[14], SVM+Cho's[14], KNN+F-test[14], KNN+Cho's[14] and our method are shown in Table 1. For MLL dataset, our method, which has 2.6–4.5% increments, gives the best classification accuracy. It is observed that our method has and the FNT classification models with the proposed mix gene selection method are better than other models for classification of microarray dataset.

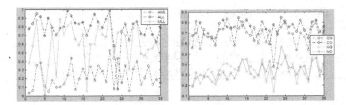

Fig. 2. In the two figures, the X axis indicates the number of selected genes and the Y axis is the mean value of the samples of each feature gene which is selected by the BW ratio method. The left picture is shown the feature gene description of the MLL and the Brain's is left.

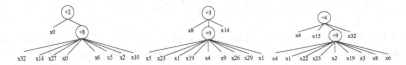

Fig. 3. An evolved best FNT for the 0 class and 1 class of MLL data classification (left), the 0 class and 2 class of MLL data (middle), the 2 class and 1 class of MLL data (right)

- Brain cancer

Firstly the BW ratio method was applied to select 35 informative genes (see Fig.3 right). The selected subset was divided into a training set of 40 samples and testing set of 10 by 5-fold cross validation technique. The Brain cancer had four classes: 0(CG), 1(CO), 2(NG), 3(NO). According to the pair-wise comparison classification strategy, there were six classifiers. And then utilize the FNT to select genes and classify. The best FNT trees obtained by the proposed method are shown from Fig. 5 to Fig. 6. It should be noted that the informative features for constructing the FNT model are formulated in the light of the procedure mentioned in the previous section. These final informative genes selected by FNT algorithm is shown in Table 4.

For comparison purpose, the classification performances of a KNN+EGSIEE[15], KNN+EGSEE[15], KNN+GS1[15], KNN+GS2[15] and our method are shown in Table 3. For Brain cancer, our method which has 4.2–10.2% increments gives the best classification accuracy. Also in our experiments the maximal number of selected genes is 19, it is smaller than others'. It is observed that the FNT classification models with the proposed mix gene selection method are better than other models for classification of micro-array dataset.

Table 1. Relative works on MLL dataset

method	Test accuracy (%)
Our method	99.3
KPCSR[14]	96.7
SVM+F-test[14]	94.8
SVM+Cho's[14]	95.5
KNN+F-test[14]	95.4
KNN+Cho's[14]	96.0

Table 2. The extracted informative genes in case of MLL dataset

0_1 class[1]	x0,x32,x14,x27,x6,x5,x2,x10
0_2 class	x8,x14,x5,x23,x19,x4,x9,x26,x29,x1
2_1 class	x4,x15,x32,x1,x22,x23,x2,x19,x3,x8,x6

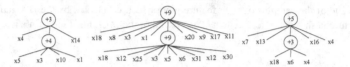

Fig. 4. An evolved best FNT for the 0 class and 1 class of Brain data classification (left), the 0 class and 2 class of Brain data (middle), the 0 class and 3 class of Brain data (right)

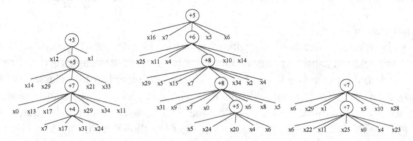

Fig. 5. an evolved best FNT for the 2 class and 1 class of Brain data classification (left), an the 3 class and 1 class of Brain data (middle), the 2 class and 3 class of Brain data (right)

Table 3. Relative works on Lymphoma dataset

method	Test accuracy (%)
Our method[2]	90.0(19)
KNN+EGSIEE[15]	84.0(19)
KNN+EGSEE[15]	78.0(10)
KNN+GS1[15]	84.0(95)
KNN+GS2[15]	80.0(92)

The number in the bracket indicates the genes which are selected

---

[1] The MLL has three classes: 0(AML), 1(ALL), 2(MLL). 0_1 class indicates the classifier which is trained by the samples with label 0 and label 1, similarly, 0_2 class is label 0 and label 2, and 2_1 class label 2 and label 1.

[2] The number in the bracket indicates the number of the final selected genes.

Table 4. The extracted informative genes in case of MLL dataset

0 _ 1 class[3]	x4,x14,x5,x10,x3,x1
0 _ 2 class	x18,x8,x3,x1,x20,x9,x17,x11,x12,x25,x3,x5,x6, x31,x12,x30
0 _ 3 class	x7,x13, x16,x4,x18,x6
2 _ 1 class	x12,x1,x14,x29,x21,x33,x0,x13,x17,x29,x34,x11, x7,x31,x24
3 _ 1 class	x16,x7,x5,x6,x25,x11,x4,x10,x14,x29,x15,x34,x2, x31,x9,x0,x8,x24,x20
2 _ 3 class	x6,x29,x1,x5,x10,x28,x22,x11,x25,x0,x4,x23

5   Conclusion

In this paper, we proposed the mix gene selection method, which was merging the BW ratio and FNT method together. The MLL and Brain cancer datasets were used for conducting all the experiments. Feature genes were first extracted by the defined relevance score technique (BW ratio method) which greatly reduces dimensionality as well as maintains the informative features. Then the FNTs were employed to classify and select more characteristic genes. Compare the results with some advanced gene selection technology; the proposed method produces the best recognition rates. In the future, we will choose more critical features to characterize the gene expression data. We also believe that FNT can have a great contribution to this study.

Acknowledgments. This research was partially supported by the Natural Science Foundation of China (61070130), the Key Project of Natural Science Foundation of Shandong Province (ZR2011FZ001), the Key Subject Research Foundation of Shandong Province and the Shandong Provincial Key Laboratory of Network Based Intelligent Computing.

References

1. Saeys, Y., Abeel, T., Van de Peer, Y.: Robust Feature Selection Using Ensemble Feature Selection Techniques. In: Daelemans, W., Goethals, B., Morik, K. (eds.) ECML PKDD 2008, Part II. LNCS (LNAI), vol. 5212, pp. 313–325. Springer, Heidelberg (2008)
2. Sandrine, D., Jane, F., Terence, P.S.: Comparison of Discrimination Methods for the Classification of Tumors Using Gene Expression Data. The American Statistical Association 97(457) (2002)
3. Hong, C., Carlotta, D.: An Evaluation of Gene Selection Methods for Multi-class Microarray Data Classification. In: Proceedings of the Second European Workshop on Data Mining and Text Mining in Bioinformatics
4. Yang, C.S., Chuang, L.Y., Li, J.C., Yang, C.H.: A Novel BPSO Approach for Gene Selection and Classification of Microarray Data. IEEE (2008)
5. Hrishikesh, M., Nitya, S., Krishna, M., Tapobrata, L.: An ANN-GA model based promoter prediction in Arabidopsis thaliana using tilling microarray data. Bioinformation 6(6), 240–243 (2011)

---

[3] The Brain cancer has four classes: 0(CG), 1(CO), 2(NG), 3(NO). 0_1 class indicates the classifier which is trained by the samples with label 0 and label 1, similarly, 0_2 class is label 0 and label 2, 0_3 class label 0 and label 3, 2_1 class label 2 and label 1, 3_1 class label 3 and label 1, 2_3 class label 2 and label 3.

6. Kohavi, R., John, G.: Wrappers for feature subset selection. Artif. Intell. 97(1-2), 273–324 (1997)

7. Liu, K.H., Xu, C.G.: A genetic programming-based approach to the classification of multiclass microarray datasets. Original Paper, Bioinformatics/btn644 25(3), 331–337 (2009)

8. Chen, Y., Peng, L., Abraham, A.: Gene Expression Profiling Using Flexible Neural Trees. In: Corchado, E., Yin, H., Botti, V., Fyfe, C. (eds.) IDEAL 2006. LNCS, vol. 4224, pp. 1121–1128. Springer, Heidelberg (2006)

9. Saeys, Y., Inza, I., Larranaga, P.: A review of feature selection techniques in bioinformatics. Bioinformatics 23(19), 2507–2517 (2007)

10. Salustowicz, R., Schmidhuber, J.: Probabilistic incremental program evolution. Evolutionary Computation 5(2), 123–141 (1997)

11. Yang, K., Cai, Z., Li, J., Lin, G.: A stable gene selection in microarray data analysis. BMC Bioinformatics 7(228) (2006)

12. Armstrong, S.A., Staunton, J.E., Silverman, L.B., Pieters, R., den Boer, M.L., Minden, M.D., Sallan, S.E., Lander, E.S., Golub, T.R., Korsmeyer, S.J.: MLL translocations specify a distinct gene expression profile that distinguishes a unique leukemia. Nature Genetics 30, 41–47 (2002)

13. Nutt, C.L., Mani, D.R., Betensky, R.A., Tamayo, P., Cairncross, J.G., Ladd, C., Pohl, U., Hartmann, C., McLaughlin, M.E., Batchelor, T.T., Black, P.M., von Deimling, A., Pomeroy, S.L., Golub, T.R., Louis, D.N.: Gene Expression-based Classification of Malignant Gliomas Correlates Better with Survival than Histological Classification. Cancer Research 63, 1602–1607 (2003)

14. Zhang, B.-L.: Cancer Classification by Kernel Principal Component Self-regression. In: Sattar, A., Kang, B.-H. (eds.) AI 2006. LNCS (LNAI), vol. 4304, pp. 719–728. Springer, Heidelberg (2006)

15. Li, G.Z., Meng, H.H., Ni, J.: Embedded Gene Selection for Imbalanced Microarray Data Analysis. IEEE (2008)

A Novel Discretization Method for Microarray-Based Cancer Classification

Ding Li[1,2], Rui Li[1,2], and Hong-Qiang Wang[2,*]

[1] Department of Automation, University of Science and Technology of China,
Hefei 230027, P.R. China
[2] Intelligent Computation Lab, Hefei Institute of Intelligent Machines, Chinese Academy of
Science, P.O.Box 1130, Hefei, Anhui 230031, P.R. China
{ld051014,hqwang126}@126.com

Abstract. In this paper, we propose a gene expression diversity-based method for gene expression discretization. By counting the numbers of samples of different classes in an open expression intervals, the method calculates class distribution diversity and then expression diversity for genes. Based on the gene expression diversity, three discretization criteria are established for discretizing gene expression levels. We evaluate the proposed method on the publicly available leukemia dataset and compare it with several previous methods.

Keywords: gene expression, gene expression diversity, gene regulation, discretization.

1   Introduction

Advent of high-throughput technology makes it possible to simultaneously measure expression levels of ten of thousands of genes in cells [1]. These data are often used to build models to predict the phenotype of cells. An increasing number of computational approaches have been developed for gene expression data analysis [2]. In such analysis, the discretization of continuous gene expression data is an important step for efficient cancer classification.

At present, a number of methods have been applied to convert continuous data to discrete ones for pattern recognition [3-9]. For example, Fayyad et al. proposed the mdlp method which is a typical non-parameter supervision discretization method [7]. Another commonly used method is the chiM method, which discretizes data based on the χ^2 distribution [10]. These methods work well in most cases, but they can not relate gene expression with gene regulation for the biological significance of gene expression discretization. In this paper, we propose a new method for gene expression discretization based on the putative 3-state regulation viewpoint [11].

The rest of the paper is organized as follows. In section Methods, we first formulate three core rules of discretization based on gene expression diversity, and describe the proposed discretization procedure in detail. In section Experimental

---

* Corresponding author.

D.-S. Huang et al. (Eds.): ICIC 2012, LNCS 7389, pp. 327–333, 2012.
© Springer-Verlag Berlin Heidelberg 2012

results, we first discuss the choices of the two parameters of the proposed method in practice, and then evaluate the proposed method on a real-world microarray data and compare with two previous methods, chiM and mdlp. Finally, the paper is concluded.

2   Methods

Gene expression levels are continuous values. However, gene regulation levels are generally assumed to be in a regulation space of 3 discrete states, i.e., down-regulatory, non-regulatory and up- regulatory states [11]. We argue that a reasonable discretization of gene expression values should consider relating gene regulation states to separated intervals of expression levels. The proposed gene expression discretization method is based on the 3-state regulation assumption, and aims at accurate cancer classification. For the convenience of presentation, we only consider binary cancer classification in this study. Assuming that the expression values of a gene in a cancer type or subtype are normally distributed in a particular expression interval, relationships between the distributions of the two cancer classes and gene regulation states could be one of the following three types: i) the two distributions are completely overlapped within a regulation state, as shown in Fig.1A; ii) the two distributions fall in two different regulation states and are clearly separated, as shown in Fig.1B; iii) the two distributions are partially overlapped in a regulation state, as shown in Fig.1C.

It can be found that only genes with the two types of expression distributions in Figs. 1B and C are useful for cancer classification. For them, the expression values can be divided into two and three discretization states, respectively. However, the genes expressed like in Fig.1A can be thought of being disabled for classification, and their expression values all should be assigned into one discretization state as noise. Based on the idea, we develop a new discretization method in the following sections.

In order to clearly depict the proposed disretizitation criteria, we first define two concepts associated with gene expression distribution. Consider binary classification problem with class 1 and class 2. For a left half-open area of gene expression, we define the diversity of class distribution within it, denoted by D, as

$$D = n_1/N_1 - n_2/N_2 \tag{1}$$

where the N_1 and N_2 represent the total numbers of samples of class 1 and class2, respectively, and n_1 and n_2 represent the numbers of samples of class 1 and class 2 having expression values in the area. Assume a gene whose expressions are divided into have n left half-open areas. Let D_i, $i=1,2,\ldots,n$, represents the corresponding n class distribution diversities, we define the diversity of expression for the gene, denoted by Δ as

$$\Delta = D_{max} - D_{min} \tag{2}$$

where D_{max} and D_{min} represent the maximum and minimum of the n class distribution diversity respectively. For D_{max} and D_{min}, we denote the right endpoints of their left half-open areas by L_{min} and L_{max}, respectively.

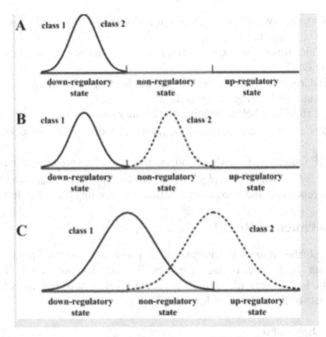

Fig. 1. The relationship between gene regulatory states and the distribution of sample category

Given two constants, $0<\alpha$, $\lambda<1$, three criteria for discretizing gene expression values can be made based on gene expression diversity as follows:

Criterion 1: if $\Delta =D_{max}-D_{min}<\alpha$, all the expression values of the gene will be grouped into one discretization state.

Criterion 2: if $\Delta =D_{max}-D_{min}>=\alpha$ and $\min(|D_{max}|,|D_{min}|)<\lambda$ where min and |.| represent the minimum and absolute functions respectively, the expression values of the gene will be discretized into two states.

Criterion 3: if $\Delta =D_{max}-D_{min}>=\alpha$ and $\min(|D_{max}|,|D_{min}|)>=\lambda$, the expression values of the gene will be discretized into three states.

The three criteria are associated with the three cases shown in Fig.1. Generally, it is assumed that a gene has three regulation states changeable to the response to external stimulus. So, for a binary classification problem, it is reasonable to divide the expression of a gene into at most three discretization states. The parameters, α and λ, are referred to as gene expression diversity threshold and class distribution diversity, respectively. The choices of the two parameters will be discussed in section Results.

Based on the three discrete criteria, a gene expression discretization procedure, especially for a binary problem can be described as follows:

Step. 1 Setting α and λ;

Step. 2 Dividing the whole range of expression values into n ($n = 50$) intervals uniformly and defining n ($n = 50$ or above) left half-open areas based on the n intervals;

Step. 3 Calcualting the class distribution diversities by Eq.(1) for each area to obtain D_{max}, D_{min}, L_{max} and L_{min};

Step. 4 Calcualting the expression diversity Δ across the n class distribution diversities by Eq.(2);

Step. 5 If Criterion 1 holds, taking the whole expression range as the only one discretiation state and ending the procedure;

Step. 6 If Criterion 2 holds, dividing the whole expression range into two intervals: $(-\infty,\ L)$ and $(L, +\infty)$, where $L=L_{max}$ if $|D_{max}|>|D_{min}|$ and L_{min} otherwise, and ending the procedure;

Step. 7 If Criterion 3 holds, dividing the whole expression range into there intervals: $(-\infty,\ L_1]$, (L_1, L_2) and $[L_2, +\infty)$, where $L_1=\min(L_{max}, L_{min})$ and $L_2=\max(L_{min}, L_{max})$ [max represents the maximum function], and ending the procedure.

3   Experimental Results

We evaluated the proposed discretization procedure on the publicly available leukemia data [12]. The dataset contains 72 samples, of which 47 are acute lymphoblastic leukemia (ALL) and 25 are acute myelogenous leukemia (AML). Each sample consists of the expression levels of 7129 genes.

3.1   The Choice of α

The proposed method uses the variable Δ to measure the diversity of gene expression. The threshold α is used to find genes whose diversities are large enough to distinguish the two classes. A too small α will bring noise to cancer classification while a too large α will run a risk of missing the genes whose expressions actually differ between the two classes. In essential, Δ means the proportion difference of the two classes, and may vary from zero to 1. So, a difference level of proportions larger than 0.4 should be statistically significant to class distinction. In this study, we prefer to use $\alpha=0.4$ for implementing the dicertizion procedure. For the justification, we have observed the random distribution of Δ on the leukemia data. We randomly shuffled the labels of the 72 samples ten times, and calculated the Δs of all the 7129 genes for each time. As a result, 71290 Δs were obtained, whose distribution is drawn in Fig.2. From Fig.2, it can be found that a random Δ seems to follow a normal distribution centered at 0.23. Further, the distribution takes on the 95% percentile of 0.37. These observations suggest that the value of $\alpha=0.4$ is acceptable for gene expression discretization.

3.2   The Choice of λ

The parameter λ is used to remove the discretization intervals in which two classes are distinguishable. This means that the magnitude of the corresponding D_{max} or D_{min} is not significantly large to specify a discretization interval. So, the value of λ should be chosen to guarantee to ignore the intervals where different classes behave the same but retain those making different classes separated. Recalling that α reflect a minimum acceptable expression diversity of a gene, we suggest setting $\lambda = \alpha/2 =0.2$ as default in practice.

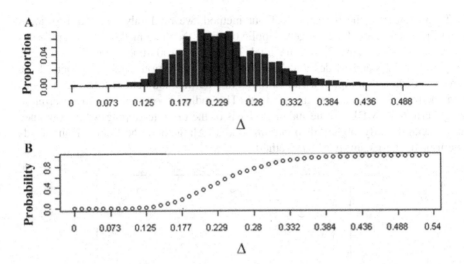

Fig. 2. Histogram (A) and cumulative probability distribution (B) of Δ

3.3   Application to the Leukemia Data

We next applied our method to the leukemia data for cancer classification. In the application, two well-known classification methods, support vector machines (SVM) and K-nearest neighborhood (KNN), were used to build classifier on the discretization results for all genes. For reliable validation, we randomly selected 50 samples (32 ALL and 18AML) from the total 72 samples as a training set for data discretization and learning classifier and the rest as a test set (15 ALL and 7 AML) for independent validation , and the random split was repeated 100 times. The following measures were used for classification evaluation: the minimum, maximum and average of classification accuracies, the average false positive rates (FPR) and the average false negative rates (FNR). For comparison, we also used two previous methods, chiM and mdlp, to analyze the data in the same setting. Table 1 compared the classification results by the three methods. From Table 1, it can be found that our method obtained the highest average classification accuracy, irrespective of the classifiers used.

Table 1. Comparison of the classification results by the three discretization methods on the leukemia data

Methods	SVM					KNN				
	ACC	max ACC	min ACC	FPR	FNR	ACC	max ACC	min ACC	FPR	FNR
Our method	**90.05**	100	77.27	12.74	0	**99.00**	100	95.45	1.06	0.87
chiM	77.27	86.22	68.18	24.92	0	87.86	100	72.73	9.04	18.85
Mdlp	80.14	90.91	68.18	22.56	0	98.95	100	95.45	0.80	1.59

Note: ACC-mean accuracy, minACC-min accuracy, maxACC-max accuracy, FPR- false positive rate, FNP- false negative rate

To demonstrate the robustness of our method, we randomly selected 100 genes from 7129 genes, and for each gene, applied each of the three methods to discretize its expression levels using 38 of the total 72 samples 100 times. Fig. 3 shows the boxplots of the standard deviations (sd) of the discretization results for each of the three methods. From Fig. 3, it can be found that our method outperforms the previous two methods with the least mean of sd. The mdlp chooses minimum description length criterion (MDL) as the judgment basis of the discrete-stoping, which obtained a sd mean slightly higher than our method's. Altogether, the comparison of sds confirms the best stability of our method.

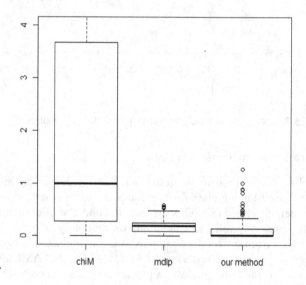

Fig. 3. Boxplots of standard deviations of the discretization results for the three methods

4   Conclusion

We have proposed a new method based on gene expression diversity for discretizing gene expression data. We evaluated it on a real-world data set, i.e., the publicly available leukemia data set, and compared with two previous methods, chiM and mdlp. The experimental results showed that our method is more stable and more robust than the previous ones for the discretization of gene expression, and can result in a better classification performance for the leukemia data. Future work will focus on applications to more real-world data sets.

Acknowledgement. This work was supported by the grants of the National Science Foundation of China, Nos. 30900321, 31071168, 60975005, 61005010, 60873012, 60973153, 61133010 and 60905023.

References

1. Berns, A.: Cancer: gene expression diagnosis. Nature 403, 491–492 (2000)
2. Borges, H.B., Nievola, J.C.: Feature Selection as a Preprocessing Step for Classification in Gene Expression Data. In: Seventh International Conference on Intelligent Systems Design and Applications, pp. 157–162 (2007), doi:10.1109/ISDA.2007.80
3. Boullé, M.: MODL: A Bayes Optimal Discretization Method for Continuous Attributes. Machine Learning, 131–165 (2006)
4. Brijs, T., Vanhoof, K.: Cost-sensitive Discretization of Numeric Attributes. In: Żytkow, J.M. (ed.) PKDD 1998. LNCS, vol. 1510, pp. 102–110. Springer, Heidelberg (1998)
5. Butterworth, R., Simovici, D.A., Santos, G.S., Ohno-Machado, L.: A Greedy Algorithm for Supervised Discretization. Journal of Biomedical Informatics, 285–292 (2004)
6. Dougherty, J., Kohavi, R., Sahami, M.: Supervised and Unsupervised Discretization of Continuous Features. In: Prieditis, A., Russell, S.J. (eds.) Proceedings of the Twelfth International Conference on Machine Learning, Tahoe City, California, pp. 194–202 (1995)
7. Fayyad, U.M., Irani, K.B.: Multi-interval Discretization of Continuous-valued Attributes for Classification Learning. In: Proceedings of the Thirteenth International Joint Conference on AI (IJCAI 1993), Chamberry, France, pp. 1022–1027 (1993)
8. Kohavi, R., Sahami, M.: Error-based and Entropy-based Discretization of Continuous Features. In: Proceedings of the Second International Conference on Knowledge Discovery and Data Mining, pp. 114–119. AAAI Press, Portland (1996)
9. Liu, H., Hissain, F., Tan, C.L., Dash, M.: Discretization: An Enabling Technique. Data Mining and Knowledge Discovery, 393–423 (2002)
10. Kerber, R.: ChiMerge: Discretization of Numeric Attributes. In: Proceedings of the Tenth National Conference on Artificial Intelligence, pp. 123–128 (1992)
11. Wang, H.Q., Huang, D.S.: Regulation Probability Method for Gene Selection. Pattern Recognition Letters 27, 116–122 (2006)
12. Golub, T.R., et al.: Molecular Classification of Cancer: Class Discovery and Class Prediction by Gene Expression Monitoring. Science 286(5439), 531–537 (1999)

Sequence-Based Prediction of Protein-Protein Interactions Using Random Tree and Genetic Algorithm

Lei Zhang[1,2]

[1] College of life Science, University of Science and Technology of China,
Hefei, Anhui 230027, China
[2] Intelligent Computing Laboratory, Institute of Intelligent Machines,
Chinese Academy of Sciences, Hefei, Anhui 230031, China
`ranyaolei@yahoo.com.cn`

Abstract. Protein-protein interactions play important roles in the course of cell functions such as metabolic pathways and genetic information processing. There are many shortcomings of traditional experiments such as tediousness and laboriousness. The machine learning methods have been developed to predict PPIs, and preliminary results have demonstrated their feasibility. Here, we introduce a sequence-based random tree and GA to infer PPI. Experimental results on *S.cerevisiae* dataset from DIP show that our novel method performs well than rotation forest, with higher accuracy, sensitivity and precision. Most importantly, our method runs faster than rotation forest.

Keywords: Protein-protein interactions, protein sequence, autocorrelation descriptor, random tree, GA.

1 Introduction

Proteins are the physical base of the biological processes. The research of protein attributes is a hot area in biological field. Protein-protein interactions contain multi-information about life, helping reveal the exact function of intrinsic biological phenomenon. With the development of experiments conditions, many PPIs have been discovered over the years and several databases have been created to store the information of these interactions. In particular, more PPIs are available from high-throughput interaction detection methods. It is a well-known fact that using high-throughput methods is expensive and time-consuming. False positives and false negatives also exist in the process of experiments. A novel method is demanded to help in the process of identifying real protein-protein interactions. Therefore, a number of computational approaches have been explored recently.

These methods [1,2], which are based on different attributes such as protein sequence, protein domain, subcellular localization, and genetic codon, can be divided into four sub-types based on: a) genome information; b) amino acids sequences; c) protein domains; d) other protein property.

D.-S. Huang et al. (Eds.): ICIC 2012, LNCS 7389, pp. 334–341, 2012.
© Springer-Verlag Berlin Heidelberg 2012

It is based on multiple data types, including the gene fusion [3], the proteins coevolution method [4], and phylogenetic profile method [5]. Sequence-based inference methods and structures-based prediction approaches have been proposed in previous research findings. As a matter of fact, the latter is more difficult than the former under later studies. The reason is that the feature means from protein structures, which are complex, diversity, and mutative, are difficult to exploit. Here we provide the details of sequence-based inference methods. The primary structure of protein, which is arrangements of amino acids, contains much useful information for basic research and application in biotechnology. The machine learning algorithms are universal method now. Support vector machine developed by Bock and Gough [6] was trained to recognize PPI based on the primary structure and the associated physicochemical properties of the proteins. Ben-Hur and Noble [7] proposed various kernel methods such as motif kernel and Pfam kernel using different data sources to predict interactions problems. Shen et al. [8] developed a novel method that combines triplet assembled information encoding with SVM to infer PPI. Xia et al. [9] used rotation forest that is an ensemble learning approach and autocorrelation descriptor to predict yeast PPIs.

Many methods of domain-based prediction PPIs are also developed over the years. Deng et al. [10] used maximum likelihood estimation(MLE) plus domains to infer protein interaction. Random decision forest framework was used by Chen et al. [11] in predicting PPIs. Lqbal et al. [12] used message-passing algorithms for the prediction of protein domain interactions from protein-protein interaction data.

In this study, we propose a novel sequence-based approach to the prediction of protein-protein interaction. As a powerful optimization tools, WEKA and GA are applied to random tree, which is a sensitive decision tree.

2   Materials and Methods

2.1   Data Sets

Database of Interacting Proteins(DIP) combines information from a variety of sources to create a single, consistent set of protein-protein interactions. Xia et al. [9] obtained the *S.cerevisiae* interaction dataset from DIP. The final data set consists of 5594 protein pairs that comprise positive set and 5594 protein pairs that comprise negative interaction set.

2.2   Extracting Sequence Features

The first priority is to extract computational information from existing bio-information. In our experiments, the universal method that each amino acid of protein sequence is converted into numerical value was used to encode protein sequence. Each residue in sequence was represented by six sequence-derived physicochemical properties, which include hydrophobicity, volumes of side chains of amino acid, polarity, polarizability, solvent-accessible surface area and net charge index of side chains of amino acid.

We formulated the problem of PPI prediction as a binary classification task, where the "interaction" class was represented by "1", and "non-interaction" class was expressed as "0". The key to the machine learning algorithm depends on that how to extract features from hidden protein sequence or structure information. There are many methods of extracting sequence features developed by in previous work such as auto covariance descriptor [13], conjoint triad [8], local descriptor [14], and autocorrelation descriptor [15, 16]. Then unequal-length vector of numerical values were obtained. It is crucial stage that how to transform unequal-length vector of protein vector into equal-length vector. In order to perform empirical studies for the PPI, the amino acid sequence information of heterogeneous length needs to be transformed into the feature vector information of homogeneous length. In previous application, Moran autocorrelation descriptor [15] was proved to be a better method for encoding features of protein sequence. It was defined as by Xia et al. [9]:

$$I(d) = \frac{1}{N-d} \sum_{j=1}^{N-d} (p_j - \bar{p})(p_{j+d} - \bar{p}) \bigg/ \frac{1}{N} \sum_{j=1}^{N} (p_j - \bar{p})^2 \tag{1}$$

$$\bar{p} = \sum_{j=}^{N} p_j \bigg/ N \tag{2}$$

where $\bar{p}$ is the average of the property of the 20 amino acids, and N is the length of the sequence, d, which ranges from 1 to 30, is the distance between one amino acid residue and its neighbors, p_j and p_{j+d} are the properties of the amino acid at positions j and $j+d$, respectively.

Thus, each protein pair is represented by a vector of 360 features. We adopted the same feature method with Xia et al. [9] to make the assessment effect more equitableness. Here, sixty percent of instances were chose as the training set. The remaining instances comprised the test set.

2.3   Random Tree and GA

Ensemble learning is a hot research topic in the field of machine learning. Integration of multiple weak classifiers should have higher classification accuracy than a single classifier. But there are still some weak points in it, i.e. the use of more basic classifier will lead to increase computational cost and storage cost. It is becoming more and more difficult to obtain divergence from a base classifier and others, with the increase of the number of individual classifier. Many empirical studies show that it can yield better performance, when we select a part of the base classifier from all. Zhou et al. [17] proposed GASEN, which is an approach that neural networks chosen as the base classifiers are effectively selected by GA, to apply to regression or classification task. The algorithm is as follows:

Input:training set S, learner L, trials ,threshold λ

Procedure:

for $t = 1$ to{

$S_t =$ bootstrap sample from S

$N_T = L(S_t)$

}

generate a population of weight vectors

evolve the population where the fitness of a weight vector w is measured as

$f(w) = 1/E_w^V$

$w^* =$ the evolved best weight vector

Output:ensemble N^*

$$N^*(x) = Ave \sum_{w_i^* > \lambda} N_T(x) \qquad \text{for regression}$$

$$N^*(x) = \arg\max_{y \in Y} \sum_{w_i^* > \lambda; N_i(x) = y} 1 \qquad \text{for classification}$$

As far as we know, there is no report discussing the application of select ensemble to predict PPIs. We presented an approach that the base classifiers of random tree were selected by GA. Trees are mathematical objects that play an important role in several areas of learning community. Decision tree uses tree model and builds training model from top and bottom according to the feature values. Because of the different root nodes, discrepant tree structures are generated, it is quite unstable. For the ensemble learning, decision tree can be chose as the base classifiers. There are many types of trees, such as C4.5, J48. In these tree structures, leaves represent class labels and branches represent conjunctions of features that lead to class labels.

Random tree on itself tends to be too weak. Thus, we want to integrate it into our method. Class for constructing a tree considers K random features at each node. Random tree does not prune. Similar to the standard random decision forest algorithm [19], training data will build a training model. The random tree randomly selects a subspace of features to focus on at each splitting node.

WEKA written in java is the popular software in machine learning community. There are many machine learning algorithms provided by WEKA. We used the random tree as the base classifier from library. Except the number of base classifier, the remaining parameters were set to the default. We used the training data to build all the basic classifiers. Here, bootstrap sampling was used to construct every random tree. Thus, all the classifiers had different with each other. In order to achieve good experimental results, a classifier, which had better performance and was discrepant compared with others, was chosen. How to select a good classifier from ensemble system can be seen as optimized problem. There are many algorithms developed in this research area, such as PSO(Particle Swarm Optimization), GA(Genetic Algorithm). GA is a search heuristic algorithm premised on the evolutionary ideas of natural selection and genetic. GA is widely applied to optimized community. There are several GA tool boxes developed by different investigators.

Here we used GAOT toolbox provided by Houck et al. [18]. The genetic operators, including select, crossover, and mutation, and the system parameters, including the crossover probability, the mutation probability and the stopping criterion, are all set to the default values of GAOT. The threshold used by GA was set to 0.01. A typical GA needs to evaluate the solution domain. In our experiment, the predictive value of each base classifier was chosen as one initial individual of population. So the number of classifiers was the size of initial population. The true labels of the training instances were used to assess evolutionary individual. Thus, every base classifier would be set a weight through their predictive outputs. The sum of weights was one. The weights that ranged from 0.01 to -0.01 would set to zero. Thus, the remaining weights, which were we wanted, were set to one. The base classifiers, which weights were set to one, would be used to predict test instances.

2.4   Evaluation Methods

Accuracy, sensitivity, precision are extensively used in classification problem. It can be used to evaluate results of inferring PPIs. For the reasonable comparison of the results, we used the same training set and test set in different algorithms. Sensitivity measures the proportion of actual PPIs which are correctly identified. Precision represents the proportion of positive inferences which are correct. Accuracy is defined as the percentage of all inferences which are correct.

3   Results and Discussion

We compared our experimental results with the rotation forest of Xia et al. [9] in Table1. The graph shows that our method has the better performance than ensemble method. Xia et al. [9] selected J48 as the base classifier and PCA as transformation method form WEKA software. Only two parameters, K and L (the number of feature subsets and the ensemble size, respectively) need to be set in previous experiments. So the remaining parameters would be set default values implemented in WEKA. Here, K was set to 12, which required the least computational cost. According the earlier reports, K, which is not too big and too small, has no great impact on our experimental results. L ranges from 1 to 100. The results show that the accuracy is changed with the different values of L: .it is a steady increase from 1 to 43, and it remains unchanged above the values.

Table 1. Chosen the best accuracy, sensitivity, precision comparison

	Our method	Rotation forest
Accuracy	94.69%	93.52%
Precision	96.82%	95.1%
Sensitivity	92.31%	91.5%
times	394.48 s	4853.53s

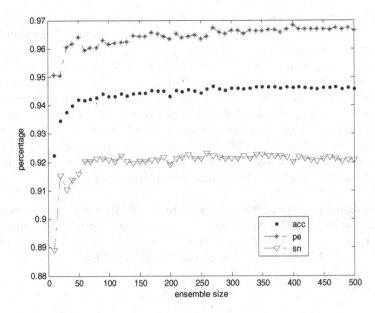

Fig. 1. The performance of our method changed with L

Our method belongs to the selective ensemble, which reveals that ensemble of the available random tree may be better than ensemble of all learners. It's also a hot research topic. In our application, the selective essence was that the base learners of high accuracy and difference with others would be chose. The results clearly show that our method has the better performance than previous studies. The accuracy is improved from 93.52% to 94.69%. In addition, the precision is 96.82% while the rotation forest has a lower value of 95.1%. The sensitivity is improved from 91.5% to 92.31%.

In Figure1, we can conclude that the predictive performance is not stable with the increase of L. Although it is implemented in the same conditions, the results have difference with others according to the number of base classifiers selected. When the L is 280, base classifiers chosen have the best performance than others. We can also find that predictive power is not more powerful with increase of base classifiers chosen. It results from the diverse difficult obtained with the increase of L. On this point the selective ensemble and traditional ensemble methods are entirely at one.

Our method has an obvious advantage that the program runs more quickly than rotation forest implemented in WEKA software. Here, we chose the random tree as our base classifiers. It's faster than J48 when building the training model. In Table1, we gave the comparison of results of time needed to generate best performance. If we used the random tree as the base classifiers of rotation forest, the bad performance was generated. So J48 is a good choice for rotation forest.

4   Conclusion

The computational methods are used to solve the biological problems. Various machine learning algorithms have been developed to resolve the difficult problems.

The crucial is that the represented and useful information must be exacted from the biological model. Here, the values of physicochemical properties of the amino acids were used to extract useful information from protein sequence. Moran autocorrelation descriptor [15], which takes into account the effect of the neighboring residues, was adopted to encode the sequence. Random tree from WEKA utilized the feature vectors to predict PPIs. It's a weak classifier that a component classifier has a lower accuracy of 81.34%. So it's a suitable base classifier in our experiment. The GA is also the powerful optimization tool. The experimental results show that our method can be applied to infer PPIs with higher accuracy, precision and sensitivity than rotation forest. In addition, our method is faster than it. Furthermore, this method can be used to study protein networks and determine protein complexes.

Acknowledgments. The author wishes to thank Dr.Hong-Jie Yu for his helpful suggestions about English writings and expressions. This work was supported by the Grants of the National Science Foundation of China, Nos. 61133010, 31071168; and Anhui Provincial Natural Science Foundation No. 1208085MF96.

References

1. Shi, M.G., Xia, J.F., Li, X.L., Huang, D.S.: Predicting Protein-protein Interactions from Sequence using Correlation Coefficient and High-quality Interaction Dataset. Amino Acids 38, 891–899 (2010)
2. Zhu, H., Bilgin, M., Bangham, R., et al.: Global Analysis of Protein Activities Using Proteome Chips. Science 293, 2101–2105 (2001)
3. Marcotte, E.M., Pellegrini, M., Ng, H.L., Rice, D.W., Yeates, T.O., Eisenberg, D.: Detecting Protein-protein Interactions from Genome Sequences. Science 285, 751–753 (1999)
4. Pazos, F., Valencia, A.: Similarity of Phylogenetic Trees as Indicator of Protein-protein Interaction. Protein Eng. 14, 609–614 (2001)
5. Pazos, F., Helmer-Citterich, M., Ausiello, G., Valencia, A.: Correlated Mutations Contain Information about Protein-protein Interaction. J. Mol. Biol. 271, 511–523 (1997)
6. Bock, J., Gough, D.: Prediticing Protein-protein Interactions from Primary Structure. Bioinformatics 17, 455–460 (2001)
7. Ben-Hur, A., Noble, W.S.: Kernel Methods for Predicting Protein-protein Interactions. Bioinformatics 21, i38–i46 (2005)
8. Shen, J.W., Zhang, J., Luo, X.M., Zhu, W.L., Yu, K.Q., Chen, K.X., Li, Y.X., Jiang, H.L.: Predicting Protein-protein Interactions based on Sequence Information. Proceedings of the National Academy of Sciences 104, 4337–4341 (2007)
9. Xia, J.F., Han, K., Huang, D.S.: Sequence-based Prediction of Protein-protein Interactions by Means of Rotation Forest and Autocorrelation Descriptor. Protein & Peptide Letters 17, 137–145 (2010)
10. Deng, M., Mehta, S., Sun, F.Z.: Inferring Domain-domain Interactions from Protein-protein Interactions. Genome Res. 12, 1540–1548 (2002)
11. Chen, X.W., Liu, M.: Prediction of Protein-protein Interactions using Random Decision Forest Framework. Bioinformatics 21, 4394–4400 (2005)
12. Lqbal, M., Freitas, A.A., Johnson, C.G., Vergassola, M.: Message-passing Algorithms for the Prediction of Protein Domain Interactions from Protein-protein Interaction Data. Bioinformatics 24, 2064–2070 (2008)

13. Guo, Y.Z., Yu, L.Z., Wen, Z.N., Li, M.L.: Using Support Vector Machine Combined with Auto Covariance to Predict Protein-protein Interactions from Protein Sequences. Nucleic Acids Research 36, 3025–3030 (2008)
14. Lo, S., Cai, C., Chen, Y., Maxey, C.M.: Effect of Training Datasets on Support Vector Machine Prediction of Protein-protein Interactions. Proteomics 5, 876–884 (2005)
15. Moran, P.A.: Notes on Continuous Stochastic Phenomena. Biometrika 37, 17–23 (1950)
16. Broto, P., Moreau, G., Vandicke, C.: Molecular Structures: Perception Autocorrelation Descriptor and Non-local Interactions. Neurocomputing 68, 66–70 (1984)
17. Zhou, Z.H., Wu, J.X., Tang, W.: Ensembling Neural Networks: Many Could be Better than All. Artificial Intelligence 137, 239–263 (2002)
18. Houck, C.R., Joines, J.A., Kay, M.G.: A Genetic Algorithm for Function Optimization: a Matlab Implementation. Technical Report: NCSU-IE-TR-95-09, North Carolina State University, Raleigh, NC (2005)
19. Breiman, L.: Random forest. Machine learning 45, 5–32 (2001)

A Two-Stage Reduction Method Based on Rough Set and Factor Analysis

Zheng Liu[1], Liying Fang[1], Mingwei Yu[2], Pu Wang[1], and Jianzhuo Yan[1]

[1] Beijing University of Technology, Beijing, China
[2] Beijing Hospital of Traditional Chinese Medicine, Beijing, China
{fangliying,wangpu,yanjianzhuo}@bjut.edu.cn,
{catherine5656,yumingwei1120}@163.com

Abstract. The performance of dimensionality reduction on multiple, relevant, and uncertain data is not satisfied by a single method. Therefore, in this paper, we proposed a two-stage reduction method based on rough set and factor analysis (RSFA). This method integrates the advantages of feature selection on treating relevant and the advantages of rough set (RS) reduction on maintaining classification power. At first, a RS reduction is used to remove superfluous and interferential attributes. Next, a factor analysis is utilized to extract common factors to replace multi-dimension attributes. Finally, the RSFA is verified by using traditional Chinese medical clinical data to predict patients' syndrome. The result shows that less attributes and more accuracy can be expected with RSFA, which is an appropriate reduction method for such problems.

Keywords: reduction, reduction, factor analysis, classification.

1 Introduction

The multiple, correlative and uncertain data is ordinary in daily life. Dimensionality reduction to this kind of data is a necessary pretreatment in the classification without enough samples [1].

Feature selection and feature extraction are two methods of dimensionality reduction. The principle component analysis (PCA), as a classical method of feature selection, can deal with the relevance of condition attributes [2-3]. However, classification power is decreased by losing information. Rough set (RS) [4] reduction is one of feature extraction methods [5], which can be used for reduction without scarifying the performance of classification especially for the uncertain data [6]. Nevertheless, the further discussion on RS reduction for relevant attributes is still in the stage of exploration as far as I know. It is concluded that, in a degree, a single method is limited to solve reduction problem. Winarski [7] integrated PCA and RS by using RS reduction to select the best principle component for PCA. However, the dependency between condition attributes and decision attributes is damaged by the linear transformation. In order to overcome the shortcomings of existing methods, a comprehensive dimensionality reduction method which integrates the advantages of feature selection and RS reduction is proposed in this paper.

D.-S. Huang et al. (Eds.): ICIC 2012, LNCS 7389, pp. 342–349, 2012.
© Springer-Verlag Berlin Heidelberg 2012

Section 2, the introduction of basic concepts and a frame of the method and algorithm are discussed. Experiments based on the Traditional Chinese Medical (TCM) clinical symptoms and results analysis are presented in section 3. Section 4 contains some conclusions and some ideas for further work.

2   Framework and Algorithms of RSFA

RS reduction can improve the clarity of knowledge system and give a better support to classification when there are uncertain data such as noisy, inaccuracy, and incomplete data.

2.1   Basic Concepts

A set $S = < U, A, V, f >$ is defined to represent an information system, where U is object sets, $f : U \times A \rightarrow V$ is the information function. If $A = C \cup D$ and $C \cap D = \phi$, the S is a decision table, where C is conditions sets, D decision sets.

Given a subset of attributes $B \subseteq A$ and a subset $X \subseteq U$, the B-lower approximation $B\_(X)$ of the set X is defined as follow:

$$B\_(X) = \{ x \in U : x \in IND(B) \wedge (x \in X) \} \tag{1}$$

Where $IND(B)$ is an indiscernibility relation on subset B. For S with $A = C \cup D$, and a subset $P \subseteq C$, a positive region $POS_P(D)$ is defined as:

$$POS_P(D) = \bigcup_{X \in IND(D)} P\_(X) \tag{2}$$

$POS_P(D)$ contains all objects in U that can be classified without an error into distinct classes defined by $IND(D)$. RS theory determines a degree of attributes' dependency based on $POS_P(D)$:

$$\gamma(P, D) = card \left[POS_P(D) \right] / card \left[U \right] \tag{3}$$

where $card\,()$ is a cardinality of a set. The $\gamma(P, D)$ is one of the most important index for rough sets. It presents a ratio of the number of objects that can be classified by attributes P absolutely to the number of all objects. It can be considered as a measure of dependency of condition subsets P to decision attributes.

2.2   Framework of RSFA

For the multiple, correlative and uncertainly data, one single method is limited to solve the complicated problem. A two-stage dimensionality reduction method based on RS attributes reduction and FA (RSFA) is proposed to obtain the reducts for the attributes sets with these characteristics in this paper. At first, RS attribute reduction is used to remove superfluous and interferential attributes without losing the power of classification. After that, FA is used to extract common factors (CF) for the correlative attributes to get the further reduction and stronger power of classification. Figure 1 shows the framework of the RSFA system.

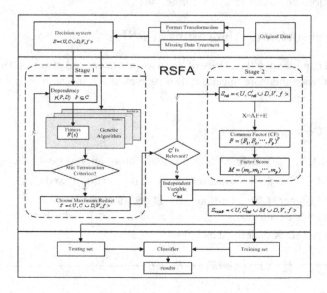

Fig. 1. Framework and process of the RSFA

In the RSFA method, a RS reduction based on genetic algorithms (GA) is chosen to implement the first stage reduction. The solutions are represented in binary as strings of 0s and 1s. After generating a population, the fitness of each individual is evaluated. The better individuals are selected from the current population based on their fitness. Then, genetic operators, crossover, mutation and inversion are used to modify the individuals to form a new population. The algorithm will be ended if the maximum amount of generations is produced, or the average fitness is not improved over the predefined amount of generations. The fitness function is defined in Eq. (4), which is proposed by Dr. Aleksander [8], use $\gamma(P,D)$ conception in RS.

$$F(s) = (1-\rho)\frac{N-L_s}{N} + \rho \times \gamma(C(s),D) \tag{4}$$

The larger the $F(s)$ of an individual is, the stronger its adaptive capacity of the individual has. Where s is the string encode candidate reduct. The $C(s)$ is a condition sets corresponding s. N is the cardinality of C. L_s is the cardinality of the selected condition subsets. In Eq. (4), the first term rewards less attributes and the second tried to ensure the classification capacity of attributes set, and the ρ is a parameter used to revise the weighing. Because the number of attributes and the classification quality are both considered in this fitness, it not only removes superfluous and interferential attributes but also can avoid losing classification capacity due to overmuch reduction.

With the acceptable classification capability, the number of attributes of the subset is various in the same generation of individuals. The subset that contains the maximum attributes is chosen to the second stage reduction because it has stronger capacity and less relevant information.

After calculating the similarity coefficient, a FA model is established by the relevant attributes subset $\{X_1, X_2, \cdots, X_k\}$ $(k \in m)$

$$
\begin{cases}
X_1 = a_{11}F_1 + a_{12}F_2 + \ldots + a_{1p}F_p + E_1 \\
X_2 = a_{21}F_1 + a_{22}F_2 + \ldots + a_{2p}F_p + E_2 \\
\ldots \\
X_k = a_{k1}F_1 + a_{k2}F_2 + \ldots + a_{kp}F_p + E_k
\end{cases}
\tag{5}
$$

In Eq. (5), $F = \left(F_1, F_2, \ldots, F_p \right)^T$ is a CF vector, $A = \left(a_{ij} \right)_{k \times p}$ is a factor loading matrix that can be used to measure the properties of the X by a linear combination of the underlying factor F.

2.3   Algorithms of RSFA

There are two algorithms used to implement the RSFA method. (a) RS reduction algorithm which consider the function in Eq. (4) as the fitness function. (b) Mining of relevant attributes algorithm based on the model represent in Eq. (5).

Algorithm 1. RS reduction algorithm based on GA
Input: An original data of $S = < U, C \cup D, V, f >$, predefine the iteration N
Output: Reduction subset s that has stronger ability and most attributes.
   Step1: Set iteration t=0 and $s = \phi$
   Step2: Generate an initial population $P(t)$ and set an individual solution $s = s_g$.
   Step3: Find all adaptable solutions
      Step3.1: Evaluate the fitness $F(s)$ of every individual in the population $P(t)$.
      Step3.2: Select the individual that has the maximal $F(s)$ value.
         If $F(s') > F(s)$ then select this individual $s = s'$
         If $F(s') < F(s)$ then discard this individual $s = s$
         If $F(s') = F(s)$ then add this individual $s = s \cup s'$
      Step3.3: Generate $P(t+1)$ through crossover and mutation operators.
      Step3.4: Judge the stop criteria. If the maximum number of generations is produced, or the average fitness is not improved over the predefined number of generations then stop; otherwise go back to step3.
   Step4: Count the number of the attributes in each selected reduction subset, and consider the subset which contains the most attributes as the reduction subset s.

Algorithm 2. Mining of relevant attributes algorithm based on FA
Input: The reduction subset s which is the output of Algorithm 1..
Output: The finally result of the dimensionality reduction.
   Step1: Calculate the proportion of missing values for each attribute. If the proportion more than 5% then deletes this attribute, that contain less information.
   Step2: Standardize the remain attributes, and calculate the similarity coefficient
   Step3: KMO test, select relevant attributes $\{X_1, X_2, \cdots, X_k\}$
   Step4: Extract the CF by PCA, and estimate the factor loading matrix
   Step5: Varimax orthogonal rotation is used on the loading matrix
   Step6: Calculated the scores on every CF by regression
   Step7: Use factor scores instead of relevant multi-variables. The finally subset result of the dimensionality reduction is consist of these factor scores and other irrelevant attributes..

3   Results and Analysis

3.1   Experiments Data

Symptom, which is mainly obtained through observation and interrogation by doctors, is relevant information to a disease. Symptom is multiple, correlative and uncertain data. A decision system $S = <U, C \cup D, V, f>$ is established in Table 1.

   In this experiment, 610 samples collected from a hospital were contained in the S. Each instance has 35 condition attributes. The attribute set C_1 contains 17 clinical symptoms and each symptom is divided into four grades marked as 0,1,2, and 3 according to the severity. The attribute set C_2 contains 18 tongue and pulse attributes which are represented in binary (0 or 1) whether or not the attribute is appeared. The syndrome is a decision attribute. I, II and III are accordingly defined as the vacuity, repletion and vacuity-repletion complication.

Table 1. Decision system S of TCM symptom and syndrome

sample	\multicolumn{8}{c	}{Attributes $C = C_1 \cup C_2$}	syndrome
	C_{11}	C_{12}	C_{1i}	C_{117}	C_{21}	C_{22}	C_{2j}	C_{218}	
	brea-thehard	weari-ness	...	asthenic fever	red tongue	Thin fur	...	slow pulse	
x_1	1	1	...	0	1	0	...	1	I
x_2	2	3	...	0	1	0	...	1	II
x_n	...	...	...	...	...	...	...	...	...
x_{610}	2	1	...	0	0	0	...	0	III

3.2   Experiments and Results

The RSFA is used to reduce the dimensions. At first, the superfluous and inter- ferential attributes are removed using the algorithm 1 mentioned in section 2.3, and the sub-set $\left\{ C' = C_1' \cup C_2' : \left(C_1' \subseteq C_1 \right) \wedge \left(C_2' \subseteq C_2 \right) \right\}$ can be obtained. Set popsize=70, Pc=0.3 and Pm=0.05. Next, a correlation test is performed for the C_1'. The result, kmo=0.76 (kmo > 0.6), indicated that it is necessary to treat the relativity to C_1' set using the algorithm 2. The finally subset result is $S_{result} = < U, C_{ind}' \cup M \cup D, V, f >$ where the C_{ind}' is the union of the C_2' and the irrelevant attributes in C_1', the M is the scores of the CF extracted from the relevant attributes.

In this experiment, three methods are used to implement the reduction for the system S, which include PCA, RS reduction, and the RSFA proposed by this paper. Based on these reduction methods, classification accuracy is calculated by LS-SVM, with the RBF kernel function, sig2=3 and gam=40. To estimate how accurately a predictive model will perform in practice, the 10-cross-validated is used in LS-SVM and repeats the experiment 5 times. It can be seen from these experiments, the accuracy results of classification based on RSFA reduction are better than based on other reduction me-thods. The dimension of the reduction subset and the average accuracy of classification based on different reduction methods are shown in Table 2.

Table 2. Dimension and average accuracy of classification based on different reduction

	un-reduction	PCA	RS	RSFA
Dimension	35	28	21	21
Accuracy rate	80.03%	78.53%	79.28%	83.27%

Compared with the performance of the PCA, higher accuracy and fewer dimensions can be got through the classification based on RS and RSFA reduction. For the latter ones, evaporative rate of the attributes [9] are 40%, more than 30% which illustrate the reduction is satisfied according to the experience [10]. Especially, for the TCM symptoms data, the accuracy of classification based on RSFA reduction is increased by 5% compared with RS reduction in the case of getting same number of attributes. Compared with the performance of PCA, the evaporative rate of attributes of RSFA is increased by 20% and the accuracy of classification based on RSFA is increased by 6%. RSFA reduction improved the classification ability, even higher than the original, not only because it removed interferential attributes but also because it held information by using CFs.

To validate the RSFA on different LS-SVM parameters, another experiment com-pared RSFA-SVM with RS-SVM is taken. The 10-cross-validated is used on each group, and the experiment is repeated 5 times. Figure 2 shows the average accuracy.

From the Figure 2, it can be seen that RSFA has a better performance than RS on different parameters. It illustrate that RSFA is more suitable to deal with the dimen-sionality reduction for relevant attributes data set.

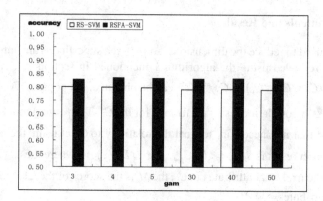

Fig. 2. Result of the comparison between RSFA and RS

Next, we tried to use some UCI datasets to verify the performance of RSFA reduction. However, the relevance between attributes (tested by KMO, P<0.4) are not statistically significant, namely, factor analysis is not appropriate for dealing with them, so the extra results have not shown here.

4   Conclusion

This paper discusses the advantages and shortcomings of single approach to dimensionality reduction. Therefore, we propose a two-stage synthetic method, named RSFA, to reduce the dimension for multiple, relevant, and uncertain data. The RSFA integrates the different advantages of RS reduction and FA. RSFA complete reduction by removing superfluous, interferential attributes and extracting common factors to replace the multi-attributes. Experiments based on the data of TCM clinical symptoms are made to verify the effectiveness. Results demonstrate better performance on dimensions and accuracy of classification based on RSFA reduction than other methods.

This paper only focused on solving the numeric variable, a correlation analysis method based on RS reduction for multi-type of data could be explored in the next researching.

Acknowledgements. This paper is supported by 2010 Program for Excellent Talents in Beijing Municipal Organization Department (2010D005015000001), the New Centaury National Hundred, Thousand and Ten Thousand Talent Project, and got the cooperation with Beijing Hospital of Traditional Chinese Medicine Affiliated to CPUMS. Special thanks have been given there.

References

1. Wang, X.Y., Yang, J., Teng, X.L., Xia, W.J., Jensen, R.: Feature Selection Based on Rough Sets and Particle Swarm Optimization. Pattern Recogn. Lett. 28, 459–471 (2007)
2. Tsai, C.F.: Feature Selection in Bankruptcy Prediction. Knowl-Based Syst. 22, 120–127 (2009)

3. Ravi, V., Pramodh, C.: Threshold Accepting Trained Principal Component Neural Network and Feature Subset Selection: Application to Bankruptcy Prediction in Banks. Applied Soft Computing 8, 1539–1548 (2008)
4. Pawlak, Z.: Rough Sets. Interantional Journal of Computer & Information Sciences 11, 341–356 (1982)
5. Chouchoulas, A., Shen, Q.: Rough Set-aided Keyword Reduction for Text Categorization. Appl. Artif. Intell. 15, 843–873 (2001)
6. Slezak, D.: Degrees of Conditional (in) Dependence: A Framework for Approximate Bayesian Networks and Examples Related to the Rough Set-based Feature Selection. Inform. Sciences 179, 197–209 (2009)
7. Swiniarski, R.W., Skowron, A.: Rough Set Methods in Feature Selection and Recognition. Pattern Recogn. Lett. 24, 833–849 (2003)
8. Vinterbo, S., Ohrn, A.: Minimal Approximate Hitting Sets and Rule Templates. Int. J. Approx. Reason. 25, 123–143 (2000)
9. Shu, H.C.: Comparative Study of Classification Method in Traditional Chinese Medicine Differentiation, vol. PhD (2008) (in Chinese)
10. Dietterich, T.G.: Machine-learning Research - Four Current Directions. Ai. Mag. 18, 97–136 (1997)

Eyebrow Segmentation Based on Binary Edge Image

Jiatao Song[1,2], Liang Wang[1] and Wei Wang[1]

[1] School of Electronic and Information Engineering,
Ningbo University of Technology, Ningbo 315016, China
[2] Zhejiang Provincial Key Laboratory of Information Network Technology,
Hangzhou 310027, China
sjt6612@163.com

Abstract. Eyebrow is one of the most salient face features. It has a lot of potential applications in face recognition, nonverbal communication, and so on. In this paper, a novel eyebrow segmentation method based on binary edge image (BEI) is proposed. Our method firstly extracts BEI from a grayscale face image, and then connections between different face components in a BEI are removed using a specially designed algorithm. After that, some eyebrow-analogue segments are extracted from a BEI based on the geometrical property of eyebrows. The fourth step is to locate eyebrows using integral projection approach. Finally the perimeter of an eyebrow block is extracted to finish the segmentation of an eyebrow. Experimental results on a set of 517 AR images with different facial expression and illumination show that a correct eyebrow segmentation rate of 93.4% is achieved, indicating that the proposed method is robust to facial expression and illumination changes.

Keywords: Eyebrow segmentation, Binary Edge Image (BEI), connection removal, illumination change, expression change.

1 Introduction

Eyebrow is the most salient and stable feature in a human face. Studies show that eyebrows play important roles in emotional expression and nonverbal communication, as well as in facial aesthetics and sexual dimorphism [1]. Sadr's research [1] suggests that for face recognition the eyebrows may be at least as influential as the eyes. They found that the absence of eyebrows in familiar faces leads to a very large and significant disruption in recognition performance. Sinha [2] also pointed out that of the different facial features, eyebrows are among the most important for recognition. Besides for face recognition, eyebrows are also used for gender classification [3].

In order to extract eyebrow feature, eyebrows should be firstly segmented from a face image. Ref. [4] locates eyebrows by means of PCA based template matching, while Ref. [5] finds a rectangular region enclosing the eyebrow based on clustering in HIS domain. Chen, et al. [6] proposed a spatial constrained sub-area K-means clustering method for eyebrow segmentation. Recently, Ding, et al. [7] proposed an eyebrow segmentation method by using color and gradient information. But their method requires eyes to be firstly located.

Although some progress has been made, the problem of automatic eyebrow segmentation is still far from being fully solved owing to its complexity. Particularly,

D.-S. Huang et al. (Eds.): ICIC 2012, LNCS 7389, pp. 350–356, 2012.
© Springer-Verlag Berlin Heidelberg 2012

when there is facial expression or illumination change in a face image, the performance of the existing methods will decrease greatly. The work described in this paper focuses on the study of eyebrow segmentation from human face images with different facial expression and lighting conditions. A novel eyebrow segmentation method based on the binary edge image (BEI) [8] is presented.

The rest of this paper is organized as follows. In section 2, the method for the extraction of BEI is described briefly. Section 3 presents our eyebrow segmentation approach. Experimental results are given in section 4. Section 5 concludes our work.

2   Extraction of Binary Edge Image (BEI)

In order to precisely locate eyes from a human face image, Song, et al. [8] proposed an approach for the extraction of BEI from a grayscale face image. Their method is based on the multi-scale analysis of wavelet transform (WT). Some BEIs are shown in Fig. 1. From this figure, we can see that for different facial expression and illumination, face components, especially eyes and eyebrows, are clearly extracted in BEIs. The shape and contour of these components are very clear. This is very helpful for the segmentation of face components.

One problem with BEIs is that sometimes pixels corresponding to different face components, such as eyebrows, eyes and nose are connected. This hampers the segmentation of face components. Thus in our work, an algorithm for the removal of connection of pixels belonging to different face components is proposed in section 3.2.

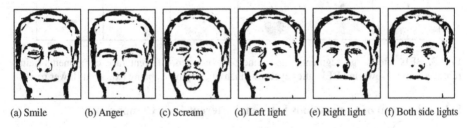

(a) Smile   (b) Anger   (c) Scream   (d) Left light   (e) Right light   (f) Both side lights

Fig. 1. Examples of BEIs with different facial expression and illumination

3   The Proposed Method

3.1   Overview of the Proposed Eyebrow Segmentation Method

Fig. 2 shows the diagram of our proposed method. The first step is to remove some very large foreground segments from a BEI, for these segments often correspond to hair and are unlikely to be eyebrow segments. The second step is to fill some holes in the segments of BEI. After that, an algorithm is designed to decrease the connection of different face components. The next step is to extract some eyebrow-analogue segments from a BEI according to some geometric properties of eyebrows. The fifth step is to detect the eyebrow segments using integral projection method and finally the contour of an eyebrow segments is extracted to finish the segmentation of eyebrows.

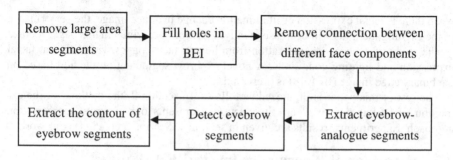

Fig. 2. Diagram of the proposed eyebrow segmentation method

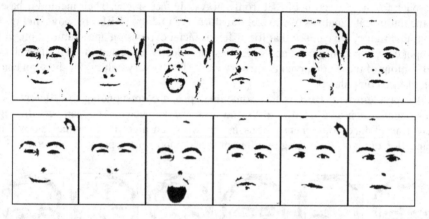

Fig. 3. BEIs after large segments and connection between different face components being removed (upper row) and BEIs after eyebrow-analogue segments being extracted (lower row)

3.2   Removal of Connections between Different Face Components

From Fig. 1 we can observe that the foreground pixels (black pixels) in BEIs may correspond to two kinds of pixels in grayscale images. The first kind is the pixels in eyebrows and iris, which often have low intensities. The second kind is the pixels on image edges, which reflect the contour of face components and have relatively larger intensities. If we remove the second kind of foreground pixels from BEIs, the connection between different face components can be decreased greatly.

Let I denote a BEI, J denote the corresponding grayscale image. The steps for connection removing include: (1) filling holes in I; (2) marking all the 8-connected components in I and extracting the perimeter pixels of each component. (3) calculating the mean intensity of the perimeter pixels of each component in J, and using it as threshold value to binarize all the pixels in this component in J. Pixels with larger intensities are segmented as background pixels.

BEIs after above operations are shown in the upper row of Fig. 3. Obviously, the connection between different face components especially that between eyebrow and nose and that between eyebrow and eye, decreases greatly.

3.3   Extraction of Eyebrow-Analogue Segments

Statistical analysis shows that the aspect ratio of the bounding rectangle of an eyebrow normally ranges from 1.5 to 7.5 when the rotation-in-plane of a face is less than 15 degrees. Thus in our method, segments in BEIs with aspect ratios of bounding rectangle outside of this range are classified as non-eyebrow blocks. Besides, segments with too small areas are viewed as noise. By removing these two kinds of segments directly from BEIs, BEIs with only eyebrow-analogue segments, such as eyebrows, eyes and mouth retained are achieved, just as shown in the lower row of Fig. 3. This greatly simplifies the segmentation of eyebrows.

3.4   Detection of Eyebrow Segments

In our method, the integral projection method and the prior knowledge about the configuration of face components on a face are employed for the detection of eyebrow segments. The prior knowledge used includes: (1) there are two eyebrows on a face; (2) eyebrows are above eyes. (3) distance between eyes and eyebrows is larger than that between eyes and mouth.

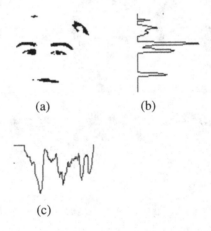

(a)   (b)

(c)

Fig. 4. Horizontal (b) and vertical (c) integral projection curves of a BEI (a)

Fig. 4 (b) and Fig. 4 (c) shows a typical horizontal and vertical integral projection curve of a BEI with only eyebrow-analogue segments remained, respectively. In the horizontal projection curve, there always exist some strong local peaks at the positions of eyebrow, eye and mouth. Based on this, some candidate horizontal eyebrows strips can be obtained. Then by projecting this strips vertically and utilizing the prior knowledge discussed above, eyebrow segments can be detected. .

After an eyebrow block is segmented, its perimeter is extracted as the contour of an eyebrow, which means the finish of eyebrow segmentation.

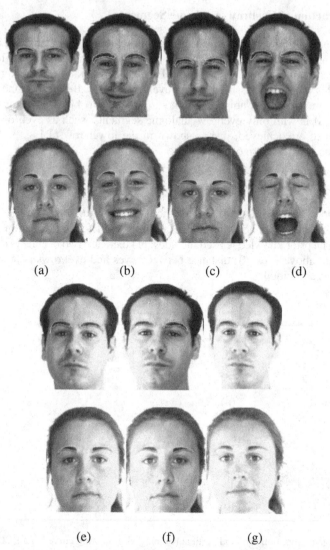

Fig. 5. Eyebrow segmentation results of AR images with different expression and illumination ((a)-(g): Neutral, smile, anger, scream, left light on, right light on and both side lights on)

4   Experimental Results and Discussion

517 AR images [9] with different facial expressions and illuminations are used for eyebrow segmentation experiments. The four different facial expressions include neutral, smile, anger and scream, and the three illumination conditions are left light on, right light on and both side lights on. Images with eyebrows being correctly segmented are shown in Fig. 5. The white contour represents the segmented shape of an eyebrow, which coincides well with the actual perimeter of an eyebrow. Further analysis shows

that of all the 517 images used, our method achieves a correctly eyebrow segmentation rate of 93.4%, indicating that our eyebrow segmentation approach is robust to expression and illumination changes.

The success of the proposed eyebrow segmentation method can be contributed to two factors. One is the high quality of BEIs. Eyebrows in BEIs with different expression and lighting are wholly extracted and noise is little, which makes eyebrows segmentation possible. The other factor is the connection removal algorithm proposed in this paper, which greatly simplifies the segmentation of eyebrows.

5   Conclusions

Eyebrow is important for face recognition, facial expression analysis, nonverbal communication, and so on. Its segmentation is the key step for eyebrow-based face image analysis system. In order to address the problem of eyebrow segmentation under varying expression and illumination conditions, a novel eyebrow segmentation method is proposed in this paper. Our method is based on the binary edge image (BEI), and includes a specially designed algorithm for the removal of connection between different face components from a BEI. Experimental results on 517 AR images with different expression and illumination show that our method achieves a successful eyebrow segmentation rate of 93.4%, indicating that the proposed method is robust to facial expression and illumination changes. Based on the extracted eyebrows, we will construct a high performance face recognition system in the future.

Acknowledgements. The work described in this paper is partially supported by a project from the National Natural Science Foundation of China (Grant No.60972163), a project from the Natural Science Foundation of Zhejiang Province of China (Grant No.Y1110086), a project from the Natural Science Foundation of Ningbo of China (Grant No.2009A610090), an open project of Zhejiang Provincial Most Important Subject of Information Processing and Automation Technology (Grant No. 201100808), an open project of Zhejiang Provincial Key Laboratory of Information Network Technology(Grant No. 201109) and two projects from the Open Fund of Mobile Network Application Technology Key Laboratory of Zhejiang Province (Grant No. MNATKL2011001, MNATKL2011003).

References

1. Sadr, J., Jarudi, L., Sinha, P.: The Role of Eyebrows in Face Recognition. Perception 32, 285–293 (2003)
2. Sinha, P., Balas, B., Ostrovsky, Y., Russell, R.: Face Recognition by Humans: Nineteen Results All Computer Vision Researchers Should Know About. Proceedings of the IEEE 94(11), 1948–1962 (2006)
3. Dong, Y., Woodard, D.L.: Eyebrow Shape-based Features for Biometric Recognition and Gender Classification: A Feasibility Study. In: 2011 International Joint Conference on Biometrics, pp. 1–8. IEEE Press, New York (2011)
4. Kapoor, A., Picard, R.W.: Real-time, Fully Automatic Upper Facial Feature Tracking. In: 5th International Conference on Automatic Face and Gesture Recognition, pp. 8–13. IEEE Press, New York (2002)

5. Lee, C.H., Kim, J.S., Park, K.H.: Automatic Human Face Location in a Complex Background Using Motion and Color Information. Pattern Recognition 29(11), 1877–1889 (1996)
6. Chen, Q., Cham, W., Lee, K.: Extracting Eyebrow Contour and Chin Contour for Face Recognition. Pattern Recognition 40, 2292–2300 (2007)
7. Ding, L., Martinez, A.M.: Features versus Context: An Approach for Precise and Detailed Detection and Delineation of Faces and Facial Features. IEEE Transactions on PAMI 32(11), 2022–2038 (2010)
8. Song, J., Chi, Z., Liu, J.: A Robust Eye Detection Method Using Combined Binary Edge and Intensity Information. Pattern Recognition 39(6), 1110–1125 (2006)
9. Martinez, A.M., Benavente, R.: The AR Face Database. CVC Technical Report #24 (1998)

X-ray Image Contrast Enhancement Using the Second Generation Curvelet Transform

Hao Li and Guanying Huo

School of Computer and Information Technology, Beijing Jiaotong University,
Beijing 100044, People's Republic of China
lihaobju@163.com

Abstract. In this paper, a novel X-ray image contrast enhancement method using the second generation curvelet transform is proposed in order to better enhance contrast and edges while remove noise. First, source images are decomposed in the curvelet transform domain. A nonlinear enhancement operator is applied to low frequency subbands to enhance global contrast. Combining with threshold denoising, the nonlinear enhancement operator is also applied to high frequency subbands to enhance edges and reduce noise. Finally, the processed coefficients are reconstructed to obtain enhanced images. Experimental results on X-ray images show that compared with histogram equalization and wavelet based contrast enhancement, the proposed method can effectively enhance contrast and edges of X-ray images while better reducing noise, thus has better visual effect.

Keywords: X-ray image, contrast enhancement, curvelet transform, nonlinear enhancement operator, denoising.

1 Introduction

X-ray is used to capture the internal body structure images which help a lot to the radiologists in recognizing the internal problems. It is the most useful imaging modality to check for the bone fractures and other related anomalies. Though there are numerous advantages of X-ray technology, but it generates low contrast images [1]. In addition to the X-ray penetration characteristics, X-ray image contrast is significantly affected by scattered radiation and the contrast characteristics of the receptor and display system. One can increase the power of X-rays for capturing images but it may harm human body. Moreover, the contrast of small objects within the body and anatomical detail is reduced by image blurring. As the X-ray images are being used for diagnostic purposes, some contrast enhancement techniques should be implemented in manual or auto-diagnose system to make the X-ray images more visual and explanatory. Image enhancement is a significant part for X-ray inspection systems.

As the reasons stated above, contrast enhancement is commonly required for the captured X-ray images. Various spatial and frequency-based techniques have been developed for image enhancement. Commonly used spatial techniques are linear stretch, histogram equalization [2], convolution mask enhancement [3], adaptive

D.-S. Huang et al. (Eds.): ICIC 2012, LNCS 7389, pp. 357–364, 2012.
© Springer-Verlag Berlin Heidelberg 2012

histogram equalization [4], etc. These conventional spatial techniques are usually simple, fast and useful for noiseless images, but they may cause noise amplification when the images have more noise. Noise not only lowers visual quality, but can cause feature extraction, analysis, and recognition algorithms to be unreliable.

Wavelet based contrast enhancement is a representative of frequency-based enhancement methods. In [5], denoising and feature enhancement are achieved by simultaneously lowering noise energy and raising feature energy in the wavelet transform domain. Compared with spatial enhancement techniques, wavelet based contrast enhancement can better remove noise while effectively enhancing contrast and edges. Despite the fact that wavelets have had a wide impact in image processing, they fail to efficiently represent objects with highly anisotropic elements such as lines or curvilinear structures (e.g. edges). The reason is that wavelets are non-geometrical and do not exploit the regularity of the edge curve [6]. In addition, separable wavelets can capture only limited directional information. For these reasons, wavelet based contrast enhancement often cause distortion at edges.

The curvelet [6] transform was developed as an answer to the weakness of the separable wavelet transform in sparsely representing lines, curves and edges. Curvelets take the form of basis elements which exhibit high directional sensitivity and are highly anisotropic. The curvelet transform has had an important success in a wide range of image processing applications including denoising, deconvolution, etc. It is expected that directionality capability of the curvelet transform results in better edge representation and enhancement in 2-D images. Therefore in this paper, the second generation curvelet transform [7] which is conceptually simpler, faster and far less redundant than the first generation is used for enhance X-ray images. A new nonlinear enhancement operator is proposed and applied to coefficients in the curvelet transform domain. Experimental results show that the proposed method is superior to both histogram equalization and wavelet based contrast enhancement in both visual effect and quantitative measurement.

2 Second Generation Curvelet Transform

The curvelet transform has gone through two major revisions. The first generation curvelet transform [6] uses a complex series of steps involving the ridgelet analysis of the radon transform of an image. The first generation curvelet transform was exceedingly slow. In order to overcome this, the second generation curvelet transform based on fast Fourier transform was proposed by [7].

2.1 Continuous-Time Curvelet Transforms

We work in two dimensions, i.e., R^2, with spatial variable x, with ω a frequency-domain variable, and with r and θ polar coordinates in the frequency-domain.

We start with a pair of windows $W(r)$ and $V(t)$, which we will call the "radial window" and "angular window", respectively. These are smooth, nonnegative and real-valued, with W taking positive real arguments and supported on $r \in (1/2, 2)$ and V taking real arguments and supported on $t \in [-1, 1]$.

For each $j \geq j_0$, we introduce the frequency window U_j defined in the Fourier domain by [6]

$$U_j(r,\theta) = 2^{-3j/4} W(2^{-j}r) V(\frac{2^{\lfloor j/2 \rfloor}\theta}{2\pi}) \tag{1}$$

where $\lfloor j/2 \rfloor$ is the integer part of $j/2$.

Define the waveform $\varphi_j(x)$ by means of its Fourier transform $\hat{\varphi}_j(\omega) = U_j(\omega)$. $\varphi_j(x)$ can be seen as a "mother" curvelet since all curvelets at scale 2^{-j} are obtained by rotations and translations of $\varphi_j(x)$. Introduce the equispaced sequence of rotation angles $\theta_l = 2\pi \cdot 2^{-\lfloor j/2 \rfloor} \cdot l$, with $l = 0, 1, \cdots$ such that $0 \leq \theta_l < 2\pi$, and the sequence of translation parameters $k = (k_1, k_2) \in Z^2$.

With these notations, we define curvelets by

$$\varphi_{j,l,k}(x) = \varphi_j(R_{\theta_l}(x - x_k^{(j,l)})) \tag{2}$$

where R_θ is the rotation by θ radians.

A curvelet coefficient is then simply the inner product between an element $f \in L^2(R^2)$ and a curvelet $\varphi_{j,l,k}$

$$c(j,l,k) := \langle f, \varphi_{j,l,k} \rangle = \int_{R^2} f(x)\overline{\varphi_{j,l,k}(x)}dx \tag{3}$$

2.2   Digital Curvelet Transforms

In the definition (1), the window U_j smoothly extracts frequencies near the dyadic corona and near the angle. Coronae and rotations are not especially adapted to Cartesian arrays. Instead, it is convenient to replace these concepts by Cartesian equivalents; here, "Cartesian coronae" based on concentric squares and shears.

Define the "Cartesian" window

$$\tilde{U}_j(\omega) := \tilde{W}_j(\omega) V_j(\omega) \tag{4}$$

$\tilde{W}_j(\omega)$ is a window of the form

$$\tilde{W}_j(\omega) = \sqrt{\Phi_{j+1}^2(\omega) - \Phi_j^2(\omega)}, \qquad j \geq 0 \tag{5}$$

where Φ is defined as the product of low-pass one dimensional windows

$$\Phi_j(\omega_1, \omega_2) = \phi(2^{-j}\omega_1)\phi(2^{-j}\omega_2) \tag{6}$$

The function ϕ obeys $0 \leq \phi \leq 1$, might be equal to 1 on $[-1/2, 1/2]$, and vanishes outside of $[-2, 2]$.

Finally, the digital curvelet transform coefficient [7] is obtained by

$$c(j,l,k) = \int \hat{f}(\omega)\tilde{U}_j(S_{\theta_l}^{-1}\omega)e^{i<S_{\theta_l}^{-T}b,\omega>}d\omega \tag{7}$$

3   Contrast Enhancement Method

First source X-ray images are decomposed in the curvelet transform domain to obtain low frequency subband and different scales of high-frequency subbands. Then in the curvelet transform domain, low frequency subband coefficients and high-frequency subbands coefficients are processed using nonlinear enhancement operator separately. Finally the processed coefficients are reconstructed to obtain the enhanced images. The nonlinear enhancement operator, low frequency subband coefficients enhancement and high-frequency subbands coefficients enhancement are the keys of the proposed method, therefore they will be described in detail as follows.

3.1   Nonlinear Enhancement Operator

Inspired by GAG enhancement operator which is used in [5], we first define a normalized nonlinear enhancement operator to enhance the second generation curvelet coefficients in each subband

$$G(x) = a*(sigm(c(x-b)) - sigm(-c(x+b))) \tag{8}$$

where $a = 1/(sigm(c(1-b)) - sigm(-c(1+b)))$, $sigm(x) = 1/(1+e^{-x})$, b is the enhancement and attenuation split point, and c is the enhancement or attenuation rate.

3.2   Low Frequency Subband Coefficients Enhancement

Low frequency subband contains the basic information and the overall contrast of the image. Therefore, the processing of low-frequency coefficients is particularly critical, and can not be ignored. In order to overcome the inflexibility of a single operator, a piecewise nonlinear operator for low frequency subband coefficients enhancement is proposed based on the nonlinear enhancement operator defined above. The piecewise nonlinear operator is defined by

$$MAG_L(C_{1,1}(i,j)) = \begin{cases} S_1 * C_{1,1}(i,j) * G_1(|C_{1,1}(i,j)|/Max) & |C_{1,1}(i,j)|/Max \geq b \\ S_2 * C_{1,1}(i,j) * G_2(|C_{1,1}(i,j)|/Max) & |C_{1,1}(i,j)|/Max < b \end{cases} \tag{9}$$

where $c_{1,1}(i,j)$ are low frequency subband coefficients with i and j are row index and column index, Max is the maximum absolute value of low frequency subband coefficients, S_1 and S_2 are gain factors, G_1 and G_2 are nonlinear enhancement operators with the same parameter b but different parameter c. In order to guarantee the continuity of piecewise operator, we have $S_2 = S_1*G_1(b)/G_2(b)$. b is calculated by $b = Mean/Max$, where $Mean$ is mean absolute value of low frequency subband coefficients. In our experiments, c_1 and c_2 are always set to 10 and 20.

3.3   High-Frequency Subbands Coefficients Enhancement

High-frequency subbands contain not only edges and details of the image, but also noise. In order to enhance edges while suppressing noise, the noise variance σ should be estimated and the threshold T for denoising should be determined. σ is estimated at the finest sub-band using MAD method [8] proposed by Donoho, and the threshold T is calculated by $\lambda \cdot \sigma$, where λ is a constant between 3 and 4. In our experiments, λ is set to 3.5. Taking into account noise, high-frequency subbands enhancement operator is a bit different from that of low frequency subband, and is defined by

$$MAG_H(C_{m,n}(i,j)) = \begin{cases} S * C_{m,n}(i,j) * G(|C_{m,n}(i,j)|/(|C_{m,n}(i,j)|+T)) & |C_{m,n}(i,j)| \geq T \\ 0 & |C_{m,n}(i,j)| < T \end{cases} \tag{10}$$

where $c_{m,n}(i,j)$ are high-frequency subbands coefficients with m and n are scale index and angel index, S is gain factor, and G is nonlinear enhancement operator. Since we need to enhance all the edges preserved after denosing and b equals 0.5 when $|C_{m,n}(i,j)|$ equals T, b of operator G is set to 0.5 here. Parameter c of G is set to 10 in all our experiments.

3.4   Contrast Evaluation Criterion for Image

The contrast of enhanced images is evaluated employing the measure function, which was proposed in [9]:

$$C_{contrast} = \frac{1}{MN} \sum_{x=1}^{M} \sum_{y=1}^{N} f'^2(x,y) - \left| \frac{1}{MN} \sum_{x=1}^{M} \sum_{y=1}^{N} f'(x,y) \right|^2 \tag{11}$$

where M and N are width and height of the original image, $f'(x,y)$ is the enhanced image. Larger the value of equation (11) is, better the contrast of the image is.

Information entropy can be used to measure richness of the information in an image, which is defined as follows:

$$E = \sum_i p_i \log(p_i) \tag{12}$$

where i is the gray value index, and p_i is the corresponding probability. Larger E is, richer the image information is.

4   Experiment Results

To demonstrate the effectiveness of our method, we test it on real X-ray images and compare it with the existing popular methods of histogram equalization (HE) and wavelet based contrast enhancement. The first test image i.e. Fig.1(a) is a low contrast X-ray of human chest to resolve the related medical issues. S_1 is set to 3 and S is set to 3.5 for enhancing Fig.1(a). Enhancement results of Fig.1(a) by different methods are

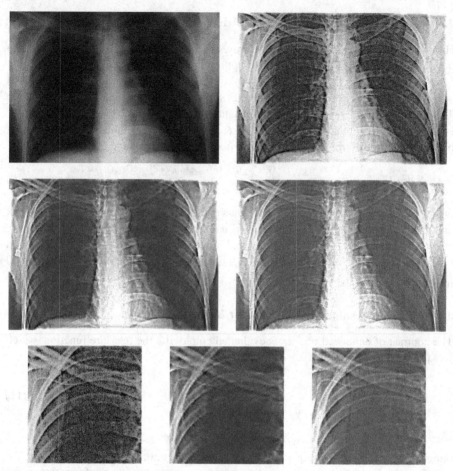

Fig. 1. Enhancement results of chest X-ray image (from left to right and top to bottom): (a) original image (b) image enhanced through HE (c) enhanced through wavelet based method (d) enhanced through proposed method (e) part of (b) (f) part of (c) (g) part of (d)

Table 1. Enhancement evaluation of $C_{contrast}$ and E for Fig.1

Method	$C_{contrast}$	E
HE	0.0821	5.4346
Wavelet	0.0886	6.8676
Curvelet	0.0995	7.3665

shown in Fig.1(b)-(g). The second test image Fig.2(a) is another low contrast X-ray capture of hysterosalpingogram. S_1 is set to 1.5 and S is set to 3.5 for enhancing Fig.2(a). The corresponding enhancement results of Fig.2(a) are displayed in Fig. 2(b)-(g). In visual analysis of both Fig.1 and Fig.2, it is observed that contrast has been enhanced to various levels by all the three methods. But histogram equalization spread noise and destroys some edge details. Wavelet based contrast enhancement can

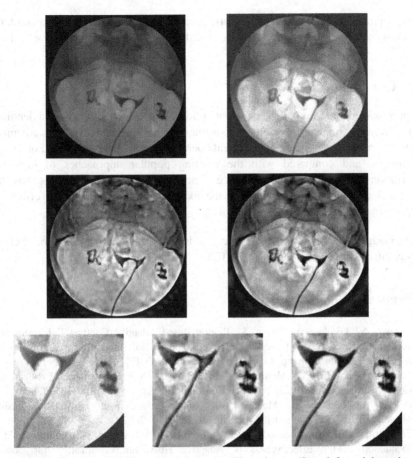

Fig. 2. Enhancement results of hysterosalpingogram X-ray image (from left to right and top to bottom): (a) original image (b) enhanced through HE (c) enhanced through wavelet based method (d) enhanced through proposed method (e) part of (b) (f) part of (c) (g) part of (d)

Table 2. Enhancement evaluation of $C_{contrast}$ and E for Fig.2

Method	$C_{contrast}$	E
HE	0.0799	5.5892
Wavelet	0.1013	7.0174
Curvelet	0.1180	7.1786

suspend noise effectively, but it causes obvious artifacts and distortion at edges. The proposed method can enhance edges more precisely while removing noise effectively.

The human visualization is not considered as benchmark for image quality, so to evaluate the performance of above mentioned methods the contrast $C_{contrast}$ and the entropy E have been calculated for the output enhanced images. Enhancement evaluation for Fig.1 is given by Table 1, and that for Fig.2 is given by Table 2. From Table 1 and Table 2, we can clearly see that the proposed method can better enhance image contrast and protect image information. Both visual effect and quantitative measurement have shown that the proposed method is superior to the existing ones.

5   Conclusions

The proposed method makes use of curvelet transform, which is superior in denoising and providing a compact representation of line-singularities, and also takes advantage of effectiveness of nonlinear enhancement operator. The method was tested on real X-ray images, and compared with the existing popular approaches of histogram equalization and wavelet based contrast enhancement. Experimental results show that the proposed technique is superior to the existing ones in both visual effect and quantitative measurement.

Acknowledgments. This research was funded by the National Natural Science Foundation of China (Grant No. 60972101).

References

1.   Kanwal, N., Girdhar, A., Gupta, S.: Region Based Adaptive Contrast Enhancement of Medical X-Ray Images. In: 2011 5th International Conference on Bioinformatics and Biomedical Engineering, pp. 1–5. IEEE Press, Wuhan (2011)
2.   Chen, S.D., Ramli, A.R.: Minimum Mean Brightness Error Bi-Histogram Eualization in Contrast Enhancement. IEEE Trans. on Consumer Electronics 49, 1310–1319 (2003)
3.   Polesel, A., Ramponi, G., Mathews, V.J.: Image Enhancement via Adaptive Unsharp Masking. IEEE Trans. on Image Processing 9, 505–510 (2002)
4.   Zimmerman, J.B., Pizer, S.M., Staab, E.V., Perry, J.R., McCartney, W., Brenton, B.C.: An Evaluation of the Effectiveness of Adaptive Histogram Equalization for Contrast Enhancement. IEEE Trans. on Medical Imaging 7, 304–312 (1988)
5.   Zong, X., Laine, A.F., Geiser, E.A.: Denoising and Contrast Enhancement via Wavelet Shrinkage and Nonlinear Adaptive Gain. In: Proc. of SPIE, vol. 2762, pp. 566–574 (1996)
6.   Candès, E.J., Donoho, D.L.: Curvelets -a Surprisingly Effective Nonadaptive Representation for Objects with Edges. Vanderbilt University Press, Nashville (2000)
7.   Candès, E.J., Demanet, L., Donoho, D.L., Ying, L.: Fast Discret Curvelet Transforms. Applied and Computational Mathematics, 1–43 (2005)
8.   Donoho, D.L.: Denoising by Soft Thresholding. IEEE Trans. on Information Theory 41, 613–627 (1995)
9.   Azriel, R., Avinash, C.K.: Digital Picture Processing. Academic Press, New York (1982)

MMW Image Blind Restoration Using Sparse ICA in Contourlet Transform Domain

Li Shang[1,2], Pin-gang Su[1,3], and Wen-jun Huai[1]

[1] Department of Electronic Information Engineering, Suzhou Vocational University,
Suzhou 215104, Jiangsu, China
[2] Department of Automation, University of Science and Technology of China,
Anhui 230026, Hefei, China
[3] State Key Lab of Millimeter Waves, Southeast University, Nanjing 210096, Jiangsu, China
{sl0930,supg,hwj}@jssvc.edu.cn

Abstract. Sparse independent component analysis (SPICA) algorithm is effective in blind separation of superimposed images, without having any priory knowledge about the image's structure and statistics. While a millimeter wave (MMW) image contains the refective information of imaging object and much unknown noise of imaging scene, so the MMW image is too high blur to be discerned. To obtain preferable MMW image, combined the advantages of contourlet sparse transform and SPICA, a new blind restoration method proposed by us of MMW images operating in the contourlet sparse transform domain is discussed in this paper. Contourlet transform can retain the better contour of an image and make this image sparser in local subspace. Here, using the low frequency band and the high frequency bands of the first layer obtained by contourlet transform as the mixed input data of SPICA, the task of MMW image restoration can be implemented. In test, the blind restoration of mixed natural images is also operated by using our method, simultaneity, using the single noise ratio (SNR) to measure the restored natural images, experimental results testify the validity of our method in doing blind separation and it is feasible to restore the MMW image using this proposed method. Further, compared with methods of contourlet transform and fast ICA, simulations again show that this MMW image restoration method proposed is indeed efficient in application.

Keywords: Millimeter wave (MMW), Independent component analysis (ICA), Sparse ICA, Contourlet sparse transform; Image restoration.

1 Introduction

Millimeter wave (MMW) imaging technology has been widely applied in different fields because of its advantages [1-2]. However, because of the lower sensitivity of the hardware imaging system, the MMW image obtained is of lower resolution, at the same time, much unknown noise is also added in the MMW image and makes the MMW image is difficult to be discerned [2-3]. In fact, the MMW image can be regard

D.-S. Huang et al. (Eds.): ICIC 2012, LNCS 7389, pp. 365–372, 2012.
© Springer-Verlag Berlin Heidelberg 2012

as a mixed image obtained by the unknown imaging object and background noise. Therefore, the blind image separation method can be explored in MMW image restoration. Independent component analysis (ICA) has been successfully used for blind source separation in many fields, especially in image processing [4]. A number of approaches such as natural gradient algorithm (NGA) [5] and FastICA [6] have been proposed. However, these iterative approaches process the sources directly, so its separation quality depends heavily on the statistical properties of the sources. To avoid the inherent defects of ICA, the assumption of separation of sparseness is developed, which is called sparse ICA (SPICA) [7-9]. This method can significantly improve accuracy and computational efficiency of the existing ICA algorithms [7-8]. The early SPICA use wavelet transform or wavelet packet transform to sparse the sources and behave preferable blind separation effect. Although this sparse decomposition can represent discontinuities at edge points, it will not see the smoothness along the contours existing in the natural images [8-9]. In recent years, the contourlet sparse transform has been used widely in image processing field [7, 10]. This method can deal with singularity in higher dimensions, especially in natural images having smooth contours. Therefore, here, the ICA operated in the contourlet sparse transform domain is discussed in separating blind images. Using the low frequency band and the high frequency bands of the first layer obtained by contourlet sparse transform as the mixed input of SPICA, the task of MMW image restoration can be implemented.

2   Sparse ICA

2.1   The Blind Source Separation Problem

In a typical blind source separation (BSS) task, M mixtures are observed, which is denoted by matrix X. Each of available signals is assumed to be generated by a linear mixture of unknown sources denoted by matrix S, where the number of sources is usually assumed to be equal to that of the mixed signals. Thus, a M dimensional vector of observed signals is generated by the product of an unknown $M \times N$ mixing matrix A and N dimensional vector of unknown source signals [7]. The task of blind separation is to estimate the mixing matrix and then recover the source signals successfully.

A typical embodiment of this problem in the context of superimposed reflections is depicted in Fig.1. The real object (a) is situated on the optical axis behind a semi-reflecting planar lens (d), inclined with respect to the optical axis. Another object (b) is partially reflected by the lens, creating a virtual image (c). The camera (f) records a superposition of the two images. Thus, the intensity of the observed mixed image is give by the following formula:

$$\begin{cases} x_1 = a_{11}s_1 + a_{12}s_2 \\ x_2 = a_{21}s_1 + a_{22}s_2 \end{cases}. \tag{1}$$

where s_1 and s_2 are the images of two source objects (a) and (b), a_{11} and a_{12} are constants and their specific values depend on the optical geometry and properties of the reflective medium. It is assumed here that the problem is spatial invariant. This reasonably good approximation of the physical conditions can be relaxed [7]. Considered the generality, the Eqn.(1) is extended to the form in matrix notation:

$$X = AS .\tag{2}$$

where $x_1, x_2, \cdots, x_m$ are the M mixed images and s_1 and s_2 are the source images represented as row vectors. The mixing matrix is usually unknown, unless side information regarding the physics of the mixing medium is available [7-8]. Under the assumption that the source is statistically independent, it is possible to recover sources s_1 and s_2 up to a permutation and multiplicative constant, by estimating the mixing matrix $\tilde{A} \approx A$, and estimating the sources by its inversion:

$$\tilde{S} = \tilde{A}^{-1}X .\tag{3}$$

This problem can be well solved using the sparse ICA method to be described in the following subsection.

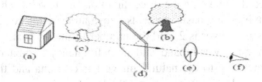

Fig. 1. The schematic diagram of the context of superimposed reflections

2.2 Sparse ICA

Supposing that the source images are sparse, the majority of pixels have a near-zero magnitude and under the assumption of statistical independence of the locations of non-zero pixels in the sources, there is a high probability that only on signal source will contribute to a given pixel in each mixture [8]. However, in reality, the source images in a separation problem are mostly non-sparse natural images, so, a sparse linear transform matrix T is needed. By applying the following transform to the mixed signals, the following formula can be obtained:

$$d_i = T(x_i) = T\left(\sum_{j=1}^{N} a_{ij}s_j\right) = \sum_{j=1}^{N} a_{ij}T(s_j) \quad i = 1, 2, \cdots, M .\tag{4}$$

Thus, the problem at hand is equivalent to separation of linearly mixed sparse sources. This blind separation in such situation can be handled by the common ICA approaches. At the same time, it is noted that the more sparse the source images are, the higher separation accuracy is [7]. Therefore, it is necessary to develop optimal sparse representation methods to improve the blind image separation quality.

3   Contourlet Sparse Decomposition

Contourlet transform is based on an efficient two dimensional non-separable filter banks and provides a flexible multi-resolution, local and directional approach for image processing [7]. It is better than wavelet or wavelet packet transform in dealing with singularity in two or higher dimensions, especially representing images with smooth contours. There are two stages in the contourlet transform [8, 10]: multi-scale analysis stage and directional analysis stage. The first stage is used to capture the point discontinuities. A Laplacian pyramid (LP) decomposes the input image into a detail sub-image and band-pass image that is difference between the input image and the prediction image. In the second stage, the band-pass image is decomposed into 2^k ($k = 1,2,\cdots,L$) wedge shape sub-image by the directional filter banks (DFB), and the detail sub-image is then decomposed by the LP for the next loop, this stage to link point discontinuities into linear structures. The whole loop can be done L_p iteratively, and the number of direction decomposition at each level can be different, which is much more flexible than the three directions in wavelet. The overall result is an image expansion by using basic elements like contour segments. Assumed that the number of the transform layer is 2, and the number of the orientation of each layer is 4. Then the low frequency sub-band and high frequency sub-bands of each layer, respectively corresponding to the natural image named Lena and the MMW image, were shown in Fig. 2.   It is clear to see that the low-pass image contains the majority of energy of the original objection, and the contour of the original image is retained well.

Using the contourlet transform, the mixing set $X = AS$ and separating system $S = A^{-1}X = WX$ can be rewritten into:

$$S_T = C_T[S] = C_T[WX] = W_T C_T[X] = W X_T. \tag{5}$$

where C_T denotes the contourlet transform and $W_T = C_T[W]$. Considering the C_T is a linear transform, then $W_T = W$. Thus, S_T and X_T are regarded as the new source signals and mixed signals. The mixing matrix A and separating matrix W can be obtained in the contourlet domain. Now, a general ICA algorithm, such as natural gradient (NA) algorithm and FastICA algorithm, can be used to train the separating matrix W_T. Here, the learning rule used is described as [7]

$$W_T(k+1) = W_T(k) + \eta(k)\left[\left(I - \varphi(S_T)\right)(S_T)^T\right]W_T(k). \tag{6}$$

where $\eta(k)$ is the learning rate, $\varphi(S_T)$ is the activation function and is closely related to the distribution of the output signal.

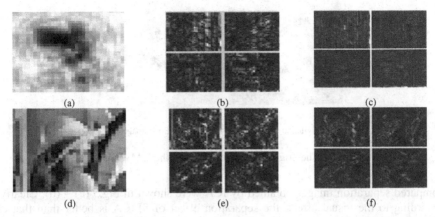

Fig. 2. The original images and the results of contourlet transform corresponding to Lena image and MMW image. (a) and (d): The low frequency sub-band. (b) and (e): The 4-orientation sub-band of the first layer. (c) and (f): The 4-orientation sub-band of the second layer.

(a) Barbara image   (b) Cameraman image   (c) Mixed image 1   (d) Mixed mage 2

Fig. 3. The original images and the corresponding mixed images

4   The Experimental Results and Analysis

4.1   Input Data

In test, two natural images with the same size of 512×512 pixels, called Barbara and Cameraman, were first used to test SPICA algorithm used in contourlet transform domain. Two original images and their mixed images were shown in Fig.3 (a) and (d), and the mixed matrix used was [-0.5444, 0.5080; -2.0102, -0.4214]. At the same time, the original toy gun image and its MMW image with the size of 41 ×41 pixels were shown in Fig.4 (a) and Fig.4 (b). In fact, in MMW imaging scene, the original object is usually unknown. The MMW image can be regarded as a mixed image being composed of the background noise and the original object. The MMW image was generated by the State Key Lab. of Millimeter Waves of Southeast University, which is our cooperation group in the research of MMW image processing.

4.2   Results of Image Restoration

The blind separation (or restoration) of mixed natural images, shown in Fig.3 (c) and Fig.3 (d), was first operated by SPICA. The corresponding separation results were shown in Fig.5 (a) ~ (b). In order to illustrate the separation validity of SPICA,

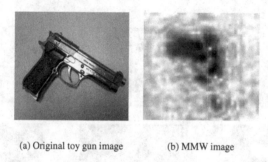

(a) Original toy gun image   (b) MMW image

Fig. 4. The original toy gun image and the MMW image

compared separation images obtained by ICA were shown in Fig.5 (c) ~ (d). Clearly, according to the visual effect, the separation effect of SPICA is better than that of ICA. Further, the single noise ratio (SNR) criterion was used to measure the quality of restored images, and calculated SNR values were listed in Table 1. The experimental data prove again that SPICA truly outperforms ICA. For the MMW image, it can be seen as a mixed image by much unknown noise and the imaging object. So, it is feasible to explore the blind separation process of the MMW image by using SPICA.

(a) Source 1   (b) Source 2   (c) Source 1   (d) Source 2

(A) Separated results of SPICA   (B) Separated results of common ICA

Fig. 5. The separated sources of natural images corresponding to ICA and SPICA

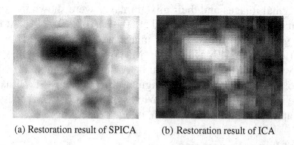

(a) Restoration result of SPICA   (b) Restoration result of ICA

Fig. 6. The separated sources of MMW image corresponding to ICA and SPICA

Table 1. SNR values of restored nature images using different algorithms

Methods \ Images	Barbara	Cameraman	Mixed image 1	Mixed image 2
ICA	29.281	18.974	19.031	11.957
SPICA	34.934	29.522		

Using SPICA to restore the MMW image, the MMW image was first transformed by contourlet. The number of decomposition layers and directions of each layer was supposed to be 2 and 4 respectively. After contourlet transform, the low frequency image and four high frequency sub-bands of each layer were obtained, as shown in Fig.2. It is well known that the low frequency image retained well the contour of the original object, and the high frequency sub-bands contained much unknown noise [10]. We used the low frequency image and the four high frequency sub-bands of the first layer as the mixed images of SPICA, the separation results, which were shown in Fig.6 (a), could be seen as the restoration of the MMW image. Meanwhile, the separation result of ICA was also shown in Fig. 6(b). Generally, in application, the original object is unknown, so it is not suitable to use SNR criterion to measure the quality of this restored MMW image. But, compared the restored MMW image (shown in Fig.6) with the MMW image (shown in Fig.4 (b)), it is distinct to see that the contour of the original imaging object can be restored, and with naked eyes, the original imaging object can be discerned without doubting. So, it can testify that the SPICA algorithm is efficient to blindly restore the original object of MMW imaging system.

5   Conclusions

In this paper, a novel MMW image restoration method using SPICA operated in the contourlet sparse transform domain is explored. In application, the MMW image is a mixed image composed of much unknown background noise and the original unknown imaging object. So, it is feasible to use blind separation method to restore the imaging object. Firstly, the mixed natural images are used to testify the validity of this SPICA. Further, the restoration task of the MMW image is discussed. Compared with general ICA algorithm, the experimental result proved that SPICA operated in the contourlet sparse transform domain is efficient and applied truly.

Acknowledgments. This work was supported by the National Nature Science Foundation of China (Grant No. 60970058), the Innovative Team Foundation of Suzhou Vocational University (Grant No. 3100125), and the "Qing Lan Project" of Jiangsu Province.

References

1. Su, P., Wang, Z., Xu, Z.: Active MMW Focal Plane Imaging System. In: Huang, D.-S., Jo, K.-H., Lee, H.-H., Kang, H.-J., Bevilacqua, V. (eds.) ICIC 2009. LNCS (LNAI), vol. 5755, pp. 875–881. Springer, Heidelberg (2009)
2. Sundareshan, M.K.: Bhattacharjee Supratik: Superresolution of Passive Millimeter-wave Images Using a Combined Maximum-likelihood Optimization and Projection-onto-convex-sets Approach. In: Proc. of SPIE Conf. on Passive Millimeter-wave Imaging Technology, Acrosense 2001, Orlando, FL, UAS, vol. 4373, pp. 105–116 (2001)

3. Cheng, P., Zhao, J.Q., Si, X.C., et al.: L-R Imaging Algorithm For Passive Millimeter Wave Based on Sparse Representation. Journal of Electronics & Information Technology 32, 1707–1711 (2010)
4. Hyvärinen, A., Karhunen, J.H., Oja, E., et al.: Independent Component Analysis. A Wiley-Interscience Publication, New York (1999)
5. Zhang, K., Chan, L.W.: ICA with Sparse Connections. Intelligent Data Engineering and Automated Learning 9, 530–537 (2006)
6. Hyvärinen, A.: Fast and Robust Fixed-point Algorithm for Independent Component Analysis. IEEE Transaction on Neural Networks 10, 626–634 (1999)
7. Liu, S.S., Fang, Y.: A Contourlet-transform Based Sparse ICA Algorithm for Blind Image Separation. Journal of Shanghai University (English Edition) 11, 464–468 (2007)
8. Zibulevsky, M., Pearlmutter, B.A.: Blind Source Separation by Sparse Decomposition. Neural Computation 13, 863–882 (2001)
9. Bronstein, A.M., Bronstein, M.M., Zibulevsky, M., et al.: Sparse ICA for Blind Separation of Transmitted and Reflected Images. International Journal of Imaging Science and Technology 15, 84–91 (2005)
10. Do, M., Vetterl, M.: The Contourlet Transform: An Efficient Directional Multiresolution Image Representation. IEEE Transactions on Image Processing 14, 2091–2106 (2003)

An Adaptive Non Local Spatial Fuzzy Image Segmentation Algorithm[*]

Hanqiang Liu[1] and Feng Zhao[2]

[1] School of computer science, Shaanxi Normal University, Xi'an, P.R. China
max.hqliu@gmail.com
[2] School of Telecommunications and Information Engineering, Xi'an University of Posts
and Telecommunications, Xi'an, P.R. China
add_zf1119@hotmail.com

Abstract. Fuzzy c-means clustering algorithm (FCM) is one of the most widely used methods for image segmentation. In order to overcome the sensitivity of FCM to noise in images, we introduce a novel non local adaptive spatial constraint term, which is defined by using the non local spatial information of pixels, into the objective function of FCM and propose an adaptive non local spatial fuzzy image segmentation algorithm (ANLS_FIS). In this method, the non-local spatial information of each pixel plays a different role in image segmentation. ANLS_FIS can effectively deal with noise while preserving the geometrical edges in the image. Experiments on synthetic and real images, especially magnetic resonance (MR) images, show that ANLS_FIS is more robust than the modified FCM algorithms with local spatial constraint.

Keywords: FCM, Image segmentation, Non-local spatial information, adaptive spatial constraint term, noisy image, Magnetic resonance (MR) image.

1 Introduction

Image segmentation [1,2] is one of the most crucial research topics in computer vision and image understanding. Fuzzy c-means clustering algorithm (FCM) [3] is one of the most widely used clustering methods for image segmentation and many new modified versions of this algorithm have been proposed [4-8].

In order to make the standard FCM more robust to noise, the local spatial information which is obtained from the neighbor window around each pixel in the image is introduced into the standard FCM. Ahmed *et al.* [8] modified the objective function of FCM by incorporating a spatial neighborhood term and applied the proposed algorithm named FCM_S to the segmentation of MR images. Subsequently, to reduce the computational complexity of FCM_S, Chen and Zhang [4] proposed two variants of

---

[*] This work is supported by the National Natural Science Foundation of China (Grant Nos. 60970067 and 61102095), Scientific Research Program Funded by Shaanxi Provincial Education Department (No. 11JK1008), Research Fund Program of Key Lab of Intelligent Perception and Image Understanding of Ministry of Education of China (No. IPIU012011008).

D.-S. Huang et al. (Eds.): ICIC 2012, LNCS 7389, pp. 373–378, 2012.
© Springer-Verlag Berlin Heidelberg 2012

FCM_S: FCM_S1 and FCM_S2. These two algorithms used the mean-filtered image and median-filtered image, respectively, to substitute the neighborhood term of the objective function of FCM_S. It is known that the number of gray levels in an image is generally much smaller than the number of pixels. Based on this fact, Szilágyi et al. [9] presented an enhanced fuzzy c-means clustering algorithm (EnFCM) to accelerate the image segmentation process. In this method, a linearly-weighted sum image is first formed from both the original image and its mean-filtered image, and then clustering is performed on the linearly-weighted sum image by using the gray level histogram instead of the pixels in the summed image. Moreover, Cai et al. [10] proposed a fast generalized fuzzy c-means clustering algorithm (FGFCM). Similar to EnFCM, FGFCM utilizes the gray level histogram of a novel non-linearly-weighted sum image to perform the image segmentation. The non-linearly-weighted sum image is formed from both the original image and the spatial coordinates and the gray level values within the neighbor window around each pixel. Recently, Krinidis and Chatzis [11] presented a fuzzy local information c-means algorithm (FLICM). This algorithm adopts a fuzzy local (both spatial and gray level) similarity measure to guarantee noise robustness and image detail preservation. FLICM is relatively independent of noise types. Furthermore, this algorithm is free of any parameter determination.

When the noise level in the image is high, the adjacent pixels of a given pixel may also have been degraded. In this case, the modified FCM algorithms with local spatial constraint can not obtain satisfactory segmentation. In this paper, we introduce a non-local spatial constraint term into the objective function of FCM and propose an adaptive non local spatial fuzzy image segmentation algorithm (ANLS_FIS). Image segmentation experiments show that ANLS_FIS is more robust than the modified FCM algorithms with local spatial constraint in noise suppression and edge preservation.

2   Adaptive Non Local Spatial Fuzzy Image Segmentation Algorithm

For images heavily contaminated by noise, the adjacent pixels of a given pixel may also have been degraded, so the modified FCM algorithms with local spatial constraint can not obtain satisfactory segmentation performances. In this paper, we first utilize the non-local mean-filtered image [12] to define a novel non local spatial constraint term. Then we introduce the spatial constraint term into the objective function of FCM and propose an adaptive non local spatial fuzzy image segmentation algorithm (ANLS_FIS). The objective function of ANLS_FIS is presented as follows

$$J_m = \sum_{k=1}^{c}\sum_{i=1}^{n} u_{ki}^m \left\| x_i - v_k \right\|^2 + \sum_{k=1}^{c}\sum_{i=1}^{n} \beta_i u_{ki}^m \left\| \overline{x_i'} - v_k \right\|^2 \tag{1}$$

where $\overline{x_i'}$ is the gray value of the ith pixel of the non-local mean-filtered image and β_i is the spatial parameter controlling the penalty effect of the spatial constraint of the ith pixel. For every pixel in an image, there are a set of pixels with a similar neighborhood configuration of it. Compared with the adjacent pixels of the pixel, it is worthwhile to use these similar pixels to obtain the spatial information of a given pixel. In detail, for the ith pixel, its spatial information $\overline{x_i'}$ is computed by the following formula

$$\overline{x_i'} = \sum\nolimits_{j \in W_i^r} w_{ij} x_j \tag{2}$$

where W_i^r denotes a $r \times r$ search window centered at the ith pixel in the noisy image. That means that the pixels in the search window are utilized to compute the spatial information of the ith pixel. The weights w_{ij} ($j \in W_i^r$) depend on the similarity between ith and jth pixels, and satisfy $0 \le w_{ij} \le 1$ and $\sum\limits_{j \in W_i^r} w_{ij} = 1$.

The similarity between the ith and jth pixels depends on the similarity of the gray level vectors $x(N_i)$ and $x(N_j)$, where N_i denotes a $s \times s$ square neighborhood centered at the ith pixel. This similarity is measured as a decreasing function of the weighted Euclidean distance $\left\| x(N_i) - x(N_j) \right\|_{2,\alpha}^2$. Therefore, the pixels with a similar gray level neighborhood to $x(N_i)$ have larger weights. These weights are defined as

$$w_{ij} = \frac{1}{Z_i} \exp\left(-\left\| x(N_i) - x(N_j) \right\|^2 \big/ h \right) \tag{3}$$

where h is called the filtering degree parameter which controls the decay of the exponential function in Eq. (3). Z_i is the normalizing constant defined as

$$Z_i = \sum_{j \in W_i^r} \exp\left(-\left\| x(N_i) - x(N_j) \right\|^2 \big/ h \right) \tag{4}$$

Intuitively, it would be better if we can adjust the spatial constraint separately for each pixel, so each pixel has its own β value. In this paper, we utilize the weights w_{ij} ($j \in W_i^r$) to determine the spatial parameter β_i for the ith pixel ($1 \le i \le n$). It is defined as

$$\beta_i = \beta_{\min} + (\beta_{\max} - \beta_{\min})(\gamma_i - \min_{1 \le l \le n}\{\gamma_l\}) \big/ (\max_{1 \le l \le n}\{\gamma_l\} - \min_{1 \le l \le n}\{\gamma_l\}) \tag{5}$$

where $[\beta_{\min}, \beta_{\max}]$ denotes the value range of β_i, and $\gamma_i = \max\limits_{j \in W_i^r}(w_{ij})$ is the maximum weight among the pixels in the window W_i^r. The value of γ_i ($1 \le i \le n$) can reflect the accuracy of $\overline{x_i'}$ estimating for the ith pixel value of the original image (non noisy image). Therefore, the larger γ_i is, the bigger the guiding effect of $\overline{x_i'}$ on the clustering of x_i is. Thus β_i obtained in Eq. (5) can make the non-local spatial constraint of each pixel able to play a different part in the pixel clustering.

When utilizing the non local spatial information of pixels, ANLS_FIS can achieve better segmentation performances, especially for the geometrical edges in the image.

By minimizing Eq. (1) using the Lagrange multiplier method, the update equations of membership function u_{ki} and cluster center v_k are given in Eqs. (6) and (7).

$$u_{ki} = \sum_{l=1}^{c} \left(\frac{\|x_i - v_k\|^2 + \beta_i \|\overline{x_i'} - v_k\|^2}{\|x_i - v_l\|^2 + \beta_i \|\overline{x_i'} - v_l\|^2} \right)^{-1/(m-1)} \tag{6}$$

$$v_k = \sum_{i=1}^{n} u_{ki}^m (x_i + \beta_i \overline{x_i'}) \bigg/ \sum_{i=1}^{n} (1 + \beta_i) u_{ki}^m \tag{7}$$

3   Experimental Results and Analysis

In order to demonstrate the performance of ANLS_FIS, we perform segmentation experiments on real images. For all the comparation methods, the fuzziness index m, the maximal iteration num T and the threshold ε are set to 2, 500 and 10^{-5}, respectively. The neighbor window size $q \times q$ of FCM_S1, FCM_S2, EnFCM FGFCM and FLICM is set to 3×3. Furthermore, we set $\beta=6$ for FCM_S1, FCM_S2 and EnFCM. According to the results and the parameter analysis presented in Ref. [10], the parameters λ_s and λ_g in FGFCM are set to 3 and 6, respectively.

Figure 1(a) presents the House image with 256×256 pixels. We artificially add the Gaussian noise of mean $M=0$ and variance $Var=0.006$ to this image, and show the noisy image in Fig. 1(b). In this section, we use this noisy image to verify the performance of FCM_S1, FCM_S2, EnFCM, FGFCM, FLICM and ANLS_FIS. The segmentation results of these six methods are shown in Fig. 1(c)-(h). In order to quantitatively assess these six methods, we adopt partition coefficient V_{pc} [13] and partition entropy V_{pe} [14] to evaluate the segmentation results. They are defined as follows:

$$V_{pc} = \sum_{i=1}^{c} \sum_{j=1}^{n} u_{ij}^2 \bigg/ n \tag{8}$$

$$V_{pe} = -\sum_{i=1}^{c} \sum_{j=1}^{n} (u_{ij} \log u_{ij}) \bigg/ n \tag{9}$$

The idea of these two validity functions is that the partition with less fuzziness means better performance. So the best clustering is achieved when the value V_{pc} is maximal and V_{pe} is minimal. It is found from Fig. 1 that ANLS_FIS is the best among these six methods for the Gaussian noisy image.

Moreover, we perform segmentation experiments on a MR image which is with 256×170 pixels and shown in Fig. 2(a). We add the Rician noise ($l=20$) to this MR image, and show the noisy image in Fig. 2(b). We adopt this noisy image to test the performance of FCM_S1, FCM_S2, EnFCM, FGFCM, FLICM and ANLS_FIS on MR images. The segmentation results on the noisy images are presented in Fig. 2(c)-(h). The segmentation result reveals that ANLS_FIS can effectively remove the noise and retain the edges in the MR image and obtains the maximum V_{pc} and minimum V_{pe} values among these six methods.

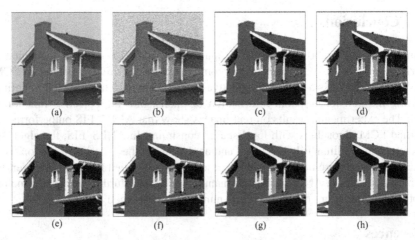

Fig. 1. Segmentation results on the "House" image corrupted by the Gaussian noise: (a) original image; (b) noisy image; (c) FCM_S1 result (Vpc=0.8165, Vpe=0.3754); (d) FCM_S2 result(Vpc=0.8071, Vpe=0.3906); (e) EnFCM result(Vpc=0.7410, Vpe=0.5018); (f) FGFCM result(Vpc=0.7398, Vpe=0.5037); (g) FLICM result (Vpc=0.8234, Vpe=0.3609); (h) ANLS_FIS result (Vpc=0.8782, Vpe=0.2609)

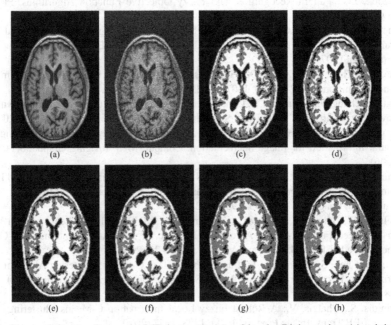

Fig. 2. Segmentation results on the MR image corrupted by the Rician noise: (a) original image; (b) noisy image; (c) FCM_S1 result (V_{pc}=0.8552, V_{pe}=0.2748); (d) FCM_S2 result (V_{pc}=0.8548, V_{pe}=0.2758); (e) EnFCM result (V_{pc}=0.7553, V_{pe}=0.4422); (f) FGFCM result (V_{pc}=0.7548, V_{pe}=0.4432); (g) FLICM result (V_{pc}=0.8500, V_{pe}=0.2793); (h) ANLS_FIS result (V_{pc}=0.8731, V_{pe}=0.2373)

4   Conclusion

In this paper, we propose an adaptive non local spatial fuzzy image segmentation algorithm (ANLS_FIS). In order to overcome the noise sensitivity of standard FCM, ANLS_FIS introduces a non local adaptive spatial constraint term, which is defined by using the non local spatial information of pixels, into the objective function of FCM. The experimental results on real images show that ANLS_FIS outperforms the modified FCM algorithms with local spatial constraint. In ANLS_FIS, the algorithm parameters are preliminarily discussed and analysized in the experiment section. How to theoretically choose these parameters deserves to be studied further. Moreover, the number of clusters in ANLS_FIS is set manually. Thus our further works also include adaptive determination for the clustering number in our method.

References

1. Pal, N.R., Pal, S.K.: A Review on Image Segmentation Techniques. Pattern Recognition 26(9), 1277–1294 (1993)
2. Zhang, H., Fritts, J.E., Goldman, S.A.: Image Segmentation Evaluation: A Survey of Unsupervised Methods. Comput. Vision and Image. Under. 110(2), 260–280 (2008)
3. Bezdek, J.C.: Pattern Recognition with Fuzzy Objective Function Algorithms. Plenum, New York (1981)
4. Chen, S.C., Zhang, D.Q.: Robust Image Segmentation Using FCM with Spatial Constraints Based on New Kernel-induced Distance Measure. IEEE Trans. Syst., Man, Cybern. 34(4), 1907–1916 (2004)
5. Fan, J.L., Zhen, W.Z., Xie, W.X.: Suppressed Fuzzy C-means Clustering Algorithm. Pattern Recognition Lett. 24(9-10), 1607–1612 (2003)
6. Zhu, L., Chung, F.L., Wang, S.T.: Generalized Fuzzy C-means Clustering Algorithm with Improved Fuzzy Partitions. IEEE Trans. Syst. Man, Cybern. B, Cybern. 39(3), 578–591 (2009)
7. Chuang, K.S., Tzeng, H.L., Chen, S., Wu, J., Chen, T.J.: Fuzzy C-means Clustering with Spatial Information for Image Segmentation. Comput. Med. Imaging Graph. 30(1), 9–15 (2006)
8. Ahmed, M.N., Yamany, S.M., Mohamed, N., Farag, A.A., Moriarty, T.: A Modified Fuzzy C-means Algorithm for Bias Field Estimation and Segmentation of MRI Data. IEEE Trans. Med. Imaging. 21(3), 193–199 (2002)
9. Szilagyi, L., Benyo, Z., Szilagyii, S., Adam, H.S.: MR Brain Image Segmentation Using An Enhanced Fuzzy C-means Algorithm. In: Proc. of 25th Annual International Conference of the IEEE EMBS, Cancun, Mexico, pp. 17–21 (2003)
10. Cai, W., Chen, S., Zhang, D.: Fast and Robust Fuzzy C-means Clustering Algorithms Incorporating Local Information for Image Segmentation. Pattern Recognition 40(3), 825–838 (2007)
11. Krinidis, S., Chatzis, V.: A Robust Fuzzy Local Information C-Means Clustering Algorithm. IEEE Trans. Image Process. 19(5), 1328–1337 (2010)
12. Buades, A., Coll, B., Morel, J.M.: A Non-local Algorithm for Image Denoising. In: Proc. IEEE Int. Conf. on Computer Vision and Pattern Recognition, vol. 2, pp. 60–65 (2005)
13. Bezdek, J.C.: Cluster Validity with Fuzzy Sets. Cybernetics and Systems 3(3), 58–73 (1973)
14. Bezdek, J.C.: Mathematical Models for Systematic and Taxonomy. In: Proc. of Eigth International Conference on Numerical Taxonomy, pp. 143–166 (1975)

MMW Image Enhancement Based on Gray Stretch Technique and SSR Theory

Wen-Jun Huai, Li Shang, and Pin-Gang Su

Department of Electronic & Information Engineering, Suzhou Vocational University,
Suzhou, 215104, Jiangsu, China
{hwj,sl0930,supg}@jssvc.edu.cn

Abstract. In order to improve the intensity and contrast qualities of Millimeter Wave (MMW) images and reduce the image noise for concealed weapons Detection, a method is proposed through combining the gray stretch technique and Retinex theory. Gray stretch is first used to preprocess the MMW image, and then Retinex theory based on Single-Scale Retinex (SSR) is used to suppress background clutters. As a result, both the image contrast and Peak Signal to Noise Ratio (PSNR) are improved efficiently. The simulation and experimental results have proved that the algorithm is not only effective in detecting MMW targets, but also has advantages of high speed and is easy for engineering applications.

Keywords: gray stretch technique, MMW image, Retinex theory, illumination image, reflectance image, image enhancement.

1   Introduction

Millimeter Wave (MMW) imaging technique is an application technology which was combined and extended from electromagnetic wave theory and optical imaging theory. It was proposed firstly in the late 1930s [1]. To the 1990s, MMW imaging technology which is applied to detect concealed weapon in the field of security inspection is becoming a hot international research spot [2]. MMW have good penetration for the clothing, buildings and other insulation materials. However, due to the wavelength of millimeter is longer than visible light and infrared light, and limited by the diffraction of optical system, there are some shortcomings such as the imaging vague, resolution of the image is low and the gray range is narrow. That is very necessary to precede the image for identify the target of MMW imaging [1, 3]. Some algorithms of classical edge detection are not very good effects such as Sober, Prewitt and Robert etc, because those are sensitive to noise. There are still some problems about threshold selection in noisy image edge detection, such as Canny algorithm and mathematical morphology algorithm. Nonlinear transform can remove image noise while preserving image detail well. Retinex algorithm is a nonlinear method which is mainly applied to reduce the non-uniform illumination caused by degraded image and focus on the interest target. It is used in the field of digital image processing [4-6]. In this paper, aiming at the characteristics of MMW imaging and the lack of general enhancement methods, extract the illumination component based on SSR method after the gray pretreated.

D.-S. Huang et al. (Eds.): ICIC 2012, LNCS 7389, pp. 379–385, 2012.
© Springer-Verlag Berlin Heidelberg 2012

2   Theory of SSR

Retinex is a mode of how the human visual system to perceive colors and brightness. It was proposed by the American physicist Edwin H Land's [7]. As is show in Retinex theory, visual imaging of an image was understood to the result of both incident components and reflected components. Incident components determine dynamic range of the image can be achieved. Reflection property, reflected intrinsic nature of the image, determined by the reflection intensity of different light waves.

SSR algorithm is improved and achieved to a center/surround Retinex [8]. It can be specific descript by the formula as follow:

$$LogR(x, y) = \log S(x, y) - \log L(x, y) \tag{1}$$

Where, $S(x, y)$ is visual imaging of an image. That is the brightness values which the human can see. $R(x, y)$ is the reflection component of the image. $L(x, y)$ is illumination component. Illumination component is usually around the original image with the convolution function. Show as type 2:

$$L(x, y) = G(x, y) * S(x, y) \tag{2}$$

In which, $S(x, y)$ is the image intensity, "*" is convolution operator. $G(x, y)$ is surrounding function. The physical meaning of convolution term can be calculated as the illumination component. It was eliminated that change on the target visual image for variation of the illumination component, though the intensity ratio of a pixel value and a weighted average of the surrounding area of point.The form of the surrounding function as follow:

$$G(x, y) = K \exp[-(x^2 + y^2)/c^2 \tag{3}$$

Where, K is the normalization factor which value is the total of visual image pixels. c is the scale constant of the Gaussian-shaped surrounding space.

3   SSR Algorithm Based on Gray-Scale Transformation

Combined with the shortcoming of classic SSR algorithm, a MMW image enhancement algorithm was proposed based on gray-scale transformation in this paper. Figure 1 shows the basic flow of the algorithm.

Fig. 1. The SSR algorithm based on gray-scale transformation

The steps of the algorithm as follow.

(1) The original image is contrasted stretching appropriate by piecewise linear transformation which is used to pretreated, to achieve rough estimates for the image brightness. The smallest error method was selected to calculate the sub-points [9].

First, statistics range is [g1, g2] in the dynamic range of gray value in the MMW images. The actual gray range is [min, max] before the image transformation. Then let a = min, b = max, c = T1, d = T2, a = g1, b = g2, the specific transformation formula can be get and shows as follows:

$$f'(x,y) = \begin{cases} k_0(f-\min)+g_1 & \min < f(x,y) < T_1 \\ f-T_1+c & T_1 < f(x,y) < T_2 \\ k_2(f-T_2)+d & T_2 < f(x,y) < \max \end{cases} \tag{4}$$

Where $k_0 = (c'-g_1)/(T_1-\min)$ is inhibition coefficient in the background area. $k_1 = (d'-c')/(T_2-T_1)$ is the remaining coefficient in the transition zone, and $k_2 = (g_2-d')/(\max-T_2)$ is drawing coefficient in the target area [10]. The gray level of the target segment in image is stretched, the transition section is keep and the background section was compressed for enhance the contrast. The target gray level of MMW image used in this paper is in the high value part of the histogram, so that the image pixels which are above the threshold was regarded as target, others pixels belong to background.

(2) The original image and convolution image was respectively transformed by logarithm operation and applied convolution function. The impact of background was eliminated after subtraction both the original image and convolution image in logarithm space, and the reflection property R of the object was obtained.

(3) According to the ranks of the expandable layer to expand the image after obtain the length L of the original image date. During Gaussian pyramid compared between image pixels in McCann99 algorithm, the original image size (col * 2^n) * (row * 2^n) was compressed into a col * row rank matrix and placed on top of the pyramid to compare firstly. Then the image after the comparison operation was extended twice, and placed in the next level for comparison operations. A way of the image edge expand was applied which selected the size of 3 x 3 kernel function. That can be largely reduced due to image distortion arising and maintain the original edge image features.

(4) The low frequency components were filtered out in the Logarithmic domain. The enhanced images after filtering out low frequency components can be obtain though the original image subtract the one after Gaussian convolution. Finally a logarithmic inverse transformation is used to the image data whose dimensions have been reduced, and output the enhanced image.

4   Experimental Results

4.1   Data Preprocessing

Experimental data come from the active millimeter wave imaging system which is the State Key Laboratory of MMW in Southeast University [11-12]. The picture 2(a) is an optical photo of pistol. The picture 2(b) is a MMW imaging of real target in clothes.

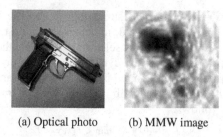

(a) Optical photo   (b) MMW image

Fig. 2. Two images of the imaging target

4.2   MMW Image Enhancement

4.2.1   Experimental of MMW Image Enhancement

As noted above, piecewise linear grayscale stretch for the image when our proposed image enhancement method was used firstly. Select the appropriate sub-point and obtained the result shows in Figure 3. As can be see, the target image and the actual shape of a pistol are some differences after stretching. But most of the background noise has been eliminated.

Fig. 3. MMW image after piecewise linear gray stretch

On the basis of the gray stretch, when different surround function of scale constant c (e.g. c = 30, 70) and different number of iterations N (i.e. N = 10, 30, 70) are selected. The corresponding results of enhancement image shows in Figure 4. The image quality has some improvement obviously as the number of iterations of the Frackle-McCann algorithm increasing. The overall tone of MMW image is tends to smooth sharply with c increases, halo phenomenon is weakly and the edge of target image is more clearly.

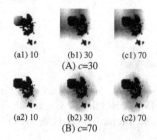

(a1) 10   (b1) 30   (c1) 70
             (A) c=30

(a2) 10   (b2) 30   (c2) 70
             (B) c=70

Fig. 4. The restore image on different scale constant c and iterations N

4.2.2   Analysis of Experimental Results

The PSNR is most commonly used as a measure of quality of reconstruction of loss compression codec. The signal is the original data, and the noise is the error introduced by compression. When comparing compression codec is used as an approximation to human perception of reconstruction quality, one reconstruction may appear to be closer to the original than another, even though a higher PSNR would indicate that the reconstruction is of higher quality [13]. But a lot of noise has been included in the MMW image, and the processed image should contain less noise, closer to the real target, so PSNR can be used to evaluate the two images different. Practically, the target image is known, and a higher PSNR would indicate that the reconstruction imaging is not similar to the original one, and a lower PSNR would show that the processed image has fewer noise, and higher quality than original MMW imaging. The PSNR formula as follow:

$$PSNR = 10 \log 10[255 * 255/MSE] \tag{5}$$

$$MSE = \frac{1}{MN} \sum_{i=1}^{M} \sum_{j=1}^{N} [I(i,j) - \hat{I}(i,j)]^2 \tag{6}$$

Where, M and N is the image matrix of rows and columns. $I(i,j)$ and $\hat{I}(i,j)$ are gray value in the row i and the column j of original image and enhanced image respectively. Change relation between the different scales constant c increases with the number of iterations N shows in figure 5.

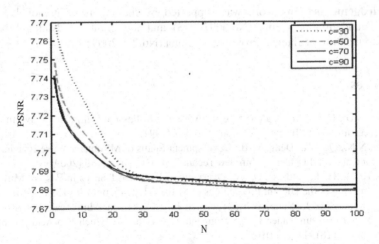

Fig. 5. The PSNR curve of MMW imaging on different value of C

It can be see from figure 5, the value of PSNR decreases rapidly with the c increasing according to the number of iterations increases. Since the original MMW image contains a large number of unknown noise, the PSNR of enhanced image is greater shows the gray scale is closer to the original MMW image, enhanced effect is poorer. Conversely, if the PSNR is smaller, it shows the gray scale is different the original MMW image largely, the noise contained in enhance image is reduced, and the result image is more similar to real target.

From figure 5, when N is between 20 and 30, the change trend of PSNR close to smooth and tends to constant 6.678db ultimately. But when c is 90, The PSNR is decrease sharply when N is less than 20. Then it tends to constant value 6.683dB in a higher position. The reason is the dynamic compression of SSR is stronger when the value of c is smaller. The gloom part detailed in image can be better enhancing. But SSR's overall tonal fidelity is higher, and the detailed in shadow is vague when the value of c is larger. By means of experimental comparison, we select the value of c equals to 70. That would allow enhanced images have lower noise, also can reduce the number of iterations and improve the real-time computing.

5   Conclusion and Future Work

In view of the insufficiency of conventional method in dealing with MMW image enhancement. In this paper, a nonlinear image enhancement algorithm based on gray-scale transformation and SSR theory is proposed. Experimental results show there are good enhancement effects to MMW image using this method. The overall contrast after processing is effectively improved. The detect target is more clearly. However, the algorithm can not completely eliminate halo phenomena. We will depth study how to integrate MMW images in the global and local characteristics in the future research, and process adaptive smoothing has been rebuilt in the brightness of image which can correct the impact due to reflection and scattering of MMW.

Acknowledgements. This work was supported by the grants of National Nature Science Foundation of China (No.60970058) and the grants of National Nature Science Foundation of Jiangsu Province of China (No. BK2009131).

References

1. Roger, A.D., Gleed, G., Anderton, R.N.: Advances in Passive Millimetre Wave Imaging. In: Proceedings of SPIE, vol. 2211, pp. 312–317 (1994)
2. Wang, N.N., Qi, J.H., Deng, W.B.: Development Status of Millimeter Wave Imaging Systems for Concealed Detection. Infrared Technology 31(3), 129–135 (2009)
3. Alan, L., Qi, H.H., Andrew, D.: An Overview of Recent Advances in Passive Millimetre Wave Imaging in the UK. In: Proceedings of SPIE, vol. 2744, pp. 146–153 (1996)
4. Wang, R.G., Zhu, J., Yang, W.T., et al.: An Improved Local Multi-scale Retinex Algorithm Based on Illuminance Image Segmentation. Tien Tzu Hsueh Pao/Acta Electronica Sinica 38(5), 1181–1186 (2010)
5. Bertalmio, M., Caselles, et al.: Issues about Retinex Theory And Contrast Enhancement. International Journal of Computer Vision 83(1), 101–119 (2009)
6. Park, Y.K., Park, S.L., Kim, J.K.: Retinex Method Based on Adaptive Smoothing for Illumination Invariant Face Recognition. Signal Processing 88(8), 1929–1945 (2008)
7. Edwin, H.L.: An Alternative Technique for The Computation of The Designator in The Retinex Theory of Color Vision. Proc. Natl. Acad., Sci. USA 83, 3078–3080 (1986)

8. Land, E.H.: The Retinex Theory of Color Vision. Scientific American 237(6), 108–128 (1977)
9. Kittler, J.: Minimum Error Thresholding. Pattern Recognition 19, 41–47 (1986)
10. Zhang, X.J., Sun, X.L.: A Research on the Piecewise Linear Transformation in Adaptive IR Image Enhancement. Electronic Science and Technology 186(3), 13–16 (2005)
11. Sun, P.G., Wang, Z.X., Xu, Z.Y.: Active MMW Focal Plane Imaging System. Journal of Suzhou Vocational University 19(1), 70–73 (2008)
12. Shang, L., Su, P.G., Zhou, C.X.: Denoising Millimeter Wave Image Using Contourlet and Sparse Coding Shrinkage. Laser & Infrared 41(9), 1049–1053 (2011)
13. Huynh-Thu, Q., Ghanbari, M.: Scope of Validity of PSNR in Image/Video Quality Assessment. Electronics Letters 44(13), 800–801 (2008)

A Study of Images Denoising Based on Two Improved Fractional Integral Marks

Changxiong Zhou[1], Tingqin Yan[1,2], Wenlin Tao[1] and Shufen Lui[1]

[1] Department of Electronic and Informational Engineering, Suzhou Vocational University,
Suzhou, Jiangsu, 215104, China
handzhou@sina.com
[2] Suzhou Key Lab. of Digital Design & Manufacturing Technology, Jiangsu 215104, China

Abstract. In this paper, applying fractional calculus Grümwald-Letnikov definition, a novel image denoising approach based on two improved fractional integral masks is proposed, Two structures of 3×3 fractional integral masks which center is the processed pixel are constructed and discussed. The denoising performance of the proposed fractional integral marks (FIM1 and FIM2) is measured using experiments according to subjective and objective standards of visual perception and SNR values. The simulation results show that SNR of FIM1 and FIM2 is prior to the mean filter and the method in Ref. [8].

Keywords: image denoising, fractional integral, Grümwald-Letnikov definition.

1 Introduction

Image processing is an important filed of information science and engineering and image denoising is one of the most fundamental image processing problems in computer vision and image processing [1, 2, 3]. The search for efficient image denosing methods still is a valid challenge, because most algorithms have not yet attained a desirable level of applicability. The most important of all is that though both edge and noise are high frequency information, the loss of edge is evident and inevitable in denoising process [3, 4]. On the other hand, fractional calculus has long history as same as integer calculus, in existing fractional calculus definition, the fractional differential operation is performed if the order is positive; oppositely, the fractional integral operation is performed if the order is negative [5-8]. Many denoising algorithms are substantially performing an integral calculation for the blocks consisting of image pixels and neighborhood pixels of image. In general, if the image is needed to process with filter, we should convolute the value of pixels of image with a mask. Accordingly, we can think of naturally that which sizes can be introduced into the mask?

The remainder of this paper is structured as follows. In Section 2, we introduce the fractional integral theory and its amplitude-frequency character. In section 3, two structures of 3×3 fractional integral masks which center is the processed pixel are

D.-S. Huang et al. (Eds.): ICIC 2012, LNCS 7389, pp. 386–392, 2012.
© Springer-Verlag Berlin Heidelberg 2012

described. In section 4, we analyze and discuss our algorithm by comparative results on noisy images. Finally, conclusions are drawn in Section 5.

2   Fractional Integral Theory and Its Amplitude-Frequency Character

Fractional calculus is essentially non-integer order calculus, and the order may be real or complex. Mathematicians have given different definitions from different perspectives, but these definitions are all not easy to understand and can not be used directly in engineering and technology. Under the Euclidean measure, Grümwald-Letnikov fractional calculus definition should be widely applied and researched. This definition is derived from integer integral definition. For $v \in (0,1)$, the differential formula is directly extended from integer $n \in Z^{+}$ to real number v, and then fractional differentiation is obtained:

$$D^{v} f(t) = \lim_{\substack{h \to 0 \\ nh = t-a}} h^{-v}(-1)^{m} \sum_{m=0}^{n} \frac{\Gamma(v+1)}{\Gamma(m+1)\Gamma(v-m+1)} f(t-mh) . \qquad (1)$$

where t is (a, b) and Γ is the function of gamma. Similarly, to make fractional integral operator, doing definition of v-order integral of $f(t)$ as follows:

$$I^{v} f(t) = \lim_{\substack{h \to 0 \\ nh = t-a}} h^{v} \sum_{m=0}^{n} \frac{\Gamma(v+m)}{\Gamma(m+1)\Gamma(v)} f(t-mh)$$

$$= h^{v} \left[f(t) + v f(t-h) + \frac{v(v+1)}{2} f(t-2h) + ... \right] \qquad (2)$$

Let $f(t) \in L^{2}(R)$ be a one-dimension signal, its Fourier transform is defined as

$$F(\omega) = \int_{R} f(t) e^{-j\omega t} dt . \qquad (3)$$

The fractional integral is $I^{v} f(t) = f^{-v}(t)$, and its Fourier transform operator can be derived:

$$I(\omega) = (j\omega)^{-v} . \qquad (4)$$

Above $I(\omega)$ represents the fractional integral operator. Its amplitude-frequency curve is shown in Fig. 1. According to fractional integral operator, as the frequency $\omega \to 0$, the amplitude frequency response of fractional integral operator $|\omega|^{-v} \to \infty$; while as $\omega \to \infty$, the amplitude frequency response $|\omega|^{-v} \to 0$. It can be illuminated from Fig.1 that fractional integral operation has the role of weaken high-frequency of signal, while keeping that of very high frequency in nonlinear manner. Meanwhile, the component with lower frequency is also strengthened, while reserving that of very low frequency in nonlinear manner. Thereby fractional integral operator can remove noise while reserving the textures and details.

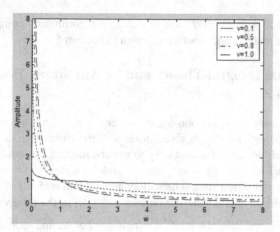

Fig. 1. Amplitude-Frequency Curve of Fractional Integral Operator

3 Construction of Fractional Integral Mask

Through formula (2), the fractional integral operation is simplified as multiplying and adding. Set $h=1$, the anterior 3 approximate backward fractional partial integral of digital image on negative x-coordinate are expressed as follows:

$$\frac{\partial^{v}}{\partial x^{v}} f(x,y) = f(x,y) + vf(x-1,y) + \frac{v(v+1)}{2} f(x-2,y) . \qquad (5)$$

The anterior 3 none-zero values of corresponding terms in formula (5) are 1, v, $\frac{v(v+1)}{2}$, which are all fractional integral masks' coefficients according to Grümwald-Letnikov fractional integral definition. To make the fractional integral mask with rotation invariant, Ref. [8] constructs eight fractional integral masks on direction 0 degrees, 45 degrees, 90 degrees, 135 degrees, 180 degrees, 225 degrees, 270 degrees and 315 degrees respectively, then does convolution filter on the above eight directions by using eight marks respectively, finally takes linear weighting sum value as approximate value of fractional integral for pixel f(x, y). The effect of center pixel f(x, y) decreases with the increased mark's size, so small mark has good denoising performance while reserving image details. Because the greatest distance between the center pixel f(x, y) and the pixels in eight marks in Ref. [8] is two pixels, so this method' SNR can not be high. On the basis of above formula 5, we propose the anterior 3 approximate forward and backward fractional partial integral are expressed as follows:

$$\frac{\partial^{v}}{\partial x^{v}}[2f(x,y)] = vf(x+1,y) + f(x,y) + vf(x-1,y) . \qquad (6)$$

Corresponding terms in formula (6), we propose two improved marks that is fractional integral mask1(FIM1) and fractional integral mask2(FIM2) shown as following Fig.2. The sizes of Mark 1 and Mark 2 are 3×3 window which center is the processed pixel f(x, y), and the coefficients of center pixel in two marks are 8 and 6 respectively, while the coefficients of other pixels are fractional integral order v. Because the distance between the center pixel f(x, y) and every pixel in window is one pixel, FIM1 and FIM2 can raise SNR.

v	v	v
v	8	v
v	v	v

v	v	v
v	6	v
v	v	v

(a) Mask1   (b) Mask 2

Fig. 2. Proposed Fractional Integral Mask

The steps of the improved fractional integral image denoising algorithm are:

(1) Overlapping the center coordinates (x, y) of fractional integral mask, the mask center's coefficient is 8(Mask1) or 6 (Mask2) with the coordinates (x, y) of the pixel f(x, y) to be filtered by fractional integral operation;

(2) Multiplying coefficients of the fractional integral masks in 3×3 sizes with the gray value of corresponding pixel, and adding all product terms respectively to obtain weighting sum;

(3) Taking the arithmetic mean of the weighting sum value in 3×3 sizes as approximate value of fractional integral for pixel f(x, y);

(4) For making every pixel in image can be filtered by fractional integral mask, translating masks one pixel by one pixel, and repeating the steps (1) to (3), thereby the approximate value of fractional integral for the whole digital image can be calculated.

4   Experiments and Comparisons

This section aims at demonstrating the denoising performance of FIM1 and FIM2 respectively. At first, we show the experiment results with four methods of the mean filter, the method in Ref.[8], the proposed FIM1 and FIM2, and then obtain the qualitative and quantitative analysis for denoising performance in various noise levels. All of the experiments were performed on images of LENA and CAMERA shown as in Fig.3 and Fig.4. Then, we also quantitatively discuss the relationship between SNR and noise level. The SNR value in this paper is defined as follows:

$$SNR = 10 \log_{10} \sum_{i=0,j=0}^{M-1,N-1} (f(i, j))^2 / \sum_{i=0,j=0}^{M-1,N-1} (f(i, j) - \hat{f}(i, j))^2 . \quad (7)$$

where f(x,y) is clear image, $\hat{f}(i, j)$ is the image after filtering. Firstly, we study the SNR performance of the mean filter and the method in Ref.[8] and FIM1 and FIM2 using images corrupted with additive, independent Gaussian noise with mean 0 and variance 0.003,0.005,0.010 respectively, wherein the fractional integral mask has a 3×3 window, the fractional integral order is 0.7.

Experiment results on Lena are shown as Fig.3 and Tab.1. Fig.3(a) is clear image of Lena; Fig.3 (b) shows Lena after being corrupted with white Gaussian noise with variance 0.010; Fig.3 (c) shows Lena after filtering with the mean filter; Fig.3 (d) shows Lena after filtering with the method in Ref.[8]; Fig.3 (e) shows Lena after filtering with FIM1; Fig.3 (f) shows Lena after filtering with FIM2. From visual sense effect of human eyes, Fig.3 shows that two proposed approaches FIM1 and FIM2 have good denoising performance for Lena compared with the mean filter. Two proposed approaches not only remove the noise well but also reserve the edge and texture details information in the image, in particular for the weak edge. In addition, compared with the method in Ref.[8], FIM1 and FIM2 do not cause the blur effect because of the distance the center pixel f(x, y) and every pixel in window being one pixel.

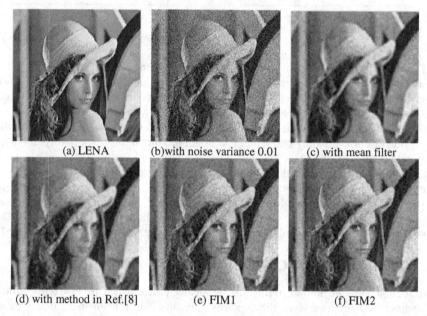

(a) LENA   (b)with noise variance 0.01   (c) with mean filter

(d) with method in Ref.[8]   (e) FIM1   (f) FIM2

Fig. 3. Noisy image of Lena and results of denoising by four methods

Table 1 shows the SNR values of four methods for the image of Lena. Data in the table 1 shows that the SNR values of the image denoised with FIM1 and FIM2 are higher than that with the mean filter and the method of Ref.[8] in different noise levels, and shows that FIM1's performance is prior to the FIM2 when noise level is less than 0.003.

Table 1. Comparison of SNR for Lena with four methods

Noise variance	Noisy image	Mean filter	Ref.[8]	FIM1	FIM2
0.001	23.94	20.17	21.27	25.23	24.61
0.003	19.17	20.01	20.91	23.58	23.45
0.005	17.00	19.89	20.59	22.36	22.50
0.010	14.01	19.52	19.85	20.36	20.86

Experiment results on Camera are shown as Fig.4 and Tab.2. Fig.4(a) is clear image of Camera; Fig.4 (b) shows Camera after being corrupted with white Gaussian noise with variance 0.010; Fig.4 (c) shows Camera after filtering with the mean filter; Fig.4 (d) shows Camera after filtering with the method in Ref.[8]; Fig.4 (e) shows Camera after filtering with FIM1; Fig.4 (f) shows Camera after filtering with FIM2. By eye's qualitative analysis for Fig.4, we observe that that two proposed approaches of FIM1 and FIM2 have good denoising performance for Camera compared with the mean filter. FIM1 and FIM2 not only remove the noise well but also reserve the edge and texture details information in the image, in particular for the weak edge. In addition, compared with the method in Ref. [8], FIM1 and FIM2 do not cause the blur effect because of the distance the center pixel f(x, y) and every pixel in window being one pixel.

| (a) CAMERA | (b) with noise variance 0.01 | (c) with mean filter |
| (d) with method in Ref.[8] | (e) FIM1 | (f) FIM2 |

Fig. 4. Noisy image of CAMERA and results of denoising by four methods

Table 2 shows the SNR values of four methods for the image of Camera. Data in the table 2 shows that the SNR values of the image denoised with FIM1 and FIM2 are higher than that with the mean filter and the method of Ref.[8] in different noise levels, and shows that FIM1's performance is prior to the FIM2 when noise level is less than 0.010.

Table 2. Comparison of SNR for Camera with four methods

Noise variance	Noisy image	Mean filter	Ref.[8]	FIM1	FIM2
0.003	19.76	16.10	16.65	20.67	20.08
0.005	17.62	16.03	16.46	20.10	19.69
0.010	14.64	15.85	16.10	18.81	18.71
0.020	11.71	15.44	15.37	16.99	17.21

5   Conclusions

In this paper, we devote to introduce factional calculus theory into the research field of digital image processing, propose a novel fractional integral image denoising algorithms which is based on Grümwald-Letnikov fractional calculus definition, and construct two fractional integral masks in 3×3 window. Simulation experiments show the availability of the improved method, that it is prior to the mean filter and the method of Ref. [8], especially when the noise deviation is less than 0.010. Furthermore, the subjective evaluations from the figure 3 and table 1, figure 4 and table 2 can be agreed upon with the objective evaluation of the above table. Certain aspects of improved method need to be studied more carefully.

Acknowledgements. This research was sponsored by the grants of Natural Science Foundation of China (No. 60970058), the grants of Natural Science Foundation of Jiangsu Province of China (No.BK2009131), Innovative Team Foundation of Suzhou Vocational University (No.3100125), Suzhou Infrastructure Construction Project of Science and Technology (No.SZS201009) and Qing Lan Project of Jiangsu Province of China.

References

1. Special issue: Fractional Signal Processing and Applications. Signal Processing 83(11), 2285–2286 (2003)
2. Tatom, F.B.: The Relationship between Fractional Calculus and Fractals. Fractals 3(1), 217–229 (1995)
3. Pu, Y., Zhou, J., Yuan, X.: Fractional Differential Mask: a Fractional Differential-based Approach for Multi-scale Texture Enhancement. IEEE Transactions on Image Processing 19(2), 491–511 (2010)
4. Oldham, K.B., Spanier, J.: The Fractional Calculus. Academic Press, New York (1974)
5. Mathieu, B., Melchior, P., Oustaloup, A.: Fractional Differentiation for Edge Detection. Signal Processing 83(11), 2421–2432 (2003)
6. Pu, Y., Wang, W.: Fractional Differential Masks of Digital Image and Their Numerical Implementation Algorithms. Acta Automatica Sinica 33(11), 1128–1135 (2007)
7. Pu, Y., Wang, W., Zhou, J., et al.: Fractional Differential Approach to Detecting Textural Features of Digital Image and its Fractional Differential Filter Implementation. Sci. China Ser. F, Inf. Sci. 51(9), 1319–1339 (2008)
8. Huang, G., Pu, Y., Chen, Q., Zhou, J.: Research on Image Denoising Based on Fractional Integral. System Engineering and Electronics 33(4), 925–932 (2011) (in Chinese)

Leaf Image Recognition Using Fourier Transform Based on Ordered Sequence

Li-Wei Yang[1,2] and Xiao-Feng Wang[2,3,*]

[1] Department of Automation, University of Science and Technology of China, Hefei, Anhui 230027, China
[2] Intelligent Computing Laboratory, Hefei Institute of Intelligent Machines, Chinese Academy of Sciences, P.O. Box 1130, Hefei, Anhui 230031, China
[3] Key Lab of Network and Intelligent Information Processing, Hefei University, Hefei, Anhui, 230601, China
lwyang.win@gmail.com, xfwang@iim.ac.cn

Abstract. There are a number of leaf recognition methods, but most of them are based on Euclidean space. In this paper, we will introduce a new description of feature for the leaf image recognition, which represents the leaf contour with the ordered sequence. For a leaf image, points on the contour represent the most important information of the leaf. Thus, by extracting serial points of the leaf contour, the unique corresponding ordered sequence can be obtained for a contour. Then, we can compute the amplitude-frequency feature by performing the Discrete Fourier transform on the ordered sequence. Since the low-frequency part of the Fourier transform represents the global information and the high-frequency part the local details, we can adopt the amplitude-frequency feature for leaf image recognition. Experimental results on the famous Swedish library and ICL library show that the proposed feature is effective for leaf image recognition.

Keywords: leaf recognition, Fourier transform, ordered sequence, amplitude-frequency.

1   Introduction

On the earth, there are a huge number of plant species. Plants are the primary food producers which sustain all other life forms including people. They are the only organisms that can convert light energy from the sun into food. Some recent work has been focused on plant image recognition [1-5]. Wang et al. [1] pointed out that plants can be basically identified by their leaves. Thus, the plant recognition methods usually depend on physical leaf features such as leaf shape [1-2] and the pattern of veins [3]. The morphological and genetic features were also employed to recognize different species of plant leaves [4-5]. Most common features are based on shape recognition. They include global and local information. Past techniques used in the

---

* Corresponding author.

D.-S. Huang et al. (Eds.): ICIC 2012, LNCS 7389, pp. 393–400, 2012.
© Springer-Verlag Berlin Heidelberg 2012

shape matching of objects are: chain code cross-correlation, Fourier descriptors and moments, statistical pattern recognition techniques, symbolic matching, syntactic and relaxation methods [6]. In order to obtain the scale, rotation and shift invariance for feature extraction based on the Euclidean distance, the time cost will be extremely large. But if the feature extraction can be performed in the frequency domain, these issues can be easily solved.

In this paper, we propose using the features extracted from frequency domain to overcome scale, rotation and shift variance problems in Euclidean space. First of all, we should obtain the points on the leaf contour, and then compute the ordered sequence with the point set. We perform the discrete Fourier transform (DFT) on the ordered sequence which is a series ordered value computed by the distance between the ordered contour points and their center point. Here, we will mainly use the linear property and cycle shift property of DFT. We make use of the linear transformation of Fourier transform to normalize the ordered sequence. One ordered sequence corresponds to a leaf contour with the one-to-one relationship. If the leaf contour was rotated, cycle shift of ordered sequence will happen. Thus we can use the cycle shift nature of Fourier transform to solve the leaf rotation problem. Finally, the nearest neighbor (K-NN) was used for recognition.

The rest of the paper is organized as follows: Section 2 introduces the Discrete Fourier Transform and its properties. In Section 3, we present our feature extraction method based on the ordered sequence. Section 4 gives the experimental results of proposed features on Swedish and ICL leaf library [14]. Finally, we will make a conclusion in Section 5.

2   Discrete Fourier Transform

In this section, we shall mainly introduce the Discrete Fourier transform and its properties.

The discrete Fourier transform (DFT) is one of the most fundamental operations in digital signal processing [7]. In cycle sequence, as a matter of fact, only limited values are meaningful, so many of its features can be replaced by limited sequence. For a discrete time series signal X:

$$X = \{x(1), x(2), ...x(N)\} \tag{1}$$

N is the length of the effective value of the ordered sequence. Its Discrete Fourier Transform can be formulated as follows:

$$x(n) = \frac{1}{N} \sum_{k=0}^{N-1} x(k) \times e^{-i \times \frac{2 \times k \times \pi \times n}{N}} \qquad 0 \leq n \leq N-1 \tag{2}$$

Discrete Fourier Transform has many good properties and two of them are important for plant leaf recognition.

I. The linear property: Since Fourier transform is a linear operation, which means that if discrete ordered sequences are multiplied by a constant $c \in R$, the corresponding frequency domain will have the same scale:

$$c \times g(x) \leftrightarrow c \times G(w) \qquad (3)$$

This property can play a very important role in data normalization especially in leaf feature normalization process.

II. The cycle shift property: If the original sequence is processed with cycle shift by k bits:

$$X(K) = \{x(k+1), x(k+2), ..., x(N), x(1), x(2), ..., x(k)\} \qquad (4)$$

So the corresponding information of every frequency should be multiplied by a complex e^{-iwk} based on w:

$$X(k) \leftrightarrow e^{-iwk} \bullet G(w) \qquad (5)$$

Because e^{-iwk} located on unit cycle, the result of Fourier transforms merely changes by phase position shifting. But the amplitude IG (w) I is stable.

$$|G(w)| = |e^{-iwk} \bullet G(w)| \qquad (6)$$

This property is useful for achieving the leaf contour rotation invariance.

3   Leaf Feature Extraction

Leaf recognition plays an important role in plant classification and its key issue lies in whether selected features are stable and have good ability to recognizing different species of leaves [8].

3.1   The Leaf Contour Extraction

There are a couple of methods for leaf contour extraction, such as Canny operator [9], eight-direction chain code method [10], gradient based edge tracing [11]. In this paper, we shall extract the leaf contour based on gradient information.

Before leaf contour extraction, the image should be pre-processed in advance. Firstly the colorful image should be transformed into gray image. Secondly the gray image should be transformed into binary image with the threshold obtained by OTSU [12]. Then, we can extract the leaf contour with eight-direction chain code. At last, we will keep only 100 points of the leaf contour by using linear interpolation on image which size is less than 128*128 pixels. You can change the number of points according to practical application. Fig.1 shows the process of our contour extraction.

Fig. 1. The Process of Contour Extraction

Fig. 2 demonstrates the extracted contours of leaves from Swedish leaf image library. It is clearly seen that 100 points can not only preserve the global information of leaf but also contain local information of leaf.

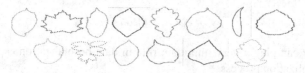

Fig. 2. Extracted contours of leaves from Swedish library

3.2   The Generation of Ordered Sequence

In our method, the computing of effective sequence is a key step since the sequence is the only representation for the leaf contour. Generally, leaf contours of same plant may have the similar sequence while contours of different plants usually have different sequences. In this way, we can distinguish the leaves by their corresponding sequence.

Here, we use the Euclidean distance between contour point and center point which is simple to compute and also accurate to express the contour information.

Firstly, the center point P_m (x_m, y_m) of the leaf contour should be computed with Eq. (7), and Fig.3 shows two leaf contours with their center point.

$$x_m = \frac{1}{n}\sum_{i=1}^{n} x_i, \quad y_m = \frac{1}{n}\sum_{i=1}^{n} y_i \tag{7}$$

Fig. 3. Leaf contours and center points

Secondly, the Euclidean distance is computed by using Eq. (8). The sequence keeps the order of the points. So the sequence and the point set are accordant.

$$d(p_i(x_i, y_i), p_j(x_j, y_j)) = \sqrt{(x_i - x_j)^2 + (y_i - y_j)^2} \tag{8}$$

The illustration of ordered sequences corresponding to 15 leaf contours in Fig.2 is shown in Fig.4. Horizontal axis represent the order of sequence. Vertical axis represents the value of sequence.

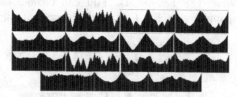

Fig. 4. The Illustration of Ordered sequences corresponding to 15 leaf contours in Fig.2

It should be emphasized that the points on the leaf contour are ordered, and the computed sequence is also ordered. Thus the relationship between sequence and the points of leaf contour is one-to-one correspondence. The normalized formula is defined by Eq. (9).

$$y = \frac{x - x_{min}}{x_{max} - x_{min}} \tag{9}$$

Through Eq. (9), the value of sequence is restricted ranging from 0 to 1 to make sure that sequence is unified. So the problem of scale variance is solved, which also means that the ordered sequence ensure the scale invariance.

The reason for why we keep the sequence ordered is that the problem of contour rotation can be solved with the cycle shift nature of DFT (as shown in Fig.5). When the leaf contour is rotated in Fig.5 (b), the corresponding sequence will circularly shift by n bits. Next, we transform it into frequency domain; the same amplitude-frequency feature can be obtained. Thus, the usage of DFT can achieve the leaf contour rotation invariance. If the sequence is disordered, the problem of contour rotation will be not solved by the nature of DFT.

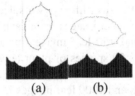

(a)   (b)

Fig. 5. The leaf contour rotation and ordered sequences. (a) Original leaf contour and ordered sequence. (b) Rotated leaf contour and ordered sequence.

3.3   Generation of Amplitude-Frequency Features

After the ordered sequence is obtained and normalized, choosing a range of frequency, we can use Fourier transform in Eq. (2) to obtain the Amplitude-Frequency Features. It can be seen from Fig.6 and Fig.7, the Amplitude-frequencies is an efficient discriminated feature for recognizing the leaves.

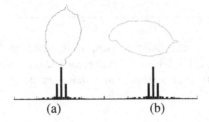

(a)   (b)

Fig. 6. Amplitude-Frequency features for the same species of leaves

The both leaf images in Fig.6 are scaled and rotated. From Fig.6, we can see that the amplitude-frequency feature is constant. That means amplitude-frequency feature can overcome the problem of scale and rotation.

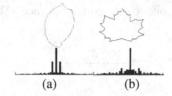

(a)   (b)

Fig. 7. Amplitude-Frequency features for the different species of leaves

The both images in Fig.7 are from different classes. From Fig.7, we can see that the amplitude-frequency feature is quite different. That means amplitude-frequency feature can be regarded as feature for classification.

4   Experimental Results

In our experiments, nearest neighbor (K-NN) was used for recognition. The parameter K is set to 1. In this paper, experiments were performed on famous Swedish library and our ICL library [13]. In Swedish library, we selected 25 leaf images from each leaf species for training, and 50 leaf images from each species for test. There are 375 leaf images in training set, and 750 leaf images in test set. In our ICL library, we select a subset which contains 50 species of plant leaves. We select 10 images from each leaf species for training, and 20 leaf images from each species for test. There are 500 leaf images in training set, and 1000 leaf images in test set.

The experimental result of Fourier descriptors and our amplitude-frequency feature by K-NN method on Swedish library and ICL library is shown in Table 1. It can be seen that the amplitude-frequency feature is more effective.

Table 1. k-NN recognition results based on amplitude-frequency features

Feature	Accuracy on Swedish Library	Accuracy on ICL Library
Amplitude frequency	**89.6%**	**91.6%**
Fourier descriptors	85.333%	80.6%

For leaves from the same plant classes, shift, scale and rotation variance may happen. In Fig.8, F1, F2, F3, F4 and F5 represent 5 leaf contours from the same class. It is obviously that shift, scale and rotation variance existed.

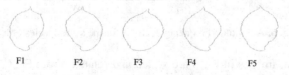

F1   F2   F3   F4   F5

Fig. 8. Leaves in same class

We select 5 images from each class, extract amplitude-frequency feature, and then calculate the inner-class variances and between-classes variances of above 5 leaf contours based on Fourier descriptors and amplitude-frequency feature. Fig.9 shows the inner-class variances based on Fourier descriptors and amplitude-frequency feature. Here, red squares represent the inner-class variance based on amplitude-frequency, and blue rhombuses represent the between-classes variances. It is obvious that inner-class variances based on amplitude-frequency are better than that based on Fourier descriptor. That is to say the amplitude-frequency feature is more effective than Fourier descriptor for recognizing leaves in the same species.

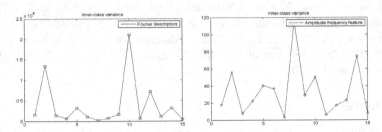

Fig. 9. Inner-class variances of Fourier descriptors and amplitude-frequency feature for each class

Table 2. Inner-class variances and between-class variances for all classes with amplitude-frequency feature and Fourier descriptors

	S_B	S_w	S_B / S_w
Amplitude-frequency	2675.0945	508.0388	5.2655
Fourier Descriptors	14327485.7251	5763918.675	2.4857

Here we compute the inner-class variances and Between-Class variances for all classes with amplitude-frequency feature and Fourier descriptors, as shown in Table 2. Inner-class variance for all classes based on amplitude-frequency is 508.0388, and between-class variance is 2675.0945.Inner-class variance based on Fourier descriptors is 5763918.675, and between-class variance is 14327485.7251. If the features can make inner variance minimal, and make between-class variance maximal, it will be more effective. From Table 2, it is obvious that the rate $S_B / S_w = 5.2655$ of amplitude-frequency is larger than $S_B / S_w = 2.4857$ of Fourier descriptors. That means amplitude-frequency feature is more effective than Fourier descriptors feature in practical classification.

5   Conclusion

In this paper, we analysis Fourier transform and its property, and then we propose a new feature for leaf recognition based on it. Firstly, the leaf image should be pre-processed into binary image, and then we extract point set to represent the contour.

Through computing the distance between point set and the center point of the set, we can obtain an ordered sequence. Next, we will transform the sequence by Fourier transform. The amplitude-frequency can be regarded as the feature. Finally, we can use this feature to do some experiments on Swedish library and our ICL library with K-NN classification. We also do some more experiments with Fourier descriptors for comparison. The experiment results show the amplitude-frequency feature is more effective than Fourier descriptors, so we can use the amplitude-frequency feature to recognize plant leaf.

Acknowledgement. This work was supported by the grants of the National Science Foundation of China, Nos. 61005010, 60975005, 60905023, 60873012, 71001072, the grant of China Postdoctoral Science Foundation, No. 20100480708, the grant of the Key Scientific Research Foundation of Education Department of Anhui Province, No. KJ2010A289, the grant of Scientific Research Foundation for Talents of Hefei University, No. 11RC05.

References

1. Wang, X.F., Huang, D.S., Du, J.X.: Classification of Plant Leaf Images with Complicated Background. Applied Mathematics and Computation 205, 916–926 (2008)
2. Wang, X.-F., Du, J.-X., Zhang, G.-J.: Recognition of Leaf Images Based on Shape Features Using a Hypersphere Classifier. In: Huang, D.-S., Zhang, X.-P., Huang, G.-B. (eds.) ICIC 2005, Part I. LNCS, vol. 3644, pp. 87–96. Springer, Heidelberg (2005)
3. Leslie, E.: Auxin Is Required for Leaf Vein Pattern in Arabidopsis. American Society of Plant Biologists 121(4), 1179–1190 (1999)
4. Soille, P.: Morphological Image Analysis Applied to Crop Field Mapping. Image and Vision Computing 18, 1025–1032 (2000)
5. Pramanik, S., Bandyopadhyay, S.K., Bhattacharyya, D., Kim, T.-H.: Identification of Plant Using Leaf Image Analysis. In: Kim, T.-H., Pal, S.K., Grosky, W.I., Pissinou, N., Shih, T.K., Ślęzak, D. (eds.) SIP/MulGraB 2010. CCIS, vol. 123, pp. 291–303. Springer, Heidelberg (2010)
6. Bhanu, B., Olivier, D.: Shape Matching of Two-Dimensional Objects. IEEE Trans. on Pattern Analysis and Machine Intelligence 6(2) (March 1984)
7. Soontom, O., Chen, Y.J.: Integer Fast Fourier Transform. IEEE Transaction on Signal Processing 50(3) (March 2002)
8. Krishna, S., Indra, G., Sangeeta, G.: SVM-BDT PNN and Fourier Moment Technique for Classification of Leaf Shape. International Journal of Signal Processing, Image Processing and Pattern Recognition 3(4) (December 2010)
9. Bao, P., Zhang, L., Wu, X.L.: Canny Edge Detection Enhancement by Scale Multiplication. IEEE Trans. on Pattern Analysis and Machine Intelligence 27(9) (2005)
10. Li, X., Shen, J.: Group Direction Difference Chain Codes for the Representation of the Border. In: Digital and Optical Shape Representation and Pattern Recognition, Orlando, FL, pp. 372–376. SPIE, Bellingham (1988)
11. Rafael, C.G., Richard, E.W.: Digital Image Processing. Prentice Hall
12. Milan, S., Vaclav, H., Roger, B.: Image Processing Analysis and Machine Vision. Thomson Learning (2008)
13. ICL Plant Leaf Images Dataset,
   http://www.intelengine.cn/English/dataset

Discriminant Graph Based Linear Embedding

Bo Li[1,2,4], Jin Liu[3,4], Wen-Yong Dong[3,4], and Wen-Sheng Zhang[4]

[1] School of Computer Science and Technology, Wuhan University of Science and Technology,
430081, Wuhan, China
[2] State Key Lab. for Novel Software Technology, Nanjing, China
[3] State Key Lab. of Software Engineering, Wuhan, China
[4] Institute of Automation, Chinese Academy of Science, Beijing, China
liberol@126.com

Abstract. LLE is a nonlinear dimensionality reduction method, which has been successfully applied to data visualization. Based on the assumption of local linearity, LLE can compute the weights between the KNN nodes using the local least reconstruction errors, which increase the computational cost. In this paper, a method titled Discriminant Graph Based Linear Embedding (DGBLE) is proposed to set the weights between the nodes in the KNN graph directly to reduce the computational expense. Moreover, label information can also be taken into account to improve the discriminant power of the original LLE. Experiments on some benchmark data show that the proposed method is feasible and effective.

Keywords: LLE, graph, linear embedding.

1   Introduction

Dimensionality reduction from the original data, which is often dictated by practical feasibility, is an important step in pattern recognition tasks. Dimensionality reduction is also an important process in exploratory data analysis, where the goal is to project the input data onto a feature space which can reflect the important inherent structure of the original data. Since there are large volumes of high dimensional data in numerous real world applications, some of which are perhaps superfluous, so extracting the most useful features from the real world data not only helps to probe into the essential structure of the data, but also contributes to accomplish the task of classification and visualization at low computational cost.

Over the past several years, the study on dimensionality reduction methods has being conducted and many useful feature extraction techniques including linear and nonlinear methods, supervised or non-supervised ones have been well developed [1]. Among them, LLE [2] is a representative nonlinear manifold learning method. Based on the assumption of the local linearity, LLE first constitutes local coordinates with the least constructed cost and then maps them to a global one. Experiments have proven that LLE is an effective method for visualization. However, some limitations are exposed when LLE is applied to pattern recognition.

D.-S. Huang et al. (Eds.): ICIC 2012, LNCS 7389, pp. 401–406, 2012.
© Springer-Verlag Berlin Heidelberg 2012

One limitation is the out-of-sample problem [3]. Because the weighted matrix of LLE is constructed on the training data, when a new data point is coming, how to generalize the results of training samples to the coming data is attracting many attentions. Another limitation lies in that the classical LLE neglects the class information, which will impair the recognition accuracy. So in this paper, we will present a discriminant graph based linear embedding method to overcome the above problems in the original LLE.

The paper is organized as follows. Section 2 describes classical LLE algorithm. Section 3 presents the proposed Discriminant Graph Based Linear Embedding algorithm. Experimental results and simulations on YALE and Palm Print data [4] sets are offered in Section 4. Then the whole paper is finished with some conclusions in Section 5.

2   Locally Linear Embedding

Let $X = [X_1, X_2, ..., X_n] \in R^{D \times n}$ be n points in a high dimensional space. The data points are well sampled from a nonlinear manifold, of which the intrinsic dimensionality is d ($d \ll D$). The goal of LLE is to map the high dimensional data into a low dimensional manifold space. Let us denote the corresponding set of n points in the embedding space as $Y = [Y_1, Y_2, ..., Y_n] \in R^{d \times n}$. The outline of LLE can be summarized as follows:

Step1: For each data point X_i, identify its k nearest neighbors by kNN criterion or ε – ball criterion;

Step2: Compute the optimal reconstruction weights which can minimize the error of linearly reconstructing X_i by its k nearest neighbors;

Step3: Compute the low-dimensional embedding Y for X that best preserves the local geometry represented by the reconstruction weights

Step1 is typically done by using Euclidean distance to define neighborhood, although more sophisticated criteria may also be used, such as Euclidean distance in kernel space or cosine distance. *Step 2* seeks the best reconstruction weights. Optimality is achieved by minimizing the local reconstruction error of X_i,

$$\varepsilon_i(W) = \arg\min \left\| X_i - \sum_{j=1}^{k} W_{ij} X_j \right\|^2 \qquad (1)$$

Step 3 computes the optimal low dimensional embedding Y based on the weight matrix W obtained from *Step 2*.

$$\varepsilon(Y) = tr\left\{ \sum_{ij} M_{ij} Y_i^T Y_j \right\} = tr\left\{ YMY^T \right\} \qquad (2)$$

So based on the weighted matrix W, a sparse, symmetric and positive semi-definite matrix M can be defined as follows.

$$M = (I - W)^T (I - W) \qquad (3)$$

3   Discriminant Graph Based Linear Embedding

In this section, we will propose a discriminant graph based linear embedding method to avoid the problems existed in the original LLE. Firstly, a linear transformation is introduced to solve the out-of-sample problem in traditional manifold learning. Second, the weights between nodes in a KNN graph are set directly by taking the label information into account, which will be described in the following.

3.1   The Weight between Two Nodes

In the original LE or LPP, the weight between two nodes is defined to be a heat kernel or simply either 1 or 0, which can not reflect their class information. In the proposed algorithm, the weight between two points is defined based on their local information and class information. The definition is stated below in details.

Shown in Fig.1 is the typical plot of W_{ij} as a function of $d^2(X_i, X_j)/\beta$. In Fig.1, S1 denotes the case that both X_i and X_j are k nearest neighbors each other and X_i, X_j have the same label; S2 denotes the case that both X_i and X_j are k nearest neighbors each other and X_i, X_j have different labels and S3 denotes the other cases. The weight W_{ij} displays the discriminant similarity between X_i and X_j. The similarity integrates the local neighborhood structure and the class information.

$$W_{ij} = \begin{cases} \exp(-\dfrac{d^2(X_i,X_j)}{\beta}) & \text{If both } X_i \text{ and } X_j \text{ are } k \\ & \text{nearest neighbors each other} \\ & \text{with the same label;} \\[2mm] \exp(-\dfrac{d^2(X_i,X_j)}{\beta})\left(1-\exp(-\dfrac{d^2(X_i,X_j)}{\beta})\right) & \text{If both } X_i \text{ and } X_j \text{ are } k \\ & \text{nearest neighbors each other} \\ & \text{with different labels;} \\[2mm] 0 & \text{otherwise} \end{cases} \qquad (4)$$

3.2   Discriminant Graph Based Linear Embedding

In order to overcome the out-of-sample problem, a linear transformation, i.e. $Y = A^T X$, is plug. Thus the objective function of the original LLE can be changed into the following form.

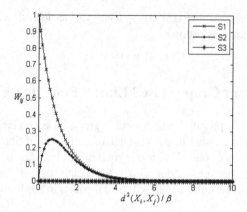

Fig. 1. Typical plot of X_j as a function of $d^2(X_i, X_j)/\beta$

$$J_1(A) = \min tr\{YMY^T\} = \min tr\{A^T XMX^T A\} \tag{5}$$

Where M can be computed by Eqn.(4) and the weights between two nodes can also obtained by Eqn.(5).Thus the outline of the proposed LPLDE can be summarized as follows.

Step1: Compute the weights between points X_i and X_j according to Eqn.(5);

Step2: Construct matrix M based on Eqn. (4);

Step3: Construct a matrix D defined to $D = XMX^T$;

Step4: Compute the d bottom generalized eigenvalues and the corresponding eigenvectors matrix V of D and obtain d -dimensional embedding $Y = V^T X$;

Step5: Adopt a suitable classifier to classify the embedding results.

4   Experiments

4.1   Experiment on Yale Face Database

The Yale face database has 165 images for 15 individuals, where each person has 11 images. The images demonstrate variations in lighting condition (left-right, center-light, right-light), facial expression (normal, happy, sad, sleepy, surprised and wink), and with or without glasses. Each image is cropped to be the size of 32×32. Fig.2 shows the cropped images of one person.

Fig. 2. Sample images of one person in Yale database

We randomly select six images as training set and the rest five images as test set for each class. The parameter k can be set to 5 when constructing the neighborhood graph. LDA[5], LPP[6] and ONPP[7] are carried out to extract features. At last we use the nearest neighbor method for classification. Shown in Table.1 is the optimal accuracy. It is found that the proposed method outperforms the other techniques.

Table 1. Performances comparison on Yale face data

Approaches	Recognition rate	Dimensions
LDA	91.15%	16
LPP	92.23%	12
ONPP	93.42%	14
DGBLE	94.16%	20

4.2 Experiments on Palmprint Data

In the PolyU palmprint database, there are 100 persons and each with 6 palmprint images. Fig.3 displays the cropped samples. The images were cropped with size of 128 by 128.

We selected the first session images as training samples and the second session images as test set. When constructing the neighborhood graph, k is set to 2. Then we use LDA, LPP and ONPP to extract the features. At last the nearest neighbor classifier is adopted to classify these features. Shown in Table.2 are performances on The PolyU palmprint database, where the proposed algorithm can gain the best results.

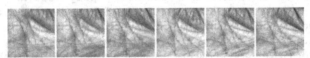

Fig. 3. The cropped sample images from PolyU palmprint

Table 2. Performances comparision on Palmprint database

Approaches	Recognition rate	Dimensions
LDA	90.33%	52
LPP	95.67%	90
ONPP	96.67%	60
DGBLE	97.33%	94

5   Conclusions

In this paper, a discriminant manifold learning method, namely Discriminant Graph Based Linear Embedding, is proposed for classification. The proposed algorithm uses the label information to construct the weights between any two nodes in the original data. So the proposed algorithm becomes more suitable for the tasks of classification. This result is validated by experiments on real-world data set.

Acknowledgments. This work was partly supported by the grants of the National Natural Science Foundation of China (61070013, U1135005, 61170305,61070009), the knowledge innovation program of the Chinese academy of sciences under grant(Y1W1031PB1), the Project for the National Basic Research 12th Five Program under Grant(0101050302) and the Science and Technology Commission of Wuhan Municipality ''Chenguang Jihua'' (201050231058), Postdoctoral Science Foundation of China(20080440073, 20100470613 & 201104173), Natural Science Foundation of Hubei Province(2010CDB03302 &2011CDC076), Project of Hubei Province(Q20121115), the Open Fund Project of State Key Lab.of Software Engineering (SKLSE2010-08-11) ,State Key Lab. of Software Novel Technology (KFKT2011B21) and Key Lab. of Shanghai Information Security Management and Technology Research(AGK2011004).

References

1. Jain, A.K., Duin, R.P.W., Mao, J.C.: Statistical Pattern Recognition: A review. IEEE Transactions on Pattern Analysis and Machine Intelligence 22, 4–37 (2000)
2. Roweis, S.T., Saul, L.K.: Nonlinear Dimensionality Reduction by Locally Linear Embedding. Science 290, 2323–2326 (2000)
3. Belkin, M., Niyogi, P.: Laplacian Eigenmaps and Spectral Techniques for Embedding and Clustering. In: Dietterich, T.G., Becker, S., Ghahramani, Z. (eds.) Advances in Neural Information Processing Systems, vol. 14, pp. 585–591. MIT Press, Cambridge (2002)
4. Biometrics Research Centre, http://www4.comp.polyu.edu.hk/~biometrics
5. Martinez, A.M., Kak, A.C.: PCA versus LDA. IEEE Trans. Pattern Analysis and Machine Intelligence 23(2), 228–233 (2001)
6. He, X., Yang, S., Hu, Y., Niyogi, P., Zhang, H.J.: Face Recognition Using Laplacianfaces. IEEE Trans. Pattern Analysis and Machine Intelligence 27(3), 328–340 (2005)
7. Kokiopoulou, E., Saad, Y.: Orthogonal Neighborhood Preserving Projections. In: The Fifth IEEE International Conference on Data Mining, pp. 1–7. IEEE Press, New York (2005)

A Performance Analysis of Omnidirectional Vision Based Simultaneous Localization and Mapping

Hayrettin Erturk[1], Gurkan Tuna[2], Tarik Veli Mumcu[3], and Kayhan Gulez[3]

[1] Yildiz Technical University, Electrical-Electronics Faculty,
Electrical Eng. Dept., Istanbul, Turkey
hayrettinerturk@gmail.com
[2] Trakya University, Department of Computer Programming, Edirne, Turkey
gurkantuna@trakya.edu.tr
[3] Yildiz Technical University, Electrical-Electronics Faculty,
Control and Automation Eng. Dept., Istanbul, Turkey
{tmumcu,gulez}@yildiz.edu.tr

Abstract. This paper presents a performance analysis of omnidirectional vision based Simultaneous Localization and Mapping (SLAM). In omnidirectional vision based SLAM; robots perform vision based SLAM using only monocular omnidirectional cameras. In this paper, we mainly investigate the use of an omnidirectional camera for Extended Kalman Filter (EKF) based SLAM. To evaluate the success of omnidirectional vision based SLAM, we have also conducted the same simulations using a laser range finder (LRF). Main contributions of this paper are the use of an omnidirectional camera to perform SLAM in the Unified System for Automation and Robot Simulation (USARSim) environment, which is controlled by MATLAB in our study. The results of USARSim simulations show that depending on the environmental conditions omnidirectional cameras can be used as an alternative to other range bearing sensors and stereo cameras.

Keywords: Omnidirectional camera, SLAM, USARSim, MATLAB.

1 Introduction

In robotic mapping, a robot explores an unknown environment and attempts to build a map of it [1]. Robots sense the landmarks in the environment via their proprioceptive and exteroceptive sensors, after calculating the range values; they create a map of the environment in a probabilistic way. Laser range finders and sonar sensors are two examples of range and bearing sensors that can be used for SLAM [2]. Researchers prefer laser range finders since they provide very accurate measurements in most cases. In addition to these sensors, vision based sensors are drawing the attention of research community since they offer exiting opportunities over other type of sensors which provide only range and bearing. A major limitation in vision based SLAM is the narrowness of the camera's field of view [3]. Another alternative is the use of omnidirectional cameras in SLAM due to their surround observation capabilities despite of their low resolutions [4], [5]. Omnidirectional camera systems are used in different type of applications such as autonomous robot navigation, surveillance and 3D reconstruction [6].

D.-S. Huang et al. (Eds.): ICIC 2012, LNCS 7389, pp. 407–414, 2012.
© Springer-Verlag Berlin Heidelberg 2012

There are different types of omnidirectional camera systems, but among these catadioptric camera systems are the most preferred ones due to their relatively low costs. A catadioptric vision system (CVS) is shown in Fig. 1. A CVS consists of a conventional camera positioned in front of a convex mirror. In a CVS, the center of the mirror is aligned with the optical axis of the camera.

Fig. 1. Catadioptric omnidirectional camera [19]

Researches show that sensor accuracy is one of the factors that directly affect the performance of SLAM [1], [2]. Omnidirectional cameras have some advantages and disadvantages that may affect SLAM performance [7]. This paper presents simulations to show the efficiency of an omnidirectional camera in SLAM.

The paper is organized as follows. Robotic mapping and Omnidirectional vision based SLAM are explained in Section 2. The simulation studies with USARSim and MATLAB are given in Section 3. Conclusions of the paper and future work are given in Section 4.

2   Omnidirectional Vision Based Simultaneous Localization and Mapping

A true omnidirectional sensor views the world through an entire sphere of view as shown in Fig. 2. The image obtained from an omnidirectional sensor can provide 360 degree information around the sensor and the direction angle information data with the principle point of camera obtained relatively accurate from the image [8]. The single viewpoint allows the construction of pure perspective images or panoramic images [9]. Panoramic sensors are different from omnidirectional sensors since they are omnidirectional only in one of the two angular dimensions [9].

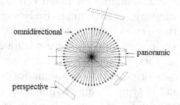

Fig. 2. Omnidirectional vision vs panoramic vision [9]

A catadioptric system uses a reflecting surface to enhance the field of view. The position, orientation, and shape of the reflecting surface are related to the viewpoint and field of view [9]. For a catadioptric system, such as the one used in our simulations, to satisfy the single viewpoint constraint (SVC), all irradiance measurements must pass through an effective viewpoint [10], [11]. In this way, geometrically correct perspective images can be constructed and planar and cylindrical perspective images reconstructed [11]. In [10], a solution of the SVC equation for hyperbolic solutions is given.

Let c denote the distance between the camera pinhole F and the effective viewpoint F′ as shown in Fig. 3, r is defined by $r = \sqrt{x^2 + y^2}$ and k is a constant.

$$(z - \frac{c}{2})^2 - r^2(\frac{k}{2} - 1) = \frac{c^2}{4}(\frac{k-2}{k}) \qquad (k \geq 2) \qquad (1)$$

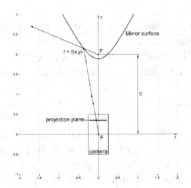

Fig. 3. Parabolic mirror and camera geometry of catadioptric cameras

The solution comprising the parameters c and k has five types of solutions: planar, conical, spherical, ellipsoidal and hyperboloidal. The omnidirectional viewing system simulated for the USARSim environment belongs to the hyperboloidal solutions, defined by (1) with k>2 and c>0 [11].

3   Simulation Studies

MATLAB and USARSim have been used in the simulation studies. USARSim is a simulation platform of robots and environments [12]. It is the basis for the RoboCup rescue virtual robot competition. USARSim consists of environmental models, many sensor models, and robot models of some experimental and commercial robots. It also includes drivers to interface with external control frameworks. We are planning to use Corobot mobile robots in our field tests but Corobot is not supported by USARsim. Hence, Pioneer P2AT robots have been used in the USARSim simulations since they support many sensor models. The USARSim MATLAB Toolbox is used to interface with USARSim and develop robot control programs in MATLAB [13]. With this toolbox, it has been possible to get sensor measurements from the USARSim environment and develop our SLAM and Foreign Object Detection algorithms in MATLAB. The toolbox also allows learning true robot position and orientation. In addition to the toolbox, USAR Screenshot Tool and FRAPS [14] have been used to get images from the Pioneer P2AT robot during the simulation.   USAR Screenshot Tool comes with USARSim MATLAB Toolbox.

 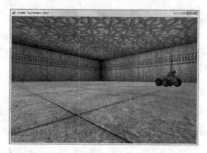

Fig. 4. (a) A Simulated omnidirectional camera, (b) A simulated P2AT with an omnidirectional camera mounted on a pillar driving autonomously in the simulation environment

A robot named "explorer" with a laser range finder and an omnidirectional camera as exteroceptive sensors has been created in USARSim. To set up the simulation of the catadioptric camera, USARBot.ini file has been edited. Though there are many map options available in USARSim, we have chosen a simple map to be able to evaluate the performances of laser range finder based SLAM and omnidirectional vision based SLAM effectively. Omnidirectional vision based SLAM requires omnidirectional cameras. An omnidirectional camera in USARSim looks upwards towards a parabolic convex mirror, and seems to capture the reflection in the mirror [12]. It depicts the complete 360° surroundings with the perspective distortion. Unreal engine used by USARSim supports only planar reflecting surfaces. The 'security camera trick' is used to simulate a non-planar mirroring surface [12]. Fig. 4 (a) shows a simulated omnidirectional camera, and Fig. 4 (b) shows a simulated P2AT robot with an omnidirectional camera mounted on a pillar in USARSim during SLAM operation.

3.1   Performing Omnidirectional Vision Based SLAM

While the robot wanders around, it needs to detect the features in the environment. The goal of omnidirectional vision based SLAM is to measure the relative distances from the robot to the features [15]. The main steps of our approach are as follows:

1. Acquiring an image from USARSim by using the USARSim Matlab Toolbox.
2. Calibrating the mirror system by detecting the center of the image and specifying external and internal boundaries of the omnidirectional image.
3. Detecting the contours of features by using an image processing technique.
4. By using the knowledge of the optics of the mirror system to calculate the distances to the features' contours.
5. Returning this information to the SLAM algorithm in MATLAB.

Since the acquired image in step 1 is obtained from an omnidirectional camera, the image has to be flipped.

image=imflipud(omnidirectional_image)

This function automatically flips the acquired image from the camera up-down.

In Step 2, the center of the omnidirectional image is found to calibrate the mirror system of the camera. To automatically calibrate the camera, we have developed a script but it is not functioning properly yet. But this is not a concern for this study since we control a P2AT robot in the simulation environment, and we do not have full control on the robot. Hence, we have set the center of the camera manually.

In Step 3, the contours of features by using an image processing technique is detected. Radial lines departing from the image center are scanned and searched for the first intensity step between white and black [16]. Before doing this, the omnidirectional image is unwrapped into a rectangular image as shown in Fig. 5. [unwrappedimage, theta] = imunwrap(image, center, angstep, Radius_max, Radius_min).

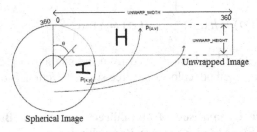

Fig. 5. Conversion of a spherical image to an unwrapped image

Unwrapped image is segmented into black and white.

Blackandwhiteimage = img2bw(unwrappedimage, BWthreshold)

The distance in pixel corresponds to the number of pixels from left image border until first black point.

distanceinpixel = getpixeldistance(blackwhiteimage , Radius_min)

For conversion of the image distances in pixels to metric distances, the optics of the camera-mirror system shown in Fig. 6 is used.

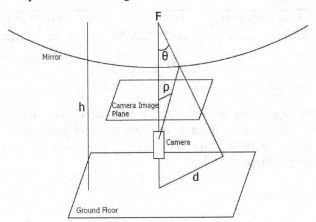

Fig. 6. The optics of the camera-mirror system

The following equation is derived by using the rectangular triangle in Fig. 6:

$$d(\theta) = h\tan(\theta) \tag{2}$$

The relation between θ, which is the distance in pixels, and the angle ρ can be approximated by a first order Taylor expansion:

$$\theta \approx \frac{1}{\alpha}\rho \tag{3}$$

Combining the equations 1 and 2 yields the equation 3. However, due to the hyperbolic shape of the mirror this formula is limited to small distances.

$$d(\rho) = h\tan\left(\frac{\rho}{\alpha}\right) d(\rho) = h\tan\left(\frac{\rho}{\alpha}\right) \tag{4}$$

The basis of our method has been derived from the method in [15]. Instead of this method, an alternative method, color-based free-space detection, is proposed in [17].

3.2   Performance Comparison of Omnidirectional Vision Based SLAM and EKF Based SLAM Using a Laser Range Finder

To evaluate the performance of omnidirectional vision for SLAM applications, we have compared the performance of omnidirectional vision based SLAM with EKF based SLAM using a LRF. The evaluation is based on calculating the differences of the estimated positions of a P2AT robot in USARSim environment with the real positions obtained from USARSim. Table 1 lists the averaged overall error bounds of the methods compared.

Table 1. Comparison of overall error bounds

Overall mean value of the error bound	Omnidirectional vision based SLAM	SLAM using a laser range finder
x	8.11 cm	3.83 cm
y	7.43 cm	2.47 cm
theta	5.21°	2.17°

Though the algorithms we use have some major drawbacks related to the implementation problems resulting from using different applications together, our simulation results provide some valuable insights into the performance of omnidirectional vision based SLAM.

4   Conclusions

This paper focuses on using omnidirectional cameras as range bearing sensors in SLAM applications, and presents a simulation study based performance analysis of omnidirectional vision based SLAM. The advantages/disadvantages of omnidirectional vision based SLAM and the design challenges related to omnidirectional vision based SLAM are investigated and shown with the results of the simulation studies in the USARSim platform. Though the results of our simulations show that omnidirectional vision based SLAM has bigger overall error bounds (global x, global y, and theta) than EKF-based SLAM using a laser range finder, omnidirectional cameras are cost effective alternatives to range bearing sensors and stereo cameras. The results of our simulations also show that the omnidirectional SLAM method proposed in this study is suitable for indoor environments since the success of this method depends mainly on the environmental conditions which can affect the camera, and the distances to features.

While our performance analysis on omnidirectional vision based SLAM provides valuable insight into design issues for omnidirectional vision based SLAM applications, they are only a first step. As a future work, field tests with autonomous robot platforms, i.e., Corobot [18], will be conducted to show the efficiency of omnidirectional vision based SLAM.

Acknowledgments. This research has been supported by Yildiz Technical University Scientific Research Projects Coordination Department. Project Number: 2010-04-02-ODAP01 and Project Number: 2010-04-02-KAP05.

References

1. Dissanayake, G., Newman, P., Clark, S., Durrant-Whyte, H.F., Csorba, M.: A Solution to The Simultaneous Localization and Map Building (SLAM) Problem. IEEE Transactions on Robotics and Automation 17(3), 229–241 (2001)
2. Williams, S.B.: Efficient Solutions to Autonomous Mapping and Navigation Problems, Ph.D. Dissertation, University of Sydney (2001)
3. Suttasupa, Y., Sudsang, A., Niparnan, N.: 3D SLAM for Omnidirectional Camera. In: Proceedings of the 2008 IEEE International Conference on Robotics and Biomimetics, pp. 828–833 (2009)
4. Kim, S., Oh, S.-Y.: SLAM in Indoor Environments Using Omnidirectional Vertical and Horizontal Line Features. Journal of Intelligent and Robotic Systems 51, 31–43 (2008)
5. Kim, J., Yoon, K.-J., Kim, J.-S., Kweon, I.: Visual SLAM by Single-Camera Catadioptric Stereo. In: Proc. of the SICE-ICASE International Joint Conference (2006)
6. Rituerto, A., Puig, L., Guerrero, J.J.: Visual SLAM with An Omnidirectional Camera. In: Proceedings of the 2010 International Conference on Pattern Recognition, pp. 348–351 (2010)
7. Burbridge, C., Spacek, L., Condell, J., Nehmzow, U.: Monocular Omnidirectional Vision based Robot Localisation and Mapping. In: Proc. of the TAROS 2008 (2008)
8. Li, M., Imou, K., Wakabayashi, K.: 3D Positioning for Mobile Robot Using Omnidirectional Vision. In: Proceedings of the 2010 International Conference on Intelligent Computing Technology and Automation, pp. 7–11 (2010)
9. Nayar, S.K.: Omnidirectional Video Camera. In: Proceedings of the DARPA Image Understanding Workshop (1997)
10. Baker, S., Nayar, S.K.: A Theory of Single-Viewpoint Catadioptric Image Formation. International Journal of Computer Vision 35, 1–22 (1999)
11. Schmits, T., Visser, A.: An Omnidirectional Camera Simulation for the USARSim World. In: Iocchi, L., Matsubara, H., Weitzenfeld, A., Zhou, C. (eds.) RoboCup 2008. LNCS, vol. 5399, pp. 296–307. Springer, Heidelberg (2009)

12. USARSim (2010), http://sourceforge.net/projects/usarsim/
13. MATLAB USARSim Toolbox (2010),
    http://robotics.mem.drexel.edu/USAR/
14. FRAPS show fps, Record Video Game Movies, screen capture software (2011),
    http://www.fraps.com/
15. Scaramuzza, D., Siegwart, R.: Appearance-Guided Monocular Omnidirectional Visual
    Odometry for Outdoor Ground Vehicles. IEEE Transactions on Robotics 24(5), 1015–
    1026 (2008)
16. Scaramuzza, D., Martinelli, A., Siegwart, R.: Appearance-based SLAM with Map Loop
    Closing Using an Omnidirectional Camera. In: Proceedings of the IEEE International
    Conference on Intelligent Robots and Systems, IROS 2006 (2006)
17. Nguyen, Q., Visser, A.: A Color Based Rangefinder for an Omnidirectional Camera. In:
    Proc. of the 2009 IEEE/RSJ International Conference on Intelligent Robots and
    Systems(IROS 2009), pp. 41–48 (2009)
18. Corobot (2010), http://robotics.coroware.com/
19. OmniAlert 360 Camera (2011), http://www.remotereality.com/
    product-components-productsmenu-116/omnialert360s-camera-
    productsmenu-92

Trajectory Estimation of a Tracked Mobile Robot Using the Sigma-Point Kalman Filter with an IMU and Optical Encoder

Xuan Vinh Ha[1], Cheolkeun Ha[1,*], and Jewon Lee[2]

[1] School of Mechanical Engineering, University of Ulsan, Republic of Korea
xuanvinhha@gmail.com, hacheol21c@yahoo.co.kr
[2] Prigent Ltd.
benzydad@hotmail.com

Abstract. Trajectory estimations of tracked mobile robots have been widely used to explore unknown environments and in military applications. In this paper, we estimate the precise trajectory of a tracked skid-steered mobile robot that contains an inertial measurement unit (IMU) and an optical encoder. For a systematic estimation, we implement a sigma-point Kalman filter (SPKF), which produces more accurate trajectory information, is easier to calculate, and requires no analytic derivations or Jacobians. The proposed SPKF compensates for the limitations of the IMU and encoder in trajectory estimation problems, as observed from our experimental results.

Keywords: Tracked mobile robots, Inertial Measurement Unit (IMU), Sensor fusion, Sigma-Point Kalman Filter (SPKF).

1 Introduction

Recently, several studies have proposed and developed various approaches to estimate the position of mobile robots. For example, one study used an extended Kalman filter with a low-cost GPS unit and inclinometer to develop a localization scheme for Ackerman steering vehicles in indoor autonomous navigation situations by estimating the positions of the vehicles and their sensor biases [1]. In another study, a navigational system, consisting of a MEMS-based digital in-plane three-axis IMU, an active beacon system, and an odometer, was utilized to obtain more precise robot position data and to monitor the robot environment in real-time [2]. In [3], a relative localization method was used to determine the navigational route of a convoy of robotic units in an indoor environment using a discrete extended Kalman filter based on low-cost laser range systems and built-in odometric sensors. Yet, discrepancies remain between the position measurements and the navigational estimations from odometric, infrared, and ultrasonic measurements and Kalman filtering techniques in skid-steered vehicles, as presented in [4]. Consequently, [5] introduced the unscented Kalman filter for a known kinematic model of robots. Then, in [6], a family of

---

* Corresponding author.

D.-S. Huang et al. (Eds.): ICIC 2012, LNCS 7389, pp. 415–422, 2012.
© Springer-Verlag Berlin Heidelberg 2012

improved derivative nonlinear Kalman filters, called sigma-point Kalman filters (SPKFs), was used and can be applied to the problem of loosely-coupled GPS/INS integration. A novel method to account for latency in GPS updates has also been developed for the SPKF.

Thus, this paper focuses on using the estimated system in order to obtain the precise trajectory of a four-wheeled, tracked, skid-steered mobile robot (Hazard Escape I) based on IMU kinematic motion and velocity constraints. It describes the details of the sensor fusion technique that combines the IMU sensor with the optical encoder. For a systematic estimation, it implements the sigma-point Kalman filter, which produces more accurate position information, is easier to calculate, and requires no analytic derivations or Jacobians. The estimated results are compared with the measurement results from the IMU and encoder sensors installed on a four-wheeled, tracked, skid-steered mobile robot, as shown in Fig. 1.

Fig. 1. The NT - Hazard Escape I Mobile Robot

This paper is organized as follows: Section 2 introduces the IMU kinematic motion model and the velocity constraints. The sigma-point Kalman filter is discussed in Section 3. The experimental setup and results are described in Section 4. Section 5 presents the conclusion.

2   IMU Kinematic Model and Velocity Constraints

2.1   IMU Kinematic Motion Model

We define a navigational reference frame $N(X,Y,Z)$ and robot body frame $B(x,y,z)$, as shown in Fig 2. Let $P_N(t)=[X_N(t),Y_N(t),Z_N(t)]^T \in \mathbb{R}^3$ and $V_N(t)=[V_x(t),V_y(t),V_z(t)]^T \in \mathbb{R}^3$ denote the position and the velocity vectors, respectively, of the IMU in the N frame. Furthermore, let $\Theta = [\phi,\theta,\psi]^T \in \mathbb{R}^3$ denote the attitude angles. We also define the IMU acceleration and the angular rate measurements in the B frame as $a_B = [a_{Bx},a_{By},a_{Bz}]^T \in \mathbb{R}^3$ and $\omega_B = [\omega_{Bx}, \omega_{By}, \omega_{Bz}]^T \in \mathbb{R}^3$, respectively. Then, it is straightforward to calculate the transformation from the B frame to the N frame, as given by the following matrix:

$$C_B^N = \begin{bmatrix} c_\theta c_\psi & -s_\psi c_\phi + c_\psi s_\phi s_\theta & s_\phi s_\psi + c_\psi s_\theta c_\phi \\ c_\theta s_\psi & c_\phi c_\psi + s_\theta s_\phi s_\psi & -s_\phi c_\psi + s_\theta c_\phi s_\psi \\ -s_\theta & c_\theta s_\phi & c_\phi c_\theta \end{bmatrix} \tag{1}$$

where $c_\theta = \cos\theta, s_\theta = \sin\theta$, and the same notation convention is used for angles ϕ and ψ.

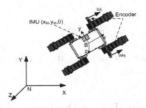

Fig. 2. An IMU kinematic model of a four-wheeled, tracked, skid-steered mobile robot

Note that the Euler angles transform according to the matrix

$$q_\Theta = \begin{bmatrix} 1 & \sin\phi\tan\theta & \cos\phi\tan\theta \\ 0 & \cos\theta & -\sin\theta \\ 0 & \sin\phi/\cos\theta & \cos\phi/\cos\theta \end{bmatrix} \tag{2}$$

Assuming that there is no moving bias, after subtracting the constant offset and local gravity vector, the accelerometer model and the gyros model can be described as [7]

$$a_B = \ddot{x}_B + w_{accel} \tag{3}$$

$$\omega_B = r_B + w_{gyros} \tag{4}$$

where the true acceleration vector, the true angular rate vector of the vehicle in the B frame, the acceleration of white noise, and the angular rate of white noise are $\ddot{x}_B = [\ddot{x}_{Bx}, \ddot{x}_{By}, \ddot{x}_{Bz}]^T$, $r_B = [r_{Bx}, r_{By}, r_{Bz}]^T$, w_{accel} and w_{gyros}, respectively. Then, the kinematic motion equation for the IMU can be simplified such that

$$\begin{aligned} \dot{P}_N &= V_N \\ \dot{V}_N &= C_B^N a_B \\ \dot{\Theta} &= q_\Theta \omega_B \end{aligned} \tag{5}$$

2.2   Velocity Constraints

The IMU coordinate is located at $(x_M, y_M, 0)$ in the B frame, as shown in Fig. 2. Using two wheel encoders in the left and right sides of the vehicle, we obtain the IMU velocity vector $v_B = [v_{Bx}, v_{By}, v_{Bz}]^T$ in the B frame. Due to the low speed of the mobile robot, the wheel slips and the longitudinal ICR location S are very small and can be neglected from the velocity constraints. Based on [8], since the four tracked wheels of our robot always contact the ground and the IMU is fixed on the robot platform, the velocity constraints for the IMU device can be simplified as

$$v_{Bz} = 0 \tag{6}$$

$$v_{Bx} = v - y_M \omega \tag{7}$$

$$v_{By} = x_M \omega \tag{8}$$

where v and ω are the velocity and angular speed of robot, respectively, as measured by two optical encoders.

3   Sigma-Point Kalman Filter Design

Now we define the state variable vector $X(k) = [P_N(k), V_N(k), \Theta(k)]^T \in \mathbb{R}^9$. We rewrite the IMU kinematic equation shown in (5) in discrete-time form such that

$$X(k) = f(X(k-1), u(k), w(k-1)) = X(k-1) + \Delta T * g(X(k-1), u(k), w(k-1)) \tag{9}$$

where the IMU input signals at the kth sampling time are $u(k) = [\ddot{x}_B(k), r_B(k)]^T$, the data sampling period is ΔT, and the nonlinear function g of the process model is given by

$$g(X(k-1), u(k), w(k-1)) = \begin{bmatrix} V_N(k) \\ C_B^N(k)(\ddot{x}_B(k) + w_{accel}(k-1)) \\ q_\Theta(r_B(k) + w_{gyros}(k-1)) \end{bmatrix} \tag{10}$$

Using the velocity constraints from (6)-(8), the IMU velocities in B frame can be described as the measurements $y(k) = [v_{Bx}(k), v_{By}(k), v_{Bz}(k)]^T \in \mathbb{R}^3$. Including the wheel encoder measurement noises and ground topography denoted by $n(k)$, the measurement model can be rewritten in discrete-time form as

$$y(k) = h(X(k)) + n(k) = \left(C_B^N\right)^T V_N + n(k) \tag{11}$$

Then, the SPKF for the system described by (9) and (11) can be implemented by concatenating the state and noise components together as $X^a(k) = [X^T(k) \quad w^T(k) \quad n^T(k)]^T$. The implementation for the SPKF is as follows [6]:

- *Initialize with:*

$$\hat{X}^a(0) = E\left[X^a(0)\right] = \left[\hat{X}^T(0) \quad 0 \quad 0\right]^T \tag{12}$$

$$P^a(0) = E\left[\left(X^a(0) - \hat{X}^a(0)\right)\left(X^a(0) - \hat{X}^a(0)\right)^T\right] = \begin{bmatrix} P(0) & 0 & 0 \\ 0 & Q & 0 \\ 0 & 0 & R \end{bmatrix} \tag{13}$$

Iterate for each time step $k \in \{1,...,\infty\}$ via

- *The sigma point calculation:*

$$\chi^a(k-1) = \left[\hat{X}^a(k-1) \quad \hat{X}^a(k-1)+c\sqrt{P^a(k-1)} \quad \hat{X}^a(k-1)-c\sqrt{P^a(k-1)} \right] \qquad (14)$$

- *Time update:*

$$\chi^x(k/k-1) = f\left(\chi^x(k-1), u(k), \chi^w(k-1)\right) \qquad (15)$$

$$\hat{X}(k/k-1) = \sum_{i=0}^{2L} W_m^{(i)} \chi_i^x(k/k-1) \qquad (16)$$

$$P(k/k-1) = \sum_{i=0}^{2L} W_C^{(i)} \left(\chi_i^x(k/k-1)-\hat{X}(k/k-1)\right)\left(\chi_i^x(k/k-1)-\hat{X}(k/k-1)\right)^T \qquad (17)$$

- *Measurement update:*

$$Y(k/k-1) = h\left(\chi^x(k/k-1), \chi^n(k/k-1)\right) \qquad (18)$$

$$\hat{y}(k/k-1) = \sum_{i=0}^{2L} W_m^{(i)} Y_i (k/k-1) \qquad (19)$$

$$S_U = \sum_{i=0}^{2L} W_C^i \left(Y_i(k/k-1)-\hat{y}(k/k-1)\right)\left(Y_i(k/k-1)-\hat{y}(k/k-1)\right)^T \qquad (20)$$

$$C_U = \sum_{i=0}^{2L} W_C^{(i)} \left(\chi^x(k/k-1)-\hat{X}(k/k-1)\right)\left(Y_i (k/k-1)-\hat{y}(k/k-1)\right)^T \qquad (21)$$

$$K = C_U / S_U \qquad (22)$$

$$\hat{X}(k) = \hat{X}(k/k-1)+K\left(y(k)-\hat{y}(k/k-1)\right) \qquad (23)$$

$$P(k) = P(k/k-1)-KS_U K^T \qquad (24)$$

where $c = \sqrt{L+\lambda}$ is a scaling parameter that determines the spread of the sigma-point.

4   Experiments

4.1   Experiment Setup

In our experimental setup, the MTI IMU and the two optical encoders (61C11-01-08-02) were installed on a mobile robot, as shown in Fig. 3. A notebook computer was used to obtain the data from the two sensors via a DSP board. During the experiment, the measurement sampling frequencies for the IMU and the optical encoders were set to 100 Hz and 5 Hz, respectively. The robot was programmed to travel in a straight line in an arbitrary direction for 3.145 meters in 15 seconds.

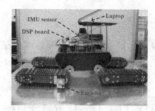

Fig. 3. The experimental setup for a mobile robot with an IMU and two optical encoders

4.2   Experiment Results

The experiment was performed in real-time to test the trajectory estimation of the mobile robot. In general, robot positions are inaccurate due to errors from the encoders, the IMU sensors, and the real-time experimental condition, among other factors. Thus, the SPKF was used to estimate the position of the mobile robot in this localization problem.

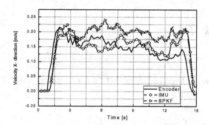

Fig. 4. The velocities from the IMU, the encoder, and the SPKF estimation along the X-direction in the N frame

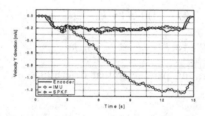

Fig. 5. The velocities from the IMU, the encoder, and the SPKF estimation along the Y-direction in the N frame

Figures 4 and 5 present graphs of the velocities from the encoder, the IMU, and the SPKF estimation along X- and Y-directions in the navigation frame. As these figures illustrate, the estimated velocities from the SPKF are more accurate than the velocities from the individual IMU or the encoder. Note that the velocity is zero at the robot's starting and ending positions (the estimated velocity along the Y-direction, especially).

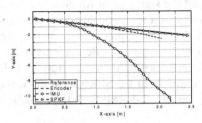

Fig. 6. The trajectories of the encoder, IMU, and SPKF estimation, as compared to a reference line

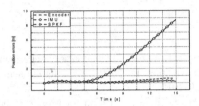

Fig. 7. The position errors of encoder, IMU and SPKF estimation

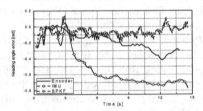

Fig. 8. The heading angle errors of the encoder, IMU, and SPKF estimation

In Fig. 6, the SPKF trajectory remains near the reference line while the IMU trajectory diverges from the reference line. Thus, the position error graphs for the SPKF and the encoder remain near zero, while the position error graph of the IMU diverges rapidly, as shown in Fig. 7. Additionally, the heading angle error of the SPKF is also close to zero, as shown in Fig. 8. Consequently, the root mean square error (RMSE) of the SPKF is the smallest (0.1711[m]) compared to that of the encoder and IMU (0.3121[m], 3.8075[m], respectively).

5   Conclusion

This paper presented an experiment to estimate the given trajectory of a tracked mobile robot using measurements of the IMU and encoder sensors. Error sources such as noise, constant offset, and bias limit the ability of these devices to obtain accurate trajectory estimations. However, based on the IMU kinematic motion and velocity constraints, we implemented the sigma-point Kalman filter in order to systematically estimate the precise trajectory of a four-wheeled, tracked, skid-steered mobile robot. The SPKF was implemented because it produces more accurate position information, is easier to calculate, and requires no analytic derivations or Jacobians. This novel systematic estimation was performed on the commercialized Hazard Escape I mobile robot, and the estimated results from the SPKF were more accurate than the sensor measurements, producing the smallest RMSE and the smallest heading angle error, among other factors.

Acknowledgements. This work was supported by the 2011 Research Fund of University of Ulsan in Korea.

References

1. Weinstein, A.J., Moore, K.L.: Pose Estimation of Ackerman Steering Vehicles for Outdoors Autonomous Navigation. In: 2010 IEEE International Conference on Industrial Technology, pp. 579–584 (2010)
2. Lee, T., Shin, J., Cho, D.: Position Estimation for Mobile Robot Using In-plane 3-axis IMU and Active Beacon. In: IEEE International Symposium on Industrial Electronics, pp. 1956–1961 (2009)
3. Espinosa, F., Santos, C., Marrón-Romera, M., Pizarro, D., Valdés, F., Dongil, F.J.: Odometry and Laser Scanner Fusion Based on a Discrete Extended Kalman Filter for Robotic Platooning Guidance. Sensor 9, 8339–8357 (2011)
4. Tamas, L., Lazea, G., Robotin, R., Marcu, C., Herle, S., Szekely, Z.: State Estimation Based on Kalman Filtering Techniques in Navigation. In: IEEE International Conference on Automation, Quality and Testing, Robotics, pp. 147–152 (2008)
5. Suliman, C., Moldoveanu, F.: Unscented Kalman Filter Position Estimation for An Autonomous Mobile Robot. Bulletin of the Transilvania University of Brasov. 3 (2010)
6. van de Merwe, R., Wan, E.A.: Sigma-Point Kalman Filters for Integrated Navigation. Institute of Navigation Annual Technical Meeting (2004)
7. Flenniken IV, W.S., Wall, J.H., Bevly, D.M.: Characterization of Various IMU Error Sources and the Effect on Navigation Performance. In: Proceedings of the 18th International Technical Meeting of the Satellite Division of The Institute of Navigation, pp. 967–978 (2005)
8. Yi, J., Wang, H., Zhang, J., Song, D., Jayasuriya, S., Liu, J.: Kinematic Modeling and Analysis of Skid-Steered Mobile Robots With Ap-plications to Low-Cost Inertial-Measurement-Unit-Based Motion estimation. IEEE Transactions on Robotics 25, 1087–1097 (2009)

Development of a Mobile Museum Guide Robot That Can Configure Spatial Formation with Visitors

Mohammad Abu Yousuf[1], Yoshinori Kobayashi[1,2], Yoshinori Kuno[1],
Akiko Yamazaki[3], and Keiichi Yamazaki[1]

[1] Saitama University, Saitama, Japan
[2] Japan Science and Technology Agency (JST), PRESTO, Kawaguchi, Japan
[3] Tokyo University of Technology, Hachioji, Japan
{yousuf,yosinori,kuno}@cv.ics.saitama-u.ac.jp,
ayamazaki@media.teu.ac.jp, BYI06561@nifty.com

Abstract. Museum guide robot is expected to establish a proper spatial formation known as "F-formation" with the visitors before starting its explanation of
any exhibit. This paper presents a model for a mobile museum guide robot that
can establish an F-formation appropriately and can employ "pause and restart"
depending on the situation. We began by observing and videotaping scenes of
actual museum galleries where human guides explain exhibits to visitors. Based
on the analysis of the video, we developed a mobile robot system that can guide
multiple visitors inside the gallery from one exhibit to another. The robot has
the capability to establish the F-formation at the beginning of explanation after
arriving near to any exhibit. The robot can also implement "pause and restart"
depending on the situation at certain moment in its talk to first elicit the visitor's attention towards the robot. Experimental results suggest the efficacy of
our proposed model.

Keywords: F-formation, human-robot interaction, mobile guide robot, pause
and restart.

1 Introduction

Museum guide robots are one of the important application areas considered in research in human robot interaction [1-2]. In the case of a museum guide robot, both
robot and visitor should have equal, direct, and exclusive access to the space where
the target exhibit exists. In our previous research, we investigated some non-verbal
actions of museum guide robots during explanation of exhibits [3]. In our previous
research, however, we considered only the situations that occur after the robot starts
its explanation. We assumed that the visitors were already in the proper position to
enjoy the explanation. However, guide robots need to bring about such situations in
order to really work as guides effectively. This paper, therefore, is concerned with the
issue of initiating explanation in a natural human-robot interaction.

D.-S. Huang et al. (Eds.): ICIC 2012, LNCS 7389, pp. 423–432, 2012.
© Springer-Verlag Berlin Heidelberg 2012

Kendon's analysis on spatial formation known as F-formation explains that "An F-formation arises whenever two or more people sustain a spatial and orientational relationship in which the space between them is one to which they have equal, direct, and exclusive access [4]." In human-human interaction, when people group themselves, they automatically form an F-formation. In human-robot interaction, if a proper F-formation is not formed during the initiation to talk, the robot should do some actions to establish an F-formation. Goodwin [5] also showed some systematic ways in which speakers obtain the attention of a recipient. When a current speaker begins to utter a sentence and if s/he finds recipients are not gazing towards him or her, the speaker can use "restart" and/or "pause" in the delivery of the utterance.

There have been several studies about robot's controlling the position of people during interaction with them. Hüttenrauch et al. found that people follow the F-formation in their interacting with robots [6]. Although their observations revealed that humans formulate O-space toward a robot, they did not study how a robot should formulate its spatial relationship. Tasaki et al. utilized Hall's proximity theory [7] to determine a combination of sensors and robot behavior along with the current distance from the interacting person [8]. Kuzuoka et al. studied the effect of body orientation and gaze in controlling the F-formation [9]. Yamaoka et al. focused on the positions and body orientations to implement their information presenting robot [10]. While this research is impressive, so far very few studies have revealed how a robot should behave to initiate interaction with multiple visitors before starting its explanation. In this paper, we focus on how a guide robot establishes an appropriate F-formation to initiate interaction with multiple visitors. We also examine how "pause and restart" plays an important role to attract the attention of visitors.

2   Interaction between Human Guide and Visitors at Museum

In order to develop our museum guide robot, we analyzed some video footage recorded at the National Japanese American Museum in Los Angeles. The videos record human guides explaining exhibits to a small group of visitors. Transcript (1) and Figs. 1(a) to 1(e) convey a typical fragment of such human guide behavior at the museum. In transcript (1), MG moves to another exhibit (Fig. 1(a)). MG clears his throat in the 3rd line. "Clearing throat" is one kind of indication that he is waiting for the visitors to come to the exhibit. In line 4, MG employs a pause of 5.0 seconds. By this time the visitors are following MG (Figs. 1(b), 1(c)). MG deploys "restart" and a "pause" of 0.2 seconds (6th line) while asking an involvement question to the visitors ("Have you all heard picture bri:des?"). At line 8, some visitors move their head vertically in response to the MG's question. From lines 10-13, V1, V2, and V3 offer verbal responses to MG's question. In lines 14 and 15, MG asks the visitors to come closer (Fig. 1(d)), an indication that they should form a proper F-formation (Fig. 1(e)).

From this analysis, two interaction patterns are derived to design our robot system:

1) Robots should have the capability to establish a proper F-formation.
2) Robots should deploy "pause and restart" depending on the situation.

Transcript (1) (Picture bride) Data:Collected at National Japanese American Museum

MG: Guide, V1,V2, V3: Visitors

		Symbols Used in the Transcript	
1	MG: Okay, let's come over here	(( ))	Vocalizations which are difficult to convey in text
2	((Guide moves to another exhibit))		
3	MG: ((clears his throat))	(5.0) (2.0)	Pauses are timed in seconds
4	(5.0) ((people follow MG))	(0.8) (0.2)	and inserted within parentheses
5	MG: Okay, so you: ah: heard of uh picture bri::des?		
6	<Have you all heard picture bri:des? (0.2) See the	,	Slight rising tone
7	picture up the::re, those are picture bri::des.	:	Stretched sound
8	(2.0) ((some people moves their head vertically))	<	Speeding up the pace of delivery
9	MG: You all heard of- you never heard of picture bri::des,	?	Final rising tone which may(or may not) indicate a question
10	V1: No:::,=		
11	V2: =(Not me.)	-	Short untimed pause within an utterance
12	(0.8)	=	Overlap
13	V3: [Uh-huh.	(did)	Guess at unclear word
14	MG: [Or if you wanna come closer this way so that other (.)	[	Simultaneous utterances
15	people could leave uh: on the other side.s	.	Stopping fall in tone, with some sense of completion

(a) Guide moves to another exhibit	(b) Visitors follow Guide	(c) Visitor moves to her appropriate position	(d) Guide asks visitors to come closer	(e) F-formation is formed (Red circle)

Fig. 1. Human guide and visitors' interaction at an actual museum

3   Mobile Museum Guide Robot System

Based on our findings, we developed a mobile museum guide robot system utilizing a humanoid robot Robovie-R Ver.3 (Overview is shown in Fig. 2). The robot can move via wheels installed on the bottom, and can move its head and arms by controlling its joints. Its head, which incorporates eye cameras and an ear microphone, moves along three axes (Yaw, Roll and Pitch) like a human head. Our system utilizes two general-purpose PCs, connected by a wired network. In our vision system, we incorporate three Logicool USB cameras and two laser range sensors. The three USB cameras are attached to a pole installed on the back of the robot, and can detect and track visitors' faces and their face directions. The two laser range sensors are attached to another long pole which is kept at a fixed position just in front of the experimental area. One of the two laser range sensor detects the ellipsoidal marker which is attached to the robot's body to obtain the position and orientation of the robot, while the other laser range sensor is used to track the position and body orientation of the visitors.

Our system consists of four software units: the face detection & tracking unit, the body tracking unit, the robot's position tracking unit and the robot control unit. During its explanation of exhibits, the robot performs predetermined bodily non-verbal actions, such as facing towards the visitors, gesturing with its hands, and pointing to the exhibits.

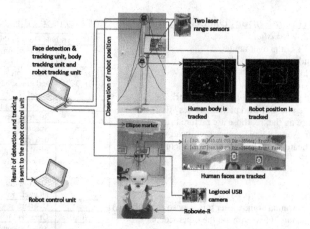

Fig. 2. System overview of our mobile guide robot

4 Proposed Modeling of Interaction

4.1 Model of F-Formation

For establishing a proper F-formation, we consider the following parameters:

- Distance between the robot and the visitors: Ranges from 90cm to 120cm (Fig 3(a)).
- Distance between the robot and the exhibits: About 110 cm, fixed in all cases for all exhibits (Fig 3(a)).
- Visitor's body orientation: Should be in the direction between the robot and the exhibit (Fig 3(b)).
- Visitor's face direction: Should be towards the robot or the exhibit.
- Robot's body orientation: The robot turns its body 30^0 towards the exhibit to explain the exhibit (Fig 3(c)).
- Robot's field of view (FOV): Set at a limit of 150^0 (Fig 3(c)).

After arriving at an appropriate position for explanation of the exhibit, the robot should examine whether or not a proper F-formation has been established.

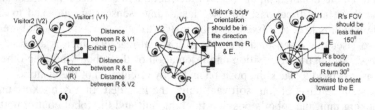

Fig. 3. Standing position of the robot and the visitors to establish an F-formation

4.2   Model of "Pause and Restart" to Achieve Mutual Gaze

The robot observes the visitors' face direction and body orientation before beginning its explanation of any exhibit. If, at the beginning of the speaker's (robot) turn, the face direction is detected as not directed toward the robot or exhibit, or body orientation is detected as not in the direction between the robot and the exhibit, the robot employs the "pause and restart" strategy. The format of "pause and restart" which is implemented in our system is as follows:

[Beginning] + [pause] + [new Beginning]
..............X_____

In this format, the solid line below the sentence structure indicates that the recipient is gazing toward the speaker, no line indicates that the recipient is gazing elsewhere, and the 'X' marks the point at which the recipient's gaze reaches the speaker. The dotted line represents the time required for the recipient to move his/her gaze from some other position to the speaker. In order to implement "pause and restart," we consider the first sentence of each script explaining the exhibits, as in the following:

Script 1: First sentence for explaining the first exhibit, "Te Nave Nave Fenua"
Robot: This is a- (2.0) This is a famous work of Gauguin.

In our proposed model, the guide robot deploys "pause and restart" depending on the situation. In all scripts, a restart with a preceding pause of 2 seconds is used.

5   Experiment with Humanoid Robot

To test the robot's effectiveness, an experiment was performed in the laboratory. Four paintings were placed in the area as shown in Fig. 4.

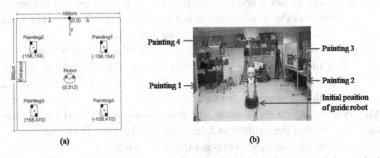

(a)   (b)

Fig. 4. (a) Overview of the experimental area. (b) Mobile guide robot and four paintings

A total of 16 graduate students (8 groups with 2 members in each group) participated in the experiment. Among the 8 groups, Group A (groups no. 1, 3, 5, and 7) participated in the sessions where the robot explained the first two paintings as the proposed guide robot system outlined above, and the remaining two paintings as a robot not equipped with the capacity to form an F-formation. On the other hand, Group B (groups no. 2, 4, 6, and 8) participated in the sessions where the robot did this in reverse order. Participants were not informed of which robot was which.

Initially, the robot with the proposed system waits in the middle of the experimental area. A schematic diagram of the main tasks performed by the robot is given in Fig. 5. The order of these main tasks is as follows:

1) When the robot finds visitors coming into its immediate vicinity, it says, "May I explain these paintings to you?" If the visitors' gaze turns toward the robot's direction for three seconds, the robot system considers the visitors to be highly interested in the exhibits (Fig. 5(a)).

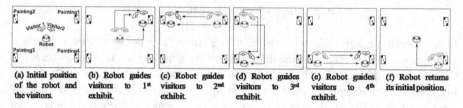

(a) Initial position of the robot and the visitors.　(b) Robot guides visitors to 1st exhibit.　(c) Robot guides visitors to 2nd exhibit.　(d) Robot guides visitors to 3rd exhibit.　(e) Robot guides visitors to 4th exhibit.　(f) Robot returns its initial position.

Fig. 5. Schematic diagram of main tasks

2) The robot then guides the visitors to the first exhibit (Fig. 5(b)).
3) After arriving at the predefined position near the first exhibit, the robot follows the following steps to establish a proper F-formation.
    i) First, the robot verifies the distance between itself and the visitors.
    ii) If the visitors are not within range, the robot turns its head towards the visitor and says to them, "Please come closer" or "Please move back a little" depending on the situation.
    iii) Next, the robot turns 30^0 clockwise to orient towards the first exhibit.
    iv) Then, the robot verifies the body orientation of the visitors.
    v) If the visitor's body orientation is not in the direction between the robot and the exhibit, the robot turns its head towards the visitor and starts its explanation using "pause and restart".
    vi) Next, the robot verifies the direction of the visitors' faces.
    vii) If the face direction is towards the robot or the exhibit, the robot begins to explain the first exhibit. If not, the robot starts its explanation with a "pause and restart".
4) After completing its explanation, the robot moves to the next exhibit and at the same time invites the visitors to follow along (Fig. 5(c), 5(d), 5(e)).
5) The robot repeats task (3) to explain the next exhibit.
6) Finally after explaining all four exhibits, the robot returns to its initial position and waits for more visitors to arrive (Fig. 5(f)).

In the experiment, the robot based on our proposed model was compared with a robot that did not employ the proposed model.

    a) Proposed Robot: Robot behaves based on the model outlined in this paper.
    b) Conventional Robot: Robot begins its explanation after finding the faces of visitors. It does not care whether or not a proper F-formation is formed, nor does it utilize the 'pause and restart' strategy. It explains the exhibits with the same preprogrammed nonverbal behaviors as the proposed robot.

We videotaped all sessions. In addition, we recorded all laser range finder and camera data so that we could obtain the exact motions of the robot and the participants for later analysis. After the experiments, we asked participants to subjectively rate the robot's effectiveness on a seven-point Likert scale, with the range: 1-very ineffective, 2-ineffective, 3-somewhat ineffective, 4-undecided, 5-somewhat effective, 6-effective, 7-very effective. The questionnaire items were as follows:

1) Did you think that the robot attended to you adequately during explanation?
2) Did you think that the robot was able to attract your attention to listen to its explanation?
3) Overall evaluation about the robot.

6   Experimental Results

We have examined the experimental results from the following three viewpoints.

1) Autonomous capability of the robot: Can the robot correctly judge the situation and behave properly according to our proposed model? (Sections 6.1, 6.2 & 6.3)
2) Effectiveness of the robot actions: Can the robot's actions make the participants form a proper F-formation and attract their attention? (Sections 6.1, 6.2, & 6.3)
3) Subjective evaluation: Do the participants prefer proposed robot? (Section 6.4)

6.1   Control of Visitors' Standing Position

We recorded the sensor data from all sessions in the experiment for analysis. As covered in section 5, the robot explained the first two paintings as the proposed robot and the remaining two as the conventional robot to Group A (groups no. 1, 3, 5, and 7) and in the reverse order to Group B (groups no. 2, 4, 6, and 8). We found from the sensor data that there were 11 cases (5 cases for Group A, 6 cases for Group B) where some visitors were not in proper position. The robot took the initiative to control visitors' standing position in all 11 cases. Thus, the success rate of robot's decisions was 100%.

In the 5 cases for Group A, out of 10 participants 6 were out of proper F-formation range, and after the robot's action 5 of these 6 moved inside the range. In the 6 cases for Group B, out of 12 participants 9 were out of proper range, and after the robot's action, 6 of these 9 moved inside the range. The total success rate was thus 73% (the robot corrected 11 out of 15).

On the other hand, when the robot explained the paintings as a conventional robot, we found from the sensor data that 17 participants in both groups (7 in Group A, 10 in Group B) were out of range. Of the total of 17 participants, only 6 moved inside the range to form a proper F-formation when the robot began its explanation. The "success rate" for the conventional robot was thus 35% (6 out of 17, although here we only use the term "success rate" for convenience since the robot did not try to correct the visitors).

6.2   Control of Visitors' Body Orientation

From the recorded sensor data, we counted the number of visitors who successfully changed their body orientation to a direction towards (meaning, oriented to the space between) the robot and the painting after the robot's employed a "pause and restart" at the beginning of its explanation. In 2 out of 8 cases in Group A and 3 out of 8 cases in Group B, the robot noticed that some visitors' body orientations were not in the direction between the robot and the painting. The robot employed "pause and restart" in all 5 cases, meaning its success rate at deploying the strategy was 100%.

In 2 cases among those of Group A, the robot noticed 2 out of 4 participants' orientations were not in the direction between the robot and the painting, while in 3 cases among those of Group B it noticed 3 out of the 6 participants were incorrectly orientated. After the robot's actions, both of the incorrectly oriented participants in Group A and 2 of the 3 in Group B changed their body orientations appropriately. The total success rate was therefore 80% (the robot corrected 4 out of 5).

On the other hand, when the robot explained the paintings as a conventional robot would, we found from the recorded sensor data that 3 out of 16 participants in Group A and 1 out of the 16 participants in Group B were not oriented towards the robot and the painting. Out of a total 4 incorrectly orientated participants, 2 changed their body orientations towards the robot and the painting just after the robot started its explanation. The conventional robot thus scored a "success rate" of 50% (2 out of 4).

Fig. 6 shows a typical example of a visitor changing his body orientation in the direction between the robot and the painting after the robot's use of "pause and restart." In Fig. 6(a), the robot noticed that the visitor's (V1) body orientation was not in the direction between the robot and the painting. It then turned its head towards V1 and employed "pause and restart" to attract the visitors' attention (Fig. 6(b)). V1 turned his body towards the robot and mutual gaze was established (Fig. 6(c)). V1 fully turned his body towards the robot (Fig. 6(d)).

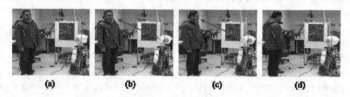

Fig. 6. A participant changing his body orientation after the robot's use of "pause and restart"

6.3   Control of Visitors' Face Direction

Here we examined those visitors whose body orientations were in the direction between the robot and the painting but whose faces were directed elsewhere. We counted the number of such visitors from the recorded sensor data. When performing as the proposed model, the robot noticed 2 cases out of 8 in Group A and 4 cases out of 8 in Group B where some visitors' face directions were not towards the robot or the painting. The robot employed "pause and restart" in all 6 cases, so its deployment success rate was 100%.

In the 2 cases in Group A, the robot noticed 3 out of 4 participants' face directions were not towards itself or the painting, while in the 4 cases in Group B it noticed 4 out of 8 participants' face were directed elsewhere. After the robot's actions, 2 of the

3 participants with faces directed elsewhere in Group A and 3 out of the 4 in Group B changed their face directions towards the robot or the painting. The total success rate was therefore 71% (the robot corrected 5 of the 7 participants).

On the other hand, when the robot explained the paintings as a conventional robot, 2 out of 16 participants in Group A and 3 out of 16 participants in Group B had faces not directed towards the robot or the painting. Out of the total of 5 whose face directions were not towards the robot or the paintings, only 1 shifted his face appropriately just after the robot started its explanation for a "success rate" of 20% (1 out of 5).

Although the number of cases is small and we cannot yet draw any definite conclusion, the results suggest that the "pause and restart" strategy can have a logical effect.

6.4   Subjective Evaluation

We conducted a subjective evaluation of the experiment among the participants, the results of which are as follows. For the question "Did you think that the robot attended to you adequately during explanation?" (Fig. 7(a)), repeated measures of ANOVA revealed a significant difference between the two conditions $(F(1,15)=15.08, p=0.0014)$. The same was true of the question "Did you think that the robot was able to attract your attention to listen to its explanation?" (Fig. 7(b)), where repeated measures of ANOVA also showed a significant difference between the two conditions $(F(1,15)=8.59, p=0.0103)$. Finally the participants' overall evaluation (Fig. 7(c)), here too repeated measures of ANOVA showed a significant difference between the two conditions $(F(1,15)=6.03, p=0.0266)$.

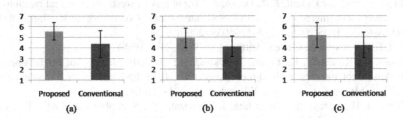

Fig. 7. Results of subjective evaluation

7   Discussion and Conclusion

In this paper, we proposed a museum guide robot that can establish a proper F-formation and employ the "pause and restart" depending on the situation. Based on our analysis of videos collected at real museums, we have found that a human guide and visitors always create an F-formation and moreover that if the visitors are not ready to listen to the presentation then the human guide may repeat sentences and/or deploy pauses during his/her explanation. Based on this, we developed a museum guide robot system that is able to move from one exhibit to another in a museum gallery, to appropriately formulate an F-formation system, and also to employ the "pause and restart" if necessary. We verified the effectiveness of our system in experiments using human participants. In future work, we plan to conduct experiments in an actual museum to further confirm the effectiveness of our robot system.

Acknowledgement. This work was supported in part by the Ministry of Education, Culture, Sports, Science and Technology under the Grant-in-Aid for Scientific Research (KAKENHI 21300316,23252001) and JST, CREST.

References

1. Nieuwenhuisen, M., Gaspers, J., Tischler, O., Behnke, S.: Intuitive Multimodal Interaction and Predictable Behavior for The Museum Tour Guide Robot Robotinho. In: Proceedings of the 10th IEEE-RAS International Conference on Humanoid Robots (Humanoids), pp. 653–658 (2010)
2. Faber, F., Bennewitz, M., Eppner, C., Görög, A., Gonsior, C., Joho, D., Schreiber, M., Behnke, S.: The Humanoid Museum Tour Guide Robotinho. In: Proceedings of the IEEE International Symposium on Robot and Human Interactive Communication (RO-MAN), pp. 891–896 (2009)
3. Yamazaki, A., Yamazaki, K., Kuno, Y., Burdelski, M., Kawashima, M., Kuzuoka, H.: Precision Timing in Human-robot Interaction: Coordination of Head Movement and Utterance. In: Proceedings of the CHI 2008, pp. 131–140. ACM Press (2008)
4. Kendon, A.: Conducting Interaction–Patterns of Behavior in Focused Encounters. Cambridge University Press (1990)
5. Goodwin, C.: Restarts, Pauses, and The Achievement of a State of Mutual Gaze at Turn Beginning. Sociological Inquiry 50(3-4), 272–302 (1980)
6. Hüttenrauch, H., Eklundh, K.S., Green, A., Topp, E.A.: Investigating spatial relationships in human-robot interactions. In: IEEE/RSJ International Conference on Intelligent Robots and Systems (IROS 2006), pp. 5052–5059 (2006)
7. Hall, E.T.: Proxemics. Current Anthopology 9, 83–108 (1968)
8. Tasaki, T., Komatani, K., Ogata, T., Okuno, H.: Spatially Mapping of Friendli-ness for Human-Robot Interaction. In: Proceedings of the IEEE/RSJ International Conference on Intelligent Robots and Systems, Edmonton, pp. 521–526 (August 2005)
9. Kuzuoka, H., Suzuki, Y., Yamashita, J., Yamazaki, K.: Reconfiguring Spatial Formation Arrangement by Robot Body Orientation. In: ACM/IEEE International Conference on Human-Robot Interaction (HRI 2010), pp. 285–292 (2010)
10. Yamaoka, F., Kanda, T., Ishiguro, H., Hagita, N.: How Close? A Model of Proximity Control for Information-presenting Robots. In: Human-Robot Interaction (HRI 2008), pp. 137–144 (2008)

A Novel Image Matting Approach Based on Naive Bayes Classifier

Zhanpeng Zhang[1,2], Qingsong Zhu[1,*], and Yaoqin Xie[1]

[1] Shenzhen Institutes of Advanced Technology, Chinese Academy of Sciences,
Shenzhen, 518055, China
[2] Sun Yat-Sen University, Guangzhou, 510006, China
{zp.zhang,qs.zhu,yq.xie}@siat.ac.cn

Abstract. Image matting is a fundamental technique used in many image and video applications. It aims to softly extract foreground from the image accurately. In this paper, we propose a new matting approach based on naive Bayes classifier to produce matting results with higher accuracy. Spatially-varying probabilistic models for the classifier are established. Confidence values are defined to make better use of the classification results. The results are then refined and combined with closed-form matting to obtain the final alpha matte. We conduct qualitative and quantitative evaluations. Results show that our method outperforms many recent algorithms.

Keywords: image matting, naive bayes classifier, foreground extraction, image segmentation.

1 Introduction

Image matting refers to the problem of extracting the opacity mask (typically called alpha matte) of the foreground, as well as the foreground and background color, from the target image. It is one of the fundamental techniques used in many image and video editing tasks. Specifically, for a pixel i with color I_i in the image, it can be model as a liner combination of the foreground color F_i and background color B_i as Eq.1.

$$I_i = \alpha_i F_i + (1 - \alpha_i) B_i \qquad (1)$$

Eq.1 is under constrained since α_i, F_i and B_i on the right-hand side are all unknown, making it a significant challenge for computer vision.

Many recent approaches require additional information from user input to build more constrains to solve the ill-posed problem. Trimaps [1] and scribbles [2] are the two most common methods, labeling some pixels which are definite foreground or background (like Fig.1b), with corresponding alpha value to be 1 or 0 respectively.

Existing matting algorithms can be classified into sampling-based or propagation-based. Sampling-based algorithms explicitly estimated the triplet (α, F, B) for every unlabeled pixel by analyzing nearby labeled pixels. These algorithms usually fit a parametric mode to the color distributions of the samples, like Bayesian Matting [1], which models the colors of the samples by oriented Gaussian distributions. However, the model assumptions may fail in some scenes. Recently, many

D.-S. Huang et al. (Eds.): ICIC 2012, LNCS 7389, pp. 433–441, 2012.
© Springer-Verlag Berlin Heidelberg 2012

non-parametric algorithms are employed, estimating the triplet directly with the sample colors under the liner model defined by Eq.1. For example, the sampling algorithms in robust matting [3] and many later approaches [4,5] assign a confidence value to a pair of foreground and background sample and choose the samples with high confidence. The confidence values are generally measured by how well the sample explains the unlabeled pixel in the linear model (Eq.1). These algorithms perform well when the true foreground and background colors are in the sample set.

For propagation-based algorithms, some affinities assumptions are made in order to derivate a constrained objective function. Poisson matting [6] deduces that the alpha matte gradient is proportional to the image gradient under the smoothness assumption. A closed form solution proposed by Levin obtains a quadratic cost function in α analytically under the color line model [2]. This approach is called as closed-form matting and has been applied in many other approaches, drawing extensive studies [7].

However, due to the color line assumption in closed-form matting, it may fail in regions with long and thin structures or holes. Many other propagation-based approaches may also fail in these situations. This is because only neighboring pixels are used in the modeling process, lacking information from further regions. Some approaches [3,4] use color information from nearby samples to fill the gap. However, gathering appropriate samples is still a challenging problem, and additional assumptions may be introduced, causing other types of artifacts.

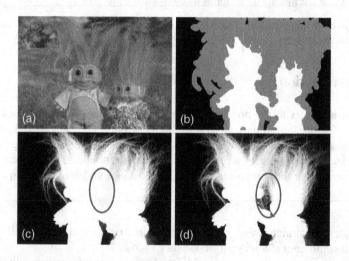

Fig. 1. (a) Original image. (b) Trimap. (c) Closed-form matting [2] result. (d) Our result (Notice the two red ellipses. The detail is missed in closed-form matting while not in the result of our method.).

In this paper, we employ naive Bayes classifier to identify the foreground and background pixels. The classification results are then refined and combined with closed-form matting. As our approach can enhance the discrimination between the foreground and the background, results show that our approach outperforms many recent algorithms and produces better alpha mattes for the images in which the closed-form matting usually fails (like Fig.1).

2   Closed-Form Matting

Levin *et al.* proposed a closed-form solution for image matting in [2]. The key assumption in closed-form matting is color line model. It assumes that in a small window, the foreground color and background color of each pixel i can be formulated as linear mixtures of two constant colors, respectively. In other words, the foreground or background colors for all pixels in a small window lie on a single line in the RGB color space.

Based on the color line model, the alpha value for pixel i in a small window w can be expressed as a liner transform of the pixel color:

$$\alpha_i = \sum_c a^c I_i^c + b, \forall i \in w \tag{2}$$

where c denotes the color channel for RGB color space. a and b are constant in w.

To derivate an alpha matte obeying the color line model (or Eq.2), the algorithm aims to find the optimal a, b and α which minimize the cost function:

$$J(\alpha, a, b) = \sum_{j \in I} \sum_{i \in w_j} (\alpha_i - \sum_c a_j^c I_i^c - b_j)^2 + \varepsilon \sum_c a_j^{c^2}) \tag{3}$$

Here wj is a small window centered at pixel j. ε is a regularization parameter while smoothing the alpha matte.

A quadratic function of α can be obtained by minimizing the cost function. Also, parameters a, b can be eliminated in the deducing process. The quadratic function is as follows:

$$J(\alpha, a, b) = \alpha^T L \alpha \tag{4}$$

pecifically, α is an N×1 vector, where N is the number of unlabeled pixels. L is typically called matting Laplacian matrix, as one of the most significant contributions of closed-form matting, drawing many further studies and applications. Formally, L is an N×N matrix with (i, j)-th element as:

$$\sum_{k|(i,j) \in w_k} (\delta_{ij} - \frac{1}{|w_k|}(1 + (I_i - \mu_k)(\Sigma_k + \frac{\varepsilon}{|w_k|} I_3)^{-1}(I_j - \mu_k))) \tag{5}$$

where δ_{ij} is the Kronecker delta. $|w_k|$ is the number of pixels in this window. Σ_k is a 3×3 covariance matrix, μ_k is a 3×1 mean vector of the colors in a window w_k, and I_3 is the 3×3 identity matrix.

Combined with constrains provided by the user (trimaps or scribbles), the objective function can be defined as:

$$J(\alpha) = \alpha^T L \alpha + \lambda (\alpha - \beta)^T D(\alpha - \beta) \tag{6}$$

Here, D is an N×N diagonal matrix whose elements are 1 for labeled pixels and 0 otherwise. β denotes the alpha value for the labeled pixels in the trimap (1 for foreground and 0 for background). λ is weighted parameter with a relatively large number (like 100). In this paper, we mainly talk about the alpha value since the foreground F and background B can be obtained easier with the estimated α (like [2]).

. Closed-form matting works well when the local region fits the color line model. To ensure this assumption holds and reduce computation cost, the windows size in Eq.5 is usually small (3×3 in Levin's implementation). That means the "propagation step" is relatively small implicitly. Over-smoothing may happen in regions with thin structures or small gaps as Fig.1c. Details will be lost in these situations. He et al. [7] improves it by introducing adaptively window sizes in different regions of the image. It is shown that larger window size can improve the matting result since large window may cover disconnected regions of the foreground/background. However, with larger window size, it is more like to break the color line model. Therefore, it is still hard to decide the appropriate windows size.

3   Matting with Naive Bayes Classifier

Different from the previous approaches, we employ naive Bayes classifier to decide whether a pixel belong to the foreground or background. The result is then "softened" by a sigmoid function. Confidence values are also computed for every alpha value. The results and the confidence values are then combined with closed-form matting, providing more accurate alpha matte.

3.1   Naive Bayes Classifier

Naive Bayes classifier [8] is a simple probabilistic classifier based on Bayes' theorem, assuming that features in a class are independent with each other. For class variable C with n features in the model, using Bayes' theorem, the probability that an instance is in class c is:

$$p(c \mid k_1, k_2 ... k_n) = \frac{p(c) p(k_1, k_2 ... k_n \mid c)}{p(k_1, k_2 ... k_n)} \tag{7}$$

where ki is the instance's value for feature i. With the assumption that every feature is independent, classification can be done by selecting the highest posterior of the classification variable with the following function:

$$\arg \max_c p(C = c) \prod_{i=1}^{n} p(K_i = k_i \mid C = c) \tag{8}$$

3.2   Classification Process

Features Selection. Color attribute is the most straight forward feature reflecting a pixel's characteristics. To provide more information of the region texture, colors of the 4-neighbor pixels are also selected. That means, for a pixel i, its feature vector $k = \{g_1, g_2, g_3, g_4, g_5\}$ where g_i is the color vector of a pixel. In our implementation, we use CIELAB color space (with every channel ranges from 0 to 255), so a 15-dimensional vector is selected for a pixel.

Classifier Parameters Estimation. We do not apply a uniform classifier probabilistic model for all the unlabeled pixels in the image. Instead, the probabilistic model is spatially-varying. Specifically, for an unlabeled pixel i, we collect other unlabeled pixels with spatial distance less than r (30 in our implementation). Pixels from this unlabeled region share a same probabilistic model. Pixels labeled by the users are selected as samples to estimate the parameters. We expect to obtain both local foreground and background characteristics, so samples are selected according to their spatial distance to the current unlabeled region. We expand form the border of the unlabeled region and collect foreground and background samples, until the numbers of the foreground and background samples are larger than that of unlabeled region, respectively. To make the features independent to some extent so as to satisfy the naive Bayes' assumption better, before parameters estimation, we apply PCA [9] to reduce the dimension of the features vector for the samples and unlabeled pixels.

Gaussian distribution is employed to model the probability of each feature of the two classes (foreground and background). The mean and variance can be computed with the collected samples. The class' prior ($p(c)$ in Eq.7) is calculated by assuming that the foreground and background are equiprobable. That means $p(c) = 0.5$ for both the two classes. With the computed parameters, unlabeled pixels can be classified as Eq.8.

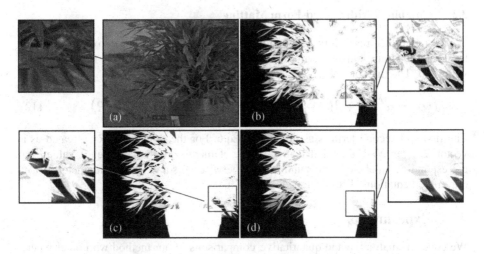

Fig. 2. (a) Original image (b) Refined result (c) Result after combined with closed-form matting (d) Closed-form matting only. Parts of the images are zoomed in for clearer distinction.

Results Refinement. With the Gaussian distribution and strong independence assumptions in the classifier, as well as the binary classification result, errors are unavoidable. To evaluate the results from the classifier, pixel-wise confidence values are computed. Also, the binary results are "softened" with a sigmoid function so as obtain values ranging from 0 to 1 indicating the opacity mask.

Instinctively, if the unlabeled pixels are more similar to the samples used to estimate the classifier, more accurate results can be obtained. To measure the similarity, we calculate D_s for every unlabeled pixel. D_s is defined as follows:

$$D_s = \min E(i, j), j \in S \qquad (9)$$

where $E(\cdot)$ denotes the Euclidean distance in the original feature space. S is the collected sample set. The confidence value is:

$$G(i) = \exp\{-\rho D_s(i)\} \qquad (10)$$

And the binary results are "softened" with a sigmoid function as:

$$R(i) = \frac{1}{1 + \exp\{-\vartheta C(i) / D_s(i)\}} \qquad (11)$$

$C(i)$ is the result of the classifier for pixel i, -1 and 1 for background and foreground, respectively. ρ and θ in the above equations are normalize parameters. In our implementation, we set $\rho = 0.5$ and $\theta = 20$. The result R is like Fig.2b. In practice, the target foreground in matting is usually connected, not several separated components. We select connected components with area less than 1% of the area of the largest one, and set the confidence values of these components to be zero.

3.3   Combined with Closed-Form Matting

We use the refined result R in Eq.11 and its corresponding confidence to define the data term as a quadratic function with the minimum at R. The data term is then combined with Eq.6. The final objective function is:

$$J(\alpha) = \alpha^T L \alpha + \lambda(\alpha - \beta)^T D(\alpha - \beta) + \gamma(\alpha - R)^T E(\alpha - R) \qquad (12)$$

The first and second term is as defined in Eq.6. For the third term, R is treated as a vector. E is an $N \times N$ diagonal matrix whose elements are the confidence values for corresponding unlabeled pixels and zeros otherwise. γ is a weighted parameter (0.1 in our implementation). The combined result is like Fig. 2c.

4   Experiments

We conduct qualitative and quantitative comparisons of our method with other recent related matting algorithms, including closed-form matting [2], learning based matting [10], large kernel matting [7] and robust matting [3]. Like closed-form matting, learning based matting also used a small window, but it employs learning techniques instead of the color line assumption. Large kernel matting improves the efficiency of closed-form matting by using larger window size. Robust matting combines the sampling-based and propagation-based algorithms, similar to our approach to some extent.

Qualitative Comparisons. We compare the result of our method and closed-form matting visually in Fig. 3a. The images and ground truth are provided by [11]. We can see that closed-form matting may fail in the regions with gaps. These regions are often

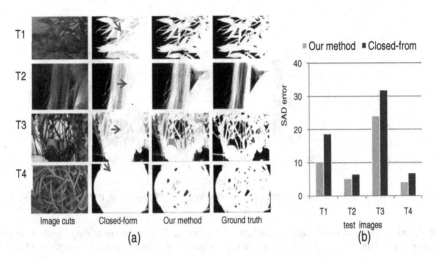

Fig. 3. (a) Qualitative comparison between closed-form matting and our method with 4 test images form [11]. Only parts of the images are shown for clearer distinction. The red arrows show the regions where closed-form matting fails while our method provides better results. (b) Quantitative comparison between our method and closed-form matting according to the SAD (sum of absolute difference) error.

recognized as definitely foreground and the details are missed. With the classification process and samples from further regions, our method can get background samples with which to indentify the background pixels and avoid this situation to some extent. However, compared to the ground truth, our method still has some artifacts. This is mainly because of the error in the classification and results refinement process. The sigmoid function simply based on the Euclidean distance in the feature space is not always effective.

Quantitative Comparisons. Fig.3b shows the quantitative comparison between our method and closed-form matting. We can see that our method can provide results with less SAD (sum of absolute difference) error. To conduct more comprehensive evaluation, we also use the matting benchmark of [11], with 8 test images and 3 different trimaps (with different sparsity level) for each of them. The average SAD errors of each method for the 3 different types of trimaps are presented in Fig.4. The comparison shows that our method is performing the best. Our method combines nearby samples and the smooth assumption for local region, providing better results. Compared to robust matting, our method does not need to find the true foreground and background color. So the accuracy can be higher in regions with color ambiguity.

Memory and Computation Cost. The memory cost in our classification process is relative small since we just establish spatially-varying probabilistic model. That means the sample size used for parameters estimation is not large (less than 3000 pixels for foreground and background respectively). We implement our algorithm in Matlab, and run it on a 3.0 GHz CPU. The classification process typically takes 20 seconds for an 800×600 image, varying with the size N of the unlabeled region in the trimap. The running time can be further reduced with parallel implementation (like GPU), since

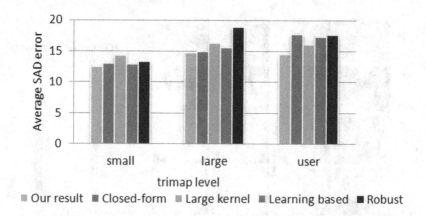

Fig. 4. Quantitative comparisons on average SAD error of alpha value in trimaps with different sparsity level (small, large, user). For the details of the sparsity level categories, refer to [11].

classification processes for different regions of pixels are independent. However, the matting Laplacian matrix needs large size of memories with the size of $N \times N$ as described in Section 2. And the time for computing the matting Laplacian matrix and solving Eq.12 in Section 3.2 is about 20 seconds for an 800×600 image, depending on N.

Limitation. Because the classification process is based on the color information, our method may fail in complex scenes or regions with foreground and background colors overlapping. Also, the naive Bayes classifier is based on a strong assumption and we used Gaussian distribution to model the probability. In some situations, these assumptions may not hold.

5   Conclusion

In this paper, a new matting approach based on naive Bayes classifier is proposed and evaluated. The binary classification results are "softened" with a sigmoid function and confidence values are computed to make better use of the results. The results are then combined with close-form matting to obtain the final alpha matte. Quantitative and qualitative comparisons between our method and other recent algorithms show that our method produce better results. However, color ambiguity or complex scenes are still challenging for our method. Future work may concentrate on providing better classification results and weaker model assumptions.

Acknowledgments. This study has been financed partially by the Projects of National Natural Science Foundation of China (Grant No. 50635030, 60932001, 61072031, 61002040), the National Basic Research (973) Program of China (Sub-grant 6 of Grant No. 2010CB732606) and the Knowledge Innovation Program of the Chinese Academy of Sciences, and was also supported by the China Scholarship Council (CSC) and China Postdoctoral Project.

Reference

1. Chuang, Y.Y., Curless, B., Salesin, D.H., et al.: A Bayesian Approach to Digital Matting. In: Proceedings of IEEE Conference on Computer Vision and Pattern Recognition (CVPR), pp. 264–271 (2001)
2. Anat, L., Dani, L., Yair, W.: A Closed Form Solution to Natural Image Matting. In: Proceedings of IEEE Conference on Computer Vision and Pattern Recognition (CVPR), pp. 61–68 (2006)
3. Wang, J., Cohen, M.F.: Optimized Color Sampling for Robust Matting. In: Proceedings of IEEE Conference on Computer Vision and Pattern Recognition (CVPR), pp. 17–22 (2007)
4. Rhemann, C., Rother, C., Gelautz, M.: Improving Color Modeling for Alpha Matting. In: Proceedings of British Machine Vision Conference, pp. 1155–1164 (2008)
5. Gastal, E.S.L., Oliveira, M.M.: Shared Sampling for Real-time Alpha Matting. Computer Graphics Forum 29, 575–584 (2010)
6. Sun, J., Jia, J., Tang, C.K., Shum, H.Y.: Poisson matting. ACM Transactions on Graphics 23(3), 315–321 (2004)
7. He, K.M., Sun, J., Tang, X.O.: Fast Matting Using Large Kernel Matting Laplacian Matrices. In: Proceedings of IEEE Conference on Computer Vision and Pattern Recognition (CVPR), pp. 2165–2172 (2010)
8. Pedro, D., Michael, P.: On the Optimality of the Simple Bayesian Classifier under Zero-one Loss. Machine Learning 29, 103–130 (1997)
9. Jolliffe, I.T.: Principal Component Analysis. Springer (1986)
10. Zheng, Y.J., Kambhamettu, C.: Learning Based Digital Matting. In: Proceedings of IEEE International Conference on Computer Vision (ICCV), pp. 889–896 (2009)
11. Rhemann, C., Rother, C., Wang, J., et al.: A Perceptually Motivated Online Benchmark for Image Matting. In: Proceedings of IEEE Conference on Computer Vision and Pattern Recognition (CVPR), pp. 1826–1833 (2009)

Detecting Insulators in the Image of Overhead Transmission Lines

Jingjing Zhao, Xingtong Liu, Jixiang Sun, and Lin Lei

School of Electronic Science and Engineering, National University of Defense Technology,
Changsha, Hunan, P.R. China, 410073
zhaojingjing63@gmail.com

Abstract. Detecting and localizing the insulators automatically are very important to intelligent inspection, which are the prerequisites for fault diagnose. A novel method for insulators detection in the image of overhead transmission lines based on lattice detection is presented in this paper. Firstly, low-level visual features of images are analyzed, feature points are generated and grouped by their appearance similarities through mean shift clustering; then a insulator lattice model consistent with the geometric relationship between candidate point clusters is proposed by voting mechanism; subsequently, performing lattice finding using an MRF model, combined with the spatial context information to localize multiple insulators jointly; Finally, extracting the minimum bounding rectangle of the target image. Since the location of each insulator is constrained by its neighbors, each of them provides knowledge about the others, the MRF model is a natural choice for inferring insulators locations while enforcing spatial lattice constraints and image likelihood constraints. The experimental results indicate that the method can effectively detect the deformed insulators of different kinds under complex background.

Keywords: insulator detection, overhead transmission lines, deformed lattice detection, texture features, MRF, intelligent inspection.

1 Introduction

As the key component of Smart Grid, it is imperative to have an effective plan to maintenance the transmission system. With the rapid spread of the transmission network, the traditional method of manual inspection is difficult to meet the growing requirements of the power grid which is rapid developed. By contrast, the technology of helicopter/UAV patrol has already become a trend for its efficiency, reliability and low-cast. Even though we can profit from the advantages of helicopter/UAV patrol, some key problems, such as recognition of potential encroachments and identifying transmission lines that need repair, were also overly dependent on manual identification. This situation led to the Low efficiency of inspection. In order to overcome various defects in the transmission line inspection, it is essential that an intelligent inspection system should be developed. Therefore, the concept of "inspection robot" is born.

The intelligent inspection system takes images of overhead transmission lines, then identifies and locates the transmission lines which are needed repair rapidly by

D.-S. Huang et al. (Eds.): ICIC 2012, LNCS 7389, pp. 442–450, 2012.
© Springer-Verlag Berlin Heidelberg 2012

analyzing these images automatically. Insulator is an important component of over-head transmission lines, which is used to prevent the live parts of transmission lines to form ground channel. The intelligent diagnosis of the insulator can be achieved through the full use of computer vision technology which refer to insulator detection, image segmentation and fault identification from the aerial images, combined with the expert system to generate strategy report. As the first step in intelligent diagnosis, detecting insulators in image accurately and automatically is of great significance, it is the basis for post-processing. Most of insulators detection methods have been pro-posed based on the infrared close range images photographed in laboratory [1], [2]. However, the backgrounds of infrared images are relatively simple, which are differ-ent from the optical images photographed in real world. For optical images, the insu-lators are often "deformed", due to variations in viewing angle, lighting, or partial occlusion, which make the detection of insulators to be difficult, therefore, there is rare research involved about that. In recent years, some scholars have suggested that the color information can be used to extract and recognize insulators [3], [4], but such method can only be applied to the detection of glass insulators, lack of versatility for detecting other types of insulators.

This paper studies on the problem of insulators detection in optical images. Essen-tially, it belongs to the field of "target detection" in computer vision. Target detection, which is widely used for the analysis of complex scenes in intelligent system, is a class of image analysis method that can automatically detect the target appears or not in the image and video based on the geometry and statistical properties of the target. When we take a photograph of the same target with different angle or distance, we may get various understanding with different appearance, but the target itself does not change, changing only happened in the appearance. The human perceptual system has the ability of extracting the invariant features of various appearances, so we can rec-ognize almost everything easily. Extraction of invariant information is one of the characteristics of human perception, but for computers, they should select and extract the invariant features to recognize the patterns, and then detect and identify targets. Therefore, how to find the invariant features of insulators image is the focus of the study in this paper.

As we discussed above, the images taken by the intelligent inspection system al-ways have complex background, we can't use the background modeling method to extract the insulators target. Since the targets are the insulators string, the appearances of the photometrical and geometrical elements remain highly correlated. It can be seen as a kind of near regular texture pattern [5] composed of deformed versions of one or more basic texture elements, thus we consider using characteristics of texture structure to detect the insulators together. Recently, some scholars have found that near-regular textures are not merely random collections of isolated texture elements, but exhibit specific geometric, topological, and statistical regularities and relations. They proposed "deformed lattice detection" theory for automatic detection of de-formed texture patterns in real-world images [6-11].

By studying on the mechanism of "deformed lattice detection" theory, this paper presents a novel method to detect the insulators in the image of overhead transmission lines from the perspective of the characteristics of texture structure. First, low-level visual invariant features of images are analyzed in order to generate high-level insulator lattice model. Second, lattice finding is performed by using an MRF model and multiple insulators are searched for jointly. Lastly, the insulator targets are located by extracting the minimum bounding rectangle. The method is robust for it combines spatial context information with image content.

2   Theory of "Deformed Lattice Detection"

"Deformed lattice detection" is a kind of structure-based texture detection method, which inform us that near regular texture can be described by a pattern element and two smallest linearly independent generating vectors t_1 and t_2 [5], and the translation subgroup of all near regular texture can be characterized by a degree-4 graphical model, where each element is a node that has four neighbors representing its own copies, offset by plus or minus t_1 and t_2. Although the "copies" may be not faithful, the appearances of the deformed elements remain highly correlated, which can be seen as deformed lattice pattern. For deformed lattice pattern, the geometric offsets of neighbors in the lattice is replaced by original terms allowing local variations of the (t_1, t_2) lattice basis vectors. The soft constraints on the geometry and appearance of deformed lattice pattern are represented as pairwise compatibility function and joint compatibility function in a degree-4 Markov Random Field (MRF) model [11]. For an image I that contains a deformed version of a true periodic pattern, the "deformed lattice detection" theory first estimates the ideal pattern element, which can be described by an appearance template T_0, and the lattice generating vectors (t_1, t_2), then infers accurate image locations ($x_0, x_1, ..., x_n$) of all elements forming the repeated pattern in image I.

The underlying topological lattice structure of near regular texture under a set of photometric and geometric deformation fields was first proposed and used by Liu et al. for texture analysis [6], [7]. Subsequently, Hays et al. developed the first deformed lattice detection algorithm for real images without pre-segmentation [8]. Then Lin and Liu developed the first deformed lattice tracking algorithm for dynamic near regular texture [9], [10]. Recently, Park et al. have proposed to formulate the detection of the underlying deformed lattice in an unsegmented image as a spatial, multi-target tracking problem [11], using Mean-shift Belief Propagation (MSBP) method [12]. Our work is partially inspired by Park et al. and we detect insulators based on the framework of their theory.

3   MRF Model of Insulators

The MRF model can be represented as an undirected graph $\mathbf{G} = (\mathbf{N}, \mathbf{E})$, where each node in $\mathbf{N}$ represents a random variable in set $\mathbf{X}$ and each edge in $\mathbf{E}$ represents a statistical dependency between random variables in $\mathbf{X}$. In the insulators detection, the random variables are the image locations of insulators, and edges in the MRF model represent two kinds of dependencies: spatial constraints between neighboring insulators and appearance consistency constraints between each candidate image patch and the reference insulator pattern element.

The meaning of introducing MRF into the work of insulators detection lies in: as a kind of man-made texture, the insulator is not independent of each other, the location of each insulator is constrained by its neighbors, so finding some of them provides knowledge about where the others may be. The result would be more effective if multiple insulators are searched for jointly, rather than one at a time. More specifically, as we know, the distance between neighboring insulators is more or less constant, which can be seen as spatial constraints in the MRF, thus we can define a pairwise compatibility function $\varphi_{i,j}(x_i, x_j)$, where the offset vector $x_i - x_j$ should be "similar" to vector t_1 or t_2. Furthermore, another piece of information that can help localize each insulator is that the image patch centered at x_i should look like the same as the ideal insulator template T_0, the difference in appearance should be small. This constraint is added to the MRF as and a joint compatibility function $\phi(x_i, y_i)$, where the y_i represents as an image likelihood measurement. The figure 1 shows the MRF model of insulators.

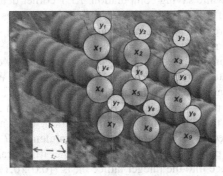

Fig. 1. This is the MRF model of insulators. The latent variables x represent locations of insulator to be inferred, while the image measurements y quantifying appearance similarity between each candidate image patch and the ideal pattern element, and the spatial neighborhood constraints provided by the lattice basis vectors (t_1, t_2).

4   Detecting Insulators in the Image of Overhead Transmission Lines

Through the above analysis, we can conclude that if we want to detect the insulators automatically in the image of overhead transmission lines, there are some problems must be focused on:

1. How to analyze the low-level visual feature to find the reference pattern element T_0 of insulator and the lattice basis vector pair (t_1, t_2) of high-level insulators lattice model;
2. How to represent the insulators lattice model by MRF, and establish the joint compatibility function and the pairwise compatibility function to detect the multiple insulators jointly;
3. How to localize the target area.

In this section, we present a method of automatically detecting insulators in optical images based on the "deformed lattice detection" theory.

4.1   Insulators Lattice Model Proposal

The effectiveness of associative detecting result depends on the insulator reference pattern element T_0 and the (t_1, t_2) basis vectors between neighbor insulators, which form a lattice that represent insulator. Therefore, how to construct the lattice model of insulators is our primary task. We view this step as a discovery process starting with low-level visual feature, and ending with a high-level insulator lattice model proposal. We achieve this lattice model proposal phase using the algorithm was used in [11].

We first study the content of insulators image, extracting the feature points which reflected the repeated structure of insulator texture. Corners, as low-level visual feature for image understanding and analysis of graphics, can effectively retain the important information while saving computation. KLT corner detection algorithm considering the change of image brightness, can find the features with high contrast of image. To reveal more repeating points locally, we apply KLT in a 50×50 pixel block and detect feature points more than 30 in every block.

To find the repeating features that can represent insulator, we cluster corners by the content of image patch through mean shift clustering [13]. We can adopt a voting mechanism to examine a cluster of feature points with similar appearance and propose the insulators lattice mode. Introduction of voting mechanism in the insulators lattice model is equivalent to adding high-level information, from a global perspective to determine a (t_1, t_2) vector pair and reference pattern element T_0. Specifically, we randomly sample three points $\{a,b,c\}$ and compute the affine transformation that maps them from image space into the integer lattice basis $\{(0,0),(0,1),(1,0)\}$, then transform all the other points from image space into their equivalent lattice positions via the same affine transform, and count those points whose lattice space coordinates are around an integer position, which can be seen as supporting votes. Random selection of sample three points $\{a,b,c\}$ multiple times, the transform with the largest number of votes   is chosen to generate insulators lattice model. The corresponding (t_1, t_2) vector pair is regard as insulators lattice generating vectors, and the reference pattern element T_0 is extracted, centered at the origin of the proposed (t_1, t_2) vector pair, with size $\min(|t_1|,|t_2|)$ by $\min(|t_1|,|t_2|)$.

4.2   Associated Detection

Since the location of each insulator is constrained by its neighbors, each of them provides knowledge about the others, the MRF model is a natural choice for inferring insulator locations while enforcing spatial lattice constraints and image likelihood constraints. As the spatial relationship between adjacent insulators is related by the offset vector t_1 or t_2, which can be represent as a pairwise compatibility function $\varphi_{i,j}(x_i,x_j)$. For each insulator, its appearance is similar as the reference pattern element, which is the foundation of joint compatibility function $\phi_i(x_i,y_i)$. When the texture structure of insulators is mapped to the MRF, we can use MSBP to solve the problem [11] [12].

The meaning of joint compatibility function $\phi_i(x_i, y_i)$ is to measure the similarity of the content between candidate image patch and the reference pattern element. As similarity measurement, the image likelihood map should not only be effective to distinguish the difference between the different targets, but also be tolerant of the noise in a class. The image likelihood map is taken as a prior density function on the image location of texture elements, and the joint compatibility function in the lattice MRF is given by:

$$\phi(x_{[i,j]}, y_{[i,j]}) = exp(-\alpha(1 - y_{[i,j]}))\tag{1}$$

$$y_{[i,j]} = NCC(T, I(x_{[i,j]}))\tag{2}$$

Where $x_{[i,j]}$ is the location of node [i, j] at the ith row and j^{th} column in the insulators lattice space, $I(x_{[i,j]})$ is an image patch centered at the location of node [i, j] and T_0 is the appearance template of reference pattern element.

The pairwise compatibility function $\varphi_{i,j}(x_i, x_j)$ specifies the spatial constraint between neighbor insulators, which can be defined by the geometric characteristics of (t_1, t_2) vector pairs in the insulators lattice. The "deformed lattice detection" theory use the normalized error below to measure the spatial consistency of two vector pairs (t_1^i, t_2^i) and (t_1^j, t_2^j):

$$E(t_1^i, t_2^i, t_1^j, t_2^j) = max(\frac{\|t_1^i - t_1^j\|_2}{\|t_1^i\|_2}, \frac{\|t_2^i - t_2^j\|_2}{\|t_2^i\|_2})\tag{3}$$

where $\| \|_2$ is L_2 vector norm. The pairwise compatibility function $\varphi(x_{[i,j]}, x_{[i,j\pm1]})$ and $\varphi(x_{[i,j]}, x_{[i\pm1,j]})$ can be defined based on the equation (3):

$$\varphi(x_{[i,j]}, x_{[i,j\pm1]}) = exp(-\beta \times h(x_{[i,j]}, x_{[i,j\pm1]})^2)\tag{4}$$

$$h(x_{[i,j]}, x_{[i,j\pm1]}) = (\overrightarrow{x^m_{[i,j]}, x^m_{[i,j\pm1]}}, \overrightarrow{x^m_{[i,j]}, x^0_{[i+1,j]}}, \overrightarrow{x^0_{[i,j]}, x^0_{[i,j\pm1]}}, \overrightarrow{x^0_{[i,j]}, x^0_{[i+1,j]}})\tag{5}$$

$$\varphi(x_{[i,j]}, x_{[i\pm1,j]}) = exp(-\beta \times v(x_{[i,j]}, x_{[i\pm1,j]})^2)\tag{6}$$

$$v(x_{[i,j]}, x_{[i\pm1,j]}) = (\overrightarrow{x^m_{[i,j]}, x^m_{[i\pm1,j]}}, \overrightarrow{x^m_{[i,j]}, x^0_{[i,j+1]}}, \overrightarrow{x^0_{[i,j]}, x^0_{[i\pm1,j]}}, \overrightarrow{x^0_{[i,j]}, x^0_{[i,j+1]}})\tag{7}$$

Where $x^m_{[i,j]}$ is the location of node (i, j) in the mth iteration. The equation (5) and the equation (7) are used to measure similarity between the assumed insulator element vector pair (t_1^m, t_2^m) and the reference vector pair (t_1, t_2). The equation (4) measures the spatial similarity between the left insulator and the right insulator, while the equation (6) measures the spatial similarity between the up insulator and the down insulator. With de definition above, we can locate the insulators with MSBP. The parameters are set to $\alpha = \beta = 5$ by experience.

(a) The input image   (b) The result of associated detection

(c) The location of the insulators

Fig. 2. The result of glass insulators detection

(a) The input image   (b) The result of associated detection

(c) The location of the insulators

Fig. 3. The result of porcelain insulators detection

4.3   Location

The purpose of insulators detection is to detect the insulator appears or not in the image automatically and localize the targets for the future work of image segmentation. Due to the influence of the image quality, there are various uncertain disturbance factors in the associated detection stage, which leads to the result of missing detection. In order to enhance the robustness of the algorithm, we analyze the geometric characteristics of the insulators region and extract the minimum bounding rectangle as the final location result.

5   Experimental Results

We have tested our proposed algorithm on real images of different kinds of insulators, there are parts of experimental results below. The detecting targets in figure 2 are suspension glass insulator strings. The background of the insulators image is waters, whose color is similar with that of glass insulators, what is more, the luminance of each insulator image patch is not exactly the same. Therefore, only from the aspect of color information, using the method of template matching is hard to overcome the influence of the interference. We combined the texture structure information with the color information to detect multiple insulators, the experimental result shows the method can overcome the impact of background and locate the insulators accurately.

The detecting targets in figure 3 are suspension porcelain insulator strings. There are not only green plants but also tower in the background. On one hand, the color of tower is similar with porcelain insulators, on the other hand, the sizes of insulators are not completely the same, due to the variations in viewing angle, there is a certain degree of "deformation" in the image. The method conquers the disadvantages and detects the insulator targets.

6   Discussion

Variations in viewing angle, lighting, and partial occlusion are the common problem in the imaging of overhead transmission lines, which make the traditional feature matching methods based on the appearance similarity to be inefficient to detect insulators. Therefore, additional constraints need to be introduced to improve the detecting results. Since the insulators are not merely random collections of isolated elements, but exhibit specific geometric, topological, and statistical regularities and relations, these spatial context information can be used to detect insulators jointly, when the local observation information are insufficient to detect the targets. We analyze the texture structure of insulator strings, and detect the targets based on the theory of "deformed lattice detection". The experimental results show that the method can overcome the disadvantage of imaging conditions and detect different kinds of insulators under a complex background of overhead transmission lines automatically. How to diagnose the fault area of the insulators is our feature work.

Acknowledgements. This work was supported by the National Natural Science Foundation of China (NO.61105031). The authors thank the Electric Power Authority of Shaoxing for providing the data and Minwoo Park for the material of "deformed lattice detection". Thanks are due to Jian Zhao and Shujin Sun for critical discussions.

Reference

1. He, H., Yao, J., Jiang, Z., Wang, X., Li, W.: Infrared Thermal Image Detecting of High Voltage Insulator Contamination Grades Based on Support Vector Machine. Automation of Electric Power Systems 29(24), 70–74 (2005)
2. Li, Z., Yao, J., Yang, Y.: Stationary Wavelet-domain Local Adaptive Denoising Method for Insulator Infrared Thermal Image. High Voltage Engineering 35(4), 833–837 (2009)
3. Lin, J., Han, J., Chen, F.: Defects Detection of Glass Insulator Based on Color Image. Power System Technology 35(1), 127–133 (2011)
4. Huang, X., Zhang, Z.: A Method to Extract Insulator Image from Aerial Image of Helicopter Patrol. Power System Technology 34(1), 194–197 (2010)
5. Liu, Y., Lin, W.C., Hays, J.: Near-regular Texture Analysis and Manipulation. ACM Transactions on Graphics 23(3), 368–376 (2004)
6. Liu, Y., Collins, R.T., Tsin, Y.: A Computational Model for Periodic Pattern Perception Based on Frieze and Wallpaper Groups. IEEE Transactions on Pattern Analysis and Machine Intelligence 26(3), 354–371 (2004)
7. Liu, Y., Tsin, Y., Lin, W.C.: The Promise and Perils of Near-regular Texture. International Journal of Computer Vision 62(1-2), 145–159 (2005)
8. Hays, J., Leordeanu, M., Efros, A.A., Liu, Y.: Discovering Texture Regularity as a Higher-Order Correspondence Problem. In: Leonardis, A., Bischof, H., Pinz, A. (eds.) ECCV 2006, Part II. LNCS, vol. 3952, pp. 522–535. Springer, Heidelberg (2006)
9. Lin, W.-C., Liu, Y.: Tracking Dynamic Near-Regular Texture Under Occlusion and Rapid Movements. In: Leonardis, A., Bischof, H., Pinz, A. (eds.) ECCV 2006, Part II. LNCS, vol. 3952, pp. 44–55. Springer, Heidelberg (2006)
10. Lin, W.C., Liu, Y.: A Lattice-based MRF Model for Dynamic Near-regular Texture Tracking. IEEE Transactions on Pattern Analysis and Machine Intelligence 29(5), 777–792 (2007)
11. Park, M., Brocklehurst, K., Collins, R.T., Liu, Y.: Deformed Lattice Detection in Real-World Images Using Mean-Shift Belief Propagation. IEEE Transactions on Pattern Analysis and Machine Intelligence 31(10), 1804–1816 (2009)
12. Park, M., Liu, Y., Collins, R.T.: Efficient Mean Shift Belief Propagation for Vision Tracking. In: Computer Vision and Pattern Recognition, Anchorage, Alaska (2008)
13. Comaniciu, D., Meer, P.: Mean shift: A Robust Approach Toward Feature Space Analysis. IEEE Transactions on Pattern Analysis and Machine Intelligence 24(5), 603–619 (2002)

Realizing Geometry Surface Modeling of Complicated Geological Object Based on Delaunay Triangulation[*]

Xiangbin Meng[1,2], Panpan Lv[1], Xin Wang[2], and Hua Chen[1]

[1] College of Science China University of Petroleumhuangdao, China
herholiness@163.com
[2] Geophysical Research Institute of Shengli Oil Field Branch CoDongying, China
mengxb_68@sina.com

Abstract. The subsurface geological structures are considerably complicated, which often appear in the form of normal fault, reverse fault, fold, pinchout and irregular body etc. In order to model geometrically the face structure of the geologic horizon and fault, some key algorithms including the Delaunay subdivision and limited Delaunay subdivision are applied to examine techniques such as curved surface intersection, division, suture, united output and so on, while the compatibility of the complicated geological structure, such as geologic horizon and fault, were maintained on geometry and topology. The analysis proposes the geometric distribution factors of geological object model for the further 3D modeling of the complicated geological object.

Keywords: Delaunay subdivision, Curved Surfaces Intersection, Curved Surfaces division, Curved Surfaces suture, Curved Surfaces united output.

1   Preface

It is an important project to build a complicated seismic model in the field of geological prospecting. The basic work is to build a geometric model of geological object. This is also one of the frontier subjects in geo-science field [1-4]. In this paper, the main issue is how to build the geometric model of the block surface by using the techniques of curved surfaces intersection[5], curved surfaces division ,curved surfaces suture and curved surfaces united output[2] on the basis of the results of the 3D reconstruction of geologic horizon, fault, etc.

2   Block Modeling Process Workflow

In this paper, we have put forward an effective block modeling process workflow, by studying the common methods of block modeling, on the basis of practical working experience (Fig.1).

According to the range of data points coordinates, we first set up the outline of a triangle. Then we add the data points to the triangle one by one using Bowyer-Watson algorithm. Finally we delete all triangles which are connected to the vertexes of the outline of triangle and then we will obtain the result of 2D Delaunay division about

---

[*] Supported by "the Fundamental Research Funds for the Central Universities"(11CX04059A).

D.-S. Huang et al. (Eds.): ICIC 2012, LNCS 7389, pp. 451–457, 2012.
© Springer-Verlag Berlin Heidelberg 2012

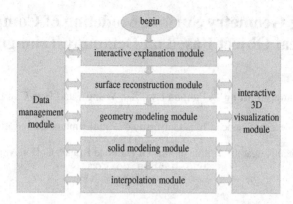

Fig. 1. Block modeling process workflow

the data points. We obtain the intersection between horizons, by making curved surface intersection in the nets of 2D Delaunay division, which is the restrictive condition. The limited segments are added into the result of 2D Delaunay division after they are standardized. After all the limited segments are added into the result, we should check the consistency of the qualification and the subdivision results in geometry and topology. If the result is affirmative, we will achieve the result of subdivision and qualification. After proceeding curved surface division, curved surface suture, curved surface united output, etc. on the basis of the result of 2D limited division, we will complete the geometric modeling of the complicated geological object surface [6-9].

3   Design and Implementation of The Modeling Algorithm about Complicated Geological Object Geometric Surface

The methods of modeling about complicated geological object geometric surface include curved surfaces intersection, curved surfaces division, curved surfaces cut, curved surfaces united output, etc. At the same time, we build the spatial topological relations of the complicated geological object in the exploration areas. That can provide reference for the next entity modeling.

4   Design and Implementation of Curved Surfaces Intersection Algorithm

Curved surfaces intersection is a basic operation for determining the topological relations between intersection and curved surface. It is the basis of curved surfaces division, curved surfaces trimming and curved surfaces transition (Fig.2).

As shown in Fig.2, First, by using the preliminary treatment of bounding box , We obtain the conditions that determine if two triangles have common point:

x1min>x2max;x2min>x1max;y1min>y2max;y2min>y1max;z1min>z2max;z2min >z1max.

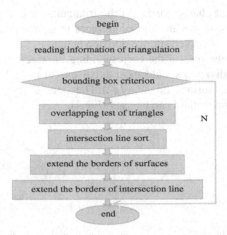

Fig. 2. Curved surfaces intersection process

Because the sampling data points are not sufficient, the geometry intersection line is different from the surface analysis results. In order to make the analysis of surfaces in line with the intersecting lines, we need to extend the borders of surfaces. In this paper we extend the surfaces by using the "boundary extension method with weighted factors". That is: by calculating the direction and the distance of trend with the method of weighted average, we can determine the terminal point of extension [5].

With the technology of surface extension with weighted factors, we can efficiently solve the two kinds of problems in the practical application (Fig.3).

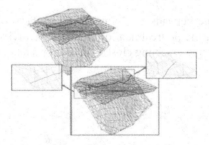

Fig. 3. Intersection lines of surfaces

For the case that the endpoints are not on the boundary of two curved surfaces, we can use the " boundary extension method with weighted factors" to extend the lines with weighted factors, with the endpoints reaching the surface boundary at the same time.

4.1   Design and Implementation of Curved Surfaces Cut Algorithm

Because the earth's crust is very uneven indeed, the intersection we get is often tortuous and irregular geometrically. The triangle spreading division technology makes full use of the adjacent characteristics in the triangulation ensuring that the surfaces after division are consistent with original surfaces in geometry and topology. Thus it can solve the problem of curved surface division better.

The surface triangulation and the corresponding intersection after pretreatment by limited Delaunay subdivision should meet either one of the following: (1) The intersection extends to the surface boundary.(2) The intersections connect end to end. Then triangle spreading method can be used (Fig.4).

Fig. 4. The recursive method of triangle spreading

For a curved surface, as long as we divide the triangulation into two parts along the existing segments or rings in the triangulation, we will finally divide all the surfaces along the intersection.

4.2   Design and Implementation of the Curved Surface Suture Sechnique

When the contour shapes of different levels are different, the suture results are prone to warping. The method of solving this problem is traversing to suture the triangle on the border, adding equipartition point (such as midpoint), and reducing the differences among the sides of all the triangles on boundary, therefore homogenizing the suturing boundaries [10].

Curved surface suture depends on the number and the geological structure of the suturing surfaces. After the pretreatment of the data, we divide the surfaces to be sutured into two types: two intersecting closed curved lines and one closed curved lines.

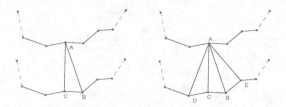

Fig. 5. Two connection points and four connection points

For type 1, the suturing method is to translate two wire frames to make their center points coincide with each other. Then we calculate the difference of the number of boundary points on the two wire frames, and determine the connection type. Finally, we can construct a triangle (The connection type shows in Fig.5).

For type 2, we discuss two cases. One case is the wire frame with two intersection points, and the other is the wire frame with four intersection points.

When there are wire frames with two intersection points, the processing approach is to structure the triangular network for the nodes except A and B with the method of the first suturing type, and to select DC or EF as the pair of initial points. Then we can get the final result by adding the triangle ACD and BEF into the triangular network (Fig.6).

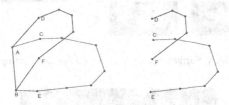

Fig. 6. Wireframe with 2 vertexes

When there are wire frames with four intersection points (Fig.7(a)), the processing approach is to select the wire frame MN-PQ(Fig.7(b)), and then construct the triangular network with MP or NQ as the initial points based on the first suturing type. Next we construct triangular network of AD-MP (Fig.7(c)) and BC-NQ (Fig.7 (d)) in the same way.

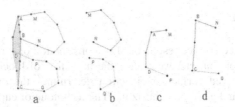

Fig. 7. Wireframe with 4 vertexes

After completing the above steps, we intensify the suturing of triangles. Finally, surface united output technique is to number geological surfaces that belong to a geological object into a triangular mesh file, and to form triangular mesh of block geometric surface.

5   Real Data Process

Above is the elaboration of the theoretical methods. Next, we apply this algorithm in the real seismic data processing. The following is the specific modeling process of a work area more than 100 square kilometers. First, model curved surfaces intersection (Fig.3) based on the result of 2D Delaunay triangulation. Second, model limited 2D Delaunay triangulation on the limited condition of first step. Third, do surface segmentation and surface cutting for the result of limited subdivision (Fig.8). Fourth, stitch the surface after segmenting the surface. Then, model surface of a single block by uniform output. Finally, output the entire block model in the whole area, and generate geometric surface model based on complicated geologic object of triangulation subdivision (Fig.9).

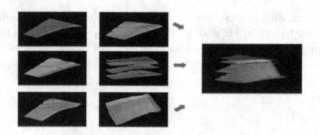

Fig. 8. Surface Segmentation And Surface Cutting

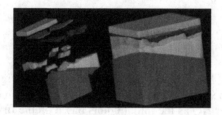

Fig. 9. Surface Uniform Output

6   Conclusion

By applying in practice and the results of 3D reconstruction of surface structure which include geological horizon, etc, we realize the modeling of block geometric surface. The technique of curved surface intersection determines the exact spatial relation between fault plane and geological horizon. The technique of curved surfaces divides geological surface into effective geological horizon and fault plane based on limited geological boundary and the condition of intersection. Surface suturing technique determines the spatial relation among different geological horizons so that the effective surface stitching algorithm can be put forward. Finally, by using the surface united output technique form the triangular mesh of block geometric surface as a basis for limited 3D Delaunay subdivision.

Of course, the geometric surface modeling of complicated geological object is a very complicated problem. Due to the limitation of our knowledge, there may be some flaws in the conclusion and methods. However, we hope this research can provide some references and discussions for improving further the geometric surface modeling of complicated geological object.

References

1. Houlding, S.W.: 3D Geoscience Modeling computer Techniques for Geological Characte-rization. Springer, Heidelberg (1994)
2. Meng, X.B., Wei, Z.Q.: Study on Seismic Imaging Block modeling Method. In: IASP 2010, 03 (2010)

3. Meng, X.B., Wei, Z.Q.: Achieving Complex Geological Object Solid Modeling Based on TIN and TEN. In: ICIC 2010, 06 (2010)
4. Wang, X.F.: 3D Geological Object Modeling Quo. School of Civil Engineering Hehai University
5. Chu, J., Wei, Z.Q.: Surface-Surface Intersection Based on Delaunay Triangulation. Journal of System Simulation 10(21), 155–158 (2009)
6. Meng, X.H., Wang, W.M.: The Geological Model and Application of Computer-Aided Design Principles. Geological Publishing House, 11 (2000)
7. Yang, Q.: Restricted Delaunay Triangulation. University of Aeronautics and Astronautics Doctoral Dissertation, Beijing (August 2001)
8. Mallet, J.-L.: Discrete Smooth Interpolation. ACM Transactions on Graphics 8(2), 121–144 (1989)
9. Mallet, J.-L.: Discrete Smooth Interpolation in Geometric Modeling. Computeraided Designer, 178–191 (1992)
10. Meng, X.H., Cai, Q., et al.: Tetrahedral Mesh Generation for Surface Delaunay Triangulation Algorithm. Journal of Engineering Graphics (1) (2006)

Researching of the Evolution of Discontent in Mass Violence Event

FanLiang Bu[1] and YuNing Zhao[2]

[1] Chinese People's Public Security University, Beijing, China
[2] Chinese People's Public Security University, Beijing, China
bufanliang@sina.com, dycppsu@163.com

Abstract. This paper uses base-Agent method to build a new model of researching the evolution of discontent in mass violence event, in which model we introduce emotion factor, relation factor, risk factor and relevance factor as the attribute value of the individual Agent, then using psychophysics' correlation theory to quantize the change of individual discontent. In the paper, by setting different initial values, we make relevant experiments to study the process of evolution of discontent of a typical mass violence event that has happened and three other different circumstances. All of what we achieved from this paper can offer help to the police to understand the inherent mechanism and law of mass violence event, also can provide guide for taking pointed measures to prevent and manage mass violence event.

Keywords: mass violence event, discontent, evolution, model.

1   Introduction

Nowadays more and more mass violence events break out. Domestic and foreign researchers have paid much more attention on these events. So far, there has been two different ways of studying mass violence events, one is to study the economic reasons or the psychology of the participants[1-2]. The other is using computer science and information technology, extracting various factors to build models, gains the regular patterns and features by simulating [3-5]. In this paper, we extract some parameters related to the emotion evolution of participants before the events to build models on the basis of the latter view[6-8]. After running considerable simulations in different settings, we get the regular pattern of the emotion evolution among participants, which provides theoretical basis to understand mass violence events.

2   Agent-Based Modeling

Agent-based modeling (ABM) is a modeling method developed from artificial intelligence in 1970s. As the basic abstract unit of the system, Agent has some simple attributes and status so it has some kind of intelligence. We define some simple rules for these Agents so that they could interact between each other and obtain corresponding system model finally. It's a bottom-up modeling method. By using Agent-based models and simulations, not only could we give answers to the problems, but also

D.-S. Huang et al. (Eds.): ICIC 2012, LNCS 7389, pp. 458–465, 2012.
© Springer-Verlag Berlin Heidelberg 2012

display all dynamic features of system evolution. We could also set up different assumptions, conditions or limited factors for the models to understand system features more adequately. More importantly, it could resolve various problems of complex system highly formally by using the interactions between Agents. Based on the characters mentioned above, the method has been widely applied in the field of computer science technology. Figure 1 shows the basic steps of ABM.

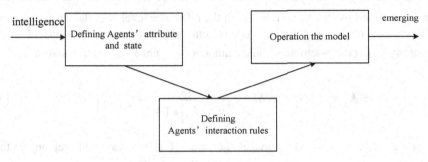

Fig. 1. Basic step of ABMS

3   The Model of Discontent Emotion Evolution

3.1   Basic Parameters of Individual Agent

In this paper, we extract four parameters from participants' individual attributes and social environment: emotion factor, relationship factor, risk factor and relevance factor.

Emotion factor is used to reflect the amount of dissatisfaction emotion of individual Agent which has a bearing on the behaviors about to happen. It is represented by Q and limited in [0, 1]. The more the dissatisfaction is, the larger the likelihood of violence is.

Relationship factor represents the individual Agent's relationship with other Agents in the crowd. The closer the relationship between two Agents is, the more proximity the relationship factors of their own are and more impact on each other's behavior they have. It's represented by R and the value is between [0, 1].

Risk factor, represented by M, refers to the rationality level of individual Agent whose valve is set between [0, 1]. The bigger the value is, the more likelihood the Agent gets excited and act violently.

Relevance factor represents the relevance level between individual Agent and the event causing the gathering. It's represented by G and its value is between [0,1]. The bigger the value is, the more relevance with the event Agent has.

So, each individual Agent in the crowd could be expressed as $I_i(Q_i, R_i, M_i, G_i)$.

3.2   Adjustment Function of Mood

According to Fechner's theory, the change of human's perceiving strength is in proportion to the change of logarithm of irritant intensity[9], as shown in equation 1.

$$\psi = K \bullet \lg \phi \tag{1}$$

ψ represents the amount of human's perception, and ϕ represents the intensity of physical irritant, while K is a constant.

We assume that there were N Agents in the crowd, and choose two individual Agents randomly I_i and I_j in a time step to analyze. Here we consider the irritant amount received by the Agent relevant to the relevance level with the event, the relationship with other Agent, other Agents' emotion and rationality status of other Agent, so we get the assemble stimulate amount of Agent I_i shown as equation 2.

$$\phi = K_1 \bullet \left(5 \cdot G_i + 3 \cdot M_j + \frac{1}{\left| R_i - R_j \right| \bullet 100} + Q_j(t) \right) \tag{2}$$

K_1 is a correction factor which makes the value of ϕ in a reasonable region. So the emotion factor of Agent I_i after evolution is expressed by equation 3.

$$Q_i(t+1) = Q_i(t) + K_2 \bullet \lg \left[K_1 \bullet \left(5 \cdot G_i + 3 \cdot M_j + \frac{1}{\left| R_i - R_j \right| \bullet 100} + Q_j(t) \right) \right] \tag{3}$$

K_2 is a correction factor which makes the value of $Q_i(t+1)$ in a reasonable region.

4 Simulation Experiment and Results Analysis

4.1 Verification Experiment

Firstly, we use a typical mass violence event that has happened as an example to verify the reasonableness of the model. According to actual situation of the event, we set the number of gathered crowd 30, containing 12 victim's relatives, 6 lawless persons, 12 bystanders. We set the Agent 1~12 with higher value of Q and G, the Agent 13~17 with higher value of M, the Agent 18~30 with lower value of Q. The specific setting of the four factors is shown in the figure 2.

Agents	Q	R	M	G	Agents	Q	R	M	G
1	0.6	0.99	0.3	0.6	16	0.34	0.60	0.5	0.08
2	0.61	0.97	0.3	0.5	17	0.35	0.58	0.5	0.09
3	0.62	0.95	0.2	0.55	18	0.36	0.56	0.5	0.10
4	0.63	0.94	0.2	0.6	19	0.10	0.38	0.1	0.04
5	0.64	0.92	0.1	0.5	20	0.11	0.36	0.2	0.05
6	0.65	0.90	0.1	0.54	21	0.12	0.34	0.1	0.06
7	0.66	0.88	0.4	0.52	22	0.13	0.32	0.1	0.07
8	0.67	0.86	0.3	0.53	23	0.14	0.30	0.2	0.08
9	0.68	0.84	0.2	0.58	24	0.15	0.28	0.05	0.09
10	0.69	0.82	0.1	0.57	25	0.16	0.26	0.05	0.10
11	0.70	0.80	0.1	0.55	26	0.09	0.24	0.1	0.11
12	0.71	0.78	0.2	0.59	27	0.08	0.22	0.1	0.12
13	0.31	0.66	0.6	0.05	28	0.07	0.20	0.05	0.13
14	0.32	0.64	0.5	0.06	29	0.06	0.18	0.05	0.14
15	0.33	0.62	0.6	0.07	30	0.05	0.16	0.05	0.15

Fig. 2. The specific setting of the four factors in the group

In order to make the thirty Agents' value of Q reasonable, we set the value of the correction factor $K_1 = 0.8, K_2 = 0.02$. Using the equation (3), we can acquire the evolution curve of discontent for the thirty Agents, as showing in the figure 3.

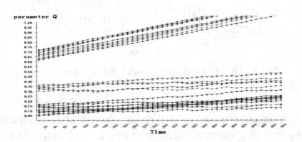

Fig. 3. The evolution curve of discontent for the thirty Agents

From the figure 3 we can know that the emotion factor Q of the victim's relatives is higher than the lawless person and bystander, also the increasing range is larger. The bystanders' discontent is increasing slowly, if the number of bystander is large, this change will lead to group violence. Because the evolution of discontent is accord with the mechanism of mass violence events, so the model presented in this paper is rational. We can use this model to research the evolution law of discontent in mass violence event.

4.2   Simulation Experiment

In this section, we design three experiments. In the first experiment, we suppose the crowd contain 18 Agents, all have low discontent, and relevancy between the Agents and the trigger event is same. We input relevant value to the model, the figure 4 shows the settings.

Under this circumstance, we get the evolution curve of every Agent's discontent emotion as the figure 5.

Agents	Q	R	M	G
1	0.15	0.35	0.25	0.15
2	0.17	0.26	0.31	0.25
3	0.13	0.15	0.26	0.24
4	0.20	0.39	0.17	0.16
5	0.14	0.46	0.13	0.11
6	0.12	0.52	0.25	0.14
7	0.10	0.43	0.24	0.10
8	0.20	0.20	0.21	0.24
9	0.16	0.07	0.17	0.22
10	0.08	0.01	0.12	0.21
11	0.09	0.76	0.10	0.23
12	0.03	0.72	0.11	0.22
13	0.05	0.88	0.08	0.18
14	0.18	0.83	0.14	0.19
15	0.21	0.90	0.03	0.17
16	0.22	0.19	0.10	0.16
17	0.06	0.99	0.22	0.11
18	0.05	0.29	0.14	0.15

Fig. 4. The settings of the first experiment

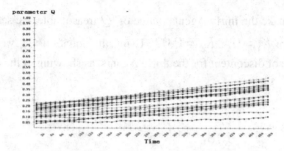

Fig. 5. The evolution curve under the first experiment

From the figure 5 we can see that the crowd's mean initial value of discontent emotion is 0.14. After experiment, most Agents' discontent emotion changes slightly, while several Agents' discontent emotion increased considerably. At the end of experiment, crowd's mean value of discontent emotion is 0.24.

On the basis of first experiment, we add four Agents with higher initial value of Q, the figure 6 shown the settings of the second experiment.

Agents	Q	R	M	G
1	0.15	0.35	0.25	0.15
2	0.17	0.26	0.31	0.25
3	0.13	0.15	0.26	0.24
4	0.20	0.39	0.17	0.16
5	0.14	0.46	0.13	0.11
6	0.12	0.52	0.25	0.14
7	0.10	0.43	0.24	0.10
8	0.20	0.20	0.21	0.24
9	0.16	0.07	0.17	0.22
10	0.08	0.01	0.12	0.21
11	0.09	0.76	0.10	0.23
12	0.03	0.72	0.11	0.22

Agents	Q	R	M	G
13	0.05	0.86	0.08	0.18
14	0.18	0.83	0.14	0.19
15	0.21	0.90	0.03	0.17
16	0.22	0.19	0.10	0.16
17	0.06	0.99	0.22	0.11
18	0.05	0.29	0.14	0.15
19	0.6	0.03	0.35	0.25
20	0.62	0.22	0.33	0.28
21	0.65	0.36	0.31	0.18
22	0.58	0.0	0.28	0.14

Fig. 6. The settings of the second experiment

Under this circumstance, we get the evolution curve of every Agent's discontent emotion as the figure 7.

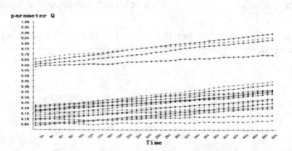

Fig. 7. The evolution curve under the second experiment

We can see from figure 7 that the crowd's mean initial value of discontent emotion is 0.14. After the second experiment, crowd's discontent emotion increases. At the 50th seconds, most Agents' value concentrate at 0.36 and 0.20, means that if the crowd includes a small number of Agents with higher value of discontent emotion, crowd's discontent emotion will be polarization. At the end, the crowd's mean value of discontent emotion is 0.28, higher than the first experiment.

On the basis of the first experiment, we add four Agents that have higher initial value of M to the crowd, the figure 8 showing the settings of the second experiment.

Under this circumstance, we get the evolution curve of every Agent's discontent emotion as the figure 9.

Agents	Q	R	M	G	Agents	Q	R	M	G
1	0.15	0.35	0.25	0.15	13	0.05	0.88	0.08	0.18
2	0.17	0.26	0.31	0.25	14	0.18	0.83	0.14	0.19
3	0.13	0.15	0.26	0.24	15	0.21	0.90	0.03	0.17
4	0.20	0.39	0.17	0.16	16	0.22	0.19	0.10	0.16
5	0.14	0.46	0.13	0.11	17	0.06	0.99	0.22	0.11
6	0.12	0.52	0.25	0.14	18	0.05	0.29	0.14	0.15
7	0.10	0.43	0.24	0.10	19	0.10	0.0	0.64	0.14
8	0.20	0.20	0.21	0.24	20	0.16	0.95	0.60	0.20
9	0.16	0.07	0.17	0.22	21	0.18	0.05	0.62	0.22
10	0.08	0.01	0.12	0.21	22	0.20	0.48	0.65	0.20
11	0.09	0.76	0.10	0.23					
12	0.03	0.72	0.11	0.22					

Fig. 8. The settings of the third experiment

Fig. 9. The evolution curve under the third experiment

We can see from figure 9 that the crowd's mean initial value of discontent emotion is 0.14. After adding four Agents that have higher initial value of risk factor, the crowd's increasing range of discontent emotion is larger than the condition under the previous experiment. At the 50th seconds, most Agents' value concentrate at 0.36 and 0.24, and the crowd's mean value of discontent emotion is 0.31, higher than the condition under the previous experiment.

4.3   Results Analysis

In the first experiment, because the crowd does not receive any stimulation and instigating from outside, the evolution of discontent emotion shows spontaneity and randomness, the average value of discontent emotion increases slightly, from 0.14 to 0.24. In the second experiment, we can see that the average value of discontent emotion increases from 0.14 to 0.28, suggesting that the evolution of discontent is influenced by the four Agents, especially, discontent emotion mainly distributes at 0.36 and 0.20 at last time step. At last experiment, we can see that the average value of discontent emotion increases from 0.14 to 0.31, and discontent emotion mainly distributes at 0.36 and 0.24 at last time step, which suggests that the Agents with high value of risking have lager influence to the crowd, also larger than influence from the Agents with high value of discontent emotion. So, if a crowd has more persons with risking, the crowd will have higher probability of conducting violence.

On the basis of curve of discontent emotion in the figure 5, 7 and 9, we draw a relation schema between crowd's average value of discontent and time under the above three experiments, shown in the figure 10.

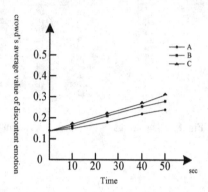

Fig. 10. Relation curve between crowd's average value of discontent emotion and time under the above three experiments

In the figure 10, A, B, C indicates relationship between crowd's value of discontent emotion and time under the above three experiments respectively. From the figure, we can see that the crowd containing high inflammatory person have larger increasing range, which suggest the high risky persons' influence to probability of conducting violence is larger than others'.

5   Conclusion

In this paper, we present a new model to research individual evolution of discontent emotion in the crowd and use a typical mass violence event that have occurred to test and verify the model's rationality. Then we design three experiments under different conditions, what we obtain from the results of the experiments is: (1) the crowd's evolution of discontent showing spontaneity and randomness when the crowd's initial discontent emotion is low; (2) when adding some persons with high value of discontent emotion, the crowd's discontent be influenced slightly, and the average value of discontent is increased; (3) when adding some persons with high value of risk, the crowd's discontent is influenced greatly, larger than the condition of adding some persons with high value of discontent, and the average value of discontent increases considerably. What we can draw from the results is: for a crowd conducting violence, instigating from the lawless persons are more dangerous than the influence from the person with high value of discontent emotion. We should pay more attention to the persons with high value of risk.

References

1. Craig, A.A., Brad, J.B.: Human Aggression. Annual Review of Psychology 53(1), 27–51 (2002)
2. Berkowitz, L., Cochran, S.T., Embree, M.C.: Physical Pain and the Goal of Aversively Stimulated Aggression. Journal of Personality and Social Psychology 40(4), 687–700 (1981)
3. Bu, F.L., Feng, P.Y.: Analysis of Agent-Based "Non-Organization and Non-Direct Interest" Collective Events. In: 2th IEEE International Conference on Emergency Management and Management Sciences, pp. 417–421. IEEE Press, Beijing (2011)
4. Bu, F.L., Sun, J.Z.: An Analysis for Agent-Based Mass Violence Event. In: 2th IEEE International Conference on Emergency Management and Management Sciences, pp. 422–425. IEEE Press, Beijing (2011)
5. Bu, F.L., Sun, J.Z.: Agent-Based Modelling and Simulation System for Mass Violence Event. In: 4th IEEE International Symposium on Computational Intelligence and Design, pp. 211–215. IEEE Press, Zhejiang (2011)
6. Bu, F.L., Zhao, Y.N.: Analysis of Mass Violence Event Based on Analytic Hierarchy Process. In: 4th IEEE International Symposium On Computational Intelligence and Design, pp. 174–177. IEEE Press, Zhejiang (2011)
7. Bu, F.L., Zhao, Y.N.: Modeling and Simulation of Mass Violence Event Based on Agent Technology. In: 2th IEEE International Conference on Emergency Management and Management Sciences, pp. 426–428. IEEE Press, Beijing (2011)
8. Bu, F.L., Zhao, Y.N.: Modeling and Warning Analysis of Mass Violence Events. In: IEEE International Conference on Automatic Control and Artificial Intelligence, pp. 2048–2052. IET Press, XiaMen (2012)
9. Narens, L., Mausfeld, R.: On the Relationship of the Psychological and the Physical in Psychophysica. Psychological Review 99(3), 467–479 (1992)

An Efficient Two-Stage Level Set Segmentation Framework for Overlapping Plant Leaf Image

Xiao-Feng Wang[1,2] and Hai Min[2,3]

[1] Key Lab of Network and Intelligent Information Processing, Department of Computer Science and Technology, Hefei University, Hefei Anhui 230022, China
[2] Intelligent Computing Lab, Hefei Institute of Intelligent Machines, Chinese Academy of Sciences, P.O.Box 1130, Hefei Anhui 230031, China
[3] Department of Automation, University of Science and Technology of China, Hefei, Anhui 230027, China
xfwang@iim.ac.cn, minhai361@gmail.com

Abstract. In this paper, an efficient two-stage segmentation framework was proposed to address the plant leaf image with overlapping phenomenon, which is built based on the leaf approximate symmetry and level set evolution theory. In the pre-segmentation stage, a straight line was manually set on the target leaf to approximate the principal leaf vein and the Local Chan-Vese (LCV) model was used on the global image region to help searching the so-called un-overlapping contour in target leaf. In the formal segmentation stage, the symmetry detection was performed based on the pre-defined approximated principal vein to obtain the narrow-band evolution region and the second initial contour. Next, the LCV model was once again used to find the complete target leaf contour in the narrow-band evolution region. Finally, experiments on some real leaf images with overlapping phenomenon have demonstrated the efficiency and robustness of the proposed segmentation framework.

Keywords: approximate symmetry, level set, overlapping leaf, principal vein, two-stage segmentation.

1   Introduction

Recently, segmenting the leaf from plant images and performing classification based on leaf features has become an efficient mean in the field of digital plant research. Plant leaf is very suitable for image processing since its shape structure is relatively stable and leaf can be easily found and collected in most of the time of four seasons [1]. It should be noted that the leaf segmentation result will directly influence the performance of leaf feature extraction and final classification. Thus, some efficient approaches have been proposed for leaf image segmentation up to now. Manh et al. [2] used deformable templates and combined the color features to segment the weed leaf. Persson et al. [3] proposed segmenting the leaf image by using the active shape models. Zheng et al. [4] proposed the mean-shift-based color segmentation method for green leaves. Wang et al. [1] proposed an automatic marker-controlled watershed

D.-S. Huang et al. (Eds.): ICIC 2012, LNCS 7389, pp. 466–474, 2012.
© Springer-Verlag Berlin Heidelberg 2012

method combined with pre-segmentation procedure and morphological operation to segment leaf images.

However, most of the proposed methods can not address the plant leaf image with overlapping phenomenon, i.e., leaves are overlapped with each other due to the original growth condition. The uncovered complete leaf in upper layer is the segmentation target. Without the prior knowledge, the segmentation result will not be satisfied since the interferential leaf in underlying layer will produce the false contour. To resolve this problem, we shall introduce a two-stage segmentation framework based on the leaf approximate symmetry. The segmentation process can be divided into two stages, i.e. pre-segmentation stage and formal segmentation stage. Here, we used our previously proposed Local Chan-Vese (LCV) model [5] as the segmentation model for both two stages. LCV model contains the controlled global term and local term. It is very suitable for two-stage image segmentation task since it can not only capture the regional image information but also segment the detailed objects.

Before segmentation, a straight line should be manually set on the target leaf in upper layer to approximate the principal vein. Then, in the pre-segmentation stage, the level set evolution driven by LCV model is performed on the global image region to help searching the so-called un-overlapping contour in target leaf. Next, the symmetry detection is implemented based on the pre-defined approximated principal vein to obtain the narrow-band evolution region and the second initial contour. Finally, in the formal segmentation stage, the LCV model is once again used to extract the complete leaf contour in the narrow-band evolution region.

The rest of this paper can be organized as follows: In Section 2, we briefly introduce the leaf approximate symmetry and LCV model. Our two-stage level set segmentation framework is presented in Section 3. In Section 4, the proposed framework is validated by some experiments on leaf images with overlapping phenomenon. Finally, some conclusive remarks are included in Section 5.

2   Background Knowledge

2.1   Leaf Approximate Symmetry

Leaf vein is the vascular bundle growing in the leaf to transfer water and nutrition. The principal vein is the longest and obvious vein located in the center of the leaf. Since the physiological function of leaf is to receive sunlight and transpire water, the shapes of both sides of the principal vein should be almost same under the same ideal conditions of sunlight and water. Fig.1 shows some basic shapes of common simple leaf. It can be seen that the shapes of leaves preserve certain symmetry although the shapes vary widely. It should be noted that practical leaf may not preserve complete symmetry due to the growth position, orientation and inclination angle of leaf. Some trivial shape difference may exist in two sides of the principal vein. So, the symmetry of leaf based on principal vein is often called approximate symmetry.

In this paper, the leaf approximate symmetry is adopted to find the complete target leaf in upper layer of the overlapping leaf image. It should be noted that we do not directly extract the principal vein but manually set a straight line on the target leaf to approximate its principal vein. The reasons are as follows: First, the principal veins are not obvious in some leaves and often connected with the lateral veins and minor veins. So, how to extract the principal veins is in itself a difficult problem in the field of leaf image processing [6]. Second, the shift and deformation may exist in the leaf contour of both sides of the principal vein due to the different conditions of sunlight and water. The approach of manual setting straight line to approximate principal vein can reduce the side-effect of shift and deformation on the final segmentation result.

Fig. 1. The basic shapes of common simple leaf

2.2   Local Chan-Vese Model

The local Chan-Vese (LCV) model was built based on the techniques of curve evolution, local statistical function and level set method. By incorporating local information into regional level set model, LCV model can not only capture the regional image information but also segment the detailed objects with the controlling balance parameters. The overall energy functional in LCV model E^{LCV} consists of three parts: global term E^{G}, local term E^{L} and regularization term E^{R}. Thus the overall energy functional can be described as:

$$E^{LCV} = \alpha \cdot E^{G} + \beta \cdot E^{L} + E^{R} \tag{1}$$

where α and β are two positive parameters which govern the tradeoff between the global term and the local term.

The global term E^{G} is directly derived from the Chan-Vese model [7]. Using the level set formulation, the boundary C is represented by the zero level set of a Lipschitz function $\phi : \Omega \rightarrow R$. The global term is stated as follows:

$$
\begin{aligned}
E^{G}(c_1, c_2, \phi) = &\iint_{\Omega} |u_0(x, y) - c_1|^2 H(\phi(x, y)) dx dy \\
&+ \iint_{\Omega} |u_0(x, y) - c_2|^2 (1 - H(\phi(x, y))) dx dy
\end{aligned}
\tag{2}
$$

where $H(z)$ is the Heaviside function.

The introduction of local term is to statistically analyze each pixel with respect to its local neighborhood. In the same manner as global term, the local term E^{L} can be reformulated in terms of the level set function $\phi(x, y)$ as follows:

$$E^L(d_1, d_2, \phi) = \int_\Omega \left| g_k * u_0(x, y) - u_0(x, y) - d_1 \right|^2 H(\phi(x, y)) dx dy$$

$$+ \int_\Omega \left| g_k * u_0(x, y) - u_0(x, y) - d_2 \right|^2 (1 - H(\phi(x, y))) dx dy$$

(3)

where g_k is a averaging convolution operator with $k \times k$ size window. d_1 and d_2 are the intensity averages of difference image $(g_k * u_0(x, y) - u_0(x, y))$ inside C and outside C, respectively.

The regularization term E^R is composed of two items. The first item is the length penalty term which is used to control the smoothness of the zero level set and further avoid the occurrence of small, isolated regions. The second item is used to force the level set function to be close to a signed distance function [8]. It plays a key role in the elimination of time-consuming re-initialization procedure.

$$E^R(\phi) = \mu \cdot L(\phi = 0) + P(\phi)$$

$$= \mu \cdot \int_\Omega \delta(\phi(x, y)) |\nabla \phi(x, y)| dx dy + \int_\Omega \frac{1}{2} (|\nabla \phi(x, y) - 1|)^2$$

(4)

where μ is the parameter which can control the penalization effect of length term: if μ is small, then smaller objects will be detected; if μ is larger, then larger objects are detected.

The minimization of E^{LCV} can be done by introducing an artificial time variable $t \geq 0$, and moving ϕ in the steepest descent direction to a steady state as follows:

$$\frac{\partial \phi}{\partial t} = \delta_\varepsilon(\phi)[-(\alpha \cdot (u_0 - c_1)^2 + \beta \cdot (g_k * u_0(x, y) - u_0(x, y) - d_1)^2)$$

$$+ (\alpha \cdot (u_0 - c_2)^2 + \beta \cdot (g_k * u_0(x, y) - u_0(x, y) - d_2)^2)]$$

$$+ [\mu \cdot \delta_\varepsilon(\phi) div(\frac{\nabla \phi}{|\nabla \phi|}) + (\nabla^2 \phi - div(\frac{\nabla \phi}{|\nabla \phi|}))]$$

(5)

For more details about the LCV model, readers can refer to the literature [5].

3   Two-Stage Segmentation Framework

In this section, we shall present and discuss the details of the two-stage level set segmentation framework. It should be noted that the final objective is to find the uncovered and complete leaf in upper layer, i.e. target leaf. The covered and incomplete leaf in underlying layer, i.e. background leaf, is regarded as the interferential object. To obtain satisfying segmentation result, we usually require that the test image should contain only one complete target leaf and one half side of the target leaf contour is un-overlapped.

3.1   The Pre-segmentation Stage

Before pre-segmentation, we should first set a straight line to approximate the principal vein for target leaf. Here, we adopt the simple two-point form to define the straight line. Two points $(x1, y1)$ and $(x2, y2)$ are manually selected at the two ends of the target leaf. Thus, the straight line equation of approximate principal vein is built as follows:

$$Ax + By + C = 0 \tag{6}$$

where $A = y1 - y2$, $B = x2 - x1$ and $C = x1 * y2 - x2 * y1$.

Next, we should set the first initial contour and initialize the level set function ϕ to a sign distance function. Benefit from the less sensitivity to the location of initial contour of LCV model, the initial contour can be placed anywhere. The parameters of LCV model should be set, which include the time-step Δt, the grid spacing h, the regularization parameter ε, the window size of averaging convolution operator k, the controlling parameter of global term α, the controlling parameter of local term β and the length controlling parameter μ.

After the level set evolution stops, the zero level sets (contours) should be extracted from the level set function ϕ. Due to the complicated background of overlapping leaf image, the extracted contours usually contain several contours with different lengths, i.e. $(C_1^*, C_2^*, \cdots, C_n^*)$, which are generated from the target leaf and the interferential leaves or branches. To provide precise results for the next formal segmentation stage, the interferential contours should be removed. According to the prior knowledge that only one target leaf exists in the test image, the longest contour C^* will be kept as the pre-segmentation result.

$$C^* = \underset{1 \leq i \leq n}{argmax} \, length(C_i^*) \tag{7}$$

Actually, the contour C^* belongs to both the target leaf in upper layer and background leaf in underlying layer. Now, it is necessary to find the contour belonging to un-overlapped half side of target leaf, which is called un-overlapping contour C_T. First, contour C^* should be further divided into two parts, i.e. C_1 and C_2 according to the approximate principal vein. For any point (x, y), it can be classified into C_1 or C_2 based on the following rule:

$$A * x + B * y + C \begin{array}{l} > 0 \\ < 0 \end{array} \begin{cases} (x, y) \in C_1 \\ (x, y) \in C_2 \end{cases} \tag{8}$$

where A, B and C are the coefficients in Eq.(6).

Then, it needs to judge which one between C_1 and C_2 is the un-overlapping contour. According to the growth characteristic of plant leaf, most of the leaf contours are smooth and bending towards the principal vein. Correspondingly, the normal directions of all leaf contour points are generally towards the principal vein. It should be noted that the overlapping contour is actually composed of two parts from the target leaf contour and background leaf contour, respectively. As a result, a sudden change of normal direction will occur at the junction of the target leaf contour and background leaf contour. Thus, we can theoretically judge which contour is the un-overlapping contour according to the continuity of normal directions of contour point. Considering that there are some tiny sawtooth on the leaf contour, the statistical analysis to the continuity of normal directions should be performed based on the points sampled from contour with the same interval in practical numerical implementation. In this way, the computations can be greatly decreased and precise statistical analysis result will be obtained.

3.2   The Formal Segmentation Stage

After obtaining the un-overlapping contour C_T, the symmetric curve C_S with respect to the approximate principal vein can be obtained. For each point $(x, y) \in C_T$, the symmetric point (x_S, y_S) on C_S can be computed as follows:

$$\begin{cases} x_S = x - 2A(Ax + By + C)/(A^2 + B^2) \\ y_S = y - 2B(Ax + By + C)/(A^2 + B^2) \end{cases} \tag{9}$$

C_T and C_S are then connected to generate a complete and closed curve. Here, we adopt the narrow-band idea from level set theory and build a narrow-band evolution region which regards C_c as its center line and has a bandwidth of $2w$. Different from the pre-segmentation stage, the second initial contour in formal segmentation stage can be automatically constructed which is a square with size of $r \times r$ centered on the center of C_T. Since the second initial contour should located in the narrow-band evolution region, the value of r should be smaller than that of $2w$.

Finally, the second levels set evolution based on LCV model can be performed in the narrow-band evolution region with the second initial contour. Here, some evolution parameters controlling the segmentation detail should be updated since the overlapping phenomenon still exists in the narrow-band evolution region. The first updated parameter is the length controlling parameter μ. A smaller value should be set for μ according to the setting rule of μ. In the same way, the controlling parameter of global term α should also be smaller than the controlling parameter of local term β. After the formal segmentation, the final obtained zero level set will be the contour of target leaf.

4   Experimental Results

In this Section, we shall present the experimental results of two-stage level set segmentation framework on some leaf images with overlapping phenomenon. The proposed segmentation framework was implemented by Matlab 7 on a computer with Intel Core 2 Duo 2.2GHz CPU, 2G RAM, and Windows XP operating system. We used the same parameters of the time-step $\Delta t = 0.1$, the grid spacing $h = 1$, $\varepsilon = 1$ (for $H_\varepsilon(z)$ and $\delta_\varepsilon(z)$), the window size of averaging convolution operator $k = 15$, the controlling parameter of local term $\beta = 1$ for all the experiments in this section. In our experiments, α has two values: 1 and 0.1 for pre-segmentation stage and formal segmentation stage. The length controlling parameter μ also has a scaling role like α. Two corresponding values, i.e. $0.01*255^2$ and $0.001*255^2$, are adopted. In conclusion, there are totally two parameters whose values need to be dynamically adjusted in our experiments.

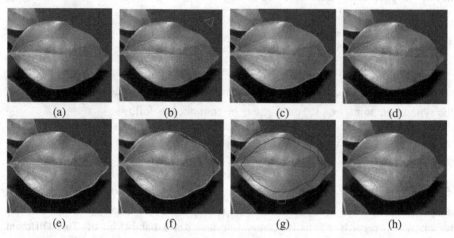

(a)   (b)   (c)   (d)

(e)   (f)   (g)   (h)

Fig. 2. The whole segmentation process of proposed two-stage level set segmentation framework on the leaf image with overlapping phenomenon. (a) Original image (b) The approximate principal vein and initial contour (c) The zero level sets after the level set evolution in pre-segmentation stage. (d) The longest contour C^* (e) C_1 (yellow line) and C_2 (blue line) (f) Center line C_c (g) The narrow-band evolution region and the second initial contour (h) The final segmentation result of formal segmentation stage. Image size=183×160.

Fig.2 demonstrates the whole process of proposed two-stage level set segmentation framework on the leaf image with overlapping phenomenon. It can be seen from Fig.2 (a) that an interferential leaf is overlapped with the target leaf at the top-left corner of image. The approximate principal vein (red straight line) and initial contour (green triangle) are shown in Fig.2 (b). The green lines in Fig.2 (c) correspond to the zero level sets after the level set evolution in pre-segmentation stage. Fig.2 (d) shows the longest contour C^* and Fig.2 (e) demonstrates the obtained C_1 (yellow line) and C_2 (blue line). By performing the statistical analysis to the continuity of normal directions, it can be judged that C_1 is the required un-overlapping contour C_T. After computing the symmetric curve C_S with respect to the approximate principal vein, C_T and C_S are then connected to generate the center line C_c (as shown in Fig.2 (f)). The blue lines in Fig.2 (g) are the boundary of narrow-band evolution region with a bandwidth of 16. The green square in Fig.2 (g) shows the second initial contour in formal segmentation stage with size of 12×12. Fig.2 (h) demonstrates the final segmentation result of formal segmentation stage. It can be seen the target leaf is segmented perfectly.

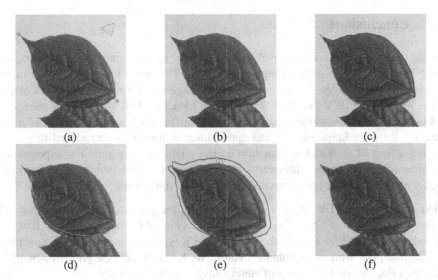

Fig. 3. The segmentation experiment of proposed framework on larger leaf image with over-lapping phenomenon. (a) The approximate principal vein and initial contour (b) The longest contour C^* (c) C_1 and C_2 (d) Center line C_c (e) The narrow-band evolution region and the second initial contour (f) The final segmentation result. Image size= 304×297.

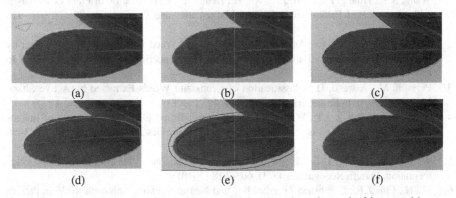

Fig. 4. The segmentation experiment of proposed framework on larger leaf image with over-lapping phenomenon. (a) The approximate principal vein and initial contour (b) The longest contour C^* (c) C_1 and C_2 (d) Center line C_c (e) The narrow-band evolution region and the second initial contour (f) The final segmentation result. Image size= 573×297.

Fig.3 and Fig.4 further show the segmentation results for two larger leaf images, respectively. The bandwidth of narrow-band evolution region in Fig.3 is 20 and the second initial contour is a square with size of 16×16. Since the size of the test leaf image in Fig.4 is larger, the corresponding bandwidth is increased to 26 and the size of the second initial contour is also increased to 20. Fig.3 (f) and Fig.4 (f) show that two target leaves are efficiently segmented from the background leaves.

5   Conclusions

In this paper, we proposed a novel two-stage level set segmentation framework for plant leaf image with overlapping phenomenon. The segmentation process can be divided into two stages, i.e. pre-segmentation stage and formal segmentation stage. Here, we used the LCV model as segmentation model for both two stages since it can not only capture the regional image information but also segment the detailed objects. Besides, the prior knowledge of leaf approximate symmetry is introduced to reduce the side-effect of shift and deformation on the segmentation result. The experiments on some real leaf images with overlapping phenomenon have demonstrated the efficiency and robustness of the proposed segmentation framework.

Acknowledgement. This work was supported by the grant of the National Natural Science Foundation of China, No. 61005010, the grant of China Postdoctoral Science Foundation, No. 20100480708, the grant of the Key Scientific Research Foundation of Education Department of Anhui Province, No. KJ2010A289, the grant of Scientific Research Foundation for Talents of Hefei University, No. 11RC05.

References

1. Wang, X.F., Huang, D.S., Du, J.X., Xu, H., Heutte, L.: Classification of Plant Leaf Images with Complicated Background. Applied Mathematics and Computation 205(2), 916–926 (2008)
2. Manh, A.G., Rabatel, G., Assemat, L., Aldon, M.J.: Weed Leaf Image Segmentation by Deformable Templates. Journal of Agricultural Engineering Research 80(2), 139–146 (2001)
3. Persson, M., Astrand, B.: Classification of Crops and Weeds Extracted by Active Shape Models. Biosystem Engineering 100(4), 484–497 (2008)
4. Zheng, L.Y., Zhang, J.T., Wang, Q.Y.: Mean-Shift-Based Color Segmentation of Images Containing Green Vegetation. Computers and Electronics in Agriculture 65(1), 93–98 (2009)
5. Wang, X.F., Huang, D.S., Xu, H.: An Efficient Local Chan-Vese Model for Image Segmentation. Pattern Recognition 43(3), 603–618 (2010)
6. Fu, H., Chi, Z.R.: Combined Thresholding and Neural Network Approach for Vein Pattern Extraction from Leaf Images. Vision, Image and Signal Processing 153(6), 881–892 (2006)
7. Chan, T.F., Vese, L.A.: Active Contours without Edges. IEEE Transactions on Image Processing 10(2), 266–277 (2001)
8. Li, C.M., Xu, C.Y., Gui, C.F., Fox, M.D.: Level Set Formulation without Re-Initialization: A New Variational Formulation. In: Proc. IEEE International Conference on Computer Vision and Pattern Recognition (CVPR 2005), pp.430–436 (2005)

Behavior Analysis of Software Systems
Based on Petri Net Slicing

Jiaying Ma[1], Wei Han[1], and Zuohua Ding[1,2,*]

[1] Lab of Intelligent Computing and Software Engineering, Zhejiang Sci-Tech University
Hangzhou, Zhejiang, 310018, P.R. China
[2] Institute of Logic and Cognition, Sun Yat-sen University
Guangzhou, Guangdong, 510275, P.R. China
sdmajiaying@yahoo.cn, hz_hanwei@163.com, zouhuading@hotmail.com

Abstract. This paper presents a new method to analyze system behavior. A software system is modeled by a Petri net, and then the Petri net is sliced into several parts based on T-invariant. It has been shown that the behavior of the slices is equivalent to the behavior of the original Petri net. Thus we can analyze the system behaviors by checking each slice. With this method, state space explosion problem in the static analysis can be solved to some extent. The Dining Philosophy problem has been used to demonstrate the benefit of our method.

Keywords: behavior analysis, petri net, model slicing, t-invariant.

1 Introduction

Traditionally, system behavior analysis can be conducted by applying static analysis techniques to the systems. These techniques include reachability based analysis techniques, symbolic model checking, flow equations, and dataflow analysis, etc. However, they may hit state space explosion problem since they have to exhaustively explore all the reachable states to detect system errors. Many techniques have been proposed to combat this explosion, including state space reductions [12], abstraction [6], and integer programming techniques [2]. With respect to these efforts, unfortunately, the explosion issue is still there, which is the main technical obstacle to transition from research to practice.

In this paper, we propose a new method to analyze the system behavior. A system is modeled by a Petri net, and the Petri net is then sliced to several slices based on T-invariants of the net, finally we check the behavior of the whole system by checking each slice. In this way, the state space explosion problem can be solved to some extent. Program slicing technique was first proposed by M. Weiser [17]. The main ideas of the slicing technique is to abstract the statements that may affect some points of interest and then divide the large and complex system model into small manageable parts. It has

---

* Corresponding author.

D.-S. Huang et al. (Eds.): ICIC 2012, LNCS 7389, pp. 475–482, 2012.
© Springer-Verlag Berlin Heidelberg 2012

been widely used in the analysis of formal models such as hierarchical state machine [8], attribute grammars [16], Object-Z specifications [4], CSP-OZ specifications [3].

This paper is structured as follows. In Section 2, we present the slicing technique based on T-invariant. In section 3, we prove the behavior equivalence of the Petri net and its slices. In Section 4, we use the dining philosophy problem as the example to illustrate our method to check system behavior. Section 5 is the discussion on the related work. Section 6 is the conclusion of the paper.

2   Petri Net Slicing Based on T-invariants

2.1   Petri Net

In this paper, we use Petri net to model software systems.

Definition 1. A Petri net is a directed bipartite graph that can be written as a tuple (P,T,F,M_0) , where P is the set of places, T is the set of transitions, $F \subset (P \times T) \cup (T \times P)$ is the set of arcs, and M_0 is the initial marking.

In our Petri net model, a place is used to denote a software state and a transition is used to denote an event. The initial marking indicates the start states.

A transition is enabled if all the input places have nonzero markings. Only enabled transitions can be fired. We assume that as soon as a transition gets enabled, it starts to fire. Firing is based on two indivisible primitives: 1) Removal of the markings from the input places and insertion of the markings in the output places (two transitions which do not share any places can be fired independently). 2) Determination of the new marking distribution function in both input and output places after firing.

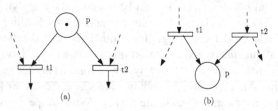

(a)   (b)

Fig. 1. The Petri net representation of conflict

Conflict is common in a Petri net, and it often leads to deadlock status. Basically, conflict is a reflection of the competitive relationship, which makes the token distribution uncertain, and thus makes the resources distribution results diverse. In a Petri nets, if a place has more than one out-transitions, but the tokens (some corresponding share resources) in the place are not enough to fire all of the enabled transitions at the same time, then only one (or some) of them can be enabled, while the rest are disenabled. Conflicts are classified as two kinds: forward conflict, which is usually called conflict, and backward conflict, which is usually called contact. For the first kind, the sharing resource is the token in place, while for the second kind conflict, the token is the capacity of post-place. Two kinds of conflicts can be described as in Fig. 1.

In the next, we will show how to divide the conflict structure into two different slices based on T-invariant, which will help us to reduce the complexity in the behavior analysis.

2.2   The Algorithm of Petri Net Slicing

Definition 2. (T-invariant) Assume T is a T-vector of Petri net N, $C:T \times S \rightarrow Z$ is the incidence matrix, $C(s_i,t_j) = W(t_j,s_i) - W(s_i,t_j)$, and θ_0 is a zero vector. If there is a T-vector satisfying $C*T = \theta_0$, then T is a T-invariant of Petri net N. Especially, if $T > \theta_0$, then T is a nonnegative invariant. If there is no other T-invariant T' satisfying $\theta_0 < T' < T$, then T is a minimal invariant.

T-invariant of a Petri net describes the locality of a system, i.e., a dynamic that begins from arbitrary initial marking M_0 can come back to the marking M_0 after a sequence of transitions.

The following algorithm (in Fig. 2) shows how a Petri net model is sliced into a set of small Petri net pieces based on T-invariant.

```
sliceset=φ
do{
       small_invariant=find_smallest_invariant(
       set_of_minimal_invariant)
       transition_connected=φ
       invariant_connected=φ
       for ∀ transition ∈ small_invariant
              transition_connected=find_ place_connected
       invariant_connected= all_transition_connected
       slice= small_invariant+invariant_connected
       sliceset=sliceset+slice
       sliceset_of_ invariant= set_of_ invariant-small_invariant
       } until(set_of_ minimal_invariant=φ )
if transition(sliceset) ≠ transition(N)
do{
       uncovered_transition_set=transition(N)-transition(sliceset)
       for ∀ transition ∈ uncovered_transition_set
              add_according_to_pre-place(post-place)
       }
end
```

Fig. 2. Petri net Slicing algorithm

First the T-invariants are computed and the minimal invariants are selected. Then the smallest invariant is chosen to construct the slice, i.e. after getting a smallest invariant, the algorithm checks the pre-places of each transition and composes a cycle as a Petri net slice.

If the invariant set becomes empty without covering all the transitions in the Petri net N, then the algorithm checks the pre-places and post-places of the transition one by one; If the pre-places (or post-places) of a transition are in some slice, then the transition is added into the slice. Accordingly, the corresponding post-places (or pre-places) are also added into the slice. For a transition, if its pre-places and post-places belong to no slices, it will be skipped and will be checked later after all uncovered transitions are checked. In this way, the algorithm guarantees that every transition in N belongs to some slice.

The modular Petri net (or Petri net pieces) obtained from the slicing algorithm is called as Petri net slice.

2.3   Complexity Analysis

Now we compute the complexity of the algorithm. The complexity of finding T-invariant is the same as that of solving the algebra equation $C * T = \theta_0$. Let the number of places and transitions in the Petri net be m and n respectively. The complexity of solving the equation is $O(n^3)$. Finding a smallest invariant is to compute the number of the invariants and then sort (bubble sort) them, thus the complexity is $O(n^2)$. Checking the pre-places and post-places of every transition is similar to locating the positive (negative, zero) number of an array, and the complexity is $O(mn)$ for m lines and n volumes. In summary, the complexity of the slicing algorithm is $O(n^3)$.

3   Justification of the Behavior Equivalence

In this section, we will prove that the original Petri net model and the Petri net slices have the same behavior.

Definition 3. (Behavior Equivalence) Two Petri net P and Q are behaviorally equivalent iff there exists an one-to-one mapping between the reachability graphs of P and Q.

We have the following observation from the algorithm:

− The change of states in Petri net is determined by the initial marking and fire sequence of transitions;
− Slicing algorithm reserves all the input and output information of every transition;
− The initial marking remains the same before or after slicing.

Hence, we can prove the following result.

Theorem 4. A Petri net and its slices obtained from T-invariant are behaviorally equivalent.

Proof. In order to prove the behavior equivalence, we only need to prove that the firing sequences remain the same after slicing.

Assume that $s_1 : t_1 t_2 t_3 \cdots t_{k-1} t_k \cdots t_{n-1} t_n$ is a firing sequence. Next we show that every t_k can fire in Petri net slices. t_k may appear in more slices, but the structures of t_k are the same (i.e. its pre-places and post-places are the same), so for the convenience, we may assume that it appears in only one slice. There are two cases.

Case 1: The pre-place of t_k is not shared.

Assume that transition t_k appears in a slice, say S, then all of its pre-places and post-places are in S, so t_k can fire.

Case 2: There is at least one shared place in the pre-places.

Let place p_a be a shared place for two transitions t_k and t_m (t_m belongs to another firing sequence, say s_2). Assume that the transition t_k is fired first and then p_a is released after the firing of sequence s_1; or t_k waits until sequence s_2 is finished and p_a is released. In both situations, the transition t_k can fire and the firing sequence s_1 can execute completely.

4   An Example: Dining Philosophy Problem

We use the dining philosophy problem as the example to show our method for behavior analysis. Dijkstra's [7] dining philosopher problem is a very well-known example for the static analysis. A group of N philosophers is sitting around a table. The philosophers alternate between thinking and eating. Initially n forks are placed on the table between each pair of philosophers. In order to eat, a philosopher must first pick up both forks next to him. The forks are put down when the philosopher finishes eating and starts thinking. This problem is interesting because of the possibility of a circular deadlock. Deadlock may occur if all philosophers pick their forks in the same order, say, the right fork followed by the left fork. In this case, there is one deadlock state corresponding to the situation in which all philosophers have picked up their right fork and are waiting for their left fork.

The Petri net for 5 philosophers is shown in Fig.3.

For every philosopher and every fork, the corresponding behavior is an independent process. If we take the first philosopher as the example, then the behavior can be described as $p_1 \xrightarrow{t_1} p_2 \xrightarrow{t_2} p_3 \xrightarrow{t_3} p_4 \xrightarrow{t_4} p_1$, and the behavior of a fork can be described as $p_{21} \xrightarrow{t_1} p_2 \xrightarrow{t_2} p_{34} \xrightarrow{t_3} p_{21}$ or ($p_{21} \xrightarrow{t_{10}} p_2 \xrightarrow{t_{11}} p_{34} \xrightarrow{t_{12}} p_{21}$).

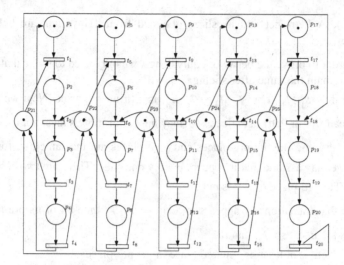

Fig. 3. Petri net model for five philosophers

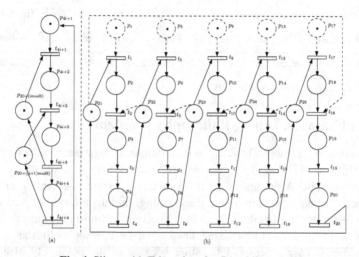

Fig. 4. Slices with T-invariant for five philosophers

After applying the T-invariant based algorithm to the Petri net model, we get 5 T-invariants:

$$T_1 = (1,1,1,1,0,0,0,0,0,0,0,0,0,0,0,0,0,0,0,0)$$
$$T_2 = (0,0,0,0,1,1,1,1,0,0,0,0,0,0,0,0,0,0,0,0)$$
$$T_3 = (0,0,0,0,0,0,0,0,1,1,1,1,0,0,0,0,0,0,0,0) \qquad (1)$$
$$T_4 = (0,0,0,0,0,0,0,0,0,0,0,0,1,1,1,1,0,0,0,0)$$
$$T_5 = (0,0,0,0,0,0,0,0,0,0,0,0,0,0,0,0,1,1,1,1)$$

The corresponding two slices are shown in the Fig. 4. Fig.4(a) describes one behavior of the dinning philosopher problem: *stop-thinking* → *get Left-Fork and Right-Fork consequently* → *eating* → *thinking*. Easy to see that the transitions in slice (a) can fire, i.e. slice (a) can execute. Fig.4(b) describes another behavior. In Fig.4(b), if transitions $t_1, t_5, t_9, t_{13}, t_{17}$ get tokens from $p_{21}, p_{22}, p_{23}, p_{24}, p_{25}$ at the same time, then transitions $t_2, t_6, t_{10}, t_{14}, t_{18}$ will be waiting for these tokens forever, i.e., the system comes to the deadlock. Hence this slice describes the deadlock behavior of the dinning philosopher problem.

Remark 1. The slices obtained with S-invariant based method [10] are related to each others, thus the execution of one slice is dependent on the execution of other slices. In other words, we can not analyze the dinning philosopher problem based on their slices.

5 Related Work

The research of Petri net slicing focuses on two aspects: one is to extract the statements that may affect the property under study, and the other is to divide a complex Petri net model into small manageable parts and then study the smaller parts.

Chang and Wang [5] presented a static slicing algorithm that slices out all sets of paths, known as concurrent sets,so that all paths within the same set should be executed concurrently. M. Llorens [11] defined a group of slicing criterion. Paths can be generated based on the slicing criterion. This method can be used for the debugging and model checking. Rakow [14][15] presented another static slicing technique to reduce the Petri net size. In his method, for a given place set, the input places, incoming transitions and outgoing transitions are added to the set iteratively to construct a subnet. The Petri nets size is reduced, thus the state space explosion problem can be solved to some extent. W. J. Lee and H. N. Kim [10] proposed a static slicing technique in which minimal S-invariants are used to partition a Petri net into concurrent units (Petri net slices) and the partitioned models can be analyzed by a compositional reachability analysis technique. They also proved that the boundedness and liveness will be retained in the partitioned models.

6 Conclusion

In this paper, a T-invariant based Petri net slicing technique is presented. The slices obtained by this technique have the equivalent behavior of the original Petri net. This technique can be used to study the behavior of a class of concurrent systems that use resource sharing as their synchronization mechanism. Instead of checking the whole system, we can check each slice to get the behavior of whole system. In the future, we will consider applying this technique to the system test.

Acknowledgments. This work is supported by the NSF under Grant No.61170015, Zhejiang NSF under Grant No.Z1090357, and Key Project of Ministry of Education of China under Grant No. 10JZD0006.

References

1. Ascher, U.M., Petzold, L.R.: Computer Methods for Ordinary Differential Equations and Differential-Algebraic Equations. Society for Industrial & Applied Mathematis, Philadelphia (1998)
2. Avrunin, G.S., Buy, U.A., Corbett, J.C., Dillon, L.K., Wileden, J.C.: Automated Analysis of Concurrent Systems With The Constrained Expression Toolset. IEEE Transactions on Software Engineering 17(11), 1204–1222 (1991)
3. Brückner, I.: Slicing CSP-OZ Specifications. In: Proc. of the 16th Nordic Workshop on Programming Theory, pp. 71–73. Uppsala University, Sweden (2004)
4. Brückner, I., Wehrheim, H.: Slicing Object-Z Specifications for Verification. In: Treharne, H., King, S., C. Henson, M., Schneider, S. (eds.) ZB 2005. LNCS, vol. 3455, pp. 414–433. Springer, Heidelberg (2005)
5. Chang, C.K., Wang, H.: A Slicing Algorithm of Concurrency Modeling Based on Petri Nets. In: Proc. of the International Conf. on Parallel Processing 1986, pp. 789–792. IEEE Computer Society Press (1986)
6. Clarke, E.M., Grumberg, O., Long, D.E.: Model Checking and Abstraction. ACM Transactions on Programming Language Systems 16(5), 1512–1542 (1994)
7. Dijkstra, E.W.: Hierarchical Ordering of Sequential Processes. Acta Informat. 2, 115–138 (1971)
8. Heimdahl, M.P.E., Whalen, M.W.: Reduction and Slicing of Hierarchical State Machines. In: Jazayeri, M. (ed.) ESEC/FSE 1997. LNCS, vol. 1301, pp. 450–467. Springer, Heidelberg (1997)
9. Lee, J.: Scheduling Analysis with Resources Share Using The Transitive Matrix Based on P-invariant. In: Proceedings of the 41st SICE Annual Conference 2002, pp. 5–7 (2002)
10. Lee, W.J., Kim, H.N.: A Slicing-Based Approach to Enhance Petri Net Reachability Analysis. Journal of Research and Practice in Information Technology 32, 131–143 (2000)
11. Llorens, M., Oliver, J., et al.: Dynamic Slicing Techniques for Petri Nets. Electronic Notes in the Theoretical Computer Science 223, 1–12 (2008)
12. Long, D.L., Clarke, L.A.: Task Interaction Graphs for Concurrency Analysis. In: Proceedings of 11th ICSE, Pittsburgh, Penn., USA, pp. 44–52 (1989)
13. Murata, T.: Petri Nets: Properties, Analysis and Applications. Proceedings of the IEEE 77, 541–580 (1989)
14. Rakow, A.: Slicing Petri Nets with an Application to Workflow Verification. In: Geffert, V., Karhumäki, J., Bertoni, A., Preneel, B., Návrat, P., Bieliková, M. (eds.) SOFSEM 2008. LNCS, vol. 4910, pp. 436–447. Springer, Heidelberg (2008)
15. Rakow, A.: Slicing Petri Nets. In: Proceeding of the Workshop on FABPWS 2007, Satellite Event, Siedlce, pp. 55–76 (2007)
16. Sloane, A.M., Holdsworth, J.: Beyond Traditional Program Slicing. In: Proc. of the International Symp. on Software Testing and Analysis, San Diego, CA, pp. 180–186 (1996)
17. Weiser, M.: Program Slicing. IEEE Transactions on Software Engineering 10, 352–357 (1984)

Aircraft Landing Scheduling Based on Semantic Agent Negotiation Mechanism

Zhao-Tong Huang, Xue-Yan Song, Ji-Zhou Sun, and Zhi-Zeng Li

School of Computer Science and Technology, Tianjin University, Tianjin, China
zhaotong@tju.edu.cn

Abstract. The aircraft landing scheduling problem (ALS) is a typical NP-hard optimization problem and exists in the Air Traffic Control for a long time. Many algorithms have been proposed to solve the problem, and most of them are centralized. With the development of the aerotechnics, the concept of free flight has been proposed. Airplanes could change their flight paths during the flight without approval from a centralized en route control. In order to support free flight, a distributed system based on Multi-Agent System is proposed in this paper. The kernel of the system is semantic agent negotiation mechanism. With the method aircrafts in the system could make landing sequence considering their own states. The Experimental results show that the proposed algorithm is able to obtain an optimal landing sequence and landing time rapidly and effectively.

Keywords: Aircraft landing scheduling problem, semantic agent negotiation, free flight.

1   Introduction

The Aircraft Landing Scheduling (ALS) problem is one of the important problems in Air Traffic Control (ATC). The problem is to determine the landing sequences and landing time for a given set of aircrafts [1]. A lot of research has been done on the optimization of ALS problem in recent years. The linear programming (LP) method has been applied in [1], [2]. Heuristic algorithms are used to solve the ALS problem owing to the powerful ability of parallel and global search [4]-[6].

In this paper we propose a distributed agent approach: Semantic Agent Negotiation Mechanism (SANM) which is based on Multi-Agent System (MAS) to solve the ALS problem. The main idea of the SANM is that aircrafts can make the landing sequence through cooperation. SANM solves the problem consider the whole condition and single aircraft, while other algorithms only consider the whole condition. Semantic knowledge is used to describe the negotiation rules in SANM. SANM has two steps: (1) every aircraft uses its own algorithm (here we will discuss the genetic algorithm) to get a landing scheduling; (2) then all aircrafts use the negotiation mechanism to optimize the scheduling and get a better result.

The remainder of this paper is organized as follows. The ALS problem in multi-runway systems, focus on minimizing the cost of the aircraft sequence, hereafter, is described in Section 2. Section 3 introduces the model of SANM. Genetic algorithm

D.-S. Huang et al. (Eds.): ICIC 2012, LNCS 7389, pp. 483–489, 2012.
© Springer-Verlag Berlin Heidelberg 2012

that used to solve the ALS problem is introduced in Section 4. The negotiation rules are proposed in Section 5. Extensive simulation study is reported in Section 6, and the paper ends with some conclusions in Section 7.

2   ALS Problem Formulation

Assume that a set of n aircrafts will land at a multi-runway airport. The objective of ALS problem is to find an optimal arrival sequence for aircrafts. In arrival sequence, time interval of adjacent aircrafts will land at the same runway should be no less than S_{ij} . S_{ij} is the required separation time between aircraft i and aircraft j . Each aircraft must land within the bounds of the time window (E_i , L_i).

Delay time of aircraft can be defined as (1). For each aircraft, extra cost caused by delay should be defined as (2).

$$d_i = \begin{cases} t_i - T_i & (L_i > t_i > T_i) \\ 0 & , else \end{cases} \quad \forall i \in (1, N) \tag{1}$$

$$c_i = \begin{cases} g_i(T_i - t_i) , & E_i < t_i < T_i \\ h_i(t_i - T_i) , & T_i < t_i < L_i \end{cases} \quad \forall i \in (1, N) \tag{2}$$

C_i is the extra cost for aircraft i. t_i stands for the time allotted that aircraft will land on the runway. g_i is the penalty cost (positive) per unit of time for landing before T_i for aircraft i ; h_i is the penalty cost (positive) per unit of time for landing after T_i for aircraft i .

Objective function of ALS problem can be described as equation (3) (4):

$$f = \min \sum_{i=1}^{N} d_i \tag{3}$$

$$f = \min \sum_{i=1}^{N} C_i \tag{4}$$

3   Semantic Agent Negotiation Mechanism (SANM)

The next-generation air transportation system should allow airplanes to change their flight paths during the flight without approval from a centralized en route control. With the development of aerotechnics and requirement of free flight, the solution of ALS problem should consider not only the macro aspect, but also the status of every aircraft. So the centralized system cannot satisfy the development tendency and a distributed system is necessary. In this paper, a solution based on Multi-Agent System is proposed: semantic agent negotiation mechanism. The method has two steps: (1) every aircraft uses algorithm (here we select genetic algorithm, discussed in section V) to get a good scheduling, (2) aircrafts use the negotiation mechanism to optimize the scheduling to satisfy their own benefits.

SANM is based on Multi-Agent System and described in Figure 1. A_i $(i \in (1, N))$ represents the aircraft agent. Controller agent and aircrafts agent use Agent Communication language (ACL) to exchange message. Genetic Algorithm, Negotiation strategy and Proposal Generation Mechanism are invoked by aircraft agent. Genetic Algorithm is used to get an initial solution. Proposal Generation Mechanism is invoked to create proposal. Negotiation strategy supplies the rule abides by aircraft agent when consulting proposal.

In order to simplify the model of SANM, we describe the model with a tuple (5).

$$M =< Ag, GA, S, I, V, R >$$ (5)

$Ag = \{A_1, ..., A_n, C\}$ is the set of aircraft agents and controller agent. GA is Genetic Algorithm. S is negotiation strategy. I is proposal generation mechanism. $V =< v_1^p, ..., v_n^p, v_c^p >$, v_i^p is the valuation of aircraft agent i for proposal p. R reserves the best scheduling at present.

The detail of the algorithm is described as follow:

(1) A_i uses GA to get a scheduling.

(2) A_i sends its scheduling to other aircraft agents and controller agent.

(3) Aircraft agents and controller agent grade the scheduling according to their own benefits. The less delay and external cost, the higher grade will be given.

(4) Calculate the score for each landing scheduling. The scheduling which gets the highest score will be selected as initial solution.

(5) Initial solution is not suitable for all aircrafts. Aircraft agent uses the proposal generation mechanism I to create the proposal for initial solution. Proposal generation mechanism will be discussed in section 5.

(6) All the aircrafts use the negotiation strategy S to decide which proposal will be accepted. V will be used in this step. Negotiation strategy will be discussed in section 5.

(7) If the terminal conditions (time constraint or negotiation suspension) are not satisfied, go to (5), otherwise reserve the scheduling in R, and then return R.

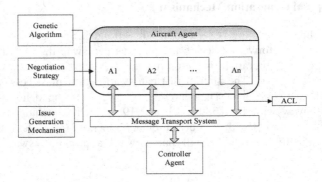

Fig. 1. SANM

4   Genetic Algorithm for ALS Problem

Genetic Algorithm (GA) is adaptive heuristic search algorithm premised on the evolutionary ideas of natural selection and genetic. In this section, we give genetic algorithm with novel constraints for ALS Problem. Chromosome coding which describes the landing scheduling is defined as Table 1. Consider that N aircrafts will land on three runways. In the first row of table 1, the sequence number of aircrafts is assigned according to the order of predicted landing time. The corresponding runways are given in the second row.

Table 1. Chromosome coding

Aircraft	1	2	3	4	5	...	N-1	N
Runway	2	1	3	1	2	...	3	2

The chromosome coding has to satisfy two constraints.

1) The load of the runways should be balanced.

In formula (6), r_i represents the number of aircrafts assigned in the *ith* runway. N and R stand for the total number of aircrafts and runways respectively. The usage rate of the runway should be in a certain range $[1/R-a, 1/R+a]$.

$$\left| \frac{r_i}{N} - \frac{1}{R} \right| \leq 0.1 \quad i \in [0, R-1] \tag{6}$$

2) Security constraint. Three or more aircrafts in continuous time can not be assigned in the same runway. Formula (7) describes the security constraint, and P_i represents the runway assigned to the aircraft i.

$$P_{i+1} \neq P_{i+2} \quad (\text{If } P_i = P_{i+1}, i \in [1, N]) \tag{7}$$

5   Proposal Generation Mechanism and Negotiation Strategy

5.1   Proposal Generation Mechanism

When a landing scheduling is send to an aircraft agent, the agent will evaluate the scheduling and put forward a set of aircrafts that can swap the runway with it. For example, in table 2, aircraft 17 is put on runway 2. A set of aircrafts it can swap with is {14, 15, 16, 18, 19}. The aircraft that can swap with it must between 13 and 20 because of time constraint. Then aircraft 17 calculates the delay and cost if it swaps the runway with 14, 15, 16, 18, and 19 respectively. Suppose it swaps the runway with aircraft 16, the value of object function (4) is less than previous scheduling. Finally aircraft 17 will put forward a proposal: swap runway with aircraft 16.

5.2   Negotiation Strategy

In our model, Strategies are adopted to direct the adaptive negotiation process. The negotiation strategies are expressed in OWL since it has a text-based representation which imposes few restrictions on protocols and (through the use of OWL schema) it has a sufficiently strict syntax to permit automated validation and processing of information in an unambiguous way. Negotiation strategies allow selections from a range of proposal provided by aircraft agents. Figure 2 shows sequence decides the proposal according to semantic descriptions.

Table 2. Example of Landing Scheduling

ID	Time(s)	Runway	ID	Time(s)	Runway	Runway1	Runway2	Runway3
1	56	3	11	346	3	3	2	1
2	113	2	12	378	2	4	5	6
3	113	1	13	400	2	8	12	7
4	122	1	14	448	1	9	13	10
5	155	2	15	467	1	14	17	11
6	156	3	16	491	3	15	20	16
7	218	3	17	500	2	19		18
8	237	1	18	653	3			
9	327	1	19	679	1			
10	332	3	20	705	2			

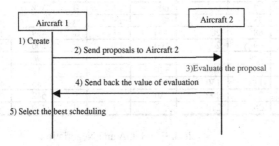

Fig. 2. Execution Step

6   Experiment

We write the program of SANM with Visual Studio 2010. To fully investigate the effectiveness of our algorithm, we compare it with the method first come first service (FCFS) on the given dataset. In table 3 we give 20 aircrafts which will land on three runways. S, L and H stand for small, middle and heavy plane respectively. Previous, earliest and latest landing time of aircrafts are given in table 3.

First of all, every aircraft works out a landing scheduling by using genetic algorithm. Total delay of each scheduling is given in Figure 3. We can see that the landing scheduling provided by aircraft 6 has the least delay time 550s. Then we will use proposal generation mechanism and negotiation strategy to optimize the initial solution. The final result is given in the third part of the Table 4.

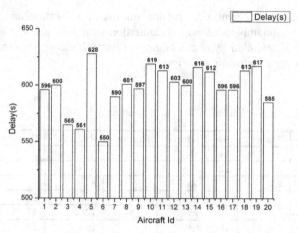

Fig. 3. Dealy time of every aircraft

Table 3. Real time data

Serial Number	Type	Time(s)	Earliest(s)	Latest(s)	Serial Number	Type	Time(s)	Earliest(s)	Latest(s)
1	S	18	0	50	11	L	343	300	380
2	H	37	0	60	12	H	398	350	440
3	H	53	10	80	13	H	487	460	520
4	H	68	20	100	14	S	489	460	520
5	L	69	20	100	15	L	500	470	540
6	S	134	100	170	16	H	596	550	640
7	S	138	100	170	17	L	619	580	670
8	S	182	150	210	18	H	635	590	680
9	S	240	200	280	19	S	642	600	680
10	H	255	210	290	20	H	719	680	760

Table 4. FCFS, GA and Negotiation

FCFS	ID	Delay(s)	GA	ID	Delay(s)	Negotiation	ID	Delay(s)	Scheduling(s)
Runway1	1	0	Runway1	2	0	Runway1	2	0	0
	4	24		4	63		5	82	114
	7	121		8	116		8	107	252
	10	78		11	29		11	20	343
	13	0		15	0		15	0	500
	16	0		17	0		17	0	619
	19	121		20	0		20	0	719
Runway2	2	0	Runway2	3	0	Runway2	3	0	10
	5	82		5	98		4	79	104
	8	107		10	0		10	0	255
	11	20		12	0		12	0	398
	14	12		13	5		13	5	492
	17	0		16	0		16	0	596
	20	0		18	55		18	55	690
Runway3	3	0	Runway3	1	0	Runway3	1	0	18
	6	86		6	0		6	0	116
	9	78		7	94		7	94	214
	12	0		9	90		9	90	312
	15	12		14	0		14	0	489
	18	0		19	0		19	0	642

In Table 5 we list the delay time of different methods which solve the ALS problem. It is obvious that our method has the better optimization performance compared with FCFS and GA.

Table 5. Data Compare

	FCFS	GA(Max)	GA(Min)	SANM
Delay(s)	741	628	550	532
Cost	13770	12960	11950	10320

7   Conclusion

In this paper, an algorithm of SANM is proposed to solve aircraft landing scheduling problem. Two techniques: Proposal Generation Mechanism and Negotiation Strategy are designed in our algorithm. The simulation results show that: this method is reasonable and it can reduce the total time delay, optimize the approach sequencing. The modeling and decision process of this methodology is easy to realize and suitable to be used in terminal area aircraft approach sequencing, it can help the controllers make a quick decision and reduce their workload if being used.

Acknowledgements. This study is supported by grants from National Natural Science Foundation of China and Civil Aviation Administration of China(61039001) and Tianjin Municipal Science and Technology Commission(11ZCKFGX04200).

References

1. Ciesielski, V., Scerri, P.: An Anytime Algorithm for Scheduling Aircraft Landing Times Using Genetic Algorithms. Australian Journal of Intelligent Information Processing System 4, 206–213 (1998)
2. Beasley, J.E., Krishnamoorthy, M., Sharaiha, Y.M., Abramson, D.: Scheduling Aircraft Landings - the Static Case. Transportation Science 34, 180–197 (2000)
3. Xu, X.H., Yao, Y.: Application of Genetic Algorithm to Aircraft Sequencing in Terminal Area. Journal of Traffic and Transportation Engineering 4, 121–126 (2004)
4. Yang, Q.: Scheduling Arrival Aircrafts on Multiple Runways Based on an Improved Genetic Algorithm. Journal of Sichuan University 38, 65–70 (2006)
5. Zhang, H., Hu, M.: Multi-runway Collaborative Scheduling Optimization of Aircraft Landing. Journal of Traffic and Transportation Engineering 9, 115–120 (2009)
6. Rong, J., Geng, S., Valasek, J., Ioerger, T.R.: Air Traffic Control Negotiation and Resolution Using an Onboard Multi-agent System. In: Proc. Digital Avionics Syst. Conf., vol. 2, pp. 7B2-1–7B2-12 (2002)
7. Pěchouček, M., Šišlák, D.: Agent-based Approach to Free-flight Planning, Control, and Simulation. IEEE Intell. Syst. 24(1), 14–17 (2009)
8. Lomuscio, R., Wooldridge, M., Jennings, N.R.: A Classification Scheme for Negotiation in Electronic Commerce. Int. Journal of Group Decision and Negotiation 12(1), 31–56 (2003)

A Real-Time Posture Simulation Method
for High-Speed Train

Huijuan Zhou[1,2], Bo Chen[2], Yong Qin[2], and Yiran Liu[1]

[1] College of Mechanical and Electronical Engineering, North China University of Technology,
Beijing, China
[2] State Key Laboratory of Rail Traffic Control and Safety, Beijing Jiaotong University,
Beijing, China
zhhjuan@sina.com, 2828900@163.com, 276356856@qq.com,
qinyong21246@126.com

Abstract. A method of real-time posture simulation aimed to high speed moving train is presented. This method decomposes the full train length in smaller train elements, so that dynamic posture simulation of train is transformed into posture simulation of each part of the train. Then according to the real-time locomotive mileage, track geometry and the center mileage, yaw angle and roll angle can be calculated for simulate the posture of each part. The whole train posture is obtained through connected each part's posture together in space-time continuum. The method is simple, accurate, efficient and practical and is used in actual system of comprehensive inspection train visualization.

Keywords: high-speed train, real time, posture, simulation.

1   Introduction

With the successful operation of Jing-Jin Intercity Railway, Beijing-Shanghai High-speed Railway and other high-speed railways, Chinese high-speed railway is now developing rapidly. Real-time posture of the high-speed moving train has an important effect on real-time train monitoring and operation safety. For the limits of monitoring equipment cost and data transmission rate, the real-time train's posture can't be obtained. However, the computer simulation technology makes it possible for real-time posture simulation of the high-speed moving train with the development in computer.

There are many methods for real-time posture simulation, such as active disturbance rejection control technique[1] and neural-predictive controller[2] in satellite attitude simulation system, the time-domain response method based on stochastic process theory of posture simulation of warship[3], the method of bodywork posture simulation based on course simulation[4], the six-degree-of-freedom movement posture simulation[5]. In train posture simulation, it's usually used to build solid model by Simpack and Adams/Rail or establish dynamic equations by Matlab/Simulink.

But Simpack and Adams/Rail is very difficult to solve the problem for multi-contacts, collision and large-scale systems[6]. Each Matlab/Simulink module is a black box for the user and its graphical user interface can't show the real-time 3D

D.-S. Huang et al. (Eds.): ICIC 2012, LNCS 7389, pp. 490–496, 2012.
© Springer-Verlag Berlin Heidelberg 2012

posture of the train[7-8]. Hence, the present simulation software does not display real-time posture of the high-speed moving train, the railway line and the surrounding buildings with the high-precision. Especially, it's impossible to realize above functions for these simulation software in Browse/Server model. This paper puts forward a method combined vehicle dynamics parameters with 3D GIS and B/S architecture to simulate real-time posture aimed to the high-speed moving train.

2   Algorithm for Real-Time Posture Simulation

For rail vehicle system, there are six kinds of basic carbody movement: lateral oscillation, bounce response, stretching vibration, yaw motion, pitch response and roll motion.

Lateral oscillation, roll and shake motion of carbody are called the lateral vibration. Because of the elastic constraints of the vehicle structure and the suspension system, the lateral oscillation and roll vibration always couple together. Bounce and pitch response happen in longitudinal vertical plane, belonged to vertical vibration. Bounce response is the vibration of the pre and post bogies with the same phase and amplitude in the vertical plane. Otherwise, if the vertical vibration direction of the pre and post bogies is opposite, the vibration is pitch response. The longitudinal stretching vibration of carbody generally appears when rail vehicle start, brake, driving and shunting, which is closely related to train dynamics[9-11]. Therefore, this paper proposed the real-time posture simulation method considered kinetic parameters such as pitch angle β, yaw angle φ and roll angle ϕ which would influence the real-time posture of the train, and didn't consider the influence of longitudinal vibration.

The method can simulate the real-time posture of high-speed moving train in 3D with vivid and continuum space-time. The flow chart of posture simulation is shown in Figure 1.

Firstly, the train is divided into multiple parts, so that posture simulation of train is transformed into posture simulation of each part of the train. And then, in the case of known mileage of the locomotive, each part's center mileage is calculated. Thirdly, according to the track geometry at the center point of each part, the real-time posture simulation method of each part is determined. Finally, real-time posture of each part of the train connected together in space-time continuum will full-dimensionally simulate the real-time posture of high-speed moving train. Specific steps are as follows:

(1) The train model is divided into n parts from head to tail (shown in Fig. 2) and posture simulation of the train is transformed into posture simulation of each part of the train.

(2) Train length is d, then each part of the train length is d/n. Train head mileage is s_0, Then the central mileage of $i(i = 0,1,2,...,n)$ part is:

$$s = s_0 - \frac{d}{2n} - \frac{d}{n}(i-1) \tag{1}$$

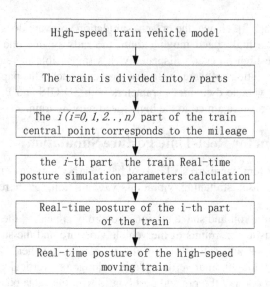

Fig. 1. Flow chart of high-speed train posture simulation

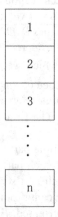

Fig. 2. Train divided into n parts

(3) In the process of high-speed train movement, real-time posture simulation method of each part of the train in real-time mileage S is as follows:

- h is the distance from the part center to the center point of left and right track. θ is the horizontal angle of left and right track center and $\theta = \arcsin v^2 / gr$ (shown in Fig. 3).Where v is design speed, g is gravity acceleration, r is track curve radius and α is track curve tangent direction angle(shown in Fig. 4).

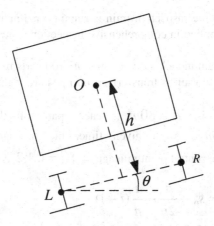

Fig. 3. Horizontal angle θ and the distance h

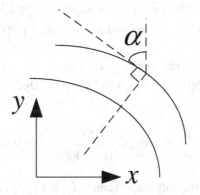

Fig. 4. Track curve tangent angle α

- $L(x_L, y_L, z_L)$ is the coordinate of the left track center point, $R(x_R, y_R, z_R)$ is the coordinate of the right track center point. The coordinate of the center O of the part is

$$O(x, y, z) = (\frac{x_L + x_R}{2} - h \bullet \sin\theta, \frac{y_L + y_R}{2}, \frac{z_L + z_R}{2} + h \bullet \cos\theta) \qquad (2)$$

- Assuming that the train wheels and tracks contact intimately and there isn't or minor pitch response, then the pitch angle $\beta = 0$, and set yaw angle $\varphi = \alpha$, roll angle $\phi = \theta$.

3   Application

In a '863' project about visualization system of high-speed comprehensive inspection train based on 3D WebGIS, the simulation method above is used to simulate the real-time posture when the inspection train is running in Jinghu high-speed railway. The system has been applied in comprehensive inspection center and the following is actual application.

(1) The inspection train model is divided into $n=100$ parts from head to tail, so that posture simulation of the train is transformed into posture simulation of each part of the train.

(2) Train length is $d = 80m$, each part of the train length is $d/n = 80/100 = 0.8m$. At a given time, the mileage of train head is $s_0 = 120km$. Then the central point mileage of $i(i = 0,1,2,...,n)$ part is:

$$s = s_0 - \frac{d}{2n} - \frac{d}{n}(i-1)$$
$$= 119.9996 - 0.0004(i-1)(km) \tag{3}$$

(3)At this moment, track curve tangent angle is $\alpha = \pi/4$ at real-time mileage $s = 119.996km(i = 10)$. Real-time posture simulation method of the part $i = 10$ is as follows:

● The distance from the part center to the center point of left and right track is $h = 2.5m$, $v = 360km/h = 100m/s$, $g = 10m/s^2$, $r = 3000m$, then the horizontal angle is

$$\theta = \arcsin \frac{v^2}{gr} = \arcsin \frac{100^2}{10 \times 3000} = 0.3398 \tag{4}$$

● The coordinate of the left track center L is $(x_L, y_L, z_L) = (12,6,5)$, the coordinate of the right track center R is $(x_R, y_R, z_R) = (13.5,6,5.5)$ $(x_R, y_R, z_R) = (13.5,6,5.5)$ and the coordinate of the center O of the part is

$$O(x, y, z) = (\frac{x_L + x_R}{2} - h \bullet \sin \theta, \frac{y_L + y_R}{2}, \frac{z_L + z_R}{2} + h \bullet \cos \theta) \tag{5}$$
$$= (11.9168, 6, 7.6071)$$

● Yaw angle $\varphi = \alpha = \pi/4$ and roll angle $\phi = \theta = 0.3398$.

Real-time dynamic posture simulation of each part of the train is obtained based above method and the simulation effect is showed in Fig.4. In this project, real-time train information including GPS, mileage and velocity is transmitted every 5 seconds from train to the ground center. The train's position and posture are drived by the accepted real-time data and updated in accordance with 24 frames every one second. In some certain, the train's position is delayed for the limit of transmission time and

Fig. 5. Actual posture simulation for high-speed comprehensive inspection train

transmission rate. So there are some other algorithms to calculate and estimate the next position and velocity according to the fore data accepted[12]. The system requires Inter Core2 multi-core CPU, 4G RAM, 1T hard disk and NVIDIA Series GeForece8800 Graphics card.

4   Conclusions

The method advanced in this paper, decomposing the full train length in smaller train element and using the same real-time posture simulation method aimed at each train element, reduces the simulation difficulty of real-time high-speed train posture and improves the simulation accuracy. At the same time, the method has the advantages of simply calculation, practical and efficient for its simulation according to curve characteristics of track and geometric features of train. Combined 3D GIS and Browse/Server, the high-speed train posture and its position, velocity can be showed in a vivid 3D platform. The actual application has proved the method accuracy and efficiency.

Acknowledgement. The authors wish to acknowledge the support and motivation provided by the 863 project NO.0912JJ0203 and Beijing college students scientific research and entrepreneurial action project.

References

1. Lian, M., Han, Z.Y., Fu, H.Y.: Application of Active Disturbances Rejection Control Technique to Satellite Attitude Simulation System. Optics and Precision Engineering 18(3), 616–622 (2010)
2. Li, Y.G., Han, J.G., Li, K., et al.: Neural-Predictive Controller and Its Application to Satellite Attitude Simulation System. Computer Measurement & Control 11(5), 359–364 (2003)
3. Qin, X.C., Mao, P., Lu, H.M.: The Simulation of One Warship's Main Motion Gestures. Computer Simulation 15(4), 59–51 (1998)
4. Dong, Z.H., Cheng, Y.Z., Fu, Q., et al.: Producing Method of Bodywork Attitude Information Based on Road Simulation. Electronic Measurement Technology 27(3), 57–67 (2002)

5. Hao, Y.N., Wang, J.Z., Wang, S.K.: A Study on a Six-Degree-of-Freedom Platform. Journal of Beijing Institute of Technology 22(3), 331–334 (2002)
6. Zhu, H.R., Chen, G., et al.: Probe into Method of Safety Analysis for Recreation Facilities Based on Virtual Prototype Technology. China Safety Science Journal 14(3), 12–15 (2004)
7. Seeking technology. MATLAB 7.0 from the Entry to the Master. Posts & Telecom Press (2006)
8. Zhou, L.N., Tang, G.J., Luo, Y.Z.: Simulation Framework for Spacecraft Attitude Dynamics and Control Based on Matlab/Simulink. Journal of System Simulation 17(10), 2517–2524 (2005)
9. Han, B.L., Chen, Z.Y., Lu, Z.J.: On-line Detection of Offset Generated by Vibration of Rolling Stock Based on Machine Vision 39(4), 787–792 (2008)
10. Jiao, S.: Railway Vehicles Elastic Vibration Analysis Method. Dalian Jiaotong University Master's Degree Thesis (2008)
11. Zhong, C.M.: Truck Center of Gravity Position of the Roll Angle Detection. Rolling Stock 11, 13–15 (1993)
12. Zhou, H.J., Qin, Y., Xing, Z.Y., et al.: Design and Implementation of Comprehensive Inspection Data Visualization System Based on 3D GIS. In: The 1st International Conference on Railway Engineering: High-speed Railway, Heavy Haul Railway and Urban Rail Transit, pp. 618–623 (2010)

High-Order Terminal Sliding-Mode Observers for Anomaly Detection

Yong Feng[1,2], Fengling Han[3], Xinghuo Yu[1], Zahir Tari[3], Lilin Li[1], and Jiankun Hu[4]

[1] School of Electrical and Computer Engineering, RMIT University,
Melbourne, VIC 3000, Australia
{yong.feng,x.yu,e88727}@rmit.edu.au
[2] Department of Electrical Engineering, Harbin Institute of Technology, Harbin 150001, China
yfeng@hit.edu.cn
[3] School of Computer Science and Information Technology, RMIT University,
Melbourne, VIC 3000, Australia
{fengling.han,zahir.tari}@rmit.edu.au
[4] School of Engineering and Information Technology, University of New South Wales at the
Australian Defence Force Academy, Canberra ACT 2600, Australia
j.hu@adfa.edu.au

Abstract. This paper proposes a high-order terminal sliding-mode observer used for the anomaly detection in TCP/IP networks. It can track the fluid-flow model representing the TCP/IP behaviors in a router level. A smooth control signal of the observer can be generated based on the high-order sliding-mode technique for estimation of the queue length dynamics in the router. The distributed anomaly in the TCP/IP network incurred by an abnormal behavior can be detected using the smooth control signal. The proposed scheme requires only the average queue length in a router for anomaly detection. The simulations are presented to verify the effectiveness of the proposed method.

Keywords: TCP/IP network model, congestion control, observers, sliding mode control.

1 Introduction

Anomaly detection aims at identifying cases when activities deviate from the normal. Anomalous events occur less frequently, however, once it occurs, the consequences could be quite dramatic. Therefore, abnormal detection has significant meaning in the security of Internet community, thus attracts outstanding researches in areas of artificial intelligence, machine learning, state machine modeling and statistical approaches, *etc.* [1]. Main anomaly detection methods include: statistical methods, expert systems, clustering, neural networks, support vector machines, outlier detection schemes, *etc.* [2]. Control theory methods, among others, have demonstrated great effectiveness in monitoring network traffic. Since a network anomaly can be considered as a perturbation in the traffic flow of the router level, an observer can be designed to estimate the network anomaly. Therefore the observer can collaborate with Active Queue Management (AQM) in the router [3] to detect and estimate anomalies [4-7].

D.-S. Huang et al. (Eds.): ICIC 2012, LNCS 7389, pp. 497–504, 2012.
© Springer-Verlag Berlin Heidelberg 2012

For avoiding congestion collapse on the Internet or restoring a congested computer network to a normal state, congestion control and network congestion avoidance techniques have been utilized widely in router level. A router in TCP/IP networks has two different functions: AQM and anomaly detection. The latter can be implemented using an observer. The anomaly detection is essential in enterprise and provider networks for diagnosing events, like attacks or failures, which severely impact performance, security, and Service Level Agreements (SLAs) [8].

Thanks to the dynamic model of the TCP/IP traffic flow [9, 10], and the contribution of control theoretic analysis of random early detection (RED) [11], in which linearized the interconnection of TCP and a bottlenecked queue. The original work of designing the AQM control system using the RED scheme opens a new area of traditional control theory. Since then, traditional control methods, Proportional (P) and Proportional-Integration (PI) controllers were first used in the queue utilization and delay in network traffic [3, 4]. Soon after, the feedback control was introduced to address the stability issue and to cope with the time-varying nature of the multiple delays in [5]. The feedback control mechanism ensures the regulation of the queue size of the congested router as well as flow rates to a prescribed level.

The construction of an observer allows the reconstruction of the system states so that the disturbance, deemed as an intrusion, can be identified in the router level. Recently following the work in observer based network anomaly estimation, some new anomaly detection methods have been proposed. Among them, sliding-mode observers [12-16] demonstrated significant potentials. In [12], an observer was proposed to detect constant anomalies for TCP/AQM networks. In [13], a new technique based on control theory and the construction of observers was developed using a simplified model describing the average dynamics of the TCP congestion window size and the queue length at the router. In [14], an unknown sliding-mode observer for anomaly detection in TCP/IP networks was proposed to distinguish false/true positives and false/true negatives in a prescribed finite time. In [15], a second-order sliding mode observer for detecting anomalies in TCP networks was investigated based on a fluid model of TCP network. Sliding-mode control (SMC) is a nonlinear control method, which alters the dynamics of a nonlinear system using a discontinuous control signal and forces the system to slide along a pre-designed sliding-mode manifold [17-19]. Sliding-mode observers apply SMC strategies so that they have some significant advantages compared to other observers, such as strong robustness, fast response, and high precision.

Following our previous work [16], this paper proposes a high-order terminal sliding-mode observer used for the anomaly detection in TCP/IP networks. The input of our observer is based on the information measurement and parameters of a router, such as the TCP/IP congestion window size and the queue length. With the help of the high-order sliding-mode technique, a smooth control signal of the observer can be generated automatically and used for estimation of the queue length dynamics which represents an anomaly in the TCP/IP network incurred in the router. Simulations are carried out and the estimation results are analyzed to validate the proposed method.

2   Simplified Dynamic Model of a TCP/IP Network

It is well known that the TCP/IP model is a descriptive framework for the Internet Protocol Suite of computer network protocols. The TCP/IP model has 5 layers: physical, data link, network, transport, and application layers. Computer network topology is the physical communication scheme used by connected devices. It is the layout pattern of interconnections of the various elements (nodes, links, peripherals, *etc.*) of a computer network. There are five basic types of network topologies and hybrid topologies, the former include bus, ring, star, mesh, and tree, the latter are a combination of two or more of the four basic topolojies.

This paper considers a TCP/IP network topology, as shown in Fig.1, where N homogeneous sources connect to a destination through a router, which links computers to the internet. Routers in TCP/IP networks play an important role during the time of the network congestion. In Fig.1, the router can include two different mechanisms: an AQM and an observer for anomaly detection. Packets are the basic unit of transmission on the Internet. AQM controls the length of the packet queues for managing the congestion by dropping the packets when necessary. The observer can detect the congestion window of the router in the TCP/IP network, and further detect the abnormal behavior in the network.

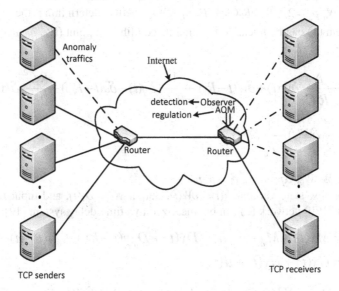

Fig. 1. TCP network topology

The TCP/IP network topology shown in Fig.1 can be described using a simplified TCP/IP model. It can be expressed by the following coupled nonlinear time-delay differential equations [4]:

$$
\begin{cases}
\dot{W}(t) = \dfrac{1}{R(t)} - \dfrac{W(t)}{2} \dfrac{W(t-R(t))}{R(t-R(t))} p(t-R(t)) \\[2mm]
\dot{q}(t) = \dfrac{W(t)}{R(t)} N - C + d(t) \\[2mm]
R(t) = \dfrac{q(t)}{C} + T_p
\end{cases}
\tag{1}
$$

where $W(t)$ is the average TCP/IP window size in packets, $q(t)$ the average queue length in packets, $R(t)$ the round-trip time in seconds, C the link capacity in packets/sec, T_p the propagation time delay in seconds, N the load factor (number of TCP/IP sessions), $p(t)$ the probability of the packet mark, and $d(t)$ the queue length dynamics.

In (1), $d(t)$ represents an additional traffic, which perturbs the normal TCP/IP network behaviors in the router level. In normal condition of the TCP/IP networks, $d(t)$ is around zero, but when an anomaly intrusion happens, it will suddenly increase. Therefore the anomaly intrusion in the TCP/IP networks can be detected by monitoring $d(t)$ based on the measurement and the parameters of the router.

The equilibrium point of system (1), (W_0, q_0, p_0), can be defined by $\dot{W} = 0$ and $\dot{q} = 0$ as: $W_0^2 p_0 = 2$, $W_0 = R_0 C/N$, $R_0 = q_0/C + T_p$. After determining the equilibrium point, system (1) can be linearized around its equilibrium point (W_0, q_0, p_0) as follows [2]:

$$
\begin{cases}
\delta\dot{W}(t) = -\dfrac{N}{R_0^2 C}(\delta W(t) + \delta W(t-R_0)) - \dfrac{1}{R_0^2 C}(\delta q(t) - \delta q(t-R_0)) - \dfrac{R_0 C^2}{2N^2}\delta p(t-R_0) \\[2mm]
\delta\dot{q}(t) = \dfrac{N}{R_0}\delta W(t) - \dfrac{1}{R_0}\delta q(t)
\end{cases}
\tag{2}
$$

where $\delta W = W - W_0$, $\delta q = q - q_0$, $\delta p = p - p_0$.

Define a new state variable $x(t) = \delta W(t)$, output $y(t) = \delta q(t)$, and input $u(t) = \delta p(t)$. Then, the TCP/IP network (2) can be linearized to a time-delay system [12]:

$$
\begin{cases}
\dot{x}(t) = Mx(t) + M_d x(t-h) + Dy(t) + D_d y(t-h) + E_d u(t-h) \\
\dot{y}(t) = Gx(t) + Hy(t) + d(t)
\end{cases}
\tag{3}
$$

where $M = M_d = -N/(R_0^2 C)$, $D = -1/(R_0^2 C)$, $D_d = 1/(R_0^2 C)$, $E_d = -R_0 C^2/(2N^2)$, $G = N/R_0$, $H = -1/R_0$. The simplified model (3) can be used for the feedback control of the router management and the TCP/IP network stability [9, 15]. In addition, this model can be utilized for the anomaly detection in the TCP/IP networks [12-16].

The objective of the paper is to design a smooth control signal of an observer of system (3) for estimating the queue length dynamics, $d(t)$, which represents an anomaly in the TCP/IP network.

3   Design of High-Order Sliding-Mode Observer

For estimating $d(t)$ in the simplified dynamic model of the TCP/IP network (3), a sliding-mode observer of the system (3) is proposed as follows:

$$\begin{cases} \dot{\hat{x}}(t) = M\hat{x}(t) + M_d\hat{x}(t-h) + Dy(t) + D_d y(t-h) + E_d u(t-h) \\ \hat{y}(t) = G\hat{x}(t) + Hy(t) + v(t) \end{cases} \tag{4}$$

where $\hat{x}(t)$ and $\hat{y}(t)$ represent the state estimation for the system states $x(t)$ and $y(t)$ respectively, and $v(t)$ is the control signal of the observer.

Define the errors between the estimations and the true states as: $e_x(t) = x(t) - \hat{x}(t)$, $e_y(t) = y(t) - \hat{y}(t)$. The error observer equation can be obtained from (3) and (4) as:

$$\begin{cases} \dot{e}_x(t) = Me_x(t) + M_d e_x(t-h) \\ \dot{e}_y(t) = Ge_x(t) + v(t) + d(t) \end{cases} \tag{5}$$

It is assumed that system (3) and its observer (4) satisfy the following assumptions.

Assumption 1. The derivative of the additional traffic signal $d(t)$ is bounded:

$$\left| \dot{d}(t) \right| \le d_m \tag{6}$$

where d_m is a positive constant.

Assumption 2. The control signal $v(t)$ in (4) satisfies the following condition:

$$Tv(t) \le v_m \tag{7}$$

where both T and v_m are also two positive constants respectively.

Theorem 1. The observer error e_y, and its derivative in system (5) will converge to zero in finite time, if a TSM manifold is chosen as (8), and the control signal $v(t)$ is designed as (9)-(12):

$$s(t) = \dot{e}_y(t) + \beta e_y^{p/q}(t) \tag{8}$$

$$v(t) = v_{eq}(t) + v_n(t) \tag{9}$$

$$v_{eq}(t) = Ge_x(t) + \beta e_y^{p/q}(t) \tag{10}$$

$$\dot{v}_n(t) + Tv_n(t) = w(t) \tag{11}$$

$$w(t) = -K\,\mathrm{sgn}(s(t)) \tag{12}$$

where $K=v_m+d_m+\eta$, $\eta>0$, both v_m and d_m are defined in (6) and (7); $\beta>0$ is a constant.

After reaching the sliding-mode surface $s(t)=0$, it can be seen from (8) that both $e_y(t)$ and $\dot{e}_y(t)$ will converge to zeros within finite time. Then the estimation of the additional traffic $d(t)$ can be obtained from (5):

$$\hat{d}(t) = -v_n(t) \tag{13}$$

Therefore $d(t)$ can be estimated using the smooth control signal of the observer.

4   Simulations

In order to evaluate the effectiveness of the proposed method for the anomaly detection in the paper, some simulations are carried out. The parameters of the system (3) for simulations are assumed as follows:

$N=60$ TCP sources; $C=3750$ packets/s; $T_p=0.2$s; $R_0=0.2045$; $q_0=17.08$; $M=M_d=-0.3824$; $D=-0.0064$; $D_d=0.0064$; $E_d=-399.5208$; $G=293.3201$; $H=-4.8887$.

The queue length dynamics $d(t)$ in the simulation is assumed to be a square waveform signal described by the function:

$$d(t) = 5[\mathrm{sgn}(\sin(0.1\pi t - 0.5\pi)) + 1] \tag{14}$$

A TSM observer is designed based on Theorem 1, and the design parameters are: $\beta=1$; $p=5$, $q=3$, $\eta=10$. The simulation results are shown in Figs.2 and 3. The estimation result of the queue length dynamics $d(t)$ in (3) is shown in Fig. 2, and the zoomed-in view of the estimation is shown in Fig.3. It can be seen that $d(t)$ can be estimated using the proposed method.

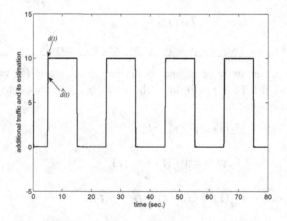

Fig. 2. Estimation of the queue length dynamics

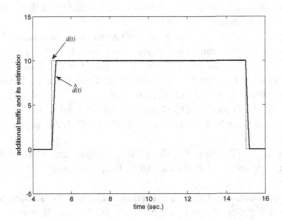

Fig. 3. Zoomed-in view of Fig.2

As aforementioned, the queue length dynamics $d(t)$ represents the additional traffic perturbing the normal TCP network behaviors in the router level. During the normal condition, there is no anomaly appears in the TCP/IP network, $d(t)$ is around zero. When an anomaly happens, $d(t)$ will rise beyond the normal range. From the simulation results, it can be seen that $d(t)$ can be estimated quickly and accurately, which means that a distributed anomaly in the TCP network can be detected using the proposed method.

5   Conclusion

This paper has proposed a high-order sliding-mode observer used for the anomaly detection in the TCP/IP networks. The observer can track the fluid-flow model representing the TCP/IP behaviors in a router. The high-order sliding-mode controller is designed for the observer and the smooth control signal of the observer can be obtained for estimation of the queue length dynamics in the router, which represents a distributed anomaly in the TCP/IP network. The proposed anomaly detection scheme requires only one signal which is the average queue length in a router. The simulation results have shown the effectiveness of the proposed anomaly detection method.

Acknowledgement. This work was supported in part by the National Natural Science Foundation of China (61074015), and also in part by ARC Linkage Project (LP100200538) of the Australian Research Council.

References

1. Labit, Y., Gouaisbaut, F., Ariba, Y.: Network Anomaly Estimation for TCP/AQM net-Works Using an Observer. In: The Third International Workshop on Feedback Control Implementation and Design in Computing Systems and Network, Annapolis, USA, pp. 63–68 (2008)

2. Steinwart, I., Hush, D., Scovel, C.: A Classification Framework for Anomaly Detection. Journal of Machine Learning Research 6, 211–232 (2005)
3. Ryu, R., Rump, C., Qiao, C.: Advances in Active Queue Management (AQM) based TCP Congestion Control. Telecommuniaction Systems 4, 317–351 (2004)
4. Ariba, Y., Gouaisbaut, F., Rahme, S., Labit, Y.: Robust Control Tools for Traffic Monitoring in TCP networks. In: 18th IEEE International Conference on Control Applications, Saint Petersburg, Russia, pp. 525–530 (2009)
5. Hollot, C., Misra, V., Towsley, D., Gong, W.: On Designing Improved Controllers for AQM Routers Supporting TCP Flows. In: IEEE INFOCOM, Anchorage, AK, USA, vol. 3, pp. 1726–1734 (2001)
6. Hollot, C., Misra, V., Towsley, D., Gong, W.: Analysis and Design of Controllers for AQM Routers Supporting TCP Flows. IEEE Trans. Automatic Control 47(6), 945–959 (2002)
7. Ariba, Y., Gouaisbaut, F., Labit, Y.: Feedback Control for Router Management and TCP/IP Network Stability. IEEE Trans. Network and Service Management 6(4), 255–266 (2009)
8. Kind, A., Stoecklin, M.P., Dimitropoulos, X.: Histogram-based Traffic Anomaly Detection. IEEE Trans. Network and Service Management 6(2), 1–12 (2009)
9. Misra, V., Gong, W., Towsley, D.: Fluid-based Analysis of a Network of AQM Routers Supporting TCP Flows with an Application to RED. In: ACM/SIGCOMM, Stockholm, Sweden, vol. 30(4), pp. 151–160 (2000)
10. Firoiu, V., Borden, M.: A Study of Active Queue Management for Congestion Control. In: IEEE INFOCOM, Tel Aviv, Israel, vol. 3, pp. 1435–1444 (2000)
11. Hollot, C., Misra, V., Towsley, D., Gong, W.: A Control Theoretic Analysis of RED. In: IEEE INFOCOM, Anchorage, AK, USA, vol. 3, pp. 1510–1519 (2001)
12. Ariba, Y., Labit, Y., Gouaisbaut, F.: Network Anomaly Estimation for TCP/AQM Networks Using an Observer. In: 3rd ACM International Workshop on Feedback Control Implementation and Design in Computing Systems and Networks, Annapolis, USA, pp. 45–50 (2008)
13. Rahme, S., Labit, Y., Gouaisbaut, F.: Sliding Mode Observer for Anomaly Detection in TCP/AQM Networks. In: The Second Inter. Conf. on Communication Theory, Reliability, and Quality of Service, Colmar, France, pp. 113–118 (2009)
14. Rahme, S., Labit, Y., Gouaisbaut, F.: An Unknown Input Sliding Observer for Anomaly Detection in TCP/IP Networks. In: 2009 International Conference on Ultra Modern Telecommunications and Workshops, St. Petersburg, Russia, pp. 1–7 (2009)
15. Rahme, S., Labit, Y., Gouaisbaut, F., Floquet, T.: Second Order Sliding Mode Observer for Anomaly Detection in TCP Networks: From Theory to Practice. In: Proc. of 49th IEEE Conf. on Decision and Control, Atlanta, GA, pp. 5120–5125 (2010)
16. Feng, Y., Han, F., Yu, X., Tari, Z., Li, L., Hu, J.: Terminal Sliding Mode Observer for Anomaly Detection in TCP/IP Networks. In: 2011 International Conference on Computer Science and Network Technology (ICCSNT), Harbin, China, vol. 1, pp. 617–620 (2011)
17. Feng, Y., Yu, X., Man, Z.: Non-singular Adaptive Terminal Sliding Mode Control of Rigid Manipulators. Automatica 38(12), 2159–2167 (2002)
18. Feng, Y., Han, X., Wang, Y., Yu, X.: Second-order Terminal Sliding Mode Control of Uncertain Multivariable Systems. International Journal of Control 80(6), 856–862 (2007)
19. Feng, Y., Zheng, J., Yu, X., Truong, N.V.: Hybrid Terminal Sliding-mode Observer Design Method for a Permanent-magnet Synchronous Motor Control System. IEEE Trans. Industrial Electronics 56(9), 3424–3431 (2009)

An Automated Bug Triage Approach: A Concept Profile and Social Network Based Developer Recommendation

Tao Zhang and Byungjeong Lee[*]

School of Computer Science, University of Seoul, Seoul, Korea
`kerryking@ieee.org, bjlee@uos.ac.kr`

Abstract. Generally speaking, the larger-scale open source development projects support both developers and users to report bugs in an open bug repository. Each report that appears in this repository must be triaged for fixing it. However, with huge amount of bugs are reported every day, the workload of developers is so high. In addition, most of bug reports were not assigned to correct developers for fixing so that these bugs need to be re-assigned to another developer. If the number of re-assignments to developers is large, the bug fixing time is increased. So "who are appropriate developers for fixing bug?" is an important question for bug triage. In this paper, we propose an automated developer recommendation approach for bug triage. The major contribution of our paper is to build the concept profile(CP) for extracting the bug concepts with topic terms from the documents produced by related bug reports, and we find the important developers with the high probability of fixing the given bug by using social network(SN). As a result, we get a ranked list of appropriate developers for bug fixing according to their expertise and fixing cost. The evaluation results show that our approach outperforms other developer recommendation methods.

Keywords: bug triage, concept profile, social network, fixing cost, re-assignment, developer recommendation.

1 Introduction

With a great amount of large-scale software projects have been developed, software maintenance becomes a challenging task due to many bugs appearing in the source code files [1]. In order to help developers track and fix these bugs for performing software maintenance well, bug tracking systems [2] have been introduced for allowing developers to serve as "testers" and report related bugs. Most well-known open source systems (e.g. Eclipse, JBoss) include an open bug repository for keeping the bug reports.

Currently, for fixing each bug in the open bug repository, the related bug report need to be arranged to a developer for fixing it. This process is called bug triage [3]. However, with a great number of bugs are reported, the workload of fixers is so high.

---

[*] Corresponding author.

D.-S. Huang et al. (Eds.): ICIC 2012, LNCS 7389, pp. 505–512, 2012.
© Springer-Verlag Berlin Heidelberg 2012

Moreover, re-assignment [4] is a serious problem as well. A lot of bug reports were not assigned to appropriate developers and need to be re-assigned to other developers. The more the number of re-assignments is, the lower the probability of bug fixing is. Meanwhile, the fixing time is increased.

In order to reduce the fixing time for submitted bugs and improve the probability of bug fixing, it is necessary to recommend appropriate developers for bug fixing. In this paper, we present an effective way of automatically recommending developers for bug fixing. This approach comprises three parts. The goal of the first part is to build the concept profile (CP). For achieving this purpose, we extract the bug concepts with topic terms from the documents deriving the corresponding bug reports. After the concept profile is built, when a new bug comes, we decide which bug concept the new bug belongs to, and then extract the corresponding developers from the concept profile. For the second part, we utilize social network (SN) [5] to find the important developers with the high probability of fixing this bug. The final part is to rank the candidates according to the developers' experience and fixing cost. We expect the proposed method can avoid the excessive number of re-assignments, improve the probability of bug fixing and reduce the bug fixing time.

The remainder of this paper is organized as follows. Section 2 presents some related work. We describe the details of our developer recommendation approach in Section 3. Section 4 shows the evaluation results on the collected bug reports from the famous open source project JBoss, and we analyze the experimental results. We summarize our work and introduce future work in Section 5.

2   Related Works

As previous work, J. Anvik et al. propose a semi-automated approach for recommending expert developers to fix the given bug [6]. Based on their work, W. Wu et al. present a bug triage approach called DREX(Developer Recommendation with K-Nearest-Neighbor Search and Expertise Ranking) [7] which collects the bug reports and comments from the open bug repository of Mozilla Firefox for recommending developers to fix given bugs. For the evaluation experiment, authors compare different ranking metrics according to the performance of bug triage. The results demonstrate DREX produces higher accuracy than traditional text categorization method when recommending ten developers for each testing bug. Specially, in the results, two metrics as Out-degree and Frequency show the best performance than others.

J. Park et al. propose a cost-aware triage algorithm, CosTRIAGE [8]. The algorithm models "developer profiles" to indicate developers' estimated costs for fixing different type of bugs. Authors apply Latent Dirichlet Allocation (LDA) to verify the bug types, and quantify each value of the developer's profile as the average time to fix the bugs in corresponding type. For a given bug report, authors can find out all developers' estimated costs by determining the most relevant topic using the LDA model. In this method, they consider the probability and cost to fix the given bug to determine which one is the most appropriate developer to resolve the bug.

Even though the purpose of our study is same as the above proposed approaches, there are some differences: first, the key idea of LDA is similar to concept profile, even so, both of the forming process and purpose are different; second, W. Wu et al. use social network metrics to rank the importance of the developers. but we find the important developers from social network according to the possibility of fixing the given bug without using the social network metrics; finally, towards candidate developers, we combine their expertise and cost of fixing historical bugs for ranking the candidates. The ranking algorithm is different with other methods.

3   A Concept Profile and Social Network Based Developer Recommendation

Our approach to recommending expert developers to fix the given bug consists of the following steps: First, we build the concept profiles which include the bug concepts with the topic terms, and the documents produced by corresponding bug reports. Next, when a new bug comes, we verify which bug concept the new bug belongs to, and we extract the developers from the documents related to this bug concept. By constructing social network, we find the important developers according to their probability of fixing the new bug. Finally, we use the ranking algorithm to recommend top-k developers for fixing the given bug.

3.1   Building the Concept Profiles

In our work, we define the concept profile (CP) as follows.

Definition 1. Concept Profile (CP)
A concept profile constitutes a pair(C, D) where C denotes a bug concept with the topic terms, and D denotes a set of documents produced by the bug reports related to the bug concept. A document includes a set of features (e.g. id, description, fixing time, comments) of corresponding bug report.

To build concept profiles, we need to address two key problems as follows.
   1. Categorizing Bugs: One effective way is to cluster the bugs in the training set. We adopt k-means clustering algorithm [9] to cluster the bug reports existing in the training set for categorizing bugs. In the clustering process, we define the similarity measure between bug reports as the distance between data points. For the similarity measure between bug reports, we use well-known cosine measure [10] to compute the textural similarity between two free-form text combining title and description of bug reports. The clustering process is an iterative process utile the error measure is a minimum value. In the other words, once the iterative process ends, each data point is close to the reference point which is selected randomly in its cluster than to any other reference point, and each reference point is the centroid of its cluster.
   2. Extracting Topic Terms: After clustered the existing bug reports in the training set, the bug concepts are determined. Now we need to extract the topic terms which

have higher frequency of appearing in the bug reports which belong to the same bug concept. We introduce normalization to transform the frequency to the weight value, we set a threshold θ_1 to determine the topic terms while the weight values of the terms are more than the threshold. Fig. 1(a) shows a bug concept with the topic terms.

We note that there are four topic terms related to the bug concept C1: 'control', 'CVS', 'repository' and 'type'. The values on the links represent the weight values which are more than the threshold θ_1. Fig. 1(b) shows an example of concept profile which includes the bug concept C1 and a set of documents D1. D1 includes the documents deriving from the clustered bug reports related to C1.

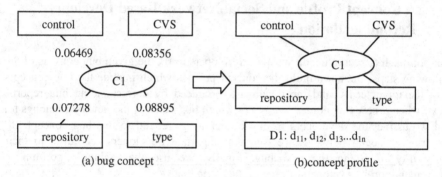

(a) bug concept   (b) concept profile

Fig. 1. An example of bug concept and concept profile. The threshold θ is set to 0.05

3.2   Retrieving Candidate Developers Using Social Network

When a new bug comes, at first, we need to determine the most relevant bug concept with the new bug. In detail, we calculate the frequency of the topic terms of each bug concept appearing in the title and description of the new bug report. Once the highest frequency is found, we identify that the new bug belongs to the corresponding bug concept due to the highest frequency of related topic terms.

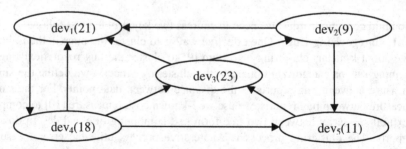

Fig. 2. An example of social network which include five nodes

Fig. 2 shows an example of a social network. There are five nodes in the social network. These nodes represent a set of developers extracted from the concept profile, and the links stand for the cooperative relationship of these developers. In detail, some nodes which launch the links represent the assignees who are arranged to fix the bugs existing the concept profile; some nodes receiving the links represent the developers who participate the comments of the bug reports fixed by the corresponding developers who lunch these links. The numeral in each node represents the number of bug reports fixed by the related developer.

Based on the number of fixed bug reports and launched links in the social network, we compute the probability of fixing the given bug for each developer in the social network as follows.

Definition 2. The probability of fixing the given bug

$$P(b_{new} \mid dev_i) = \frac{nb_{devi}}{\sum_{k=1}^{N} nb_{devk}} \times \frac{nl_{devi}}{\sum_{k=1}^{N} nl_{devk}}$$

where

- nb_{devi} stands for the number of bug reports fixed by developer dev_i.
- N is the total number of developers(or nodes) in the social network.
- nl_{devi} represents the number of links which are lunched by developer dev_i.

When getting the probability of fixing the given bug for each developer, we set a threshold θ_2 to determine the important developers with the high probability of fixing the new bug. If the probability is more than θ_2, we add the corresponding developer to the candidate list.

3.3   Ranking Developers for Recommending to Fix the Given Bug

For a list of candidate developers, we need to rank these developers for finding the most appropriate developers to fix the given bug. We define a ranking algorithm to get the ranking score as follows.

Definition 3. Developer Ranking Algorithm

$$RScore(dev_i) = \alpha \times \frac{E(dev_i)}{\sum_{k=1}^{M} E(dev_k)} + (1-\alpha) \times \frac{C(dev_i)}{\sum_{k=1}^{M} C(dev_k)}$$

where

- $E(dev_i)$ represents the developer's experience, which is defined by the ratio of the number of fixed bug reports by dev_i to the number of bug reports assigned to dev_i.
- $C(dev_i)$ stands for the fixing cost for the candidate developer dev_i, which is defined by the inverse of the average fixing time of all bug reports which are fixed by dev_i.
- α is a weight vector, where $0 \leq \alpha \leq 1$.
- M represents the total number of candidate developers when a new bug comes.

By utilizing the ranking algorithm which combines the candidate's experience and the fixing cost of historical bug reports, we get a ranked list of candidate developers.

4   Experimental Results

4.1   Setup

For the evaluating experiment, we collect 836 bug reports labeled with "resolved" from Jan 2010 to Dec 2011 in the JBoss open bug repository. We find that 2,325 developers participated in the fixing and commenting activities. We divide the data to training set and testing set. The training set includes 736 bug reports and the testing set includes 100 bug reports. We execute the pre-process(e.g. stemming, stop words removal and tokenization)[11] to all bug reports, and cluster the bug reports in the training set. As a result, we get 91 bug concepts and 915 topic terms when the threshold θ_1 is set to 0.05. In addition, we find 651 active developers after removing the inactive developers who did not work more than 30 days.

When a new bug from the testing set comes, we verify the bug type of this new bug, and utilize the social network and developer ranking algorithm for recommending the appropriate developers to fix the given bug. In this experiment, we set the threshold θ_2 to 0.05 and the weight vector α to 0.6.

We adopt Precision-Recall measure[12] to evaluate the performance of different developer recommendation approaches. It is defined as follows.

Definition 4. Precision-Recall Measure

$$Precsion(b_{new}) = \frac{\left| DEV(dev, b_{new}) \cap RDEV(dev, b_{new}) \right|}{\left| DEV(dev, b_{new}) \right|}$$

$$Recall(b_{new}) = \frac{\left| DEV(dev, b_{new}) \cap RDEV(dev, b_{new}) \right|}{\left| RDEV(dev, b_{new}) \right|}$$

where

- DEV(dev, b_{new}) denotes all recommended developers for the new bug report b_{new}.
- RDEV(dev, b_{new}) represents the real developers who participate in the fixing and commenting activities to the new bug report.

4.2   Results and Discussion

In this section, we apply SVM based recommendation, DERX with Frequency, DERX with Degree and our approach(Recommendation 1) to same data sets. In addition, we also compare our method without using social network and ranking algorithm(Recommendation 2). Fig. 3 shows the comparison results. From left to right side of the curves, six data points represents the number of recommended developers from 1 to 6.

We can see that the uppermost curve shows the highest precision and recall than others, while employing our recommendation approach. On the other hand, the performance of DREX with Frequency and DREX with Degree is better than SVM-based recommendation method. In our opinion, the major reason is lack of ranking process for SVM-based recommendation method. For Recommendation 2, that is the first phase of our method without using social network and ranking algorithm, we note this method shows the worst performance.

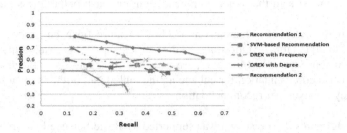

Fig. 3. Performance of our approach with different developer recommendation methods

In order to analyze the major reason of this result, we compare the different processes as each developer recommendation methods. Table. 1 gives the comparison results. We note that both of COSTRIAGE and our recommendation approach use concept profile(or LDA) to extract the bug concepts(or bug types) and corresponding topic terms. Comparing with BR(Similar bug reports retrieval), concept profile(or LDA) resolves over-specialization due to introducing bug types. There are only DERX and our method(Recommendation-1) utilize the social network, but, for our method, we did not use the social network metrics to rank developers. Moreover, except for SVM-based recommendation and Recommendation-2, other studies also utilize the ranking algorithm for recommending top-k appropriate developers to fix the new bug. However, our study combines developers' experience and fixing cost as the factors of ranking algorithm, it is different from other methods. In summary, our approach includes concept profile, social network and developer ranking algorithm to recommend experienced developers for bug triage. So we think this is the major reason that the performance of our method is better than others and it is expected to improve the probability of bug fixing.

Table 1. The comparison results of different processes as each developer recommendation method(BR: Similar bug reports retrieval; Concept: Concept profile or LDA; SN: Social network; Ranking: Developer ranking algorithm)

Criticism	Recommendation-1	Recommendaiton-2	SVM-based	DERX	COSTRIAGE
BR	N/A	N/A	Y	Y	N/A
Concept	Y	Y	N/A	N/A	Y
SN	Y	N	N	Y	N
Ranking	Y	N	N	Y	Y

5   Conclusion and Future Work

In this paper, we propose a new idea for recommending appropriate developers for bug triage. We utilize concept profile to extract the bug concepts and corresponding topic terms, construct social network to find the important developers with the high probability of fixing the given bug, and rank the candidate developers according to their experience and fixing cost. The experimental results on JBoss open bug repository demonstrated that our recommendation approach outperforms other recent studies.

In the future, we will examine our method to more open bug repositories(e.g. Eclipse, Mozilla) for demonstrating the efficiency and availability of our recommendation approach. Moreover, we plan to investigate other factors affecting the developer recommendation for enhancing the ranking algorithm so that improving the accuracy of developer recommendation.

Acknowledgments. This research was supported by Basic Science Research Program through the National Research Foundation of Korea(NRF) funded by the Ministry of Education, Science and Technology(No. 2011-0026461).

References

1. Kagdi, H., Gethers, M., Poshyvanyk, D., Hammad, M.: Assigning Change Request to Software Developers. Journal of Software: Evolution and Process 24, 3–33 (2012)
2. Jalbert, N., Weimer, W.: Automated Duplicate Detection for Bug Tracking System. In: International Conference on Dependable System & Networks, pp. 52–61 (2008)
3. Matter, D., Kuhn, A., Nierstrasz, O.: Assigning Bug Reports Using a Vocabulary-based Expertise Model of Developers. In: 6th IEEE International Working Conference on Mining Software Repositories, pp. 131–140 (2009)
4. Jeong, G., Kim, S., Zimmermann, T.: Improving Bug Triage with Bug Tossing Graphs. In: 7th Joint Meeting of the European Software Engineering Conference and the ACM Symposium on the Foundations of Software Engineering, pp. 111–120 (2009)
5. Chen, I.X., Yang, C.Z., Lu, T.K., Jaygarl, H.: Implicit Social Network Model for Predicting and Tracking the Location of Faults. In: Annual IEEE International Computer Software and Applications Conference, pp. 136–143 (2008)
6. Anvik, J., Hiew, L., Murphy, G.C.: Who Should Fix This bug? In: 28th International Conference on Software Engineering, pp. 361–370 (2006)
7. Wu, W., Zhang, W., Yang, Y., Wang, Q.: DREX: Developer Recommendation with K-Nearest-Neighbor Search and Expertise Ranking. In: 18th Asia-Pacific Software Engineering Conference, pp. 389–396 (2011)
8. Park, J., Lee, M., Kim, J., Hwang, S., Kim, S.: CosTRIAGE: A Cost-Aware Triage Algorithm for Bug Reporting Systems. In: 25th AAAI Conference on Artificial Intelligence, pp. 139–144 (2011)
9. Kanungo, T., Mount, D.M., Netanyahu, N.S., Piatko, C.D., Silverman, R., Wu, A.Y.: An Efficient K-means Clustering Algorithm: Analysis and Implementation. IEEE Transaction on Pattern Analysis and Implementation 24, 881–892 (2002)
10. Tata, S., Patel, J.M.: Estimating The Selectivity of TF-IDF based Cosine Similarity Predicates. ACM SIGMOD Record 36, 75–80 (2007)
11. Wang, X., Zhang, L., Xie, T., Anvik, J., Sun, J.: An Approach to Detecting Duplicate Bug Reports Using Natural Language and Execution Information. In: 30th International Conference on Software Engineering, pp. 461–470 (2008)
12. Wu, R., Zhang, H., Kim, S., Cheung, S.C.: Relink: Recovering Links between Bugs and Changes. In: Joint Meeting of the European Software Engineering Conference and the ACM Symposium on the Foundations of Software Engineering, pp. 15–25 (2011)

A New Method for Filtering IDS False Positives with Semi-supervised Classification

Minghua Zhang and Haibin Mei

Information College, Shanghai Ocean University, Shanghai, China
mhzhang@shou.edu.cn

Abstract. Constructing alert classifiers is an efficient way to filter IDS false positives. Classifiers built with supervised classification technique require large amounts of labeled training alerts which are difficult and expensive to prepare. This paper proposes to use semi-supervised learning technique to build alert classification model to reduce the number of needed labeled training alerts. Experiments conducted on the DARPA 1999 dataset have demonstrated that the semi-supervised alert classification model can improve the classification performance dramatically, especially when the labeled alert training dataset is small. As a result, the feasibility of deploying alert classifier for filtering false positives is enhanced.

Keywords: intrusion detection system, false positive, semi-supervised learning, EM algorithm.

1   Introduction

Filtering false positives is an important and basic work in intrusion detection system (IDS). Recently, some researchers have observed that building alert classifier (or classification model) based on machine learning can do the job trick and reduce false positives dramatically [1-3]. However, current alert classifiers are built on the supervised learning techniques, which require large amounts of labeled training data. In reality, such alert training data are very difficult and expensive to obtain. As a consequence, practicability of these methods has been greatly limited.

This paper has made a study on the applicability of semi-supervised learning to build intrusion alert classification model. The goal of semi-supervised learning is making use of large number of unlabeled data that are easy to achieve to improve the quality of machine learning. For classification, semi-supervised learning is a special classification technique [4], which learns from both labeled data and unlabeled data. This makes it very suitable for building an accurate alert classification model when labeled alert training data are scarce.

2   Related Work

In order to filtering false positives generated by IDS, existing solutions include alert correlation [5] and modeling false positives by frequent or periodic alert sequences [6]. Recently, methods based on classification have also been presented. The most famous

D.-S. Huang et al. (Eds.): ICIC 2012, LNCS 7389, pp. 513–519, 2012.
© Springer-Verlag Berlin Heidelberg 2012

work is done by Pietrazek [1], in which he constructed an adaptive learner called ALAC to classify alerts based on RIPPER algorithm. A Naïve Bayesian (NB) classifier is proposed in [3] and used in a multi-tier intrusion detection system. Moon et al. [2] applied data mining technique to the classification and developed an alert classifier using decision tree. This method can classify alerts automatically. But the above methods are all based on the supervised learning and the main problem is that obtaining a large, representative, training dataset that is fully labeled is difficult, time consuming and expensive. The power of semi-supervised learning for building classification model has been demonstrated in many applications. In the field of network security, semi-supervised learning has also been used [7]. However, to the best of our knowledge, this is the first practice to apply semi-supervised learning to intrusion alert classification.

3 Construction of Alert Classification Model

The alert contains many inherent properties, some are redundant, and others are very specific or general. To improve classification efficiency, we firstly adopt the information gain method [8] to select a subset of inherent properties as classification features.

3.1 Alert Generative Model

In this paper, we choose generative models-based approach [9], one of often-used semi-supervised learning methods. Alert generative model is a probability model which explicitly states how the alerts are generated. Alert data can be generated by a mixture model, which is parameterized by θ. The mixture model consists of mixture components and each component, which is parameterized by a disjoint subset of θ, corresponds to a class $c_j \in C$, where $C = \{c_1, c_2, ..., c_{|C|}\}$ represents the classes of alert data (this paper only considers two class, C={"true positive", "false positive"}). Formally, every alert object x_i is created in two steps. First, a mixture component is selected according to the class probabilities $P(c_j|\theta)$. Then, this component is used to generate an alert object according to its own distribution model $P(x_i|c_j;\theta)$. The likelihood of creating alert object x_i is equation (1).

$$P(x_i \mid \theta) = \sum_{j=1}^{|C|} P(c_j \mid \theta) P(x_i \mid c_j; \theta) \tag{1}$$

where $P(c_j|\theta)$ is the jth mixture component. Every alert object has a class label y_i. If alert object x_i was generated by mixture component c_j then let $y_i = c_j$. $P(x_i \mid c_j; \theta)$ represents the probability distribution to generate an alert object. Assume that $X = <a_1, a_2, ..., a_n>$ is a feature vector extracted from a raw alert A, the second term $P(x_i \mid c_j; \theta)$ in equation (1) becomes:

$$P(x_i \mid c_j; \theta) = P(< a_{x_i,1}, ..., a_{x_i,n} > \mid c_j; \theta) \tag{2}$$

where $a_{x_i,k}$ is the value of the kth feature a_k of alert object x_i. Since the features are conditionally independent of other features in the same alert when the class label is given, the equation can be further expressed as equation (3).

$$P(x_i \mid c_j; \theta) = P(< a_{x_i,1}, ..., a_{x_i,n} > \mid c_j; \theta) = \prod_{k=1}^{n} P(a_{x_i,k} \mid c_j; \theta) \tag{3}$$

Let $P(a_{x_i,k} \mid c_j; \theta)$ in equation (3) be parameter $\theta_{a_k \mid c_j}$ and the mixture component distribution $P(c_j \mid \theta)$ be parameter $\theta_{c_j} : c_j \in C$, the complete model parameter set can be described as $\theta = \{ \theta_{a_k \mid c_j} : a_k \in \{a_1, a_2, ..., a_n\}, c_j \in C; \ \theta_{c_j} : c_j \in C \}$. With equation (1) and (3), the alert generative model $P(x_i \mid \theta)$ is shown in equation (4).

$$P(x_i \mid \theta) = \sum_{j=1}^{|C|} P(c_j \mid \theta) \prod_{k=1}^{n} P(a_{x_i,k} \mid c_j; \theta) \tag{4}$$

3.2 Alert Classification Based on Generative Model

Suppose the estimated of parameters θ is $\hat{\theta}$, for a given alert object x_i, the probability that x_i belongs to category c_j can be calculated by equation (5).

$$P(y = c_j \mid x_i; \hat{\theta}) = \frac{P(x_i \mid c_j; \hat{\theta}) P(c_j \mid \hat{\theta})}{P(x_i \mid \hat{\theta})} \tag{5}$$

Using equation (1) and (3), equation (5) finally becomes:

$$P(y = c_j \mid x_i; \hat{\theta}) = \frac{P(c_j \mid \hat{\theta}) \prod_{k=1}^{|a_k|} P(a_{x_i,k} \mid c_j; \hat{\theta})}{\sum_{g=1}^{|C|} P(c_g \mid \hat{\theta}) \prod_{k=1}^{|a_k|} P(a_{x_i,k} \mid c_g; \hat{\theta})} \tag{6}$$

For an unlabeled alert object x_i, its estimated class y_i is the one which obtains the maximum value of the posterior probability, that is $y_i = \arg\max_{j=1,...,|C|}(P(y = c_j \mid x_i; \hat{\theta}))$.

Suppose the labeled alert training data is $D_l = \{< x_1, y_1 >, < x_2, y_2 >, ..., < x_{|D_l|}, y_{|D_l|} >\}$, and use the MAP estimate, we can find that $\hat{\theta} = \arg\max_{\theta} P(\theta \mid D_l)$. All estimated parameters in $\hat{\theta}$ that result from maximization are the familiar ratios of empirical counts. The estimated probability of $\hat{\theta}_{a_k \mid c_j}$ is the number of the alerts whose attribute value a_k is $a_{k,v}$ and class label is c_j divided by the total number of alerts whose category are class c_j. The calculation equation of $\hat{\theta}_{a_k \mid c_j}$ is in (7).

$$\hat{\theta}_{a_k | c_j} \equiv P(a_{k,v} \mid c_j; \hat{\theta}) = \frac{\sum_{i=1}^{|D_i|} S_{ij} N(a_{k,v}, x_i)}{\sum_{m=1}^{|a_k|} \sum_{i=1}^{|D_i|} S_{ij} N(a_{k,m}, x_i)} \tag{7}$$

where S_{ij} is determined by the label of the ith alert object, satisfying $S_{ij} = \begin{cases} 1, & y_i = c_j \\ 0, & y_i \neq c_j \end{cases}$,

the function $N(a_{k,v}, x_i)$ is defined as $N(a_{k,v}, x_i) = \begin{cases} 1, & a_{x_i,k} = a_{k,v} \\ 0, & a_{x_i,k} \neq a_{k,v} \end{cases}$. After applying the

Laplace smoothing operation, equation (7) becomes:

$$\hat{\theta}_{a_k | c_j} \equiv P(a_{k,v} \mid c_j; \hat{\theta}) = \frac{1 + \sum_{i=1}^{|D_i|} S_{ij} N(a_{k,v}, x_i)}{|a_k| + \sum_{m=1}^{|a_k|} \sum_{i=1}^{|D_i|} S_{ij} N(a_{k,m}, x_i)} \tag{8}$$

In the same way, the class prior probabilities $\hat{\theta}_{c_j}$ can be estimated as follows.

$$\hat{\theta}_{c_j} \equiv P(c_j \mid \hat{\theta}) = \frac{1 + \sum_{i=1}^{|D_i|} S_{ij}}{|C| + |D_i|} \tag{9}$$

This paper also proposes an EM algorithm, which utilizes both labeled alert training data and unlabeled alert data to improve the accuracy of parameter estimates. We firstly give some relevant definitions.

Definition 1. Weak labeled dataset. Let unlabeled alert dataset $D_u = \{x_1^u, x_2^u, ..., x_{|D_u|}^u\}$, then its weak labeled dataset is $D_p = \{<x_1^u, y_1^u> ..., <x_{|D_u|}^u, y_{|D_u|}^u>\}$, where y_i^u is the predicted label for x_i^u, where i=1,2,...,|D_u|.

Definition 2. Weighted labeled dataset. It consists of triples $<x_i, y_i, p_i>$, where y_i is the label of x_i, p_i represents the probability that x_i is assigned the label y_i ($0 < p_i \leq 1$). For the labeled dataset $D_l = \{<x_1, y_1>, <x_2, y_2>, ..., <x_{|D_l|}, y_{|D_l|}>\}$, each element is assigned a maximum weight 1. For the weak labeled dataset D_p, its weighted labeled dataset is $D_p^w = \{<x_1^u, y_1^u, p_1^u>, ..., <x_{|D_u|}^u, y_{|D_u|}^u, p_{|D_u|}^u>\}$, where p_i satisfies $0.5 \leq p_i \leq 1$, i=1,2,...,|D_u|.

Definition 3. Expanded labeled dataset. It is a subset of weighted labeled dataset, in which the weight is equal or bigger than a given threshold value ρ.

The EM algorithm mainly consists of three steps. First, based on the limited amount of labeled alert training dataset D_l, the parameters of generative model are estimated, and a NB classification model is built. Second, the weight labeled dataset of the weak labeled dataset D_p^w is created by using the NB classification model and the expanded labeled dataset D_s^w constructed based on D_p^w. Finally, the optimal classification model $\hat{\theta} = \arg\max_\theta P(D_t \mid \theta)P(\theta)$ is rebuilt using the expanded training dataset D_t^w. The algorithm iterates the above three steps until the class members in D_t^w do not change much.

4   Experiments

4.1   Experimental Dataset and Preprocessing

The DARPA 1999 [10] dataset is a collection of data files including tcpdump files, BSM and NT audit data files, and directory listings files. We adopt Snort to detect attacks and generate alerts by reading the inside tcpdump data files in five weeks. First, we use the type, temporal and spatial characteristic of alert to identify and eliminate the redundant alerts, more details about the redundant elimination method refer to [11]. Then, the alerts meeting the following criteria are labeled as true alerts: (a) matching the source IP address, (b) matching the destination IP address, and (c) alert time stamp in the time window in which the attack has occurred. All remaining alerts are labeled as false alerts.

4.2   Results and Analysis

In order to evaluate the performance of the proposed semi-supervised classification method, we conduct experiments to compare it with two typical supervised methods, the traditional NB and the rule based RIPPER. Fig. 1 illustrates the classification accuracy of the three methods with only 20 training labeled alerts. Our approach shows a better classification accuracy (86.7%), which is respectively more than 10 percent and 7 percent higher than that with the NB(76%) and the RIPPER(79%) method.

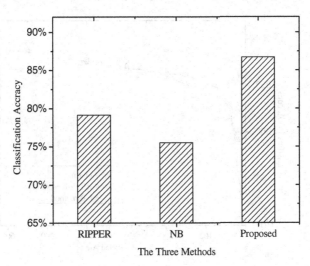

Fig. 1. Classification accuracy with only 20 labeled alerts

Fig. 2 compares the number of labeled alerts required by different methods when classification accuracy achieves more than 88%. The proposed method only need 69 labeled alerts, while NB and RIPPER methods need more than 530 and 409 alerts respectively. With varying number of labeled alerts, we compare the classification accuracy with the NB and RIPPER methods. Results in Fig. 3 show that the proposed method performs significantly better than the other methods even with a very small fraction of labeled alerts.

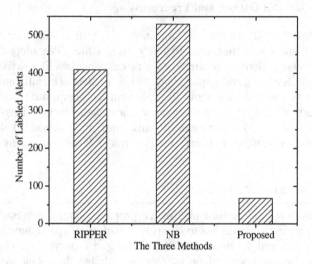

Fig. 2. Number of labeled alerts needed for achieving 88% classification accuracy

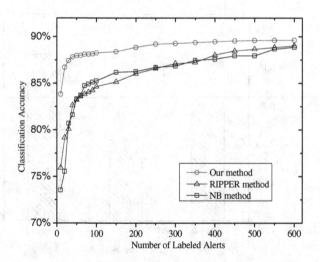

Fig. 3. The comparison of three methods on classification accuracy

5   Conclusion

This paper proposes a novel approach to alert classification for filtering IDS false positives. An alert generative model and an EM algorithm based on the generative model is designed. Using the intrusion detection test datasets, experiments are performed to compare the approach with two supervised methods based on Naïve Bayes and RIPPER. Experimental results show that the proposed method can significantly improve alert classification accuracy to limited sized labeled alert data. With this approach, manual effort is substantially reduced, and the feasibility of deploying alert classifier is enhanced. So it is more suitable to the real network environment.

References

1. Pietraszek, T.: Using Adaptive Alert Classification to Reduce False Positives in Intrusion Detection. In: Jonsson, E., Valdes, A., Almgren, M. (eds.) RAID 2004. LNCS, vol. 3224, pp. 102–124. Springer, Heidelberg (2004)
2. Shin, M.S., Kim, E.H., Ryu, K.H.: False Alarm Classification Model for Network-Based Intrusion Detection System. In: Yang, Z.R., Yin, H., Everson, R.M. (eds.) IDEAL 2004. LNCS, vol. 3177, pp. 259–265. Springer, Heidelberg (2004)
3. Bildoy, J., Clausen, S., Klausen, T.E.: Classifying Alerts in Multi-tier Intrusion Detection Systems. Master, Agder University College, Arendal (2004)
4. Chapelle, O., Scholkopf, B., Zien, A.: Semi-supervised Learning. MIT Press, Cambridge (2006)
5. Fatima, L.S., Mezrioui, A.: Improving the Quality of Alerts with Correlation in Intrusion Detection. International Journal of Computer Science and Network Security 7(12) (2007)
6. Nehinbe, J.O.: Automated Method for Reducing False Positives. In: Proc. of 2010 Int. Conf. on Intelligent Systems, Modelling and Simulation, pp. 54–59 (2010)
7. Li, H., Hu, Z., Wu, Y., Wu, F.: Behavior Modeling and Abnormality Detection Based on Demi-supervised Learning Method. Journal of Software 18(3), 527–537 (2007)
8. Liu, H., Yu, L.: Toward Integrating Feature Selection Algorithms for Classification and Cluster. IEEE Transactions on Knowledge and Data Engineering 17(3), 491–502 (2005)
9. Nigam, K.: Using Unlabeled Data to Improve Text Classification. PhD Thesis, Carnegie Mellon University, Pittsburgh, PA, USA (2001)
10. Lippmann, R., Haines, J.W., Fried, D.J., Korba, J., Das, K.: The 1999 DARPA Off-line Intrusion Detection Evaluation. Computer Networks 4(4), 579–595 (2000)
11. Gong, J., Mei, H.B., Ding, Y., Wei, D.H.: A multi-feature Correlation Redundance Elimination of Intrusion Event. Journal of Southeast University 35(3), 366–371 (2005)

Dual-form Elliptic Curves Simple Hardware Implementation

Jianxin Wang and Xingjun Wang

Department of Electronic and Engineering,
Beijing Tsinghua University, Beijing 100084, China
wangjianxin09@mails.tsinghua.edu.cn

Abstract. Performing standard Weierstrass-form curves' operations based on Edwards-form curves' addition law, the overall security of elliptic curves can be strengthened while remain compatible with existing ECC system. We present a simplified algorithm for finding such dual-form elliptic curves over prime field F_p with $p \equiv 3 \bmod 4$. Using the generated curves, algorithms for implementing dual-form operations on affine, projective and twisted coordinates are further discussed and optimized for the case of Weierstrass-form operations. The algorithms are implemented on FPGA and show competitive time and area performance both in Edwards form and Weierstrass form.

Keywords: Elliptic curve, Edwards curve, birational equivalence, FPGA.

1 Introduction

Elliptic curve cryptosystems (ECC) was proposed by Koblitz [1] and Miller [2] in the mid-1980s, showing comparable level of security with shorter key sizes and computationally more efficient algorithms[3]. However, as discussed in [4,5], side-channel attacks, which recover the private key using side channel information such as computing time and power consumption, are potential threats to existing Weierstrass-form ECC system.

Edwards curves was recently introduced in [6], and Bernstein and Lange further proposed twisted Edwards curves [7]. Edwards curves have unified addition law, which have the native ability to countermeasure some kinds of side-channel attacks. Furthermore, Bernstein et al in [8] suggest that Edwards-form group formula need fewer operations than fastest known Weierstrass-form formula. Because of super timing and security performance of Edwards curves, work has been done to try to perform Weierstrass curves' operations based on Edwards-form addition law [12].

As discussed in [7-9], only certain simplified Weierstrass-form can be transformed to Edwards-form. However, performing standard Weierstrass-form curves' operations based on Edwards-form curves' addition law, the overall security of elliptic curves can be strengthened while remain compatible with existing ECC system. This paper will focus on F_p with $p \equiv 3 \bmod 4$ and discuss finding such dual-form curves and optimizing algorithm for hardware implementation. For the sake of simplicity and

D.-S. Huang et al. (Eds.): ICIC 2012, LNCS 7389, pp. 520–527, 2012.
© Springer-Verlag Berlin Heidelberg 2012

without confusion, we just take Edwards-form curves for twisted Edwards curves and take Weierstrass-form curves for simplified Weierstrass curves.

The organization of this paper is as follows. Section 2 reviews Weierstrass-form and Edwards-form curves, defining the main parameters and discussing the group law. In Section 3, birational equivalence between Weierstrass and Edwards form is investigated and simplified algorithm to generate dual-form curves is further proposed. The algorithms for hardware implementation are discussed and optimized in Section 4. Finally Section 5 concludes the paper.

2   Elliptic Curves Arithmetic

2.1   Weierstrass-Form Elliptic Curves

Let $p \geq 5$ be a prime and F_p be the finite field of order p. For $a, b \in F_p$, an elliptic curve E can be written in the simplified Weierstrass form as (see [10]):

$$E(F_p): y^2 = x^3 + ax + b \tag{1}$$

2.2   Montgomery-Form Elliptic Curves

Montgomery originally introduced this form of curves for speeding up the Pollard and Elliptic curve methods of integer factorization [11]. The Montgomery-form elliptic curves over F_p is defined as follow:

$$M_{A,B}: BY^2 = X^3 + AX + X$$
$$A \in F_p \setminus \{\pm 2\}, B \in F_p \setminus \{0\} \tag{2}$$

The transformability from Weierstrass-form to Montgomery-form is thoroughly discussed in [12], which is very important for our dual-form implementation. We just show the result as (3).

$$v^2 = t^3 + (3 - A^2)/(3B^2) t + (2A^3 - 9A)/(27B^3) \tag{3}$$

2.3   Edwards-Form Elliptic Curves

Edwards in [6] presented a unified addition law for the curves $x^2 + y^2 = c^2(1 + x^2 y^2)$ over a non-binary field k (including F_p). Bernstein and Lange in [7] further generalized the Edwards curves and introduced the twisted Edwards curves over the prime field F_p, which is described as follow with coefficients a and d over F_p and non-zero:

$$E_{E,a,d}: ax^2 + y^2 = 1 + dx^2 y^2 \tag{4}$$

Points on $E_{E,a,d}$ can be added by the unified addition law as (5), even if the two points are the equal. The unified addition law is very important for the anti-side-channel implementation as we discuss in the following section.

$$(x_1, y_1) + (x_2, y_2) = \left(\frac{x_1 y_2 + x_2 y_1}{1 + d x_1 x_2 y_1 y_2}, \frac{y_1 y_2 - a x_1 x_2}{1 - d x_1 x_2 y_1 y_2} \right) \tag{5}$$

Theorem 3.2 in [7] gives that every twisted Edwards curve over F_p is birationally equivalent over F_p to a Montgomery curve, and [6] showed that a significant number of elliptic curves over GF(p) (roughly 1/4 of isomorphism classes of elliptic curves) are birationally equivalent to a twisted Edwards curve. And the twisted Edwards curve $E_{E,a,d}$ is birationally equivalent to the Montgomery curve $M_{A,B}$, especially over F_p with $p \equiv 3 \bmod 4$, where

$$A = 2(a+d)/(a-d), B = 4/(a-d) \tag{6}$$

3   Birational Equivalence

3.1   Birational Transformation

Based on the work of [7] and [12], we can now deduce the birationally mapping between twisted Edwards form and simplified Weierstrass form. Substitute (6) into (3), gives(with $a_w = -(a^2 + 14ad + d^2)/48$ and $b_w = -(a^3 - 33a^2 d - 33ad^2 + d^3)/864$):

$$W(F_p) : v^2 = t^3 + a_w t + b_w \tag{7}$$

The transformation and reverse transformation from $E_{E,a,d}$ to $W(F_p)$ is shown as (8) and (9).

$$t = ((5a-d) + (a-5d)y)/(12(1-y)), v = (a-d)(1+y)/(4x(1-y)) \tag{8}$$

Here just omitting the special cases for simplicity. Although these would bring in two more inversion, which is the most expensive operation, when incorporated with projective or inverted twisted Edwards coordinates, the additional inversion can be eliminated.

$$x = (6t - (a+d))/(6v), y = (12t + d - 5a)/(12t + a - 5d) \tag{9}$$

As we can deduce from projective twisted Edwards coordinates and inverted twisted Edwards coordinates in [6,7], we just need about 8 extra field multiplications to transform the point in simplified Weierstrass form to twisted Edwards form, performing point operations in twisted Edwards form and then transform back.

3.2   Birational Equivalence

Reference [12] gives the transformability between Weierstrass-form and Montgomery-form curves, and theorem 3.4 in [7] shows that for a prime F_p with $p \equiv 3 \bmod 4$, every Montgomery-form curve over F_p is birationally equivalent over

F_p to an Edwards-form curve. In this section, we discuss the birational equivalence between Edwards-form and Weierstrass-form curves over F_p with $p \equiv 3 \bmod 4$.

As mentioned in section 6 of [7], if a is a square in F_p and d is nonsquare, twisted Edwards curves have unified and complete addition law. Without loss of generality, we focus our discussing on these kinds of twisted Edwards curves. Taking these into consideration, we can generalize theorem 1 in [12] as follows:

Corollary. If a is a square in F_p and d is nonsquare:

Table 1. Criterion

$p \equiv 3 \bmod 4((-1/p) = -1)$		
d	$a-d$	$8 \nmid\!\# E_{E,a,d}$
QNR	QNR	ND
QNR	QNR	D
QR	QNR	D
QR	QR	D

with QR: quadratic residue; QNR: quadratic non-residue; D: divisible by 8; ND: non-divisible by 8

Proof: With $p \equiv 3 \bmod 4$, Montgomery-form curve is birationally equivalent over F_p to an Edwards-form curve. As the theorem 1 of [12] indicated, if and only if $A+2$ is quadratic non-residue and $A-2$ is quadratic residue, $\# E_{E,a,d}$ is non-divisible by 8. With (8), we get $4a/(a-d)$ is quadratic non-residue and $4d/(a-d)$ is quadratic residue. Because a is a square in F_p, then a should be quadratic residue. For the product of a non-residue and a (nonzero) residue is a non-residue and $4a/(a-d)$ is quadratic non-residue, then $1/(a-d)$ is quadratic non-residue. Also the inversion of quadratic non-residue is quadratic non-residue. So, $(a-d)$ is quadratic non-residue. As the same again for $4d/(a-d)$, we get d is quadratic non-residue.

Table 2. Algorithm 1

Algorithm 1 Generate dual-form elliptic curves whose Edwards-form has complete addition law
INPUT A prime $p \equiv 3 \bmod 4$.
OUTPUT A Weierstrass-form elliptic curve with the Edwards-form which has complete addition law and with cofactor 4.
1. Generate a and d, and put $E_{E,a,d}$.
2. Check a is a square, d and $(a-d)$ is quadratic non-residue. If not, go to step 1;
3.Set $a_w = -(a^2 + 14ad + d^2)/48$, $b_w = -(a^3 - 33a^2d - 33ad^2 + d^3)/864$, get $W(F_p)$
4.Compute $\# E_{W(F_p)}$ and check $\# E_{W(F_p)} = 4l$, for some prime l. Go to Step 1 if not.
5. Check other security tests, and output the parameters if it passes all tests.

With this conclusion, we can simplified the algorithms to generate Elliptic curves with cofactor 4 discussed in [12] for $p \equiv 3 \bmod 4$, as shown in table 2.

As discussed in [2], we can apply all security tests used for finding Weierstrass-form curves to check the generated dual-form curves. So the security of Edwards-form curves is guaranteed and equal to that of Weierstrass-form curves. One generated group for field F_p with $p = 2^{256} - 2^{224} - 2^{96} + 2^{64} - 1$ is shown in table 3.

Table 3. Group parameter Example

Generated curve from algorithm 1
p =0xFFFFFFFEFFFFFFFFFFFFFFFFFFFFFFFFFFFFFFFF FF00000000FFFFFFFFFFFFFFFF
a_E =0xFFFFFFFEFFFFFFFFFFFFFFFFFFFFFFFFFFFFFE140000010400000004100000020000001
d_E =0x168D9
n =0x3FFFFFFFBFFFFFFFFFFFFFFFFFFFFFFFFFDF36BCC0EA4B456F51C94F1D2F7FF825*4
a_W =0x3C4CFFFFD951FFFFDB098000129B4A45C6D726869E29A3DBEDDFF2D8096672AC
b_W =0x8BD877D54CD8B264E09D5141B88BE222352167ABEDBCC61F861B33B600699623

4   Hardware Implementation Issue and Result

In this section, we discuss and optimize the algorithms for hardware implementation, based on the curves on table 3. Firstly, we optimize the montgomery multiplication.

4.1   Montgomery Multiplication and Group Law

Montgomery multiplication [13] is key operation for optimizing the field multiplication on F_p. But as we all know, the Montgomery transformation $x \to \tilde{x} = xR \bmod p$ and the reverse transformation $\tilde{x} \to x = \tilde{x}R^{-1} \bmod p$ [11] is required for montgomery multiplication. Integrating the dual-form transform into projective and inverted twisted Edwards coordinates' operations, we can further eliminate the montgomery transformation overhead, which further simplify the hardware control logic.

As transform from $W(F_p)$ to $E_{E,a,d}$ in projective or inverted twisted Edwards coordinates in [7] indicate, the transform just need one field multiplication. If we omit $x \to \tilde{x} = xR \bmod p$, all of three coordinates just introduce exactly one extra coefficient $c = R^{-1}$. And the same for the reverse transform, both inversions will just eliminate the introduced extra coefficient. Also notice that, if we transform parameter of $E_{E,a,d}$ as $a' \leftarrow a' = aR \bmod p$ and $b' \leftarrow b' = bR \bmod p$, the Edwards-form group law in projective coordinates will just get the same extra coefficient for all three coordinates. Combined the transform and reverse transform from $W(F_p)$ to $E_{E,a,d}$, we can totally eliminate the montgomery transform and reverse transform.

4.2   Implementation Results and Performance Comparison

Based on the curves generated by Algorithm 1, we implement the dual-form elliptic curves processor on FPGA for verification purposes. We design and compare the performance on affine, projective and inverted coordinates respectively. All of following results are based on a Xilinx FPGA (Spartan 3 xc3s5000).

Affine Coordinates. Weierstrass-form, Edwards-form and dual-form processor were implemented separately and time-area performance was compared in table 4. From table 4 we can see that the point multiplication time in dual forms for $W(F_p)$ is 27% slower then Edwards form and the area is about 35% larger.

Table 4. Affine Coordinates Result

Form	Weierstrass	Edwards	Dual-Form
AREA/Number of Slices	11,339	12,357	16,928
TIME/Clock cycles(W(Fp))	533,378	1,140,554	1,452,226

Also it is apparent that for affine coordinate, point multiplication in Weierstrass form is twice as fast as Edwards-form and Dual-form, with area about 50% smaller. The overall performance shows that dual-form is not quite useful in affine coordinate.

Projective and Inverted Coordinates. Using Montgomery multiplication algorithm discussed previously, we compared the performance of Edwards form and dual form in table 5 for projective and inverted coordinates.

Table 5. Projective/Inverted Coordinates Result

Form	Projective coordinates		Inverted coordinates	
	Edwards	Dual-Form	Edwards	Dual-Form
AREA/Number of Slices	14,773	18,755	16,609	20,207
TIME/Clock cycles(W(Fp))	104,538	104,731	97,443	97,552

Table 5 show that when integrate the dual-form operation into projective or inverted twisted coordinates, we get almost 30% boost in area, while keeping the operating time about the same. Also we can see that, inverted coordinates is trading 7% larger in area for 7% faster in time.

Compared with Existing Result. We also compare the implementation with reported result in 256-bit length prime field [14-17] in table 6. From table 6 we can see that our dual-form implementations have better time-area performance, trading almost 30% boost in size for 35% decrease in clock cycles, while at the same time enhancing the overall security.

Table 6. Result Comparision

ECC system	Platform	Max freq. (MHz)	Clock cycles	Area
McIvor et al. [17]	XC2VP125-7	34.46	151,360	15,755 slices 256 mults
Sakiyama et al. [14]	XC3S5000-5	40	-	27,597 slices
dual-form projetive	XC3S5000-5	34.75	104,731	18,755 slices 64 mults
dual-form inverted	XC3S5000-5	35.41	97,552	20,207 slices 64 mults

5   Conclusions

In this paper, we discuss the issues relating to dual-from, i.e. the twisted Edwards form and simplified Weierstrass form, elliptic curves over F_p with $p \equiv 3 \mod 4$, from curves generating to hardware implementation. We generalize the theorem of birational equivalence and optimize the algorithm for generating such curves. With generated curves, algorithm for hardware implementation is optimized and compared with reported result.

Affine coordinate in Edwards form is not quite suitable for dual form, for both timing and area performance are not quite satisfied. Projective and inverted twisted coordinates are designed and optimized for Weierstrass-form operations and both show comparable timing and area performance. Compared with reported result, both have better timing and area performance, trading almost 30% boost in size for 35% faster in clock cycles. And dual-form inverted implementation also outperforms the projective one in timing, which is 7% faster.

Future work could be incorporating more countermeasures of side-channel attacks into the processor to enhance the overall security levels.

Acknowledgments. The authors would like to express their thanks to Dr. Wang and the members of Department of Electronic and Engineering for their support, assistance and inspiration. Moreover, special thanks to anonymous reviewers for their contributions to improve the manuscript.

References

1. Koblitz, N.: Elliptic Curve Cryptosystems. Mathematics of Computation 48, 203–209 (1987)
2. Miller, V.S.: Use of Elliptic Curves in Cryptography. In: Williams, H.C. (ed.) CRYPTO 1985. LNCS, vol. 218, pp. 417–426. Springer, Heidelberg (1986)
3. Lenstra, A.K., Verhul, E.R.: Selecting Cryptographic Key Sizes. J. Cryptol. 14, 255–293 (2001)
4. Kocher, P.C.: Timing Attacks on Implementations of Diffie-Hellman, RSA, DSS, and Other Systems. In: Koblitz, N. (ed.) CRYPTO 1996. LNCS, vol. 1109, pp. 104–113. Springer, Heidelberg (1996)
5. Biehl, I., Meyer, B., Müller, V.: Differential Fault Attacks on Elliptic Curve Cryptosystems. In: Bellare, M. (ed.) CRYPTO 2000. LNCS, vol. 1880, pp. 131–146. Springer, Heidelberg (2000)
6. Edwards, H.M.: A Normal Form for Elliptic Curves. Bulletin of the American Mathematical Society 44(3), 393–422 (2007)
7. Bernstein, D.J., Birkner, P., Joye, M., Lange, T., Peters, C.: Twisted Edwards Curves. In: Vaudenay, S. (ed.) AFRICACRYPT 2008. LNCS, vol. 5023, pp. 389–405. Springer, Heidelberg (2008)
8. Bernstein, D.J., Lange, T.: Faster Addition and Doubling on Elliptic Curves. In: Kurosawa, K. (ed.) ASIACRYPT 2007. LNCS, vol. 4833, pp. 29–50. Springer, Heidelberg (2007)
9. Verneuil, V.: Elliptic Curve Cryptography on Standard Curves Using the Edwards Addition Law (2011) (not published yet)

10. Hankerson, D.R., Vanstone, S.A., Menezes, A.J.: Guide to Elliptic Curve Cryptography. Springer (2004)

11. Montgomery, P.L.: Speeding the Pollard and Elliptic Curve Methods of Factorizations. Math. Comp. 48, 243–264 (1987)

12. Okeya, K., Kurumatani, H., Sakurai, K.: Elliptic Curves with the Montgomery-Form and Their Cryptographic Applications. In: Imai, H., Zheng, Y. (eds.) PKC 2000. LNCS, vol. 1751, pp. 238–257. Springer, Heidelberg (2000)

13. Koc, C.K., Acar, T., Kaliski, B.S.: Analyzing and Comparing Montgomery Multiplication Algorithms. IEEE Micro 16(3), 26–33 (1996)

14. Sakiyama, K., Mentens, N., Batina, L., Preneel, B., Verbauwhede, I.: Reconfigurable Modular Arithmetic Logic Unit Supporting High-performance RSA and ECC over GF(p). International Journal of Electronics 94(5), 501–514 (2007)

15. Kocabas, U., Fan, J., Verbauwhede, I.: Implementation of Binary Edwards Curves for very-Constrained Devices. In: 21st IEEE International Conference on Application-specific Systems Architectures and Processors, ASAP 2010, pp. 185–191 (2010)

16. McIvor, C., McLoone, M., McCanny, J.: Hardware Elliptic Curve Cryptographic Processor over GF(p). IEEE Trans. Circuits and Systems I 53(9), 1946–1957 (2006)

17. Chatterjee, A., Gupta, I.S.: FPGA Implementation of Extended Reconfigurable Binary Edwards Curve Based Processor. In: 2012 International Conference on Computing, Networking and Communications (ICNC), pp. 211–215 (2012)

Genetic Based Auto-design of Fuzzy Controllers for Vector Controlled Induction Motor Drives

Moulay Rachid Douiri and Mohamed Cherkaoui

Mohammadia Engineering School, Department of Electrical Engineering, Avenue Ibn Sina, 765, Agdal-Rabat, Morocco
douirirachid@hotmail.com

Abstract. This work presents the Genetic algorithm based auto-design of fuzzy logic controller for speed controller of the indirect field oriented controlled induction motor drives, to automate and at the same time to optimize the fuzzy controller design process. To do this, the normalization parameters, membership functions and decision table are converted into binary bit string. This optimization requires a predefined performance index. The task of such a design algorithm is the modification of the existing knowledge and at the same time, the investigation of new feasible structures.

Keywords: fuzzy logic controller, genetic algorithms, indirect field oriented, induction motor.

1 Introduction

Field oriented control (FOC) or vector control (VC) proposed by Blaschke [1] and Hasse [2] has become an industry standard for control of induction machines in high performance drive applications. Control of induction motor using the principle of field orientation gives control characteristics similar to that of a separately excited dc machine. Orientation is possible along mutual flux or stator flux or the rotor flux; however, orientation of the stator current space vector with respect to the rotor flux alone gives natural decoupling between the torque and flux producing components of the stator current space vector. VC induction motor drive outperforms the dc drive because of higher transient current capability, increased speed range and lower rotor inertia. It is due to this reason that modern high performance drive application is moving towards using induction machine as the drive element.

Today, the new trends in this field now involve the application of modern non-linear control techniques to further enhance the performance of such controllers as well as optimizing drive operation based on a specific requirement [3], [4], [5]. The research underlying this paper involves the development of a novel synthesis methodology to automate and at the same time, to optimize the performance of fuzzy controllers based on a predefined objective function for any particular application. It also aims, in particular, to design an optimal fuzzy controller for induction motor drives with indirect field oriented control. Two intelligent techniques were used in this paper namely fuzzy logic and genetic algorithms.

D.-S. Huang et al. (Eds.): ICIC 2012, LNCS 7389, pp. 528–537, 2012.
© Springer-Verlag Berlin Heidelberg 2012

Fuzzy logic, first developed by Zadeh (1965) [6], is an approach for handling complex problems using reasoning that is approximate as opposed to precise, formally deduced logic. The key difference between fuzzy logic and probability theory is that the former is interested in capturing partial truths, that is, how to reason about things that are not wholly true or false, while the latter is concerned with making predictions about events based on a partial state of knowledge [6], [7]. Fuzzy logic is derived from fuzzy set theory whereby subjects in a set have degrees of membership, described by membership functions, and where each subject can belong to one or more fuzzy sets. Membership in a fuzzy set is denoted by a membership value between 0 and 1; it can be thought of as the possibility of association of a particular subject with a particular set, as opposed to the probabilistic likelihood of an event occurring. Fuzzy logic is useful in situations where vagueness exists, there are no clear cut definitions, and results cannot be categorized as "true" or "false" outcomes [6], [7]. The fuzzy logic controller (FLC) to be investigated is the Mamdani's type [8], although there exist other types, for example, the Sugeno's [9] and the Yamakawa's [10].

The underlying principles of GAs were first published by Holland in 1962 [12]. The mathematical framework was developed in the late 1960's, and presented in Holland's pioneering book in 1975. Genetic Algorithms are search algorithms which are based on the genetic processes of biological evolution. They work with a population of individuals, each representing a possible solution to a given problem. Each individual is assigned a fitness score according to how well it solves the given problem. For instance, the fitness score might be a performance index for a dosed loop control system. In nature, this is equivalent to assessing how effective an organism is at competing for resources. The highly adapted individuals will have relatively large numbers of offspring's. Poorly performing ones will produce few or even no offspring at all. The combination of selected individuals produces super fit offspring's, whose fitness's are greater than that of the parents. In this way, the individuals evolve to become more and better suited to their environment [12], [13].

This paper is organized as follows: The principle of indirect field oriented control is presented in the second part, the fuzzy logic speed controller in section three, the genetic algorithm optimization based auto-design of fuzzy speed controllers in the fourth section, the five part is devoted to illustrate the simulation performance of this control approach, a conclusion and reference list at the end.

2   Indirect Field Oriented Control

The field orientation concept implies that the currents supplied to the machine should be oriented in phase and in quadrature to the rotor flux vector ψ_{qdr}. This can be achieved by selecting ω_e to be the instantaneous speed of ψ_{qdr} and locking the phase of the reference system such that the rotor flux is aligned with the d-axis, resulting in the mathematical constraint [1]. At any instant, d electrical axis is in angular position θ_e relative to α axis. The angle θ_e is the result of the sum of both rotor angular and slip angular positions, as follows:

$$\begin{cases} \theta_e = \theta_r + \theta_{sl} \\ \omega_e t = \omega_r t + \omega_{sl} t = (\omega_r + \omega_{sl})t \end{cases} \tag{1}$$

where: ω_r and θ_r are the position and rotor angular velocity; θ_{sl} and ω_{sl} are the position and sliding angular velocity.

$$\psi_r = \psi_{rd}, \qquad \psi_{qr} = 0 \tag{2}$$

The main equations of indirect vector control,

$$\omega_{sl} = \frac{L_m}{\widehat{\Psi}_r} \cdot \frac{R_r}{L_r} \cdot i_{qs} \tag{3}$$

$$\frac{L_r}{R_r} \cdot \frac{d\widehat{\Psi}_r}{dt} + \widehat{\Psi}_r = L_m i_{ds} \tag{4}$$

$$\Gamma_{em} = \frac{3}{4} \cdot p \cdot \frac{L_m}{L_r} \cdot i_{qs} \cdot \widehat{\Psi}_r \tag{5}$$

3   Fuzzy Speed Controller

A conventional PI controller can be described by:

$$\Gamma_{em}^* = k_p e + k_i \int_0^t e(t)dt \tag{6}$$

where k_p and k_i are the proportional and the integral gain coefficients and $e = \omega_r^* - \omega_r$ is the speed error between the command speed ω_r^* and the actual motor speed ω_r. If the above integral equation is converted into a differential equation by taking the derivative with respect to time, the equivalent equation will be:

$$\dot{\Gamma}_{em}^* = k_p \dot{e} + k_i e \tag{7}$$

The PI controller (7) can be written in a fuzzy rule form as follows:

$$\text{If } e(k) \text{ is } LV_e, \text{ and } \Delta e(k) \text{ is } LV_{\dot{e}}, \text{ then } \Delta\Gamma_{em}^*(k) \text{ is } LV_{\dot{\Gamma}_{em}} \tag{8}$$

with LV: linguistic variable

The most significant variables entering the fuzzy logic speed controller have been selected as the speed error e and its change $\dot{e}$, the output this controller is $\dot{\Gamma}_{em}^*$. The equation input/output controller FLC written at time k:

$$e(k) = \omega_r^*(k) - \omega_r(k) \tag{9}$$

$$\dot{e}(k) = e(k) - e(k-1) \tag{10}$$

$$\Gamma_{em}^*(k) = \Gamma_{em}^*(k-1) + \dot{\Gamma}_{em}^*(k) \tag{11}$$

The principle of this strategy is shown in Fig. 1.

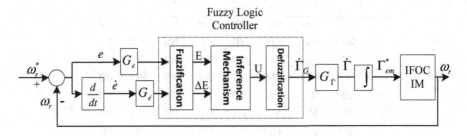

Fig. 1. Basic structure of the fuzzy logic controller for indirect field oriented control

The fuzzy sets are characterized by standard designations: *NB* (negative big), *NM* (negative medium), *NS* (negative small), *AZ* (approximate zero), *PS* (positive small), *PM* (positive medium) and *PB* (positive big). Fuzzy distribution is symmetric, and non-equidistant in our choice. We have chosen also in our application the triangular-shaped membership function. In order to design a universal FLC, we can transform the values range in standard ranges. Therefore, the input and output gains are introduced:

$$G_e = \frac{e_G(k)}{e(k)}, \qquad G_{\dot{e}} = \frac{\dot{e}_G(k)}{\dot{e}(k)}, \qquad G_{\dot{\Gamma}} = \frac{\dot{\Gamma}(k)}{\dot{\Gamma}_G(k)} \tag{12}$$

From behavior study of the system closed-loop speed based on experience, we can establish the command rules which connect output with inputs [14], [7]. As we have seen, there are seven fuzzy sets, which imply forty-nine possible combinations of these inputs, in which forty-nine rules. They can be presented in a matrix called matrix inference shown in the Table 1.

We choose min-max inference method, for each rule, we obtain the partial membership function by relation (13) [11]:

$$\mu_{R_i}(\dot{\Gamma}_G) = \min\left(\mu_{c_i}, \mu_{O_i}(\dot{\Gamma}_G)\right) \quad i = 1, 2 ... m \tag{13}$$

where μ_{Ci} is a membership factor assigned to each rule R_i; $\mu_{Oi}(\dot{\Gamma}_G)$ is the membership function related in operation imposed by rule R_i.

The resulting membership function is then given by [11], [6]:

$$\mu(\dot{\Gamma}_G) = \max\left(\mu_{R_1}(\dot{\Gamma}_G), \mu_{R_2}(\dot{\Gamma}_G), ..., \mu_{R_m}(\dot{\Gamma}_G)\right) \tag{14}$$

The defuzzification process employs the center of gravity method. As a result, the control increment is obtained by the following formula [11], [6]:

$$\dot{\Gamma}_G = \frac{\int \dot{\Gamma}_G . \mu(\dot{\Gamma}_G) d(\dot{\Gamma}_G)}{\int \mu(\dot{\Gamma}_G) d(\dot{\Gamma}_G)} \qquad (15)$$

Table 1. The fuzzy linguistic rule table

$\dot{\Gamma}$		e					
	NB	NM	NS	ZE	PS	PM	PB
NB	NB	NB	NB	NB	NM	NS	AZ
NM	NB	NB	NB	NM	NS	AZ	PS
NS	NB	NB	NM	NS	AZ	PS	PM
$\dot{e}$ AZ	NB	NM	NS	AZ	PS	PM	PB
PS	NM	NS	AZ	PS	PM	PB	PB
PM	NS	AZ	PS	PM	PB	PB	PB
PB	AZ	PS	PM	PB	PB	PB	PB

4   Genetic Algorithms Based Fuzzy Logic Controller

The genetic algorithm is applied to automate and optimize the fuzzy controller design process. This optimization requires a predefined objective function. Moreover, the normalization parameters, membership functions and decision table are converted into binary bit string constructed by cascading.

An individual bit length is 597 bit and composed of three gene block:

Gene block 1 (Normalization factors): Determine the proper domain of the control surface, which represents 10 bit normalization factors for speed error, 10 bit normalization for speed error derivative, 10 bit denormalization factor for control output.

Gene block 2 (Membership functions): To have complete freedom in partitioning the state space, asymmetrical membership functions should be chosen that consequently suggest three different design parameters, i.e. M_1, M_2, and M_3 for each membership functions (Fig. 2). Let us consider seven membership functions for each

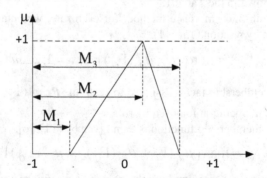

Fig. 2. Membership function parameters

variable for a controller with two inputs, 42 parameters are required to define the entire set of membership functions, where every parameter is encoded with 10-bit resolution.

Gene block 3 (Decision table): Every consequent part of a fuzzy rule should be encoded in a binary form. Since every consequent can take on only one of seven different values based on Table 1, every consequent can be represented by only three bits i.e. 3×49 parameters = 147 bits will represent the entire decision table (Fig. 3).

Normalization factors			Membership functions parameters					Decision table				
G_e	$G_{\dot{e}}$	$G_{\dot{r}}$	M_1	M_2	M_3	....	M_{42}	R_1	R_2	R_3	....	R_{49}

Fig. 3. Bit-string representation of entire controller

Each individual represents a possible solution to the problem; a particular fitness function is required for the evaluation of the individuals [12], [13], [15]. In this way, for every particular chromosome (i.e. each solution), the fitness function returns a single numerical value, which indicates the quality of that solution. In the context of optimization it is the performance index of the closed loop system that becomes the fitness function. Our goal is to have a response speed with a short rise time, small overshoot, and near-zero steady state error. In this respect, a multiple objective function is required:

$$J = \int_0^t |e| dt + 4 \underbrace{\int_0^t \delta(\frac{dz}{dt}).|z^* - z(t)| dt}_{(2)} + 0.5 \underbrace{\int_0^t |e| t dt}_{(3)} \tag{16}$$
$$\underbrace{}_{(1)}$$

(1)   Measure of a fast dynamic response;

(2) The penalty on the multiple overshoot of the response, where $\delta(dz/dt)$ detects the instances that overshoots (or undershoots) occur:

$$\int_{0^-}^{0^+} \delta(\frac{dz}{dt}) = \begin{cases} 1 & if & \dfrac{dz}{dt} = 0 \\ 0 & if & \dfrac{dz}{dt} \neq 0 \end{cases} \tag{17}$$

and $|z^*-z(t)|$ determines the response deviation from the desired value;

(3) Measure the steady state error.

The genetic algorithm with the free parameters shown in Table 2 was able to find the near-optimum solution with a population of 44 individuals, in almost 358 generations Fig. 5. This is due to the large number of design parameters involved in concurrent optimization. The principle of genetic algorithms based fuzzy logic controller is shown in Fig. 4.

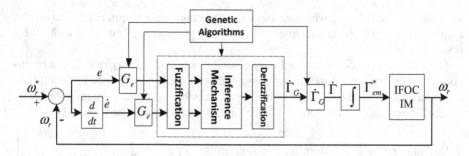

Fig. 4. Basic structure of the genetic algorithms based fuzzy logic controller for indirect field oriented control

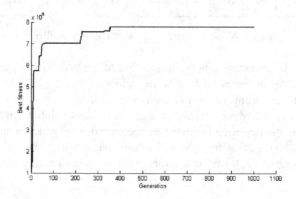

Fig. 5. Speed of convergence

The optimization algorithm and the motor drive response are then verified under loading and unloading conditions. A speed command of 50 *rad/s* at 0.02 *s* is given to the drive system, the full load is applied at 0.2 *s*; then load is completely removed at 0.4 *s*. and then accelerated further to 100 *rad/s*, full load is applied at 0.8 *s*; then load is completely removed at 1 *s*. Later, after speed reversal of -50 *rad/s* at 1.2 *s*, full load is applied at 1.4 *s* and the load is fully removed at 1.6 *s*. Fig. 5 shows the speed optimization result and response of the drive system.

The FLC-GA speed response (Fig. 6) shows that the drive can follow the low command speed very quickly and smoothly without overshoot, no steady-state error and rapid rejection of disturbances, with a low dropout speed (Fig. 7 and Table 3). The current responses are sinusoidal and balanced, well as the decoupling between the flux and torque is verified (Figs. 8 and 9).

Table 2. Genetic algorithm parameters

GA property	Value	GA property	Value/Method
Number of generations	358	Selection method	Roulette wheel
No of chromosomes in each generation	44	Crossover method	Double-point
No of genes in each chromosome	3	Crossover probability	0.8
Chromosome length	597	Mutation rate	0.05

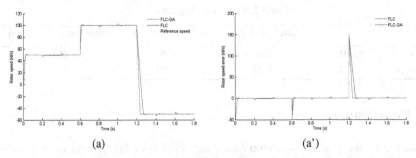

(a)   (a')

Fig. 6. Rotor speed acceleration and reversal: (a) FLC and FLC-GA (a') speed error

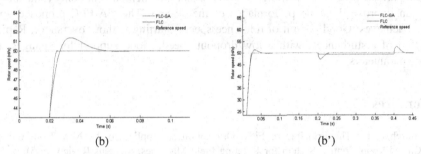

(b)   (b')

Fig. 7. Zoom speed: (b) starting transient performance and overshoot (b') response due to load and unload change

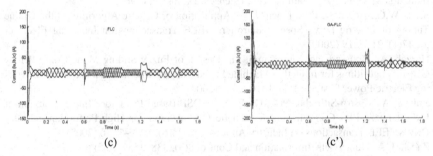

(c)   (c')

Fig. 8. Three phase stator current: (c) FLC (c') FLC-GA

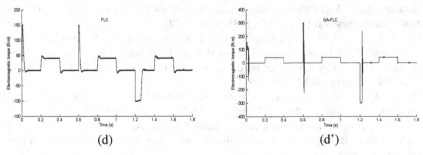

(d)   (d')

Fig. 9. Electromagnetic torque response: (d) FLC (d') FLC-GA

Table 2. Summary of results

	Rise time (s)	Overshoot (%)	Settling time (%)	Steady state error (%)
FLC	0.03	2.9	0.08	0.7
FLC-GA	0.02	0.05	0.001	0.4

5   Conclusions

This work uses the genetic algorithm based auto-design of fuzzy logic controller as the speed controller of the indirect field oriented controlled induction motor drives. By comparison with FLC controller, it testifies that this method is not only robust, but also can improve dynamic performance of the system. The GA-FLC proposed approach achieves: Good pursuit of reference speed; Starting without overshoot; Rapid rejection of disturbances, with a low dropout speed; Good support for changes in engine parameters.

References

1. Blaschke, F.: The Principle of Field Orientation as Applied to the New Trans-vector Closed-Loop Control System for Rotating Field Machines. Siemens Review 34(5), 217–219 (1972)
2. Hasse, K.: On the Dynamics of Speed Control of a Static AC Drive with a Squirrel Cage Induction Machine. Dissertation, Tech. Hochsch. Darmstadt (1969)
3. Silva, W.G., Acarnley, P.P., Finch, J.W.: Application of Genetic Algorithm to the Online Tuning of Electric Drive Speed Controllers. IEEE Transactions on Industrial Electronics 47(1), 217–219 (2000)
4. Hazzab, A., Bousserhane, I.K., Kamli, M.: Design of Fuzzy Sliding Mode Controller by Genetic Algorithms for Induction Machine Speed Control. International Journal of Emerging Electric Power Systems 1(2), 1016–1027 (2004)
5. Rubaai, A., Castro-Sitiriche, M.J., Ofoli, A.R.: DSP-Based laboratory Implementation of Hybrid Fuzzy-PID Controller using Genetic Optimization for High-Performance Motor Drives. IEEE Transactions on Industry Applications 44(6), 1977–1986 (2008)
6. Zadeh, L.A.: Fuzzy Sets. Information and Control 8(3), 338–353 (1965)
7. Zhao, Z.-Y., Tomizuka, M., Isaka, S.: Fuzzy Gain Scheduling of PID Controllers. IEEE Transactions on Systems Man and Cybernetics 23(5), 1392–1398 (1993)
8. Mamdani, E.H.: Application of Fuzzy Logic Algorithms for Control of Simple Dynamic Plant. Proceedings of IEEE 121(12), 1585–1588 (1974)
9. Takagi, T., Sugeno, M.: Fuzzy Identification of Systems and its Applications to Modeling and Control. IEEE Transactions on Systems Man and Cybernetics 15(1), 116–132 (1985)
10. Yamakawa, T.: Fuzzy Controller Hardware System. In: Proceedings of 2nd IFSA Congress, Tokyo, Japan, pp. 827–830 (1987)
11. Chen, S.-M.: A Fuzzy Approach for Rule-Based Systems Based on Fuzzy Logics. IEEE Transactions on Systems Man and Cybernetics 26(5), 769–778 (1996)
12. Holland, J.: Adaptation in Natural and Artificial Systems: An Introductory Analysis with Applications to Biology Control and Artificial Intelligence. University of Michigan Press, Ann Arbor (1975)

13. Goldberg, D.E.: Genetic Algorithms in Search Optimization and Machine Learning. Addison-Wesley, Reading (1989)
14. Uddin, M.N., Radwan, T.S., Rahman, M.A.: Performance of Fuzzy-Logic-Based Indirect Vector Control for Induction Motor Drive. Transactions on Industry Applications 38(5), 1219–1225 (2002)
15. Cardoso, F.D.S., Martins, J.F., Pires, V.F.: A Comparative Study of a PI, Neural Network and Fuzzy Genetic Approach Controllers for an AC-Drive. In: 5th IEEE International Workshop on Advanced Motion Control, AMC 1998, Coimbra, pp. 375–380 (1998)

Appendix: Induction Motor Parameters

Rated power = $7.5Kw$, Rated voltage = $220V$, Rated frequency = $60Hz$, $R_r = 0.17\Omega$, $R_s = 0.15\Omega$, $L_r = 0.035H$, $L_s = 0.035H$, $L_m = 0.0338H$, $J = 0.14Kg.m^2$.

Very Short Term Load Forecasting
for Macau Power System

Chong Yin Fok and Mang I Vai

University of Macau, Department of Electrical and Computer Engineering, Macau, China
kelvin.fok@cem-macau.com,
fstmiv@umac.mo

Abstract. This paper presents the implementation of very short time load fore-casting (VSLF) for Macau power system with the forecasting period ranging from several minutes to 8 hours. The methodology adopted is the hybrid model with ANN-based and similar days methods included weather information va-riables, which are seldom considered as input variables of VSLF in other litera-tures. It is shown that weather information is one of influence factors of the VSLF for a small city like Macau and the MAPE of VSLF for 15-minutes to 3-hours ahead load is 0.96% and 0.85% for Jan 2011 and Jul 2011 respectively. In this work, the author also utilizes the result of VSLF to adjust a day ahead short term load forecasting (STLF) result, by this approach, it is demonstrated that MAPE of STLF can be reduced by 20% for the data of Jul 2011.

Keywords: very short term load forecasting, hybrid model.

1 Introduction

Short term load forecasting (STLF) is essential to daily dispatch planning and opera-tion: unit commitment, economic dispatch, load flow analysis and even resource man-agement. Economic dispatch is getting more and more concern due to the cost of fuel oil and purchased electricity from other electric utilities dramatically increases, Accu-rate STLF can help dispatcher to deploy the suitable and cheaper generation source and can save the cost in the degree of hundred thousand Patacas (Macau Currency, 1USD ≈ 8 Patacas) per day for Macau case! On the other hand, network security is also crucial in Macau due to less endurance on electricity outage in Macau flourishing tour and gambling industry. Accurate STLF can also provide the good balance of the trade-off between economic dispatch and network security.

In order to improve the accuracy of STLF, very short time load forecasting (VSLF) is introduced in this paper to adjust the result of STLF. Furthermore, VSLF can also help dispatcher to forecast ranging from 15-minutes up to few-hours ahead peak load for dispatching the most suitable unit for coming peak load to fulfill the network secu-rity as well as saving the generation cost.

In the past several decades, lot of efforts was put in developing the short and very short term load forecasting model with different approaches: Time series methods and statistical methods, such as moving average [1,5], exponential smoothing methods

D.-S. Huang et al. (Eds.): ICIC 2012, LNCS 7389, pp. 538–546, 2012.
© Springer-Verlag Berlin Heidelberg 2012

[2,5], linear regression models [3-5], moving average (ARMA) models [5-7], Kalman filtering method [5], are difficult to model the non-linear relationship between input variables and forecasted load. On the other hand, lot of literatures focus on machine learning technique support vector machines (SVM) [9-13] and intelligent algorithm ANN [8,14], which are good candidates of load forecasting model due to they can model complex and non-linear relationship without knowledge of detail analytical model between input variables and forecasted load. Hybrid models with different stages of forecasting models and forecasted load adjustment [2, 15-16] or combination of different models [17] are adopted for improving the accuracy of load forecasting. Among VSLF models, only few works consider weather information involved in the input variables selection due to lagging weather effect on very short term load [17-19]. Refer to one of STLF ANN model [20], hourly weather information is selected to be the inputs of STLF to forecast one-hour ahead load. In this work, by examining high correlation between hourly weather information and coming 15 minutes to few hours load for a small city like Macau, the author utilizes the hourly weather information to forecast very short time load ranging from 15 minutes to few hours or even several minutes by interpolating methods as the load within 15 minutes can be approximated to be monotonic linear relationship.

In this work, the author aims at developing the hybrid forecasting model to forecast 15-minutes (can be several minutes by interpolation) up to 8-hours ahead load for Macau system and the result of VSLF can also help to improve a day ahead STLF accuracy. The content of the paper is organized as follows: First of all, introduction is given in section 1. In section 2, load characteristic of Macau power system is discussed for the choice of forecasting model. Section 3 presents the selection of forecasting model and the applied strategy of hybrid model of VSLF. Section 4 shows the result of VSLF. Section 5 discusses how to use the VSLF result to modify a day ahead STLF for Macau power system. In the last section, conclusion is drawn.

2   Load Characteristic of Macau

First of all, load characteristic of Macau is analyzed for choosing suitable input variables and models. They come from two main sources: statistical analysis of historical load and the information from dispatcher experience:

i)   The load incremental rate of Macau power system in summer and winter is shown in fig. 1 and fig. 2
ii)   Dispatcher experience:
1. Sudden raining and sunshine can affect load change for coming few couple minutes and few hours.
2. The load characteristic is similar for the similar days and the load trend won't change abruptly in recent days.
3. For the coming day ahead forecasting, even though the forecasted maximum temperature is same as that of historical day, the load level can vary a lot due to accumulative effect of weather and different residents' custom in previous recent days during different period.

To deal with the above observations of the characteristic of load profile and experience from dispatchers: for 1), the author tries to examine the relationship between

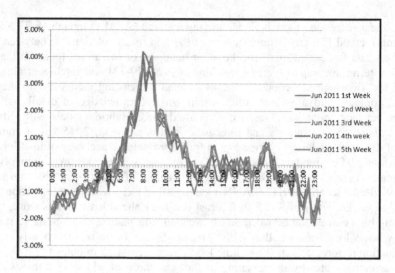

Fig. 1. Similar load incremental rate for the days in Jun 2011

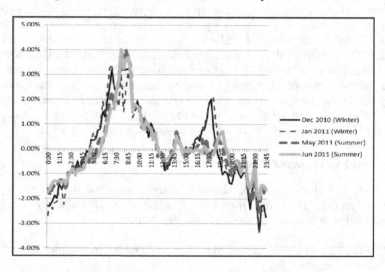

Fig. 2. Similar load incremental rate of winter 2010 and summer 2011

Table 1. Correlation analysis of current load and weather information at current and previous hours

Correlation	Temperature (-3h)	Temperature (-2h)	Temperature (-1h)	Temperature (current)
Load	0.839	0.855	**0.860**	0.846
Correlation	Dew point (-3h)	Dew point (-2h)	Dew point (-1h)	Dew point (current)
Load	0.833	0.838	**0.841**	0.835

hourly weather information and coming hourly load (Table 1), for 2) the VSLF model includes the similar days and average incremental rate of recent 3 days methods, for 3) it is necessary to utilize the trend of load from VSLF to adjust the day ahead STLF, which will be discussed in section 5.

- Among different hourly weather information obtained from Macau observatory, temperature and dew point temperature are chosen to be the variables, which have highest correlation with the load. The next step is to examine how long the weather information is used to forecast the load. It can be shown in table 1 that current weather information is suitable to forecast the load within three hours.
- Only historical hourly weather information is used as input variables of VSLF without future hourly forecasted weather information involved, since weather forecast is another source contributed to the error of load forecasting and there is also lack of completed forecasted hourly weather information from observatory.

3   Methodology

Based on the previous discussion, ANN, similar day method and average incremental rate of recent 3 days method are chosen to be the individual forecasters of VSLF, which are categorized into two groups with and without weather information input variables and is summarized in table 2.

ANN: The incremental rate instead of load level is used for the ANN inputs due to incremental rate is more stable than load level when the weather change rapidly [18].

Similar days approach: Euclidean norm [21-22] examines the similarity of two highest correlation inputs from historical data. In this work, two-steps Euclidean norm is implemented for the reason that temperature is the dominant input than others, the 10 most similar days in term of temperature are filtered out first and then the 3 most similar days are obtained by evaluating Euclidean norm of other inputs.

Recent 3 days approach: recent days have strong relationship in term of load trend due to residents' behavior and enterprises' consumption is similar. The number of days used for the average of the load incremental rate is determined by MAPE of data of previous months. It is selected to be 3 days in this work.

Table 2. Individual forecasters of VSLF model

	without weather information			with weather information	
	ANN	Similar days	Recent 3 days	ANN*	Similar days
Input	The last 5 load incremental rate	The last 2 hour load level + current load level	The average of load incremental rate of last 3 recent days	The last 3 load incremental rate + current hour temperature and dew point temperature	The current hour temperature + dew point temperature + current load level

*for ANN, the current hourly weather information is only suitable to forecast load up to 3 hour based on correlation analysis

Furthermore, many literatures described that using a combined forecasting method [17, 23] can enhance the accuracy of the load forecast. In this work, hybrid model is also adopted by combining different individual forecasters running in parallel, which can also be extended by including more suitable individual forecasters in parallel later on. By examining the previous load forecasting MAPE by different individual forecasters, two possible approaches are as follows:

Method 1) Weighting is assigned between different forecasters to give the final result [17].

Or

Method 2) Individual forecaster with the smallest MAPE of last hour is used for the forecasting model for next few hours.

Ambiguous combined effect on load forecasting of different forecasters (method 1) is avoided in this work. Instead, the reason of the choice of method 2 for the hybrid model is that within several hours in one day, the characteristic of the trend of previous load is similar to that of forecasting load, same method has lowest MAPE result for a specific continuous hours in specific day, which is examined by the large sample size of historical data shown in table 4. Furthermore, the probability of same individual forecaster applied for previous 1 hour and next few hours with lowest MAPE should be very close to 50% and 33.33% for two and three forecasters respectively if two MAPE have no relationship, after examining the recent historical data in large sample size in table 3, it is more than 50% and 33.33%, so we use the previous hour MAPE by each individual forecaster to evaluate which method is used for next 15 minutes and few hours forecasting. Based on table 3, the author prefers to use 3 forecasters with previous 1 hour MAPE evaluation to forecast next 3 hour load, since 39.5% is most far away from its reference value 33.33% (for 3 forecasters) compared with other cases.

Table 3. The probability of lowest MAPE for previous and forecasted load when same method applied

	Same method applied, Lowest MAPE for previous 1 hr and next forecasted 1 hr	Same method applied, Lowest MAPE for previous 1 hr and next forecasted 3 hr
Hybrid model (with 2 forecasters)	52.3%	52.6%
Hybrid model (with 3 forecasters)	37.7%	39.5%

Table 4. Methods with lowest MAPE in forecasted day

	0	1	2	3	4	5	6	7	...	21	22	23
4 Jul	A*	B	B	B	B	B	C	C		B	C	B
5 Jul	C	C	B	B	B	B	A	A		C	B	C
6 Jul	A	B	A	C	C	B	A	C		B	A	A
7 Jul	B	C	B	B	B	B	A	A		A	A	A
...												
25Jul	C	B	A	A	A	B	C	C		C	C	C

*Individual forecaster with the lowest MAPE applied for every hour per day (A: ANN, B: similar days, C: recent)

By method 2, there is also a mechanism for the model to store the times of different forecasters utilized within a day for further analyzing the characteristic of the load during different period of a day. As a result, further improving the forecasted load by using different models during a day [17-18, 24] can be accomplished, which can be further analyzed and done in future works.

The flowchart of the hybrid model is shown in fig. 3. There are many developed individual forecasters to fit different load characteristic of Macau for a specific day or hour. Two or three most suitable forecasters among them are chosen to be the candidates and run in parallel in hybrid model, the system also evaluates the MAPE of the rest of the forecasters in off-line mode. Once the system discovers that the individual forecaster in hybrid model with large MAPE for recent period, this forecaster need to be improved or replaced by other off-line forecasters that had better performance in the mentioned period.

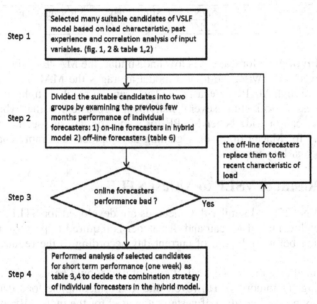

Fig. 3. The flowchart of the hybrid model

4   VSLF Result

The ANN model was trained by Nov and Dec 2010, May and Jun 2011 data and forecast the load for Jan and Jul for verification. The result of MAPE of hybrid model for 3 hours (table 5) and 8 hours (table 7) is as follows and the result of 8-hours ahead load forecasting for Jul is utilized to adjust the STLF result in the next section. Furthermore, MAPE can be improved with hourly weather information involved in the individual forecaster as shown in table 6.

Table 5. MAPE of individual forecasters and hybrid model.

Hybrid model (3 forecasters)	ANN	Similar days	Recent 3 days	Hybrid
MAPE (3 hours) Jan	2.01%	1.02%	0.58%	0.955%
MAPE (3 hours) Jul	0.99%	0.81%	0.82%	0.853%

Table 6. MAPE of individual forecasters with and without weather information inputs.

Different individual forecasters	ANN		Similar days		Recent 3 days
Input with/without weather info	weather info	w/o weather info	weather info	w/o weather info	w/o weather info
MAPE(3 hours) Jul	0.99%	1.10%	0.810%	0.813%	0.823%

Table 7. MAPE for 8th hour load of individual forecasters and hybrid model.

Hybrid model (3 forecasters)	ANN	Similar days	Recent 3 days	Hybrid
MAPE (8th hour) Jul	2.31%	1.73%	1.86%	1.87%

If the MAPE of three forecasters is low and similar, the MAPE of the hybrid model can be improved, otherwise, the hybrid model averages the MAPE of the individual forecasters. The high MAPE forecaster will worsen the result of hybrid model; however, it is necessary to include several individual forecasters in the hybrid model as each forecaster can provide better MAPE for a specific day or specific period of the day, especially for Macau flourishing gaming development and rapid change of load characteristic year by year.

5   The Result of VSLF to Adjust STLF

Currently, ANN [25] and similar-day methods are developed for STLF, which is applied in dispatch center of Macau and dispatcher is required to perform the next day load forecasting before 5:30p.m. of current day according to the requirement of the regulator.

For ANN methods, the actual load of current day should be used for the inputs of load forecasting for tomorrow. However, due to the incomplete load data of current day after 5:30p.m., the load data of yesterday is used for the inputs. By availability of VSLF in this work, the actual load of current day and forecasted load after 5:30p.m. can be utilized as inputs to enhance the accuracy of next day forecasted load as shown in table 8.

For the similar day methods, it is recognized by other literatures [18] and by dispatcher experience that the load level can vary a lot among different similar days in term of weather information due to recent residents and enterprises' custom as well as the accumulative effect of previous consecutive days' weather information. That information is inherited in the result of VSLF in the current day. As a result, we can use the VSLF forecast load at 00:00 of next day to adjust the next day STLF by (1) and (2). The adjusted result is shown in the table 8 and the MAPE is reduced by 12% and 20% for the current two approaches by this work.

Result from a day-ahead load STLF forecasting from 0:00-23:45: L_1, L_2, L_3......L_{96} Result from 8th hour ahead load VSLF forecasting at 0:00: S_1. Adjusted STLF by VSLF result from 0:00-23:45: $L_{adj,1}$, $L_{adj,2}$, $L_{adj,3}$......$L_{adj,96}$. Adjusted coefficient: α_1, α_2, α_3...... α_{96}

$$L_{adj,1} = L_1 + \alpha_1(S_1 - L_1) \tag{1}$$

$$L_{adj,i} = L_i + \alpha_i(S_1 - L_1) \tag{2}$$

where $0 < \alpha_1 \leq 1$, $0 < \alpha_i \leq 1$ if $|L_1 - S_1|/L_1 > 0.03$, $\alpha_1 = 0$, $\alpha_i = 0$ if $|L_1 - S_1|/L_1 \leq 0.03$

Table 8. The MAPE of adjusted result of STLF by VSLF of this work

	Reference: ANN (actual load from 0-24 of current day as inputs)	This work: ANN (actual load from 0-17 of current day plus VSLF result of 17-24 of cur-rent day as inputs)	Original approach: ANN (actual load from 0-24 of yesterday as inputs)
MAPE (ANN) Jul	2.25%	2.37% (-12%)	2.72%
		This work: Similar days (with VSLF adjustment)	Original approach: Similar days
MAPE(Similar days) Jul		2.95% **(-20%)**	3.72%

6   Conclusion

In this work, the author developed the VSLF hybrid model to forecast the load ranging from 15 minutes or even several minutes up to few hours by utilizing hourly weather information as input variables, which is seldom to be considered in other VSLF literatures. For small power system, with weather information involvement, the MAPE of forecasters can be improved. The strategy of hybrid model is also discussed and the MAPE of VSLF for 15-minutes to 3-hours ahead load is 0.96% and 0.85% for Jan 2011 and Jul 2011 respectively. Furthermore, due to limited input variables and time constraints of performing current STLF, the result of VSLF is utilized for facilitating to perform STLF to reduce the MAPE by 12% to 20% for July data.

Acknowledgement. I would like to express my special gratitude to the great support of my bosses Benjamin Yue, Calvin Ho, dispatch center, my company Companhia de Electricidade de Macau, the feedback of dispatchers and the opinion of previous STLF project author Cheuk Fung Wong.

References

1. Box, G.E., Jenkins, P.G.M.: Time Series Analysis - Forecasting and Control. Holden-Day (1970)
2. Song, K.B., Ha, S.K., Park, J.W., Kweon, D.J., Kim, K.H.: Hybrid Load Forecasting Method with Analysis of Temperature Sensitivities. IEEE Trans. Power Syst. 21(2), 869–876 (2006)

3. Papalexopoulos, A.D., Hesterberg, T.C.: A Regression-based Approach to Short-term Load forecasting. IEEE Trans. Power Syst. 5(4), 1535–1550 (1990)
4. Haida, T., Muto, S.: Regression Based Peak Load Forecasting Using a Transformation Technique. IEEE Trans. Power Syst. 9(4), 1788–1794 (1994)
5. Moghram, I., Rahman, S.: Analysis and Evaluation of Five Short-term Load Forecasting Techniques. IEEE Trans. Power Syst. 4(4), 1484–1491 (1989)
6. Amjady, N.: Short-term Hourly Load Forecasting Using Time-series Modeling with Peak Load Estimation Capability. IEEE Trans. Power Syst. 15(3), 498–505 (2001)
7. Galiana, F.D.: Short-term Load Forecasting. Proceedings of IEEE 75(12), 1558–1572 (1987)
8. Swarup, K.S., Satish, B.: Integrated ANN Approach to Forecast Load. IEEE Computer Applications in Power 15, 46–51 (2002)
9. Cortes, C., Vapnik, V.: Support-vector Network. Mach. Learn. 20, 273–297 (1995)
10. Cristianini, N., Tylor, J.S.: An Introduction to Support Vector Machines and Other Kernel-Based Learning Methods. Cambridge Univ. Press, Cambridge (2000)
11. Chen, B.J., Chang, M.W., Lin, C.J.: Load Forecasting Using Support Vector Machines: A study on EUNITE competition 2001. IEEE Trans. Power Syst. 19(4), 1821–1830 (2004)
12. Pang, S.L., Mu, G., Wang, X.Q., Jin, P., Ma, J.G.: Short-term Load Forecasting Method Based on Load Regularity Analysis for Supporting Vector Machines. Proc. Northeast Dianli Univ. 26(4) (2006)
13. Hong, W.C., Cheng, Y.L., Hung, W.M., Dong, Y.C.: Electric Load Forecasting by SVR with Chaotic Ant Swarm Optimization. In: IEEE CIS (2010)
14. Hippert, H.S., Pedreira, C.E., Souza, R.C.: Neural Networks for Short-Term Load Forecasting: A Review and Evaluation. IEEE Trans. Power Syst. 16(1) (2001)
15. Senjyu, T., Mandal, P., Uezato, K., Funabashi, T.: Next Day Load Curve Forecasting Using Hybrid Correction Method. IEEE Trans. Power Syst. 20(1), 102–109 (2005)
16. Wang, Y., Xia, Q., Kang, C.Q.: Secondary Forecasting Based on Deviation Analysis for Short Term Load Forecasting. IEEE Trans. Power Syst. 26(2) (2011)
17. Daneshi, H., Daneshi, A.: Real Time Load Forecast in Power System. In: DRPT 2008, Nanjing (2008)
18. Charytoniuk, W., Chen, M.S.: Very Short-Term Load Forecasting Using Artificial Neural Network. IEEE Trans. Power Syst. 15(1) (2000)
19. Ding, Q., Lu, J.G., Liao, H.Q.: A Practical Super Short Term Load Forecast Method and its Implementations. In: IEEE PES, Power Systems Conference and Exposition (2004)
20. Zainab, H.O., Awad, M.L., Mahmoud, T.K.: Neural Network Based Approach for Short-Term Load Forecasting. In: Power Systems Conference and Exposition (2009)
21. Jain, A., Srinivas, E., Rauta, R.: Short Term Load Forecasting using Fuzzy Adaptive Inference and Similarity. In: World Congress on Nature & Biologically Inspired Computing (2009)
22. Senjyu, T., Uezato, T., Higa, P.: Future Load Curve Shaping Based on Similarity Using Fuzzy Logic Approach. IEE Proceedings of Generation, Transmission, Distribution 145(4), 375–380 (1998)
23. Chen, Q., Milligan, J., Germain, E.H., Raub, R., Shamsollahi, P., Cheung, K.W.: Implementation and Performance Analysis of Very Short Term Load Fore-caster – based on the Electronic Dispatch Project in ISO New England. In: 2001 Large Engineering Systems Conference on Power Engineering, LESCOPE 2001 (2001)
24. Chen, D.G., York, M.: Neural Network Based Very Short Term Load Prediction. In: Transmission and Distribution Conference and Exposition, T&D. IEEE/PES (2008)
25. Wong, C.F.: ANN STLF for a Rapidly Expanding Power System. In: CEPSI (2006)

A Method for the Enhancement of the Detection Power and Energy Savings against False Data Injection Attacks in Wireless Sensor Networks

Su Man Nam and Tae Ho Cho

College of Information and Communication Engineering, Sungkyunkwan University, Suwon
440-746, Republic of Korea
{smnam,taecho}@ece.skku.ac.kr

Abstract. Malicious attackers spread various attacks to destroy the system of the sensor network. False report injection attacks occur on the application layer and drain the energy resources of each node. Statistical en-route filtering (SEF) is proposed to detect and drop false reports in intermediate nodes during the forwarding process. In this work, we propose a security method to improve the detection power and energy savings using four types of keys. The performance of the proposed method was evaluated and compared to that of SEF against the attack. Our experimental results reveal that our method improves detection power and energy savings by up to 25% and 9%, respectively.

1 Introduction

Wireless sensor networks (WSNs) supply economically feasible technologies for applications that use wireless communication [1-2]. WSNs are comprised of a number of sensor nodes and a base station in a sensor field. The sensor nodes operate sensing, computing, and wireless communication modules. The base station provides information for many uses across the network infrastructure. The nodes are vulnerable to being captured and compromised due to limited resources [3]. Malicious attackers use various methods to destroy the sensor network.

The source nodes usually transmit legitimate reports as events occur in the sensor network. When a legitimate report is generated by a source node, intermediate nodes forward the report toward the base station. On the other hand, when a malicious attacker captures a node, the node becomes a compromised node. When a false report is generated by the compromised node, intermediate nodes transmit the false report to the base station. During the attack, the intermediate nodes drain their energy resources due to transmitting and receiving the false report. In addition, false alarms are caused as the false report arrives at the base station. That is, false report injection attacks harm the entire network.

Ye et al. [4] proposed statistical *en route* filtering (SEF) to detect a false report injection attack in the sensor network. In SEF, when a real event occurs, each of the detecting nodes generates message authentication codes (MACs). A center-of-stimulus (CoS) node collects the MAC and produces a report. When the report is

D.-S. Huang et al. (Eds.): ICIC 2012, LNCS 7389, pp. 547–554, 2012.
© Springer-Verlag Berlin Heidelberg 2012

forwarded, each node statistically verifies the MACs. If a forged MAC is detected, the node drops the report. Thus, SEF decreases the energy consumption of each node by dropping the false report while forwarding processes.

In this work, we propose a method to improve detection power and energy consumption. The proposed method uses four types of keys: a new individual key, a pairwise key, a new cluster key, and a group key. Each node has four keys to encrypt and exchange MACs and reports between two nodes. The new individual key and the cluster key filter out false reports within a cluster region early. The proposed method improves the detection power and energy savings of each node against false report injection attacks.

The remainder of this paper is organized as follows. SEF and the motivation for the study are described in Section 2. The proposed method is introduced in Section 3, and the optimization results are presented in Section 4. Finally, conclusions and future work are discussed in Section 5.

2   Background and Motivation

2.1   Statistical En-route Filtering (SEF)

SEF is proposed to prevent false report injection attacks in the sensor network. SEF aims at early detection and elimination of false reports, with low computation and communication overhead. SEF has four phases: (1) key assignment, (2) report generation, (3) en-route filtering, and (4) base station verification. Phase (1) is initially performed, and phases (2), (3), and (4) are repetitively performed as events occur. In the key assignment phase, each node stores a small number of keys of a partition (PID) from the base station which maintains a global key pool divided into many partitions before sensor nodes are deployed. In the report generation phase, one of the detecting nodes is elected as the center-of-stimulus (CoS) node when an event occurs. Its neighboring nodes select a key to submit message authentication code (MAC) including event information to the CoS node. After collecting MACs from the neighboring nodes, the CoS node produces a report and forwards it to the base station. In the en-route phase, intermediate nodes verify MACs of the report using keys of the intermediate nodes while forwarding processes between nodes. Finally, during a base station verification, the base station again identifies MACs of the report through keys of the global key pool when the report arrives at the base station.

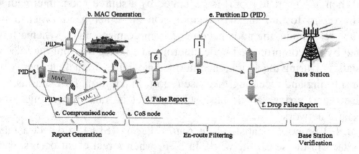

Fig. 1. Filtering of false reports

Fig. 1 shows phase report generation, en-route filtering, and base station verification, and false reports via intermediate nodes forwarded from the CoS node to the base station. When an event occurs within a region including a number of nodes, the CoS node (Fig. 1-a) is elected. It then forwards a broadcast to its neighbors, and the neighboring nodes generate MACs 1 and 4 (Fig. 1-b) using PIDs 1 and 4 submit to submit it if the neighbors detect the same event. A compromised node (Fig. 1-c) also submits a fake MAC to the CoS node. After collecting MACs, the CoS node forwards a false report (Fig. 1-d) to the base station because a fake MAC is included. When node A receives the false report, the false report is transmitted to node B of the next hop because the MACs of the false report and the PID (Fig. 1-e) of node B are different. When node B receives the false report, the node verifies the MAC_1 of the report through the PID 1 of node B. If the PID is legitimate, node B transmits the false report to node C. On the other hand, when the false report arrives at node C, it is dropped (Fig. 1-f) because its MAC_3 is forged by the compromised node. If a legitimate report is generated, the report safely arrives at the base station, and the base station verifies the report using the global key pool. Therefore, SEF statistically filters out false reports through keys of PIDs of intermediate nodes while a forged MAC that occurs from a compromised node is forwarded via PID keys.

2.2   Motivation for the Study

In SEF, the detection power of false reports is affected by key authentication of intermediate nodes. If a false report arrives at the base station without sanctions, the base station intermediate nodes transmits the false report and are drained even when the base station drops the false report [5]. Thus, the lifetime of the sensor network is decreased due to low detection power.

In this paper, we propose a method to achieve both improved detection power and energy consumption for reducing damage from the false report. Our proposed method effectively detects false reports using four types of keys [6]. When a compromised node generates a fake MAC, a CoS node verifies NC of the compromised node through its NC. After dropping the fake MACs, the CoS node encrypts a report using NG and the report is transmitted while maintaining secure paths using a pairwise key (PK) until the base station. Therefore, we enhance the detection power of the false report including a forged MAC through NC in a CoS node and conserve the energy resources of intermediate nodes by preventing the inflow of false reports. The proposed method detects and drops the false report for both the improved detection power and the energy consumption in a CoS node.

3   Proposed Method

3.1   Assumptions

We assume a static sensor network. Sensor nodes are fixed after they are deployed. The sensor network is comprised of a base station and a large number of sensor nodes, e.g., the Berkeley MICA2 motes [7]. The initial paths of the topology are established through directed diffusion [8], and minimum cost forwarding algorithms [9] after distribution of the sensor nodes. It is further assumed that every node forwards data packets toward the base station along their path. An attacker generates a false report injection using compromised nodes. The resulting false report is forwarded from a compromised node to the base station before it is filtered out.

3.2   Overview

In this paper, we effectively detect false report injection attacks that occur on the application layer using four types of keys – an individual key shared between each node and the base station, a pairwise key shared between a node and another node, a cluster key shared between a node and a CoS node, and a group key shared by all nodes. When an event occurs in a range, nodes that detect the event transmit event data, including each MAC and NC, to a CoS node. After collecting event data from its neighbors, the CoS node verifies MACs through its NC, selects legitimate MACs, and encrypts a report. All intermediate nodes maintain secure paths through the PK while forwarding processes into the base station. Therefore, we improve both detection power and energy savings in the sensor network by preventing false report injection attacks.

3.3   Keys Explanation

- *New Individual Key (NI)*: Each sensor node has a unique key associated with the base station for one-on-one communications. This key is used to generate a MAC when an event occurs. For example, when a node detects a real event, the node produces a MAC after encrypting event information through its NI, and it transmits event data including the MAC to a CoS node. When MACs arrive at the base station, the base station verifies the MACs through its NIs to detect forged MACs. That is, the NI is used to encrypt and decrypt event information in the node and the base station.
- *Pairwise Key (PK)*: Sender and receiver nodes have the same keys to check conditions. The PKs of two nodes are verified before transmitting a report. For example, a report generated by a CoS node is forwarded via multiple intermediate nodes toward the base station. The intermediate nodes maintain secure paths through PK against an adversary node that captures the report information. That is, PK retains secure paths between all pairs of nodes.
- *New Cluster Key (NC):* Nodes that are located within a cluster range have a common key. This key detects forged MACs generated by compromised nodes in a CoS node. For example, a node that detects an event transmits event data including its NK to the CoS node. After collecting event data, the CoS node verifies the NK of the event data through its NK. The CoS node then forwards a report including the MACs to the base station. On the other hand, if a compromised node is located in the same cluster range, the compromised node also transmits event data including the forged MAC and NC to the CoS node. The CoS node then verifies the NC of the event data through its NC, and drops the forged MAC. That is, NC is used to detect forged MACs that are generated from the compromised node.
- *New Group Key (NG):* Every node and the base station have a key. This key is used to encrypt or decrypt reports in the CoS node or the base station. For example, the CoS node encrypts a report through NG for preventing information extraction by an adversary node. If the adversary node captures the report while

forwarding processes, it is difficult to decrypt the report without NG. That is, NG prevents information extraction against the adversary node if the report is captured.

3.4   The Detection of False Report Injection Attack

Fig. 2. False MAC Detection using NC shows the detection processes of a forged MAC in a CoS node when a real event occurs within a cluster region. In the region shown, there are four nodes including a compromised node. There is also a node (Fig. 2-a) that is outside of the region. When a real event (Fig. 2-b) occurs within the region, a CoS node (Fig. 2-c) is elected to generate an event report. The nodes of its neighbors generate MACs after they encrypt event information through NI. The neighboring nodes transmits each event data including NC and MAC to the CoS node. A compromised node (Fig. 2-d) transmits forged event data (Fig. 2-e) including a forged NC and MAC to the CoS node during this time. The outside node also sends event data. After collecting event data, the CoS node filters out both the forged MAC and the MAC of another region by verifying the NC. The CoS node forwards a report with legitimate MACs into the base station. Our proposed method reduces the flow of false reports using early detection within a cluster region. The proposed method decreases communications traffic between intermediate nodes.

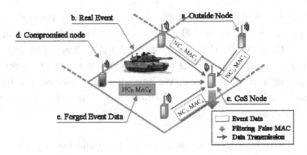

Fig. 2. False MAC Detection using NC

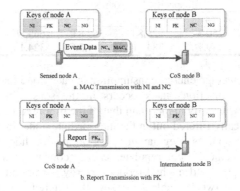

Fig. 3. Transmission of MAC or Report

Fig. 3 shows keys used for each node between two nodes according to a MAC and report. In Fig. 3-a, node A detects an event and encrypts event information using its NI. It then transmits event data including its NC_A and MAC_A to the CoS node B. After the CoS node B verifies NC_A through the NC of the CoS node, it drops the forged MAC. The CoS node B then collects verified MACs and forwards a report to the base station. When the report arrives at the base station, the base station again verifies the MACs of the report through its global key pool. If a forged MAC is detected, the base station drops the forged MAC. Thus, NI and NC are used to detect false reports in the CoS node or the base station. In Fig. 3-b, the CoS node A encrypts a report using NK and the encrypted report is forwarded to intermediate node B at the base station. The CoS node verifies the PK of intermediate node B and checks the condition of the intermediate node during this time. If intermediate node B is inserted by an adversary and has no key, the CoS node forwards the report to another node after verifying the PK of the intermediate node. When the report arrives at the base station, it encrypts the report through its NG. Thus, because PK checks the conditions of the intermediate nodes, the secure paths of each node are maintained between two intermediate nodes. Therefore, we are able to improve both the detection power and energy consumption using four types of keys against false report injection attack.

4   Performance Results

The performance of the proposed method was evaluated through simulations as compared to SEF. A field size of $500 \times 500m^2$ is used and the base station is located in the middle of the field. It takes 16.25 and 12.5µJ to transmit/receive a byte, and each MAC generation consumes 15µJ [4]. Moreover, the size of an original report is 24 bytes, and size of a MAC is 1 byte. We have decided to transmit false reports from 20 compromised nodes in the field. A false report per 10 legitimate reports is generated by the compromised nodes.

Table 1. Performance against false report injection attacks

	a. Average energy consumptions of each node for specific hop counts					b. Ratio of filtered reports
	2	4	6	8	10	
SEF	776.92	250.37	528.83	357.83	649.86	76.60%
P.M	232.60	449.76	384.11	319.50	324.16	100.00 %

Table 1-a shows the average energy consumption in each node for specific hop counts when a false report injection attack occurs in the sensor network. The energy consumption of SEF is usually higher than that of the proposed method (P.M) because false reports are statically filtered out in the intermediate node. In addition, the energy consumption of each node is irregular in terms of specific hop count. On the other hand, the proposed method better regulates energy consumption due to hop counts than SEF. Thus, the proposed method reduces energy consumption and prolongs the lifetime of the sensor network. Table 1-b shows the filtered report ratios of SEF and the proposed method. In SEF, the number of false reports that arrive at the base station drops in 25% of cases and the filtered report rate is about 75% for intermediate

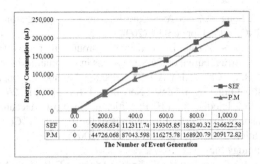

Fig. 4. Energy Consumption of the Sensor Network

nodes. On the other hand, the proposed method initially detects all forged MACs through CK in a CoS node. Therefore, the proposed method enhances the detection power of the false report more than SEF because the CoS node drops all the forged MACs.

Fig. 4 illustrates the total energy consumption of SEF and the proposed method after 1,000 events. When 200 events have occurred, the proposed method conserves energy resources by up to 9% more than SEF. The proposed method enhances detection power by 25% and reduces energy consumption in the sensor network by 9% because it detects all forged MACs in the CoS node and prevents false report inflows.

5   Conclusion and Future Work

We use four types of keys to detect attacks and to plan countermeasures against false reports. Each node has four keys: a NK for encrypting event information, a PK for maintaining secure paths, an NC for detecting the false report, and an NG for encrypting a report. The effectiveness of the proposed method is evaluated and compared to that of SEF while generating a false report. The proposed method improves the detection level by about 25% and reduces energy consumption by about 9%. The proposed method effectively detects false report injection attacks using four types of keys. In future work, we will run additional simulation scenarios and perform experiments to test the robustness of our method against other types of attacks.

Acknowledgment. This work was supported by National Research Foundation of Korea Grant funded by the Korean Government (No. 2012-0002475).

References

1. Akyildiz, I.F., Su, W., Sankarasubramaniam, Y., Cayirci, E.: A Survey on Sensor Networks. IEEE Communications Magazine 40, 102–114 (2002)
2. Chan, H., Perrig, A.: Security and Privacy in Sensor Networks. Computer 36, 103–105 (2003)

3. Przydatek, B.D., Song, A.P.: SIA: Secure Information Aggregation in Sensor Networks. In: Proc. of CCNC, vol. 23, pp. 63–98 (2004)
4. Ye, F., Luo, H., Lu, S., Zhang, L.: Statistical En-route Filtering of Injected False Data in Sensor Networks. IEEE Journal on Selected Areas in Communications 23, 839–850 (2005)
5. Sun, C.I., Lee, H.Y., Cho, T.H.: A Path Selection Method for Improving the Detection Power of Statistical Filtering in Sensor Networks. J. Inf. Sci. Eng. 25, 1163–1175 (2009)
6. Zhu, S., Setia, S., Jajodia, S.: LEAP: Efficient Security Mechanisms for Large-scale Distributed Sensor Networks. In: Proceedings of the 10th ACM Conference on Computer and Communications Security, pp. 62–72. ACM (2003)
7. Crossbow technology Inc., http://www.xbow.com
8. Intanagonwiwat, C., Govindan, R., Estrin, D.: Directed Diffusion: A Scalable and Robust Communication Paradigm for Sensor Networks. In: MOBICOM, pp. 56–67. ACM (2000)
9. Ye, F., Chen, A., Lu, S., Zhang, L.: A Scalable Solution to Minimum Cost Forwarding in Large Sensor Networks. In: Proceedings of the Tenth International Conference on Computer Communications and Networks, pp. 304–309 (2001)

Virtual Cluster Tree Based Distributed Data Classification Strategy Using Locally Linear Embedding in Wireless Sensor Network

Xin Song[1,2], Cuirong Wang[1], Cong Wang[1], and Xi Hu[1]

[1] School of Information Science and Engineering, Northeastern University, 110819,
Shenyang, China
[2] The Key Laboratory of Complex System and Intelligence Science, Institute of Automation,
Chinese Academy of Sciences, 100190, Beijing, China
neuqsong@sohu.com

Abstract. With recent advances in wireless communication and low cost, low power sensors are enabling the deployment of large-scale and collaborative wireless sensor network (WSN), which performs different tasks using the subsets formation of sensor on specific requirement of event monitoring. For implementing the differentiated monitoring task of the heterogeneous sensor network in the same geographical region, efficient classification of the various sensed data becomes a critical task due to stringent constraint on network resources, frequent link and indeterminate variations in sensor readings. In this paper, we present a virtual cluster tree based distributed data classification strategy using locally linear embedding (LLE). The strategy can realize the structure representation of original sensed data by LLE to meet the classification for forming the virtual cluster tree of the various monitoring task. The theoretical analysis and experimental results show that the proposed strategy can effectively reduce the energy consumption using LLE based classifier.

Keywords: wireless sensor network, distributed data classification, virtual cluster tree, locally linear embedding, energy efficient.

1 Introduction

One important class of share resources technologies is to build a virtual cluster tree, which use the connecting sensor nodes subset to observe the same phenomenon. The existence of such virtual cluster tree capability implies that the sensors are the multiplexing device if they have sensed different occurrence. Then, another key issue is how to dynamically determine which node should join which virtual cluster tree of implementing same task. A simple approach is to design a classifier in sensor network for determining the classed of newly arrived sensed data from monitoring region. However, conventional centralized solution is not directly applicable in the new type of heterogeneous WSN environment. To address the above challenges, in this paper, we present a virtual cluster tree based distributed data classification strategy using locally linear embedding in energy-constrained sensor network. Our approach performs event

D.-S. Huang et al. (Eds.): ICIC 2012, LNCS 7389, pp. 555–562, 2012.
© Springer-Verlag Berlin Heidelberg 2012

monitoring in the built virtual cluster tree and implements classification of sensed data, utilizing a locally linear embedding algorithm for distinguishing task, in particular, an adaptive and bottom-up manner.

The rest of this paper is organized as follows: in section 2, we briefly review some closely related works. The proposed virtual cluster tree based distributed data classification strategy is derived and discussed in section 3. The validity analysis and performance evaluation of the approach is presented in section 4. Finally, the conclusions and future work directions are described in section 5.

2   Related Work

The sensor data collection and object count using data mining technology have been extensively studied in the literature. Ref. [1] proposed a distributed data mining model for WSN. The model poses a three-layer MLP (multilayer perceptron) for data aggregation in the clustered sensor network. Input layer and the first hidden layer are between cluster members, and then output layer and the second hidden layer are between cluster head nodes. In WSN the construction of a classifier might be necessary if a WSN is intended to classify phenomena in its sensing environment. For avoiding the costly data transfer of sensor node, the possibility of in-network classification tree construction by the SPRINT (Scalable PaRallelizable Induction of decision Tree) algorithms family was evaluated in Ref. [2]. Concept resource-aware framework on data mining online is a very important for monitoring continuously, data aggregation and in-network processing. The system implementation was introduced by applying data monitoring technologies with light weight classification (LWClass), light weight frequent item (LWF) and light weight clustering (LWCluster) [3]. In order to decrease the traffic load in a tunable manner, a new lightweight classification algorithm in WSN was proposed in Ref. [4]. By this algorithm, node classification is adaptive to topology changes and has no constraint on routing protocols and hardware. Because WSN provides service to the users, in Ref. [5], a new classification method based on SOM (Self-Organizing feature Map) for information fusion of WSN to satisfy user's requirements was proposed. SOM is powerful in processing magnanimous clustering information. The advantage is that such classification and disposal method according with particular characters of WSN not only can obtain more exact and more general information in application but also can assist in intelligence decision-making later. Decision tree is one of the most important models for data classification application. For a classification of application, the data collection behavior of sensor nodes and indicating the metrics can suit to the evaluation of particular application classes [6]. Several model for classifications exist, for example neural network, fuzzy decision, statistical models, virtual machine, genetic models and decision trees [7-11]. For example, in Ref. [12], a novel decision-tree-based hierarchical distributed classification approach in WSN was proposed. In the algorithm the classifiers are iteratively enhanced by combining strategically generated pseudo data and new local data, eventually converging to a global classifier for the whole network. Rajkamal R. and Vanaja Ranjan P. use artificial neural network, which is conventional feed-forward back propagation network, to classify the packets based on nature of the data to overcome the traffic in WSN [13]. Yet efficient classification and energy saving for WSN have not been sufficiently addressed and researched. Through the distributed signal sensing and classification performed by collaborated sensor is proven to be beneficial to increasing the modulation classification reliability [14]. Better resource efficiency was not achieved by collaborating and sharing of sensor nodes in the same geographical region. A novel distributed data classification of WSN applications is now presented to facilitate the multi-function sensor collaboration and to form logically separating multipurpose sensor network using nonlinear data classifier.

3   Virtual Cluster Tree Based Distributed Data Classification Using LLE in WSN

Imposing some structure within the network to effectively achieve the application objectives is an attractive option for the self-organization of large-scale WSNs. Cluster based organization, and arranging clusters in form of a tree simplifies many higher-level functions and distributed application deployments [15]. We use EADEEG (an energy-aware data gathering protocol for wireless sensor networks) protocol to form a cluster [16], then virtual cluster tree was built by connecting CH nodes observing the same phenomenon, which was identified using locally linear embedding (LLE) algorithm [17]. The virtual cluster tree based distributed data classification strategy includes two phases: the virtual cluster tree formation and classifier generation using LLE.

3.1   The Network Model

The network model of the strategy was assumed that a set of N energy-constrained heterogeneous sensor nodes were randomly deployed in M*M two-dimensional field. Each node covers an area of the monitoring region and is responsible for collecting environment data. All sensor nodes are not mobile and unaware of their location. The immobile Sink node is only and considered to be a powerful node endowed with enhanced communication and computation capabilities and no energy constraints. Sink node received the messages from the cluster head nodes using cluster-based hierarchical routing approach in WSN. A subset of cluster head (CH) nodes is selected to facilitate communication functions with a certain task. The communication radius of sensor nodes is more than a multiple of the cluster radius for implementing the direct communication between cluster head nodes.

3.2   The Virtual Cluster Tree Formation

After the distributed cluster was formed by using EADEEG protocol, the network topology was built by multi-hop communication between the CH nodes. The virtual cluster tree can be formed by providing logical connectivity among these collaborative communication CH nodes (the CH nodes can communicate directly each other). Fig.1 illustrates the ideal cluster detecting two different events happened. Two monitoring events are located around clusters CH3, CH4, CH15, CH6, CH8, CH9 and CH10. CH nodes can be grouped into different virtual cluster tree based on the phenomenon they track (e.g., rockslides vs. animal crossing) or the task they perform.

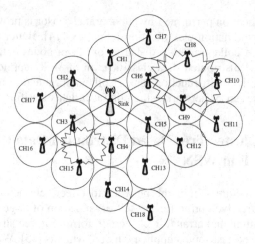

Fig. 1. The ideal clusters detecting two different events happened

Assume that a cluster member node in cluster CH10 first detects the interested event. It sends a virtual cluster tree formation message towards the root node (sink node). Firstly, the message is forwarded to its own CH10. This is the first time that CH10 is receiving this interested event information therefore; it marks its self as detecting the event happened. It then forwards the message to its parent CH9, CH9 caches the event information about the new virtual cluster tree and forwards the message to its parent CH6. Finally, the message is forwarded to the root node (sink). Meanwhile other nodes in cluster CH8 will also detect the interested event. CH8 also marks itself as detecting the event. The message is then forwarded to its parent CH6. However, CH6 is already aware of the interested event and has already informed its parent sink node. Therefore, CH6 will not forward the message any further. CH6 keeps track that CH6 is also in the interested event. Similarly, cluster CH15, CH3 and CH4 can form another logical tree. The virtual trees formed were shown in Fig.2.

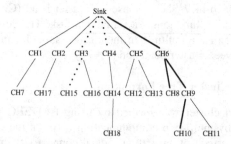

Fig. 2. Virtual cluster tree that connect CH members with special task

If these trees monitor the same interested event, they belong to the same virtual cluster tree. If not, they can be considered as two separate virtual cluster trees. Our strategy can also enable multiple logical structures to communicate with each other because each virtual tree is guaranteed to meet at the root sink node.

3.3   The Classifier Generation Using Locally Linear Embedding Algorithm

One technique is to implement the distributed data classification and enforce different parts of the sensor network to be active for different application at different times. In general, the energy consumption of computation is significantly smaller than that of data transmission and the information representation of monitoring environment are formed by processing large numbers of sensory inputs. Note the monitoring data of the sensors with multi-function has multi-dimensional and nonlinear feature. Therefore, our approach utilizes LLE algorithm as the basis of classification for classifier generation. The LLE algorithm can establish the mapping relationship between the observed data and the corresponding low-dimensional data. Here is a brief description of LLE algorithm. That is, the main principle of LLE is to obtain the corresponding low-dimensional data $Y(Y \subset R^d)$ of the training set $X(X \subset R^N, N \gg d)$. Here is a brief description of LLE algorithm.

Step1. Discover the adjacency information. For each x_i find its n nearest neighbors in the dataset $x_j(j = 1, 2, \cdots, n)$.

Step2. Construct the approximation matrix. Suppose each data point and its neighbors to lie on or close to a locally linear patch. We characterize the local geometry of these patches by linear coefficients that reconstruct each data point from its neighbors. Reconstruction errors are measured by the cost function as function (1):

$$E(W) = \sum_i \left| x_i - \sum_j^n W_{ij} x_j \right|^2 \tag{1}$$

The weigh W_{ij} summarize the contribution of the jth data point to the ith reconstruction. The weigh W_{ij} were computed by minimizing the cost function subject to the constraints that $\sum_j W_{ij} = 1$ for each i. Each data point x_i is reconstructed only from its neighbors, enforcing $W_{ij} = 0$ if x_j does not belong to neighbors set of x_i.

Step3. Map to embedded coordinates. Each high-dimensional observation x_i is mapped to a low-dimensional vector y_i by minimizing the cost function (2).

$$\phi(y) = \sum_i \left| y_i - \sum_j^n W_{ij} y_j \right|^2 \tag{2}$$

The optimal embedding is found by computing the bottom $d + 1$ eigenvectors of matrix $M = (I - W)^T (I - W)$. The bottom eigenvector of this matrix, which we discard, is the unit vector with all equal components.

We let the sensor nodes in WSN be organized by multiple virtual cluster trees. A leaf node builds the classifier with the local sensed training data and sends the interested event to the parent. The intermediate CH node periodically checks if there is any new class from children. If yes, it will mark and combine them with its local classifier for forming a new virtual cluster tree. The base station will build the global classifier. The processing model of distributed data classification in WSN was shown in Fig. 3.

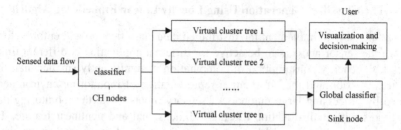

Fig. 3. The processing model of distributed data classification

4   Experiment Results and Performance Evaluation

In the experimental scenario, 200 sensor nodes are randomly distributed be-
tween $(x = 0, y = 0)$ and $(x = 400, y = 400)$ with the sink node at location $(x = 200, y = 200)$.
Each node begins with only 6J of energy and an unlimited amount of data to send to
the sink. The simulation time is 1000 seconds. Fig.4 shows the nodes deployment and
the cluster trees that are formed in nodes detecting the same phenomenon.

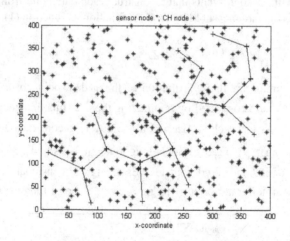

Fig. 4. Virtual cluster tree by nodes detecting the same phenomenon

We supposed that two interested event occurred within two different monitoring
regions. For homogeneous data, they were buffered in the CH node as training data
for data classification. For heterogeneous data, the CH node executed the LLE based
classifier to recognize for forwarding new marker to parent node. We investigate the
energy consumption, which is one of the most important concerns in WSN. Fig.5
shows the total energy dissipation of nodes obtained with separate non-cooperation
WSN and two virtual cluster trees formed in implementing different phenomenon
monitoring. From the figure, we find that the energy consumption of our strategy is
much lower than the conventional approach. Due to enhanced data classification

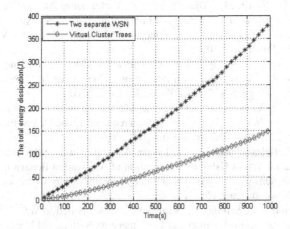

Fig. 5. The total energy dissipation of two strategies

function of CH node, the effective dimensionality of the data may be significantly lower than the total number of sensor measurements while decreasing the communication requirements.

5   Conclusion

In this paper, we have proposed a novel virtual cluster tree based distributed data classification using locally linear embedding in wireless sensor network. The strategy can realize the structure representation of original sensed data by LLE to meet the classification for forming the virtual cluster tree of the various monitoring task. Virtual cluster tree is an emerging concept that supports collaborative, resource efficient, and multipurpose wireless sensor networks. The experimental results show that the proposed strategy can maintain very low communication overhead and decrease the energy dissipation. We are also interested in evaluating the performance with real wireless sensor network.

Acknowledgment. The research work was supported by open research fund of Key Laboratory of Complex System and Intelligence Science, Institute of Automation, Chinese Academy of Sciences under grant No.20100106.

References

1. Hong, Y., Xu, S., Wu, H.: Study on Distributed Data Mining Model in Wireless Sensor Networks. In: International Conference on Intelligent Computing and Integrated Systems, pp. 866–869. IEEE Press, Guilin (2010)
2. Lantow, B.: Applying Distributed Classification Algorithm to Wireless Sensor Networks- A Brief View into the Application of the SPRINT Algorithm Family. In: 7th International Conference on Networking, pp. 52–59. IEEE Press, Cancun (2008)

3. Parenrenq, J.M., Syarif, M.I., Djanali, S., et al.: Performance Analysis of Resource-aware Framework Classification, Clustering and Frequent Items in Wireless Sensor Networks. In: 2011 International Conference on e-Education, Entertainment and e-Management, pp. 117–120. IEEE Press, Bali (2011)

4. Tezcan, N., Wang, W.: A Lightweight Classification Algorithm for Energy Conservation in Wireless Sensor Network. In: 14th International Conference on Computer Communications and Networks, pp. 87–92. IEEE Press, San Diego (2005)

5. Zhao, C., Wang, Y.: A New Classification Method on Information Fusion of Wireless Sensor Networks. In: 2008 International Conference on Embedded Software and Systems Symposia, pp. 231–236. IEEE Press, Chengdu (2008)

6. Boyd, A.W.F., Balasubramaniam, D., Dearle, A., et al.: On the Selection of Connectivity-Based Metrics for WSNs Using a Classification of Application Behavior. In: International Conference on Sensor Networks, Ubiquitous, and Trustworthy Computing, pp. 268–275. IEEE Press, Newport Beach (2010)

7. Costa, N., Pereira, A., Serodio, C.: Virtual Machines Applied to WSN's the State-of-the-art and Classification. In: International Conference on System and Networks Communications, pp. 50–57. IEEE Press, Cap Esterel (2007)

8. Gök, S., Yazici, A., Cosar, A., et al.: Fuzzy Decision Fusion for Single Target Classification in Wireless Sensor Networks. In: International Conference on Fuzzy Systems, pp. 1–8. IEEE Press, Barcelona (2010)

9. Imran, N., Khan, A.: Identifier Based Graph Neuron: A Light Weight Event Classification Scheme for WSN. In: Wong, K.W., Mendis, B.S.U., Bouzerdoum, A. (eds.) ICONIP 2010, Part II. LNCS, vol. 6444, pp. 300–309. Springer, Heidelberg (2010)

10. Imran, N., Khan, A.: A Single Shot Associated Memory Based Classification Scheme for WSN. In: Liu, D., Zhang, H., Polycarpou, M., Alippi, C., He, H. (eds.) ISNN 2011, Part III. LNCS, vol. 6677, pp. 94–103. Springer, Heidelberg (2011)

11. Yan, Z., Ansari, N., Wei, S.: Optimal Decision Fusion Based Automatic Modulation Classification by Using Wireless Sensor Networks in Multipath Fading Channel. In: Proceedings of the 2011 IEEE Global Communications Conference (GLOBECOM), pp. 1–5. IEEE Press, Kathmandu (2011)

12. Cheng, X., Xu, J., Pei, J., Liu, J.: Hierarchical Distributed Data Classification in Wireless Sensor Networks. Computer Communications 33, 1404–1413 (2010)

13. Rajkamal, R., Vanaja Ranjan, P.: Packet Classification for Network Processors in WSN Traffic Using ANN. In: The IEEE International Conference on Industrial Informatics, pp. 707–710. IEEE Press, Daejeon (2008)

14. Xu, J., Su, W., Zhou, M.: Distributed Automatic Modulation Classification with Multiple Sensors. IEEE Sensors Journal 10(11), 1779–1785 (2010)

15. Dilum Bandara, H.M.N., Jayasumana, A.P., Illangasekare, T.H.: Cluster Tree Based Self Organization of Virtual Sensor Networks. In: IEEE Globecom Workshops (GLOBECOM), pp. 1–6. IEEE Press, New Orleans (2008)

16. Liu, M., Cao, J., Chen, G., et al.: EADEEG: An Energy-Aware Data Gathering Protocol for Wireless Sensor Networks. Journal of Software 18(5), 1092–1109 (2007) (in Chinese)

17. Roweis, S.T., Saul, L.K.: Nonlinear Dimensionality Reduction by Locally Linear Embedding. Science 290(22), 2323–2326 (2000)

Fault Detection and Isolation in Wheeled Mobile Robot

Ngoc Bach Hoang[1], Hee-Jun Kang[2,*], and Young-Shick Ro[2]

[1] Graduate School of Electrical Engineering, University of Ulsan,
680-749, Ulsan, South Korea
hoangngocbach@gmail.com
[2] School of Electrical Engineering, University of Ulsan,
680-749, Ulsan, South Korea
{hjkang,ysro}@ulsan.ac.kr

Abstract. This paper presents a fault detection and isolation scheme for wheeled mobile robots. A nonlinear observer is designed based on the mobile robot dynamic model. The fault is detected when at least one of the residuals exceeds its corresponding threshold. After that, three observers are activated to isolate three types of faults: right wheel fault, left wheel fault, and the other changed dynamic faults.

Keywords: wheeled mobile robots, nonlinear observer, fault detection, fault isolation.

1 Introduction

Over last two decades, several researches have been investigated about WMR fault detection and fault tolerant control [1, 2]. Until now, most of papers in field of fault diagnosis for mobile robots only use kinematics model with nonholonomic constraints and neglect robot dynamic model to design robot controller. As proposed in [3], robot dynamic model needs to be considered. Recently some approaches have been presented in which mobile robot dynamics is taken into account [3, 4]. On the other hand, in the robot manipulator diagnosis field, faults are successfully detected and isolated based on manipulator dynamic models [5, 6]. However, those fault diagnosis design techniques have never been applied to the mobile robot system. Therefore, we intend to apply them to the mobile robot system by taking its dynamics into account.

In this paper, we present a fault detection and isolation scheme for wheeled mobile robots. This work is about the extension of the results in [5, 8] for wheeled mobile robots. First, a nonlinear observer is designed based on the WMR dynamics. The fault is detected by comparing the error state observer with its corresponding derived threshold. Second, three observers are constructed for three types of faults: right wheel fault, left wheel fault, and all the other dynamic faults. Finally, the computer simulation has been performed and its results show that the developed scheme is very effective for the fault detection and isolation for the WMR.

This paper is organized as follows. In section 2, the robot kinematics and dynamics are described, and the fault diagnosis problem is formulated. In section 3, the fault

---

* Corresponding author.

D.-S. Huang et al. (Eds.): ICIC 2012, LNCS 7389, pp. 563–569, 2012.
© Springer-Verlag Berlin Heidelberg 2012

detection and isolation scheme are discussed. The simulation results are shown in section 4. Section 5 has some conclusions and future works.

2   Problem Formulation

I_c : Moment of inertia of the mobile platform about P_c; I_m : Moment of inertia of wheel about its diameter; I_w : Moment of inertia of wheel about its rotation axle; m_c : Mass of the mobile platform without wheels; m_w : Mass of each wheel; O-xy : The world coordinates system; P-XY: The coordinate system fixed to mobile platform; P_c: Center of mass of WMR without wheels; d : Distance from P to P_c; b : Distance from each wheel to P; r_R : Radius of right wheel; r_L : Radius of left wheel;

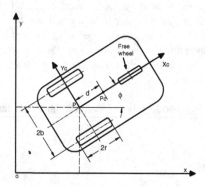

Fig. 1. Differential-driven WMR

The configuration of the WMR can be described by five generalized coordinates:

$$q_0 = [x, y, \phi, \theta_r, \theta_l]^T$$

The kinematic and dynamic models of WMR are

$$\dot{q}_0 = S(\dot{q}_0)\dot{q} \tag{1}$$

$$M\ddot{q} + V(\dot{q}) + F(\dot{q}) + \tau_d = \tau \tag{2}$$

where: $S(q_0) = \begin{bmatrix} \dfrac{r_R}{2}\cos\phi & \dfrac{r_R}{2}\sin\phi & \dfrac{r_R}{2b} & 1 & 0 \\ \dfrac{r_L}{2}\cos\phi & \dfrac{r_L}{2}\sin\phi & -\dfrac{r_L}{2b} & 0 & 1 \end{bmatrix}^T$

$M = \begin{bmatrix} I_w + (mb^2 + I)\dfrac{r_R^2}{4b^2} & (mb^2 - I)\dfrac{r_R r_L}{4b^2} \\ (mb^2 - I)\dfrac{r_R r_L}{4b^2} & I_w + (mb^2 + I)\dfrac{r_L^2}{4b^2} \end{bmatrix}$, $V(\dot{q}) = \begin{bmatrix} \dfrac{m_c d r_R r_L}{4b^2}(r_R \dot{q}_1 \dot{q}_2 - r_L \dot{q}_2^2) \\ \dfrac{m_c d r_R r_L}{4b^2}(r_L \dot{q}_1 \dot{q}_2 - r_R \dot{q}_1^2) \end{bmatrix}$

$F(\dot{q}) = S^T(\dot{q}_0)F_0(\dot{q}_0)$ With $I = (I_c + m_c d^2) + 2(I_w + m_w b^2)$ and $m = m_c + 2m_w$

In the presence of fault, the WMR dynamic model can be represented as:

$$\ddot{q} = M^{-1}(\tau - V(\dot{q}) - F(\dot{q}) - \tau_d) + \beta(t - T)\psi(q,\dot{q},\tau) \tag{3}$$

The term $\beta(t - T)\psi(q,\dot{q},\tau)$ is the fault function. The fault time profile is modeled by: $\beta_i(t-T) = 1 - e^{-\alpha_i(t-T)}$ if $(t \geq T)$ and $\beta_i(t-T) = 0$ if $(t < T)$ (i=1,2). The modeling uncertainty is assumed to be bounded: $\left| M^{-1}(F(\dot{q}) + \tau_d) \right| \leq \eta$

3   Fault Detection and Isolation Scheme

3.1   Fault Detection

Let $x = \dot{q}^T(t)$. The dynamic model of eq.(2) can be rewritten as

$$\dot{x} = M^{-1}(\tau - V(\dot{q}) - F(\dot{q}) - \tau_d) + \beta(t - T)\psi(q,\dot{q},\tau) \tag{4}$$

The following estimated model is considered

$$\dot{\hat{x}} = M^{-1}(\tau - V(\dot{q})) + A(\hat{x} - x) \tag{5}$$

where $\hat{x} \in R^2$, $A = diag(-a_1, -a_2)$ with $a_1, a_2 > 0$ are the estimation velocity vector of x and the stable matrix, respectively. From eqs.(4) and (5), we get the error dynamic:

$$\dot{e} = A(\hat{x} - x) - M^{-1}(F(\dot{q}) + \tau_d) + \beta(t - T)\psi(q,\dot{q},\tau) \tag{6}$$

The threshold bound can be chosen as

$$\|e\| \leq \int_0^t exp(A(t-\tau))\|M^{-1}\|\eta d\tau = e_M \tag{7}$$

Fault detection decision scheme: The decision on the occurrence of fault is made when at least one of the residual e exceeds its corresponding threshold bound e_M.

3.2   Fault Isolation

In this paper, we classify all that reasons into three types of faults: Right wheel fault $^1\psi$: right wheel radius change; Left wheel fault $^2\psi$: left wheel radius change; the other faults $^3\psi$: all remaining faults which make the dynamics change. Each fault function is described by $^s\psi(\tau,\dot{q}) = {}^s\theta_{ij}\, {}^s g_j^T(\tau,\dot{q})$, $s = \{1,2,3\}$ where $^s\theta_{ij}$ (i =1, 2; j =1... m) is an unknown matrix assumed to belong to a known compact set $^s\Omega_{ij} \in R^{2 \times m}$. $^s g_j^T(\tau,\dot{q}) \in R^m$ is a known smooth vector field. More details about fault functions see appendix 1. To isolate after a fault is detected, three isolation estimators are activated [8]. Each estimator has the following form as

$$\dot{\hat{x}}_s = A_s(\hat{x}_s - x) + M^{-1}(\tau - V(\dot{q})) + {}^s\hat{\theta}_{ij}\, {}^s g_j^T(\tau,\dot{q}) \tag{8}$$

where ${}^s\hat{\theta}_{ij}$ (i =1, 2; j =1... m) is the estimate of fault parameter ${}^s\theta_{ij}$, $A_s = diag(-a_1,-a_2)$ where $a_1,a_2 > 0$ is a stable matrix. The online updating law for ${}^s\hat{\theta}_{ij}$ is derived by using Lyapunov synthesis approach, with the projection operator restricting ${}^s\hat{\theta}_{ij}$ to the corresponding known set ${}^s\Omega_{ij} \in R^{2\times m}$ [8].

$$ {}^s\dot{\hat{\theta}}_{ij} = P_{{}^s\Omega_{ij}}\{ \gamma_{ij}{}^s g_i e_i \} \tag{9}$$

where γ_{ij} are the positive-definite learning rate, ${}^s g_i$ are the corresponding smooth vector field and $e_i = x_i - \hat{x}_i$ ($i = 1,2$) are the error state. The error dynamics is given

$$ \dot{e} = Ae - M^{-1}(F(\dot{q}) + \tau_d) + \beta(t - T)\psi(q,\dot{q},\tau)-{}^s\hat{\theta}_{ij}{}^s g_j^T(\tau,\dot{q}) \tag{10}$$

Thus, each element of state estimation error is given by

$$ e_i = exp(-a_i(t-T_d))e_i(T_d) - \int_{T_d}^{t} exp(-a_i(t-\tau)){}^s\hat{\theta}_{ij}{}^s g_j^T(\tau,\dot{q})d\tau $$
$$ + \int_{T_d}^{t} exp(-a_i(t-\tau))(1-e^{-\alpha_i(t-\tau)}){}^s\theta_{ij}{}^s g_j^T(\tau,\dot{q})d\tau - \int_{T_d}^{t} exp(-a_i(t-\tau))\sum_{l=1}^{2}mv_{il}(f_l + \tau_l)d\tau \tag{11}$$

The following functions are defined to use later for deriving the threshold as

$$ {}^1 h_j = \begin{cases} {}^s g_j(\tau,\dot{q}) & ({}^s g_j(\tau,\dot{q}) \geq 0) \\ 0 & ({}^s g_j(\tau,\dot{q}) < 0) \end{cases} {}^2 h_j = \begin{cases} 0 & ({}^s g_j(\tau,\dot{q}) \geq 0) \\ {}^s g_j(\tau,\dot{q}) & ({}^s g_j(\tau,\dot{q}) < 0) \end{cases} \tag{12}$$

Because ${}^s\theta_{ij} \in {}^s\Omega_{ij} \in R^{2\times m}$, there exist two values ${}^m\theta_{ij}, {}^M\theta_{ij}$ satisfy the inequality:

$$ {}^m\theta_{ij} \leq {}^s\theta_{ij} \leq {}^M\theta_{ij} < 0 \quad or \quad 0 \leq {}^m\theta_{ij} \leq {}^s\theta_{ij} \leq {}^M\theta_{ij} \tag{13}$$

In the incipient fault case, $\bar{\alpha}$ is the lower bound of fault evolution rate which satisfies $\alpha_i \geq \bar{\alpha}_i$, see[8]. So that the following inequality can be established

$$ 0 \leq 1 - e^{-\bar{\alpha}_i(t-T_d)} \leq 1 - e^{-\alpha_i(t-T_d)} \leq 1 \tag{14}$$

From eqs(14) and (15) if ${}^M\theta_{ij} \geq {}^m\theta_{ij} \geq 0$ we have (16.a)if $0 > {}^M\theta_{ij} \geq {}^m\theta_{ij}$, we have (16.b)

$$ m_{ij} = (1 - e^{-\bar{\alpha}_i(t-T_d)}){}^m\theta_{ij} \leq (1 - e^{-\alpha_i(t-T_d)}){}^s\theta_{ij} \leq {}^M\theta_{ij} = M_{ij} \tag{15}$$

$$ m_{ij} = {}^m\theta_{ij} \leq (1 - e^{-\alpha_i(t-T_d)}){}^s\theta_{ij} \leq (1 - e^{-\bar{\alpha}_i(t-T_d)}){}^M\theta_{ij} = M_{ij} \tag{16.b}$$

Now, let's combine eqs(13) and (16)

$$ m_{ij}{}^1 h_j + M_{ij}{}^2 h_j \leq (1 - e^{-\alpha_i(t-T_d)}){}^s\theta_{ij}{}^s g_j \leq M_{ij}{}^1 h_j + m_{ij}{}^2 h_j \tag{17}$$

Thus, the following thresholds can be chosen for isolation decision

$$
^U e_i = exp(-a_i(t - T_d))e_i(T_d) - \int_{T_d}^{t} exp(-a_i(t - \tau))^s \hat{\theta}_{ij}{}^s g_j^T(\tau, \dot{q})d\tau
$$

$$
+ \int_{T_d}^{t} exp(-a_i(t - \tau))(M_{ij}{}^1 h_j + m_{ij}{}^2 h_j)d\tau + \int_{T_d}^{t} exp(-a_i(t - \tau))\eta_i d\tau \tag{18}
$$

$$
^L e_i = exp(-a_i(t - T_d))e_i(T_d) - \int_{T_d}^{t} exp(-a_i(t - \tau))^s \hat{\theta}_{ij}{}^s g_j^T(\tau, \dot{q})d\tau
$$

$$
+ \int_{T_d}^{t} exp(-a_i(t - \tau))(m_{ij}{}^1 h_j + M_{ij}{}^2 h_j)d\tau - \int_{T_d}^{t} exp(-a_i(t - \tau))\eta_i d\tau \tag{19}
$$

Fault isolation scheme: If one of the state estimation errors (e_i) exceeds its threshold $^U e_i, {}^L e_i$, for some time t > T_d, then the occurrence of that fault is excluded.

4   Simulation Results

The WMR parameters: I_c = 0.85(Kgm²); I_m = 0.0061(Kgm²); I_w = 0.012(Kgm²); m_c = 29(Kg); m_w = 3.54(Kg); d = 0.053(m); b = 0.149(m); r_R = 0.08(m); r_L =0.08(m). The values of $F(\dot{q}), \tau_d : |F(\dot{q}) + \tau_d| \leq 0.05$ .Three observers will be used to isolate faults: *O1*: right wheel fault, *O2*: left wheel fault, *O3*: other fault. First, right wheel radius suddenly changes from 0.08(m) to 0.07(m) at T = 10s. The fault is detected at T_d = 10.1s as shown in Fig.2.a,b. The outputs of three observers are shown in: O1 (2.c, 2.d); O2 (2.e, 2.f); O3 (2.g, 2.h). The right wheel fault is successfully isolated at T_{iso} = 10.2s. Second, left wheel radius changes from 0.08(m) to 0.07(m) at T = 10s. The fault is detected at T_d = 10.1s as shows in Fig.3.a, 3.b. The left wheel fault is successfully isolated at T_{iso} = 10.15s by observing O1 (3.c, 3.d); O2 (3.e, 3.f); O3 (3.g, 3.h). Third, a fault occurs at T = 10s and results 2kg loss in mass of platform. The fault is detected at T_d = 10.1s as shown in Fig.4.a, 4.b. The fault is isolated at T_{iso} = 10.22s by observing O1 (4.c, 4.d); O2 (4.e, 4.f); O3 (4.g, 4.h).

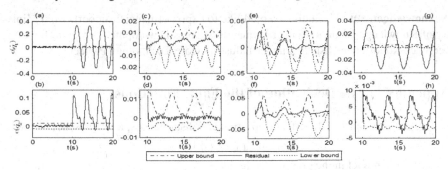

Fig. 2. Fault detection and Isolation in case of right wheel fault

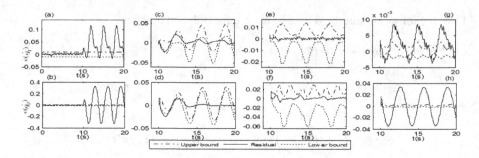

Fig. 3. Fault detection and Isolation in case of left wheel fault

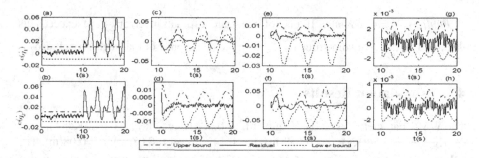

Fig. 4. Fault detection and Isolation in case of the other faults

5   Conclusions and Future Works

In this paper, the fault detection and isolation schemes have been proposed for wheeled mobile robot. By using three observers, three types of faults (right wheel fault, left wheel fault, and the other faults change robot dynamics) are successfully isolated after a fault is detected. For future research, we will find a suitable fault accommodation control scheme and test the proposed algorithm in a real mobile robot.

Acknowledgement. The authors would like to express financial supports from Ministry of Knowledge Economy under Human Resources Development Program for Convergence Robot Specialists and under Industrial Core Technology Project.

Appendix 1: The Derivation of the Fault Functions

We temporarily remove the effect of friction and disturbances, in the absence of fault:
$$\ddot{q} = M^{-1}(\tau - V(\dot{q}))$$
The dynamic model in the presence of fault:
$$\ddot{q} = M^{-1}(\tau - V) - ((M + \Delta M)M)^{-1}\Delta M(\tau - V) - (M + \Delta M)^{-1}\Delta V$$

From (30) and (31), the fault function will have the following form:

$$\psi(\tau,\dot{q}) = -((M+\Delta M)M)^{-1}\Delta M(\tau-V)-(M+\Delta M)^{-1}\Delta V$$

Right wheel fault: $\Delta V = \dfrac{\Delta r_R}{r_R}V + \dfrac{m_c d(r_R+\Delta r_R)r_L}{4b^2}\Delta r_R\begin{bmatrix}\dot{q}_1\dot{q}_2 \\ -\dot{q}_1^2\end{bmatrix}$

$${}^1\psi(\tau,\dot{q}) = \begin{bmatrix}{}^1\theta_{11} & .. & {}^1\theta_{16} \\ {}^1\theta_{21} & .. & {}^1\theta_{26}\end{bmatrix}\begin{bmatrix}\tau_1-V_1 & \tau_2-V_2 & V_1 & V_2 & \dot{q}_1^2 & \dot{q}_1\dot{q}_2\end{bmatrix}^T = {}^1\theta_{ij}{}^1g_j^T(\tau,\dot{q})$$

Left wheel fault: $\Delta V = \dfrac{\Delta r_L}{r_L}V + \dfrac{m_c d r_R(r_L+\Delta r_L)}{4b^2}\Delta r_L\begin{bmatrix}-\dot{q}_2^2 \\ \dot{q}_1\dot{q}_2\end{bmatrix}$

$${}^2\psi(\tau,\dot{q}) = \begin{bmatrix}{}^2\theta_{11} & .. & {}^2\theta_{16} \\ {}^2\theta_{21} & .. & {}^2\theta_{26}\end{bmatrix}\begin{bmatrix}\tau_1-V_1 & \tau_2-V_2 & V_1 & V_2 & \dot{q}_1\dot{q}_2 & \dot{q}_2^2\end{bmatrix}^T = {}^2\theta_{ij}{}^2g_j^T(\tau,\dot{q})$$

The other fault: $\Delta V = \Delta nV$

$${}^3\psi(\tau,\dot{q}) = \begin{bmatrix}{}^3\theta_{11} & .. & {}^3\theta_{14} \\ {}^3\theta_{21} & .. & {}^3\theta_{24}\end{bmatrix}\begin{bmatrix}\tau_1-V_1 & \tau_2-V_2 & V_1 & V_2\end{bmatrix}^T = {}^3\theta_{ij}{}^3g_j^T(\tau,\dot{q})$$

References

1. Ji, M., Sarkar, N.: Supervisory Fault Adaptive Control of a Mobile Robot and Its Application in Sensor-Fault Accommodation. IEEE Transactions on Robotics 23(1), 174–178 (2007)
2. Meng, J.: Hybrid Fault Adaptive Control of a Wheeled Mobile Robot. IEEE/ASME Transactions on Mechatronics 8(2), 226–233 (2003)
3. Dongkyoung, C.: Tracking Control of Differential-Drive Wheeled Mobile Robots Using a Backstepping-Like Feedback Linearization. IEEE Transactions on Systems, Man and Cybernetics, Part A: Systems and Humans 40(6), 1285–1295 (2010)
4. Fierro, R., Lewis, F.L.: Control of a Nonholonomic Mobile Robot: Backstepping Kinematics into Dynamics. Journal of Robotic Systems, 149–163 (1997)
5. Huang, S.N., Kiang, T.K.: Fault Detection, Isolation, and Accommodation Control in Robotic Systems. IEEE Transactions on Automation Science and Engineering 5(3), 480–489 (2008)
6. Vemuri, A.T., Polycarpou, M.M., Diakourtis, S.A.: Neural Network Based Fault Detection in Robotic Manipulators. IEEE Transactions on Robotics and Automation 14(2), 342–348 (1998)
7. Vemuri, A.T., Polycarpou, M.M.: Neural-network-based Robust Fault Diagnosis in Robotic Systems. IEEE Transactions on Neural Networks 8(6), 1410–1420 (1997)
8. Zhang, X.D., Parisini, T., Polycarpou, M.M.: Adaptive Fault-tolerant Control of Nonlinear Uncertain Systems: an Information-based Diagnostic Approach. IEEE Transactions on Automatic Control 49(8), 1259–1274 (2004)

The Research on Mapping from DAML+OIL Ontology to Basic-Element and Complex-Element of Extenics

Wen Bin[1,2]

[1] School of Mechanical Electric & Information Engineering, China University of Mining &
Technology, Beijing, China
[2] School of Information Science and Technology, Yunnan Normal University, Kunming, China
wenbin-315@163.com

Abstract. DAML+OIL is an important ontology language. Basic-element and
complex-element are the formalization basis of Extenics and used to solve the
contradiction problems. We firstly introduced the DAML+OIL and the basic-
element and complex-element. Secondly, we analyzed the element of
DAML+OIL and the basic-element and complex-element, then gave the
corresponding relation of them. Finally, we gave the mapping rules from
DAML+OIL to the basic-element and complex-element. We studied how to
transform DAML+OIL ontology to the basic-element and complex-element in
syntax, and gave a way to solve the contradiction problem of the knowledge
domain.

Keywords: DAML+OIL, Ontology, Extenics, Basic-element, Mapping rules.

Introduction

For AI community,there are many definitions of an ontology[1].An ontology is an
explicit specification of a conceptualisation[2]. An ontology consits of the description
of concepts,the desciption of properties of each concept and a set of individual
instances of classes in a domain of discourse[1,3,4,5].OIL(Ontology Interference
Layer) is the production of the On-To-Knowledge plan. OIL unifies three important
aspects: formal semantics and efficient reasoning support as provided by Description
Logics, epistemologically rich modeling primitives as provided by the Frame system,
and a standard proposal for syntactical exchange notations as provided by the Web
languages(XML and RDF(S))[6]. DAML(DARPA Agent Markup Language)is the
web language which is created by DARPA. DAML allowed users to mark the semantic information on their data, and it let the computer can "understand" the marked
information resource[7].A joint committee with two organizations combined DAML
and OIL, and named it DAML+OIL In December 2000[8]. DAML+OIL is a standard
description language of Semantic Web[9][10]. DAML+OIL is set up based on the
description logics, which only can represent the static knowledge by concepts and
roles, and it is difficult to represent the dynamic knowledge by description logic. So,
DAML+OIL can't describe the essential change of thing which is caused by the internal attribute change, and it is difficult to solve the contradiction problem by
DAML+OIL.

D.-S. Huang et al. (Eds.): ICIC 2012, LNCS 7389, pp. 570–578, 2012.
© Springer-Verlag Berlin Heidelberg 2012

In Extenics, we research the possibility of extending the things and the rules and methods of development and innovation with the formalization models, and use them to solve the contradiction problems[11,12].We use the matter-element, affair-element, relation-element and complex-element to describe the concepts and their relationships. In a certain level, there is a similarity between the ontology and the basic-element and complex-element.If the ontology can be mapped to the basic-element and complex-element, it will help us to use Extenics engineering to solve contradiction problems of the domain relvant the ontology.The Mapping from OWL lite to complex-elments is researched in article[14],but OWL Lite is the restrict ontolgoy language.DAML+OIL is an important ontology language of the Semantic Web. If the DAML+OIL ontology can be mapped to the basic-element and complex-element, it is useful to solve contradiction problems of the Semantic Web. In this paper, we researched the mapping from DAML+OIL to the basic-element and complex-element in syntax, and gave the corresponding relation between their elements and their mapping rules.

1   DAML+OIL

DAML+OIL is an ontology description language, and it describe the domain structure which include classes and attributes. The basis of DAML+OIL is the RDF triple collection. DAML+OIL use its vocabulary to express the RDF triple means.The elements of DAML+OIL language are show in table1,table2 and table3[5,8,9,10,13].

Table 1. Property elements

DAML+OIL element	statement
daml:ObjectProperty	To state the object property
daml:ObjectProperty	To state the datatype property
rdfs:subPropertyOf	To define the level relation of properties
daml:toClass	The universal quantification restriction
daml:hasClass	The existential quantification restriction
daml:hasValue	A speceial case of hasClass using enumeration
daml:maxCardinality	The maxcardinality restriction
daml:minCardinality	The mincardinality restriction
daml:cardinality	The cardinality restriction
rdfs:domain	To state the property domain
rdfs:range	To state the property range
daml:samePropertyAs	To assert one property is equivalent to another property
daml:inverseOf	To assert one property is inverse to another property
daml:TransitiveProperty	To assert one property is transitive
daml:UniqueProperty	To assert one property is unique

Table 2. Instance elements

DAML+OIL element	statement
daml:sameIndividualAs	To state that objects are same
daml:differentIndividualFrom	To state that objects are distinct

Table 3. Class elements

DAML+OIL element	statement
daml:Class	To define an object class
rdfs:subClassOf	To define the level relation of classes
daml:sameClassAs	To assert one class is equal to another class
daml:disjointWith	To assert one class is disjoint with another class
daml:intersectionOf	To define the intersection of classes
daml:unionOf	To define the union of classes
daml:complementOf	To define the complement of class

2   Basic-Element and Complex-Element of Extenics

Matter-element and affair-element are the basic concepts of Extenics. Extension transformation is the basic tool to solve the contradiction problems. Extension analysis is the basis for exploring the extension transformation[11,12]. Using these we can analyze the extension possibilities of things from a qualitative point of view.

2.1   Basic-Element

Matter-element describes the things,affair-element describes the events,relation-element describes the relationship between the matter-elements or affair-elements. Matter-element, affair-element and relation-element are collectively called basic-element, it is described in detail in article[12].

Given an object class $\{O\}$, for any $O \in \{O\}$, there is $v_i = c_i(O) \in V_i$ on the property of c_i $(i = 1, 2, ..., n)$, the basic-element set

$$\{B\} = (\{O\}, C, V) = \begin{bmatrix} \{O\}, c_1, \overline{V}_1 \\ c_2, V_2 \\ \dots, \dots \\ c_n, V_n \end{bmatrix} \qquad (1)$$

is called class basic-element [14,15].

2.2   Complex-Element

The compounded form of matter-element, affair-element and relationship-element are collectively called complex-element.Complex-element has seven forms:matter-element composites with matter-element, matter-element composites with affair-element, matter-element composites with relation-element, etc. The relevant description of all complex-element in detail is in the article[11].

Given an object class $\{O_{cm}\}$, for any $O_{cm} \in \{O_{cm}\}$, there is $v_{cmi} = c_{cmi}(O_{cm}) \in V_{cmi}$ on the property c_{cmi} $(i = 1, 2, ..., n)$, the complex-element set

$$\{CM\} = (\{O_{cm}\}, C_{cm}, V_{cm}) = \begin{bmatrix} \{O_{cm}\}, c_{cm1}, V_{cm1} \\ c_{cm2}, V_{cm2} \\ \dots, \dots \\ c_{cmn}, V_{cmn} \end{bmatrix} \tag{2}$$

is called class complex-element[14,15].

3   Mapping from DAML+OIL to the Basic-Element and Complex-Element

3.1   Corresponding Relation between DAML+OIL and the Basic-Element and Complex-Element

Table 4. The corresponding relation between DAML+OIL element and basic-element and complex-element

DAML+OIL element	Basic-element,Complex-element
Class	Class basic-element or class complex-element
ObjectProperty	Propery of basic-element or complex-element
DatatypeProperty	Propery of basic-element or complex-element
Datatype Range of one Property	Text
Object Range of one Property	Class basic-element or class complex-element
Individual	Basic-element or complex-element

3.2   Mapping Rules of Class

(1)daml:Class According whether it has the object property or not, it is mapped to class basic-element or class complex-element. If it has the object property, it is mapped to class complex-element, otherwise it is mapped to class basic-element.

(2)rdfs:subClassOf This is mapped to the logical relation $\subseteq$ of classes.For example,

<daml:Class rdf:ID="Man">
    <rdfs:subClassOf rdf:resource="#Person"/>
</daml:Class >

Its mapping is $\{CM_{Man}\} \subseteq \{CM_{Person}\}$.

(3)daml:sameClassAs If there is daml:sameClassAs between class A and class B,its mapping is $\{CM_A\}=\{CM_B\}$.For example,

<daml:Class rdf:ID="HumanBeing">
    <daml:sameClassAs rdf:resource="#Person"/>
</daml:Class>

Its mapping is $\{CM_{HumanBeing}\}=\{CM_{Person}\}$.

(4) daml:disjointWith If there is daml:disjointWith between class A and class B,its mapping is $\{CM_A\} \cap \{CM_B\}=\varnothing$.

(5)daml:intersectionOf This is mapped to ∧ of Extenics,For example,
```
<daml:Class rdf:ID="Woman">
  <daml:intersectionOf rdf:parseType="daml:collection">
    <daml:Class rdf:about="#Person"/>
    <daml:Class rdf:about="#Female"/>
  </daml:intersectionOf>
</daml:Class>
```
Its mapping is $\{CM_{Woman}\}=\{CM_{Person}\}\wedge\{CM_{Female}\}$.

(6)daml:unionOf This is mapped to ∨ of Extenics,For example,
```
<daml:Class rdf:ID="Parents">
  <daml:unionOf parseType="daml:collection">
  <Class rdf:resource="#Mother"/>
  <Class rdf:resource="#Father"/>
  </ daml:unionOf>
</daml:Class>
```
Its mapping is $\{CM_{Parent}\}=\{CM_{Mather}\}\vee\{CM_{Father}\}$.

(7)daml:complementOf This is mapped to ¬ of Extenics, For example,
```
<complementOf>
       <Class>
              <Class rdf:resource="#Man"/>
       </Class>
</complementOf>
```
Its mapping is $\neg\{CM_{Man}\}$.

3.3   Mapping Rules of Property

(1)daml:ObjectProperty This is mapped to the property, whose range is class basic-element or class complex-element, of class complex-element.

(2)daml:DatatypeProperty This is mapped to the property, whose range is text,of class basic-element or class complex-element.

(3)daml:hasValue This is mapped to the all range of the property of class basic-element or class complex-element.

(4)rdfs:domain This is mapped to the domain of the property of class basic-element or class complex-element.

(5)rdfs:range This is mapped to the range of the property of class basic-element or class complex-element.

(6)daml:samePropertyAs This is mapped to the equal property of the property of class basic-element or class complex-element.

(7)daml:inverseOf This is mapped to the inverse property of the property of class basic-element or class complex-element.

3.4   Mapping Rules of Instance

Individual of instance is mapped to the individual, that is basic-element or complex-element, of some class basic-element or class complex-element.For example,

<daml:Class rdf:ID="Person">
<Person rdf:ID="Wenbin">

Person is mapped to class basic-element $\{B_{Person}\}=(O_{Person}, name, V_{name})$, where V_{name} is the collection of all person name. Wenbin is mapped "to basic-element $B_{Wenbin}=(O_{Person}, name, "Wenbin")$, and $B_{Wenbin} \in \{B_{Person}\}$.

Daml:sameIndividualAs is mapped to that the basic-elements or complex-elements are equal. Daml:differentIndividualFrom is mapped to that the basic-elements or complex-elements are distinct.

4   Example of the Mapping from DAML+OIL Ontology to Basic-Element and Complex-Element

As basic-element and complex-element have the extended property, which include the divergence, the relevance, the implication and the extension. We can analyze the extension of basic-element and complex-element to get many approaches to solve the contradiction[12]. After transforming DAML+OIL ontology to basic-element and complex-element, and extendedly analyzing the intrinsic property of the concept and relation which are represented by basic-element and complex-element, the contradiction problems can be solved. The following example illustrates this.

A petroleum company set up the information ontology of its petrol sales with DAML+OIL, the class and property of the information ontology show in Fig 1, and the instance of the information ontology shows in Fig 2.

Using the above mapping rules, the DAML+OIL ontology in Fig.1 and Fig.2 is mapped to basic-element and complex-element as follows.

```
<daml:Class rdf:ID="ProvinceSalesRegion"/>
   <rdfs:label>ProvinceSalesRegion</rdfs:label>
</daml:Class>
<daml:Class rdf:ID="PetrolProduct">
   <rdfs:label>PetrolProduct</rdfs:label>
</daml:Class>
<daml:Class rdf:ID="CitySalesRegion ">
   <rdfs:label>CitySaleRegion</rdfs:label>
   <daml:subClassOf rdf:resource="ProvinceSalesRegion"/>
</daml:Class>
<daml:DatatypeProperty rdf:ID="AreaCode">
   <rdfs:label>AreaCode</rdfs:label>
   <rdfs:range rdf:resource="http://www.w3.org/2001/XMLSchema#string"/>
   <rdfs:domain rdf:resource="ProvinceSalesRegion"/>
</daml:DatatpyeProperty>
<daml:DatatpyeProperty rdf:ID="Region">
   <rdfs:label>Region</rdfs:label>
   <rdfs:range rdf:resource="http://www.w3.org/2001/XMLSchema#string"/>
   <rdfs:domain rdf:resource="ProvinceSalesRegion"/>
</daml:DatatypeProperty>
<daml:DatatpyeProperty rdf:ID="SalesVolume">
   <rdfs:label>SalesVolume</rdfs:label>
   <rdfs:range rdf:resource="http://www.w3.org/2001/XMLSchema#string"/>
   <rdfs:domain rdf:resource="ProvincialSalesRegion"/>
</daml:DatatpyeProperty>
```

Fig. 1. Definition of class and property

```
<CitySalesRegion rdf:ID="Kunming">
   <Region xml:lang="en">Kunming</Region>
   <AreaCode xml:lang="en">0871</AreaCode>
   <SalesVolume xml:lang="en">4 million tons</SalesVolume >
   <daml:sameIndividualAs>
      <CitySalesRegion rdf:ID="Kunming">
      <daml:sameIndividualAs rdf:resource="KunmingCity"/>
      <Region xml:lang="en">KunmingCity</Region>
      <AreaCode xml:lang="en">0871</AreaCode>
      <SalesVolume xml:lang="en">4 million tons</SalesVolume>
      </CitySalesRegion>
   </daml:sameIndividualAs>
</CitySalesRegion>
<ProvinceSalesRegion rdf:ID="Yunnan">
   <Region xml:lang="en">Yunnan</Region>
   <AreaCode xml:lang="en">087</AreaCode>
   <SalesVolume xml:lang="en">10 million tons</SalesVolume>
   <daml:sameIndividualAs>
      <ProvinceSalesRegion rdf:ID="Yunnan">
      <daml:sameIndividualAs rdf:resource="YunnanProvince"/>
      <Region xml:lang="en">YunnanProvince</Region>
      <AreaCode xml:lang="en">087</AreaCode>
      <SalesVolume xml:lang="en">10 million tons</SalesVolume>
      </ProvinceSalesRegion>
   </daml:sameIndividualAs>
</ProvinceSalesRegion>
<CitySalesRegion rdf:ID="Zhaotong">
   <Region xml:lang="en">Zhaotong</Region>
   <AreaCode xml:lang="en">0870</AreaCode>
   <SalesVolume xml:lang="en">2 million tons</SalesVolume>
   <daml:sameIndividualAs>
      <CitySalesRegion rdf:ID="Zhaotong">
      <daml:sameIndividualAs rdf:resource="ZhaotongCity"/>
      <Region xml:lang="en">ZhaotongCity</Region>
      <AreaCode xml:lang="en">0870</AreaCode>
      <SalesVolume xml:lang="en">2 million tons</SalesVolume>
      </CitySalesRegion>
   </daml:sameIndividualAs>
</CitySalesRegion>
```

Fig. 2. Definition of instance

$$\{CM_{ProvinceSalesRegion}\} = \begin{bmatrix} O_{ProvinceSalesRegion}, Name, V_{ProvinceSalesRegionName} \\ Region, V_{Region} \\ AreaCode, V_{AreaCode} \\ SalesVolume, V_{SalesVolume} \end{bmatrix} \tag{3}$$

Where, $V_{ProvinceSalesRegionName}$ is the set of all province sales region names. $V_{ProvinceSalesRegionName}, V_{Region}, V_{AreaCode}, V_{SalesVolume}$ is the description text respectively.

$$\{CM_{CitySalesRegion}\} = \begin{bmatrix} O_{CitySalesRegion}, Name, V_{CitySalesRegionName} \\ Region, V_{Region} \\ AreaCode, V_{AreaCode} \\ SalesVolume, V_{SalesVolume} \end{bmatrix} \tag{4}$$

Where, $V_{CtitySalesRegionName}$ is the set of all city sales region names, $V_{CitySalesRegionName}$, $V_{Region}, V_{AreaCode}, V_{SalesVolume}$ is the description text respectively, and $\{CM_{CitySalesRegion}\} \subseteq \{CM_{ProvinceSalesRegion}\}$.

$$B_{Yunnan} = \begin{bmatrix} O_{Yunnan}, Name, \text{"Yunnan"} \\ Region, \text{"Yunnan"} \\ AreaCode, \text{"087"} \\ SalesVolume, \text{"10 million tons"} \end{bmatrix} \tag{5}$$

$$B_{YunnanProvince}= \begin{bmatrix} O_{YunnanProvince}, & Name, & \text{"YunnanProvince"} \\ & Region, & \text{"YunnanProvince"} \\ & AreaCode, & \text{"087"} \\ & SalesVolume, & \text{"10 million tons"} \end{bmatrix} \quad (6)$$

$$B_{Kunming}= \begin{bmatrix} O_{Kunming}, & Name, & \text{"Kunming"} \\ & Region, & \text{"Kunming"} \\ & AreaCode, & \text{"0871"} \\ & SalesVolume, & \text{"4 million tons"} \end{bmatrix} \quad (7)$$

$$B_{KunmingCity}= \begin{bmatrix} O_{KunmingCity}, & Name, & \text{"KunmingCity"} \\ & Region, & \text{"KunmingCity"} \\ & AreaCode, & \text{"0871"} \\ & SalesVolume, & \text{"4 million tons"} \end{bmatrix} \quad (8)$$

$$B_{Zhaotong}= \begin{bmatrix} O_{Zhaotong}, & Name, & \text{"Zhaotong"} \\ & Region, & \text{"Zhaotong"} \\ & AreaCode, & \text{"0870"} \\ & SalesVolume, & \text{"2 million tons"} \end{bmatrix} \quad (9)$$

$$B_{ZhaotongCity}= \begin{bmatrix} O_{ZhaotongCity}, & Name, & \text{"ZhaotongCity"} \\ & Region, & \text{"ZhaotongCity"} \\ & AreaCode, & \text{"0870"} \\ & SalesVolume, & \text{"2 million tons"} \end{bmatrix} \quad (10)$$

$B_{Kunming}, B_{KunmingCity}, B_{Zhaotong}, B_{ZhaotongCity} \in \{CM_{CitySalesRegion}\} \subseteq \{CM_{ProvinceSalesRegion}\}$
$B_{Yunnan}, B_{YunnanProvince} \in \{CM_{ProvinceSalesRegion}\}$
$B_{Kunming}=B_{KunmingCity}, B_{Zhaotong}=B_{ZhaotongCity}, B_{Yunnan}= B_{YunnanProvince}.$

If you find the sales volume of some region is too low and the petrol is surplus, and the sales volume of other region is too large and the petrol is shortage, we can extendedly analyze the property of the above basic-element and complex-element, and find the way to solve the problem of the petrol shortage and overstock.

5   Conclusion

DAML+OIL and the basic-element and complex-element can describe the concepts and their relationship in some domain. But DAML+OIL only can describe the static concept, the basic-element and complex-element which can be with the parameter variables can describe the dynamic things and it conforms to the reality. DAML+OIL can't solve the contradiction problems, and the basic-element and complex-element have the extension ability. In this paper, we researched the mapping from DAML+OIL to the basic-element and complex-element in syntax, we firstly analyzed the corresponding relation from DAML+OIL to the basic-element and complex-element, then gave the mapping rules from DAML+OIL to the basic-element and complex-element. In the future, we will study their mapping in semantic and the application of conversion from DAML+OIL to the Extenics.

References

1. Noy, N.F., McGuinness, D.L.: Ontology Development 101: A Guide to Creating Your First Ontology. Stanford Knowledge Systems Laboratory Technical Report KSL-01-05 and Stanford Medical Informatics Technical Report SMI-2001—0880 (2001)
2. Gruber, T.R.: A Translation Approach to Portable Ontology Specifications. Journal of Knowledge Acquisition 5, 199–220 (1993)
3. Yu, L.C., Wu, C.H., Jang, F.L.: Psychiatric Document Retrieval Using a Discourse-Aware Model. Journal of Artificial Intelligence 173, 817–829 (2009)
4. Yu, L.C., Chan, C.L., Lin, C.C., Lin, I.C.: Mining Association Language Patterns Using a Distributional Semantic Model for Negative Life Event Classification. Journal of Biomedical Informatics 44, 509–518 (2011)
5. Lai, Y.S., Wang, R.J., Hsu, W.T.: A DAML+OIL-Compliant Chinese Lexical Ontology. In: 19th International Conference on Computational Linguistics, Taipei (2002)
6. Fensel, D., et al.: OIL: An Ontology Infrastructure for the Semantic Web. Journal of IEEE Intelligent Systems 16, 38–45 (2001)
7. The DARPA Agent Markup Language (DAML) (EB/OL), http://www.Daml.org
8. Yi, Q.W., Li, S.P.: Semantic Web Language DAML+OIL and its Initiatory Application. Journal of Computer Science 30, 139–141 (2003)
9. Mcguinness, D.L., Fikes, R., Hendler, J., Stein, L.A.: DAML+OIL:An Ontology Language for The Semantic Web. Journal of IEEE Intelligent Systems 17, 72–80 (2002)
10. Horrocks, I., Patel-Schneider, P.F., van Harmelen, F.: Reviewing The Design of DAML + OIL: An Ontology Language for The Semantic Web. In: Proc. of the 18th National Conference on Artificial Intelligence, pp. 792–797 (2002)
11. Cai, W., Yang, C.Y., He, B.: Extension Logic Initial. Science Press, Beijing (2003)
12. Yang, C.Y., Cai, W.: Extension Engineering. Science Press, Beijing (2007)
13. van Harmelen, F., et al.: Reference Description of the DAML+OIL Ontology Markup Language (2001), http://www.daml.org/2001/03/reference.html
14. Liu, Z.M., Li, W.H., Tan, J.X.: Research on Mapping OWL Ontology to Complex-Elements. Journal of Guangdong University of Technology 26, 78–83 (2009)
15. Tan, J.X., Li, W.H., Liu, Z.M.: Research on Database Retrieval for Complex-Element. Journal of Guangdong University of Technology 25, 57–61 (2008)

The Study of Codeswitching in Advertisements

Wang Hua

Department of Basics,Chinese People's Armed Police Force Academy, Langfang, Hebei
065000, China
wanghuahuawang@126.com

Abstract. Codeswitching has become one of the most common language phenomena in these days. Researchers have launched various studies on codeswitching from different perspectives and thus achieved different findings. The Chinese-English codeswitching in Chinese advertising discourse can be categorized into three types: insertional codeswitching, alternational codeswitching and diglossia codeswitching. Moreover, the insertional codeswtiching can be further divided into letter insertion, lexical and phrasal insertion, clausal insertion, and discourse insertion. Chinese-English codeswitching in ads is a normal linguistic and social phenomenon. The purpose of advertisers' applying Chinese-English codeswitching is mainly to realize their final commercial goals. Good Chinese-English codeswitching in ads can achieve good results, while overuse and misuse of it may leave bad impressions on consumers. Therefore, proper guidance should be offered and other actions should be taken immediately in order to address the problems in Chinese-English codeswitching in ads and to maintain a healthy development of Chinese advertising, which will facilitate the harmonious coexistence between English and Chinese in Chinese advertising, thus contributing to the final social harmony.

Keywords: codswitching, advertiding, reasons, problems, solutions.

1   Introduction

With the introduction of economic reform and open-door policy, China develops closer ties with the west, especially the English-speaking countries for the political, economic and cultural purposes, which unavoidably leads to the language contact. As one of the outcomes of language contact, Chinese-English codeswitching appears constantly in people's daily communications. A number of scholars have launched various researches on codeswitching from different viewpoints, but little attention has been paid to the Chinese-English codeswitching in advertising. This paper attempts to further study the phenomenon of Chinese-English codeswitching in advertising. What types could the Chinese-English codeswitching in ads be classified into? What might be the reasons for which the advertisers make use of the Chinese-English codeswitching during their composition of the advertising discourse? What problems are existing in the Chinese-English codeswitching in ads?

2   Codeswitching and Advertising

It is Hans Vogt [1] who introduced code-switching for the first time as a psychological phenomenon in 1954: code-switching in itself is perhaps not a linguistic

D.-S. Huang et al. (Eds.): ICIC 2012, LNCS 7389, pp. 579–585, 2012.
© Springer-Verlag Berlin Heidelberg 2012

phenomenon, but rather a psychological one, and its causes are obviously extra-linguistic. Up to the present, a variety of definitions have been proposed by different researchers [2][3][4] based on their research focuses and the data they use. We prefer to choose a generally accepted definition: codeswitching is the alternate use of two or more languages within the same conversation or utterance.

A French advertising agent Robert says that what we are breathing today are oxygen, nitrogen and advertising; we are swimming in the ocean of advertising. It is no exaggeration. As consumers, we are exposed everyday to hundreds and even thousands of commercial messages in the form of newspaper ads, publicity, event sponsorships or TV commercials. No doubt, advertising is getting increasingly important and has become an essential part of our life. It is clear from this definition that advertising, by its very nature, is a kind communication with the purpose of convincing the target audience about something (ideas, goods or services), especially persuading them into the action of purchasing.

3   The Types of Chinese-English Codeswitching in Advertising

Three kinds of codeswitching in Chinese advertising are insertional codeswitching, alternational codeswitching and the diglossia codeswitching [5].

3.1   Insertional Codeswitching in Ads

Insertional codeswitching occurs when the language items from English are put into the grammatical framework of a discourse made up of Chinese. According to the linguistic structure of the English items, the classification of insertional codeswiching is shown in Table 1.

Table 1. The Classification of Insertional Codeswitching

Insertional codeswitching			
Letter insertion	Lexical and phrasal insertion	Clausal insertion	Discourse insertion

As the name indicates, letter insertion refers to the codeswitching in which an English letter is embedded into the Chinese advertising. If the pronunciations of the inserted letters get associated with some Chinese elements, the effect may be more powerful especially when an English pronunciation is punned into a certain Chinese syntactic frame. Lexical and phrasal insertion refers to the English elements at levels of lexicon and phrase that are interposed in the Chinese discourse. In fact, such insertion may be regarded as the so-called intrasentential codeswitching in Poplack's [6] division. It involves switching languages at sentential boundaries and occurs within the same sentence or sentence fragment. The clausal insertion includes the codeswitching in which one or more English sentences are put into the Chinese discourse. Different from lexical and phrasal insertion, it is similar to the so-called intersentential codeswitching which involves switches of codes from one language to the other between sentences. It is a fact that not all the Chinese-English codeswitching are the sheer lexical, phrasal or the complete clausal insertion. Usually, in most ads, there is a mixture of the above-mentioned two or even three types of insertions. Discourse insertion refers to the codeswitching in which a discourse from English as an entire unit is inserted or embedded into another discourse from Chinese.

3.2   Alternational Codeswiching in Ads

Alternational codeswitching means that two or more languages, English and Chinese in the present study, appear alternately and symmetrically in the same clause or discourse.

3.3   Diglossia in Ads

Diglossia originally is a term used to describe a situation existing in a speech community where people use two very different varieties of language to perform different functions. But in this section, we borrow diglossia to analyze the ads in which both Chinese and its English equivalent are employed to express the same meaning. Apart from the complete diglossia, there exist many ads in which only part of the Chinese elements is accompanied by the English translation. We call this form of codeswitching partial diglossia. Comparatively speaking, partial diglossia is more popular than complete diglossia. And it appears most frequently in the switching of company names, product brand names and slogans.

4   The Reasons for the Codeswitching in Advertising

4.1   Codeswitching as Adaptation to the Linguistic Reality

The adaptation to linguistic reality in the Chinese–English codeswitching means the adaptation to the linguistic existence and linguistic properties of both Chinese and English.

Every language possesses its own linguistic existence that other languages do not share. It is the same with the language of Chinese and English. By reasons of dissimilar geographical positions, natural environment, culture and social practice and the like, the linguistic elements and structures of the two languages that are employed to express the same objective can not be in complete agreement. Consequently, when the linguistic existence possessed uniquely by either English or Chinese is needed, it is no surprise that the language users including advertisers will resort to codeswitching to fill the lexical gap between the two languages so that the communication could be continued smoothly.

This specific adaptation mainly involves the codeswitching triggered by differences in the semantic features contained by a particular Chinese expression and its English equivalent.

Linguistically speaking, different languages can express the same meaning. But when it comes to the actual use, this usually can not be true due to the complexity of meaning. Sometimes, though a word has a counterpart in other languages, there still exist subtle semantic differences between them. Once the language users become aware of the subtlety, they will possibly switch to another language for a certain purpose. As to the advertising discourse, there are many instances of codeswitching resulted from the advertisers' desire to adapt to the linguistic features of Chinese and English.

4.2   Codeswitching as Adaptation to the Social Factors

Being a mirror of the society, language use is always closely tied up with a complex social world. The social world is so complex that there is no principled limit to the range of social factors to which the linguistic choices are inter-adaptable. Nonetheless, just as Verschueren [7] remarks, phenomenon of the utmost importance in the relationship between linguistic choices and the social world are the setting-, institution-, or community-specific communicative norms that have to be observed. The social settings and norms can regulate the content and means of the actualization of a speech event greatly.

With regard to the Chinese advertising, the social world to which the codeswitching adapts is from two aspects. One is the adaptation to the social environment on the macro-level which concerns the widespread of English in China. And the other is the adaptation to a certain social convention or custom in terms of the micro-level.

4.3   Codeswitching as Adaptation to the Psychological Motivations

Psychological motivations do work powerfully on the communicators' language choice making since their motives or intentions influence or even decide not only what to say but also how to say, for instance, switching from one code to another.

Given that advertising is a sort of one-way communication from the advertisers to the consumers, the psychological motivations in this study refer to the advertisers' spontaneous motives or intentions behind their performing of Chinese-English codeswitching in editing the advertising discourse. Usually, the advertisers make use of codeswitching, which is itself a communicative strategy as has been indicated earlier, to satisfy different motives or intentions, including their assumptions on the expectations of the addressee. Once the assumptions are in conformity with the real situations, various functions will be fulfilled in the form of a series of communicative strategies and the codeswitching will prove to be a success.

The Chinese-English codeswitching as adaptation to psychological motivations in advertising is rather complicated, as the analysis shows: codeswitching as attention seeking strategy, codeswitching as foreign flavor gaining strategy, codeswitching as authenticity-keeping strategy, codeswitching as solidarity-building strategy, codeswitching as addressee selection strategy, codeswitching as convention strategy and codeswitching as decoration strategy. However, it must be made clear that the adaptation to psychological motivations is rather complicated and the above list is far from being complete, there is still more to be identified and discussed.

5   Problems in the Code Switched Ads and Possible Solutions

There is no doubt that the Chinese-English codeswitching, as a communicative strategy, can function to be much more persuasive if it is used by the advertisers rationally and efficiently. But this is by no means an easy task. The advertisers have to be very careful with the employment of English in their work, otherwise they may make mistakes which will in the end destroy their work and also cause economic loss for the companies.

5.1   Problems

The author finds there does exist some problems in the code switched Chinese advertising which are illustrated as follows:

First of all, it seems that in some ads the English code is abused or overused so much that it makes the customers, especially those who are less proficient in English, feel hard to catch the main information conveyed.

The second problem lies in the advertisers' misjudgments on public reactions to the code switched ads. In China, some advertisers normally overestimate the power of English and show a blind trust on codeswitching without considering the specific context. Take "Da Bao SOD Mi", a cosmetic well known by Chinese, for instance, SOD here is a kind of chemical ingredient which is helpful to prevent the skin from premature senility. The reason why the advertisers adopt "SOD" into the brand name as an important constituent of the ad is probably based on the psychological motivation of implying the hi-technology to the consumers. But unfortunately, the spelling of "SOD" is happen to be identical with that of "sod", a swear word used mainly to express anger, annoyance or contempt. So how will the English native speakers react to such cosmetic if it is promoted in the international market?

The excessive craziness for the foreign flavor gives rise to another serious problem. If it is acceptable for those produced in the west to have English names, so that the English name will enable the perfect match of the name and content, then what do you think of the English brand names for some commodities which are manufactured for the domestic market? As a Chinese, sometimes, we get so confused by a surge of English brand names in the ads, such as "Canadian Garden", "Monte Carlo Villa" and "Manhattan Square" ect., that we can not help wondering whether we are living in China. Predictably, such ads will leave no good impressions on the consumers and may even arouse their antipathy, let alone motivate them to buy.

Besides, the problems stemming from the advertisers' less proficiency in English also spread all over the code switched ads. Misspellings, grammatical mistakes and inappropriate translations are not uncommon. Translation problems in the diglossia ads or partial diglossia ads appear even more frequent. The author once read a piece of diglossia ad in which "Bing Tang Yan Wo"is translated as "Bird's nest". Such a nest is made of straw and mud, and how can it be nourishing and nutritious?

5.2   Solutions

In spite of various problems mentioned above, we have no reasons to reject all the code switched ads like throwing the baby out with the bath water. On the contrary, as an inevitable result of language contacts and social development, the codeswitching in Chinese advertising does reflect the real life and has very much potential impact onto the social language. Confronted with the negative influences, what we should do is not to be annoyed with the language of code switched ads, and argue that it does nothing good but destructing the purity of Chinese. In fact, it is not the language itself but the people who abuse the language deserve the blame. So, instead of forbidding the occurrence of English in Chinese ads, more considered guidance should be provided so that such codeswitching can be used properly.

First, efforts should be made to help people realize that codeswitching is a natural result of language contacts, also an inevitable and essential language phenomenon. The quality of a product actually has nothing to do with the language in which it is advertised. As the advertiser, they should set up good professional ethics and adopt the language of English into their ads in a proper way. Meanwhile, both the cultural factors and acceptability of the addressees should be taken into considerations by them when editing the ads. As the consumer, however, they should develop rational consuming concepts but not merely pursue fashion and go after westernization blindly.

Second, as to those linguistic-related errors, they are very easy to be avoided by the advertisers through careful review of the translated version, or seeking help from an English expert.

Last but not least, a more efficient monitoring system should be established. On the one hand, the government should take measures to protect the consumers from being misled by the appearance of English code in some ads. On the other hand, the consumers themselves should keep an eye on and defend their rights against some false code switched ads when it is necessary.

Certainly, there are many other different ways to solve the problems existing in the code switched ads. But anyhow, a balanced relationship between the embedded English and the matrix Chinese is a must for the healthy development of Chinese advertising. And we still need make more efforts to facilitate the harmonious coexistence of the two languages and thus contribute to the final social harmony.

6   Conclusion

The Chinese-English codeswitching in ads can be classified into three types including the insertional codeswitching, alternational codeswitching, and diglossia codeswitching. The reason for which the advertisers switch codes between Chinese and English is that they need adapt to the contextual elements including the linguistic reality, social factors and psychological motivations, so as to realize their communicative purposes. But the boundaries among the adaptation to these three elements are not definitely clear-cut because there exists overlapping in some cases. Furthermore, the author has found some problems from the advertisers' application of Chinese-English codeswitching which involves English abuse or overuse, linguistic errors and so on. To fight against the negative influences, reasonable guidance should be provided such as, setting up good professional ethics and rational consuming concepts, establishing efficient monitory system and advocating correct English ect., but not simply prohibiting the occurrence of Chinese-English codeswitching in ads.

Against the backdrop of growing economic globalization, it is predictable that Chinese-English codeswitching in Chinese ads will become increasingly intense in the not too distant future. Therefore, for language policy makers, instead of forbidding the occurrence of English in fields such as advertising, they should look for ways in which people can prepare for their encounter with this phenomenon. Nothing is more effective than education in making a population critically alert and empowered to deal with the ongoing realities of language contact.

Acknowledgments. I'm indebted to a number of people for their invaluable assistance in the preparation and completion of this paper writing, especially my teachers, friends and family members whose loving consideration and helps are the source of my strength.

References

1. Vogt, H.: Language Contacts in Word, pp. 365–374 (1954)
2. Di Pietro, R.: Code-switching as A Verbal Strategy Among Bilinguals. In: Current Themes in Linguistics: Bilingualism, Experimental Linguistics and Language Typologies. Hemisphere Publishing (1977)
3. Valde's Fallis, G.: Social Interaction and Code-switching Patterns: A Case Study of Spanish English alternation. In: Bilingualism in the Bicentennial and Beyond. Bilingual Press (1976)
4. Scotton, C., Ury, W.: Bilingual Strategies: The Social Functions of Code-Switching. Linguistics (193), 5–20 (1977)
5. Muysken, P.: Code-switching and Grammatical Theory. In: Two Languages: Cross-disciplinary Perspectives on Code-switching. Cambridge University Press (1995)
6. Poplack, S.: Sometimes I Start A Sentence in Spanish Y Ternimo En Espanol: Towards A Typology of Codeswitching. Linguistics, 581–618 (1980)
7. Verschueren, J.: Understanding Pragmatics, Arnold (1999)

Powered Grid Scheduling by Ant Algorithm

Feifei Liu[1,2] and Xiaoshe Dong[1]

[1] Department of Computer Science and Technology, Xi'an Jiaotong University, Xi'an, China
[2] Engineering University of Armed Police Force of China, Xi'an, China
liuff.wj@stu.xjtu.edu.cn, xsdong@mail.xjtu.edu.cn

Abstract. The grid environment has the characteristics of distribution, dynamic and heterogeneous. How to schedule jobs is one of most important issues of computing grid. To address this problem, this paper presents a novel reputation-based ant algorithm in the computing grid scheduling. The reputation is a comprehensive measure and used to reflect the ability of compute node or network for a long-running stability. The reputation-based ant algorithm introduce reputation index both in tasks and resources to the local and global pheromone. Experimental results show that the reputation-based ant algorithm outperforms Round Robin, Min-Min, and reputation based Min-Min in makespan and system load balancing.

Keywords: improved ACS, reputation, grid scheduling, SimGrid.

1   Introduction

Grid computing is now being researched by more institutions and scholars as a new dynamic, multi-virtual organization domain heterogeneous resource sharing and collaboration processing technology. At the same time, grid scheduling is getting more attention as an important function module of the grid.

Traditional algorithms of distributed grid scheduling aim to better complete task-resource mapping. The dynamic, distributed and heterogeneous characteristics of the grid environment decide that the instability and non-credit factors in the grid environment and these factors may be caused by the inherent instability of system resources or behavior of some malicious viruses. Therefore, paying more attention to reputation of physical resource in job scheduling will be brings better quality of grid service.

In this article, we regard the reputation of running environment which user's job expect as a request QoS, and deal the reputation mapping between user job and physical resource. An improved ant algorithm is being used to solve the resource allocation problem with the special QoS constrains. The reputation mapping relation is reflected in the pheromone initialization and update process of ant colony algorithm.

This paper is organized as follows. Section 2 will discuss the background and related work of grid job scheduling. In section 3, an improved ant scheduling algorithm is introduced. In section 4, we present and discuss the simulation architecture and show the experimental results. We conclude this study in section 5.

D.-S. Huang et al. (Eds.): ICIC 2012, LNCS 7389, pp. 586–593, 2012.
© Springer-Verlag Berlin Heidelberg 2012

2   Related Works

Early grid scheduling system often focused on how best to complete the mapping from tasks to resources. But with the continuous development of grid technology, as well as the proposed Open Grid Services Architecture, Quality of Service (QoS) has became an important factor which should be considered in job scheduling. Generally speaking, QoS is a comprehensive metric to measure user's satisfaction with the service. It describes certain performance characteristics of a service and these performance features are visible to the user.

As an earlier research, Qos was firstly introduced into grid job scheduling in [1]. The work presented by Zhiang et al. [2] considers multi-dimensional QoS indicators in the grid scheduling and proposes user satisfaction function with more detailed description. There are also a number of works which are based on special QoS metrics. These indicators usually have a specific influence to efficient completion of user jobs. In [3], based on the grid trust model and trust utilization functions, a computational service scheduling problem based on trust QoS enhancement is proposed. Although this method can effectively reduce the risk of failure of user job execution, complicated relationship of mutual trust model also brings more complexity for resource allocation and job scheduling.

Since various features of the grid itself, the traditional grid scheduling more likely use heuristics algorithm to solve the resource allocation problem. The related work introduced by [4-6] indicate that heuristic algorithm also applies to the Market-based job scheduling strategy. With the introduction of the concept of QoS, the authors of [7] not only proposes a hierarchical structure of grid QoS, but also apply these researches to improve on Min-Min algorithm.

Ant algorithm was first introduced by Dorigo in [8]. The work presented by Z. Xu et al. [9] apply ant colony algorithm to the grid scheduling and consider the reward and punishment factors in pheromone update. A Balanced Ant Colony Optimization (BACO) algorithm presented in [10] has a major contributions to balance the entire system load while trying to minimize the makespan of a given set of jobs. The scheme proposed in [11] is more focused on giving better throughput with a controlled cost. In [12], a modified ant algorithm combined with local search is proposed which takes into consideration the free time of the resources and the execution time of the jobs to achieve better resource utilization. Unlike those methods mentioned above, we will focus on how to solve the job schedulig based reputation by ant algorithm in this article.

3   Improved Ant Algorithm

3.1   The Scheduling Model

Resources in grid environment are heterogeneous, which includes not only high-performance cluster but also the ordinary household hosts, and most of them run jobs which are communication-intensive or computing-intensive.

This study abstracts and encapsulates a variety of heterogeneous resources in a grid environment. Grid physical resources can be simplified by $G = < Node_i, Network >$, where $node_i, i \in 1..M$ is the set of servers hosts. Each node can be characterized by a set of independent parameters, including of computing power and the reputation value, which can express as:

$$Node = < Computation\ Power, \mathrm{Re}\ putation > \tag{1}$$

The core attribute of nodes is the computing power. In addition, the long-running stability is also an important attribute of nodes, which may be reflected in **reputation of node.**

Based on the definition of the node, we can describe the grid links and networks with the aim of statistical analysis of communication overhead. In essence, the network is a weighted undirected graph, each vertex represents a mesh node, each edge represents a data link, and the weight of edge is depended on many factors (e.g. bandwidth and latency). We can describe the network as:

$$Network = \{(Node_i, Node_j, Bandwidth, Latency, \mathrm{Re}\ putation) \mid i, j \in M\} \tag{2}$$

where **reputation of network** is a important value which reflect the stability of network between distinct nodes.

In the Service-Oriented grid environment, how to improve the QoS of the grid and satisfy the user's request is much more urgent to be resolve. Generally, the user only concerns whether their computation goals have been achieved. From user's viewpoint, how to manage the gird and complete the job is insignificant. At the same time, some specific requirements for the QoS are also requested by user. So user's request can be described like the formula (3).

$$Request = < Task, Input, QoS_{request} > \tag{3}$$

In our study, we take the reputation as the mainly QoS indicator. That is to say, users of grid hope that their tasks are dispatched to those nodes owned a reputation value which exceeds a baseline. So we can describe the QoS as:

$$Qos = < \mathrm{Re}\ putation_{node}, \mathrm{Re}\ putation_{network} > \tag{4}$$

3.2   The Algorithm Improvement

Let $R = \{r_1, r_2...r_m\}$ denotes M heterogeneous hosts in the grid and $T = \{t_1, t_2...t_n\}$ denotes N independent tasks in the grid. In order to apply ant algorithm, we should first map the ant colony to scheduling systems. Ant is on behalf of a mapping solution of task-resource. An ant goes from the task i to node j denotes that task i is scheduled to service host j and executed on it. The ACS algorithm based reputation is described in Table 1.

Table 1. The ACS algorithm based reputation

```
Initiate  α,β,ρ,C_e,C_p
While NC < NC_MAX do
    Initialize Allowed
    for all ants  a_k in  Allowed_k
        for all tasks  t_i in T,
            find the largest value  τ_ij  in  τ
            update the local pheromone
        end for
        Wait for all the tasks to finish, when t_i finishes
        update the local pheromone
        update the reputation of node
    end for
update the global pheromone
end while
```

In this process, pheromone denotes the weight of the path from task to resource. The higher the pheromone, the greater the weight, also the possibility of the corresponding allocation solution is higher. The pheromone matrix is as follows:

$$\tau = \left\{ \begin{matrix} \tau_{11} & \cdots & \pi_{1m} \\ \vdots & & \vdots \\ \tau_{n1} & \cdots & \tau_{nm} \end{matrix} \right\} \tag{5}$$

where every τ_{ij} represents the degree of match between the node i and task j. The initial value of τ_{ij} can be calculated by:

$$\tau_{ij}^{init} = (TT_{ij} + WT_{ij} + ET_{ij} + \text{Re } putation\_Match)^{-1} \tag{6}$$

In formula (6), ET_{ij} represents the expected execution time for task t_i to execute on node h_j given that h_j is idle, TT_{ij} denotes the expected transmission time for task t_i to transmit to node h_j, and WT_{ij} is the time that task t_i need to wait before node h_j begin to execute it. ET_{ij} and TT_{ij} can be calculated respectively by:

$$ET_{ij} = \frac{Computation\ size}{Computing\ Power} \tag{7}$$

$$TT_{ij} = \frac{Communication\ size}{Bandwidth} + Latency \tag{8}$$

The smaller ET_{ij}, TT_{ij}, and WT_{ij} can be, the higher possibility allocating the task to this resource is. Similarly, the task is hopeful to assigned to host in which the task and resource reputation value is more closer.

Stochastic transmission probability P_{ij}^k denotes the probability of the kth ant scheduling job i to host j, P_{ij}^k can be calculate by formula (9), where $allowed_k$ denotes the tasks that kth ant can schedule next step.

$$P_{ij}^k = \begin{cases} \dfrac{(\tau_{ij})^\alpha (\eta_{ij})^\beta}{\sum\limits_{r=1}^{m} (\tau_{ir})^\alpha (\eta_{ir})^\beta} , i \in allowed_k \\ 0 \qquad\qquad\qquad otherwise \end{cases} \qquad (9)$$

Visibility factor η_{ij} is given by:

$$\eta_{ij} = (\frac{1}{Bandwidth} + Latency + \frac{1}{Power} + \mathrm{Re}\, putation_j)^{-1} \qquad (10)$$

The rule of job scheduling and state transfer is as follows: the kth ant which just scheduled job s selects job t that will be scheduled next step and the corresponding execution node r through the rule given in (11).

$$<t,r> = \begin{cases} \arg\max\limits_{\substack{t \in allowed_k, \\ r \in \{1,2...m\}}} \{[\tau_{tr}]^\alpha \cdot [\eta_{tr}]^\beta\} & if\ q < q_0 \\ <T,R> & otherwise\ use\ (9) \end{cases} \qquad (11)$$

In formula (11), q is a random number uniformly distributed in [0,1], q_0 is a parameter ($0 \le q_0 \le 1$), $< T, R >$ is a random variable selected from the probability distribution given in (9).

Local pheromone update formula is as $\tau_{ij} = (1-\rho)\tau_{ij} + \Delta\tau_{ij}, 0 < \rho < 1$. After the job is scheduled and before it is executed, update the pheromone by:

$$\Delta\tau_{ij} = K = (TT_{ij} + ET_{ij} + \mathrm{Re}\, putation\_Match)^{-1} \qquad (12)$$

After the task execution succeed, the reputation value of the node would update, meanwhile, the pheromone would update as follows: $\Delta\tau_{ij} = C_e \times K$, which C_e is the encourage factor. If job execution failed, the reputation value of the node would update, meanwhile, the pheromone would update as follows: $\Delta\tau_{ij} = C_p \times K$, which C_p is punish factor.

4   Experimental Test

Based on SimGrid simulation library, this paper developed Round-Robin, Min-Min, Reputation based Min-Min, and Reputation based ACS scheduling algorithm. The ant algorithm improved in this paper is based on the implementation of the Ant Colony

System. ACS uses the pseudo-random-proportional rule to replace state transition rule for decreasing computation time of selecting paths and update the pheromone on the optimal path only. It is proved that it helps ants search the optimal path.

4.1 Simulation Architecture

There are many kind of projects related to grid simulator, such as SimGrid, GridSim, ChicSim, GSSIM, Alea etc. They are especially powerful for development and testing of new grid scheduling algorithms. SimGrid[13] is a toolkit that provides core functionalities for the simulation of distributed applications in heterogeneous distributed environments. The specific goal of the project is to facilitate research in the area of distributed and parallel application scheduling on distributed computing platforms ranging from simple network of workstations to Computational Grids. It provides high-level user interfaces for researchers to use in either C or Java. It seeks a compromise between execution speed and simulation accuracy because the main performance matrix in the gird domain is makespan of every application. Our experiment is based on SimGrid.

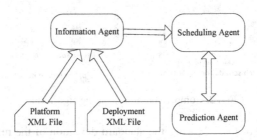

Fig. 1. Simulation Architecture

The Simulation Architecture is illustrated in Fig.1. It consists of a scheduling agent that implements different scheduling strategies in the computing grid. The deployment and platform scheme of the system which include tasks, resources and their properties are all loaded form the XML files. Information Agent is responsible for collecting these data and storage them, which is very useful for the scheduling agent to retrieve. It is the prediction agent's responsible to provide TT, WT and CT.

4.2 Implementation Environment

Job attributes: We suppose that there are 200 independent tasks. We only consider one-dimensional QoS: reputation. Reputation is a random value in [0, 1]. The task computing size is a random value in [0, 100000000], the task transmission size is also a random valued in [0,100]. This fully guarantees the heterogeneity of tasks.

   Resource attributes: We suppose that there are 15 resources. Resource computing power, network bandwidth and delay value save as a configuration file that comes with the simulator. The reputation is also a random value in [0, 1].

Table 2. Simulation Parameter

PARAMETER	NUM OF ANTS	NC_MAX	α	β	ρ	C_e	C_p
VALUES	20	200	0.5	0.5	0.1	1.1	0.8

Parameter Setup: There are lots of parameters to setup in the simulator. We set them just as [11] in Table 2.

4.3   Experiment Results

Makespan: As can be seen from the Fig. 2(a), RACS performs the best.

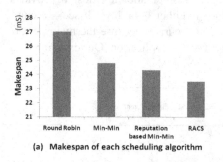 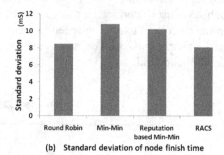

(a) Makespan of each scheduling algorithm   (b)   Standard deviation of node finish time

Fig. 2. Comparison of scheduling algorithms

Standard deviation of load: We use the standard deviation of the resource finish time as load balance metrics, which is defined as follows:

$$\delta = \sqrt{\frac{1}{N} \sum_{i=1}^{M} (f(t)_i - \overline{f(t)})^2}$$ (13)

where $f(t)_i$ it the time when node i finishes all the tasks that dispatched to it, $\overline{f(t)}$ is the average finish time of all nodes in the grid system.

If the standard deviation value of an algorithm is small, it means that the difference of each load is small. The small standard deviation tells that the load of the entire system is balanced. The lower value the standard deviation has, the more load balanced the system is.

As can be seen from Fig. 2(b), Reputation based Min-Min performs well in terms of makespan, but it is not good in terms of load balancing. But the reputation based ACS outperforms others three algorithm both in makespan and load balancing.

5   Conclusion And Future Work

In this paper, we proposed a novel QoS metric, named reputation. In contract, value of Reputation is a simplified description about the stability and credibility of variety of grid resources. In the context of the independent batch mode job scheduling problems, we analyzed the current ant colony system algorithm, combined it with the reputation mechanism, and proposed an improved ant system algorithm. Its main contribution is introducing reputation mechanism to the pheromone change, taking into account both the node computing power and the network link capacity. We use SimGrid simulator to simulate the heterogeneous resources and algorithm. The experiment results show that the reputation based ACS not only outperforms other three algorithms in makespan, but also in system load balancing.

This study aimed at the independent task scheduling, in a real grid environment, there are always communication and data dependencies between tasks, meanwhile, due to the dynamic nature of grid resources, the availability of the resource increasing cause for concern. Our future research will focus on resource availability and relative task scheduling in the complex grid environment.

References

1. He, X.S., Sun, X.H., Laszewski, G.V.: QoS Guided Min-min Heuristic for Grid Task Scheduling. Journal of Computer Science and Technology 18, 442–451 (2003)
2. Wu, Z.A., Luo, J.Z., Dong, F.: Measurement Model of Grid QoS and Multi-dimensional QoS Scheduling. In: Shen, W., Luo, J., Lin, Z., Barthès, J.-P.A., Hao, Q. (eds.) CSCWD. LNCS, vol. 4402, pp. 509–519. Springer, Heidelberg (2007)
3. Zhang, W.Z., Fang, B.X., et al.: A Trust-QoS Enhanced Grid Service Scheduling. Chinese Journal of Computers 7, 1157–1166 (2006)
4. Sonmez, O.O., Gursoy, A.: A Novel Economic-Based Scheduling Heuristic for Computational Grids. International Journal of High Performance Computing Applications 21(1), 21–29 (2007)
5. Kumar, S., Dutta, K., Mookerjee, V.: Maximizing Business Value by Optimal Assignment of Jobs to Resources in Grid Computing. European Journal of Operational Research 194, 856–872 (2009)
6. Vanderstera, D.C., Dimopoulosb, N.J., et al.: Resource Allocation on Computational Grids Using a Utility Model and the Knapsack Problem. Future Generation Computer Systems 25(1), 35–50 (2009)
7. Wu, Z.A., Luo, J.Z., Song, A.B.: QoS-Based Grid Resource Management. Journal of Software 17(11), 2264–2276 (2006)
8. Dorigo, M., Maniezzo, V., Colorni, A.: The Ant System: Optimization by a Colony of Cooperating Agents. IEEE Transactions on Systems, Man, and Cybernetics-Part B 26(1), 29–41 (1996)
9. Xu, Z., Hou, X., Sun, J.: Ant Algorithm-Based Task Scheduling in Grid Computing. In: Canadian Conference on Electrical and Computer Engineering, IEEE CCECE (2003)
10. Chang, R.S., Changa, J.S., Lina, P.S.: An Ant Algorithm for Balanced Job Scheduling in Grids. Future Generation Computer Systems 25, 20–27 (2009)
11. Sathish, K., Reddy, A.R.M.: Enhanced Ant Algorithm Based Load Balanced Task Scheduling in Grid Computing. International Journal of Computer Science and Network Security 8(10), 219–223 (2008)
12. Kousalya, K., Balasubramanie, P.: Ant Algorithm for Grid Scheduling Powered by Local Search. International Journal of Open Problems Computational Mathematics 1, 223–240 (2008)
13. Casanova, H., Legrand, A., Quinson, M.: SimGrid: A Generic Framework for Large-Scale Distributed Experiments. In: Tenth International Conference on Computer Modeling and Simulation (uksim 2008) 126–131 (2008)

An Efficient Palmprint Based Recognition System Using 1D-DCT Features

G.S. Badrinath, Kamlesh Tiwari, and Phalguni Gupta

Department of Computer Science and Engineering,
Indian Institute of Technology Kanpur
Kanpur 208016, India
{badri,ktiwari,pg}@cse.iitk.ac.in

Abstract. This paper makes use of one dimensional Discrete Cosine Transform (DCT) to design an efficient palmprint based recognition system. It extracts the palmprint from the hand images which are acquired using a flat bed scanner at low resolution. It uses new techniques to correct the non-uniform brightness of the palmprint and to extract features using difference of 1D-DCT coefficients of overlapping rectangular blocks of variable size and variable orientation. Features of two palmprints are matched using Hamming distance while nearest neighbor approach is used for classification. The system has been tested on three databases, viz. IITK, CASIA and PolyU databases and is found to be better than the well known palmprint systems.

Keywords: Biometrics, Image Enhancement, Palmprint, 1D-DCT, Decidability index, ROC curve.

1 Introduction

Biometrics helps to provide the identity of the user based on his/her physiological or behavioral characteristics. Palmprint is the region between wrist and fingers and has features like principle lines, wrinkles, datum points, delta point, ridges, minutiae points, singular points and texture pattern which can be considered as biometric characteristics [17]. Compared to other biometric systems, palmprint has many advantages. Features of the human hand are relatively stable and unique. It needs very less co-operation from users for data acquisition. Collection of data is non-intrusive. Low cost devices are sufficient to acquire good quality of data. The system uses low resolution images but provides high accuracy. Compared to the fingerprint, a palm provides larger surface area so that more features can be extracted. Because of the use of lower resolution imaging sensor to acquire palmprint, the computation is much faster at the preprocessing and feature extraction stages. System based on hand features is found to be most acceptable to users. Furthermore, palmprint also serves as a reliable human identifier because the print patterns are found to be unique even in mono-zygotic twins [9].

There exist palmprint recognition systems which are based on Gabor filter [16], Ordinal code [14], Local binary pattern [15], 2D-DCT [6,7]. The system in [6] retains less than 50% of 2D-DCT coefficients of palmprint image as its features and it has

D.-S. Huang et al. (Eds.): ICIC 2012, LNCS 7389, pp. 594–601, 2012.
© Springer-Verlag Berlin Heidelberg 2012

used partial PolyU database. For the identification mode, this system performs with CRR of 99%. However, the recognition rates on this partial database are not promising.

This paper proposes an efficient technique to extract palmprint features. The proposed technique to generate palmprint features uses the variation of 1D-DCT coefficients of local rectangular blocks of variable size and variable orientation to represent the palmprint. The variable size rectangular blocks are oriented at 45° for capturing characteristics of both horizontal and vertical direction. The binary features of query and enrolled palmprints are matched using Hamming distance. Classification is done using nearest neighborhood approach. The performance of the system is studied using IITK, CASIA [1] and PolyU [2] databases where the hand images are obtained under different environmental circumstances.

The rest of the paper is organized as follows: Section 2 describes the use of the Discrete Cosine Transform (DCT) in the proposed system to extract features of the palmprint. Next section presents the proposed system. Performance of the system has been discussed in Section 4. Conclusion is presented in the last section.

2   Discrete Cosine Transform

The objective of any palmprint feature extraction technique is to obtain good inter-class separation in minimum time. Features should be obtained from the extracted, enhanced and normalized palmprint for recognition. Discrete Cosine Transform (DCT) [5] has variance distribution. Further, DCT is also source independent, which means that the DCT basis vectors are fixed and independent of data. Thus, DCT is more suitable for systems where the reference database is dynamic over the life of the system. Moreover, DCT has increased tolerance to variation of illumination [13]. One can use DCT to extract palmprint features.

DCT [5] is a real valued transformation. There exist variants of DCT such as DCT-I to DCT-VIII. But DCT-II has certain advantages over other DCT types. It can be applied for real sequence of any positive length. It has strong energy compaction property [12]. It has been used efficiently for image compression in JPEG, MPEG and for pattern recognition problems like biometrics [8, 11]. As a result, DCT-II has been considered for feature extraction. In [3], it is shown that the 1D-DCT coefficients, CT_0, CT_1, CT_2 \ldots CT_{N-1} of length N using DCT-II [1] can be generated as

$$CT_n = \frac{2}{N} w(k) \sum_{n=0}^{N-1} x_n \cos\left[\frac{\pi}{N}\left(n + \frac{1}{2}\right)k\right], \qquad k = 0, \dots, N - 1 \qquad (1)$$

And

$$x_n = \sum_{k=0}^{N-1} w(k) CT_k \cos\left[\frac{\pi}{N}\left(n + \frac{1}{2}\right)k\right], \qquad k = 0, \dots, N - 1 \qquad (2)$$

---

[1]   Henceforth, DCT-II is refereed as DCT.

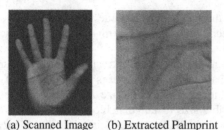

(a) Scanned Image   (b) Extracted Palmprint   (c) Enhanced Palmprint

Fig. 1. Palmprint Extraction Process

where x_n is a real sequence of length N and

$$w(k) = \begin{cases} \dfrac{1}{\sqrt{2}} & if\ k = 0 \\ 1 & Otherwise \end{cases} \tag{3}$$

3   Proposed System

This section discusses the proposed palmprint based system. Like other biometrics, it consists of five major tasks. Hand image has been acquired with the help of a low cost flat bed scanner. Desired area of palm, termed as palmprint or region of interest (ROI), has been extracted from the hand image and subsequently enhanced to improve its texture as described in [4]. The enhanced palmprint is segmented into oriented   overlapping rectangular blocks and features are extracted using the difference of 1D-DCT coefficients of adjacent blocks. Features of two palmprints are matched using Hamming distance and decision is made based on a threshold.

3.1   Feature Extraction

Conventional image transformation is done by applying 2D-DCT on whole image. But if the image has lot of textures or randomness, the transformation of larger image results in severe ringing artifacts around edges and the 2D-DCT coefficients may not represent the image effectively [10]. Since the palmprint is texture in nature, the 2D-DCT coefficients obtained from transforming whole palmprint may not represent properly and it results in poor performance. This paper considers a non-conventional approach of applying 1D-DCT on rectangular blocks of variable size and variable orientation for extracting features. This non-conventional method can take the advantages of varying the block size either to large size or small size and varying the orientation to any direction according to the regional properties such that performance is optimized. Further, the system based on fixed block is a special case of the proposed non-conventional method of rectangular blocks.

The enhanced palmprint is segmented into overlapped rectangular blocks of variable size with orientation at particular direction θ. The oriented rectangular block is of size $W \times H$ where W is the width and H is the height of the block. The rectangular block as shown in Fig. 2 is the basic structure used to extract features of the palmprint in the proposed system. The figure also shows the schematic diagram for segmenting the palmprint and other parameters that can be used for feature extraction.

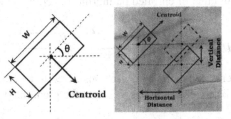

Fig. 2. Overlapping Rectangular Blocks at an angle θ and other parameters

The segmented block is averaged across its height to obtain 1D-intensity signal of width W. Formally, the rectangular block R of width W and height H is averaged across width to obtain 1D-intensity signal $\mathbf{R}$ of width W as

$$\mathbf{R}_j = \sum_{k=1}^{H} R_{j,k} \qquad j = 1,2,\dots,W \tag{4}$$

Averaging helps to smooth the image; thus, it reduces the noise. This 1D intensity signal $\mathbf{R}$ is windowed using Hanning window and then is subjected to 1D-DCT. Since the signal type of the palmprint is not known, Hanning window is chosen. The difference of 1D-DCT coefficients from each vertically adjacent block is computed. That is, if $\mathbf{R1}$ and $\mathbf{R2}$ are the 1D-DCT coefficients of vertically adjacent rectangular blocks of width W, the difference of 1D-DCT coefficients $\mathbf{D}$ of these vertically adjacent blocks are computed as

$$D_j = \mathbf{R1}_j - \mathbf{R2}_j \qquad j =1,2,\dots,W \tag{5}$$

Further, the difference of 1D-DCT coefficients $\mathbf{D}$ obtained from each of the vertically adjacent rectangular blocks are binarised using zero crossing to obtain a binary vector $\mathbf{B}$ as

$$\mathbf{B}_j = \begin{cases} 1 & if\ \mathbf{D}_j > 0 \\ 1 & Otherwise \end{cases} \qquad j = 1,2,\dots,W \tag{6}$$

The resulted W-bit binary vector $\mathbf{B}$ is considered as sub-feature vector and sub-feature vectors from all adjacent blocks create the feature vector of palmprint.

3.2   Matching

The extracted feature vector from the query palmprint is matched with that of each enrolled palmprint stored in the database for verification or identification. In this paper, nearest neighbor approach is used for matching. If both the query and the enrolled palmprints are from same class, then it is a genuine (Non-False) match; otherwise, it is a imposer (False) match. The distance between the feature vectors is computed using Hamming distance.

Formally, let $L = \{l_{i,j}\}$ and $E = \{e_{i,j}\}$ be binarised matrix of sub-feature vectors from query and enrolled palmprints. Hamming distance between feature vector of query palmprint L and that of enrolled palmprint E is computed as

$$HD = \frac{\sum_{ni=1}^{m} \sum_{j=1}^{n} \left(\sum_{k=1}^{W} \frac{l_{i,j}(k) \otimes e_{i,j}(k)}{W} \right)}{m \times n} \tag{7}$$

The query and the enrolled palmprints are considered to be matched if the normalized Hamming distance between L and E is less than the predefined threshold Th.

4   Experimental Results

The proposed system has been tested on three hand image databases. These databases are obtained from The Indian Institute of Technology Kanpur (IITK), The Chinese Academy of Sciences Institute of Automation (CASIA) [2], and The Hong Kong Polytechnic University (PolyU) [1].

Database of the IITK consists of 549 hand images obtained from 150 users corresponding to 183 different palms with the help of low cost flat bed scanner at spatial resolution of 200 dpi with 256 gray levels. Three images per palm are collected in the same session. The palmprint region is extracted and normalized to size of the 448×448 pixels and then enhanced. One image per palm is considered for training and remaining two images are used for testing. The CASIA database [1] contains 5,239 hand images captured from 301 subjects corresponding to 602 palms. For each subject, around 8 images from left hand and 8 images from right hand are collected. All images collected using CMOS and 256 gray-levels. The device is pegs free; so user is free to place his hand facing the scanner. The palmprint region is extracted and normalised to size of 256×256 pixels and then enhanced. Two images per palm are considered for training while remaining images are used for testing. The PolyU database [2] consists of 7752 grayscale images from 193 users corresponding to 386 different palms. Around 17 images per palm are collected in two sessions. The images are collected using CCD at spatial resolution of 75 dots per inch and 256 gray-levels. Images are captured placing pegs. Palmprints are normalised to size of 176×176 pixels and then enhanced. The database is classified into training set and testing set. Six images per palm are considered for training and remaining images are used for testing.

The performance of the proposed system is measured for both identification and verification mode. For identification mode, the Correct Recognition Rate (CRR) of the system is measured as

$$CRR = \frac{N_1}{N_2} \times 100 \tag{8}$$

Where N_1 denotes the number of correct (Non-False) recognition of palmprints and N_{21} is the total number of palmprints in the testing set. For verification mode Equal Error Rate (EER) is used to measure accuracy.

The performance of the proposed system is compared with the best known systems proposed in [6, 7, 14, 16] for all three databases. The systems in [6, 7] have used 2D-DCT to extract palmprint features. It is reported that systems in [6] and [7] are evaluated using 500 images and 198 images of PolyU database respectively. However, experiments for evaluating the performance of the systems in [6, 7, 14, 16] are carried out on the same set of extracted and enhanced palmprint images used for the proposed system.

CRR and ERR of all the systems are given in Table 1. It can be observed that CRR and EER of the proposed system are better than the best known systems [6, 7, 14, 16]. Further, the proposed system performs better with IITK database which has the better quality images as compared to CASIA and PolyU databases. It should also be noted that it performs with 100% CRR for CASIA which has the worst quality images among the three databases.

Table 1. Performance Comparison of the Proposed System

	IITK			CASIA			PolyU		
	CRR%	ERR%	DI	CRR%	ERR%	DI	CRR%	ERR%	DI
Palm-code [16]	100.00	5.210	1.800	99.619	3.673	3.073	99.916	0.533	5.580
Ordinal code [14]	100.00	1.188	2.129	99.843	1.754	3.396	100.00	0.070	6.678
Kipsang et.al.,[6]	92.769	10.234	0.9281	95.326	6.768	0.644	85.84	20.11	0.798
Dale et. al., [7]	89.097	1.025	1.025	93.544	7.712	2.563	90.90	11.90	2.269
Proposed	**100.00**	**6.490**	**6.490**	**100.0**	**1.004**	**5.526**	**100.0**	**0.033**	**7.750**

The Receiver Operating Characteristics (ROC) curve which is used as graphical depiction of performance of a system for verification mode is generated by plotting FAR against FRR at different thresholds. Fig 4 shows the ROC curves of various systems for IITK, CASIA and PolyU databases. It is found to have minimum FRR at low FAR. Hence, from Table. 1 and Fig. 3, it can be inferred that the proposed system performs better than the systems in [6, 7, 14, 16].

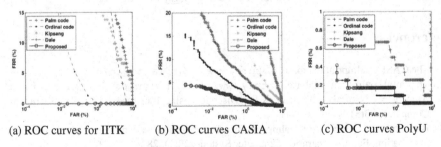

(a) ROC curves for IITK   (b) ROC curves CASIA   (c) ROC curves PolyU

Fig. 3. Comparing ROC Curves

Since EER is near to zero, significant conclusion cannot be drawn. So, decidability index DI which gives the measure of the separability of genuine and imposter matching scores, is used. Decidability index is defined as

$$DI = \frac{|\mu_1 - \mu_2|}{\sqrt{\frac{\sigma_1^2 + \sigma_2^2}{2}}} \tag{9}$$

where μ_1 and μ_2 are the means, and σ_1 and σ_2 are standard deviations of genuine and imposter scores respectively. DIs for various systems for all three databases are shown in Table 1. It can be observed that the proposed system has the highest DI compared to other systems for the corresponding databases.

5   Conclusion

This paper has presented an palmprint based recognition system using 1D-DCT coefficients of local region. Palmprints are extracted from hand images with the help of low cost flat bed scanner at IITK, which is free of constraints (pegs). So user is free to place hand independent of orientation and translation on scanner bed. Extracted palmprint is found to be robust to translation and orientation of placement of hand on scanner bed. A non-conventional technique to extract palmprint features based on variation of 1D-DCT coefficients of adjacent rectangular blocks of variable size and variable orientation has been proposed and experimentally evaluated. Hamming distance is used as classifier. The system has been tested using IITK database of 549, CASIA database of 5239 and PolyU database of 7752 hand images, which are acquired using flat bed scanner, CCD and CMOS respectively.

Experimentally optimum parameters of the proposed system are determined to achieve CRR of 100% for all three databases. Also it is observed that EERs are 0.00%, 1.00% and 0.03% for IITK, CASIA and PolyU databases respectively. The system has been compared with the best known systems in [14, 16] and those in [6, 7] which have used 2D-DCT to extract palmprint features. The proposed system is found to perform better than all these systems.

Acknowledgement. Authors acknowledge the support provided by the Department of Information Technology, Government of India, to carry out this work.

References

1. The CASIA palmprint database, http://www.cbsr.ia.ac.cn/
2. The PolyU palmprint database, http://www.comp.polyu.edu.hk/biometrics
3. Ahmed, N., Natarajan, T., Rao, K.R.: Discrete cosine transform. IEEE Transaction on Computers 23(3), 90–93 (1974)
4. Badrinath, G., Gupta, P.: Palmprint based recognition system using phase diference information. Future Generation Computer Systems 28(1), 287–305 (2012)
5. Britanak, V., Yip, P.C., Rao, K.R.: Discrete Cosine and Sine Transforms: General Properties, Fast Algorithms and Integer Approximations. Academic Press (2006)
6. Choge, H.K., Oyama, T., Karungaru, S., Tsuge, S., Fukumi, M.: Palmprint Recognition Based on Local DCT Feature Extraction. In: Leung, C.S., Lee, M., Chan, J.H. (eds.) ICONIP 2009, Part I. LNCS, vol. 5863, pp. 639–648. Springer, Heidelberg (2009)

7. Dale, M.P., Joshi, M.A., Gilda, N.: Texture based palmprint identification using DCT features. In: International Conference on Advances in Pattern Recognition, pp. 221–224 (2009)
8. Hafed, Z.M., Levine, M.D.: Face recognition using the discrete cosine transform. International Journal of Computer Vision 43(3), 167–188 (2001)
9. Kong, A., Zhang, D., Lu, G.: A study of identical twins' palmprints for personal authentication. Pattern Recognition 39(11), 2149–2156 (2006)
10. Mao, X., He, Y.: Image subjective quality with variable block size coding. In: International Video Coding and Video Processing Workshop, pp. 26–28 (2008)
11. Monro, D., Rakshit, S., Zhang, D.: DCT-based iris recognition. IEEE Transactions on Pattern Analysis and Machine Intelligence 29(4), 586–595 (2007)
12. Rao, K.R., Yip, P.: Discrete Cosine Transform: Algorithms, Advantages, Applications. Academic Press (1990)
13. Schwerin, B., Paliwal, K.: Local-DCT features for facial recognition. In: International Conference on Signal Processing and Communication Systems, pp. 1–6 (2008)
14. Sun, Z., Tan, T., Wang, Y., Li, S.Z.: Ordinal palmprint representation for personal identification. In: Computer Vision and Pattern Recognition, pp. 279–284 (2005)
15. Wang, X., Gong, H., Zhang, H., Li, B., Zhuang, Z.: Palmprint identification using boosting local binary pattern. In: International Conference on Pattern Recognition, pp. 503–506 (2006)
16. Zhang, D., Kong, A.W., You, J., Wong, M.: Online palmprint identification. IEEE Transactions on Pattern Analysis and Machine Intelligence 25(9), 1041–1050 (2003)
17. Zhang, D.D.: Palmprint Authentication. International Series on Biometrics. Springer-Verlag New York, Inc., Secaucus (2004)

An Efficient Algorithm for De-duplication
of Demographic Data

Vandana Dixit Kaushik[1], Amit Bendale[2], Aditya Nigam[2], and Phalguni Gupta[2]

[1] Department of Computer Science & Engineering, Harcourt Butler Technological Institute,
Kanpur 208002, India
vandanadixitk@yahoo.com
[2] Department of Computer Science & Engineering, Indian Institute of Technology Kanpur,
Kanpur 208016, India
{bendale,naditya,pg}@iitk.ac.in

Abstract. This paper proposes an efficient algorithm to de-duplicate based on demographic information which contains two name strings, viz. *GivenName* and *Surname*, of individuals. The algorithm consists of two stages - enrolment and de-duplication. In both stages, all name strings are reduced to generic name strings with the help of phonetic based reduction rules. Thus there may be several name strings having same generic name and also there may be many individuals having the same name. The generic name with all name strings and their Ids forms a bin. At the enrolment stage, a database with demographic information is efficiently created which is an array of bins and each bin is a singly linked list. At the de-duplication stage, name strings are reduced and all neighbouring bins of the reduced name strings are used to determine the top k best matches. In order to see the performance of the proposed algorithm, we have considered a large demographic database of 4,85,136 individuals. It has been observed that the phonetic reduction rules could reduce both the name strings by more than 90%. Experimental results reveal that there is very high hit rate against a low penetration rate.

Keywords: De-duplication, Demographic Data, Edit Distance, Levenshtein Distance, Phonetics.

1 Introduction

Individual recognition has gained great importance recently for personal as well as national security. Techniques of recognition can be broadly classified on the basis of the data used into two categories: biometric and demographic. Biometric data makes uses of physical characteristics like face, fingerprint, iris, etc. for recognition. Demographic data, on the other hand, is the data in the form of text/string which can be used for identification. For example, in electoral department, any voter can be identified uniquely with the help of his name, his father's name, his date of birth and his permanent residential address (all in the form of text), the combination of which is very less likely to be matching with the other voter.

D.-S. Huang et al. (Eds.): ICIC 2012, LNCS 7389, pp. 602–609, 2012.
© Springer-Verlag Berlin Heidelberg 2012

Demographic data in general refers to selected population characteristics as used in government, marketing or opinion research. Commonly-used demographics include race, age, income, disabilities, mobility (in terms of travel time to work or number of vehicles available), educational attainment, home ownership, employment status, and even location.

In computing, data de-duplication is a process to eliminate redundant data to improve storage utilization. In the de-duplication process, duplicate data is deleted, leaving only one copy of the data to be stored, along with references to the unique copy of data. In this paper, by demographic data of an individual, we mean a record consisting of two string fields viz. GivenName and Surname of the individual. demographic deduplication process determines whether there exists any set of records in the demographic database which are matched with the record of a query data within some tolerable range. For any new demographic data, a negative background search is performed to obtain all its close matches. To accomplish this task, every string (i.e. first name, and last name of the query individual) is considered to be found in the database if any of the following criteria satisfies.

- Identical strings in the database
- String having similar phonetics but difference with alphabets.
- String which can be constructed from new string by few transformations like insertion, deletion or substitution.

Data de-duplication has been approached previously using machine learning approaches ([6], [8], [9]). In these approaches, a labeled set of duplicates and non-duplicates is provided to learn the de-duplication function and is used to predict on testing set. These approaches depend heavily on providing an exhaustive and diverse set of duplicates and non-duplicates that bring out the subtlety of the de-duplication function.

In this paper, we propose an efficient algorithm for de-duplication of demographic data. We have considered GivenName and Surname as demographic data. Thus, beside Id of each individual, each record in the database contains two string fields: GivenName and Surname. The proposed de-duplication search is based on the approximate string matching like edit-distance. Next section discusses the method of edit distance which has been used in this paper for de-duplication. Section 3 describes the proposed algorithm. Experimental results have been analyzed in the next section. Finally conclusions are given in the last section.

2   Edit Distance

Edit-distance is used commonly for *approximate string matching* [3]. In information theory and computer science, the edit distance between two strings of characters is the number of operations required to transform one of them into the other. There are several ways to define an edit distance depending on which edit operations are allowed: replace, delete, insert, transpose [1, 2, 4, 5, 7]. In this paper we have used Levenshtein distance to measure the amount of difference between two sequences (i.e. an edit distance). The Levenshtein distance between two strings is the minimum number of edits that are required to transform one string into the other, with the help of edit operations like insertion, deletion, or substitution of a single character. This edit distance is normalized by dividing the maximum of the length of two name strings. Cost of computing this distance is proportional to the product of the two string lengths. Let A and B be two name strings of size m and n respectively. The complete algorithm which computes the normalized edit distance between A and B, NED(A,B), is described in Algorithm 1.

```
Algorithm 1: Normalised_Edit_Distance(A,B)
Require: String A    of length m, String B   of length n
Ensure: normalised Levenshtein Edit-distance of A    and B
1. Create 2D array d0..m, 0..n
// for all i and j, di,j  holds the Levenshtein distance
between the first i characters of A and the first j char-
acters of B, note that d   has (m + 1) x (n + 1) values
2. for i = 0 to m do
3.    di,0  := i   //distance of null substring of B
from A1..i
4. end for
5. for j = 0 to n do
6.    d0,j  := j   //distance of null substring of A from
B1..:j
7. end for
8. for j = 1 to n do
9.      for i = 1 to m do
10.        if Ai == Bj then
11.            di,j := di-1, j-1 //no editing required
12.        else
13.            di,j  :=   min(di-1,j, di,j-1, di-1,j-1)+1
//deletion, insertion, substitution
14.        end if
15.     end for
16. end for
17.  N_ED(A,B)= d_mn/max(|A|,|B|)
18. return N_ED(A,B)   // Normalized edit distance
```

3   Proposed Algorithm

In this section we have proposed an efficient demographic de-duplication algorithm. It consists of two major components: (1) Enrolment of demographic data of an individual in the database and (2) Searching the database for a query demographic data to find potential duplicates. By the term of potential duplicates, we mean the duplicates with some tolerable distance. Each demographic data is represented by two string fields viz. *GivenName* and *Surname*. It has been observed that one can write or pronounce a name in several ways and as a result, the database size becomes unnecessarily very large which leads to a very large amount of searching time. In order to reduce the search space in the database, each name is reduced to its generic name.

Phonetic reduction rules to convert a name string into a generic name string are based on the above said rules. There may be several name strings which are converted into a single generic name. A bin can be defined by this generic name which contains the set of all such name strings. Further, one can observe that there may be several individuals having same name but different Ids. So, one can represent the bin in the form of a singly linked list where each node contains a name string with different Ids.

Thus, a demographic database consists of an array of records where each record is a singly linked list with a header node marked by the reduced name. Each node in the linked list has several Id fields, one name string field and a link field. At enrolment stage, the reduced name of a name string is used to get the desired bin (i.e. singly linked list) where the name string is to be inserted. If there exists a field in a node of the list having the same name string then the Id of the name string is inserted in one of the empty Id fields; otherwise a new node with the name string and its Id is inserted in the appropriate position of the linked list. The entire scheme to reduce the name strings is depicted in Figure 1.

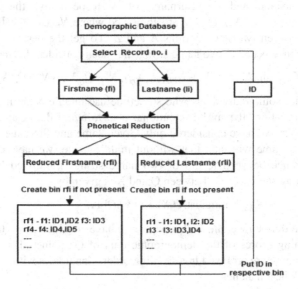

Fig. 1. Pre-processing - Creation of bins

At the time of verification or looking for the existence of duplicate entries with some tolerable range, each name string of the query demographic data (Given-Name(F), Surname(S)) is reduced and all bins having reduced names whose normalized edit distance(N_{ED}) from the reduced name of the name string of the query data is within the tolerable range are selected for further processing. It can be noted that edit distance enables to capture two similar strings that cannot be identified simply by using the phonetical rules of reduction. It can effectively detect all those duplications where there are some modifications in a string and do not conform to phonetical rules. All name strings in these selected bins having the normalized edit distance from the name string of the query data less than the tolerable range are considered as existence of the name string of the query data in the database. Ids of all these name strings are the probable candidates of the duplicates with some tolerable range.

These probable candidates are considered for obtaining top k best matches. Let Q and X be the demographic data of a query and of a probable candidate in the database respectively. One could have obtained sum of the normalized edit distances of Given Name and Surname between Q and X to find its matching score. But it has been observed that individuals need not be consistent in providing their demographic information. By inconsistency we mean, any individual may interchange the strings of his name. This may be due to make an attempt to fraud the environment or to make use of different styles of writing his name etc. In order to tackle such type of scenarios, normalized edit distances between every pair of name strings have been considered.

More clearly, let Q_{rf} and Q_{rs} be the generic form of GivenName and the Surname of a query individual. Further, it is assumed that X is a record in the database with whom the normalized edit distances are to be computed. If X_{rf} and X_{rs} be the generic form of GivenName and the Surname of X respectively, the $N_{ED}(Q_{rf},X_{rf})$, $N_{ED}(Q_{rs},X_{rs})$, $N_{ED}(Q_{rf},X_{rs})$, $N_{ED}(Q_{rs},X_{rf})$ are computed where $N_{ED}(A,B)$ is the normalized edit distance between two name strings A and B. To get the best match, minimum normalized distance between two name strings has been considered. That means,

$$minDist(Q,X) = min(N_{ED}(Q_{rf},X_{rf})+N_{ED}(Q_{rs},X_{rs}), N_{ED}(Q_{rf},X_{rs})+N_{ED}(Q_{rs},X_{rf})) \qquad (1)$$

Further, there are some individuals who sometime use middle name in providing demographic information. It may be intentional also to avoid duplication in the database. In this paper, we have considered only GivenName and Surname. But to distinguish between a name with and that without middle name, we have considered the factor of difference between length of two name strings multiplied by a constant. Thus the matching score SQ,X   between Q and X is given by

$$S_{Q,X} = minDist(Q,X) + w * \|Q|-|X\| \qquad (2)$$

where w is a pre-defined weight. In this paper, we have considered $w = 0.01$.

These matching scores of the demographic data of Q against all probable candidates in the database are arranged in ascending order. Candidates with top k scores are selected as top k best matches.

4   Experimental Results

In order to analyze the performance of the proposed algorithm, we have used our database consisting of 4,85,136 individuals, each having his GivenName and Surname as his demographic data. The database size is large enough and is sufficient to prove its robustness and efficiency. At the enrollment stage, name strings are reduced at two stages- one with GivenName and other with Surname. It has been seen that percentage of reduction is 90.94% in case of GivenName and is 94.40% in case of Surname string. Number of distinct bins containing reduced name strings is found to be 43941 and 27184 with respect to GivenName and Surname respectively. Further, maximum number of nodes in a bin of GivenName is 16 and that of Surname is 15.

Detailed statistics of both type of bins is given in Table 1 and Table 2. We have not shown those bins having negligible densities. These tables indicate that there is a lot of redundancies of names if exact spellings and phonetics are ignored.

Table 1. Bin Size Statistics

Name String	Number of Bins	Minimum Bin size	Maximum Bin size	Average Bin size	Standard deviation	Reduction Percentage
Given Name	43941	1	16	1.278	0.808	90.94%
Surname	27184	1	15	1.288	0.797	94.40%

Table 2. Bin Density

Bin Size	1	2	3	4	5
GivenName	36461	5090	1288	532	270
Bin Density %	83	12	3	1	1
Surname	22349	3222	9321	371	152
Bin Density %	82	12	3	1	1

To analyze the performance of the proposed de-duplication algorithm, 500 query records are chosen at random from above mentioned database to create gallery demographic database (DB). An edit-distance of i ($i = 1$ to 6) units is artificially induced in each full name (consisting of first and last name) of DB, and inserted in query dataset DB_i, alongwith the corresponding ID. In this manner, six query datasets DB_1, DB_2, DB_3, DB_4, DB_5, DB_6 are created, each having 500 queries. Thus, the dissimilarity of query datasets and gallery dataset DB increases from DB_1 to DB_6. A random sample of consonants from the name are randomly substituted or augmented with another consonant or deleted to make the modified name at the desired edit-distance from original one. Matching of queries from these size sets is carried out with DB and top-k matches for each query are reported.

If the correct ID for a query is found in the top-k matches, it is reported as a hit otherwise it is considered as a miss. The parameter t_1 is the relative tolerance threshold for normalized edit-distance between generic form of GivenName string of a query data and that of a record in the database, whereas t_2 is the threshold for Surname. For different tolerance thresholds t_1, t_2, hit rates for DB_1, DB_2, DB_3, DB_4, DB_5, DB_6 with $k = 5$ are found and shown in Figure 2 and Table 3. From the figure we can see that matching performance degrades with an increase in dissimilarity between query and gallery datasets.

Table 3. Hitrates with k = 5 for varying thresholds

Dataset	$t_1=0.4, t_2=0.4$	$t_1=0.5, t_2=0.5$	$t_1=0.6, t_2=0.6$
DB_1	99.0	99.0	98.0
DB_2	92.2	97.2	97.6
DB_3	82.6	92.8	95.2
DB_4	74.6	89.8	93.8
DB_5	66.8	83.0	90.4
DB_6	56.8	79.6	88.2

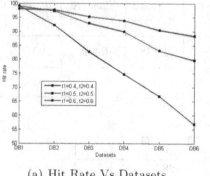   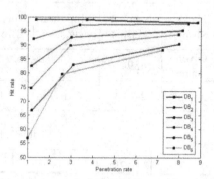

(a) Hit Rate Vs Datasets   (b) Penetration Rate Vs Hit Rate

Fig. 2. Performance Analysis

Penetration rate for a given query is the average fraction of total database (in %) scanned to get top k best matches. Penetration is directly proportional to threshold values t_1, t_2. It can be seen from the Figure 2(b) that increasing penetration rate generally increases the hit rate for all the datasets. It can be inferred that hit rate improves greatly by increasing thresholds, but at the cost of increased penetration.

5   Conclusion

This paper has proposed an efficient de-duplication algorithm for demographic data. It is a challenging task when two name strings with phonetical and other random differences (within the tolerable range) are referring to the same individual. Simple phonetic reduction and exact matching cannot identify potential duplicates due to their rigidity. Thus, this paper has reduced each name string to a generic name string so that the name string is put in a bin with the reduced name string. Finally, name string of a query data is searched in the corresponding bin and its neighbors to compute the top k best matches. We have used a large database of 4,85,136 demographic data. It has been observed that there is a reduction of 91.6% in case of GivenName and more than 94% reduction in case of Surname. In order to test the performance of the algorithm, a test set of 500 demographic data has been used. It has been found that it has achieved 94% average accuracy over DB_1 through DB_6 for top 5 best matches, using best thresholds.

Acknowledgement. Authors acknowledge the support provided by the Department of Information Technology, Government of India, to carry out this work.

References

1. Jaro, M.: Advances in Record-linkage Methodology as Applied to Matching the 1985 Census
2. Levenshtein, V.: Binary Codes Capable of Correcting Deletions, Insertions, and Reversals. Soviet Physics Doklady 10, 707–710 (1966)
3. Navarro, G.: A Guided Tour to Approximate String Matching. ACM Computing Surveys (CSUR) 33(1), 31–88 (2001)
4. Oommen, B., Loke, R.: Pattern Recognition of Strings with Substitutions, Insertions, Deletions and Generalized Transpositions. Pattern Recognition 30(5), 789–800 (1997)
5. Sankoff, D., Kruskal, J.B. (eds.): Time Warps, String Edits, and Macromolecules: The Theory and Practice of Sequence Comparison. Addison-Wesley Publication, Reading (1983)
6. Sarawagi, S., Bhamidipaty, A.: Interactive Deduplication Using Active Learning. In: Proceedings of the Eighth International Conference on Knowledge Discovery and Data Mining (ACM SIGKDD), pp. 269–278. ACM (2002)
7. Ukkonen, E.: On-line Construction of Suffix Trees. Algorithmica 14(3), 249–260 (1995)
8. Winkler, W.: Matching and Record Linkage. Wiley Online Library (1993)
9. Winkler, W.: The State of Record Linkage and Current Research Problems. Statistical Research Division, US Census Bureau, Citeseer (1999)

A Transportation Model with Interval Type-2 Fuzzy Demands and Supplies

Juan C. Figueroa-García[1] and Germán Hernández[2]

[1] Universidad Distrital Francisco José de Caldas, Bogotá – Colombia
[2] Universidad Nacional de Colombia, Sede Bogotá – Colombia
jcfigueroag@udistrital.edu.co,
gjhernandezp@gmail.com

Abstract. This paper presents a basic transportation model (TM) where its demands and supplies are defined as Interval Type-2 Fuzzy sets (IT2FS). This kind of constraints involves uncertainty to the membership function of a fuzzy set, so we called this model as Interval Type-2 Transportation Model (IT2TM). Using convex optimization techniques, a global solution of this problem can be-found. To do so, we define a general model for IT2TM and then we present an application example to illustrate how the algorithm works.

1 Introduction and Motivation

Transportation problems are interesting ones due to its applicability to logistics, production and assignment scenarios. There are some possible situations where the demands of the customers and availabilities of the suppliers are neither constant nor well defined. In this way, Fuzzy Linear Programming (FLP) appears as an important tool to find solutions of uncertain problems using Type-1 and Type-2 fuzzy sets.

Optimization with IT2FS is a recent approach that allows us to deal with linguistic uncertainty of a Type-1 fuzzy set. The main scope of this paper is to define a TM model with IT2FS, compute its optimal solutions by using classical algorithms, and finally use a Type-reduction strategy for obtaining a crisp solution of the problem.

Some fuzzy uncertain optimization methods have been proposed by Figueroa [1], [2], [3] and [4] who solved LP problems with uncertain fuzzy constraints. The problem addressed in this paper is the problem of having both uncertain demands and supplies, which can be extended to VRP, TSP and related problems.

The paper is divided as follows. In Section 1, a brief introduction and motivation is presented; Section 2 presents the IT2TM model; in Section 3, an application example is introduced and Section 4 presents the concluding remarks of the study.

2 The Uncertain TM Model

FLP problems have well known solution procedures. For further information about FLP problems, see Klir & Yuan in [5], Lai & Hwang in [6], Kacprzyk & Orlovski in [7], Pandiant in [8] and Zimmermann in [9] and [10]. In this way, the classical TM with crisp parameters is:

D.-S. Huang et al. (Eds.): ICIC 2012, LNCS 7389, pp. 610–617, 2012.
© Springer-Verlag Berlin Heidelberg 2012

$$\min z = c_{ij} x_{ij} \tag{1}$$

$$s.t.$$

$$\sum_{i=1}^{m} x_{ij} \leqslant a_j \quad \forall \ j \in \mathbb{N}_n \tag{2}$$

$$\sum_{j=1}^{n} x_{ij} \geqslant d_i \quad \forall \ i \in \mathbb{N}_m \tag{3}$$

Index Sets

$\mathbb{N}_m$ is the set of all "i" resources, $i \in \mathbb{N}_m$, $\mathbb{N}_m = 1, 2, \cdots, m$.

$\mathbb{N}_n$ is the set of all "j" products, $j \in \mathbb{N}_n$, $\mathbb{N}_n = 1, 2, \cdots, n$.

Decision Variables

x_{ij} = Quantity of product to be shipped from the supplier "i" to the customer "j".

Parameters

a_j = Quantity of product available by the supplier "j".

b_i = Quantity of product required by the customer "i".

In this problem, all their parameters are crisp sets, and they have no uncertainty sources involved. A common issue here is that the demands of the customers and the available units that each supplier has, can be represented by fuzzy numbers, so the TM becomes a more complex one.

Now, the uncertain TM called IT2TM uses two partial orders: $\precsim$ and $\succsim$ with different membership functions. This implies that its supplies and demands are defined as IT2FS, so an Interval Type-2 fuzzy parameter is defined as follows:

Definition 1 (Uncertain constraint). *Let* $(\tilde{b}_j)$ *a constraint of an IT2FLP. The membership function which represents the fuzzy space[1]* $\operatorname{supp}(\tilde{b}_j)$ *is:*

$$\tilde{b}_j = \int_{b_j \in \mathbb{R}} \left[\int_{u \in J_{b_j}} 1/u \right] / b_j, \ J_{b_j} \subseteq [0,1] \tag{4}$$

The difference between a Type-1 and Type-2 fuzzy sets lies in the Footprint of Uncertainty (FOU) of a Type-2 fuzzy set, which can be understood as uncertainty in the membership grades of a fuzzy set, which is an interval itself, as defined in (4). A graphical representation of an IT2FS is shown in Figure 1.

---

[1] A Fuzzy Space is defined by the interval $b_j \subseteq \tilde{b}_j$.

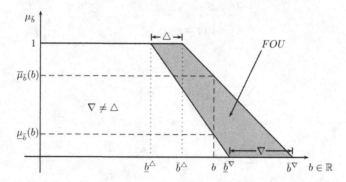

Fig. 1. Interval Type-2 fuzzy constraint with Joint Uncertain Δ & ∇

This Figure depicts a linear IT2FS, note that its FOU is weighted by 1, this means that its primary membership functions J_b are decomposed into *Lower* and *Upper* primary membership functions namely $\overline{\mu}_{\tilde{b}}$ and $\underline{\mu}_{\tilde{b}}$ respectively. In particular, each value b has an interval-valued membership grade modeled by following interval:

$$\mu_{\tilde{b}}(b) \in \left[\underline{\mu}_{\tilde{b}}(b), \overline{\mu}_{\tilde{b}}(b) \right]; b \in \mathbb{R} \tag{5}$$

Definition 2. (Uncertain supplies and demands). *Consider the demands* (d_i) *and supplies* (a_j) *of the crisp TM. Their uncertain counterparts according to the Definition 1 are defined as linear IT2FS, namely* $(\tilde{d}_i)$ *and* $(\tilde{a}_j)$.

The transportation model with uncertain demands and supplies $(\tilde{d}_i)$ and $(\tilde{a}_j)$, is

$$\min z = c_{ij} x_{ij} \tag{6}$$

s.t.

$$\sum_{i=1}^{m} x_{ij} \precsim \tilde{a}_j \quad \forall\ j \in \mathbb{N}_n \tag{7}$$

$$\sum_{j=1}^{n} x_{ij} \succsim \tilde{d}_i \quad \forall\ i \in \mathbb{N}_m \tag{8}$$

where $c_{ij}, x_{ij} \in \mathbb{R}^{n,m}$, $\tilde{a}_j$ and $\tilde{d}_i$ are IT2FS whose supports are defined over the real numbers $\mathbb{R}$, $\precsim$ and $\succsim$ are Type-2 fuzzy partial orders.

In this model, $\tilde{a}_j$ is a linear IT2FS with parameters $\underline{a}_j^\Delta, \overline{a}_j^\Delta, \underline{a}_j^\nabla$ and $\overline{a}_j^\nabla$, and $\tilde{d}_i$ is a linear IT2FS with parameters $\underline{d}_i^\Delta, \overline{d}_i^\Delta, \underline{d}_i^\nabla$ and $\overline{d}_i^\nabla$, analogous to Figure 1.

This problem can be solved through the proposals of Figueroa [1-3] and Figueroa and Hernández [4] by computing a set of optimal solutions called $\tilde{Z}$ and applying a Type-reduction strategy to find an optimal solution through the Zimmermann's method. In next section we provide more details about its computation.

2.1   Solution Procedure

Firstly, we need to compute a fuzzy set of optimal solutions $\tilde{Z}$ through the boundaries of each fuzzy supply and demand, in the following way:

$$\overline{d}_i^\nabla, \overline{a}_j^\nabla \to \underline{z}^\Delta \tag{9}$$

$$\overline{d}_i^\Delta, \overline{a}_j^\Delta \to \underline{z}^\nabla \tag{10}$$

$$\underline{d}_i^\nabla, \underline{a}_j^\nabla \to \overline{z}^\Delta \tag{11}$$

$$\underline{d}_i^\Delta, \underline{a}_j^\Delta \to \overline{z}^\nabla \tag{12}$$

Then, we define the following auxiliary variables:

Definition   3. *Consider an uncertain supply $\tilde{a}_j$ defined as IFS with linear membership function. Then Δ_j is defined as the distance between $\underline{a}_j^\Delta$ and $\overline{a}_j^\Delta$, $\Delta_j = \overline{a}_j^\Delta - \underline{a}_j^\Delta$ and ∇_j is defined as the distance between $\underline{a}_j^\nabla$ and $\overline{a}_j^\nabla$, $\nabla_j = \overline{a}_j^\nabla - \underline{a}_j^\nabla$.*
Similarly, for an uncertain demand $\tilde{d}_i$, Δ_i is defined as the distance between $\underline{d}_i^\Delta$ and $\overline{d}_i^\Delta$, $\Delta_i = \underline{d}_i^\Delta - \overline{d}_i^\Delta$ and ∇_i is defined as the distance between $\underline{d}_i^\nabla$ and $\overline{d}_i^\nabla$, $\nabla_i = \underline{d}_i^\nabla - \overline{d}_i^\nabla$.

First, we describe the Zimmermann's Soft Constraints method as follows

Algorithm 1. [Zimmermann's Soft Constraints method]

1. Calculate an inferior bound called *Z Minimum* ($\underline{z}$) by solving a LP model with $\underline{b}$.
2. Calculate a superior bound called *Z Maximum* ($\overline{z}$) by solving a LP model with $\overline{b}$.
3. Define a Fuzzy Set $\tilde{z}(x^*)$ with bounds $\underline{z}$ and $\overline{z}$, and linear membership function. This set represents the degree that any feasible solution has regarding the optimization objective:

4. If the objective is to minimize, then its membership function[2] is:

$$\mu_{\tilde{z}}(x;\underline{z},\overline{z}) = \begin{cases} 1, & c'x \leqslant \underline{z} \\ \dfrac{\overline{z}-c'x}{\overline{z}-\underline{z}}, & \underline{z} \leqslant c'x \leqslant \overline{z} \\ 0, & c'x \geqslant \overline{z} \end{cases} \tag{13}$$

— Thus, solve the following LP model[3]:

$$\max\{\alpha\} \tag{14}$$

$$s.t.$$

$$c'x + \alpha(\overline{z}-\underline{z}) = \overline{z} \tag{15}$$

$$Ax - \alpha(\overline{b}-\underline{b}) \geqslant \underline{b} \tag{16}$$

$$x \geqslant 0$$

where α is the overall satisfaction degree of all fuzzy sets.

Now, Figueroa [1-3] proposed a method for solving IT2FLP problems which uses Δ and ∇ as auxiliary variables with weights C^Δ and C^∇ respectively to find an optimal fuzzy set enclosed into the FOU of $\tilde{Z}$ and then solve it by using the Soft Constraints model. Its description is presented next.

Algorithm 2. [Figueroa's method for IT2FLP]

— Calculate an inferior boundary called Z *minimum* ($\underline{z}$) by using $\overline{b}^\nabla + \nabla$ as a frontier of the model, where ∇ is an auxiliary set of variables weighted by C^∇ which represents the lower uncertainty interval subject to the following statement:

$$\nabla \leqslant \underline{b}^\nabla - \overline{b}^\nabla \tag{17}$$

— Calculate a superior boundary called Z *maximum* ($\overline{z}$) by using $\overline{b}^\Delta + \Delta$ as a frontier of the model, where Δ is an auxiliary set of variables weighted by C^Δ which represents the upper uncertainty interval subject to the following statement:

$$\Delta \leqslant \underline{b}^\Delta - \overline{b}^\Delta \tag{18}$$

Find the final solution using the third and subsequent steps of the Algorithm 1.

In this algorithm, Δ and ∇ operate as Type-reducers, this means that for each uncertain $\tilde{d}_i, \tilde{a}_j$, we obtain a fuzzy set embedded on its FOU where Δ_i^* and ∇_i^*

---

[2] For a maximization problem, its membership function is defined as the complement of $\mu_{\tilde{z}}(x;\underline{z},\overline{z})$

[3] In the same way, a minimization problem is defined as the complement of a maximization problem.

leads to $\overline{d}_i$ and $\underline{d}_i$, and Δ_j^* and ∇_j^* leads to $\underline{a}_j$ and $\overline{a}_j$ in the Zimmermann's method.

$$\Delta_j^* \to \underline{a}_j \tag{19}$$

$$\nabla_j^* \to \overline{a}_j \tag{20}$$

$$\Delta_i^* \to \overline{d}_i \tag{21}$$

$$\nabla_i^* \to \underline{d}_i \tag{22}$$

Then we apply the Algorithm 1, finding an optimal solution in terms of α.

It is important to note that the analyst needs a solution in terms of $x \in \mathbb{R}$, so the uncertain problem must be Type-reduced by using the Algorithm 2 and then a real valued solution is found by using the Algorithm 1. This means that we come from an IT2FLP problem to an FLP and finally to a crisp solution embedded into $\tilde{Z}$.

3   Application Example

We present an application of our proposal to a small example, where all its demands and supplies are IT2FS, so its optimal solution is given by an α-cut. The obtained solution is a set enclosed into $\tilde{Z}$, where its type-reduced solution is   provided by the Algorithm 2. All parameters of $\tilde{a}_j$ and $\tilde{b}_i$ are shown in following matrices.

$$\overline{d}_i^\nabla = \begin{bmatrix} 10 \\ 11 \\ 12 \end{bmatrix} \overline{d}_i^\Delta = \begin{bmatrix} 13 \\ 14 \\ 18 \end{bmatrix} \underline{d}_i^\nabla = \begin{bmatrix} 12 \\ 13 \\ 15 \end{bmatrix} \underline{d}_i^\Delta = \begin{bmatrix} 16 \\ 17 \\ 21 \end{bmatrix}$$

$$\overline{a}_j^\nabla = \begin{bmatrix} 20 \\ 32 \\ 24 \end{bmatrix} \overline{a}_j^\Delta = \begin{bmatrix} 14 \\ 25 \\ 18 \end{bmatrix} \underline{a}_j^\nabla = \begin{bmatrix} 24 \\ 37 \\ 29 \end{bmatrix} \underline{a}_j^\Delta = \begin{bmatrix} 16 \\ 30 \\ 23 \end{bmatrix}$$

$$C_{ij} = \begin{bmatrix} 2 \\ 3 \\ 2 \\ 4 \\ 1 \\ 3 \\ 2 \\ 4 \\ 2 \end{bmatrix} C^\Delta = \begin{bmatrix} 1 \\ 0.5 \\ 1 \\ 1 \\ 0.5 \\ 0.5 \end{bmatrix} \Delta = \begin{bmatrix} 4 \\ 4 \\ 6 \\ 8 \\ 7 \\ 6 \end{bmatrix} C^\nabla = \begin{bmatrix} 1 \\ 0.5 \\ 1 \\ 1 \\ 0.5 \\ 1 \end{bmatrix} \nabla = \begin{bmatrix} 3 \\ 3 \\ 6 \\ 6 \\ 7 \\ 6 \end{bmatrix}$$

This example is composed by three suppliers and three customers which parameters are IT2FS, so we apply the Algorithm 2 to find a crisp solution of the problem. The obtained fuzzy set $\tilde{Z}$ is defined by the following boundaries:

$$\overline{z}^\Delta = 76 \quad \overline{z}^\nabla = 91 \quad \underline{z}^\Delta = 56 \quad \underline{z}^\nabla = 69$$

The values of $\underline{z}^*$ and $\overline{z}^*$ are 65.5 and 79 respectively. After applying the Zimmermann's method we obtain a crisp solution of $\alpha^* = 0.1053$ and $z^* = 77.57$. The obtained solution in terms of x is:

$$x_{11}^* = 2.1053, \; x_{13}^* = 18.3158, \; x_{22}^* = 14.3158, \; x_{31}^* = 11.2105$$
$$\Delta_{i=1}^* = 3, \Delta_{i=2}^* = 3, \Delta_{i=3}^* = 6, \nabla_{i=1}^* = 4, \nabla_{i=2}^* = 4, \nabla_{i=3}^* = 6$$

The resultant set of optimal solutions is displayed next.

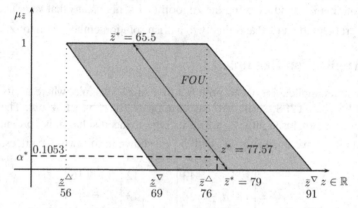

Fig. 2. Fuzzy Set $\tilde{z}$ embedded into the FOU of $\tilde{Z}$

These results indicate that the analyst should send x_{ij}^* units from its suppliers to customers as shown above. A global satisfaction degree of $\alpha^* = 0.1053$ is computed over the Type-reduced fuzzy set of optimal solutions $\tilde{z}$ which is embedded into the FOU of $\tilde{Z}$ as shown in Figure 2.

In this problem, all suppliers availability and customers demands are considered as IT2FS. We apply both Algorithm 1 and 2 to find a crisp solution of the problem which is what the analyst needs to make a decision about how many units has to send getting an optimal minimum transportation cost.

In this way, the IT2TM can be handled with classical optimization techniques and the analyst can involve linguistic uncertainty provided by the experts of the system to find an operation point namely α which satisfies both customers and suppliers.

4   Concluding Remarks

The presented model is a tool which deals with uncertainty of a fuzzy set involving the perception of different experts into IT2FS. Its computation and solution are explained through a small example. Although the example presented in this paper is small, it is intended to be easy for implementation and checking its results by readers.

The presented IT2TM deals with uncertainty coming from the opinion and perception of the experts about the availability of suppliers and customers demands, getting an optimal solution which allows us make decisions under fuzzy uncertainty.

Finally, the presented model can be extended to other scenarios and models as TSP, VRP and its extensions, by applying a similar reasoning than the presented here.

4.1   Further Topics

The theory of Generalized Interval Type-2 Fuzzy Sets (GT2 FS) can be considered as a new optimization focus. In this field an additional degree of freedom should be considered in the modeling process, this feature is the secondary membership function $f_x(u)/u$ which induces to new directions.

References

1. Figueroa, J.C.: Linear Programming with Interval Type-2 Fuzzy Right Hand Side Parameters. In: 2008 Annual Meeting of the IEEE North American Fuzzy Information Processing Society, NAFIPS (2008)
2. Figueroa, J.C.: Solving Fuzzy Linear Programming Problems with Interval Type-2 RHS. In: 2009 Conference on Systems, Man and Cybernetics. IEEE (2009)
3. Figueroa, J.C.: Interval Type-2 Fuzzy Linear Programming: Uncertain constraints. In: IEEE Symposium Series on Computational Intelligence, pp. 1–6. IEEE (2011)
4. Figueroa-García, J.C., Hernandez, G.: Computing Optimal Solutions of a Linear Programming Problem with Interval Type-2 Fuzzy Constraints. In: Corchado, E., Snášel, V., Abraham, A., Woźniak, M., Graña, M., Cho, S.-B. (eds.) HAIS 2012, Part I. LNCS, vol. 7208, pp. 567–576. Springer, Heidelberg (2012)
5. Klir, G.J., Yuan, B.: Fuzzy Sets and Fuzzy Logic: Theory and Applications. Prentice Hall (1995)
6. Lai, Y.J., Hwang, C.: Fuzzy Mathematical Programming. Springer (1992)
7. Kacprzyk, J., Orlovski, S.A.: Optimization Models Using Fuzzy Sets and Possibility Theory. Kluwer Academic Press (1987)
8. Pandiant, M.V.: Application of Fuzzy Linear Programming in Production Planning. Fuzzy Optimization and Decision Making 2(3), 229 (2003)
9. Zimmermann, H.J.: Fuzzy programming and Linear Programming with Several Objective Functions. Fuzzy Sets and Systems 1(1), 45–55 (1978)
10. Zimmermann, H.J., Fullér, R.: Fuzzy Reasoning for Solving Fuzzy Mathematical Programming Problems. Fuzzy Sets and Systems 60(1), 121–133 (1993)

Unstructured Scene Object Localization Algorithm Based on Sparse Overcomplete Representation[*]

Peng Lu, Yuhe Tang, Eryan Chen, Huige Shi, and Shanshan Zhang

School of Electrical Engineering, Zhengzhou University, Zhengzhou, China
lupeng@zzu.edu.cn

Abstract. Unstructured scene has many uncertainties and unpredictable states. It brings difficulties to the object localization, which is pixel-based processing. The method of analog visual information processing is an effective way to solve the problem mentioned above. Sparse overcomplete representation is an image representation model which is more in line with visual mechanism. However, the overcomplete representation not only increases the combinatorial search difficulty of sparse decomposition, but also changes the symmetry between input and code space. Furthermore, it makes the model solution and calculation method complicated. In order to solve the afore mentioned problem and effectively use this model to achieve automatic image object localization, this paper takes the unstructured scenes object localization as the background. Firstly, the overcomplete representation computational model which is based on energy model and score matching method is established. Then an automatic object localization method based on the neuronal response and dynamic threshold strategy is proposed and applied to the movement object localization. On this basis, the error analysis is done. Experimental results show that the method can achieve the movement object localization.

Keywords: Unstructured scenes, sparse overcomplete representation, object localization, score matching.

1   Introduction

Unstructured scene has many uncertainties and unpredictable states [1], which leads to the pixel values in a fixed position change constantly. It brings difficulties to the method pixel-based processing. Data processing of high-dimensional heterogeneous in the unstructured scene involves complex computational process and needs an overwhelming computational amount. Thus calculation method with efficient processing of unstructured scene is needed. Excitedly, visual computing method has advantages in environmental perception and cognitive abilities [2]. It is an important way to solve the above problems. Therefore, simulating the visual perception mechanism plays an important role in solving the moving object localization in unstructured scenes.

[*] This work is supported by National Natural Science Foundation of China (No.60841004 & 60971110 &61172152) and Zhengzhou Science and Technology Development Project (112PPTGY219-8).

D.-S. Huang et al. (Eds.): ICIC 2012, LNCS 7389, pp. 618–625, 2012.
© Springer-Verlag Berlin Heidelberg 2012

Recently, the sparse coding is commonly used as the visual computing model [3]. This model assumes that the code space is equal to input space dimension [4]. However, the image sparse representation based on overcomplete mechanism [5] meets the below three properties: sparse, divisibility and the shift invariance when it is used to represent the image data. The simulation of overcomplete mechanism plays an important role in solving image processing problems, such as object localization.

However, the overcomplete representation increases the combinatorial search difficulty of sparse decomposition, meanwhile changes the symmetry between input space and code space, thus makes the model solution and calculation method complicated[6]. In order to solve the afore mentioned problem and effectively use the overcomplete representation model to achieve image object localization, this paper takes object localization of unstructured scenes as its problem background. Firstly, we design the algorithm to obtain overcomplete receptive fields of simple cells in primary visual cortex from unstructured scenes sequences based on sparse overcomplete representation model. Secondly, we design the algorithm to achieve object localization. Finally, the effectiveness of the model and algorithm is tested through the unstructured scenes experiments. The results show that this method can achieve the movement object localization.

2   Computational Model

The basic sparse coding model is:

$$I(x, y) = \sum_{i=1}^{m} A_i(x, y) s_i \tag{1}$$

Where $I(x, y)$ is a $\sqrt{n} \times \sqrt{n}$ dimensional image, $A_i(x, y)$ is a basis function with n-dimensional column vector. m is the number of basis functions, s_i is the response coefficient. If $m > n$, and A is full rank, then A is overcomplete.

From the physiological point, formula (1) is modeling the respond process of cells of visual cortex. Basis function set A simulates the receptive fields of visual cortex. Increase m to meet $m > n$, A would contains more basic characteristics of the joint space, such as localization, orientation and frequency. We can clearly see that sparse overcomplete representation would enhance the robustness of system.

However,if $m > n$, two problems will be caused. The first one is that sparse decomposition about $I(x, y)$ based on overcomplete needs combination search [7].So the approximation algorithm such as relaxation optimization algorithm [8] and greedy tracking algorithm [9] to solve this problem is needed. But the calculation of approximation algorithm is very complicated. The second problem is that overcomplete sparse coding model doesn't like the complete ICA model, in which row vector $A^{-1} = W$ is defined as the receptive field.Therefore; a new method must be adopted to solve the problem of model solution.

In this paper, we estimated the receptive field through overcomplete sparse coding model based on energy model [10]. Meanwhile, we defined an objective function to measure the sparseness, and estimated the optimal receptive field W through maximizing the sparseness. First, we used the non-normalized log-likelihood function to define the energy model:

$$\log L(x; w_1, ..., w_n) = -\log Z(W) + \sum_{i=1}^{n} G_i(w_i^T x)$$ (2)

Where x is a single observation data, $Z(W)$ is the normalization constant of w_i, $\int L(x; w_1, ..., w_n) dx = 1$, $Z(W) = \int \prod_{i=1}^{n} \exp(G_i(w_i^T x)) dx$. The classic log cosh function G, which is used to replace the logarithm probability density of s_i, is defined as:

$$G_i(u) = -\alpha_i \log \cosh(u)$$ (3)

Where α_i are parameters that are estimated following with w_i.

Maximization likelihood estimation needs to calculate $Z(W)$ firstly. However, under situation of overcomplete, the calculation is complicated. Therefore, score matching [11] is used to directly estimate the linear receptive field w_k. The log density function of the data vector is defined as:

$$\log p(\mathbf{x}) = \sum_{k=1}^{m} \alpha_k G(\mathbf{w}_k^T \mathbf{x}) + Z(\mathbf{w}_1, ..., \mathbf{w}_n, \alpha_1, ..., \alpha_n)$$ (4)

Where n is the dimension of the data, m is the number of components, which is the number of the receptive fields. And the number of components m is larger than the dimension of the data n. x is the single sample data vector, that is ,an image patch, where the vector $\mathbf{w}_k = (w_{k1}, ..., w_{kn})$ is constrained to the unit norm.

Use the gradient of the log-density function of data vector to define the score function. The model score function is defined as;

$$\psi(\mathbf{x}; W, \alpha_1, ..., \alpha_m) = \nabla_x \log p(\mathbf{x}; W, \alpha_1, ..., \alpha_m)$$
$$= \sum_{k=1}^{m} \alpha_k \mathbf{w}_k g(\mathbf{w}_k^T \mathbf{x})$$ (5)

Where g is the first derivative of G.

We used the square of distance between model score function and data score function to get the objective function:

$$J = \sum_{K=1}^{m} \alpha_k \frac{1}{T} \sum_{t=1}^{T} g'(\mathbf{w}_k^T \mathbf{x}(t)) + \frac{1}{2} \sum_{j,k=1}^{m} \alpha_j \alpha_k \mathbf{w}_j^T \mathbf{w}_k \frac{1}{T} \sum_{t=1}^{T} g(\mathbf{w}_k^T \mathbf{x}(t)) g(\mathbf{w}_j^T \mathbf{x}(t))$$ (6)

Where T is the number of samples, denoted by $x(1), x(2), ..., x(T)$.

We used the gradient descent algorithm to make the objective function minimization:

$$w(t) = w(t-1) - \eta(t) \frac{\partial J(w)}{\partial w} \Big|_{w=w(t-1)}$$ (7)

Where $\eta(t)$ is the step of the negative gradient direction, which is the learning rate.

3   Algorithm

3.1   Learning Algorithm of Receptive Field Overcomplete Set

Input: sample images

  Output: the receptive field W

  Steps:

  Step1: random sampling to the images to obtain the training samples $X = \{x_1, x_2, ..., x_n\}$;

  Step2: remove DC component of the x_i, whiten the samples X by the principal component analysis (PCA) method, and project X into whitenization space $Z = VX$

  Step3: use unit row vector to initialize the W_s;

  Step4: calculate $J(w(0))$, move along the negative gradient direction, $t = 1, 2....$, update W according to the formula (7), and normalize the unit vector; meanwhile update parameters α;

  Step5: If the result is convergent or the number of iterations reaches the set value, then stop iteration, otherwise, return to step 4.

  Step6: Project W_s back into the original image space $W = W_s V$.

Then, we can obtain overcomplete receptive field of simple cell use this way. According to the properties of competitive response and visual sparse [12], only a small amount of neurons is activated to constitute the internal representation of image. So the object locating algorithm (OLA) selects N neurons which have larger response to achieve image object localization.

3.2   Image Locating Algorithm

Input: test image I_1 and its corresponding background image I_2

  Output: results of object locating

  Steps:

  Step1: sequentially sampling image I_1 and I_2, and obtain the image patches $X_1, X_2, ..., X_n$;

  Step2: remove DC component of the X_i; calculate the neurons response according to the formula $S_i = WX_i$

  Step3: retain N larger responses in each group S_i;

  Step4: obtain the perception results S_g and S_d of I_1 and I_2;

  Step5: $h = |S_{di} - S_{gi}|$, set dynamic thresholds $\delta = \dfrac{\sum\limits_{i=1}^{n} |S_{di} - S_{gi}|}{n}$, if $h \geq \delta$, then X_i is the object patch. Store the coordinates of X_i;

  Step6: display the results of object locating.

4   Experiments and Data Analysis

We selected the unstructured traffic images as test object of the above algorithm.

4.1   Data Preprocessing and Learning the Receptive Field

We select 14 images as the training images for learning the receptive fields. Using algorithm 1 for learning the receptive field. We can get 50000 image patches using 16 * 16 window. Every patch is converted to a column vector. Input data set X 256*50000 is acquired. We preprocess the data just as algorithm 1 and retain 128 principal components after dimension reduction. Then, a representation with 512 receptive fields is estimated based on the energy model and score matching. The resulting receptive fields are shown in Fig.1. The number of receptive fields is four times when compared to the PCA dimensions. These results are exactly similar with the characteristics that found in the simple cell receptive fields of V1 area.

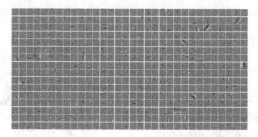

Fig. 1. Overcomplete set of receptive fields

4.2   Experiment of Image Object Localization

We select background image and test images of 512*512. Using algorithm 2 for content-aware. Firstly, sampling the test images sequentially, starting from the top left-hand corner apex of the image and adopting16*16 pixels space sub window to sample, we can get the first image patch. And then shifting the window to right 16 pixels, we can get the second image patch. Repeat the above step until we gathered all patches of first line. Then let 16 pixels shift downward from the apex coordinate position and image patches of the second line are collected. At last, we could sample the entire image.

Response of neurons caused by one image patch is shown in Fig.2 (a). N neurons with larger response are shown in Fig.2 (b). The number N is determined by experiment.

The first experiment selected images with complex background. The images are taken from video supervision sequence, we do the grayscale processing. Fig.3 (a) is the test images. The corresponding results are shown in Fig.3 (b). The area of rectangular box is the target area marked by OLA algorithm.

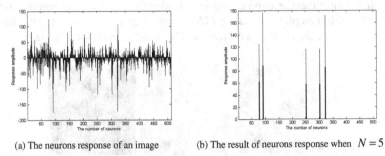

(a) The neurons response of an image   (b) The result of neurons response when $N = 5$

Fig. 2. The neurons response aroused by one image patch

(a)Test image   (b) the result of targeting

Fig. 3. Test results of the first experiments

The second experiment selected vehicle image at high-speed. The test image is shown in Fig.4. The corresponding results are shown in Fig.5. The area of rectangular box is the target area marked by OLA algorithm.

(a)one car   (b)many cars

Fig. 4. Two test images

(a) the test results when $N = 5$ (b) the test results when $N = 1$

Fig. 5. Test results of the second experiments

The third experiment is comparison test. Under the same experimental conditions, respectively with Template Matching (TM) method and the edge detection based on wavelet method for the contrast experiment. The results can be seen in Fig.6.

(a)OLA algorithm   (b) IM algorithm   (c) edge detection algorithm

Fig. 6. Contrast experiment

4.3   Results Analysis

As can be seen in Fig.6, OLA algorithm achieves the object localization accurately; the TM and edge detection methods all have erroneous judgment. 68 images were chosen for other experiments. Statistics results of three algorithms are shown in Table 1.

Table 1. The statistics results of three algorithms

algorithms	image (numbers)	located correctly (numbers)	located error (numbers)	the accurate rates
OLA	68	65	3	95.59%
TM	68	52	16	76.47%
edge detection	68	42	26	61.76%

It can be seen from the Table 1, the accurate rate of object localization based on OLA algorithm has improved greatly compared with TM and edge detection methods.

From the above experiments, we can see that the OLA algorithm can achieve the object localization under different conditions. However, there are still small amount of error patches, from Table 1, we know that the error have little effect to results. The edge of object exist error is attributed to vehicle and pedestrian in close distance, the error of leaves is caused by the swing. The error patches in Fig.5 (a) less than Fig.5 (b) shows that keeping N neurons have more strong anti-jamming than retaining neuron with largest response.

Experiment results show that OLA methods based on sparse overcomplete representation is effective for image object localization. It has a low error rate and high accuracy rate under unstructured background. And has anti-jamming and relatively strong robustness to leaves swing.

5   Conclusion

By simulating visual information processing mechanism and method, we established the image sparse overcomplete representation model, and then put forward OLA algorithm. The algorithm solved the problem of the increasing of sparse decomposition search difficulty and is effectively used to realize image object localization. The algorithm has an important theoretical and practical application value to traffic flow statistics. The furthermore work is to solve the problem of error image patches caused by leaves swing and shadow.

References

1. Rodriguez, M., Ali, S., Kanade, T.: Tracking in Unstructured Crowded Scenes. In: IEEE 12th International Conference on Computer Vision, pp. 1389–1396 (2009)
2. Lu, P., Li, Y.Q.: Fault-image Detection Algorithm Based on Visual Perception. Chinese Journal of Scientific Instrument 31(9), 1997–2002 (2010)
3. Zheng, M., Bu, J.J.: Graph Regularized Sparse coding for Image Rrepresentation. IEEE Transactions on Image Processing 20(5), 1327–1336 (2011)
4. Geisler, W.S.: Visual Perception and the Statistical Properties of Natural Scenes. Annual Review of Psychology 59, 167–192 (2008)
5. Thang, N.D., Rasheed, T.: Content-based Facial Image Retrieval Using Constrained Independent Component Analysis. Information Sciences 181(15), 3162–3174 (2011)
6. Tseng, P.: Further Results on Stable Recovery of Sparse Overcomplete Representations in the Presence of Noise. IEEE Trans. Inf. Theory 55(2), 888–899 (2009)
7. Cai, Z.M., Lai, J.Z.: An Overcomplete Learned Dictionary-Based Image Denoising Method. Acta Electronica Sinica 37(2), 347–350 (2009)
8. Mohimani, H., Babaie-Zadeh, M., Jutten, C.: A fast approach for overcomplete sparse decomposition based on smoothed L0-norm. IEEE Trans. Signal, Processing 57(1), 289–301 (2009)
9. Needell, D., Vershynin, R.: Uniform Uncertainty Principle and Signal Recovery via Regularized Orthogonal Matching Pursuit. Foundations of Computational Mathematics 9(3), 317–334 (2009)
10. Valgaerts, L., Bruhn, A.: Dense Versus Sparse Approaches for Estimating the Fundamental Matrix. Int. J. Comput Vision 96(2), 212–234 (2012)
11. Joder, C., Essid, S., Richard, G.: A Conditional Random Field Framework for Robust and Scalable Audio-to-score Matching. IEEE Trans. Audio Speech Lang. Process. 19(8), 2385–2397 (2011)
12. Tiesinga, P.H., Buia, C.I.: Spatial Attention in Area V4 is Mediated by Circuits in Primary Visual Cortex. Neural Networks 22(8), 1039–1054 (2009)

Solving the Distribution Center Location Problem Based on Multi-swarm Cooperative Particle Swarm Optimizer

Xianghua Chu[1], Qiang Lu[1,*], Ben Niu[2,*], and Teresa Wu[3]

[1] Shenzhen Graduate School, Harbin Institute of Technology Shenzhen 518055, China
[2] College of Management, Shenzhen University Shenzhen 518060, China
[3] School of Computing, Informatics, Decision Systems Engineering,
Arizona State University Tempe, USA
{Drniuben,qiang.lu.home}@gmail.com

Abstract. The discrete location of distribution center is a NP-hard issue and has been studying for many years. Inspired by the phenomenon of symbiosis in natural ecosystems, multi-swarm cooperative particle swarm optimizer (MCPSO) is proposed to solve the location problem. In MCPSO, the whole population is divided into several sub-swarms, which keeps a well balance of the exploration and exploitation in MCPSO. By competition and collaboration of the individuals in MCPSO the optimal location solution is obtained. The experimental results demonstrated that the MCPSO achieves rapid convergence rate and better solutions.

Keywords: discrete location, distribution center, improved PSO.

1 Introduction

Determination of optimal distribution center location is an abiding challenge for researchers and professionals in the logistics sector. The location of distribution center, however, is a NP-hard problem due to its various variables and complex constraints, which makes it hard to be solved within polynomial time. Some mathematical methods have been developed to cope with the location problem, such as Gravity Method (GM) [1], Lagrangian Relaxation Algorithm (LRA) [2-4], Branch and Bound Method (BBM) [5].

In recent years, with the rapid development of heuristic stochastic optimization algorithms, more and more techniques have been introduced to solve the location problem of distribution center, for instance, Genetic Algorithm [6-7], Simulated Annealing [8], Neural Network [9], etc.

Based on the simulation of simplified social models, particle swarm optimization (PSO) is first introduced in [10] as a novel evolutionary computation tool that has been applied in many engineering fields. Nevertheless, PSO suffers premature convergence as a result of weak exploitation and exploration.

---

* Corresponding authors.

D.-S. Huang et al. (Eds.): ICIC 2012, LNCS 7389, pp. 626–633, 2012.
© Springer-Verlag Berlin Heidelberg 2012

In this paper, the distribution center location problem is studied with an improved PSO, Multi-swarm Cooperative PSO (MCPSO) [11].The Section 2 describes the location problem modeling. Section 3 provides a brief introduction of PSO. Section 4 presents MCPSO algorithm. Section 5 introduces MCPSO-based location algorithm and experimental study followed by conclusions in Section 6.

2   The Discrete Location Model for Distribution Centers

The formulation extends the *Median-based* location model to include turnover cost and setup cost. The goal is to select a certain number of sites out of a finite set of potential sites to locate distribution centers while minimize total cost of the supply network. Before formulating the objective function, the issue is concentrated on the following situations.

(1) The decision for distribution center location is based on a set of candidate sites.
(2) The capacity of factory is capable of satisfying the demand of ending customers.
(3) The potential distribution centers have capacity limitation.
(4) The demand of customer is predictable.
(5) Each customer only supplied by one distribution center and each distribution center satisfies at least one customer.
(6) Transportation cost is directly proportional to the quantity of transportation.
(7) Total cost includes turnover cost, setup cost and transportation cost.

Some relevant notations are explained as follows: K is a set of factories, I is a set of candidate sites, J represents the demand notes, the unit cost of shipping from factory k to distribution center i is given by c_{ki}, and that cost from distribution center i to customer j is given by x_{ij}. The turnover cost in site i is given by g_i. Letting f_i be the fixed facility cost of site i and pq denotes the predefined number of established distribution centers.

To formulate our problem the following decision variables are introduced: for any $i \in I$, $j \in J$, y_{ij}, z_i are a binary decision variables, $y_{ij} = 1$ if demand note j is supplied by facility i, and 0 otherwise, $z_i = 1$ if a facility is located at candidate i, and 0 otherwise. The transportation volume that from factory k to distribution center i and from distribution center i to customer j is denoted as w_{ki} and x_{ij}, respectively, t_{ij} denotes whether center i covers demand note j, it take on the value 1 if distribution center i provides goods with customer j, otherwise it is 0. The formulation for the above discrete location problem is

$$\min U = \sum_{k=1}^{l}\sum_{i=1}^{m} c_{ki} w_{ki} + \sum_{i=1}^{m}\sum_{j=1}^{n} h_{ij} x_{ij} + \sum_{k=1}^{l}\sum_{i=1}^{m} g_i w_{ki} + \sum_{i=1}^{m} z_i f_i \qquad (1)$$

subject to:

$$\sum_{i \in I} w_{ki} \leq A_k , \quad k \in K \tag{2}$$

$$\sum_{k \in K} w_{ki} \leq M_i , \quad i \in I \tag{3}$$

$$\sum_{i \in I} z_i \leq pq \tag{4}$$

$$\sum_{i \in I} x_{ij} \geq D_j , \quad j \in J \tag{5}$$

$$\sum_{i \in I} y_{ij} = 1, \quad j \in J \tag{6}$$

$$\sum_{j \in J} x_{ij} = \sum_{k \in K} w_{ki} , \quad i \in I \tag{7}$$

$$t_{ij} \in \{0, 1\}, \quad \forall i \in I, \ \forall j \in J \tag{8}$$

$$w_{ki} \geq 0, \quad x_{ij} \geq 0, \quad k \in K, \ i \in I, \ j \in J \tag{9}$$

Constraints (2) limit the supply of factories within their capacities. Constraints (3) ensure that each distribution center receives input that within its processing capacity. Constraint (4) states that the selected sites should not exceed the predefined number. Constraints (5) stipulate that each demand note is satisfied. Constraints (6) ensure that each customer is supplied by only one distribution center. Constraints (7) state that the input goods of each distribution center equals to its output, (8) and (9) are constraints for decision variables.

3   Standard PSO

PSO is a population-based, stochastic optimization algorithm based on the idea of a swarm moving over a given landscape. In PSO algorithm, each member of the population is called a "particle", and each particle flies around in the d-dimensional search space with a velocity, which is constantly updated by the particle's own experience and the experience of the whole swarm. The Standard PSO (SPSO) can be described as follows:

$$v_i^d(t+1) = w \times v_i^d(t) + c_1 \times r_1 \times (pbest_i^d - x_i^d(t)) + c_2 \times r_2 \times (gbest^d - x_i^d(t)) \tag{10}$$

$$x_i^d(t+1) = x_i^d(t) + v_i^d(t+1) \tag{11}$$

For $i = 1,\ldots,s$ and $d = 1,\ldots,n$, where s is the population size, n is the problem dimension, c_1 and c_2 are acceleration coefficients, $v_i^d(t)$ is particle velocity, $x_i^d(t)$ is particle i's current position, r_1 and r_2 is uniformly distributed function in the interval [0,1], w is the inertia weight which provides the necessary diversity to the swarm by changing the momentum of particles, $pbest_i^d$ is the location of the best solution vector found by i, and $gbest^d$ represents the best solution found so far in the population.

In SPSO, there is only one swarm for evolution that the communication among particles is so simple, thus the whole population is easy to be trapped into the local optima. It is necessary to improve the convergence of SPSO, especially where exists a large number of sub-optima.

4   Multi-swarm Cooperative PSO

In ecosystems, many species have developed cooperative interactions with other species to improve their survival which is called symbiosis. The phenomenon of symbiosis can be found in all forms of life, from simple cells to birds and mammals. Inspired by this phenomenon, a master–slave mode is incorporated into the SPSO, and the multi-swarm (species) cooperative optimizer (MCPSO) is thus developed [11,12].

In MCPSO, a population consists of one master swarm and several slave swarms. The slave swarms mainly explore the feasible space, while the master swarm focuses on exploitation. The master-slave communication is showed in Fig. 1.

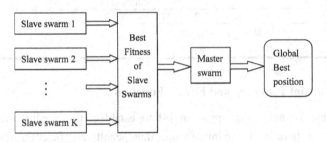

Fig. 1. The master-slave model

Each slave swarm executes a single PSO or its variants within MCPSO. When all slave swarms are ready with the new generations, each slave swarm then sends their best individual to the master swarm, and the best fitness of all received individuals would be selected and employed for evolution according to the following equations:

$$v_{id}^M (t+1) = wv_{id}^M (t) + c_1 * r_1 (p_{id}^M (t) - x_{id}^M (t)) + c_2 * r_2 (p_{gd}^M (t) - x_{id}^M (t))$$

$$+c_3 * r_3 (p_{gd}^S (t) - x_{id}^M (t)) \tag{12}$$

$$x_{id}^M (t+1) = x_{id}^M (t) + v_{id}^M (t+1) \tag{13}$$

where $p_{id}^M (t)$ is the best previous position of particle i in master swarm, $p_{gd}^M (t)$ is the best previous position of the master swarm, $p_{gd}^S (t)$ is the best previous position in the slave swarms, c_3 is acceleration coefficient which determines the communication intensity that the slave swarms impact the master swarm.

5   MCPSO-Based Location Algorithm

5.1   Particle Coding

The key to implement MCPSO-based location is to develop an efficient particle representation method. In the above location model, there are two types of variables, the binary variables (e.g. z_i and y_{ij}) and the numeric variables (e.g. w_{ki} and x_{ij}).

For the binary decision variables, discrete binary method is convenient to code the particle. However, it may be inefficient for the numeric variables, e.g., for a company with l factories, m candidate sites and n demand notes, then the potential problem scale of w_{ki} and x_{ij} is $l \times m$ and $m \times n$, respectively. Apparently, the algorithm efficiency would deteriorate dramatically with the increase of problem scale. Besides, the representation of floating number by binary method is lack of accuracy. Therefore, a hybrid parallel coding method is employed as shown in Table 1.

Table 1. Particle coding structure

Variable	Candidate				
	1	2	3	......	m
y_{ij}	00...0	$y_{21}y_{21}{\cdots}y_{2n}$	$y_{31}y_{31}{\cdots}y_{3n}$	......	00...0
w_{ki}	0	w_{k2}	w_{k3}		0

5.2   Constraint Handling and Fitness Function

We make use of annealing approach [8] to establish the penalty factor τ, i.e., $\tau = \tau_0/(1+t)$, where τ_0 is the initial value, then penalty coefficient is given by (14):

$$r = \frac{(1+t)}{(2 \times \tau_0)} \tag{14}$$

According to (14), an infeasible solution may not be penalized enough at the beginning, while it may be penalized seriously in the lately stage, hence, population likely to concentrate on local exploitation.

With the proposed penalty function method, the location model problem can be transformed into an unconstrained one, and the fitness function is built as (15):

$$fitness = U + \frac{1+t}{2\tau_0} \times \sum_{i \in A} d_q^2(x) \tag{15}$$

where $d_q(x)$ is a real-valued function:

$$d_q(x) = \max\{0, nc_1, nc_2, nc_3, nc_4, nc_5\} \quad q = 1, 2, \cdots, 1+l+2m+n \tag{16}$$

where nc_{ii}, $ii = 1, 2, 3, 4, 5$, is given as follows: $nc_1 = \sum_{i \in I} w_{ki} - A_k$, $k \in K$;

$nc_2 = \sum_{k \in K} w_{ki} - M_i$, $i \in I$; $nc_3 = \sum_{i \in I} z_i - p$; $nc_4 = D_j - \sum_{i \in I} x_{ij}$, $j \in J$;

$nc_5 = | \sum_{j \in J} x_{ij} - \sum_{k \in K} w_{ki} | - \varepsilon$, $i \in I$.

5.3   Experimental Study and Discussion

A company has established a factory in southern China, and it aims to build no more than 4 distribution centers to serve 15 demand notes. The number of candidate sites is 10. The detailed data and parameter settings are given in [13].

Experiments are conducted to compare MCPSO with SPSO. When conducting MCPSO-based algorithm, one master swarm and three slave swarms are set up in a population. For master swarm, $c_1 = c_2 = c_3 = 1.333$, and for slave swarms the coefficients are set at 2. The population size for each slave swarm is 50 and the maximum generation is set at 100. Comparing with MCPSO, the population size of SPSO is set at 200. Besides, SPSO has the same parameters settings as well as MCPSO. 50 runs for each algorithm are carried out. The evolution and convergence of MCPSO and SPSO are shown in Fig. 2, and the results are presented in Table 2.

Fig. 2 (a) and (b) illustrate how the algorithms evolve and converge during the evolutions. In Fig. 2 (a), although the master swarm gets inferior values at the beginning, its fitness figure decreases steeply and overtakes the slave swarms at about 18th iteration, then the master swarm continuously evolves. With collaboration between the master swarms and the slave swarms, MCPSO gain a superior performance. According to Fig. 2 (b), SPSO is trapping into local optima in the 9th iteration, and the whole algorithm gets stagnated that no better solution is found.

Comparing with SPSO, as shown in Table 2, MCPSO generates a feasible solution that without any penalty cost, which has practical values in the facility location.

Comparison is also conducted between the proposed algorithm and GA. GA is carried out with the population size of 50, for maximum of 100 iterations, the Roulette Wheel Selection is used to select the best individuals. The probability of crossover and mutation is set at 0.4 and 0.05, respectively. The results in Table 3 are the average of minimum values and standard deviations for 30 trials.

In Table 3, the MCPSO-based algorithm achieves the best mean result and standard deviation over the other algorithms. SPSO gains the second place in the average cost, while its standard deviation is worse than GA. This phenomenon is probably attributed to its premature convergence and stochastic instability.

In addition, comparing with some traditional location methods, such as [1], [14], the proposed MCPSO-based algorithm not only demonstrates the turnover of each selected location center, but also presents additional valued information (e.g. service coverage, transportation volume) for the decision makers to optimize supply network.

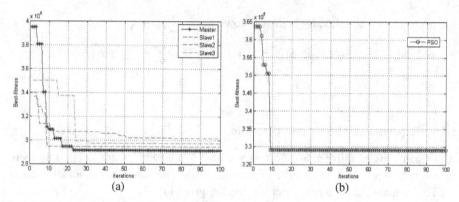

Fig. 2. The median convergence characteristics: (a) MCPSO-based algorithm; (2) SPSO

Table 2. Solution results

Algorithm	Decision	Coverage of each facility	Total cost	Penalty cost
MCPSO	5	3,5,11,12,14		
	6	2,4,6,7	291279	0
	8	1,8,9,10,13,15		
SPSO	5	3,4,9,11,14		
	8	2,5,6,12,15	329090	1333.3
	10	1,7,8,10,13		

Table 3. The results of MCPSO, SPSO and GA

Algorithm	MCPSO	SPSO	GA
Average cost	293565 ± 5993	316582 ± 18562	328962 ± 10256

6   Conclusion

In this paper we considered an extension of the traditional *Median*-based distribution center location model in terms of fixed cost, turnover cost and transportation cost, which is highly applicable in distribution center location. A MCPSO-based algorithm is proposed to solve the NP-hard location model. Due to the discrete characteristics of this location model, a hybrid parallel coding method is developed and the SA factor is integrated into penalty coefficient. A real-world location problem is employed to assess to the performance of MCPSO-based algorithm. In the experimental simulation, MCPSO-based location algorithm shows a superior result and overtakes that of GA and SPSO algorithm.

There are still several problems remaining to be investigated, such as how to make the swarms more adaptive in evolution, and how to apply the algorithms of this kind to more real-world engineering problems.

Acknowledgements. This work is supported by the National Natural Science Foundation of China (Grants No. 71171064, No.71001072), Science and Technology Project of Shenzhen (Grant No. JC201005280492A), The Natural Science Foundation of Guangdong Province (Grant no. 9451806001002294).

References

1. Plastria, F.: Facility Location: A Survey of Application and Method, pp. 25–80. Springer, New York (1995)
2. Beasley, J.E.: Lagrangean Heuristics for Location Problems. European Journal of Operational Research 65, 383–399 (1993)
3. Agar, M., Salhi, S.: Lagrangean Heuristics Applied to a Variety of Large Capacitated Plant Location Problem. Journal of the Operational Research Society 49, 1072–1084 (1998)
4. Aykin, T.: Lagrangian Relaxation Based Approaches to Capacitated Hub-and-spoke Network Design Problem. European Journal of Operational Research 79, 501–523 (1994)
5. Land, A.H., Doig, A.G.: An Automatic Method of Solving Discrete Programming Problems. Econometrica 28, 497–520 (1960)
6. Topcuoglu, H., Corut, F., Ermis, M., Yilmaz, G.: Solving the Uncapacitated Hub Location Using Genetic Algorithms. Computers and Operations Research 32, 967–984 (2005)
7. Qian, J., Pang, X.H., et al.: An Improved Genetic Algorithm for Allocation Optimization of Distribution Centers. Journal of Shanghai Jiaotong University 9, 73–76 (2004)
8. Abdinnour-Helm, S.: Using Simulated Annealing to Solve the P-hub Median Problem. Int. Journal of Physical Distribution and Logistics Management 31, 203–220 (2001)
9. Smith, K.A., Krishnamoorthy, M., Palaniswami, M.: Neural versus Traditional Approaches to the Location of Interacting Hub Facilities. Location Science 4, 155–171 (1996)
10. Eberhart, R., Kenedy, J.: Particle Swarm Optimization. In: Proceedings of IEEE Int. Conf. Neural Networks, Piscataway, pp. 1114–1121 (1995)
11. Niu, B., Zhu, Y.L., He, X.X.: MCPSO: A Multi-Swarm Cooperative Particle Swarm Optimizer. Applied Mathematics and Computation 185, 1050–1062 (2007)
12. Niu, B., Xue, B., Li, L., Chai, Y.: Symbiotic Multi-swarm PSO for Portfolio Optimization. In: Huang, D.-S., Jo, K.-H., Lee, H.-H., Kang, H.-J., Bevilacqua, V. (eds.) ICIC 2009. LNCS (LNAI), vol. 5755, pp. 776–784. Springer, Heidelberg (2009)
13. https://docs.google.com/open?id=0B0LeEyhqhWOGbVdTeGRHR3VWT2c
14. Huo, H.: Research on Distribution Center Location Problems. Logistic Science Technology 27, 50–52 (2004)

Improved Bacterial Foraging Optimization with Social Cooperation and Adaptive Step Size

Xiaohui Yan[1,2,*], Yunlong Zhu[1], Hanning Chen[1], and Hao Zhang[1,2]

[1] Key Laboratory of Industrial Informatics, Shenyang Institute of Automation,
Chinese Academy of Sciences, 110016, Shenyang, China
[2] Graduate School of the Chinese Academy of Sciences, 100039, Beijing, China
{yanxiaohui,ylzhu,chenhanning,zhanghao}@sia.cn

Abstract. This paper proposed an Improved Bacterial Foraging Optimization (IBFO) algorithm to enhance the optimization ability of original Bacterial Foraging Optimization. In the new algorithm, Social cooperation is introduced to guide the bacteria tumbling towards better directions. Meanwhile, adaptive step size is employed in chemotaxis process. The new algorithm is tested on a set of benchmark functions. Canonical BFO, Particle Swarm Optimization and Genetic Algorithm are employed for comparison. Experiment results show that the IBFO algorithm offers significant improvements over the original BFO algorithm and is a competitive optimizer for numerical optimization.

Keywords: bacterial foraging optimization, social cooperation, adaptive search strategies.

1 Introduction

Bacterial Foraging Optimization (BFO) is a novel swarm intelligence algorithm first proposed by Passino in 2002 [1]. It is inspired by the foraging and chemotactic behaviors of bacteria. Recently, BFO algorithm and its variants have been used for many numerical optimization [2] or engineering optimization problems [3-4].

However, original BFO algorithm has some weaknesses. First, the tumble angles are generated randomly. Useful information can't be shared between bacteria. Second, the step size in the original BFO is a constant. If the step size is large at the end stage, it is hard to converge to the optimal point. In this paper, we proposed an Improved Bacterial Foraging Optimization (IBFO). Two adaptive strategies are used in IBFO to improve its optimization ability. First, social cooperation is introduced to enhance the information sharing between bacteria. Then, adaptive step size is employed, which could make the bacteria use different search step sizes in different stages.

The rest of the paper is organized as followed. Section 2 introduces the original BFO algorithm. In section 3, the proposed IBFO algorithm is described in detail. In Section 4, the IBFO algorithm is tested on a set of benchmark functions compared with several other algorithms. Results are presented and discussed. Finally, conclusions are drawn in Section 5.

---

* This research is partially supported by National Natural Science Foundation of China 61174164, supported by National Natural Science Foundation of China 61003208 and supported by National Natural Science Foundation of China 61105067.

D.-S. Huang et al. (Eds.): ICIC 2012, LNCS 7389, pp. 634–640, 2012.
© Springer-Verlag Berlin Heidelberg 2012

2   Original Bacterial Foraging Optimization

The *E. coli* bacterium is one of the earliest bacterium which has been researched. It has a plasma membrane, cell wall, and several flagella which are randomly distributed around its cell wall 1. By the rotation of the flagella, *E. coli* can "tumble" or "run" in the nutrient solution. By simulating the foraging process of bacteria, Passino proposed the BFO algorithm. The main steps of BFO are explained as followed.

2.1   Chemotaxis

In BFO, the position updating which simulates the chemotaxis procedure is used the Eq. (1) as followed. θ^t is the position of the bacterium in the *t*th chemotaxis step. $C(i)$ presents the step size. $\Phi(i)$ is a randomly produced unit vector which stands for the tumble angle.

$$\theta_i^{t+1} = \theta_i^t + C(i)\phi(i) \tag{1}$$

In each chemotactic step, the bacterium generated a tumble direction firstly. Then the bacterium moves in the direction using Eq. (1). If the nutrient concentration in the new position is higher than the last position, it will run one more step in the same direction. This procedure continues until the nutrient get worse or the maximum run step is reached. The maximum run step is controlled by a parameter called N_s.

2.2   Reproduction

For every N_c times of chemotactic steps, a reproduction step is taken in the bacteria population. The bacteria are sorted in descending order by their nutrient values. Bacteria in the first half of the population will split into two. Bacteria in the residual half of the population die. By this operator, individuals with higher nutrient are survived and reproduced, which guarantees the potential optimal areas are searched more carefully.

2.3   Eliminate and Dispersal

After every N_{re} times of reproduction steps, an eliminate-dispersal event happens. For each bacterium, it will be moved to a random place according to a certain probability, known as P_e. This operator enhances the diversity of the algorithm.

3   Improved Bacterial Foraging Optimization

3.1   Social Cooperation

As we known, swarm intelligence is emerged by the cooperation of simple individuals [5]. However, the social cooperation hasn't been used in original BFO algorithm. In chemotactic steps, the tumble directions are generated randomly. Information carried by the bacteria in rich-nutrient positions is not utilized. In our IBFO, the tumble direc-tions are generated using Eq. (2). Where θ_{gbest} is the global best of the population found so far. $\theta_{i,\,pbest}$ is the ith bacterium's personal historical best. The tumble direction is then normalized as unit vector and the position updating is still using the Eq. (1).

$$\Delta_i = (\theta_{gbest} - \theta_i) + (\theta_{i,pbest} - \theta_i) \tag{2}$$

The direction generating equation is similar with the velocity updating equation of PSO algorithm [6]. They all used the global best and personal best. However, they are not the same. First, there is no inertia term in Eq. (2). In chemotactic steps of bacteria, inertia term will enlarge the difference of between θ_{gbest}, $\theta_{i,\,pbest}$ and the current position tremendously. Second, there are no learning factors in Eq. (2) as the direction will be normalized to unit vector. By social cooperation, the bacteria will move to better areas with higher probability as good information is fully utilized.

3.2   Adaptive Step Size

As it mentioned above, the constant step size will make the population hard to converge to the optimal point. In an intelligence optimization algorithm, it is important to balance its exploration ability and exploitation ability. Generally, in the early stage of an algorithm, we should enhance the exploration ability to search all the areas. In the later stage of the algorithm, we should enhance the exploitation ability to search the good areas intensively.

There are various step size varying strategies [7-8]. In IBFO, we use the decreasing step size. The step size will decrease with the iteration, as shown in Eq. (3). C_s is the initial step size. C_e is the ending step size. *NowEva* is the current function evaluations count. *TotEva* is the total function evaluations. In the early stage of IBFO algorithm, we use larger step size to guarantee the exploration ability. And at the end stage, small-er step size is used to make sure the algorithm can converge to the optimal point.

$$C = C_s - (C_s - C_e) \times NowEva \, / \, TotEva \tag{3}$$

4   Experiments

In this section, we tested the optimization ability of IBFO algorithm on six benchmark functions. Original BFO, PSO and Genetic Algorithm (GA) were employed for comparison.

4.1   Benchmark Functions

The six benchmark functions are listed in Table 1. They are widely adopted by other re-searchers to test their algorithms in many works [9-10]. Dimension of all functions are 20.

To compare algorithms fairly, we use number of function evaluations (FEs) as a measure criterion in this paper. It is also used in many other works [11-12]. All algo-rithms were terminated after 60,000 function evaluations.

4.2 Parameter Settings for the Involved Algorithms

The population sizes S of all algorithms are 50. In original BFO and IBFO algorithm, the parameters are set as followed: N_c=50, N_s=4, N_{re}=4, N_{ed}=10, P_e=0.25, C=0.1, S_r=S/2=25. The initial step size of IBFO C_s=0.1(Ub-Lb), ending step C_e=0.00001(Ub-Lb) where Lb and Ub refer the lower bound and upper bound of the variables of the problems. In PSO algorithm, ω decreased from 0.9 to 0.7. $C1$=$C2$=2.0 [13]. V_{min}=0.1×Lb, V_{max}=0.1×Ub. In GA, P_c is 0.95 and P_m is 0.1.

Table 1. Benchmark functions used in the experiment

Function	Formulation	Variable ranges	f(x*)
Sphere	$f_1(x) = \sum_{i=1}^{D} x_i^2$	[-5.12, 5.12]	0
Rosenbrock	$f_2(x) = \sum_{i=1}^{D-1}\left(100(x_i^2 - x_{i+1})^2 + (1-x_i)^2\right)$	[−15, 15]	0
Rastrigin	$f_3(x) = \sum_{i=1}^{D}\left(x_i^2 - 10\cos(2\pi x_i) + 10\right)$	[-10, 10]	0
Ackley	$f_4(x) = 20 + e - 20e^{\left(-0.2\sqrt{\frac{1}{D}\sum_{i=1}^{D} x_i^2}\right)} - e^{\left(\frac{1}{D}\sum_{i=1}^{D}\cos(2\pi x_i)\right)}$	[−32.768, 32.768]	0
Griewank	$f_5(x) = \frac{1}{4000}\left(\sum_{i=1}^{D} x_i^2\right) - \left(\prod_{i=1}^{D}\cos(\frac{x_i}{\sqrt{i}})\right) + 1$	[−600, 600]	0
Schwefel2.22	$f_6(x) = \sum_{i=1}^{n}|x_i| + \prod_{i=1}^{n}|x_i|$	[-10,10]	0

4.3 Experiment Results and Statistical Analysis

The results of IBFO, BFO PSO and GA on the benchmark functions are listed in Table 2. Best values of them on each function are marked as bold. Convergence plots of the algorithms on these functions are shown in Fig. 1.

It is clear from table 2 that IBFO obtained the best values on five of all six functions. PSO is best on the rest one. BFO and GA performed worst. On Rosenbrock and Rastrigin functions, all algorithms didn't performed well. However, the results of IBFO are a little better than that of PSO. On Ackley function, BFO, PSO and GA all performed badly while only IBFO obtained remarkable results. IBFO converged fast at the end stage and seemed was able to continue improving its result. On Griewank function, it got a rank of 2. However, it is only a little worse than PSO. On Schwefel2.22, it performed better than the other three algorithms, too. Overall, IBFO shows significant improvement over the original BFO algorithm. And its optimization ability is better than the classic PSO and GA algorithms on most functions.

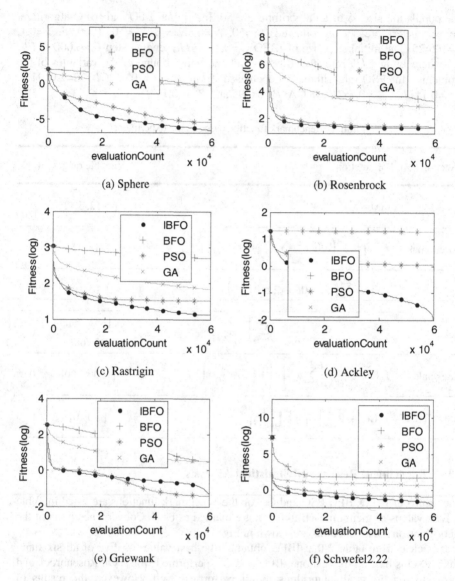

(a) Sphere

(b) Rosenbrock

(c) Rastrigin

(d) Ackley

(e) Griewank

(f) Schwefel2.22

Fig. 1. Convergence plots of IBFO, BFO, PSO and GA algorithms on functions

Table 2. Results obtained by the IBFO, BFO, PSO and GA algorithms

Function		IBFO	BFO	PSO	GA
f_1	Mean	**3.16004e-007**	6.17635e-001	4.52322e-006	6.61625e-001
	Std	**1.50001e-007**	1.94300e-001	3.34529e-006	2.34433e-001
f_2	Mean	**2.05631e+001**	2.02447e+003	2.49117e+001	6.56345e+002
	Std	**1.71408e+001**	8.62087e+002	2.12081e+001	4.55687e+002
f_3	Mean	**1.32458e+001**	4.94922e+002	3.35838e+001	6.80152e+001
	Std	**5.41958e+000**	6.74151e+001	1.07949e+001	1.77908e+001
f_4	Mean	**1.00880e-002**	1.95483e+001	1.10071e+000	1.86919e+001
	Std	**1.47724e-003**	3.71818e-001	9.14650e-001	1.30767e-000
f_5	Mean	4.96128e-002	3.00705e+000	**3.89587e-002**	3.11937e+000
	Std	3.97896e-002	4.5062e-001	**3.13277e-002**	8.55760e-001
f_6	Mean	**8.51024e-003**	5.43634e+001	1.39941e-001	4.92696e+000
	Std	**1.71202e-003**	1.20626e+001	1.87068e-001	1.09457e+000

5   Conclusions

This paper analyzes the shortages of original BFO algorithm. To overcome its short-ages, an Improved Bacterial Foraging Optimization (IBFO) algorithm is proposed. Social cooperation and adaptive step size strategies are used in IBFO. In the chemo-tactic steps, the tumble angles are no longer generated randomly. Instead, they are produced using the information of the bacteria's global best, the bacterium' personal best and its current position. The step size in chemotactic processes decreases linearly with iterations too, which could balance its exploration ability and exploitation ability.

To test the optimization ability of IBFO algorithm, it is tested on a set of bench-mark functions compared with original BFO, PSO and GA algorithms. The results show that IBFO algorithm performed best on five functions of all six. On the rest one, it is only a little worse than PSO. In general, the proposed IBFO algorithm offers significant improvements over original BFO, and is a competitive algorithm for opti-mization compared other algorithms.

References

1. Passino, K.M.: Biomimicry of Bacterial Foraging for Distributed Optimization and Con-trol. IEEE Control Systems Magazine 22, 52–67 (2002)
2. Chen, H., Zhu, Y., Hu, K.: Cooperative Bacterial Foraging Optimization. Discrete Dynam-ics in Nature and Society, Article ID 815247, 17 pages (2009)

3. Wu, C., Zhang, N., Jiang, J., Yang, J., Liang, Y.: Improved Bacterial Foraging Algorithms and Their Applications to Job Shop Scheduling Problems. In: Beliczynski, B., Dzielinski, A., Iwanowski, M., Ribeiro, B. (eds.) ICANNGA 2007, Part I. LNCS, vol. 4431, pp. 562–569. Springer, Heidelberg (2007)
4. Majhi, R., Panda, G., Majhi, B., Sahoo, G.: Efficient Prediction of Stock Market Indices Using Adaptive Bacterial Foraging Optimization (ABFO) and BFO Based Techniques. Expert Systems with Applications 36(6), 10097–10104 (2009)
5. Eberhart, R.C., Shi, Y., Kennedy, J.: Swarm Intelligence. Morgan Kaufmann (2001)
6. Shi, Y., Eberhart, R.C.: A Modified Particle Swarm Optimizer. In: Proceedings of the IEEE Congress on Evolutionary Computation, Piscataway, pp. 303–308 (1998)
7. Chen, H., Zhu, Y., Hu, K.: Self-Adaptation in Bacterial Foraging Optimization Algorithm. In: Proceedings of the 3rd International Conference on Intelligent System & Knowledge Engineering, Xiamen, China, pp. 1026–1031 (2008)
8. Zhou, B., Gao, L., Dai, Y.: Gradient Methods with Adaptive Step-Sizes. Computational Optimization and Applications 35(1), 69–86 (2006)
9. Karaboga, D., Basturk, B.: A Powerful and Efficient Algorithm for Numerical Function Optimization: Artificial Bee Colony (ABC) Algorithm. Journal of Global Optimization 39(3), 459–471 (2007)
10. Zou, W., Zhu, Y., Chen, H., Sui, X.: A Clustering Approach Using Cooperative Artificial Bee Colony Algorithm. Discrete Dynamics in Nature and Society, Article ID 459796, 16 pages (2010)
11. Yan, X., Zhu, Y., Zou, W.: A Hybrid Artificial Bee Colony Algorithm for Numerical Function Optimization. In: 11th International Conference on Hybrid Intelligent Systems, pp. 127–132 (2011)
12. Liang, J.J., Qin, A.K., Suganthan, P.N., Baskar, S.: Comprehensive Learning Particle Swarm Optimizer for Global Optimization of Multimodal Functions. IEEE Transcations on Evolutionary Computing 10, 281–295 (2006)
13. Shi, Y., Eberhart, R.C.: Empirical Study of Particle Swarm Optimization. In: Proceedings of the IEEE Congress on Evolutionary Computation, Piscataway, NJ, USA, vol. 3, pp. 1945–1950 (1999)

Root Growth Model for Simulation of Plant Root System and Numerical Function Optimization[*]

Hao Zhang[1,2], Yunlong Zhu[1], and Hanning Chen[1]

[1] Key Laboratory of Industrial Informatics, Shenyang Institute of Automation of Chinese
Academy of Sciences 110016, Shenyang, China
[2] Graduate School of the Chinese Academy of Sciences 100039, Beijing, China
{zhanghao,ylzhu,chenhanning}@sia.cn

Abstract. This paper presents the study of modelling root growth behaviours in the soil. The purpose of the study is to investigate a novel biologically inspired methodology for optimization of numerical function. A mathematical framework is designed to model root growth patterns. Under this framework, the interactions between the soil and root growth are investigated. A novel approach called "root growth algorithm" (RGA) is derived in the framework and simulation studies are undertaken to evaluate this algorithm. The simulation results show that the proposed model can reflect the root growth behaviours and the numerical results also demonstrate RGA is a powerful search and optimization technique for numerical function optimization.

Keywords: Root growth, simulation, numerical function optimization, modelling.

1   Introduction

Nature ecosystems have been the rich source of mechanisms for designing computational systems to solve difficult engineering and computer science problems. Modeling is an important tool for the comprehension of complex systems such as nature ecosystems and the model inspired from nature ecosystem is instantiated as an optimizer for numerical function and engineering optimization.

Computational root growth models or "virtual root system of plant" are increasingly seen as a useful tool for comprehending complex relationships between plant physiology, root system development, and the resulting plant form. Therefore accurate root growth models for simulation of root-soil interactions are of major significance. In the field of modeling root system of plant, there already exists a variety of elaborate approaches [1-3]. However, the complexity of the root–soil system requires an accurate and detailed description not only of each subsystem, but also of their mutual linkage and influence.

---

[*] This research is partially supported by National Natural Science Foundation of China 61174164, supported by National Natural Science Foundation of China 61003208 and supported by National Natural Science Foundation of China 61105067.

D.-S. Huang et al. (Eds.): ICIC 2012, LNCS 7389, pp. 641–648, 2012.
© Springer-Verlag Berlin Heidelberg 2012

In pursuit of finding solution to the optimization problems many researchers have been drawing inspiration from the nature [4]. A lot of such biologically inspired algorithms have been developed namely Genetic Algorithm (GA) [5], Particle Swarm Optimization (PSO) [6], Artificial Immune System (AIS) [7] and Artificial Bee Colony (ABC) [8]. Plant growth simulation algorithm (PGSA) is a powerful evolutionary algorithm simulating plant growth that has been proposed for solving integer programming [9]. These algorithms with their stochastic means are well equipped to handle such problems. However, the algorithms inspired from plant root models are very limited.

This study presents a root growth model which can simulate plant root and formulated as an optimization algorithm for numerical function optimization. This paper is structured as follows: root growth model is presented in Section 2. The self-adaptive growth of root hairs is simulated in Section 3. In Section 4, experimental settings and experimental results are given. Finally, Section 5 concludes the paper.

2   Proposed Root Growth Model

2.1   Description for Growth Behavior of Root System

At the end of the 1960s, A. Lindenmayer introduced Transformational-generative grammar into biology and developed a variant of a formal grammar, namely L-Systems, most famously used to model the growth processes of plant development. L-Systems are based on simple rewriting rules and branching rules and successfully make a formal description for the plant growth. We use L-Systems to describe growth behavior of root system as follows:

- A seed germinates in the soil, partly becoming stems of plant above the earth's surface. The other part grows down, becoming plant root system. New root hairs grow from the root tips.
- More new root hairs grow from the root tips of old root hairs. The behavior of root system which is repeated is called as branching of the root tips.
- Most root hairs and root tips are similar to each other. The entire root system of plant has self-similar structure. The root system of each plant is composed of numerous root hairs and root tips with similar structure.

2.2   Plant Morphology

The uneven concentration of nutrients in the soil makes root hairs growing towards different directions. This characteristic of root growth relates to the morphogenesis model in biological theory. When the rhizome of root system grows, three or four growing points with different rotation directions will be generated at each root tip. The rotation diversifies the growth direction of the root tip. Root tips from which root hairs germinate contain undifferentiated cells. These cells are considered as fluid bags in which there are homogeneous chemical constituents. One of chemical constituents is a version of the growth hormone, called as morphactin. The morphactin concentration determines whether cells start to divide. When cells start to divide, root hairs appear.

With regard to the process of root growth, there have been the following conclusions in biology:

- If root system of plant have more than one root tip, which root tips could germinate root hairs depends on their morphactin concentration. The probability of new root hairs germinating is higher from root tips with larger morphactin concentration than root tips with less ones.
- The morphactin concentration in cells is not static, but depends on its surroundings. After new root hairs germinate and grow, the morphactin concentration will be reallocated among new root tips in line with the new concentration of nutrients in the soil.

In order to simulate the above process, it is assumed that the sum of the morphactin concentration of is constant (considered as 1) in the morphactin state space of the multi-cellular closed system. If there are n root tips x_i ($i=1, 2, \cdots, n$) which are D-dimensional vectors, the morphactin concentration of any cell is defined as E_i ($i=1, 2, \cdots, n$). The morphactin concentration of each root tip can be expressed as:

$$E_i = \frac{1/f(x_i)}{\sum_{i=1}^{n} 1/f(x_i)} \tag{1}$$

where $f(*)$ is objective function which represents the spatial distribution of nutrients in the soil. In expression (1), the morphactin concentration of each root tip is determined by the relative position of each point and environmental information (objective function value) at this position. When new root hairs germinate, the morphactin concentration may be changed.

2.3 Branching of the Root Tips

There are four rules for branching of the root as follow:

- Plant growth begins from a seed.
- In each cycle of growth process, some excellent root tips which have larger morphactin concentration values are selected to branch.
- The distance should not be close between the selected root tips in order to make spatial distribution of the root system wider and increase the diversity of the fitness values.
- If the number of the root tips selected equals the predefined value, the loop of the selection process terminates.

In order to produce a new growing point from the old root tip in memory, the proposed model uses the following expression:

$$pg_{ij} = \begin{cases} x_{ij} + (2 \times \delta_{ij} - 1) & j = k \\ x_{ij} & j \neq k \end{cases} \tag{2}$$

where $k \in \{1, 2, \cdots, D\}$ are randomly chosen indexes and $j \in \{1, 2, \cdots, D\}$. pg_i ($i=1, 2, \cdots, S$) are S new growing points. δ_{ij} is a random number between [-1, 1].

2.4   Root Hair Growth

Root hair growth depends on its growth angle and growth length. The growth angle is a vector for measuring the growth direction of root hair. The growth angle of each root hair φ_i (i=1, 2, $\cdots$, n) which is produced randomly can be expressed as:

$$(\phi_1, \phi_2, \cdots, \phi_D) = rand(D) \tag{3}$$

$$\varphi_i = \frac{(\phi_1, \phi_2, \cdots, \phi_D)}{\sqrt{\phi_1^2 + \phi_2^2 + \cdots + \phi_D^2}} \tag{4}$$

The growth length of each root hair is defined as δ_i (i=1, 2, $\cdots$, n) which is an important parameter in the root growth model. Some strategies of tuning the parameter can produce multiple versions of the root growth model. After growing, a new root tip is produced by the following expression:

$$x_i = x_i + \delta_i \varphi_i \tag{5}$$

2.5   Root Growth Algorithm

The root growth model proposed is instantiated as root growth algorithm (RGA) for simulation of root system of plant and numerical function optimization. The threshold of the distance between root tips and the growth length of each root hair are important parameters for RGA. The flowchart of the RGA is presented in Figure 1.

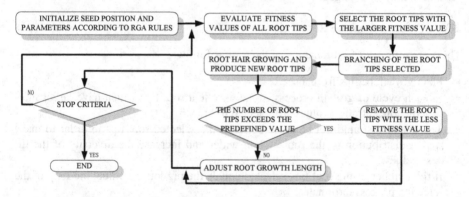

Fig. 1. The flowchart of the RGA

3   Simulating Self-adaptive Growth of Root Hairs

Simulating self-adaptive growth of root hairs is important for learning relationships between the nutrient concentration of the soil and the growth length of root hairs. Rastrigin function is used as soil environment in computer. Rastrigin function is presented in Section 4.1 and the setting of the parameters is the same as one in Section

4.2. The growth length of each root hair δ_i ($i = 1, 2, \cdots, n$) is the master variable for self-adaptive growth of root hairs. The growth length of all root hairs change as their fitness values change. The pseudocode and the formula of the way of self-adaptive growth are listed in Table 1 in which τ is a nonnegative number. Figure 6 shows the simulation of the self-adaptive growth way. The roots of plant grow towards the same direction with the high concentration of nutrients. From the view point of the optimization, the self-adaptive growth way can make all points moving toward the area with the largest fitness value and then these points move around back and forth for intensive search. So RGA using the way of self-adaptive growth can obtain better optimal solution easily and quickly.

Table 1. Pseudocode for self-adaptive growth way

Set: δ_i = Initial value
FOR each generation
   FOR each point x_i
     Set: $\delta_i = |E_i| / |E_i + \tau|$
   END FOR
END FOR

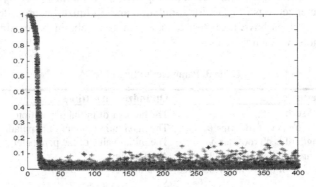

Fig. 2. The simulation of self-adaptive growth way

4   Numerical Experiments for Optimization

4.1   Benchmark Functions

The set of benchmark functions contains five functions that are commonly used in evolutionary computation literature [10–12] to show solution quality and convergence rate. The five function are Sphere function, Rosenbrock function, Ackley function, Griewank function and Rastrigin function. The first two functions are unimodal problems and the remaining functions are multimodal. The functions are listed below. The dimensions, initialization ranges, global optimum, and the criterion of each function are listed in Table 2.

Table 2. Parameters of the benchmark functions

Function	Dimensions	Initial range	Minimum
Sphere	30	$[-100, 100]^D$	0
Rosenbrock	30	$[-30, 30]^D$	0
Ackley	30	$[-32.768, 32.768]^D$	0
Griewank	30	$[-600, 600]^D$	0
Rastrigin	30	$[-10,10]^D$	0

4.2   Settings

For the experiments, the value of the common parameter, total evaluation number, used in each algorithm was chosen to be the same. The maximal number of fitness function evaluations was 200,000 for all functions.

For GA, a binary coded standard GA were employed. The rate of single point crossover operation was 0.8. Mutation rate was 0.01. The selection method was stochastic uniform sampling technique. Generation gap is the proportion of the population to be replaced. Chosen generation gap value in experiments was 0.9. For DE, F was set to 0.5. Crossover rate was chosen to be 0.9 as recommended in [13]. For PSO, inertia weight was 0.6 and cognitive and social components were both set to1.8 [14].

The parameter settings of RGA for simulation and optimization are listed in Table 3. All parameter values have been tested many times to obtain better simulation and optimal solution and then were used.

Table 3. Parameter setting of RGA

Simulation settings	Optimization settings	Value
The number of seed	The number of initial population	1
The maximum number of root tips	The maximum number of population	100
The initial length of each root fair δ_i	The initial value of the parameter δ_i	1
The distance threshold between root tips	A parameter	1
Branching number of each root tip selected	A parameter	4
S	S	4
α	α	1.2
β	β	0.8
τ	τ	0.25~1.5
λ	λ	200

4.3   Numerical Results and Comparison

The GA, DE, PSO and RGA algorithms are compared on five functions described in the previous section and are listed in Table 2. Each of the experiments in this section was repeated 30 times. The mean values, the standard deviation, the minimum values and the maximum values produced by the algorithms have been recorded in Table 4. The best values obtained by the four algorithms for each function are marked as bold. As shown in Table 4, RGA algorithm can obtain better performance than the other three algorithms on Sphere, Rosenbrock, Rastrigin, and Griewank functions while PSO algorithm shows better performance on Ackley. On Sphere function, PSO and

RGA obtained satisfying results. However, RGA algorithm showed the best performance. Rosenbrock function is a unimodal non-separable function. On Rosenbrock
function, RGA algorithm showed the best performance and PSO was a little worse
than RGA. GA and DE cannot obtain better performance on this function. Ackley and
Griewank functions are two multimodal non-separable functions. On these two functions, the results obtained by RGA and PSO were significantly better than GA and DE.
On Ackley function, RGA showed a better converge performance than PSO, but on
Griewank function, RGA converged better. PSO algorithm converged fast at the beginning and trapped in the local optimum soon on Griewank function. Rastrigin function is a multimodal variable-separable function. The convergence profile of RGA
was significantly better than the other three algorithms and GA, DE and PSO also
trapped in the local optimum soon on Rastrigin function.

Overall, RGA algorithm can obtain good performance on most functions with high-
er-dimension, especially on the multimodal variable-separable functions. RGA based
on root growth model is appropriate for numerical function optimization.

Table 4. Comparison of results obtained by GA, DE, PSO and RGA

Function		GA	DE	PSO	RGA
Sphere	Mean	0.92505	22.6724	2.25409e-008	**1.09852e-008**
	Std	1.29811	28.6836	1.76046e-008	**1.61596e-008**
	Min	0.50220	3.69461	1.06366e-008	**5.90561e-024**
	Max	3.19916	55.6673	4.27632e-008	**2.95410e-008**
Rosenbrock	Mean	4.30021e+003	11859.6	36.0147	**21.2087**
	Std	638.258	5221.57	37.1823	**0.808404**
	Min	3.66262e+003	7170.21	**12.1717**	20.3453
	Max	4.93916e+003	17486.4	78.8579	**21.9481**
Ackley	Mean	4.90392	12.8558	**6.44018e-007**	0.532827
	Std	0.42856	1.06169	**4.68484e-007**	0.460087
	Min	4.57918	11.6323	**2.42885e-007**	0.00157042
	Max	5.38955	13.5347	**1.1589e-006**	0.80068
Griewank	Mean	1.81473	1.38865	0.0338252	**0.0180136**
	Std	0.20679	0.290332	0.0483038	**0.0214339**
	Min	1.58355	1.05381	4.98359e-006	**3.53763e-007**
	Max	1.98206	1.57037	0.0891463	**0.0417191**
Rastrigin	Mean	87.1626	156.074	28.5221	**4.28574e-004**
	Std	12.4927	70.4707	9.02802	**9.69091e-005**
	Min	79.1588	96.8458	21.8891	**3.35807e-004**
	Max	101.558	234.011	38.8034	**5.29153e-004**

5   Conclusion

In this paper, we present a root growth model based on root growth behaviours in the
soil. By using this model as a computational metaphor, we propose a novel algorithm
called "root growth algorithm" (RGA) for simulation of plant root system and

numerical function optimization. The self-adaptive growth of root hairs, are simulated and he characteristics of root growth are showed in the form of images. The numerical results obtained from the proposed algorithm have been compared with those obtained from GA, DE and PSO. It is seen from the comparison that RGA performs better than GA, DE and PSO. RGA can optimize numerical function better.

References

1. Gerwitz, A., Page, E.R.: An Empirical Mathematical Model to Describe Plant Root Systems. Journal of Applied Ecology 11(2), 773–781 (1974)
2. Hodge, A.: Root Decisions. Plant, Cell and Environment 32(6), 628–640 (2009)
3. Leitner, D., Klepsch, S., Bodner, G., Schnepf, A.: A Dynamic Root System Growth Model Based on L-Systems. Plant Soil 332, 177–192 (2010)
4. Eberhart, R.C., Shi, Y., Kennedy, J.: Swarm Intelligence. Morgan Kaufmann (2001)
5. Holland, J.H.: Adaptation in Natural and Artificial Systems. University of Michigan Press, Ann Arbor (1975)
6. Kennedy, J., Eberhart, R.C.: Particle Swarm Optimization. In: 1995 IEEE International Conference on Neural Networks, vol. 4, pp. 1942–1948. IEEE Press, New York (1995)
7. Castro, D.L.N., Zuben, V.F.J.: Artificial Immune Systems, Part I. Basic Theory and Applications, Technical Report Rt Dca 01/99, Feec/Unicamp, Brazil (1999)
8. Karaboga, D., Basturk, B.: On the Performance of Artificial Bee Colony (ABC) Algorithm. Applied Soft Computing 8(1), 687–697 (2008)
9. Cai, W., Yang, W., Chen, X.: A Global Optimization Algorithm Based on Plant Growth Theory: Plant Growth Optimization. In: International Conference on Intelligent Computation Technology and Automation (ICICTA), vol. 1, pp. 1194–1199 (2008)
10. Krink, T., Vestertroem, J.S., Riget, J.: Particle Swarm Optimization with Spatial Particle Extension. In: Proceedings of the IEEE Congress on Evolutionary Computation, Honolulu, Hawaii, pp. 1474–1479. IEEE Press, New York (2002)
11. Shi, Y., Ebrehart, R.C.: A Modified Particle Swarm Optimizer. In: Proceeding of the 1998 IEEE International Conference on Evolutionary Computation, Piscataway, NJ, pp. 69–73 (1998)
12. Shi, Y., Eberhart, R.C.: Empirical Study of Particle Swarm Optimization. In: Proceedings of the 1999 IEEE Congress on Evolutionary Computation, Piscataway, NJ, pp. 1945–1950. IEEE Press, New York (1999)
13. Corne, D., Dorigo, M., Glover, F.: New Ideas in Optimization. McGraw-Hill (1999)
14. Vesterstrom, J., Thomsen, R.: A Comparative Study of Differential Evolution Particle Swarm Optimization and Evolutionary Algorithms on Numerical Benchmark Problems. In: IEEE Congress on Evolutionary Computation (CEC 2004), pp. 1980–1987. IEEE Press, New York (2004)

Bacterial-Inspired Algorithms for Engineering Optimization

Ben Niu[1,2,3,4,*], Jingwen Wang[1], Hong Wang[1], and Lijing Tan[5,*]

[1] College of Management, Shenzhen University, Shenzhen 518060, China
[2] Hefei Institute of Intelligent Machines, Chinese Academy of Sciences, Hefei 230031, China
[3] e-Business Technology Institute, The University of Hongkong, Hongkong, China
[4] Institute for Cultural Industries, Shenzhen University, Shenzhen 518060, China
[5] Management School, Jinan University, Guangzhou 510632, China
drniuben@gmail.com

Abstract. Bio-inspired optimization techniques using analogy of swarming principles and social behavior in nature have been adopted to solve a variety of problems. In this paper, Bacterial foraging optimization (BFO) was employed to achieve high-quality solutions to engineering optimization problems. Two modifications of BFO, BFO with linear decreasing chemotaxis step (BFO-LDC) and BFO with non-linear decreasing chemotaxis step (BFO-NDC) were proposed to further improve the performance of the original algorithm. In order to illustrate the efficiency of the proposed method (BFO-LDC and BFO-NDC) for engineering problem, an engineering design problem was selected as testing functions, and the performance is compared against some state-of-the-art approaches. The experimental results demonstrated that the modified BFOs are of greater efficiency and can be used as general approach for engineering problems.

Keywords: Engineering problem, constrained handling, optimization, bacterial foraging, linear decreasing chemotaxis, non-linear decreasing chemotaxis.

1 Introduction

In 2002, a new optimization algorithm based on foraging behavior of bacteria was introduced by Passino [1]. Although lots of significant research articles published so far have focused on analysis of the foraging behavior and self adaptability properties of BFO as an unconstrained optimizer, till date, little such analysis has been done for solving the constrained engineering optimization problems.

Constrained Optimization Problems arise in numerous applications [2]. Yet constraint handing remains to be one of the most difficult parts encountered in practical engineering design optimization. Real-world limitations frequently introduce multiple, non-linear and non-trivial constraints on a design so that feasible solutions would be restricted to a small subset of the design space. In general, engineering optimization problem can be defined as follows:

---

[*] Corresponding authors.

D.-S. Huang et al. (Eds.): ICIC 2012, LNCS 7389, pp. 649–656, 2012.
© Springer-Verlag Berlin Heidelberg 2012

$$\text{Minimize } f(X), X = \{x_1, x_2, \ldots, x_n\} \in R^n$$

$$\text{subject to: } g_i(X) \leq 0, i = 1, 2, \ldots, p \tag{1}$$

$$\text{and } \quad h_i(X) = 0, i = 1, 2, \ldots, m$$

$$\text{where } \quad l_i \leq x_i \leq u_i, i = 1, 2, \ldots, n$$

The objective function f is defined on an n-dimensional search space in R^n. Constraints g_i and equality constraints h_i are called active constraints.

Many approaches, especially evolution algorithms, have received a lot of attention regarding their potential for solving constrained engineering optimization problems over the past several years. The society and civilization simulation proposed by Ray & Liew [3], made use of intra- and intersociety interactions within a formal society and civilization model to solve engineering optimization problems. Belegundu [4] used Various mathematical programming methods for optimal design of structural systems. C.Coello [5] proposed an evolution strategy which is called Simple multi-membered Evolution Strategy (SMES) to solve global nonlinear optimization problems. However, most of the work is centered on some algorithms such as Particle Swarm Optimization (PSO), Genetic Algorithm (GA) or other evolutionary algorithms. In this paper, the performance of two Bacterial-inspired Algorithms proposed by Niu B. et al. [6] −BFO-LDC and BFO-NDC in solving constrained engineering problems was investigated and compared with results from other methods.

The paper is organized as follows: the description of the two improved algorithms-BFO-LDC and BFO-NDC is provided in section 2. Then, the process of solving constrained engineering optimization problems by two bacterial-inspired algorithms will be presented in section 3, together with constraint handling, pseudo-code of algorithm, experimental settings and the results. Finally, the paper ends with conclusions and ideas for future work reported in section 4.

2   Modified Bacterial Foraging Optimization

Bacterial Foraging Optimization (BFO) algorithm has been applied to model E. coli foraging behavior and optimization problems. Bacteria have the tendency to migrate toward the nutrient-rich areas and this behavior is termed 'chemotaxis'. In the original BFO, each bacterium updated its position of chemotaxis procedure is the key step. However, the chemotaxis step length C is a constant. If this rule holds true, balancing between the global and local search ability will be difficult and as a result, the accuracy and searching speed will be affected. To solve this problem, after solving multi-objective optimization [7] and portfolio optimization by symbiotic multi-swarm PSO [8], Niu B. et al. [6]proposed a simple scheme to modulate the chemotaxis step size to improve its convergence behavior. In the following section, two novel variants of original BFO were proposed, i.e. BFO-LDC that employed linear variation and BFO-NDC with non-linear variation of chemotaxis step length.

2.1   BFO with Linear Decreasing Chemotaxis Step (BFO-LDC)

In this method, a linearly varying chemotaxis step length over iterations is used. The chemotaxis step length starts with a high value C_{max} and linearly decreases to C_{min} at the maximal number of iterations. The mathematical representations of the BFO method are given as shown in:

$$C_j = C_{min} + \frac{iter_{max} - iter}{iter_{max}}(C_{max} - C_{min}) \tag{2}$$

where $iter_{max}$ is the maximal number of iterations, j is the current number of iterations, jth is the chemotaxis step. With $C_{max} = C_{min}$, the system becomes a special case of fixed chemotaxis step length, as the original proposed BFO algorithm. From hereafter, this BFO algorithm will be referred to as bacterial foraging optimizer with linear decreasing chemotaxis (BFO-LDC).

2.2   BFO with Non-linear Decreasing Chemotaxis Step (BFO-NDC)

Different from the linearly decreasing strategy, a nonlinear function modulated chemotaxis step adaptation with time is used to improve the performance of BFO algorithm. The key element is the determination of the chemotaxis step length as a nonlinear function of the present iteration number at each time step. The proposed adaptation of C is given as:

$$C_j = C_{min} + \exp(-\lambda \times (\frac{iter}{iter_{max}})^2) \times (C_{max} - C_{min}) \tag{3}$$

where $iter_{max}$, $iter$ is the same as linearly decreasing strategy. λ, the nonlinear modulation index. The system starts with a high initial chemotaxis step length (C_{max}) which should allow it to explore new search areas aggressively and then decreases it gradually according to (4) following different paths for different values of λ to reach C_{min} at $iter = iter_{max}$.

3   Solving Engineering Optimization Problems by Bacterial-Inspired Algorithms

3.1   Constraint Handling-Penalty Function Approach

Penalty function approach is the most widely used method for problems with evolutionary algorithm. This approach converts the constrained problem to an unconstrained one by introducing a penalty term into the original objective function to penalize constraint violations. Through this, unconstrained optimization algorithm, the violations can be minimized. If the penalty values are high, the minimization algorithms are likely to be trapped in local minima. On the other hand, if penalty values are low, feasible optimal solutions can hardly be detected.

Penalty functions can be grouped into two main categories: stationary and non-stationary. Stationary penalty functions use fixed penalty values throughout the minimization while non-stationary penalty functions use dynamically modified penalty values. In this paper, the stationary one is adopted.

The basic form of the penalty function is as follows:

$$F(x) = f(x) + \Sigma M \times G(x) + \Sigma M \times H(x)$$

$$G(x) = \max[0, g(x)] \times a \tag{4}$$

$$H(x) = |h(x)| \times b$$

$f(x)$ is the original objective function of the constrained optimization problem, $F(x)$ is the new objective function , M is a constant called penalty factor. An initial value of M has to be provided by the user and there are no rigorous ways of finding this initial value. This getting the fittest penalty factor is the most difficult part in using the method we may face. $G(x)$ and $H(x)$ are functions of the constraints, $g(x)$ and $h(x)$ are the original constrains. Finally, a and b are both constant, the value is one or two. In our experiments, a stationary assignment penalty function was used, which means M is fixed.

3.2   BFOs for Engineering Optimization

We illustrate the pseudo-code of the modified BFOs. The main features of BFO-LDC and BFO-NDC can be summarized in Table 1.

3.3   Settings

Bacterial foraging optimization algorithm includes six parameters. S is population size, N_c is chemotactic step, N_{re} is the number of reproduction steps to be taken, N_{ed} is the number of elimination-dispersal events, and for each elimination-dispersal event each bacterium in the population is subjected to elimination-dispersal with probability P_{ed} , N_s is the largest number of step along the random search direction. For the engineering problem in the experiment, we choose $S = 50$, $N_c = 1000$, $N_{re} = 5$, $N_{ed} = 2$, $N_s = 4$,and $P_{ed} = 0.25$ .The chemotaxis step length starts with a high value $C_{start} = 0.2$ and linearly decreases to $C_{min} = 0.01$ in BFO-LDC , and $C_{start} = 0.6$, $C_{min} = 0.05$ in BFO-NDC. In penalty function, $M = 50000, a = b = 1$.

Table 1. The pseudo-code of BFO-LDC and BFO-NDC

INITIALIZE.

Set parameters S , N_s , N_c , N_{re} , N_{ed} , P_{ed} , θ^i , C_{start} , C_{min} , λ , $C(i)$

The chemotaxis step start length C_{start} and it decreases linearly to the chemotaxis step length C_{min} , nonlinear modulation index (λ).

Constrain handing by Penalty Function Approach as penalty function 4.

WHILE (the termination conditions are not met)

FOR ($l=1:N_{ed}$) Elimination-Dispersal loop

FOR ($k=1:N_{re}$) Reproduction loop

FOR ($j=1:N_c$) Chemotaxis loop

    For each bacterium i

      Tumble: Generate a random vector $\Delta(i)\in R^n$.

      Move: Let

$$\theta^i(j+1,k,l) = \theta^i(j,k,l) + c(i)\frac{\Delta(i)}{\sqrt{\Delta^T(i)\Delta(i)}}$$

      This results in a step of the start size C_{start} in the direction of the tumble for bacteria i.

      Swim: Let $m=0$ (counter for swim length).

        While ($m<N_s$) $m=m+1$.

          If $J(i,j+1,k,l)<J_{last}$, let $J_{last}=J(i,j+1,k,l)$;

          Calculate the new $J(i,j+1,k,l)$ using $\theta^i(j+1,k,l)$

          Else

          let $m=N_s$.

          END

      END

END FOR Chemotaxis loop end

    Select highest J^i_{health} bacteria and reproduce.

END FOR Reproduction loop end

    With probability P_{ed} , eliminates and disperse each bacteria.

END FOR Elimination and Dispersal loop end

END WHILE

3.4   Experiments and Results

In this section, a well-known engineering problem---*A tension/compression spring design*---from the real-world optimization literature is used. This problem posed a challenge for constraint-handling engineering problem and is good measurements for testing the ability of the proposed algorithms.

The constrained optimization problem is the minimization of the weight of spring. It consists of minimizing the weight of a tension/compression spring subject to constraints on shear stress, surge frequency and minimum deflection. The design variables are the mean coil diameter $D(= x_1)$, the wire diameter $d(= x_2)$ and the number of active coils $N(= x_3)$, and the problem involves four nonlinear inequality constrains. The best feasible solution found by BFO-LDC and BFO-NDC is $f(x^*) = 0.01273838$ and $f(x^*) = 0.01275064$ respectively. The problem has been studied by Ray and Liew [4], CDE [9], Belegundu [3], Arora [10] and Mahdavi et al. [11]. The problem can be described as follows:

Minimize:

$$f(x) = (2\sqrt{2}x_1 + x_2) \times l \tag{5}$$

Subject to:

$$g_1(x) = \frac{\sqrt{2}x_1 + x_2}{\sqrt{2}x_1^2 + 2x_1 x_2} P - \sigma \le 0$$

$$g_2(x) = \frac{x_2}{\sqrt{2}x_1^2 + 2x_1 x_2} P - \sigma \le 0$$

$$g_3(x) = \frac{1}{\sqrt{2}x_2 + x_1} P - \sigma \le 0$$

where $0 \le x_1 \le 1$ and $0 \le x_2 \le 1$, $l = 100cm$, $P = 2KN/cm^2$, and $\sigma = 2KN/cm^2$.

Five independent runs are performed in MATLAB 7.0. And then we measured the quality of results and the robustness of BFO-LDC and BFO-NDC (the standard deviation values) by compared with other methods. The statistical results are summarized in Table 2. As it can be seen, both BFO-LDC and BFO-NDC outperformed the two compared approaches proposed by Belegundu and Mahdavi et al. Our method and Arora's showed similar performances. Although Ray and Liew and CDE gave better results than BFO-LDC or BFO-NDC, our method showed better performance in terms of robustness. When the computational cost (i.e., the number of fitness functions evaluations---FFEs) is concerned, it could be noted that the results of CDE requires 240,000 FFEs, while those of BFO-LDC and BFO-NDC were obtained after only 50,000 FFEs. So we can conclude that the computational cost of BFO-LDC and BFO-NDC is less than that of CDE.

Table 2. Comparing of tension/compression spring design problem results of BFO-LDC and BFO-NDC with respect to the other state-of-the-art algorithms

Design problem	Best	Mean	Worst	SD
BFO-LDC	0.01273838	0.01275498	0.01277329	1.3774e-005
BFO-NDC	0.01275064	0.01277656	0.01279454	1.9135e-005
BFO	0.01273111	0.01276133	0.01281023	2.9571e-005
Ray and Liew	0.012669249	0.012922669	0.016717272	5.9e-04
CDE	0.0126702	0.012703	0.012790	2.7e-05
Belegundu	0.012833	NA	NA	NA
Arora	0.012730	NA	NA	NA
Mahdavi et al.	0.0128874	NA	NA	NA

4   Conclusions and Future Work

It is well known that nearly all engineering optimization problems in the real world would involve multiple, non-linear and non-trivial constraints due to many limitations. From an engineering standpoint, a better, faster, cheaper solution is always desired.

In this paper, two novel variants of the original BFO were proposed: BFO-LDC that employed linear variation and BFO-NDC with non-linear variation of chemotaxis step length. The experimental results showed that the proposed BFO-LDC and BFO-NDC algorithms are effective methods for handling constraints of engineering problems.

Future works should focus on comparing the two proposed methods BFO-LDC and BFO-NDC with PSO, GA and other algorithms and effort should be made to optimize the performance of them. In addition, studying of the applications in more complex practical optimization problems in engineering is necessary for thorough investigation of the properties and performances of BFO-LDC and BFO-NDC. Meanwhile, we are also setting about to explore other proposed methods for chemotaxis step to improve the performance of BFO.

Acknowledgements. This work is supported by National Natural Science Foundation of China (Grant No.71001072, 60905039, China Postdoctoral Science Foundation (Grant No. 20100480705), Science and Technology Project of Shenzhen (Grant No. JC201005280492A), The Natural Science Foundation of Guangdong Province (Grant no. 9451806001002294).

References

1. Passino, K.M.: Biomimicry of Bacterial Foraging for Distributed Optimization and Control. IEEE Control Systems Magazine 22(3), 52–67 (2002)
2. Michalewicz, Z., Schoenauer, M.: Evolutionary Algorithms for Constrained Parameter Optimization Problems. Evolutionary Computation 4(1), 1–32 (1996)
3. Ray, T., Liew, K.M.: Society and Civilization: An Optimization Algorithm Based on the Simulation of Social Behavior. IEEE Transactions on Evolutionary Computation 7(4), 386–396 (2003)

4. Belegundu, A.D.: A Study of Mathematical Programming Methods for Structural Optimization. Science and Engineering 43(12), 383 (1983)
5. Coello, C.A.C.: Constraint-handling in Genetic Algorithms Through The Use of Dominance-based Tournament Selection. Advanced Engineering Informatics 16, 193–203 (2002)
6. Niu, B., Fan, Y., Wang, H., Li, L., Wang, X.F.: Novel Bacterial Foraging Optimization with Time-varying Chemotaxis Step. International Journal of Artificial Intelligence 7(A11), 257–273 (2011)
7. Niu, B., Wang, H., Tan, L.J., Xu, J.: Multi-objective Optimization Using BFO Algorithm. In: Huang, D.-S., Gan, Y., Premaratne, P., Han, K. (eds.) ICIC 2011. LNCS (LNBI), vol. 6840, pp. 582–587. Springer, Heidelberg (2012)
8. Niu, B., Xue, B., Li, L., Chai, Y.: Symbiotic Multi-swarm PSO for Portfolio Optimization. In: Huang, D.-S., Jo, K.-H., Lee, H.-H., Kang, H.-J., Bevilacqua, V. (eds.) ICIC 2009. LNCS, vol. 5755, pp. 776–784. Springer, Heidelberg (2009)
9. Amirjanov, A.: The Development of a Changing Range Genetic Algorithm. Computer Methods in Applied Mechanics and Engineering 195, 2495–2508 (2006)
10. Arora, J.S.: Introduction to Optimum Design. McGraw-Hill, New York (1989)
11. Mahdavi, M., Fesanghary, M., Damangir, E.: An Improved Harmony Search Algorithm for Solving Optimization Problems. Applied Mathematics and Computation 188(2), 1567–1579 (2007)

Multiobjective Dynamic Multi-Swarm Particle Swarm Optimization for Environmental/Economic Dispatch Problem

Jane-Jing Liang[1], Wei-Xing Zhang[1], Bo-Yang Qu[2], and Tie-Jun Chen[1]

[1] School of Electrical Engineering, Zhengzhou Univerisity, Zhengzhou, China
[2] School of Electric and Information Engineering,
Zhongyuan University of Technology, Zhengzhou, China
{liangjing,tchen}@zzu.edu.cn,
boystarboy@163.com, e070088@e.ntu.edu.sg

Abstract. This paper presents a new multiobjective particle swarm optimization (MOPSO) technique to solve environmental/economic dispatch (EED) problem. The EED problem is a non-linear constrained multiobjective optimization problem. The Multi-objective Dynamic Multi-Swarm Particle Swarm Optimizer (DMS-MO-PSO) proposed employs novel pbest and lbest updating criteria which are more suitable for solving multi-objective problems. In this work, the standard IEEE 30-bus six-generator test system is used and simulation results showed that the proposed approach is efficient and confirms its potential to solve the multiobjective EED problem.

Keywords: environmental/economic dispatch, particle warm optimization, multi-objective optimization, evolutionary algorithm.

1 Introduction

The passage of the clean air act amendments in 1990 has forced the utilities to reduce their SO_2 and NO_X emissions by 40 percent from 1980 levels [1].Therefore, not only cost but also emission objective have to be considered. Environmental/Economic dispatch (EED) is a multi-objective problem having conflicting objectives, as the minimization of cost and minimization of the pollution. This leads to a trade-off analysis to define admissible dispatch policies for any demand level [2].

In the past decade, the meta-heuristic optimization methods have been used to solve EED problems primarily due to their nice feature of population-based search [3]. With the development of multi-objective evolutionary, a number of effective multi-objective evolutionary search strategies [4]such as the novel Nondominated Sorting Genetic Algorithm (NSGA) [5], Niched Pareto Genetic Algorithm (NPGA)[6], Strength Pareto Evolutionary Algorithm (SPEA)[7] and NSGA-II [8] have been successful used to solve EED problem.

Particle swarm optimization (PSO) is an effective search method that based on swarm intelligence. Although PSO has been applied to many areas [9]-[13], few

D.-S. Huang et al. (Eds.): ICIC 2012, LNCS 7389, pp. 657–664, 2012.
© Springer-Verlag Berlin Heidelberg 2012

efficient works have been reported to implement MOPSO for solving EED problems. In this paper, the Multiobjective Dynamic Multi-Swarm Particle Swarm Optimizer is proposed for Environmental/Economic Dispatch Problem. The DMS-MO-PSO [14] algorithm uses an external archive and a novel pbest and lbest updating strategy to solve the EED problem.

The rest of the paper is organized as follows. In section 2, we present the problem formulation. Section 3 and 4 gives a brief description about the DMS-MO-PSO algorithm and constrain method used respectively. The experimental results and discussions are provided in Section 5 and Section 6 concludes the paper.

2   Problem Formulation

2.1   Problem Objectives

Objective 1: The total fuel cost $F(P_G)$ can be expressed as

$$F(P_G) = \sum_{i=1}^{N} (a_i + b_i * P_{G_i} + c_i * P_{G_i}^2) \tag{1}$$

Where N is the number of generators, a_i, b_i, c_i are the cost coefficients of the *ith* generator, and P_{G_i} is the real power output of the *ith* generator. P_G is the vector of real power outputs and defined as

$$P_G = [P_{G1}, P_{G2}, \cdots, P_{G_N}]^T \tag{2}$$

Objective 2: The total emission $E(P_G)$ can be expressed as

$$E(P_G) = \sum_{1}^{N} 10^{-2} (\alpha_i + \beta_i P_{G_i} + \gamma_i P_{G_i}^2) + \xi_i \exp(\lambda_i P_{G_i}) \tag{3}$$

Where $\alpha_i, \beta_i, \gamma_i, \xi_i, \lambda_i$ are coefficients of the *ith* generator emission characteristics.

2.2   Problem Constraints

Constraint 1: Generation capacity constraint
For stable system operation, real power output of each generator is restricted by lower and upper limits as follows:

$$P_{G_i}^{min} \leq P_{G_i} \leq P_{G_i}^{max}, i = 1, \cdots, N \tag{4}$$

Where $P_{G_i}^{min}$ and $P_{G_i}^{max}$ are the minimum and maximum power generated by the *ith* generator.

Constraint 2: Power balance constraint

The total power generation must cover the total demand P_D and the real power loss in transmission lines P_{loss}. This relation can be expressed as

$$\sum_{i=1}^{N} P_{G_i} - P_D - P_{loss} = 0 \tag{5}$$

As a matter of fact, the power loss in transmission lines can be calculated by different methods such as B matrix loss formula method. The system loss formula can be expressed as follows:

$$P_{loss} = \sum_{i=1}^{N} \sum_{j=1}^{N} P_{G_i} B_{ij} P_{G_i} + \sum_{i=1}^{N} P_{G_i} B_{i0} + B_{00} \tag{6}$$

where B_{ij} is the transmission loss coefficient, B_{i0} is the ith element of the loss coefficient vector. B_{00} is the loss coefficient constant. B-coefficients are needed, they are shown as follows:

$$B_{ij} = \begin{bmatrix} 0.0218 & 0.0107 & -0.00036 & -0.0011 & 0.00055 & 0.0033 \\ 0.0107 & 0.01704 & -0.0001 & -0.00179 & 0.00026 & 0.0028 \\ -0.0004 & -0.0002 & 0.02459 & -0.01328 & -0.0118 & -0.0079 \\ -0.0011 & -0.00179 & -0.01328 & 0.0265 & 0.0098 & 0.0045 \\ 0.00055 & 0.00026 & -0.0118 & 0.0098 & 0.0216 & -0.0001 \\ 0.0033 & 0.0028 & -0.00792 & 0.0045 & -0.00012 & 0.02978 \end{bmatrix}$$

$$B_{i0} = \begin{bmatrix} 0.010731e-3 & 1.7704e-3 & -4.0645e-3 & 3.8453e-3 & 1.3832e-3 & 5.5503e-3 \end{bmatrix}$$

$$B_{00} = 0.0014$$

2.3   The Standard IEEE 30-Bus 6-Generator Test System [15].

In this paper the standard IEEE 30-bus 6-generator system is used. The power system's total demand is 2.834 p.u. The system parameters are listed in Table1, including fuel cost and emission coefficients and each generator's limit.

Table 1. Parameters of the standard IEEE 30-bus six-generator test

	G_1	G_2	G_3	G_4	G_5	G_6
a	10	10	20	10	20	10
b	200	150	180	100	180	150
c	100	120	40	60	40	100
α	4.091	2.543	4.258	5.326	4.258	6.131
β	-5.554	-6.047	-5.094	-3.55	-5.094	-5.555
γ	6.49	5.638	4.586	3.38	4.586	5.151
ξ	2.0e-4	5.0e-4	1.0e-6	2.0e-3	1.0e-6	1.0e-5
λ	2.857	3.333	8.000	2.000	8.000	6.667
P_i^{min}	0.05	0.05	0.05	0.05	0.05	0.05
P_i^{max}	0.50	0.60	1.00	1.20	1.00	0.60

3   DMS-MO-PSO

The dynamic multi-swarm particle swarm optimizer (DMS-PSO) [16-17] is con-structed based on the local version of PSO. In the proposed approach, the population was divided into small sized swarms and each swarm uses its own members to search the space. Every R generations, the population is regrouped randomly and starts searching using a new configuration of swarms. R is called regrouping period. In this way, the good information obtained by each swarm is exchanged among the swarms and the diversity of the population is increased simultaneously.

Base on DMS-PSO, an extended version for multi-objective optimization is pro-posed in [14]. An external archive is added to keep a historical record of the non-dominated solutions obtained during the search process. The maximum size of the archive N_{max} is predefined. The technique of updating the external archive is similar to the NSGAII. Different from other multi-objective PSOs, in DMS-MO-PSO, lbest are chosen from the best non-dominated solutions set in the external archive.

4   Constraints Handling

To solving multi-objective optimization problems with constraints, we need to adopt certain constraint handling method. In this paper, the superiority of feasible solutions technique is used [18]. In this technique, feasible solutions always dominate infeasi-ble. Thus in order to simplify the constraints handling mechanism, we modify the objective functions of EED problems as follows:

$$\underset{P_G}{Minimize} \quad [F'(P_G), E'(P_G)] \tag{8}$$

$$V(P_G) = g(P_G) + |h(P_G)|$$

$$if \quad V(P_G) = 0, F(P_G)' = F(P_G), \qquad E(P_G)' = E(P_G)$$

$$if \quad V(P_G) > 0, F(P_G)' = \max(F(P_G)) + V, \quad E(P_G)' = \max(E(P_G)) + V$$

Here max $(F(P_G))$ and   max$(E(P_G))$ are the maximum F and E value found so far.

5   Experimental Results and Discussions

5.1   Experimental Setup

To demonstrate the effectiveness of the proposed approach, two cases have been con-sidered which are listed as follows:

Case 1: The generation capacity and the power balance constraints without consi-dering P_{loss}.

Case 2: The generation capacity and the power balance constraints with consider-ing P_{loss}.

The results obtained by proposed approach are compared with FCPSO [6], NSGA [5], NPGA [7], SPEA [10], MOPSO [19], IMOPSO [19]. As reported in literature, for all the six compared algorithms, the size of the archive is set as 50, the population size is also selected as 50 and the maximum generation number is 7000. For DMS-MO-PSO, the sub-swarm number is set to 5 and in each sub-swarm there are five particles. Thus the population size is 25, which is just a half of others. And the maximum number of iteration is set to 3000, which means we only use about 21% fitness evaluations for DMS-MO-PSO.

5.2   Experimental Results

Fig. 1 shows the pareto-optimal front of the proposed approach of case 1. Table 2 and 3 show the two non-dominated solutions that represent the best cost and best emission for case 1. It can be seen from these tables DMS-MO-PSO is able to generate better or similar performance with much less number of function evaluations.

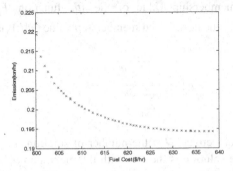

Fig. 1. DMS-MO-PSO Pareto-optimal front in case 1

Table 2. The result of best cost in case 1

	NPGA [6]	NSGA [5]	SPEA [7]	FCPSO [10]	MOPSO [19]	IMPSOD E[19]	DMS-MO-PSO
P1	0.1080	0.1567	0.1099	0.1070	0.1018	0.1060	0.1102
P2	0.3284	0.2870	0.3186	0.2897	0.2900	0.3063	0.2998
P3	0.5386	0.4671	0.5400	0.5250	0.5404	0.5209	0.5250
P4	1.0067	1.0467	0.9903	1.0150	1.0175	1.0140	1.0169
P5	0.4949	0.5037	0.5336	0.5300	0.5292	0.5262	0.5231
P6	0.3574	0.3729	0.36507	0.3673	0.3548	0.3602	0.3590
Cost	600.259	600.572	600.22	600.131	600.142	600.118	**600.1116**
Emission	0.22116	0.22282	0.2206	0.22226	0.22282	0.2220	0.2222

Table 3. The result of best emission in case 1

	NPGA [6]	NSGA [5]	SPEA [7]	FCPSO [10]	MOPSO [19]	IMPSO E [19]	DMS-MO-PSO
P1	0.4002	0.4394	0.4240	0.4097	0.3993	0.4018	0.3964
P2	0.4474	0.4511	0.4577	0.4550	0.4536	0.4583	0.4572
P3	0.5166	0.5105	0.5301	0.5363	0.5478	0.5426	0.5318
P4	0.3688	0.3871	0.3721	0.3842	0.3801	0.3844	0.4095
P5	0.5751	0.5553	0.5311	0.5348	0.5408	0.5415	0.5367
P6	0.5259	0.4905	0.5180	0.5140	0.5124	0.5052	0.5025
Cost	639.182	639.231	640.42	638.357	637.969	637.775	635.4356
Emission	0.19433	0.19436	0.1942	0.1942	0.19421	0.1942	**0.1942**

Beside best cost and best emission, the most important criterion is to select the best compromise solution from the final non-dominated solutions. The best solution will be extracted by using a fuzzy-based mechanism [20] that defined as follows:

$$
\mu_i = \begin{cases} 1 & if \quad F_i \leq F_i^{min} \\ \dfrac{F_i^{max} - F_i}{F_i^{max} - F_i^{min}} & if \quad F_i^{min} \leq F_i \leq F_i^{max} \\ 0 & if \quad F_i \geq F_i^{max} \end{cases} \tag{9}
$$

where F_i is the *ith* objective function of a solution in the Pareto-optimal. F_i^{max} and F_i^{min} are the maximum and minimum values of the objective function, μ_i stands for the membership value of the *ith* function(F_i). For each non-dominated solution k, the normalized membership value $\mu[k]$ is calculate using

$$
\mu[k] = \frac{\sum\limits_{i=1}^{M} \mu_i[k]}{\sum\limits_{j=1}^{N} \sum\limits_{i=1}^{M} \mu_i[j]} \tag{10}
$$

M is the number of objectives, N is the number of solutions in pareto-optimal front. The best compromise solution is the one with the maximum $\mu[k]$. The best compromise solutions of these six algorithms for case 1 are calculated listed in the Table 4.

Table 4. Compromise solution in case 1

	NPGA [5]	NSGA [4]	SPEA [6]	MOPSO [19]	IMPSODE [19]	DMS-MO-PSO
P1	0.2696	0.2571	0.2623	0.2484	0.2445	0.2415
P2	0.3673	0.3774	0.3765	0.3779	0.3771	0.3741
P3	0.5594	0.5381	0.5428	0.5409	0.5416	0.4974
P4	0.6496	0.6872	0.6838	0.7117	0.7190	0.7270
P5	0.5396	0.5404	0.5381	0.5324	0.5229	0.5631
P6	0.4486	0.4337	0.4305	0.4224	0.4346	0.4273
Cost	612.127	610.067	610.300	608.737	608.414	608.1710
Emission	0.19941	0.20060	0.2004	0.2015	0.2018	0.2022

From Table 4, it is clear that using the proposed DMS-MO-PSO method can give us a stable result with much less number of function evaluations. The better performance is due to the high diversity and fast converge speed generated by the dynamic multiple swarms.

Fig. 2 shows the pareto-optimal front obtained by the proposed approach for case 2. The best cost, best emission, and compromise solutions found by DMS-MO-PSO for case 2 are presented in Table 5.

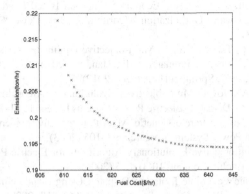

Fig. 2. DMS-MO-PSO Pareto-optimal front in case 2

Table 5. Results achieved for case 2

	Best Cost	Best Emission	Compromise Solution
P1	0.1214	0.3891	0.2423
P2	0.3012	0.4565	0.3672
P3	0.6203	0.5505	0.5806
P4	0.9592	0.4173	0.7496
P5	0.5157	0.5392	0.5234
P6	0.3528	0.5127	0.4031
Cost	608.8074	641.6982	614.1244
Emission	0.2186	0.1942	0.2033

6   Conclusions

In this work, a novel DMS-MO-PSO has been employed to solve the environmental / economic dispatch problem. The proposed algorithm is compared with a number of algorithms in the literature. Due to the high diversity and fast converge speed generated by the dynamic multiple swarms; DMS-MO-PSO is able to generate stable and satisfactory performance over multi-objective problems. From the experimental results, we can observe that the proposed algorithm performs well on the EED problems.

Acknowledgment. This research is partially supported by National Natural Science Foundation of China (Grant NO. 60905039) and Postdoctoral Science Foundation of China (Grants 20100480859).

References

1. IEEE Current Operating Problems Working Group, Potential Impacts of Clean Air Regulations on System Operations, pp. 647–656 (1995)
2. Zahavi, J., Eisenberg, L.: An Application of the Economic-environmental Power Dispatch. IEEE Trans. Syst., Man, Cybernet., 523–530 (1977)

3. Wang, L.F., Singh, C.: Stochastic Economic Emission Based Load Dispatch Through a Modified Particle Swarm Optimization Algorithm. Electric Power Systems Research 78, 1466–1476 (2008)
4. Niu, B., Wang, H., Tan, L., Xu, J.: Multi-objective Optimization Using BFO Algorithm. In: Huang, D.-S., Gan, Y., Premaratne, P., Han, K. (eds.) ICIC 2011. LNCS (LNBI), vol. 6840, pp. 582–587. Springer, Heidelberg (2012)
5. Abido, M.A.: A Novel Multiobjective Evolutionary Algorithm for Environmental/Economic Power Dispatch. Electric Power Systems Research, 71–81 (2003)
6. Abido, M.A.: A Niched Pareto Genetic Algorithm for Environmental/Economic Power Dispatch. Electric Power Systems Research, 97–105 (2003)
7. Abido, M.A.: Multiobjective Evolutionary Algorithms for Electric Power Dispatch Problem. IEEE Trans. Evolut. Comput. 10(3), 315 (2006)
8. Basu, M.: Dynamic Economic Emission Dispatch Using Nondominated Sorting Genetic Algorithm-II. Electric Power Energy Systems 30(2), 140–210 (2008)
9. Niu, B., Xue, B., Li, L.: Symbiotic Multi-swarm PSO for Portfolio Optimization. In: Huang, D.-S., Jo, K.-H., Lee, H.-H., Kang, H.-J., Bevilacqua, V. (eds.) ICIC 2009. LNCS (LNAI), vol. 5755, pp. 776–784. Springer, Heidelberg (2009)
10. Shubham, A., Panigrahi, B.K., Tiwari, M.K.: Multiobjective Particle Swarm Algorithm with Fuzzy Clustering for Electrical Power Dispatch. IEEE Trans. Evolutionary Computation, 529–541 (2008)
11. Lu, S., Sun, C., Lu, Z.: An Improved Quantum-behaved Particle Swarm Optimization Method for Short-term Combined Economic Emission Hydrothermal Scheduling. Energy Conversion and Management, 561–571 (2010)
12. Wang, L., Singh, C.: Environmental/Economic Power Dispatch Using a Fuzzified Multiobjective Particle Swarm Optimization. Electric Power Systems Research, 1654–1664 (2007)
13. Cai, J., Ma, X., Li, Q., Li, L., Peng, H.: A Multi-objective Chaotic Particle Swarm Optimization for Environmental/economic Dispatch. Energy Conversion and Management, 1318–1325 (2009)
14. Liang, J.J., Qu, B.Y., Suganthan, P.N.: Dynamic Multi-Swarm Particle Swarm Optimization for Multi-Objective Optimization Problems. Has been accepted by IEEE Congress on Evolutionary Computation (2012)
15. Hemamalini, S., Sishaj, P.S.: Emission Constrained Economic Dispatch with Valve-Point Effect using Particle Swarm Optimization. In: TENCON 2008 – 2008 IEEE Region 10 Conference, pp. 1–6 (2008)
16. Liang, J.J., Suganthan, P.N.: Dynamic Multi-Swarm Particle Swarm Optimizer with Local Search. In: Proceedings of IEEE Congress on Evolutionary Computation (CEC 2005), vol. 1, pp. 522–528 (2005)
17. Liang, J.J., Suganthan, P.N.: Dynamic Multi-Swarm Particle Swarm Optimizer. In: Proceedings of IEEE International Swarm Intelligence Symposium (SIS 2005), pp. 124–129 (2005)
18. Deb, K.: An Efficient Constraint Handling Method for Genetic Algorithms. Computer Methods in Applied Mechanics and Engineering 186, 311–338 (2000)
19. Wu, Y.L., Xu, L.Q., Zhang, J.: Multiobjective Particle Swarm Optimization Based on Differential Eevolution for Environmental/Economic Dispatch problem. In: Control and Decision Conference (CCDC), Chinese, pp. 1498–1503 (2011)
20. Abido, M.A.: Environmental/Economic Power Dispatch Using Multiobjective Evolutionary Algorithms. IEEE Trans. Power Systems, 1529–1537 (2003)

Author Index